清华版双语教学用书

非线性规划
（第3版）

Nonlinear Programming
(Third Edition)

Dimitri P. Bertsekas

清华大学出版社
北　京

北京市版权局著作权合同登记号　图字：01-2017-6580

Nonlinear Programming 3rd edition by Dimitri P. Bertsekas

© Athena Scientific, Belmont, MA, USA

本书封面贴有清华大学出版社防伪标签，无标签者不得销售。
版权所有，侵权必究。举报：010-62782989，beiqinquan@tup.tsinghua.edu.cn。

图书在版编目（CIP）数据

非线性规划（第 3 版）= Nonlinear Programming 3rd：英文/（美）伯特瑟卡斯（Dimitri P. Bertsekas）著. —北京：清华大学出版社，2018（2022.12重印）
（清华版双语教学用书）
ISBN 978-7-302-48234-5

Ⅰ. ①非… Ⅱ. ①伯… Ⅲ. ①非线性规划 – 高等学校 – 教材 – 英文 Ⅳ. ①O221.2

中国版本图书馆 CIP 数据核字（2018）第 033805 号

责任编辑：王一玲
封面设计：常雪影
责任校对：白　蕾
责任印制：沈　露

出版发行：清华大学出版社
　　　网　　址：http://www.tup.com.cn, http://www.wqbook.com
　　　地　　址：北京清华大学学研大厦 A 座　　　邮　编：100084
　　　社 总 机：010-83470000　　　　　　　　　邮　购：010-62786544
　　　投稿与读者服务：010-62776969, c-service@tup.tsinghua.edu.cn
　　　质量反馈：010-62772015, zhiliang@tup.tsinghua.edu.cn
　　　课件下载：http://www.tup.com.cn, 010-83470236
印 装 者：北京九州迅驰传媒文化有限公司
经　　销：全国新华书店
开　　本：185mm×260mm　　　印　张：55.25　　　字　数：1208 千字
版　　次：2018 年 6 月第 1 版　　　　　　　　　印　次：2022 年 12 月第 8 次印刷
定　　价：169.00 元

产品编号：075271-01

影印版序

非线性规划是具有非线性约束条件或目标函数的数学规划，是运筹学的一个重要分支，在工程设计、经济管理、科学研究、军事指挥等很多方面有着广泛的应用。

本书系统深入地介绍了非线性规划的理论和方法，是研究并掌握相关理论和方法的重要教材及学术著作。

本书作者 Dimitri P. Bertsekas 曾在希腊雅典国立技术大学机械和电子工程系学习，并在麻省理工学院获得系统科学专业博士学位。Bertsekas 教授先后在斯坦福大学、伊利诺伊大学执教，自 1979 年开始执教于麻省理工学院，其研究领域广泛，包括优化理论、控制理论、大规模计算和数据通信网络等，目前已著有 14 本教材和研究专著。Bertsekas 教授凭借其与合作者在运筹优化和计算科学交叉领域的合作专著 *Neuro-Dynamic Programming* 的杰出贡献而获得 1997 年 INFORMS 奖，2001 年他被选为美国国家工程院院士。近年来，他又获得优化、控制理论及应用领域的众多国际奖项。

本书涵盖非线性规划的主要内容，包括无约束优化、凸优化、拉格朗日乘子理论和算法、对偶理论及方法等，包含了大量的实际应用案例。本书从无约束优化问题入手，通过直观分析和严格证明给出了无约束优化问题的最优性条件，并讨论了梯度法、牛顿法、共轭方向法等基本实用算法。进而，本书将无约束优化问题的最优性条件和算法推广到具有凸集约束的优化问题中，进一步讨论了处理约束问题的可行方向法、条件梯度法、梯度投影法、双度量投影法、近似算法、流形次优化方法、坐标块下降法等。拉格朗日乘子理论和算法是非线性规划的核心内容之一，也是本书的重点。本书中的第 4 章和第 5 章详尽阐述了这方面的内容。首先从等式约束优化问题最优解的必要条件入手，给出了拉格朗日乘子理论最基本的形式，然后给出了等式约束优化问题最优解的充分条件以及不等式约束优化问题的充分条件和必要条件。拉格朗日乘子算法的引入则基于将约束优化问题转化为无约束优化问题和求解最优性条件对应的方程组的两个角度展开，分别讨论了障碍函数法、惩罚函数法、序贯二次规划法、拉格朗日法和原始对偶内点法等。本书的另一个重点是对偶性理论和方法。第 6 章从几何的角度阐述了拉格朗日对偶理论和 Fenchel 对偶理论，并讨论了离散优化及分支定界、拉格朗日松弛方法；第 7 章则详细讨论了求解对偶问题的理论和方法，包括对偶上升方法、近似增强拉格朗日方法、乘子交替方向方法、基于次梯度的优化方法及分解方法等。

本书在第 2 版基础上进行了必要扩充与完善，在无约束优化、凸集上优化、对偶方法内容上进行了部分扩充，将无约束优化问题写成两章，并补充了增量方法、分布式异步算法内容。本书将深层次的优化理论分析与实用的计算方法密切结合，以解决各种不同类型的优化问题，与其他关于优化理论和方法的书籍对比，具有如下几个特点：一是内容完整，自成体系。附录部分提供了关于矩阵分析、凸分析和线性搜索等内容的数学基础知识，阅读本书时读者也不需要提前掌握线性规划、网络优化等其他相关知识。二是层次清晰，由浅入深，理论性较强。对于理论性的定理命题，都首先给出直观的解释或者进行启发式的思想引导，最后给出严格的数学证明。全书内容按照从无约束优化问题到约束优化问题、从拉格朗日乘子理论到具体算法、从对偶理论到其求解方法的顺序安排，便于读者掌握。三是分析视角独特，特色鲜明。书中采用大量图片对抽象问题进行直观解释和说明，例如从几何角度对次梯度和次微分概念、对偶性理论进行分析和说明，同时各章节最后对非线性规划相关理论方法的最新进展进行注释，并在多处对线性规划和非线性规划的联系进行了深入的分析和对比。

  本书可以作为高年级本科生、研究生运筹优化类课程的教材或者相关研究者、工程师的工具参考书。本书作为清华大学自动化系研究生教学多年选用的优秀教材，培养了研究生从形象几何直观到严谨数学推导再到实际应用案例的思维模式，在分析和解决实际科学技术及工程问题能力方面发挥了不可替代的作用。

宋士吉 教授
清华大学自动化系
2018 年 1 月

ABOUT THE AUTHOR

Dimitri Bertsekas studied Mechanical and Electrical Engineering at the National Technical University of Athens, Greece, and obtained his Ph.D. in system science from the Massachusetts Institute of Technology. He has held faculty positions with the Engineering-Economic Systems Department, Stanford University, and the Electrical Engineering Department of the University of Illinois, Urbana. Since 1979 he has been teaching at the Electrical Engineering and Computer Science Department of the Massachusetts Institute of Technology (M.I.T.), where he is currently the McAfee Professor of Engineering.

His teaching and research spans several fields, including deterministic optimization, dynamic programming and stochastic control, large-scale and distributed computation, and data communication networks. He has authored or coauthored numerous research papers and sixteen books, several of which are currently used as textbooks in MIT classes, including "Dynamic Programming and Optimal Control," "Data Networks," "Introduction to Probability," "Convex Optimization Theory," "Convex Optimization Algorithms," as well as the present book.

Professor Bertsekas was awarded the INFORMS 1997 Prize for Research Excellence in the Interface Between Operations Research and Computer Science for his book "Neuro-Dynamic Programming" (co-authored with John Tsitsiklis), the 2001 AACC John R. Ragazzini Education Award, the 2009 INFORMS Expository Writing Award, the 2014 AACC Richard Bellman Heritage Award, the 2014 Khachiyan Prize for Life-Time Accomplishments in Optimization, and the MOS/SIAM 2015 George B. Dantzig Prize. In 2001, he was elected to the United States National Academy of Engineering for "pioneering contributions to fundamental research, practice and education of optimization/control theory, and especially its application to data communication networks."

ATHENA SCIENTIFIC
OPTIMIZATION AND COMPUTATION SERIES

1. Nonlinear Programming, 3rd Edition, by Dimitri P. Bertsekas, 2016, ISBN 1-886529-05-1, 880 pages
2. Dynamic Programming and Optimal Control, Two-Volume Set, by Dimitri P. Bertsekas, 2017, ISBN 1-886529-08-6, 1270 pages
3. Convex Optimization Algorithms, by Dimitri P. Bertsekas, 2015, ISBN 978-1-886529-28-1, 576 pages
4. Abstract Dynamic Programming, by Dimitri P. Bertsekas, 2013, ISBN 978-1-886529-42-7, 256 pages
5. Convex Optimization Theory, by Dimitri P. Bertsekas, 2009, ISBN 978-1-886529-31-1, 256 pages
6. Introduction to Probability, 2nd Edition, by Dimitri P. Bertsekas and John N. Tsitsiklis, 2008, ISBN 978-1-886529-23-6, 544 pages
7. Convex Analysis and Optimization, by Dimitri P. Bertsekas, Angelia Nedić, and Asuman E. Ozdaglar, 2003, ISBN 1-886529-45-0, 560 pages
8. Network Optimization: Continuous and Discrete Models, by Dimitri P. Bertsekas, 1998, ISBN 1-886529-02-7, 608 pages
9. Network Flows and Monotropic Optimization, by R. Tyrrell Rockafellar, 1998, ISBN 1-886529-06-X, 634 pages
10. Introduction to Linear Optimization, by Dimitris Bertsimas and John N. Tsitsiklis, 1997, ISBN 1-886529-19-1, 608 pages
11. Parallel and Distributed Computation: Numerical Methods, by Dimitri P. Bertsekas and John N. Tsitsiklis, 1997, ISBN 1-886529-01-9, 718 pages
12. Neuro-Dynamic Programming, by Dimitri P. Bertsekas and John N. Tsitsiklis, 1996, ISBN 1-886529-10-8, 512 pages
13. Constrained Optimization and Lagrange Multiplier Methods, by Dimitri P. Bertsekas, 1996, ISBN 1-886529-04-3, 410 pages
14. Stochastic Optimal Control: The Discrete-Time Case, by Dimitri P. Bertsekas and Steven E. Shreve, 1996, ISBN 1-886529-03-5, 330 pages

Contents

1. Unconstrained Optimization: Basic Methods . . . . p. 1
  1.1. Optimality Conditions . . . . p. 5
    1.1.1. Variational Ideas . . . . p. 5
    1.1.2. Main Optimality Conditions . . . . p. 15
  1.2. Gradient Methods – Convergence . . . . p. 28
    1.2.1. Descent Directions and Stepsize Rules . . . . p. 28
    1.2.2. Convergence Results . . . . p. 49
  1.3. Gradient Methods – Rate of Convergence . . . . p. 67
    1.3.1. The Local Analysis Approach . . . . p. 69
    1.3.2. The Role of the Condition Number . . . . p. 70
    1.3.3. Convergence Rate Results . . . . p. 82
  1.4. Newton's Method and Variations . . . . p. 95
    1.4.1. Modified Cholesky Factorization . . . . p. 101
    1.4.2. Trust Region Methods . . . . p. 103
    1.4.3. Variants of Newton's Method . . . . p. 105
    1.4.4. Least Squares and the Gauss-Newton Method . . . . . . p. 107
  1.5. Notes and Sources . . . . p. 117

2. Unconstrained Optimization: Additional Methods . . p. 119
  2.1. Conjugate Direction Methods . . . . p. 120
    2.1.1. The Conjugate Gradient Method . . . . p. 125
    2.1.2. Convergence Rate of Conjugate Gradient Method . . . . p. 132
  2.2. Quasi-Newton Methods . . . . p. 138
  2.3. Nonderivative Methods . . . . p. 148
    2.3.1. Coordinate Descent . . . . p. 149
    2.3.2. Direct Search Methods . . . . p. 154
  2.4. Incremental Methods . . . . p. 158
    2.4.1. Incremental Gradient Methods . . . . p. 161
    2.4.2. Incremental Aggregated Gradient Methods . . . . p. 172
    2.4.3. Incremental Gauss-Newton Methods . . . . p. 178
    2.4.3. Incremental Newton Methods . . . . p. 185
  2.5. Distributed Asynchronous Algorithms . . . . p. 194

2.5.1. Totally and Partially Asynchronous Algorithms	p. 197
2.5.2. Totally Asynchronous Convergence	p. 198
2.5.3. Partially Asynchronous Gradient-Like Algorithms	p. 203
2.5.4. Convergence Rate of Asynchronous Algorithms	p. 204
2.6. Discrete-Time Optimal Control Problems	p. 210
2.6.1. Gradient and Conjugate Gradient Methods for Optimal Control	p. 221
2.6.2. Newton's Method for Optimal Control	p. 222
2.7. Solving Nonlinear Programming Problems - Some Practical Guidelines	p. 227
2.8. Notes and Sources	p. 232

3. Optimization Over a Convex Set   p. 235

3.1. Constrained Optimization Problems	p. 236
3.1.1. Necessary and Sufficient Conditions for Optimality	p. 236
3.1.2. Existence of Optimal Solutions	p. 246
3.2. Feasible Directions - Conditional Gradient Method	p. 257
3.2.1. Descent Directions and Stepsize Rules	p. 257
3.2.2. The Conditional Gradient Method	p. 262
3.3. Gradient Projection Methods	p. 272
3.3.1. Feasible Directions and Stepsize Rules Based on Projection	p. 272
3.3.2. Convergence Analysis	p. 283
3.4. Two-Metric Projection Methods	p. 292
3.5. Manifold Suboptimization Methods	p. 298
3.6. Proximal Algorithms	p. 307
3.6.1. Rate of Convergence	p. 312
3.6.2. Variants of the Proximal Algorithm	p. 318
3.7. Block Coordinate Descent Methods	p. 323
3.7.1. Variants of Coordinate Descent	p. 327
3.8. Network Optimization Algorithms	p. 331
3.9. Notes and Sources	p. 338

4. Lagrange Multiplier Theory   p. 343

4.1. Necessary Conditions for Equality Constraints	p. 345
4.1.1. The Penalty Approach	p. 349
4.1.2. The Elimination Approach	p. 352
4.1.3. The Lagrangian Function	p. 356
4.2. Sufficient Conditions and Sensitivity Analysis	p. 364
4.2.1. The Augmented Lagrangian Approach	p. 365
4.2.2. The Feasible Direction Approach	p. 369
4.2.3. Sensitivity	p. 370
4.3. Inequality Constraints	p. 376
4.3.1. Karush-Kuhn-Tucker Necessary Conditions	p. 378

Contents

- 4.3.2. Sufficient Conditions and Sensitivity . . . . p. 383
- 4.3.3. Fritz John Optimality Conditions . . . . p. 386
- 4.3.4. Constraint Qualifications and Pseudonormality . . . . . p. 392
- 4.3.5. Abstract Set Constraints and the Tangent Cone . . . . p. 399
- 4.3.6. Abstract Set Constraints, Equality, and Inequality
  Constraints . . . . p. 415
- 4.4. Linear Constraints and Duality . . . . p. 429
  - 4.4.1. Convex Cost Function and Linear Constraints . . . . . p. 429
  - 4.4.2. Duality Theory: A Simple Form for Linear
    Constraints . . . . p. 432
- 4.5. Notes and Sources . . . . p. 441

5. Lagrange Multiplier Algorithms . . . . p. 445

- 5.1. Barrier and Interior Point Methods . . . . p. 446
  - 5.1.1. Path Following Methods for Linear Programming . . . p. 450
  - 5.1.2. Primal-Dual Methods for Linear Programming . . . . . p. 458
- 5.2. Penalty and Augmented Lagrangian Methods . . . . p. 469
  - 5.2.1. The Quadratic Penalty Function Method . . . . p. 471
  - 5.2.2. Multiplier Methods – Main Ideas . . . . p. 479
  - 5.2.3. Convergence Analysis of Multiplier Methods . . . . . p. 488
  - 5.2.4. Duality and Second Order Multiplier Methods . . . . . p. 492
  - 5.2.5. Nonquadratic Augmented Lagrangians - The Exponential
    Method of Multipliers . . . . p. 494
- 5.3. Exact Penalties – Sequential Quadratic Programming . . . . p. 502
  - 5.3.1. Nondifferentiable Exact Penalty Functions . . . . . . p. 503
  - 5.3.2. Sequential Quadratic Programming . . . . p. 513
  - 5.3.3. Differentiable Exact Penalty Functions . . . . p. 520
- 5.4. Lagrangian Methods . . . . p. 527
  - 5.4.1. First-Order Lagrangian Methods . . . . p. 528
  - 5.4.2. Newton-Like Methods for Equality Constraints . . . . p. 535
  - 5.4.3. Global Convergence . . . . p. 545
  - 5.4.4. A Comparison of Various Methods . . . . p. 548
- 5.5. Notes and Sources . . . . p. 550

6. Duality and Convex Programming . . . . p. 553

- 6.1. Duality and Dual Problems . . . . p. 554
  - 6.1.1. Geometric Multipliers . . . . p. 556
  - 6.1.2. The Weak Duality Theorem . . . . p. 561
  - 6.1.3. Primal and Dual Optimal Solutions . . . . p. 566
  - 6.1.4. Treatment of Equality Constraints . . . . p. 568
  - 6.1.5. Separable Problems and their Geometry . . . . p. 570
  - 6.1.6. Additional Issues About Duality . . . . p. 575
- 6.2. Convex Cost – Linear Constraints . . . . p. 582
- 6.3. Convex Cost – Convex Constraints . . . . p. 589

6.4. Conjugate Functions and Fenchel Duality	p. 598
6.4.1. Conic Programming	p. 604
6.4.2. Monotropic Programming	p. 612
6.4.3. Network Optimization	p. 617
6.4.4. Games and the Minimax Theorem	p. 620
6.4.5. The Primal Function and Sensitivity Analysis	p. 623
6.5. Discrete Optimization and Duality	p. 630
6.5.1. Examples of Discrete Optimization Problems	p. 631
6.5.2. Branch-and-Bound	p. 639
6.5.3. Lagrangian Relaxation	p. 648
6.6. Notes and Sources	p. 660

7. Dual Methods . . . . p. 663

7.1. Dual Derivatives and Subgradients	p. 666
7.2. Dual Ascent Methods for Differentiable Dual Problems	p. 673
7.2.1. Coordinate Ascent for Quadratic Programming	p. 673
7.2.2. Separable Problems and Primal Strict Convexity	p. 675
7.2.3. Partitioning and Dual Strict Concavity	p. 677
7.3. Proximal and Augmented Lagrangian Methods	p. 682
7.3.1. The Method of Multipliers as a Dual	
Proximal Algorithm	p. 682
7.3.2. Entropy Minimization and Exponential	
Method of Multipliers	p. 686
7.3.3. Incremental Augmented Lagrangian Methods	p. 687
7.4. Alternating Direction Methods of Multipliers	p. 691
7.4.1. ADMM Applied to Separable Problems	p. 699
7.4.2. Connections Between Augmented Lagrangian-	
Related Methods	p. 703
7.5. Subgradient-Based Optimization Methods	p. 709
7.5.1. Subgradient Methods	p. 709
7.5.2. Approximate and Incremental Subgradient Methods	p. 714
7.5.3. Cutting Plane Methods	p. 717
7.5.4. Ascent and Approximate Ascent Methods	p. 724
7.6. Decomposition Methods	p. 735
7.6.1. Lagrangian Relaxation of the Coupling Constraints	p. 736
7.6.2. Decomposition by Right-Hand Side Allocation	p. 739
7.7. Notes and Sources	p. 742

Appendix A: Mathematical Background . . . . p. 745

A.1. Vectors and Matrices	p. 746
A.2. Norms, Sequences, Limits, and Continuity	p. 749
A.3. Square Matrices and Eigenvalues	p. 757
A.4. Symmetric and Positive Definite Matrices	p. 760
A.5. Derivatives	p. 765

A.6. Convergence Theorems . . . . p. 770

Appendix B: Convex Analysis . . . . **p. 783**

B.1. Convex Sets and Functions . . . . p. 783
B.2. Hyperplanes . . . . p. 793
B.3. Cones and Polyhedral Convexity . . . . p. 796
B.4. Extreme Points and Linear Programming . . . . p. 798
B.5. Differentiability Issues . . . . p. 803

Appendix C: Line Search Methods . . . . **p. 809**

C.1. Cubic Interpolation . . . . p. 809
C.2. Quadratic Interpolation . . . . p. 810
C.3. The Golden Section Method . . . . p. 812

Appendix D: Implementation of Newton's Method . . . **p. 815**

D.1. Cholesky Factorization . . . . p. 815
D.2. Application to a Modified Newton Method . . . . . . p. 817

References . . . . **p. 821**

Index . . . . **p. 857**

Preface to the First Edition

Nonlinear programming is a mature field that has experienced major developments in the last ten years. The first such development is the merging of linear and nonlinear programming algorithms through the use of interior point methods. This has resulted in a profound rethinking of how we solve linear programming problems, and in a major reassessment of how we treat constraints in nonlinear programming. A second development, less visible but still important, is the increased emphasis on large-scale problems, and the associated algorithms that take advantage of problem structure as well as parallel hardware. A third development has been the extensive use of iterative unconstrained optimization to solve the difficult least squares problems arising in the training of neural networks. As a result, simple gradient-like methods and stepsize rules have attained increased importance.

The purpose of this book is to provide an up-to-date, comprehensive, and rigorous account of nonlinear programming at the beginning graduate student level. In addition to the classical topics, such as descent algorithms, Lagrange multiplier theory, and duality, some of the important recent developments are covered: interior point methods for linear and nonlinear programs, major aspects of large-scale optimization, and least squares problems and neural network training.

A further noteworthy feature of the book is that it treats Lagrange multipliers and duality using two different and complementary approaches: a variational approach based on the implicit function theorem, and a convex analysis approach based on geometrical arguments. The former approach applies to a broader class of problems, while the latter is more elegant and more powerful for the convex programs to which it applies.

The chapter-by-chapter description of the book follows:

Chapter 1: This chapter covers unconstrained optimization: main concepts, optimality conditions, and algorithms. The material is classic, but there are discussions of topics frequently left untreated, such as the behavior of algorithms for singular problems, neural network training, and discrete-time optimal control.

Chapter 2: This chapter treats constrained optimization over a convex set without the use of Lagrange multipliers. I prefer to cover this material before dealing with the complex machinery of Lagrange multipliers because I have found that students absorb easily algorithms such as conditional gradient, gradient projection, and coordinate descent, which can be viewed as natural extensions of unconstrained descent algorithms. This chapter contains also a treatment of the affine scaling method for linear programming.

Chapter 3: This chapter gives a detailed treatment of Lagrange multipliers, the associated necessary and sufficient conditions, and sensitivity analysis. The first three sections deal with nonlinear equality and inequality constraints. The last section deals with linear constraints and develops a simple form of duality theory for linearly constrained problems with differentiable cost, including linear and quadratic programming.

Chapter 4: This chapter treats constrained optimization algorithms that use penalties and Lagrange multipliers, including barrier, augmented Lagrangian, sequential quadratic programming, and primal-dual interior point methods for linear programming. The treatment is extensive, and borrows from my 1982 research monograph on Lagrange multiplier methods.

Chapter 5: This chapter provides an in-depth coverage of duality theory (Lagrange and Fenchel). The treatment is totally geometric, and everything is explained in terms of intuitive figures.

Chapter 6: This chapter deals with large-scale optimization methods based on duality. Some material is borrowed from my Parallel and Distributed Algorithms book (coauthored by John Tsitsiklis), but there is also an extensive treatment of nondifferentiable optimization, including subgradient, ϵ-subgradient, and cutting plane methods. Decomposition methods such as Dantzig-Wolfe and Benders are also discussed.

Appendixes: Four appendixes are given. The first gives a summary of calculus, analysis, and linear algebra results used in the text. The second is a fairly extensive account of convexity theory, including proofs of the basic polyhedral convexity results on extreme points and Farkas' lemma, as well the basic facts about subgradients. The third appendix covers one-dimensional minimization methods. The last appendix discusses an implementation of Newton's method for unconstrained optimization.

Inevitably, some coverage compromises had to be made. The subject of nonlinear optimization has grown so much that leaving out a number of important topics could not be avoided. For example, a discussion of variational inequalities, a deeper treatment of optimality conditions, and a more detailed development of Quasi-Newton methods are not provided. Also, a larger number of sample applications would have been desirable. I hope that instructors will supplement the book with the type of practical examples that their students are most familiar with.

The book was developed through a first-year graduate course that I taught at the Univ. of Illinois and at M.I.T. over a period of 20 years. The mathematical prerequisites are matrix-vector algebra and advanced calculus, including a good understanding of convergence concepts. A course in analysis and/or linear algebra should also be very helpful, and would provide the mathematical maturity needed to follow and to appreciate the mathematical reasoning used in the book. Some of the sections in the book may be ommited at first reading without loss of continuity. These sections have been marked by a star. The rule followed here is that the material discussed in a starred section is not used in a non-starred section.

The book can be used to teach several different types of courses.

(a) A two-quarter course that covers most sections of every chapter.

(b) A one-semester course that covers Chapter 1 except for Section 1.9, Chapter 2 except for Sections 2.4 and 2.5, Chapter 3 except for Section 3.4, Chapter 4 except for parts of Sections 4.2 and 4.3, the first three sections of Chapter 5, and a selection from Section 5.4 and Chapter 6. This is the course I usually teach at MIT.

(c) A one-semester course that covers most of Chapters 1, 2, and 3, and selected algorithms from Chapter 4. I have taught this type of course several times. It is less demanding of the students because it does not require the machinery of convex analysis, yet it still provides a fairly powerful version of duality theory (Section 3.4).

(d) A one-quarter course that covers selected parts of Chapters 1, 2, 3, and 4. This is a less comprehensive version of (c) above.

(e) A one-quarter course on convex analysis and optimization that starts with Appendix B and covers Sections 1.1, 2.1, 3.4, and Chapter 5.

There is a very extensive literature on nonlinear programming and to give a complete bibliography and a historical account of the research that led to the present form of the subject would have been impossible. I thus have not attempted to compile a comprehensive list of original contributions to the field. I have cited sources that I have used extensively, that provide important extensions to the material of the book, that survey important topics, or that are particularly well suited for further reading. I have also cited selectively a few sources that are historically significant, but the reference list is far from exhaustive in this respect. Generally, to aid researchers in the field, I have preferred to cite surveys and textbooks for subjects that are relatively mature, and to give a larger number of references for relatively recent developments.

Finally, I would like to express my thanks to a number of individuals for their contributions to the book. My conceptual understanding of the subject was formed at Stanford University while I interacted with David Luenberger and I taught using his books. This experience had a lasting influence on my thinking. My research collaboration with several colleagues, particularly Joe

Dunn, Eli Gafni, Paul Tseng, and John Tsitsiklis, were very useful and are reflected in the book. I appreciate the suggestions and insights of a number of people, particularly David Castanon, Joe Dunn, Terry Rockafellar, Paul Tseng, and John Tsitsiklis. I am thankful to the many students and collaborators whose comments led to corrections and clarifications. Steve Patek, Serap Savari, and Cynara Wu were particularly helpful in this respect. David Logan, Steve Patek, and Lakis Polymenakos helped me to generate the graph of the cover, which depicts the cost function of a simple neural network training problem. My wife Joanna cheered me up with her presence and humor during the long hours of writing, as she has with her companionship of over 30 years. I dedicate this book to her with my love.

Dimitri P. Bertsekas
November, 1995

Preface to the Second Edition

The second edition has expanded by about 130 pages the coverage of the original. Nearly 40% of the new material represents miscellaneous additions scattered throughout the text. The remainder deals with three new topics. These are:

(a) A new section in Chapter 3 that focuses on a simple but far-reaching treatment of Fritz John necessary conditions and constraint qualifications, and also includes semi-infinite programming.

(b) A new section in Chapter 5 on the use of duality and Lagrangian relaxation for solving discrete optimization problems. This section describes several motivating applications, and provides a connecting link between continuous and discrete optimization.

(c) A new section in Chapter 6 on approximate and incremental subgradient methods. This material is the subject of ongoing joint research with Angelia Nedić, but it was thought sufficiently significant to be included in summary here.

One of the aims of the revision was to highlight the connections of nonlinear programming with other branches of optimization, such as linear programming, network optimization, and discrete/integer optimization. This should provide some additional flexibility for using the book in the classroom. In addition, the presentation was improved, the mathematical background material of the appendixes has been expanded, the exercises were reorganized, and a substantial number of new exercises were added.

A new internet-based feature was added to the book, which significantly extends its scope and coverage. Many of the theoretical exercises, quite a few of them new, have been solved in detail and their solutions have been posted in the book's www page

http://www.athenasc.com/nonlinbook.html

These exercises have been marked with the symbol **WWW**

The book's www page also contains links to additional resources, such as computer codes and my lecture slides from my MIT Nonlinear Programming class.

I would like to express my thanks to the many colleagues who contributed suggestions for improvement of the second edition. I would like to thank particularly Angelia Nedić for her extensive help with the internet-posted solutions of the theoretical exercises.

Dimitri P. Bertsekas
June, 1999

Preface to the Third Edition

The third edition of the book is a thoroughly rewritten version of the 1999 second edition. New material was included, some of the old material was discarded, and a large portion of the remainder was reorganized or revised. The total number of pages has increased by about 10 percent.

Aside from incremental improvements, the changes aim to bring the book up-to-date with recent research progress, and in harmony with the major developments in convex optimization theory and algorithms that have occurred in the meantime. These developments were documented in three of my books: the 2015 book "Convex Optimization Algorithms," the 2009 book "Convex Optimization Theory," and the 2003 book "Convex Analysis and Optimization" (coauthored with Angelia Nedić and Asuman Ozdaglar). A major difference is that these books have dealt primarily with convex, possibly nondifferentiable, optimization problems and rely on convex analysis, while the present book focuses primarily on algorithms for possibly nonconvex differentiable problems, and relies on calculus and variational analysis.

Having written several interrelated optimization books, I have come to see nonlinear programming and its associated duality theory as the lynchpin that holds together deterministic optimization. I have consequently set as an objective for the present book to integrate the contents of my books, together with internet-accessible material, so that they complement each other and form a unified whole. I have thus provided bridges to my other works with extensive references to generalizations, discussions, and elaborations of the analysis given here, and I have used throughout fairly consistent notation and mathematical level.

Another connecting link of my books is that they all share the same style: they rely on rigorous analysis, but they also aim at an intuitive exposition that makes use of geometric visualization. This stems from my belief that success in the practice of optimization strongly depends on the intuitive (as well as the analytical) understanding of the underlying theory and algorithms.

Some of the more prominent new features of the present edition are:

(a) An expanded coverage of incremental methods and their connections to stochastic gradient methods, based in part on my 2000 joint work with Angelia Nedić; see Section 2.4 and Section 7.3.2.

(b) A discussion of asynchronous distributed algorithms based in large part on my 1989 "Parallel and Distributed Computation" book (coauthored with John Tsitsiklis); see Section 2.5.

(c) A discussion of the proximal algorithm and its variations in Section 3.6, and the relation with the method of multipliers in Section 7.3.

(d) A substantial coverage of the alternating direction method of multipliers (ADMM) in Section 7.4, with a discussion of its many applications and variations, as well as references to my 1989 "Parallel and Distributed Computation" and 2015 "Convex Optimization Algorithms" books.

(e) A fairly detailed treatment of conic programming problems in Section 6.4.1.

(f) A discussion of the question of existence of solutions in constrained optimization, based on my 2007 joint work with Paul Tseng [BeT07], which contains further analysis; see Section 3.1.2.

(g) Additional material on network flow problems in Section 3.8 and 6.4.3, and their extensions to monotropic programming in Section 6.4.2, with references to my 1998 "Network Optimization" book.

(h) An expansion of the material of Chapter 4 on Lagrange multiplier theory, using a strengthened version of the Fritz John conditions, and the notion of pseudonormality, based on my 2002 joint work with Asuman Ozdaglar.

(i) An expansion of the material of Chapter 5 on Lagrange multiplier algorithms, with references to my 1982 "Constrained Optimization and Lagrange Multiplier Methods" book.

The book contains a few new exercises. As in the second edition, many of the theoretical exercises have been solved in detail and their solutions have been posted in the book's internet site

http://www.athenasc.com/nonlinbook.html

These exercises have been marked with the symbols **WWW**. Many other exercises contain detailed hints and/or references to internet-accessible sources. The book's internet site also contains links to additional resources, such as many additional solved exercises from my convex optimization books, computer codes, my lecture slides from MIT Nonlinear Programming classes, and full course contents from the MIT OpenCourseWare (OCW) site.

I would like to express my thanks to the many colleagues who contributed suggestions for improvement of the third edition. In particular, let me note with appreciation my principal collaborators on nonlinear programming topics since the 1999 second edition: Angelia Nedić, Asuman Ozdaglar, Paul Tseng, Mengdi Wang, and Huizhen (Janey) Yu.

Dimitri P. Bertsekas
June, 2016

1
Unconstrained Optimization: Basic Methods

Contents

1.1. Optimality Conditions . . . . p. 5
   1.1.1. Variational Ideas . . . . p. 5
   1.1.2. Main Optimality Conditions . . . . p. 15
1.2. Gradient Methods – Convergence . . . . p. 28
   1.2.1. Descent Directions and Stepsize Rules . . . . p. 28
   1.2.2. Convergence Results . . . . p. 49
1.3. Gradient Methods – Rate of Convergence . . . . p. 67
   1.3.1. The Local Analysis Approach . . . . p. 69
   1.3.2. The Role of the Condition Number . . . . p. 70
   1.3.3. Convergence Rate Results . . . . p. 82
1.4. Newton's Method and Variations . . . . p. 95
   1.4.1. Modified Cholesky Factorization . . . . p. 101
   1.4.2. Trust Region Methods . . . . p. 103
   1.4.3. Variants of Newton's Method . . . . p. 105
   1.4.4. Least Squares and the Gauss-Newton Method . . . . p. 107
1.5. Notes and Sources . . . . p. 117

Mathematical models of optimization can be generally represented by a *constraint set* X and a *cost function* f that maps elements of X into real numbers. The set X consists of the available decisions x and the cost $f(x)$ is a scalar measure of undesirability of choosing decision x. We want to find an optimal decision, i.e., an $x^* \in X$ such that

$$f(x^*) \leq f(x), \qquad \forall\ x \in X.$$

In this book we focus on the case where each decision x is an n-dimensional vector; that is, x is an n-tuple of real numbers $(x_1, \ldots, x_n)$. Thus the constraint set X is a subset of $\Re^n$, the n-dimensional Euclidean space. (We refer to Appendix A for an account of our terminology and notational conventions.)

The optimization problem just stated is very broad and contains as special cases several important classes of problems that have widely differing structures. Our focus will be on nonlinear programming problems, so let us provide some orientation about the character of these problems and their relations with other types of optimization problems.

Continuous and Discrete Problems

Perhaps the most important characteristic of an optimization problem is whether it is *continuous* or *discrete*. Continuous problems are those where the constraint set X is infinite and has a "continuous" character. Typical examples of continuous problems are those where there are no constraints, i.e., where $X = \Re^n$, or where X is specified by some equations and inequalities involving continuous functions. Generally, continuous problems are analyzed using the mathematics of calculus and convexity.

Discrete problems are basically those that are not continuous, usually because of finiteness of the constraint set X. Typical examples arise in scheduling, route planning, and matching, among many others. An important type of discrete problems is *integer programming*, where there is a constraint that the optimization variables must take only integer values from some range (such as 0 or 1). Discrete problems are addressed with combinatorial and discrete mathematics, and other special methodology, some of which relates to continuous problems.

Nonlinear programming, the case where either the cost function f is nonlinear or the constraint set X is specified by nonlinear equations and inequalities, lies squarely within the continuous problem category. Several other important types of optimization problems have more of a hybrid character, but are strongly connected with nonlinear programming.

In particular, *linear programming* problems, the case where f is linear and X is a polyhedral set specified by linear inequality constraints, have many of the characteristics of continuous problems. However, they also have in part a combinatorial structure: according to a fundamental theorem [Prop. B.20(c) in Appendix B], optimal solutions of a linear program can be found by searching among the (finite) set of extreme points of X. Thus the search for an optimum can be confined within this finite set, and indeed one of the most popular methods for linear programming, the simplex method, is based on this idea. We note, however, that other important linear programming methods, such as the interior point methods to be discussed in Section 5.1, and some of the duality-based methods in Chapters 6 and 7, rely on the continuous structure of linear programs and are based on nonlinear programming ideas.

Another major class of problems with a strongly hybrid character is *network optimization*. Here the constraint set X is a polyhedral set that is defined in terms of a graph consisting of nodes and directed arcs. The salient feature of this constraint set is that its extreme points have *integer components*, something that is not true for general polyhedral sets. As a result, important combinatorial or integer programming problems, such as for example some matching and shortest path problems, can be embedded and solved within a continuous network optimization framework.

Our objective in this book is to focus on nonlinear programming problems, their continuous character, and the associated mathematical analysis. However, we will maintain a view to other broad classes of problems that have in part a discrete character. In particular, we will consider extensively those aspects of linear programming that bear a close relation to nonlinear programming methodology, such as interior point methods and polyhedral convexity (see Section 5.1, and Sections B.3 and B.4 in Appendix B).

We will also discuss various aspects of network optimization problems that relate to both their continuous and their discrete character in Sections 3.1, 3.8, and 6.4.3. A far more extensive treatment, which straddles the boundary between continuous and discrete optimization, can be found in the author's network optimization textbook [Ber98].

Finally, we will discuss some of the major methods for integer programming and combinatorial optimization, such as branch-and-bound and Lagrangian relaxation. These methods rely on duality and the solution of continuous optimization subproblems (see Sections 6.5 and 7.5).

Let us also note that there is a methodological division within the class of continuous problems. On one hand we have problems where the cost function f is a differentiable, or even twice differentiable. This allows a calculus-based analysis, which will be the primary approach in our analysis of Chapters 1-5. On the other hand we have problems where f is nondifferentiable but is convex. This requires a line of analysis that relies on convexity (rather than differentiability). It will be our primary approach in Chapters 6 and 7. Of course differentiability can also play an important role within the context of convex problems. Moreover, nondifferentiable (convex or nonconvex) problems can often be fruitfully converted to differentiable ones by using smoothing transformations, as we will explain later (see Section 2.7).

Unconstrained Differentiable Optimization: An Outline

In this chapter and the next, we focus on unconstrained differentiable nonlinear programming problems. These are problems where f is at least once continuously differentiable and where $X = \Re^n$:

$$\text{minimize } f(x)$$
$$\text{subject to } x \in \Re^n. \tag{UP}$$

The first and second derivatives of f play an important role in the characterization of optimal solutions via necessary and sufficient conditions, which are the main subject of Section 1.1. The first and second derivatives are also central in numerical algorithms for computing approximately optimal solutions. There is a broad range of such algorithms, with a rich theory, which is discussed in Sections 1.2-1.4, and in Chapter 2.

In Chapter 2, we will also discuss two types of special problem structures. The first involves an *additive cost function*,

$$f(x) = \sum_{i=1}^{m} f_i(x),$$

where the component functions f_i are differentiable. Many problems of interest, arising in signal processing, machine learning, and neural network training have this form. For such problems, an incremental algorithmic approach is often used, which involves sequential steps along the gradients of the functions f_i, with intermediate adjustment of x after processing each f_i. We will discuss algorithms of this type and their applications in Section 2.4. These methods include incremental versions of the gradient, Newton, and Gauss-Newton methods, discussed in Chapter 1. In many important contexts, some of the components f_i are nondifferentiable. Incremental methods for problems of this kind will also be developed in Chapter 7. In Section 2.5, we will also discuss various methods in a distributed asynchronous computation setting, involving multiple processors and communication delays between the processors.

The second type of special structure that we will discuss is *optimal control problems*, which involve a discrete-time dynamic system (see Section 2.6). These are problems of potentially very large dimension, whose structure can be exploited for the convenient implementation of gradient and Newton-like methods. An important characteristic of these problems is that the gradient and Newton directions can be computed economically, using the dynamic system structure.

Both additive cost and optimal control problems arise also in constrained settings, and on occasion we will pause to discuss constrained variants in subsequent chapters. A third type of special structure that arises primarily in a constrained setting is *network optimization problems*.

We will discuss these problems in Sections 3.1 and 3.8, after the development of the relevant constrained optimization algorithms.

Our analysis in this chapter will focus on explaining the basic properties of the various methods, and primarily their convergence and rate of convergence properties. Many of these properties can be adequately and intuitively explained using a quadratic problem. The rationale is that behavior of an algorithm for a positive definite quadratic cost function is typically a correct predictor of its behavior for a twice differentiable cost function in the neighborhood of a minimum where the Hessian matrix is positive definite. Since the gradient is zero at that minimum, the positive definite quadratic term dominates the other terms in the series expansion of f, and the asymptotic behavior of the method does not depend on terms of order higher than two. This line of analysis underlies some of the most widely used unconstrained optimization methods, such as Newton, Gauss-Newton, quasi-Newton, and conjugate direction methods. However, the rationale for these methods is weakened when the Hessian is singular at the minimum, since in this case third and higher order terms may become significant. Then it may be best to use first order methods and analysis that relies primarily on the first order differentiability of the cost function. Consistent with this idea, we will discuss both first and second order methods, in this and later chapters, and explain the circumstances under which each type of method is most suitable.

1.1 OPTIMALITY CONDITIONS

1.1.1 Variational Ideas

The main ideas underlying optimality conditions in nonlinear programming usually admit simple explanations although their detailed proofs are sometimes tedious. For this reason, we will first discuss informally these ideas in the present subsection, and leave detailed statements of results and proofs for the next subsection.

Local and Global Minima

A vector x^* is an *unconstrained local minimum* of a function $f : \Re^n \mapsto \Re$ if it is no worse than its neighbors; that is, if there exists an $\epsilon > 0$ such that†

$$f(x^*) \leq f(x), \qquad \forall\, x \in \Re^n \text{ with } \|x - x^*\| < \epsilon.$$

---

† Unless stated otherwise, we use the standard Euclidean norm $\|x\| = \sqrt{x'x}$. Appendix A describes in detail our mathematical notation and terminology.

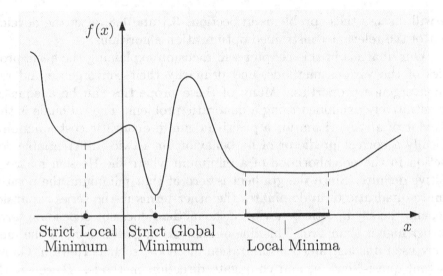

Figure 1.1.1. Unconstrained local and global minima in one dimension.

A vector x^* is an *unconstrained global minimum* of f if it is no worse than all other vectors; that is,

$$f(x^*) \leq f(x), \quad \forall\, x \in \Re^n.$$

The unconstrained local or global minimum x^* is said to be *strict* if the corresponding inequality above is strict for $x \neq x^*$. Figure 1.1.1 illustrates these definitions.

The definitions of local and global minima can be extended to the case where minimization of f is subject to a constraint set $X \subset \Re^n$, the points of which are called *feasible*. In particular, we say that x^* is a *local minimum of f over X* if $x^* \in X$ and there is an $\epsilon > 0$ such that

$$f(x^*) \leq f(x), \quad \forall\, x \in X \text{ with } \|x - x^*\| < \epsilon;$$

see Fig. 1.1.2. The definitions of a global and a strict minimum of f over X are analogous.

Local and global *maxima* are similarly defined. In particular, x^* is an unconstrained local (global) maximum of f, if x^* is an unconstrained local (global) minimum of the function $-f$.

Necessary Conditions for Optimality

If the cost function is differentiable, we can use gradients to compare the cost of a vector with the cost of its close neighbors. In particular, we consider small variations Δx from a given vector x^*, which approximately, up to first order, yield a cost variation

$$f(x^* + \Delta x) - f(x^*) \approx \nabla f(x^*)'\Delta x,$$

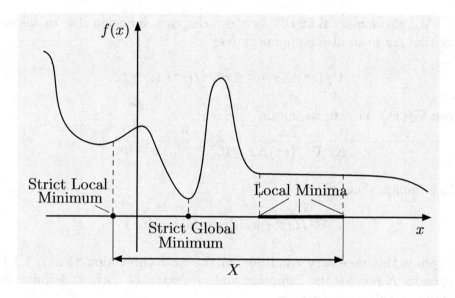

Figure 1.1.2. Local and global minima of f over the constraint set X.

and, up to second order, yield a cost variation

$$f(x^* + \Delta x) - f(x^*) \approx \nabla f(x^*)' \Delta x + \tfrac{1}{2} \Delta x' \nabla^2 f(x^*) \Delta x.$$

We expect that if x^* is an unconstrained local minimum, the first order cost variation due to a small variation Δx is nonnegative:

$$\nabla f(x^*)' \Delta x = \sum_{i=1}^{n} \frac{\partial f(x^*)}{\partial x_i} \Delta x_i \geq 0.$$

In particular, by taking Δx to be positive and negative multiples of the unit coordinate vectors (all coordinates equal to zero except for one which is equal to unity), we obtain $\partial f(x^*)/\partial x_i \geq 0$ and $\partial f(x^*)/\partial x_i \leq 0$, respectively, for all $i = 1, \ldots, n$. Equivalently, we have the necessary condition

$$\nabla f(x^*) = 0,$$

[originally formulated by Fermat in 1637 in the short treatise "Methodus as Disquirendam Maximam et Minimam" without proof (of course!)]. This condition is proved formally in Prop. 1.1.1, given in the next subsection.

The idea that at a local minimum x^*, the condition $\nabla f(x^*)' \Delta x \geq 0$ should hold for small variations Δx applies more broadly, including for problems with convex constraint sets X, when it takes the form

$$\nabla f(x^*)'(x - x^*) \geq 0, \qquad \forall \, x \in X.$$

This condition will be shown in Prop. 1.1.2 for the case of a convex cost function. In Chapter 3, it will become the basis for constrained versions of the computational methods of the present chapter.

We also expect that the second order cost variation due to a small variation Δx must also be nonnegative:

$$\nabla f(x^*)'\Delta x + \tfrac{1}{2}\Delta x' \nabla^2 f(x^*)\Delta x \geq 0.$$

Since $\nabla f(x^*)'\Delta x = 0$, we obtain

$$\Delta x' \nabla^2 f(x^*) \Delta x \geq 0, \qquad \forall\ \Delta x \in \Re^n,$$

which implies that

$$\nabla^2 f(x^*) : \text{positive semidefinite}.$$

We prove this necessary condition in the next subsection (Prop. 1.1.1). Appendix A reviews the definition and properties of positive definite and positive semidefinite matrices.

In what follows, we refer to a vector x^* satisfying the condition $\nabla f(x^*) = 0$ as a *stationary point*.

The Case of a Convex Cost Function

Convexity plays a very important role in nonlinear programming.† One reason is that when the cost function f is convex, there is no distinction between local and global minima; every local minimum is also global. The idea is illustrated in Fig. 1.1.3 and the formal proof is given in Prop. 1.1.2.

Another important fact is that the first order condition $\nabla f(x^*) = 0$ is also sufficient for optimality if f is convex. This is established in Prop. 1.1.3. The proof is based on a basic property of a convex function f: the linear approximation at a point x^* based on the gradient, i.e.,

$$f(x^*) + \nabla f(x^*)'(x - x^*),$$

underestimates $f(x)$, so if $\nabla f(x^*) = 0$, then $f(x^*) \leq f(x)$ for all x (see Prop. B.3 in Appendix B).

Sufficient Conditions for Optimality

If f is not convex, the first and second order necessary conditions can fail to guarantee local optimality of x^*. This is illustrated in Fig. 1.1.4. However, by strengthening the second order condition we obtain sufficient conditions

---

† The theory of convex sets and functions, particularly as it relates to optimization theory, is reviewed in Appendix B and is discussed extensively in the author's books [BNO03] and [Ber09].

Sec. 1.1   Optimality Conditions

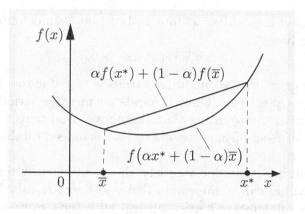

Figure 1.1.3. Illustration of why a local minimum of a convex function must also global. Suppose that f is convex and that x^* is not a global minimum, so that there exists $\bar{x}$ with $f(\bar{x}) < f(x^*)$. By convexity, for all $\alpha \in (0, 1)$,

$$f\big(\alpha x^* + (1-\alpha)\bar{x}\big) \leq \alpha f(x^*) + (1-\alpha)f(\bar{x}) < f(x^*).$$

Thus, f has value strictly lower than $f(x^*)$ at every point on the line segment connecting x^* with $\bar{x}$, except x^*. Therefore x^* cannot be a local minimum.

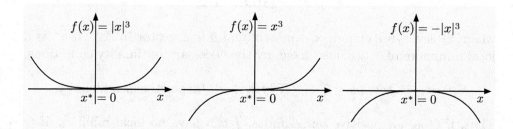

Figure 1.1.4. Illustration of the first order necessary optimality condition of zero slope $[\nabla f(x^*) = 0]$ and the second order necessary optimality condition of nonnegative curvature $[\nabla^2 f(x^*) \geq 0]$ for functions of one variable. The first order condition is satisfied not only by local minima, but also by local maxima and "inflection" points, such as the one on the middle figure above. In some cases [e.g. $f(x) = x^3$ and $f(x) = -|x|^3$] the second order condition is also satisfied by local maxima and inflection points. If the function f is convex, the condition $\nabla f(x^*) = 0$ is necessary and sufficient for global optimality of x^*.

for optimality. In particular, consider a vector x^* that satisfies the first order necessary optimality condition

$$\nabla f(x^*) = 0, \tag{1.1}$$

and also satisfies the following strengthened form of the second order necessary optimality condition

$$\nabla^2 f(x^*) : \text{positive definite}, \tag{1.2}$$

(i.e., the Hessian is positive definite rather than semidefinite). Then, for all $\Delta x \neq 0$ we have

$$\Delta x' \nabla^2 f(x^*) \Delta x > 0,$$

implying that at x^* the second order variation of f due to a small nonzero variation Δx is positive. Thus, f tends to increase strictly with small excursions from x^*, suggesting that the above conditions (1.1) and (1.2) are sufficient for local optimality of x^*. This is indeed established in Prop. 1.1.5.

Local minima that don't satisfy the positive definiteness sufficient condition (1.2) are called *singular*; otherwise they are called *nonsingular*. Singular local minima are harder to deal with for two reasons. First, in the absence of convexity of f, their optimality cannot be ascertained using easily verifiable sufficiency conditions. Second, in their neighborhood, the behavior of the most commonly used optimization algorithms tends to be slow and/or erratic, as we will see in the subsequent sections.

Quadratic Cost Functions

Consider the quadratic function

$$f(x) = \tfrac{1}{2} x' Q x - b' x,$$

where Q is a symmetric $n \times n$ matrix and b is a vector in $\Re^n$. If x^* is a local minimum of f, we must have, by the necessary optimality conditions,

$$\nabla f(x^*) = Q x^* - b = 0, \qquad \nabla^2 f(x^*) = Q : \text{positive semidefinite}.$$

Thus, if Q is not positive semidefinite, f can have no local minima. If Q is positive semidefinite, f is convex [Prop. B.4(d) of Appendix B], so any vector x^* satisfying the first order condition $\nabla f(x^*) = Q x^* - b = 0$ is a global minimum of f. On the other hand there may not exist a solution of the equation $\nabla f(x^*) = Q x^* - b = 0$ if Q is singular. If, however, Q is positive definite (and hence invertible, by Prop. A.20 of Appendix A), the equation $Q x^* - b = 0$ can be solved uniquely and the vector $x^* = Q^{-1} b$ is the unique global minimum. This is consistent with Prop. 1.1.3(a) to be given shortly, which asserts that strictly convex functions can have at most one global minimum [f is strictly convex if and only if Q is positive definite; Prop. B.4(d) of Appendix B]. Figure 1.1.5 illustrates the various special cases considered.

Quadratic cost functions are important in nonlinear programming because they arise frequently in applications, but they are also important for another reason. From the second order expansion around a local minimum x^*,

$$f(x) = f(x^*) + \tfrac{1}{2}(x - x^*)' \nabla^2 f(x^*)(x - x^*) + o(\|x - x^*\|^2),$$

Sec. 1.1 Optimality Conditions

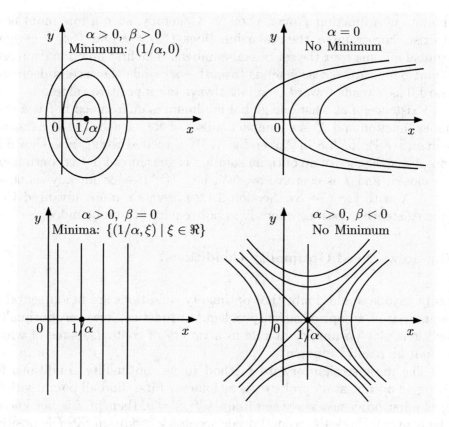

Figure 1.1.5. Illustration of the level sets $\{x \mid f(x) \leq c\}$ of the quadratic cost function $f : \Re^2 \mapsto \Re$ given by

$$f(x,y) = \tfrac{1}{2}\left(\alpha x^2 + \beta y^2\right) - x$$

for various values of α and β.

it is seen that a nonquadratic cost function can be approximated well by a quadratic function if x^* nonsingular [$\nabla^2 f(x^*)$: positive definite]. This means that we can carry out much of our analysis and experimentation with algorithms using positive definite quadratic functions and expect that the conclusions will largely carry over to more general cost functions near convergence to such local minima. However, for local minima near which the Hessian matrix either does not exist or is singular, the higher than second order terms in the series expansion are not negligible and an algorithmic analysis based on quadratic cost functions will likely be seriously flawed.

Existence of Optimal Solutions

In many cases it is useful to know that there exists at least one global minimum of a function f over a set X. Generally, such a minimum need not exist. For example, the scalar functions $f(x) = x$ and $f(x) = e^x$ have no global minima over the set of real numbers. The first function decreases without bound to $-\infty$ as x tends toward $-\infty$, while the second decreases toward 0 as x tends toward $-\infty$ but always takes positive values.

Existence of at least one global minimum is guaranteed if f is a continuous function and X is a compact subset of $\Re^n$. This is the *Weierstrass theorem*; see Prop. A.8 in Appendix A. By a related result, also shown in Prop. A.8, existence of an optimal solution is guaranteed if f is continuous, X is closed, and f is coercive over X, i.e., $f(x^k) \to \infty$ for any sequence $\{x^k\} \subset X$ with $\|x^k\| \to \infty$. Section 3.1.2 presents a more advanced view of the existence question, where X is not required to be bounded.

Why do we Need Optimality Conditions?

Hardly anyone would doubt that optimality conditions are fundamental to the analysis of an optimization problem. In practice, however, optimality conditions play an important role in a variety of contexts, some of which may not be readily apparent.

The most straightforward method to use optimality conditions for solving an optimization problem, is as follows: First, find all points satisfying the first order necessary condition $\nabla f(x) = 0$; then (if f is not known to be convex), check the second order necessary condition ($\nabla^2 f$ is positive semidefinite) for each of these points, filtering out those that do not satisfy it; finally for the remaining candidates, check if $\nabla^2 f$ is positive definite, in which case we are sure that they are strict local minima.

A slightly different alternative is to find all points satisfying the necessary conditions, and to declare as global minimum the one with smallest cost value. However, here it is essential to know that a global minimum exists. As an example, for the one-dimensional function

$$f(x) = x^2 - x^4,$$

the points satisfying the necessary condition

$$\nabla f(x) = 2x - 4x^3 = 0$$

are 0, $1/\sqrt{2}$, and $-1/\sqrt{2}$, and of these, 0 gives the smallest cost value. Nonetheless, we cannot declare 0 as the global minimum, because we don't know if a global minimum exists. Indeed, in this example none of the points 0, $1/\sqrt{2}$, and $-1/\sqrt{2}$ is a global minimum, because f decreases to $-\infty$ as $|x| \to \infty$, and has no global minimum. Here is an example where the approach can be applied.

Example 1.1.1 (Arithmetic-Geometric Mean Inequality)

We want to show the following classical inequality [due to Cauchy (1821)]:

$$(x_1 x_2 \cdots x_n)^{1/n} \leq \frac{\sum_{i=1}^n x_i}{n}$$

for any set of positive numbers x_i, $i = 1, \ldots, n$. By making the change of variables

$$y_i = \ln(x_i), \qquad i = 1, \ldots, n,$$

we have $x_i = e^{y_i}$, so that this inequality is equivalently written as

$$e^{\frac{y_1 + \cdots + y_n}{n}} \leq \frac{e^{y_1} + \cdots + e^{y_n}}{n},$$

which must be shown for all scalars $y_1, \ldots, y_n$. Note that with this transformation, the nonnegativity requirements on the variables have been eliminated.

One approach to proving the above inequality is to minimize the function

$$\frac{e^{y_1} + \cdots + e^{y_n}}{n} - e^{\frac{y_1 + \cdots + y_n}{n}},$$

and to show that its minimal value is 0. An alternative, which works better if we use optimality conditions, is to minimize instead

$$e^{y_1} + \cdots + e^{y_n},$$

over all $y = (y_1, \ldots, y_n)$ such that

$$y_1 + \cdots + y_n = s$$

for an arbitrary scalar s, and to show that the optimal value is no less than $ne^{s/n}$.

To this end, we use an elimination technique, a common device to convert constrained optimization problems to unconstrained ones. In particular, we use the constraint $y_1 + \cdots + y_n = s$ to eliminate the variable y_n, thereby obtaining the equivalent *unconstrained* problem of minimizing

$$g(y_1, \ldots, y_{n-1}) = e^{y_1} + \cdots + e^{y_{n-1}} + e^{s - y_1 - \cdots - y_{n-1}},$$

over $y_1, \ldots, y_{n-1}$. The necessary conditions $\partial g / \partial y_i = 0$ yield the system of equations

$$e^{y_i} = e^{s - y_1 - \cdots - y_{n-1}}, \qquad i = 1, \ldots, n-1,$$

or

$$y_i = s - y_1 - \cdots - y_{n-1}, \qquad i = 1, \ldots, n-1.$$

This system has only one solution: $y_i^* = s/n$ for all i. The solution must be the unique global minimum if we can show that that there exists a global minimum. Indeed, it can be seen that the function $g(y_1, \ldots, y_{n-1})$ is coercive, so it has an unconstrained global minimum (Prop. A.8, in Appendix A). Therefore, $(s/n, \ldots, s/n)$ is this minimum. Thus the optimal value of

$$e^{y_1} + \cdots + e^{y_n}$$

is $ne^{s/n}$, which as argued earlier, is sufficient to show the arithmetic-geometric mean inequality.

It is important to realize, however, that except under very favorable circumstances, using optimality conditions to obtain a solution as described above does *not* work. The reason is that solving for x the system of equations $\nabla f(x) = 0$ is usually nontrivial; algorithmically, it is typically as difficult as solving the original optimization problem.

The principal context in which optimality conditions become useful will not become apparent until we consider iterative optimization algorithms in subsequent sections. We will see that optimality conditions often provide the basis for the development and the analysis of algorithms. In particular, algorithms recognize solutions by checking whether they satisfy various optimality conditions and terminate when such conditions hold approximately. Furthermore, the behavior of various algorithms in the neighborhood of a local minimum often depends on whether various optimality conditions are satisfied at that minimum. Thus, for example, sufficiency conditions play a key role in assertions regarding the speed of convergence of various algorithms.

There is one other important context, prominently arising in microeconomic theory, where optimality conditions provide the basis for analysis. Here one is interested primarily not in finding an optimal solution, but rather in how the optimal solution is affected by changes in the problem data. For example, an economist may be interested in how the prices of some raw materials will affect the availability of certain goods that are produced by using these raw materials; the assumption here is that the amounts produced are the variables of a profit optimization problem, which is solved by the corresponding producers. This type of reasoning is known as *sensitivity analysis*, and is discussed next.

Sensitivity

Suppose that we want to quantify the variation of the optimal solution as a vector of parameters changes. In particular, consider the optimization problem

$$\text{minimize } f(x, a)$$
$$\text{subject to } x \in \Re^n,$$

where $f : \Re^{m+n} \mapsto \Re$ is a twice continuously differentiable function involving the m-dimensional parameter vector a. Let $x(a)$ denote the global minimum corresponding to a, assuming for the moment that it exists, is

Sec. 1.1 Optimality Conditions

unique, and it is differentiable as a function of a. By the first order necessary condition we have

$$\nabla_x f(x(a), a) = 0, \qquad \forall\, a \in \Re^m,$$

and by differentiating this relation with respect to a, we obtain

$$\nabla x(a) \nabla^2_{xx} f(x(a), a) + \nabla^2_{xa} f(x(a), a) = 0,$$

where the elements of the $m \times n$ gradient matrix $\nabla x(a)$ are the first partial derivatives of the components of $x(a)$ with respect to the different components of a. Assuming that the inverse below exists, we have

$$\nabla x(a) = -\nabla^2_{xa} f(x(a), a) \left(\nabla^2_{xx} f(x(a), a) \right)^{-1}, \qquad (1.3)$$

which gives the first order variation of the components of the optimal x with respect to the components of a.

For the preceding analysis to be precise, we must be sure that $x(a)$ exists and is differentiable with respect to a. The principal analytical framework for this is the implicit function theorem (Prop. A.25 in Appendix A). With the aid of this theorem, we can define $x(a)$ in some sphere around a minimum $\bar{x} = x(\bar{a})$ corresponding to a nominal parameter value $\bar{a}$, assuming that the Hessian matrix $\nabla^2_{xx} f(\bar{x}, \bar{a})$ is positive definite. Thus, the preceding development and the formula (1.3) for the matrix $\nabla x(a)$ can be justified provided the nominal local minimum $\bar{x}$ is nonsingular.

We postpone further discussion of sensitivity analysis for Sections 4.2.3, 4.3.2, and 4.3.6, where we will show constrained versions of the expression (1.3) for $\nabla x(a)$.

1.1.2 Main Optimality Conditions

We now provide formal statements and proofs of the optimality conditions discussed in the preceding section.

Proposition 1.1.1: (Necessary Optimality Conditions) Let x^* be an unconstrained local minimum of $f : \Re^n \mapsto \Re$, and assume that f is continuously differentiable in an open set S containing x^*. Then

$$\nabla f(x^*) = 0. \qquad \text{(First Order Necessary Condition)}$$

If in addition f is twice continuously differentiable within S, then

$$\nabla^2 f(x^*) : \text{positive semidefinite}. \qquad \text{(Second Order Necessary Condition)}$$

Proof: Fix some $d \in \Re^n$. Then, using the chain rule to differentiate the function $g(\alpha) = f(x^* + \alpha d)$ of the scalar α, we have

$$0 \leq \lim_{\alpha \downarrow 0} \frac{f(x^* + \alpha d) - f(x^*)}{\alpha} = \frac{dg(0)}{d\alpha} = d' \nabla f(x^*),$$

where the inequality follows from the assumption that x^* is a local minimum. Since d is arbitrary, the same inequality holds with d replaced by $-d$. Therefore, $d' \nabla f(x^*) = 0$ for all $d \in \Re^n$, which shows that $\nabla f(x^*) = 0$.

Assume that f is twice continuously differentiable, and let d be any vector in $\Re^n$. For all $\alpha \in \Re$, the second order expansion yields

$$f(x^* + \alpha d) - f(x^*) = \alpha \nabla f(x^*)' d + \frac{\alpha^2}{2} d' \nabla^2 f(x^*) d + o(\alpha^2).$$

Using the condition $\nabla f(x^*) = 0$ and the local optimality of x^*, we see that there is a sufficiently small $\epsilon > 0$ such that for all α with $\alpha \in (0, \epsilon)$,

$$0 \leq \frac{f(x^* + \alpha d) - f(x^*)}{\alpha^2} = \tfrac{1}{2} d' \nabla^2 f(x^*) d + \frac{o(\alpha^2)}{\alpha^2}.$$

Taking the limit as $\alpha \to 0$ and using the fact $\lim_{\alpha \to 0} o(\alpha^2)/\alpha^2 = 0$, we obtain $d' \nabla^2 f(x^*) d \geq 0$, showing that $\nabla^2 f(x^*)$ is positive semidefinite. **Q.E.D.**

The Convex Case

We will now consider the case where both the cost function f and the constraint set X are convex. The following proposition shows that a local minimum of f over X is also a global minimum over X. The proposition also deals with the cases where f is strictly convex and where it is strongly convex (see Appendix B for the definition and properties of strictly and strongly convex functions).

> **Proposition 1.1.2:** If X is a convex subset of $\Re^n$ and $f : \Re^n \mapsto \Re$ is convex over X, then a local minimum of f over X is also a global minimum. If in addition f is strictly convex over X, then f has at most one global minimum over X. Moreover, if f is strongly convex and X is closed, then f has a unique global minimum over X.

Proof: Assume, to arrive at a contradiction, that x is a local minimum of f but not a global minimum. Then there exists some $y \neq x$ such that $f(y) < f(x)$. Using the convexity of f, we have

$$f(\alpha x + (1 - \alpha) y) \leq \alpha f(x) + (1 - \alpha) f(y) < f(x), \qquad \forall \, \alpha \in [0, 1).$$

Sec. 1.1 Optimality Conditions

This contradicts the assumption that x is a local minimum.

Assume, to arrive at a contradiction, that f is strictly convex, and two distinct global minima x and y exist. Then their average $(x+y)/2$ must belong to X, since X is convex, and the value of f at the average must be smaller than $\bigl(f(x)+f(y)\bigr)/2$, by the strict convexity of f. Since x and y are global minima, we obtain a contradiction.

A strongly convex function is coercive, so it has at least one minimum over the closed set X by Prop. A.8 in Appendix A. It is also strictly convex by Prop. B.5(a) of Appendix B, so the minimum is unique. **Q.E.D.**

The following proposition provides a simple necessary and sufficient condition for optimality; see Fig. 1.1.6.

Proposition 1.1.3: (Convex Case - Necessary and Sufficient Conditions) Let X be a convex set and let $f : \Re^n \mapsto \Re$ be a convex function over X.

(a) If f is continuously differentiable, then
$$\nabla f(x^*)'(x - x^*) \geq 0, \qquad \forall\, x \in X,$$
is a necessary and sufficient condition for a vector $x^* \in X$ to be a global minimum of f over X.

(b) If X is open and f is continuously differentiable over X, then $\nabla f(x^*) = 0$ is a necessary and sufficient condition for a vector $x^* \in X$ to be a global minimum of f over X.

Proof: (a) Using the convexity of f and Prop. B.3(a) of Appendix B, we have
$$f(x) \geq f(x^*) + \nabla f(x^*)'(x - x^*), \qquad \forall\, x \in X.$$
If the condition $\nabla f(x^*)'(x-x^*) \geq 0$ holds for all $x \in X$, then $f(x) \geq f(x^*)$ for all $x \in X$, so x^* minimizes f over X.

Conversely, assume to arrive at a contradiction that x^* minimizes f over X and that $\nabla f(x^*)'(x - x^*) < 0$ for some $x \in X$. Then, we have
$$\lim_{\alpha \downarrow 0} \frac{f\bigl(x^* + \alpha(x - x^*)\bigr) - f(x^*)}{\alpha} = \nabla f(x^*)'(x - x^*) < 0,$$
so $f\bigl(x^* + \alpha(x - x^*)\bigr)$ decreases strictly for sufficiently small $\alpha > 0$, contradicting the optimality of x^*.

(b) If $\nabla f(x^*) = 0$, the optimality of x^* follows as a special case of part (a). Conversely, assume to arrive at a contradiction that x^* minimizes f over X and that $\nabla f(x^*) \neq 0$. Then, since X is open and $x^* \in X$, there must

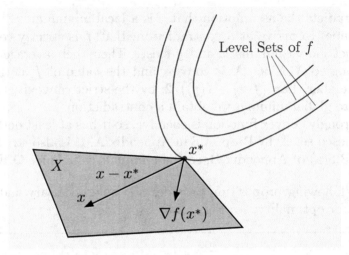

Figure 1.1.6. Illustration of the necessary and sufficient condition of Prop. 1.1.3(a) for x^* to minimize f over X. The gradient $\nabla f(x^*)$ makes an angle less or equal to $\pi/2$ with any vector of the form $x - x^*$, $x \in X$.

exist an open ball centered at x^* that is contained in X. Thus for some $x \in X$ we have
$$\nabla f(x^*)'(x - x^*) < 0.$$
The proof now proceeds as in part (a). **Q.E.D.**

The following is an illustration of the use of the preceding optimality conditions.

Example 1.1.2 (Arithmetic-Geometric Mean Inequality Revisited)

As an application of the preceding proposition, let us provide an alternative optimization-based proof of the arithmetic-geometric mean inequality, given in Example 1.1.1. We argued there that showing the inequality is equivalent to showing that for an arbitrary scalar s, the minimum value of the convex function
$$g(y) = e^{y_1} + \cdots + e^{y_n},$$
over all $y = (y_1, \ldots, y_n)$ such that
$$y_1 + \cdots + y_n = s$$
is no less than $ne^{s/n}$. We verified this by showing that the symmetric solution
$$y_1^* = \cdots = y_n^* = s/n$$
is optimal, via conversion of the problem to an unconstrained problem through elimination of one of the variables. Alternatively, we can prove the optimality

Sec. 1.1   Optimality Conditions

of this solution by checking the sufficiency condition of Prop. 1.1.3(a) (the cost function and constraint are easily verified to be convex). This condition is written as

$$\nabla g(y^*)'(y - y^*) = (e^{s/n}, \ldots, e^{s/n})' \begin{pmatrix} y_1 - s/n \\ \vdots \\ y_n - s/n \end{pmatrix} \geq 0,$$

or

$$e^{s/n}(y_1 + \cdots + y_n - s) \geq 0$$

for all y with $y_1 + \cdots + y_n = s$, which is evidently true.

Let us use Prop. 1.1.2 and the optimality condition of Prop. 1.1.3(a) to prove a basic theorem of analysis and optimization, which is illustrated in Fig. 1.1.7 and will be used frequently in this book.

Proposition 1.1.4: (Projection Theorem) Let X be a closed convex set and let $\|\cdot\|$ be the Euclidean norm.

(a) For every $x \in \Re^n$, there exists a unique vector that minimizes $\|y - x\|$ over all $y \in X$. This vector is called the *projection of x on X*, and is denoted by $[x]^+$, i.e.,

$$[x]^+ = \arg\min_{y \in X} \|y - x\|.$$

(b) Given some $x \in \Re^n$, a vector $x^* \in X$ is equal to $[x]^+$ if and only if

$$(y - x^*)'(x - x^*) \leq 0, \qquad \forall\, y \in X. \tag{1.4}$$

(c) The mapping $f : \Re^n \mapsto X$ defined by $f(x) = [x]^+$ is continuous and nonexpansive, i.e.,

$$\|[x]^+ - [y]^+\| \leq \|x - y\|, \qquad \forall\, x, y \in \Re^n.$$

Proof: (a) Given x, minimizing $\|y - x\|$ over $y \in X$ is equivalent to minimizing the convex and differentiable function

$$f(y) = \tfrac{1}{2}\|y - x\|^2$$

over X. Since, f is strongly convex, existence and uniqueness of the minimizing vector follows from Prop. 1.1.2.

(b) By Prop. 1.1.3, x^* minimizes f over X if and only if

$$\nabla f(x^*)'(y - x^*) \geq 0, \qquad \forall\, y \in X.$$

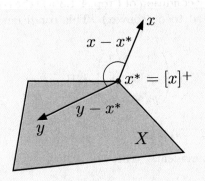

Figure 1.1.7. Illustration of the condition (1.4) of the projection theorem. For x^* to to be the projection of x on X, the vector $x - x^*$ should make an angle greater or equal to $\pi/2$ with any vector of the form $y - x^*$, $y \in X$.

Since $\nabla f(x^*) = x^* - x$, this condition is equivalent to Eq. (1.4).

(c) Let x and y be elements of $\Re^n$. From part (b), we have

$$\big(w - [x]^+\big)'\big(x - [x]^+\big) \leq 0, \qquad \forall\, w \in X.$$

Since $[y]^+ \in X$, we can use $w = [y]^+$ in the preceding relation and obtain

$$\big([y]^+ - [x]^+\big)'\big(x - [x]^+\big) \leq 0.$$

Exchanging the roles of x and y, we also obtain

$$\big([x]^+ - [y]^+\big)'\big(y - [y]^+\big) \leq 0.$$

Adding these two inequalities, we have

$$\big([y]^+ - [x]^+\big)'\big(x - [x]^+ - y + [y]^+\big) \leq 0.$$

By rearranging and by using the Schwarz inequality, we obtain

$$\big\|[y]^+ - [x]^+\big\|^2 \leq \big([y]^+ - [x]^+\big)'(y - x) \leq \big\|[y]^+ - [x]^+\big\| \cdot \|y - x\|,$$

showing that $[\cdot]^+$ is nonexpansive and *a fortiori* continuous. **Q.E.D.**

Sufficient Conditions without Convexity

In the absence of convexity, we have the following sufficiency conditions for local optimality.

Sec. 1.1 Optimality Conditions

> **Proposition 1.1.5: (Second Order Sufficient Optimality Conditions)** Let $f : \Re^n \mapsto \Re$ be twice continuously differentiable over an open set S. Suppose that a vector $x^* \in S$ satisfies the conditions
>
> $$\nabla f(x^*) = 0, \qquad \nabla^2 f(x^*) : \text{positive definite.}$$
>
> Then, x^* is a strict unconstrained local minimum of f. In particular, there exist scalars $\gamma > 0$ and $\epsilon > 0$ such that
>
> $$f(x) \geq f(x^*) + \frac{\gamma}{2}\|x - x^*\|^2, \qquad \forall\, x \text{ with } \|x - x^*\| < \epsilon. \qquad (1.5)$$

Proof: Denote by λ the smallest eigenvalue of $\nabla^2 f(x^*)$. By Prop. A.20(b) of Appendix A, λ is positive since $\nabla^2 f(x^*)$ is positive definite. Furthermore, by Prop. A.18(b) of Appendix A,

$$d'\nabla^2 f(x^*)d \geq \lambda \|d\|^2, \qquad \forall\, d \in \Re^n.$$

Using this relation, the hypothesis $\nabla f(x^*) = 0$, and a second order expansion, we have for all d

$$\begin{aligned}
f(x^* + d) - f(x^*) &= \nabla f(x^*)'d + \tfrac{1}{2}d'\nabla^2 f(x^*)d + o(\|d\|^2) \\
&\geq \frac{\lambda}{2}\|d\|^2 + o(\|d\|^2) \\
&= \left(\frac{\lambda}{2} + \frac{o(\|d\|^2)}{\|d\|^2}\right)\|d\|^2.
\end{aligned}$$

Choose any $\epsilon > 0$ and $\gamma > 0$ such that

$$\frac{\lambda}{2} + \frac{o(\|d\|^2)}{\|d\|^2} \geq \frac{\gamma}{2}, \qquad \forall\, d \text{ with } \|d\| < \epsilon.$$

Then Eq. (1.5) is satisfied. **Q.E.D.**

EXERCISES

1.1.1

For each value of the scalar β, find the set of all stationary points of the following function of the two variables x and y

$$f(x, y) = x^2 + y^2 + \beta xy + x + 2y.$$

Which of these stationary points are global minima?

1.1.2

In each of the following problems fully justify your answer using optimality conditions.

(a) Show that the 2-dimensional function $f(x,y) = (x^2 - 4)^2 + y^2$ has two global minima and one stationary point, which is neither a local maximum nor a local minimum.

(b) Find all local minima of the 2-dimensional function $f(x,y) = \frac{1}{2}x^2 + x \cos y$.

(c) Find all local minima and all local maxima of the 2-dimensional function $f(x,y) = \sin x + \sin y + \sin(x+y)$ within the set
$$\{(x,y) \mid 0 < x < 2\pi,\ 0 < y < 2\pi\}.$$

(d) Show that the 2-dimensional function $f(x,y) = (y-x^2)^2 - x^2$ has only one stationary point, which is neither a local maximum nor a local minimum.

(e) Consider the minimization of the function f in part (d) subject to no constraint on x and the constraint $-1 \le y \le 1$ on y. Show that there exists at least one global minimum and find all global minima.

1.1.3 [Hes75]

Let $f : \Re^n \mapsto \Re$ be a differentiable function. Suppose that a point x^* is a local minimum of f along every line that passes through x^*; that is, the function
$$g(\alpha) = f(x^* + \alpha d)$$
is minimized at $\alpha = 0$ for all $d \in \Re^n$.

(a) Show that $\nabla f(x^*) = 0$.

(b) Show by example that x^* need not be a local minimum of f. *Hint*: Consider the function of two variables
$$f(y,z) = (z - py^2)(z - qy^2),$$
where $0 < p < q$; see Fig. 1.1.8. Show that $(0,0)$ is a local minimum of f along every line that passes through $(0,0)$. Moreover, if $p < m < q$, then
$$f(y, my^2) < 0 \qquad \forall\ y \ne 0,$$
while $f(0,0) = 0$, so $(0,0)$ is not a local minimum of f.

1.1.4

Use optimality conditions to show that for all $x > 0$ we have
$$\frac{1}{x} + x \ge 2.$$

Sec. 1.1   Optimality Conditions   23

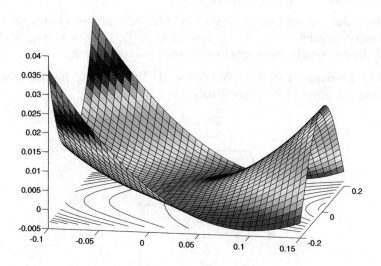

Figure 1.1.8. Three-dimensional graph of the function $f(y,z) = (z-py^2)(z-qy^2)$ for $p = 1$ and $q = 4$ (cf. Exercise 1.1.3). The origin is a local minimum with respect to every line that passes through it, but is not a local minimum of f.

1.1.5

Find the rectangular parallelepiped of unit volume that has the minimum surface area. *Hint*: By eliminating one of the dimensions, show that the problem is equivalent to the minimization over $x > 0$ and $y > 0$ of

$$f(x,y) = xy + \frac{1}{x} + \frac{1}{y}.$$

Show that the sets $\{(x,y) \mid f(x,y) \leq \gamma,\ x > 0,\ y > 0\}$ are compact for all scalars γ.

1.1.6 (The Weber Point of a Set of Points)

We want to find a point x in the plane whose sum of weighted distances from a given set of points $y_1, \ldots, y_m$ is minimized. Mathematically, the problem is

$$\text{minimize} \quad \sum_{i=1}^{m} w_i \|x - y_i\|$$
$$\text{subject to} \quad x \in \Re^n,$$

where $w_1, \ldots, w_m$ are given positive scalars.

(a) Show that there exists a global minimum for this problem and that it can be realized by means of the mechanical model shown in Fig. 1.1.9.

(b) Is the optimal solution always unique?

(c) Show that an optimal solution minimizes the potential energy of the mechanical model of Fig. 1.1.9, defined as $\sum_{i=1}^{m} w_i h_i$, where h_i is the height of the ith weight, measured from some reference level.

Note: This problem stems from Weber's work [Web29], which is generally viewed as the starting point of location theory.

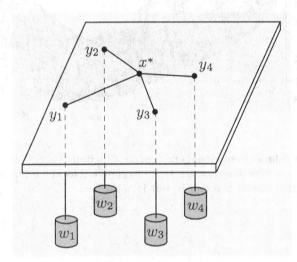

Figure 1.1.9. Mechanical model (known as the Varignon frame) associated with the Weber problem (Exercise 1.1.6). It consists of a board with a hole drilled at each of the given points y_i. Through each hole, a string is passed with the corresponding weight w_i attached. The other ends of the strings are tied with a knot as shown. In the absence of friction or tangled strings, the forces at the knot reach equilibrium when the knot is located at an optimal solution x^*.

1.1.7 (Fermat-Torricelli-Viviani Problem)

Given a triangle in the plane, consider the problem of finding a point whose sum of distances from the vertices of the triangle is minimal. Show that such a point is either a vertex, or else it is such that each side of the triangle is seen from that point at a 120 degree angle (this is known as the Torricelli point). *Note*: This problem, whose detailed history is traced in [BMS99], was suggested by Fermat to Torricelli who solved it. Viviani also solved the problem a little later and proved the following generalization: Suppose that x_i, $i = 1, \ldots, m$, are points in the plane, and x is a point in their convex hull such that $x \neq x_i$ for all i, and the angles $x_i x x_{i+1}$, $i < m$, and $x_m x x_1$ are all equal to $2\pi/m$. Then x minimizes $\sum_{i=1}^{m} \|z - x_i\|$ over all z in the plane (show this as an exercise by using sufficient optimality conditions; compare with the preceding exercise). Fermat is credited with being the first to study systematically optimization problems in geometry.

1.1.8 (Diffraction Law in Optics)

Let p and q be two points on the plane that lie on opposite sides of a horizontal axis. Assume that the speed of light from p and from q to the horizontal axis is v and w, respectively, and that light reaches a point from other points along paths of minimum travel time. Find the path that a ray of light would follow from p to q.

1.1.9 (www)

Let $f : \Re^n \mapsto \Re$ be a twice continuously differentiable function that satisfies

$$m\|y\|^2 \leq y'\nabla^2 f(x)y \leq M\|y\|^2, \qquad \forall\, x, y \in \Re^n,$$

where m and M are some positive scalars. Show that f has a unique global minimum x^*, which satisfies

$$\frac{1}{2M}\|\nabla f(x)\|^2 \leq f(x) - f(x^*) \leq \frac{1}{2m}\|\nabla f(x)\|^2, \qquad \forall\, x \in \Re^n,$$

and

$$\frac{m}{2}\|x - x^*\|^2 \leq f(x) - f(x^*) \leq \frac{M}{2}\|x - x^*\|^2, \qquad \forall\, x \in \Re^n.$$

Hint: Use a second order expansion and the relation

$$\min_{y \in \Re^n} \left\{ \nabla f(x)'(y - x) + \frac{\alpha}{2}\|y - x\|^2 \right\} = -\frac{1}{2\alpha}\|\nabla f(x)\|^2, \qquad \forall\, \alpha > 0.$$

1.1.10 (Nonconvex Level Sets [Dun87])

Let $f : \Re^2 \mapsto \Re$ be the function

$$f(x) = x_2^2 - ax_2\|x\|^2 + \|x\|^4,$$

where $0 < a < 2$ (see Fig. 1.1.10). Show that $f(x) > 0$ for all $x \neq 0$, so that the origin is the unique global minimum. Show also that there exists a $\bar{\gamma} > 0$ such that for all $\gamma \in (0, \bar{\gamma}]$, the level set $L_\gamma = \{x \mid f(x) \leq \gamma\}$ is not convex. *Hint*: Show that for $\gamma \in (0, \bar{\gamma}]$, there is a $p > 0$ and a $q > 0$ such that the vectors $(-p, q)$ and (p, q) belong to L_γ, but $(0, q)$ does not belong to L_γ.

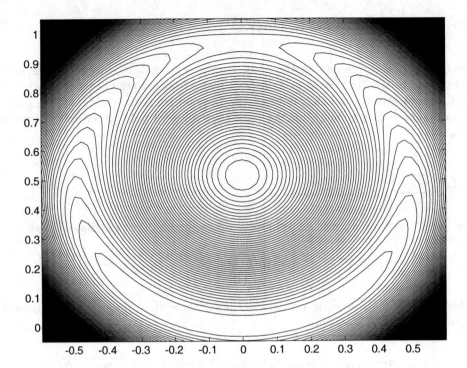

Figure 1.1.10. Level sets of the function f of Exercise 1.1.10 for the case where $a = 1.98$. The unique global minimum is the origin, but the level sets of f are nonconvex.

1.1.11 (Singular Strict Local Minima [Dun87]) (www)

Show that if x^* is a nonsingular strict local minimum of a twice continuously differentiable function $f : \Re^n \mapsto \Re$, then x^* is an isolated stationary point; that is, there is a sphere centered at x^* such that x^* is the only stationary point of f within that sphere. Use the following example function $f : \Re \mapsto \Re$ to show that this need not be true if x^* is a singular strict local minimum:

$$f(x) = \begin{cases} x^2 \left(\sqrt{2} - \sin\left(\frac{5\pi}{6} - \sqrt{3} \ln\left(x^2 \right) \right) \right) & \text{if } x \neq 0, \\ 0 & \text{if } x = 0. \end{cases}$$

In particular, show that $x^* = 0$ is the unique (singular) global minimum, while the sequence $\{x^k\}$ of nonsingular local minima, where

$$x^k = e^{\frac{(1-8k)\pi}{8\sqrt{3}}},$$

converges to x^* (cf. Fig. 1.1.11). Verify also that there is a sequence of nonsingular local maxima that converges to x^*.

Sec. 1.1   Optimality Conditions   27

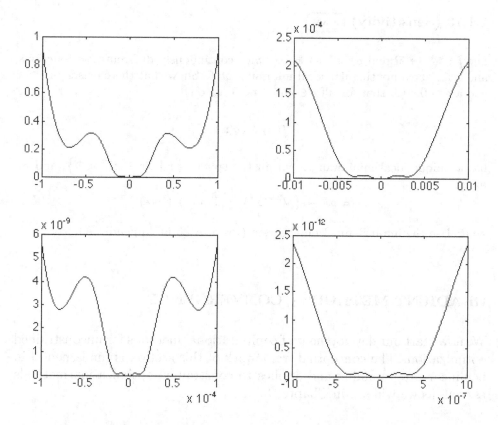

Figure 1.1.11. Illustration of the function f of Exercise 1.1.11 in progressively finer scale. The point $x^* = 0$ is the unique (singular) global minimum, but there are sequences of nonsingular local minima and local maxima that converge to x^*.

1.1.12 (Stability) (www)

We are often interested in whether optimal solutions change radically when the problem data are slightly perturbed. This issue is addressed by *stability analysis*, to be contrasted with sensitivity analysis, which deals with *how much* optimal solutions change when problem data change. An unconstrained local minimum x^* of a function f is said to be *locally stable* if there exists a $\delta > 0$ such that all sequences $\{x^k\}$ with

$$f(x^k) \to f(x^*), \qquad \|x^k - x^*\| < \delta, \quad \forall \, k \geq 0,$$

converge to x^*. Suppose that f is a continuous function and let x^* be a local minimum of f.

(a) Show that x^* is locally stable if and only if x^* is a strict local minimum.

(b) Let g be a continuous function. Show that if x^* is locally stable, there exists a $\delta > 0$ such that for all sufficiently small $\epsilon > 0$, the function $f(x) + \epsilon g(x)$ has an unconstrained local minimum x_ϵ that lies within the sphere centered at x^* with radius δ. Furthermore, $x_\epsilon \to x^*$ as $\epsilon \to 0$.

1.1.13 (Sensitivity) (www)

Let $f : \Re^n \mapsto \Re$ and $g : \Re^n \mapsto \Re$ be twice continuously differentiable functions, and let x^* be a nonsingular local minimum of f. Show that there exists an $\bar{\epsilon} > 0$ and a $\delta > 0$ such that for all $\epsilon \in [0, \bar{\epsilon})$ the function

$$f(x) + \epsilon g(x)$$

has a unique local minimum x_ϵ within the sphere $\{x \mid \|x - x^*\| < \delta\}$, and we have

$$x_\epsilon = x^* - \epsilon \big(\nabla^2 f(x^*)\big)^{-1} \nabla g(x^*) + o(\epsilon).$$

Hint: Use the implicit function theorem (Prop. A.25 in Appendix A).

1.2 GRADIENT METHODS – CONVERGENCE

We now start our development of computational methods for unconstrained optimization. The conceptual framework of this section is fundamental in nonlinear programming and applies to constrained optimization methods as well, as we will see in Chapter 3.

1.2.1 Descent Directions and Stepsize Rules

As in the case of optimality conditions, the main ideas of unconstrained optimization methods have simple geometrical explanations, but the corresponding convergence analysis is often complex. Thus, for pedagogical reasons, we first discuss informally the methods and their behavior in the present subsection, and we substantiate our conclusions with rigorous analysis in Section 1.2.2.

Consider the problem of unconstrained minimization of a continuously differentiable function $f : \Re^n \mapsto \Re$. Most of the interesting algorithms for this problem rely on an important idea, called *iterative descent* that works as follows: We start at some point x^0 (an initial guess) and successively generate vectors $x^1, x^2, \ldots$, such that f is decreased at each iteration, i.e.,

$$f(x^{k+1}) < f(x^k), \qquad k = 0, 1, \ldots,$$

(cf. Fig. 1.2.1). In doing so, we successively improve our current solution estimate and we hope to decrease f all the way to its minimum. In this section, we introduce a general class of algorithms based on iterative descent, and we analyze their convergence to local minima. In Section 1.3 we examine their rate of convergence properties.

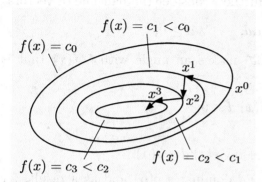

Figure 1.2.1. Iterative descent for minimizing a function f. Each vector in the generated sequence has a lower cost than its predecessor.

Gradient Methods

Given a vector $x \in \Re^n$ with $\nabla f(x) \neq 0$, consider the half line of vectors

$$x_\alpha = x - \alpha \nabla f(x), \qquad \forall\, \alpha \geq 0.$$

From the first order expansion around x we have

$$f(x_\alpha) = f(x) + \nabla f(x)'(x_\alpha - x) + o(\|x_\alpha - x\|)$$
$$= f(x) - \alpha \|\nabla f(x)\|^2 + o(\alpha \|\nabla f(x)\|),$$

so we can write

$$f(x_\alpha) = f(x) - \alpha \|\nabla f(x)\|^2 + o(\alpha).$$

The term

$$\alpha \|\nabla f(x)\|^2$$

dominates $o(\alpha)$ for α near zero, so for positive but sufficiently small α, $f(x_\alpha)$ is smaller than $f(x)$ as illustrated in Fig. 1.2.2.

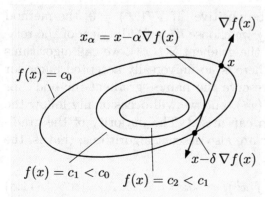

Figure 1.2.2. If $\nabla f(x) \neq 0$, there is an interval $(0, \delta)$ of stepsizes such that

$$f(x - \alpha \nabla f(x)) < f(x)$$

for all $\alpha \in (0, \delta)$.

Carrying this idea one step further, consider the half line of vectors

$$x_\alpha = x + \alpha d, \qquad \forall\ \alpha \geq 0,$$

where the direction vector $d \in \Re^n$ makes an angle with $\nabla f(x)$ that is greater than 90 degrees, i.e.,

$$\nabla f(x)'d < 0.$$

Again we have

$$f(x_\alpha) = f(x) + \alpha \nabla f(x)'d + o(\alpha).$$

For α near zero, the term $\alpha \nabla f(x)'d$ dominates $o(\alpha)$ and as a result, for positive but sufficiently small α, $f(x+\alpha d)$ is smaller than $f(x)$ as illustrated in Fig. 1.2.3.

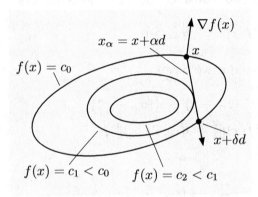

Figure 1.2.3. If the direction d makes an angle with $\nabla f(x)$ that is greater than 90 degrees, i.e., $\nabla f(x)'d < 0$, there is an interval $(0,\delta)$ of stepsizes such that

$$f(x+\alpha d) < f(x)$$

for all $\alpha \in (0,\delta)$.

The preceding observations form the basis for the broad and important class of algorithms

$$x^{k+1} = x^k + \alpha^k d^k, \qquad k = 0, 1, \ldots, \tag{1.6}$$

where, if $\nabla f(x^k) \neq 0$, the direction d^k is chosen so that

$$\nabla f(x^k)' d^k < 0, \tag{1.7}$$

and the stepsize α^k is chosen to be positive. If $\nabla f(x^k) = 0$, the method stops, i.e., $x^{k+1} = x^k$ (equivalently we choose $d^k = 0$). In view of the relation (1.7) of the direction d^k and the gradient $\nabla f(x^k)$, we call algorithms of this type *gradient methods*. [There is no universally accepted name for these algorithms; some authors reserve the name "gradient method" for the special case where $d^k = -\nabla f(x^k)$, and we will occasionally follow the same practice, when no confusion can arise.] The majority of the gradient methods that we will consider are also descent algorithms; that is, the stepsize α^k is selected so that

$$f(x^k + \alpha^k d^k) < f(x^k), \qquad k = 0, 1, \ldots. \tag{1.8}$$

Sec. 1.2   Gradient Methods – Convergence

However, there are some exceptions, which will be described shortly.

There is a large variety of possibilities for choosing the direction d^k and the stepsize α^k in a gradient method. Indeed there is no single gradient method that can be recommended for all or even most problems. Otherwise said, given any one of the numerous methods and variations thereof that we will discuss, there are interesting types of problems for which this method is well-suited. Our principal analytical aim is to develop a few guiding principles for understanding the performance of broad classes of methods and for appreciating the practical contexts in which their use is most appropriate.

Selecting the Descent Direction

Many gradient methods are specified in the form

$$x^{k+1} = x^k - \alpha^k D^k \nabla f(x^k), \tag{1.9}$$

where D^k is a positive definite symmetric matrix. Since $d^k = -D^k \nabla f(x^k)$, the descent condition $\nabla f(x^k)'d^k < 0$ is written as

$$\nabla f(x^k)' D^k \nabla f(x^k) > 0,$$

and holds thanks to the positive definiteness of D^k.

Here are some examples of choices of the matrix D^k, resulting in methods that are widely used:

Steepest Descent

$$D^k = I, \qquad k = 0, 1, \ldots,$$

where I is the $n \times n$ identity matrix. This is the simplest choice but it often leads to slow convergence, as we will see in Section 1.3. The difficulty is illustrated in Fig. 1.2.4 and motivates the methods of the subsequent examples. The name "steepest descent" is derived from an interesting property of the (normalized) negative gradient direction

$$d^k = -\frac{\nabla f(x^k)}{\|\nabla f(x^k)\|}.$$

Among all directions $d \in \Re^n$ that are normalized so that $\|d\| = 1$, it is the one that minimizes the slope $\nabla f(x^k)'d$ of the cost $f(x^k + \alpha d)$ along the direction d at $\alpha = 0$. Indeed, by the Schwarz inequality (Prop. A.2 in Appendix A), we have for all d with $\|d\| = 1$,

$$\nabla f(x^k)'d \geq -\|\nabla f(x^k)\| \cdot \|d\| = -\|\nabla f(x^k)\|,$$

and it is seen that equality is attained above for d equal to $-\nabla f(x^k)/\|\nabla f(x^k)\|$.

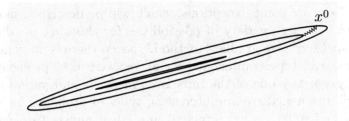

Figure 1.2.4. Slow convergence of the steepest descent method

$$x^{k+1} = x^k - \alpha^k \nabla f(x^k)$$

when the equal cost surfaces of f are very "elongated." The difficulty is that the gradient direction is almost orthogonal to the direction that leads to the minimum. As a result the method is zig-zagging without making fast progress.

Newton's Method

$$D^k = \left(\nabla^2 f(x^k)\right)^{-1}, \qquad k = 0, 1, \ldots,$$

provided $\nabla^2 f(x^k)$ is positive definite. If $\nabla^2 f(x^k)$ is not positive definite, some modification is necessary as will be explained in Section 1.4. The idea in Newton's method is to minimize at each iteration the quadratic approximation of f around the current point x^k given by

$$f^k(x) = f(x^k) + \nabla f(x^k)'(x - x^k) + \tfrac{1}{2}(x - x^k)' \nabla^2 f(x^k)(x - x^k),$$

(see Fig. 1.2.5). By setting the derivative of $f^k(x)$ to zero,

$$\nabla f(x^k) + \nabla^2 f(x^k)(x - x^k) = 0,$$

we obtain the next iterate x^{k+1} as the minimum of $f^k(x)$:

$$x^{k+1} = x^k - \left(\nabla^2 f(x^k)\right)^{-1} \nabla f(x^k).$$

This is the pure Newton iteration. It is the special case of the more general iteration

$$x^{k+1} = x^k - \alpha^k \left(\nabla^2 f(x^k)\right)^{-1} \nabla f(x^k),$$

where the stepsize $\alpha^k = 1$. Note that Newton's method finds the global minimum of a positive definite quadratic function in a single iteration (assuming $\alpha^k = 1$). For other cost functions, Newton's method typically converges very fast asymptotically and does not exhibit the zig-zagging behavior of steepest descent, as we will show in Section 1.4. For this reason many other methods try to emulate Newton's method. Some examples will be given shortly.

Sec. 1.2 Gradient Methods – Convergence

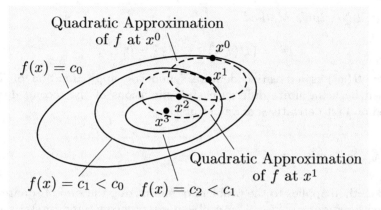

Figure 1.2.5. Illustration of the fast convergence rate of Newton's method with a stepsize $\alpha^k = 1$. Given x^k, the method obtains x^{k+1} as the minimum of a quadratic approximation of f based on a second order expansion around x^k.

Diagonally Scaled Steepest Descent

$$D^k = \begin{pmatrix} d_1^k & 0 & 0 & \cdots & 0 & 0 & 0 \\ 0 & d_2^k & 0 & \cdots & 0 & 0 & 0 \\ \vdots & \vdots & \vdots & \ddots & \vdots & \vdots & \\ 0 & 0 & 0 & \cdots & 0 & d_{n-1}^k & 0 \\ 0 & 0 & 0 & \cdots & 0 & 0 & d_n^k \end{pmatrix}, \qquad k = 0, 1, \ldots,$$

where d_i^k are positive scalars, thus ensuring that D^k is positive definite. A popular choice, resulting in a method known as a *diagonal approximation to Newton's method*, is to take d_i^k to be an approximation to the inverted second partial derivative of f with respect to x_i, i.e.,

$$d_i^k \approx \left(\frac{\partial^2 f(x^k)}{(\partial x_i)^2} \right)^{-1}$$

(making sure of course that $d_i^k > 0$).

Modified Newton's Method

$$D^k = \left(\nabla^2 f(x^0) \right)^{-1}, \qquad k = 0, 1, \ldots,$$

provided $\nabla^2 f(x^0)$ is positive definite. This method is the same as Newton's method except that to economize on overhead, the Hessian matrix is not recalculated at each iteration. A related method is obtained when the Hessian is recomputed every $p > 1$ iterations.

Discretized Newton's Method

$$D^k = \left(H(x^k)\right)^{-1}, \qquad k = 0, 1, \ldots,$$

where $H(x^k)$ is a positive definite symmetric approximation of $\nabla^2 f(x^k)$, formed by using finite difference approximations of the second derivatives, based on first derivatives or values of f.

Gauss-Newton Method

This method applies to the problem of minimizing the sum of squares of real-valued functions $g_1, \ldots, g_m$, a problem often encountered in statistical data analysis and in the context of neural network training (see Section 1.4.4). By denoting $g = (g_1, \ldots, g_m)$, the problem is written as

$$\text{minimize } f(x) = \tfrac{1}{2}\|g(x)\|^2 = \tfrac{1}{2}\sum_{i=1}^m \bigl(g_i(x)\bigr)^2$$
$$\text{subject to } x \in \Re^n.$$

We choose

$$D^k = \left(\nabla g(x^k)\nabla g(x^k)'\right)^{-1}, \qquad k = 0, 1, \ldots,$$

assuming the matrix $\nabla g(x^k)\nabla g(x^k)'$ is invertible. The latter matrix is always positive semidefinite, and it is positive definite and hence invertible if and only if the matrix $\nabla g(x^k)$ has rank n (Prop. A.20 in Appendix A). Since

$$\nabla f(x^k) = \nabla g(x^k) g(x^k),$$

the Gauss-Newton method takes the form

$$x^{k+1} = x^k - \alpha^k \left(\nabla g(x^k)\nabla g(x^k)'\right)^{-1} \nabla g(x^k) g(x^k). \tag{1.10}$$

We will see in Section 1.4.4 that the Gauss-Newton method may be viewed as an approximation to Newton's method, particularly when the optimal value of $\|g(x)\|^2$ is small.

Other choices of D^k yield the class of *Quasi-Newton methods* discussed in Section 2.2. There are also some interesting descent methods where the direction d^k is not usually expressed as $d^k = -D^k \nabla f(x^k)$. Important examples are the *conjugate gradient method* and the *coordinate descent methods*, which are discussed in Sections 2.1.1 and 2.3.1, respectively.

Stepsize Selection

There are a number of rules for choosing the stepsize α^k in a gradient method. We give some that are used widely in practice:

Minimization Rule

Here α^k is such that the cost function is minimized along the direction d^k, i.e., α^k satisfies

$$f(x^k + \alpha^k d^k) = \min_{\alpha \geq 0} f(x^k + \alpha d^k). \qquad (1.11)$$

Limited Minimization Rule

This is a version of the minimization rule, which is more easily implemented in many cases. A fixed scalar $s > 0$ is selected and α^k is chosen to yield the greatest cost reduction over all stepsizes in the interval $[0, s]$, i.e.,

$$f(x^k + \alpha^k d^k) = \min_{\alpha \in [0, s]} f(x^k + \alpha d^k).$$

The minimization and limited minimization rules must typically be implemented with the aid of one-dimensional line search algorithms (see Appendix C). In general, the minimizing stepsize cannot be computed exactly, and in practice, the line search is stopped once a stepsize α^k satisfying some termination criterion is obtained. Some stopping criteria are discussed in Exercise 1.2.15.

Generally, compared with other stepsize rules, the minimization rules tend to require more function and/or gradient evaluations per iteration. However, their use tends to reduce the number of required iterations for practical convergence, because of the greater cost reduction per iteration that they achieve. The minimization rules are also favored in cases where the structure of the problem can be exploited to economize on the associated computations. A prominent example arises when the cost function has the form

$$f(x) = h(Ax),$$

where A is a matrix such that the calculation of the vector $y = Ax$ for a given x is far more expensive than the calculation of $h(y)$ and its gradient and Hessian (assuming it exists). In this case, calculation of values, first, and second derivatives of the function $g(\alpha) \equiv f(x + \alpha d) = h(Ax + \alpha Ad)$ requires just two expensive operations: the one-time calculation of the matrix-vector products Ax and Ad.

Successive Stepsize Reduction – Armijo Rule

To avoid the often considerable computation associated with the line minimization rules, it is natural to consider rules based on successive stepsize reduction. In the simplest rule of this type an initial stepsize s is chosen, and if the corresponding vector $x^k + sd^k$ does not yield an improved value of f, i.e., $f(x^k + sd^k) \geq f(x^k)$, the stepsize is reduced, perhaps repeatedly, by a certain factor, until the value of f is improved. While this method often works in practice, it is theoretically unsound because the cost improvement obtained at each iteration may not be substantial enough to guarantee convergence to a minimum. This is illustrated in Fig. 1.2.6.

The Armijo rule is essentially the successive reduction rule just described, suitably modified to eliminate the theoretical convergence difficulty shown in Fig. 1.2.6. Here, fixed scalars s, β, and σ, with $0 < \beta < 1$, and $0 < \sigma < 1$ are chosen, and we set $\alpha^k = \beta^{m_k} s$, where m_k is the first nonnegative integer m for which

$$f(x^k) - f(x^k + \beta^m s d^k) \geq -\sigma \beta^m s \nabla f(x^k)' d^k. \quad (1.12)$$

In other words, the stepsizes $\beta^m s$, $m = 0, 1, \ldots$, are tried successively until the above inequality is satisfied for $m = m_k$. Thus, the cost improvement must not be just positive; it must be sufficiently large as per the test (1.12). Figure 1.2.7 illustrates the rule.

Usually σ is chosen close to zero, for example, $\sigma \in [10^{-5}, 10^{-1}]$. The reduction factor β is usually chosen from $1/2$ to $1/10$ depending on the confidence we have on the quality of the initial stepsize s. We can always take $s = 1$ and multiply the direction d^k by a scaling factor. Many methods, such as Newton-like methods, incorporate some type of implicit scaling of the direction d^k, which makes $s = 1$ a good stepsize choice (see the discussion on rate of convergence in Section 1.3). If a suitable scaling factor for d^k is not known, one may use various ad hoc schemes to determine one. For example, a simple possibility is based on quadratic interpolation of the function

$$g(\alpha) = f(x^k + \alpha d^k),$$

which is the cost along the direction d^k, viewed as a function of the stepsize α. In this scheme, we select some stepsize $\bar{\alpha}$, evaluate $g(\bar{\alpha})$, and perform the quadratic interpolation of g on the basis of $g(0) = f(x^k)$, $dg(0)/d\alpha = \nabla f(x^k)' d^k$, and $g(\bar{\alpha})$. If $\hat{\alpha}$ minimizes the quadratic interpolation, we replace d^k by $\hat{d}^k = \hat{\alpha} d^k$, and we use an initial stepsize $s = 1$. Of course some safeguards are needed when implementing heuristics of this type; for example if $g(\alpha)$ is linear or concave in the interval $[0, \bar{\alpha}]$, the quadratic interpolation scheme just described will fail; see also Exercise 1.2.15.

Constant Stepsize

Here a fixed stepsize $s > 0$ is selected and

$$\alpha^k = s, \quad k = 0, 1, \ldots.$$

The constant stepsize rule is very simple. However, if the stepsize is too large, divergence will occur, while if the stepsize is too small, the rate of convergence may be very slow. Thus, the constant stepsize rule is useful only for problems

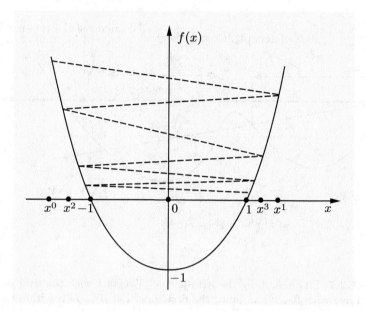

Figure 1.2.6. Example of failure of the successive stepsize reduction rule for the one-dimensional function

$$f(x) = \begin{cases} \dfrac{3(1-x)^2}{4} - 2(1-x) & \text{if } x > 1, \\ \dfrac{3(1+x)^2}{4} - 2(1+x) & \text{if } x < -1, \\ x^2 - 1, & \text{if } -1 \leq x \leq 1. \end{cases}$$

The gradient of f is given by

$$\nabla f(x) = \begin{cases} \dfrac{3x}{2} + \dfrac{1}{2} & \text{if } x > 1, \\ \dfrac{3x}{2} - \dfrac{1}{2} & \text{if } x < -1, \\ 2x, & \text{if } -1 \leq x \leq 1. \end{cases}$$

It is seen that f is strictly convex, continuously differentiable, and is minimized at $x^* = 0$. Furthermore, $f(x) < f(\tilde{x})$ if and only if $|x| < |\tilde{x}|$. For $x > 1$, we have

$$x - \nabla f(x) = x - \frac{3x}{2} - \frac{1}{2} = -\left(1 + \frac{x-1}{2}\right),$$

from which it can be verified that $|x - \nabla f(x)| < |x|$, so that $f\big(x - \nabla f(x)\big) < f(x)$ and $x - \nabla f(x) < -1$. Similarly, for $x < -1$, we have $f\big(x - \nabla f(x)\big) < f(x)$ and $x - \nabla f(x) > 1$. Consider now the steepest descent iteration where the stepsize is successively reduced from an initial stepsize $s = 1$ until descent is obtained. Let the starting point satisfy $|x^0| > 1$. From the preceding equations, it follows that $f\big(x^0 - \nabla f(x^0)\big) < f(x^0)$ and the stepsize $s = 1$ will be accepted by the method. Thus, the next point is $x^1 = x^0 - \nabla f(x^0)$, which satisfies $|x^1| > 1$. By repeating the preceding argument, we see that the generated sequence $\{x^k\}$ satisfies $|x^k| > 1$ for all k, and cannot converge to the unique stationary point $x^* = 0$. In fact, it can be shown that $\{x^k\}$ will have two limit points, $\bar{x} = 1$ and $\bar{x} = -1$, for every x^0 with $|x^0| > 1$.

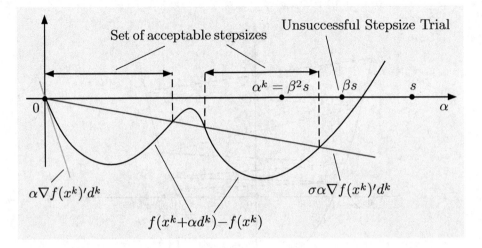

Figure 1.2.7. Line search by the Armijo rule. We start with the trial stepsize s and continue with $\beta s, \beta^2 s, \ldots$, until the first time that $\beta^m s$ falls within the set of stepsizes α satisfying the inequality

$$f(x^k) - f(x^k + \alpha d^k) \geq -\sigma \alpha \nabla f(x^k)' d^k.$$

While this set need not be an interval, it will always contain an interval of the form $[0, \delta]$ with $\delta > 0$, provided $\nabla f(x^k)' d^k < 0$. For this reason the stepsize α^k chosen by the Armijo rule is well defined and will be found after a finite number of trial evaluations of f at the points $(x^k + sd^k), (x^k + \beta s d^k), \ldots$

where an appropriate constant stepsize value is known or can be determined fairly easily.

For the case where f is convex, there are methods that attempt to determine automatically an appropriate value of stepsize; see Exercise 1.2.19 and also [Ber15a], Section 6.1. In these methods an initial value of stepsize is selected, and using the results of the computation over several iterations, the stepsize is reduced to a level that eventually stays constant. Still the convergence of a gradient method using a constant or eventually constant stepsize requires that the gradient ∇f satisfies a Lipschitz condition (see the subsequent Prop. 1.2.2 and the discussion that follows it). By contrast the line minimization rules and the Armijo rule do not require this restriction.

Diminishing Stepsize

Here the stepsize converges to zero,

$$\alpha^k \to 0.$$

This stepsize rule is different than the preceding ones in that it does not guarantee descent at each iteration, although descent becomes more likely as the stepsize diminishes. One difficulty with a diminishing stepsize is that it may become so small that substantial progress cannot be maintained, even when far from a stationary point. For this reason, we require that

$$\sum_{k=0}^{\infty} \alpha^k = \infty.$$

The last condition guarantees that $\{x^k\}$ does not converge to a nonstationary point. Indeed, if $x^k \to \bar{x}$, then for any large indexes m and n ($m > n$) we have

$$x^m \approx x^n \approx \bar{x}, \qquad x^m \approx x^n - \left(\sum_{k=n}^{m-1} \alpha^k\right) \nabla f(\bar{x}),$$

which is a contradiction when $\bar{x}$ is nonstationary and $\sum_{k=n}^{m-1} \alpha^k$ can be made arbitrarily large. Generally, the diminishing stepsize rule has good theoretical convergence properties (see Prop. 1.2.3, and Exercises 1.2.12 and 1.2.13). The associated convergence rate tends to be slow, so this stepsize rule is used primarily in situations where slow convergence is inevitable; for example, in singular problems or when the gradient is calculated with error (see the discussion later in this section).

The preceding stepsize rules are based on cost function reduction (or eventual cost function reduction, in the case of a diminishing stepsize). There are also some other rules, often called *nonmonotonic*, which do not explicitly try to enforce cost function descent and have achieved some success, but are based on ideas that we will not discuss in this book; see [BaB88], [GLL91], [Ray93], [Ray97], [BMR00], [DHS06].

Convergence Issues

Let us now delineate the type of convergence issues that we would like to clarify. We will first discuss informally these issues, and we will state and prove the associated convergence results in Section 1.2.2.

Given a gradient method, ideally we would like the generated sequence $\{x^k\}$ to converge to a global minimum. Unfortunately, however, this is too much to expect, at least when f is not convex, because of the presence of local minima that are not global. Indeed a gradient method is guided downhill by the form of f near the current iterate, while being oblivious to the global structure of f, and thus, can easily get attracted to any type of minimum, global or not. Furthermore, if a gradient method starts or lands at any stationary point, including a local maximum, it stops at that point. Thus, the most we can expect from a gradient method is that it converges to a stationary point. Such a point is a global minimum if f is convex, but this need not be so for nonconvex problems. It must therefore be recognized that gradient methods can be quite inadequate, particularly if little is known about the location and/or other properties of global minima. For such problems one should either try an often difficult and frustrating process of running a gradient method from multiple starting points, or else resort to a fundamentally different approach.

Generally, depending on the nature of the cost function f, the sequence $\{x^k\}$ generated by a gradient method need not have a limit point; in fact $\{x^k\}$ is typically unbounded if f has no local minima. If, however, we know that the level set $\{x \mid f(x) \leq f(x^0)\}$ is bounded, and the stepsize is chosen to enforce descent at each iteration, then the sequence $\{x^k\}$ must be bounded since it belongs to this level set. It must then have at least one limit point; this is because every bounded sequence has at least one limit point (see Prop. A.5 of Appendix A).

Even if $\{x^k\}$ is bounded, convergence to a single limit point may not be easy to guarantee. However, it can be shown that local minima, which are isolated stationary points (they are unique stationary points within some open sphere), tend to attract most types of gradient methods, i.e., once a gradient method gets sufficiently close to such a local minimum, it converges to it. This is the subject of a simple and remarkably powerful result, the *capture theorem*, which is given in the next subsection (Prop. 1.2.4).

Another single limit convergence result for the steepest descent method is given in Exercise 1.2.12 for the case where f is convex and the constant or diminishing stepsize rule is used. Generally, if there are multiple global minima, it is possible that $\{x^k\}$ has multiple limit points (see Exercise 1.2.17 from [Zou76]; also [Gon00]).

We now address the question whether each limit point of a sequence $\{x^k\}$ generated by a gradient method is a stationary point. From the first order expansion

$$f(x^{k+1}) = f(x^k) + \alpha^k \nabla f(x^k)'d^k + o(\alpha^k),$$

we see that if the slope of f at x^k along the direction d^k, which is $\nabla f(x^k)'d^k$, has "substantial" magnitude, the rate of progress of the method will also tend to be substantial. If on the other hand, the directions d^k tend to become asymptotically orthogonal to the gradient direction,

$$\frac{\nabla f(x^k)'d^k}{\|\nabla f(x^k)\|\|d^k\|} \to 0,$$

as x^k approaches a nonstationary point, there is a chance that the method will get "stuck" near that point. To ensure that this does not happen, we consider rather technical conditions on the directions d^k, which are either naturally satisfied or can be easily enforced in most algorithms of interest.

One such condition for the case where

$$d^k = -D^k \nabla f(x^k),$$

is to assume that the eigenvalues of the positive definite symmetric matrix D^k are bounded above and bounded away from zero, i.e., for some positive

Sec. 1.2 Gradient Methods – Convergence

scalars c_1 and c_2, we have

$$c_1\|z\|^2 \leq z'D^k z \leq c_2\|z\|^2, \qquad \forall\, z \in \Re^n,\ k=0,1,\ldots \qquad (1.13)$$

It can then be seen that

$$\left|\nabla f(x^k)'d^k\right| = \left|\nabla f(x^k)'D^k \nabla f(x^k)\right| \geq c_1 \left\|\nabla f(x^k)\right\|^2.$$

It follows that as long as $\nabla f(x^k)$ does not tend to zero, $\nabla f(x^k)$ and d^k cannot become asymptotically orthogonal.

We will introduce another "nonorthogonality" type of condition, which is more general than the "bounded eigenvalues" condition (1.13). Let us consider the sequence $\{x^k, d^k\}$ generated by a given gradient method. We say that the direction sequence $\{d^k\}$ is *gradient related* to $\{x^k\}$ if the following property can be shown:

For any subsequence $\{x^k\}_{k \in \mathcal{K}}$ that converges to a nonstationary point, the corresponding subsequence $\{d^k\}_{k \in \mathcal{K}}$ is bounded and satisfies

$$\limsup_{k \to \infty,\, k \in \mathcal{K}} \nabla f(x^k)'d^k < 0. \qquad (1.14)$$

In particular, if $\{d^k\}$ is gradient related to $\{x^k\}$, it follows that if a subsequence $\{\nabla f(x^k)\}_{k \in \mathcal{K}}$ tends to a nonzero vector, the corresponding subsequence of directions d^k is bounded and does not tend to be orthogonal to $\nabla f(x^k)$. Roughly, this means that d^k does not become "too small" or "too large" relative to $\nabla f(x^k)$, and that the angle between d^k and $\nabla f(x^k)$ does not get "too close" to 90 degrees.

We can often guarantee *a priori* that $\{d^k\}$ is gradient related. In particular, if $d^k = -D^k \nabla f(x^k)$ and the eigenvalues of D^k are bounded as in the "bounded eigenvalues" condition (1.13), it can be seen that $\{d^k\}$ is gradient related. [The boundedness requirement of $\{d^k\}_{k \in \mathcal{K}}$ holds in view of the relation

$$\|d^k\|^2 = \left|\nabla f(x^k)'(D^k)^2 \nabla f(x^k)\right| \leq c_2^2 \left\|\nabla f(x^k)\right\|^2,$$

which follows from Eq. (1.13), since c_2 is no less than the largest eigenvalue of D^k, and the eigenvalues of $(D^k)^2$ are equal to the squares of the corresponding eigenvalues of D^k (Props. A.18 and A.13 in Appendix A).]

Two other examples of conditions that, if satisfied for some scalars $c_1 > 0$, $c_2 > 0$, $p_1 \geq 0$, $p_2 \geq 0$, and all k, guarantee that $\{d^k\}$ is gradient related are

(a)
$$c_1 \left\|\nabla f(x^k)\right\|^{p_1} \leq -\nabla f(x^k)'d^k, \qquad \|d^k\| \leq c_2 \left\|\nabla f(x^k)\right\|^{p_2}.$$

(b)
$$d^k = -D^k \nabla f(x^k),$$

with D^k a positive definite symmetric matrix satisfying

$$c_1 \|\nabla f(x^k)\|^{p_1} \|z\|^2 \leq z'D^k z \leq c_2 \|\nabla f(x^k)\|^{p_2} \|z\|^2, \qquad \forall\ z \in \Re^n.$$

This condition generalizes the "bounded eigenvalues" condition (1.13), which is obtained for $p_1 = p_2 = 0$.

An important convergence result is that if $\{d^k\}$ is gradient related and the minimization rule, or the limited minimization rule, or the Armijo rule is used, then all limit points of $\{x^k\}$ are stationary. This is shown in Prop. 1.2.1, given in the next subsection. When a constant stepsize is used, convergence can be proved assuming that the stepsize is sufficiently small and that f satisfies some further conditions (cf. Prop. 1.2.2).

There is a common line of proof for these convergence results. The main idea is that the cost function is improved at each iteration and that, based on our assumptions, the improvement is "substantial" near a nonstationary point, i.e., it is bounded away from zero. We then argue that the algorithm cannot approach a nonstationary point, since in this case the total cost improvement would accumulate to infinity.

Termination of Gradient Methods

Generally, gradient methods are not finitely convergent, so it is necessary to have criteria for terminating the iterations with some assurance that we are reasonably close to at least a local minimum. A typical approach is to stop the computation when the norm of the gradient becomes sufficiently small, i.e., when a point x^k is obtained with

$$\|\nabla f(x^k)\| \leq \epsilon,$$

where ϵ is a small positive scalar. Unfortunately, it is not known a priori how small one should take ϵ in order to guarantee that the final point x^k is a "good" approximation to a stationary point. The appropriate value of ϵ depends on how the problem is scaled. In particular, if f is multiplied by some scalar, the appropriate value of ϵ is also multiplied by the same scalar. It is possible to correct this difficulty by replacing the criterion $\|\nabla f(x^k)\| \leq \epsilon$ with

$$\frac{\|\nabla f(x^k)\|}{\|\nabla f(x^0)\|} \leq \epsilon.$$

Still, however, the gradient norm $\|\nabla f(x^k)\|$ depends on all the components of the gradient, and depending on how the optimization variables are scaled, the preceding termination criterion may not work well. In particular, some components of the gradient may be naturally much smaller than others, thus requiring a smaller value of ϵ than the other components.

Sec. 1.2   Gradient Methods – Convergence

Assuming that the direction d^k captures the relative scaling of the optimization variables, it may be appropriate to terminate computation when the norm of the direction d^k becomes sufficiently small, i.e.,

$$\|d^k\| \leq \epsilon.$$

Still the appropriate value of ϵ may not be easy to guess, and it may be necessary to experiment prior to settling on a reasonable termination criterion for a given problem. Sometimes, other problem-dependent criteria are used, in addition to or in place of $\|\nabla f(x^k)\| \leq \epsilon$ and $\|d^k\| \leq \epsilon$.

When $\nabla^2 f(x)$ is positive definite, the condition $\|\nabla f(x^k)\| \leq \epsilon$ yields bounds on the distance from local minima. In particular, if x^* is a local minimum of f and there exists $m > 0$ such that for all x in a sphere S centered at x^* we have

$$m\|z\|^2 \leq z'\nabla^2 f(x)z, \qquad \forall\, z \in \Re^n,$$

then every $x \in S$ satisfying $\|\nabla f(x)\| \leq \epsilon$ also satisfies

$$\|x - x^*\| \leq \frac{\epsilon}{m}, \qquad f(x) - f(x^*) \leq \frac{\epsilon^2}{m},$$

(see Exercise 1.2.9).

In the absence of positive definiteness conditions on $\nabla^2 f(x)$, it may be very difficult to infer the proximity of the current iterate to the optimal solution set by just using the gradient norm. We will return to this point when we will discuss singular local minima in the next section.

Spacer Steps

Often, optimization problems are solved with complex descent algorithms in which the rule used to determine the next point may depend on several previous points or on the iteration index k. Some of the conjugate direction algorithms discussed in Section 2.1 are of this type. Other algorithms consist of a combination of different methods and switch from one method to the other in a manner that may either be prespecified or may depend on the progress of the algorithm. Such combinations are usually introduced in order to improve speed of convergence or reliability. However, their convergence analysis can become extremely complicated.

It is thus often valuable to know that if in such algorithms one inserts, perhaps irregularly but infinitely often, an iteration of a convergent algorithm such as the gradient methods of this section, then the theoretical convergence properties of the overall algorithm are quite satisfactory. Such an iteration is known as a *spacer step*. The related convergence result is given in Prop. 1.2.5. The only requirement imposed on the iterations of the algorithm other than the spacer steps is that they do not increase the cost; these iterations, however, need not strictly decrease the cost, and this allows for flexibility in the design of algorithms.

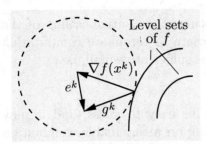

Figure 1.2.8. Illustration of the descent property of the direction $g^k = \nabla f(x^k) + e^k$ of the steepest descent method with error. If the error e^k has smaller norm than the gradient $\nabla f(x^k)$, then g^k lies strictly within the sphere centered at $\nabla f(x^k)$ with radius $\|\nabla f(x^k)\|$, and thus makes an angle less than 90 degrees with $\nabla f(x^k)$.

Gradient Methods with Random and Nonrandom Errors

Frequently in optimization problems the gradient $\nabla f(x^k)$ is not computed exactly. Instead, one has available

$$g^k = \nabla f(x^k) + e^k,$$

where e^k is an uncontrollable error vector. There are several potential sources of error; roundoff error, and discretization error due to finite difference approximations to the gradient are two possibilities, but there are others that will be discussed in more detail later in Chapter 2, in the context of incremental and asynchronous algorithms. Let us focus for concreteness on the steepest descent method with errors,

$$x^{k+1} = x^k - \alpha^k g^k,$$

and let us consider several qualitatively different cases:

(a) e^k **is small relative to the gradient**, i.e.,

$$\|e^k\| < \|\nabla f(x^k)\|, \qquad \forall\ k.$$

Then, assuming $\nabla f(x^k) \neq 0$, $-g^k$ is a direction of cost improvement, i.e., $\nabla f(x^k)'g^k > 0$. This is illustrated in Fig. 1.2.8, and is verified by the calculation

$$\begin{aligned}
\nabla f(x^k)'g^k &= \|\nabla f(x^k)\|^2 + \nabla f(x^k)'e^k \\
&\geq \|\nabla f(x^k)\|^2 - \|\nabla f(x^k)\|\,\|e^k\| \\
&= \|\nabla f(x^k)\|\bigl(\|\nabla f(x^k)\| - \|e^k\|\bigr) \\
&> 0.
\end{aligned} \qquad (1.15)$$

In this case convergence results that are analogous to Props. 1.2.3 and 1.2.4 can be shown.

(b) $\{e^k\}$ **is bounded**, i.e.,

$$\|e^k\| \leq \delta, \qquad \forall\ k,$$

Sec. 1.2   Gradient Methods – Convergence   45

where δ is some scalar. Then by the preceding calculation (1.15), the method operates like a descent method within the region

$$\{x \mid \|\nabla f(x)\| > \delta\}.$$

In the complementary region where $\|\nabla f(x)\| \leq \delta$, the method can behave quite unpredictably. For example, if the errors e^k are constant, say $e^k \equiv e$, then since $g^k = \nabla f(x^k) + e$, the method will essentially be trying to minimize $f(x) + e'x$ and will typically converge to a point $\bar{x}$ with $\nabla f(\bar{x}) = -e$. If the errors e^k vary substantially, the method will tend to oscillate within the region where $\|\nabla f(x)\| \leq \delta$ (see Exercise 1.2.16 and also Exercise 1.3.5 in the next section). The precise behavior will depend on the precise nature of the errors, and on whether a constant or a diminishing stepsize is used (see also the following cases).

(c) $\{e^k\}$ **is proportional to the stepsize**, i.e.,

$$\|e^k\| \leq \alpha^k q, \quad \forall\, k,$$

where q is some scalar. If the stepsize is constant, we come under case (b), while if the stepsize is diminishing, the behavior described in case (b) applies, but with $\delta \to 0$, so the method will tend to converge to a stationary point of f. Important situations where the condition $\|e^k\| \leq \alpha^k q$ holds will be encountered in the context of incremental methods in Section 2.4. A more general condition under which similar behavior occurs is

$$\|e^k\| \leq \alpha^k \left(q + p\|\nabla f(x^k)\|\right), \quad \forall\, k,$$

where q and p are some scalars. Generally, under this condition and with a diminishing stepsize, the convergence behavior is similar to the case where there are no errors; see the following Prop. 1.2.3 (also Exercise 1.2.20, whose solution is posted online).

(d) $\{e^k\}$ **are independent zero mean random vectors with finite variance**. An important special case where such errors arise is when f is of the form

$$f(x) = E_w\{F(x, w)\}, \qquad (1.16)$$

where $F : \Re^{m+n} \to \Re$ is some function, w is a random vector in $\Re^m$, and $E_w\{\cdot\}$ denotes expected value. Under very mild assumptions it can be shown that if F is continuously differentiable, the same is true of f and furthermore,

$$\nabla f(x) = E_w\{\nabla_x F(x, w)\}.$$

Often an approximation g^k to $\nabla f(x^k)$ is computed by simulation or by using a limited number of samples of $\nabla F(x, w)$, with potentially substantial error resulting. In an extreme case, we have

$$g^k = \nabla_x F(x^k, w^k),$$

where w^k is a single sample value corresponding to x^k. Then the error

$$e^k = \nabla_x F(x^k, w^k) - \nabla f(x^k) = \nabla_x F(x^k, w^k) - E_w\{\nabla_x F(x^k, w)\}$$

need not diminish with $\|\nabla f(x^k)\|$, but has zero mean, and under appropriate conditions, its effects are "averaged out." What is roughly happening here is that the descent condition $\nabla f(x^k)'g^k > 0$ holds *on the average* at nonstationary points x^k. It is still possible that for some sample values of e^k, the direction g^k is "bad", but with a diminishing stepsize, the occasional use of a bad direction cannot deteriorate the cost enough for the method to oscillate, given that on the average the method uses "good" directions (see also the discussion of incremental gradient methods in Section 2.4.1). The detailed analysis of gradient methods with random errors (also called *stochastic gradient* methods) is beyond the scope of this text, and properly belongs to the algorithmic field of *stochastic approximation* (see e.g. [BeT89], [BeT96], [BeT00], [KuC78], [KuY97], [LPW92], [Pfl96], [PoT73a], [Pol87], [TBA86]). We mention one representative convergence result from [BeT00], which parallels the following Prop. 1.2.3 that deals with a gradient method without errors: if in the iteration

$$x^{k+1} = x^k - \alpha^k\bigl(\nabla f(x^k) + e^k\bigr)$$

the random variables $e^0, e^1, \ldots$ are independent, zero mean, with bounded variance, the stepsize is diminishing and satisfies

$$\alpha^k \to 0, \qquad \sum_{k=0}^{\infty} \alpha^k = \infty, \qquad \sum_{k=0}^{\infty} (\alpha^k)^2 < \infty,$$

and the gradient ∇f is Lipschitz continuous, then with probability one, we either have $f(x^k) \to -\infty$ or else $\nabla f(x^k) \to 0$. Furthermore, every limit point of $\{x^k\}$ is a stationary point of f.

The Role of Convergence Analysis

The following subsection gives a number of mathematical propositions relating to the convergence properties of gradient methods. The meaning of these propositions is usually quite intuitive but their statement often requires complicated mathematical assumptions. Furthermore, their proof often involves tedious ϵ-δ arguments, so at first sight students may wonder whether "we really have to go through all this."

When Euclid was faced with a similar question from king Ptolemy of Alexandria, he replied that "there is no royal road to geometry." In our case, however, the answer is not so simple because we are not dealing with a purely mathematical subject such as geometry that may be developed without regard for its practical application. In the eyes of most people, the value of an analysis or algorithm in nonlinear programming is judged primarily by its practical utility in solving various types of problems. It is therefore important to give some thought to the interface between convergence analysis and its practical application. To this end it is useful to consider two extreme viewpoints; most workers in the field find themselves somewhere between the two.

In the first viewpoint, convergence analysis is considered primarily a mathematical subject. The properties of an algorithm are quantified to the extent possible through mathematical statements. General and broadly applicable assertions, and simple and elegant proofs are at a premium here. The rationale is that simple statements and proofs are more readily understood at an intuitive level, and general statements apply not only to the problems at hand but also to other problems that are likely to appear in the future. On the negative side, one may remark that simplicity is not always compatible with relevance, and broad applicability is often achieved through assumptions that are hard to verify or appreciate.

The second viewpoint largely discounts the role of mathematical analysis. The rationale here is that the validity and the properties of an algorithm for a given class of problems must be verified through practical experimentation anyway, so if an algorithm looks promising on intuitive grounds, why bother with a convergence analysis. Furthermore, there are a number of important practical questions that are hard to address analytically, such as roundoff error, multiple local minima, and a variety of finite termination and approximation issues. The main criticism of this viewpoint is that mathematical analysis often reveals (and explains) fundamental flaws of algorithms that experimentation may miss. These flaws often point the way to better algorithms or modified algorithms that are tailored to the type of practical problem at hand. Similarly, analysis may be more effective than experimentation in delineating the types of problems for which particular algorithms are well-suited.

Our own mathematical approach is tempered by practical concerns, but we note that the balance between theory and practice in nonlinear programming is particularly delicate, subjective, and problem dependent. Aside from the fact that the mathematical proofs themselves often provide valuable insight into algorithms, here are some of our reasons for insisting on a rigorous convergence analysis:

(a) We want to delineate the range of applicability of various methods. In particular, we want to know for what type of cost function (once or twice differentiable, convex or nonconvex, with singular or nonsingular minima) each algorithm is best suited. If the cost function violates the assumptions under which a given algorithm can be proved to converge, it is reasonable to suspect that the algorithm is unsuitable for this cost function.

(b) We want to understand the qualitative behavior of various methods. For example, we want to know whether convergence of the method depends on the availability of a good starting point, whether the iterates x^k or just the function values $f(x^k)$ are guaranteed to converge, etc. This information, supplemented by theoretical examples and counterexamples, may guide the computational experimentation.

(c) We want to provide guidelines for choosing a few algorithms for further experimentation out of the often bewildering array of candidate algorithms that are applicable for the solution of a given type of problem. One of the principal means for this is the rate of convergence analysis to be given in Section 1.3. Note here that while an algorithm may provably converge, in practice it may be entirely inappropriate for a given problem because it converges very slowly. Experience has shown that without a good understanding of the rate of convergence properties of algorithms it may be difficult to exclude bad candidates from consideration without costly experimentation.

At the same time one should be aware of some of the limitations of the mathematical results that we will provide. For example, some of the assumptions under which an algorithm will be proved convergent may be hard to verify for a given type of problem. Furthermore, our convergence rate analysis of Section 1.3 is largely asymptotic; that is, it applies near the eventual limit of the generated sequence. It is possible, that an algorithm has a good asymptotic rate of convergence but it works poorly in practice for a given type of problem because it is very slow in its initial phase. Finally, some lines of mathematical analyses can be so dry and lacking in intuition that they are of little use to all but a small number of theorists.

There is still another viewpoint, which is worth addressing because it is often adopted by the casual user of nonlinear programming algorithms. This user is interested in a particular application of nonlinear programming in his/her special field, and is counting on an existing code or package to solve the problem (several such packages are commercially or publicly available). Since the package will do most of the work, the user may hope that a superficial acquaintance with the properties of the algorithms underlying the package will suffice. This hope is sometimes realized but unfortunately in many cases it is not. There are a number of reasons for this. First, there are many packages implementing a lot of different methods, and to choose the right package, one needs to have insight into the suitability of different methods for the special features of the application at hand. Second, to use a package one must often know how to suitably formulate the

Sec. 1.2 Gradient Methods – Convergence

problem, how to set various parameters (e.g. termination criteria, stepsize parameters, etc.), and how to interpret the results of the computation (particularly when things don't work out as hoped initially, which is often the case). For this, one needs considerable insight into the inner workings of the algorithm underlying the package. Finally, for a challenging practical optimization problem (e.g. one of large dimension), it may be essential to exploit its special structure, and packages often do not have this capability. As a result the user may have to modify the package or write an altogether new code that is tailored to the application at hand. Both of these require an intimate understanding of the convergence properties and other characteristics of the relevant nonlinear programming algorithms.

1.2.2 Convergence Results

We now provide an analysis of the convergence behavior of gradient methods. The following proposition is the main convergence result.

Proposition 1.2.1: (Stationarity of Limit Points for Gradient Methods) Let $\{x^k\}$ be a sequence generated by a gradient method $x^{k+1} = x^k + \alpha^k d^k$, and assume that $\{d^k\}$ is gradient related [cf. Eq. (1.14)] and α^k is chosen by the minimization rule, or the limited minimization rule, or the Armijo rule. Then every limit point of $\{x^k\}$ is a stationary point.

Proof: Consider first the Armijo rule and let $\bar{x}$ be a limit point of $\{x^k\}$. Since $\{f(x^k)\}$ is monotonically nonincreasing, $\{f(x^k)\}$ either converges to a finite value or diverges to $-\infty$. Since f is continuous, $f(\bar{x})$ is a limit point of $\{f(x^k)\}$, so it follows that the entire sequence $\{f(x^k)\}$ converges to $f(\bar{x})$, and

$$f(x^k) - f(x^{k+1}) \to 0. \tag{1.17}$$

Moreover, by the definition of the Armijo rule, we have

$$f(x^k) - f(x^{k+1}) \geq -\sigma \alpha^k \nabla f(x^k)' d^k, \tag{1.18}$$

so the right-hand side in the above relation tends to 0.

Let $\{x^k\}_{\mathcal{K}}$ be a subsequence converging to $\bar{x}$, and assume to arrive at a contradiction that $\bar{x}$ is nonstationary. Since $\{d^k\}$ is gradient related, we have

$$\limsup_{\substack{k \to \infty \\ k \in \mathcal{K}}} \nabla f(x^k)' d^k < 0,$$

and therefore from Eqs. (1.17) and (1.18),

$$\{\alpha^k\}_{\mathcal{K}} \to 0.$$

Hence, by the definition of the Armijo rule, we must have for some index $\bar{k} \geq 0$

$$f(x^k) - f\big(x^k + (\alpha^k/\beta)d^k\big) < -\sigma(\alpha^k/\beta)\nabla f(x^k)'d^k, \qquad \forall\; k \in \mathcal{K},\; k \geq \bar{k}, \tag{1.19}$$

i.e., the initial stepsize s will be reduced at least once for all $k \in \mathcal{K}$, $k \geq \bar{k}$. Since $\{d^k\}$ is gradient related, $\{d^k\}_\mathcal{K}$ is bounded, so there exists a subsequence $\{d^k\}_{\bar{\mathcal{K}}}$ of $\{d^k\}_\mathcal{K}$ such that

$$\{d^k\}_{\bar{\mathcal{K}}} \to \bar{d},$$

where $\bar{d}$ is some vector. From Eq. (1.19), we have

$$\frac{f(x^k) - f(x^k + \bar{\alpha}^k d^k)}{\bar{\alpha}^k} < -\sigma \nabla f(x^k)'d^k, \qquad \forall\; k \in \bar{\mathcal{K}},\; k \geq \bar{k}, \tag{1.20}$$

where $\bar{\alpha}^k = \alpha^k/\beta$. By using the mean value theorem, this relation is written as

$$-\nabla f(x^k + \tilde{\alpha}^k d^k)'d^k < -\sigma \nabla f(x^k)'d^k, \qquad \forall\; k \in \bar{\mathcal{K}},\; k \geq \bar{k},$$

where $\tilde{\alpha}^k$ is a scalar in the interval $[0, \bar{\alpha}^k]$. Taking limits in the above relation we obtain

$$-\nabla f(\bar{x})'\bar{d} \leq -\sigma \nabla f(\bar{x})'\bar{d}$$

or

$$0 \leq (1 - \sigma)\nabla f(\bar{x})'\bar{d}.$$

Since $\sigma < 1$, it follows that

$$0 \leq \nabla f(\bar{x})'\bar{d}, \tag{1.21}$$

which contradicts the assumption that $\{d^k\}$ is gradient related. This proves the result for the Armijo rule.

Consider next the minimization rule, and let $\{x^k\}_\mathcal{K}$ converge to $\bar{x}$ with $\nabla f(\bar{x}) \neq 0$. Again we have that $\{f(x^k)\}$ decreases monotonically to $f(\bar{x})$. Let $\tilde{x}^{k+1}$ be the point generated from x^k via the Armijo rule, and let $\tilde{\alpha}^k$ be the corresponding stepsize. We have

$$f(x^k) - f(x^{k+1}) \geq f(x^k) - f(\tilde{x}^{k+1}) \geq -\sigma \tilde{\alpha}^k \nabla f(x^k)'d^k. \tag{1.22}$$

By repeating the arguments of the earlier proof following Eq. (1.18), replacing α^k by $\tilde{\alpha}^k$, we can obtain a contradiction. In particular, we have

$$\{\tilde{\alpha}^k\}_\mathcal{K} \to 0,$$

and by the definition of the Armijo rule, we have for some index $\bar{k} \geq 0$

$$f(x^k) - f\big(x^k + (\tilde{\alpha}^k/\beta)d^k\big) < -\sigma(\tilde{\alpha}^k/\beta)\nabla f(x^k)'d^k, \qquad \forall\; k \in \mathcal{K},\; k \geq \bar{k},$$

Sec. 1.2   Gradient Methods – Convergence   51

[cf. Eq. (1.19)]. Proceeding as earlier, we obtain Eqs. (1.20) and (1.21) (with $\bar{\alpha}^k = \tilde{\alpha}^k/\beta$), and a contradiction of Eq. (1.21).

The line of argument just used [cf. Eq. (1.22)] establishes that any stepsize rule that gives a larger reduction in cost at each iteration than the Armijo rule inherits the convergence properties of the latter. This also proves the proposition for the limited minimization rule. **Q.E.D.**

The next proposition establishes, among other things, convergence for the case of a constant stepsize. The idea is that if the rate of growth of the gradient of f is bounded from above (i.e., the curvature of f is bounded), then one can construct a quadratic function that majorizes f; see Fig. 1.2.9. Given x^k and d^k, an appropriate constant stepsize α^k can then be obtained within an interval around the scalar $\bar{\alpha}^k$ that minimizes this quadratic function along the direction d^k. The proposition requires that for some constant $L > 0$, we have

$$\|\nabla f(x) - \nabla f(y)\| \leq L\|x - y\|, \quad \forall\ x, y \in \Re^n, \tag{1.23}$$

which insures the boundedness of the curvature of f in every direction. This is called *Lipschitz continuity* of ∇f.

Proposition 1.2.2: (Constant Stepsize) Let $\{x^k\}$ be a sequence generated by a gradient method $x^{k+1} = x^k + \alpha^k d^k$, where $\{d^k\}$ is gradient related. Assume that the Lipschitz condition (1.23) holds, and that for all k we have $d^k \neq 0$ and

$$\epsilon \leq \alpha^k \leq (2 - \epsilon)\bar{\alpha}^k, \tag{1.24}$$

where

$$\bar{\alpha}^k = \frac{|\nabla f(x^k)'d^k|}{L\|d^k\|^2},$$

and $\epsilon \in (0, 1]$ is a fixed scalar. Then every limit point of $\{x^k\}$ is a stationary point of f.

Proof: By using the descent lemma (Prop. A.24 of Appendix A), we obtain

$$\begin{aligned}f(x^k) - f(x^k + \alpha^k d^k) &\geq -\alpha^k \nabla f(x^k)'d^k - \tfrac{1}{2}(\alpha^k)^2 L\|d^k\|^2 \\ &= \alpha^k\bigl(|\nabla f(x^k)'d^k| - \tfrac{1}{2}\alpha^k L\|d^k\|^2\bigr).\end{aligned} \tag{1.25}$$

The right-hand side of Eq. (1.24) yields

$$|\nabla f(x^k)'d^k| - \tfrac{1}{2}\alpha^k L\|d^k\|^2 \geq \tfrac{1}{2}\epsilon|\nabla f(x^k)'d^k|.$$

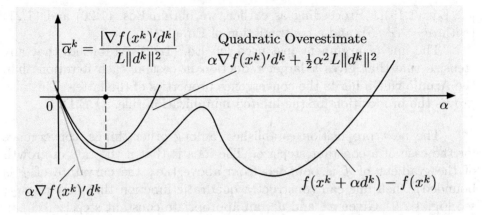

Figure 1.2.9. The idea of the proof of Prop. 1.2.2. Given x^k and the descent direction d^k, the cost difference $f(x^k + \alpha d^k) - f(x^k)$ is majorized by

$$\alpha \nabla f(x^k)'d^k + \tfrac{1}{2}\alpha^2 L \|d^k\|^2$$

(see the proof of Prop. 1.2.2). Minimization of this function over α yields the stepsize

$$\bar{\alpha}^k = \frac{|\nabla f(x^k)'d^k|}{L\|d^k\|^2},$$

which reduces the cost function f as well (see the proof of Prop. 1.2.2).

Using this relation together with the condition $\alpha^k \geq \epsilon$ in the inequality (1.25), we obtain the following bound on the cost improvement obtained at iteration k:

$$f(x^k) - f(x^k + \alpha^k d^k) \geq \tfrac{1}{2}\epsilon^2 |\nabla f(x^k)'d^k|.$$

Now if a subsequence $\{x^k\}_\mathcal{K}$ converges to a nonstationary point $\bar{x}$, we must have, as in the proof of Prop. 1.2.1, $f(x^k) - f(x^{k+1}) \to 0$, and the preceding relation implies that $|\nabla f(x^k)'d^k| \to 0$. This contradicts the assumption that $\{d^k\}$ is gradient related. Hence, every limit point of $\{x^k\}$ is stationary. **Q.E.D.**

In the case of steepest descent $[d^k = -\nabla f(x^k)]$, the condition (1.24) on the stepsize becomes

$$\epsilon \leq \alpha^k \leq \frac{2-\epsilon}{L}.$$

Thus a constant stepsize roughly in the middle of the interval $[0, 2/L]$ guarantees convergence. This is a classical convergence result.

Note that the existence of an ϵ satisfying Eq. (1.24) is guaranteed by standard conditions that imply the gradient related assumption. In particular, if $\{d^k\}$ is such that there exist positive scalars c_1, c_2 such that for all k,

$$c_1\|\nabla f(x^k)\|^2 \leq -\nabla f(x^k)'d^k, \qquad \|d^k\|^2 \leq c_2\|\nabla f(x^k)\|^2, \qquad (1.26)$$

then Eq. (1.24) is satisfied if for all k we have

$$\epsilon \leq \alpha^k \leq \frac{c_1(2-\epsilon)}{Lc_2}. \qquad (1.27)$$

Furthermore, if d^k has the form

$$d^k = -D^k \nabla f(x^k)$$

with D^k positive definite symmetric and having eigenvalues in an interval $[\gamma, \Gamma]$, the condition (1.26) can be seen to hold with

$$c_1 = \gamma, \qquad c_2 = \Gamma^2.$$

Exercise 1.2.3 provides an example showing that the Lipschitz condition (1.23) is essential for the validity of Prop. 1.2.2. This condition requires roughly that the "curvature" of f is no more than L at all points and in all directions. In particular, it is possible to show that this condition is satisfied, if f is twice differentiable and the eigenvalues of the Hessian $\nabla^2 f$ are bounded over $\Re^n$ by L. Unfortunately, however, it is generally difficult to obtain an estimate of L, so in most cases the range of stepsizes that guarantee convergence [cf. Eq. (1.24) or (1.27)] is unknown, and experimentation may be necessary to obtain appropriate stepsize values.

Even worse, many types of cost function f, while twice differentiable, have Hessian $\nabla^2 f$ that is unbounded over $\Re^n$ [this is so for any function $f(x)$ that grows faster than a quadratic as $x \to \infty$, such as $f(x) = \|x\|^3$]. Fortunately, the Lipschitz condition can be significantly weakened, as shown in Exercise 1.2.5. In particular, it is sufficient that it holds for all x, y in the level set

$$\{x \mid f(x) \leq f(x^0)\},$$

in which case, however, the range of stepsizes that guarantee convergence depends on the starting point x^0. Let us also note that, for the case where f is convex, there are methods that attempt to determine automatically a constant stepsize [one such method is given in Exercise 1.2.19 (with solution posted online), and another is described in [Ber15a], Section 6.1].

The Lipschitz continuity condition also essentially guarantees convergence for a diminishing stepsize, as shown by the following proposition.

Proposition 1.2.3: (Diminishing Stepsize) Let $\{x^k\}$ be a sequence generated by a gradient method $x^{k+1} = x^k + \alpha^k d^k$. Assume that the Lipschitz condition (1.23) holds, and that there exist positive scalars c_1, c_2 such that for all k we have

$$c_1 \|\nabla f(x^k)\|^2 \leq -\nabla f(x^k)'d^k, \qquad \|d^k\|^2 \leq c_2 \|\nabla f(x^k)\|^2. \qquad (1.28)$$

Suppose also that

$$\alpha^k \to 0, \qquad \sum_{k=0}^{\infty} \alpha^k = \infty.$$

Then either $f(x^k) \to -\infty$ or else $\{f(x^k)\}$ converges to a finite value and $\nabla f(x^k) \to 0$. Furthermore, every limit point of $\{x^k\}$ is a stationary point of f.

Proof: Combining Eqs. (1.25) and (1.28), we have

$$f(x^{k+1}) \leq f(x^k) + \alpha^k \left(\tfrac{1}{2}\alpha^k L \|d^k\|^2 - |\nabla f(x^k)'d^k| \right)$$
$$\leq f(x^k) - \alpha^k \left(c_1 - \tfrac{1}{2}\alpha^k c_2 L \right) \|\nabla f(x^k)\|^2.$$

Since the linear term in α^k dominates the quadratic term in α^k for sufficiently small α^k, and $\alpha^k \to 0$, we have for some positive constant c and all k greater than some index $\bar{k}$,

$$f(x^{k+1}) \leq f(x^k) - \alpha^k c \|\nabla f(x^k)\|^2. \qquad (1.29)$$

From this relation, we see that for $k \geq \bar{k}$, $\{f(x^k)\}$ is monotonically decreasing, so either $f(x^k) \to -\infty$ or $\{f(x^k)\}$ converges to a finite value. In the latter case, by adding Eq. (1.29) over all $k \geq \bar{k}$, we obtain

$$c \sum_{k=\bar{k}}^{\infty} \alpha^k \|\nabla f(x^k)\|^2 \leq f(x^{\bar{k}}) - \lim_{k \to \infty} f(x^k) < \infty.$$

We see that there cannot exist an $\epsilon > 0$ such that

$$\|\nabla f(x^k)\|^2 > \epsilon$$

for all k greater than some $\hat{k}$, since this would contradict the assumption $\sum_{k=0}^{\infty} \alpha^k = \infty$. Therefore, we must have

$$\liminf_{k \to \infty} \|\nabla f(x^k)\| = 0.$$

Sec. 1.2   Gradient Methods – Convergence   55

To show that $\nabla f(x^k) \to 0$, assume the contrary; that is,

$$\limsup_{k\to\infty} \|\nabla f(x^k)\| \geq \epsilon > 0. \tag{1.30}$$

Let $\{m_j\}$ and $\{n_j\}$ be sequences of indexes such that

$$m_j < n_j < m_{j+1},$$

$$\frac{\epsilon}{3} < \|\nabla f(x^k)\| \qquad \text{for } m_j \leq k < n_j, \tag{1.31}$$

$$\|\nabla f(x^k)\| \leq \frac{\epsilon}{3} \qquad \text{for } n_j \leq k < m_{j+1}. \tag{1.32}$$

Let also $\bar{j}$ be sufficiently large so that

$$\sum_{k=m_{\bar{j}}}^{\infty} \alpha^k \|\nabla f(x^k)\|^2 < \frac{\epsilon^2}{9L\sqrt{c_2}}. \tag{1.33}$$

For any $j \geq \bar{j}$ and any m with $m_j \leq m \leq n_j - 1$, we have

$$\|\nabla f(x^{n_j}) - \nabla f(x^m)\| \leq \sum_{k=m}^{n_j-1} \|\nabla f(x^{k+1}) - \nabla f(x^k)\|$$

$$\leq L \sum_{k=m}^{n_j-1} \|x^{k+1} - x^k\|$$

$$= L \sum_{k=m}^{n_j-1} \alpha^k \|d^k\|$$

$$\leq L\sqrt{c_2} \sum_{k=m}^{n_j-1} \alpha^k \|\nabla f(x^k)\|$$

$$\leq \frac{3L\sqrt{c_2}}{\epsilon} \sum_{k=m}^{n_j-1} \alpha^k \|\nabla f(x^k)\|^2$$

$$\leq \frac{3L\sqrt{c_2}}{\epsilon} \frac{\epsilon^2}{9L\sqrt{c_2}}$$

$$= \frac{\epsilon}{3},$$

where the last two inequalities follow using Eqs. (1.31) and (1.33). Thus

$$\|\nabla f(x^m)\| \leq \|\nabla f(x^{n_j})\| + \frac{\epsilon}{3} \leq \frac{2\epsilon}{3}, \qquad \forall \, j \geq \bar{j},\ m_j \leq m \leq n_j - 1.$$

Thus, using also Eq. (1.32), we have for all $m \geq m_{\bar{j}}$

$$\|\nabla f(x^m)\| \leq \frac{2\epsilon}{3}.$$

This contradicts Eq. (1.30), implying that $\lim_{k\to\infty} \nabla f(x^k) = 0$.

Finally, if $\bar{x}$ is a limit point of x^k, then $f(x^k)$ converges to the finite value $f(\bar{x})$. Thus we have $\nabla f(x^k) \to 0$, implying that $\nabla f(\bar{x}) = 0$. **Q.E.D.**

Under the assumptions of the preceding proposition, descent is not guaranteed in the initial iterations. However, if the stepsizes are all sufficiently small [e.g., they satisfy the right-hand side inequality of Eq. (1.27)], descent is guaranteed at all iterations. In this case, it is sufficient that the Lipschitz condition $\|\nabla f(x) - \nabla f(y)\| \leq L\|x - y\|$ holds for all x, y in the set $\{x \mid f(x) \leq f(x^0)\}$ (see Exercise 1.2.5); otherwise the Lipschitz condition must hold over a set larger than $\{x \mid f(x) \leq f(x^0)\}$ to guarantee convergence (see Exercise 1.2.14 for an example).

The following proposition explains to some extent why sequences generated by gradient methods tend in practice to have unique limit points. It essentially states that local minima which are "isolated" tend to attract gradient methods: once the method gets close enough to such a minimum it remains close and converges to it.

Proposition 1.2.4: (Capture Theorem) Let f be continuously differentiable and let $\{x^k\}$ be a sequence satisfying $f(x^{k+1}) \leq f(x^k)$ for all k and generated by a gradient method $x^{k+1} = x^k + \alpha^k d^k$, which is convergent in the sense that every limit point of sequences that it generates is a stationary point of f. Assume that there exist scalars $s > 0$ and $c > 0$ such that for all k there holds

$$\alpha^k \leq s, \qquad \|d^k\| \leq c\|\nabla f(x^k)\|.$$

Let x^* be a local minimum of f, which is the only stationary point of f within some open set. Then there exists an open set S containing x^* such that if $x^{\bar{k}} \in S$ for some $\bar{k} \geq 0$, then $x^k \in S$ for all $k \geq \bar{k}$ and $\{x^k\} \to x^*$. Furthermore, given any scalar $\bar{\epsilon} > 0$, the set S can be chosen so that $\|x - x^*\| < \bar{\epsilon}$ for all $x \in S$.

Proof: Suppose that $\rho > 0$ is such that

$$f(x^*) < f(x), \qquad \forall\ x \text{ with } \|x - x^*\| \leq \rho.$$

Define for $t \in [0, \rho]$

$$\phi(t) = \min_{\{x \mid t \leq \|x - x^*\| \leq \rho\}} f(x) - f(x^*),$$

Sec. 1.2   Gradient Methods – Convergence   57

and note that ϕ is a monotonically nondecreasing function of t, and that $\phi(t) > 0$ for all $t \in (0, \rho]$. Given any $\epsilon \in (0, \rho]$, let $r \in (0, \epsilon]$ be such that

$$\|x - x^*\| < r \quad \Rightarrow \quad \|x - x^*\| + sc\|\nabla f(x)\| < \epsilon. \tag{1.34}$$

Consider the open set

$$S = \{x \mid \|x - x^*\| < \epsilon,\ f(x) < f(x^*) + \phi(r)\}.$$

We claim that if $x^k \in S$ for some k, then $x^{k+1} \in S$.

Indeed if $x^k \in S$, from the definition of ϕ and S we have

$$\phi(\|x^k - x^*\|) \le f(x^k) - f(x^*) < \phi(r).$$

Since ϕ is monotonically nondecreasing, the above relation implies that $\|x^k - x^*\| < r$, so that by Eq. (1.34),

$$\|x^k - x^*\| + sc\|\nabla f(x^k)\| < \epsilon.$$

We also have by using the hypotheses $\alpha^k \le s$ and $\|d^k\| \le c\|\nabla f(x^k)\|$

$$\|x^{k+1} - x^*\| \le \|x^k - x^*\| + \|\alpha^k d^k\| \le \|x^k - x^*\| + sc\|\nabla f(x^k)\|,$$

so from the last two relations it follows that $\|x^{k+1} - x^*\| < \epsilon$. Since $f(x^{k+1}) < f(x^k)$, we also obtain $f(x^{k+1}) - f(x^*) < \phi(r)$, so we conclude that $x^{k+1} \in S$.

By using induction it follows that if $x^{\bar{k}} \in S$ for some $\bar{k}$, we have $x^k \in S$ for all $k \ge \bar{k}$. Let $\bar{S}$ be the closure of S. Since $\bar{S}$ is compact, the sequence $\{x^k\}$ will have at least one limit point, which by assumption must be a stationary point of f. Now the only stationary point of f within $\bar{S}$ is the point x^* (since we have $\|x - x^*\| \le \rho$ for all $x \in \bar{S}$). Hence $x^k \to x^*$. Finally given any $\bar{\epsilon} > 0$, we can choose $\epsilon \le \bar{\epsilon}$ in which case we have $\|x - x^*\| < \bar{\epsilon}$ for all $x \in S$. **Q.E.D.**

Note that in the preceding proposition, the conditions $f(x^{k+1}) \le f(x^k)$ and $\alpha^k \le s$ are satisfied for the Armijo rule and the limited minimization rule. They are also satisfied for a constant and a diminishing stepsize under conditions that guarantee descent at each iteration (see the proofs of Props. 1.2.2 and 1.2.3). The condition $\|d^k\| \le c\|\nabla f(x^k)\|$ is satisfied if $d^k = -D^k \nabla f(x^k)$ with the eigenvalues of D^k bounded from above.

Finally, we state a result that deals with the convergence of algorithms involving a combination of different methods. It shows that for convergence it is enough to insert, perhaps irregularly but infinitely often, an iteration of a convergent gradient algorithm, provided that the other iterations do not degrade the value of the cost function. The proof is similar to the one of Prop. 1.2.1, and is left for the reader.

Proposition 1.2.5: (Convergence for Spacer Steps) Consider a sequence $\{x^k\}$ such that

$$f(x^{k+1}) \leq f(x^k), \qquad k = 0, 1, \ldots$$

Assume that there exists an infinite set $\mathcal{K}$ of integers for which

$$x^{k+1} = x^k + \alpha^k d^k, \qquad \forall\ k \in \mathcal{K},$$

where $\{d^k\}_\mathcal{K}$ is gradient related and α^k is chosen by the minimization rule, or the limited minimization rule, or the Armijo rule. Then every limit point of the subsequence $\{x^k\}_\mathcal{K}$ is a stationary point.

EXERCISES

1.2.1

Consider the problem of minimizing the function of two variables $f(x, y) = 3x^2 + y^4$.

(a) Apply one iteration of the steepest descent method with $(1, -2)$ as the starting point and with the stepsize chosen by the Armijo rule with $s = 1$, $\sigma = 0.1$, and $\beta = 0.5$.

(b) Repeat (a) using $s = 1$, $\sigma = 0.1$, $\beta = 0.1$ instead. How does the cost of the new iterate compare to that obtained in (a)? Comment on the tradeoffs involved in the choice of β.

(c) Apply one iteration of Newton's method with the same starting point and stepsize rule as in (a). How does the cost of the new iterate compare to that obtained in (a)? How about the amount of work involved in finding the new iterate?

1.2.2

Describe the behavior of the steepest descent method with constant stepsize s for the function $f(x) = \|x\|^{2+\beta}$, where $\beta \geq 0$. For which values of s and x^0 does

Sec. 1.2 Gradient Methods – Convergence

the method converge to $x^* = 0$. Relate your answer to the assumptions of Prop. 1.2.2.

1.2.3

Consider the function $f : \Re^n \mapsto \Re$ given by

$$f(x) = \|x\|^{3/2},$$

and the method of steepest descent with a constant stepsize. Show that for this function, the Lipschitz condition $\|\nabla f(x) - \nabla f(y)\| \leq L\|x - y\|$ for all x and y is not satisfied for any L. Furthermore, for any value of constant stepsize, the method either converges in a *finite* number of iterations to the minimizing point $x^* = 0$ or else it does not converge to x^*.

1.2.4

Apply the steepest descent method with constant stepsize α to the function f of Exercise 1.1.11. Show that the gradient ∇f satisfies the Lipschitz condition

$$\|\nabla f(x) - \nabla f(y)\| \leq L\|x - y\|, \qquad \forall\, x, y \in \Re,$$

for some constant L. Write a computer program to verify that the method is a descent method for $\alpha \in (0, 2/L)$. Do you expect to get in the limit the global minimum $x^* = 0$?

1.2.5 (www)

Suppose that the Lipschitz condition

$$\|\nabla f(x) - \nabla f(y)\| \leq L\|x - y\|, \qquad \forall\, x, y \in \Re^n,$$

[cf. Eq. (1.23)] is replaced by the condition that for every bounded set $A \subset \Re^n$, there exists some constant L such that

$$\|\nabla f(x) - \nabla f(y)\| \leq L\|x - y\|, \qquad \forall\, x, y \in A. \tag{1.35}$$

Show that:

(a) Condition (1.35) is always satisfied if the level sets $\{x \mid f(x) \leq c\}$, $c \in \Re$, are bounded, and f is twice continuously differentiable.

(b) The convergence result of Prop. 1.2.2 remains valid provided that the level set

$$A = \{x \mid f(x) \leq f(x^0)\}$$

is bounded and the stepsize is allowed to depend on the choice of the initial vector x^0. *Hint:* The key idea is to show that x^k stays in the set A, and

to use a stepsize α^k that depends on the constant L corresponding to this set. Let
$$R = \max\{\|x\| \mid x \in A\},$$
$$G = \max\{\|\nabla f(x)\| \mid x \in A\},$$
and
$$B = \{x \mid \|x\| \leq R + 2G\}.$$

Using condition (1.35), there exists some constant L such that $\|\nabla f(x) - \nabla f(y)\| \leq L\|x - y\|$, for all $x, y \in B$. Suppose the stepsize α^k satisfies
$$0 < \epsilon \leq \alpha^k \leq (2 - \epsilon)\gamma^k \min\{1, 1/L\},$$
where
$$\gamma^k = \frac{|\nabla f(x^k)' d^k|}{\|d^k\|^2}.$$

Let $\beta^k = \alpha^k(\gamma^k - L\alpha^k/2)$, which can be seen to satisfy $\beta^k \geq \epsilon^2 \gamma^k/2$ by our choice of α^k. Show by induction on k that with such a choice of stepsize, we have $x^k \in A$ and
$$f(x^{k+1}) \leq f(x^k) - \beta^k \|d^k\|^2, \qquad \forall\, k \geq 0.$$

1.2.6

Suppose that f is quadratic and of the form $f(x) = \frac{1}{2} x' Q x - b'x$, where Q is positive definite and symmetric.

(a) Show that the Lipschitz condition $\|\nabla f(x) - \nabla f(y)\| \leq L\|x - y\|$ is satisfied with L equal to the maximal eigenvalue of Q.

(b) Consider the gradient method $x^{k+1} = x^k - sD\nabla f(x^k)$, where D is positive definite and symmetric. Show that the method converges to $x^* = Q^{-1}b$ for every starting point x^0 if and only if $s \in (0, 2/\bar{L})$, where $\bar{L}$ is the maximum eigenvalue of $D^{1/2}QD^{1/2}$.

1.2.7

An electrical engineer wants to maximize the current I between two points A and B of a complex network by adjusting the values x_1 and x_2 of two variable resistors, where $0 \leq x_1 \leq R_1$, $0 \leq x_2 \leq R_2$, and R_1, R_2 are given. The engineer does not have an adequate mathematical model of the network and decides to adopt the following procedure. She keeps the value x_2 of the second resistor fixed and adjusts the value of the first resistor until the current I is maximized. She then keeps the value x_1 of the first resistor fixed and adjusts the value of the second resistor until the current I is maximized. She then repeats the procedure until no further progress can be made. She knows *a priori* that during this procedure, the values x_1 and x_2 can never reach their extreme values 0, R_1, and R_2. Explain whether there is a sound theoretical basis for the engineer's procedure. *Hint:* Consider how the steepest descent method works for two-dimensional problems.

1.2.8

Consider the gradient method $x^{k+1} = x^k + \alpha^k d^k$, where α^k is chosen by the Armijo rule or the line minimization rule and

$$d^k = - \begin{bmatrix} 0 \\ \vdots \\ 0 \\ \frac{\partial f(x^k)}{\partial x_i} \\ 0 \\ \vdots \\ 0 \end{bmatrix},$$

where i is the index for which $\left|\partial f(x^k)/\partial x_j\right|$ is maximized over $j = 1, \ldots, n$. Show that every limit point of $\{x^k\}$ is stationary.

1.2.9 (www)

Let f be twice continuously differentiable. Suppose that x^* is a local minimum such that for all x in an open sphere S centered at x^*, we have, for some $m > 0$,

$$m\|d\|^2 \leq d' \nabla^2 f(x) d, \qquad \forall\, d \in \Re^n.$$

Show that for every $x \in S$, we have

$$\|x - x^*\| \leq \frac{\|\nabla f(x)\|}{m}, \qquad f(x) - f(x^*) \leq \frac{\|\nabla f(x)\|^2}{2m}.$$

Hint: Use the relation

$$\nabla f(y) = \nabla f(x) + \int_0^1 \nabla^2 f\big(x + t(y - x)\big)(y - x) dt.$$

See also Exercise 1.1.9.

1.2.10 (Alternative Assumptions for Convergence) (www)

Consider the gradient method $x^{k+1} = x^k + \alpha^k d^k$. Instead of $\{d^k\}$ being gradient related, assume *one* of the following two conditions:

(i) It can be shown that for any subsequence $\{x^k\}_{k \in \mathcal{K}}$ that converges to a nonstationary point, the corresponding subsequence $\{d^k\}_{k \in \mathcal{K}}$ is bounded and satisfies

$$\liminf_{k \to \infty,\, k \in \mathcal{K}} \nabla f(x^k)' d^k < 0.$$

(ii) α^k is chosen by the minimization rule, and for some $c > 0$ and all k, we have

$$\left|\nabla f(x^k)' d^k\right| \geq c \|\nabla f(x^k)\| \, \|d^k\|.$$

Show that the conclusion of Prop. 1.2.1 holds.

1.2.11 (Behavior of Steepest Descent Near a Saddle Point)

Let $f(x) = (1/2)x'Qx$, where Q is symmetric, invertible, and has at least one negative eigenvalue. Consider the steepest descent method with constant stepsize and show that unless the starting point x^0 belongs to the subspace spanned by the eigenvectors of Q corresponding to the nonnegative eigenvalues, the generated sequence $\{x^k\}$ diverges.

1.2.12 (Convergence of Steepest Descent to a Single Limit) (www)

Consider the steepest descent method $x^{k+1} = x^k - \alpha^k \nabla f(x^k)$, and assume that f is convex, has at least one minimizing point, and for all x, y, and some $L > 0$, satisfies
$$\|\nabla f(x) - \nabla f(y)\| \le L \|x - y\|.$$
Show that $\{x^k\}$ converges to a minimizing point of f under each of the following two stepsize rule conditions:

(i) For some $\epsilon > 0$, we have
$$\epsilon \le \alpha^k \le \frac{2 - \epsilon}{L}, \qquad \forall\, k.$$

(ii) $\alpha^k \to 0$ and $\sum_{k=0}^{\infty} \alpha^k = \infty$.

Note: This result is due to [BGI95], who also show convergence to a single limit for a variant of the Armijo rule.

1.2.13 (Steepest Descent with Diminishing Stepsize [CoL94]) (www)

Consider the steepest descent method $x^{k+1} = x^k - \alpha^k \nabla f(x^k)$, and assume that the function f is convex.

(a) Use the convexity of f to show that for any $y \in \Re^n$, we have
$$\|x^{k+1} - y\|^2 \le \|x^k - y\|^2 - 2\alpha^k \left(f(x^k) - f(y)\right) + \left(\alpha^k \|\nabla f(x^k)\|\right)^2.$$

(b) Assume that
$$\sum_{k=0}^{\infty} \alpha^k = \infty, \qquad \alpha^k \|\nabla f(x^k)\|^2 \to 0.$$
Show that $\liminf_{k \to \infty} f(x^k) = \inf_{x \in \Re^n} f(x)$. *Hint*: Argue by contradiction. Assume that for some $\delta > 0$, there exists y with $f(y) < f(x^k) - \delta$ for all k sufficiently large. Use part (a).

(c) Assume that
$$\alpha^k = \frac{s^k}{\|\nabla f(x^k)\|},$$

where
$$\sum_{k=0}^{\infty} s^k = \infty, \qquad \sum_{k=0}^{\infty} (s^k)^2 < \infty.$$

Show that $\liminf_{k\to\infty} f(x^k) = \inf_{x\in\Re^n} f(x)$, and that if f has at least one global minimum, then $\{x^k\}$ converges to some global minimum. *Hint*: In part (a), set y to some x^* such that $f(x^*) < f(x^k)$ for all k (if no such x^* exists, we are done). Show that the relation
$$\|x^{k+1} - x^*\|^2 \le \|x^k - x^*\|^2 + (s^k)^2$$
implies that $\{x^k\}$ is bounded and hence also that $\{\nabla f(x^k)\}$ is bounded. Use part (b).

1.2.14 (Divergence with Diminishing Stepsize)

Consider the one-dimensional function
$$f(x) = \frac{2}{3}|x|^3 + \frac{1}{2}x^2$$
and the method of steepest descent with stepsize $\alpha^k = \gamma/(k+1)$, where γ is a positive scalar.

(a) Show that for $\gamma = 1$ and $|x^0| \ge 1$ the method diverges. In particular, show that $|x^k| \ge k+1$ for all k.

(b) Characterize as best as you can the set of pairs (γ, x^0) for which the method converges to $x^* = 0$.

(c) How do you reconcile the results of (a) and (b) with Prop. 1.2.3.

1.2.15 (Wolfe Conditions for Line Search Accuracy)

There are several criteria for implementing approximately the minimization rule in a gradient method. An example of such a criterion is that α^k satisfies simultaneously
$$f(x^k) - f(x^k + \alpha^k d^k) \ge -\sigma \alpha^k \nabla f(x^k)' d^k, \tag{1.36}$$
$$\nabla f(x^k + \alpha^k d^k)' d^k \ge \beta \nabla f(x^k)' d^k, \tag{1.37}$$
where α and β are some scalars with $\sigma \in (0, 1/2)$ and $\beta \in (\sigma, 1)$. If α^k is indeed a minimizing stepsize, then $\nabla f(x^k + \alpha^k d^k)' d^k = dg(\alpha^k)/d\alpha = 0$, where g is the function $g(\alpha) = f(x^k + \alpha d^k)$, so Eq. (1.37) is in effect a test on the accuracy of the minimization (see Fig. 1.2.10).

(a) Show that if conditions (1.36) and (1.37) are satisfied by a gradient method at each iteration and the direction sequence is gradient related, then all limit points of the generated sequence $\{x^k\}$ are stationary points of f.

(b) Assume that there is a scalar b such that $f(x) \ge b$. Show that there exists an interval $[c_1, c_2]$ with $0 < c_1 < c_2$, such that every $\alpha \in [c_1, c_2]$ satisfies Eqs. (1.36) and (1.37).

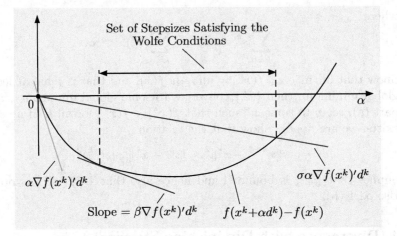

Figure 1.2.10. Illustration of the stepsize selection criterion based on the Wolfe conditions.

1.2.16 (Steepest Descent with Errors) (www)

Consider the steepest descent method $x^{k+1} = x^k - \alpha^k \left(\nabla f(x^k) + e^k \right)$, where e^k is an error satisfying $\|e^k\| \leq \delta$ for all k. Assume that ∇f is Lipschitz continuous. Show that for any $\delta' > \delta$, there exists a range of positive stepsizes $[\underline{\alpha}, \bar{\alpha}]$ such that if $\alpha^k \in [\underline{\alpha}, \bar{\alpha}]$ for all sufficiently large k, then either $f(x^k) \to -\infty$ or $\left\| \nabla f(x^k) \right\| < \delta'$ for infinitely many values of k. *Hint:* Use the reasoning of Prop. 1.2.2.

1.2.17 (Multiple Limit Points for Steepest Descent [Zou76])

Consider the two-dimensional function

$$f(x) = \begin{cases} (r-1)^2 - \frac{1}{2}(r-1)^2 \cos\left(\frac{1}{r-1} - \phi\right) & \text{if } r \neq 1, \\ 0 & \text{if } r = 1, \end{cases}$$

where

$$r = \sqrt{x_1^2 + x_2^2}, \qquad \phi = \arctan(x_1/x_2).$$

This function is minimized at each point of the circle where $r = 1$. Consider a nonoptimal starting point and the method of steepest descent where x^{k+1} is set equal to the first local minimum along the line $\left\{ x^k - \alpha \nabla f(x^k) \mid \alpha \geq 0 \right\}$. Show that this method follows a spiral path that comes arbitrarily close to every point of the circle of optimal points. *Note:* For another example of convergence to multiple limit points, which involves a convex differentiable cost function, see [Gon00]. In this example the steepest descent method with the exact line minimization rule produces a sequence with four limit points.

Sec. 1.2   Gradient Methods – Convergence   65

1.2.18 (Simplified Steepest Descent) (www)

(a) Consider the unconstrained minimization of a function f of the form

$$f(x) = F(x, g(x)),$$

where $g : \Re^m \to \Re^n$ is continuously differentiable and $F(x, y)$ is a continuously differentiable function of the two arguments $x \in \Re^n$ and $y \in \Re^m$. It is sometimes convenient to approximate the gradient of $F(x, g(x))$ by neglecting the dependence on g. This leads to the method

$$x^{k+1} = x^k - \alpha^k \nabla_x F(x^k, g(x^k)),$$

where α^k is chosen by the minimization rule or the Armijo rule on the function f. (Such a method makes sense when $\nabla_x F$ is much easier to compute than $\nabla g \nabla_y F$.) Show that if there exists $\gamma \in (0, 1)$ such that

$$\|\nabla g(x) \nabla_y F(x, g(x))\| \leq \gamma \|\nabla_x F(x, g(x))\|, \quad \forall \, x \in \Re^n,$$

then the method is convergent in the sense that all limit points of the sequences that it generates are stationary points of f.

(b) Consider the constrained minimization problem

$$\text{minimize } f(x, y)$$
$$\text{subject to } h(x, y) = 0$$

where $f : \Re^{n+m} \to \Re$ and $h : \Re^{n+m} \to \Re^m$ are continuously differentiable functions of the two arguments $x \in \Re^n$ and $y \in \Re^m$. Consider also a method of the form

$$x^{k+1} = x^k - \alpha^k \nabla_x f(x^k, y^k),$$

where y^k is a solution of $h(x^k, y) = 0$, viewed as a system of m equations in the unknown vector y, and α^k is chosen by the minimization rule or the Armijo rule. Formulate conditions that guarantee that this method is convergent.

1.2.19 (A Stepsize Reduction Rule for Convex Problems) (www)

Suppose that the cost function f is convex, and consider a gradient method $x^{k+1} = x^k + \alpha^k d^k$ where the assumptions of Prop. 1.2.3 are satisfied, except that the stepsize α^k is determined by the following rule:

$$\alpha^{k+1} = \begin{cases} \alpha^k & \text{if } \nabla f(x^{k+1})' d^k \leq 0, \\ \beta \alpha^k & \text{otherwise,} \end{cases}$$

where $\beta \in (0,1)$ is a fixed scalar and α^0 is any positive scalar.

(a) Show that the stepsize is reduced after iteration k if and only if the interval I^k connecting x^k and x^{k+1} contains in its interior all the vectors $\bar{x}^k$ that minimize $f(x)$ over $x \in I^k$.

(b) Show that the stepsize will be constant after a finite number of iterations.

(c) Show that either $f(x^k) \to -\infty$ or else $\{f(x^k)\}$ converges to a finite value and $\nabla f(x^k) \to 0$. Furthermore, every limit point of $\{x^k\}$ is a global minimum of f.

1.2.20 (Convergence of Gradient Method with Errors [BeT96], [BeT00]) (www)

Let $\{x^k\}$ be a sequence generated by the gradient method with errors

$$x^{k+1} = x^k + \alpha^k (d^k + e^k),$$

where the following hold:

(1) ∇f satisfies the Lipschitz assumption of Prop. 1.2.3.

(2) d^k satisfies

$$c_1 \|\nabla f(x^k)\|^2 \leq -\nabla f(x^k)' d^k, \quad \|d^k\| \leq c_2 (1 + \|\nabla f(x^k)\|), \quad \forall \, k,$$

where c_1 and c_2 are some scalars.

(3) The stepsizes α^k satisfy

$$\sum_{k=0}^{\infty} \alpha^k = \infty, \qquad \sum_{k=0}^{\infty} (\alpha^k)^2 < \infty.$$

(4) The errors e^k satisfy

$$\|e^k\| \leq \alpha^k (q + p\|\nabla f(x^k)\|), \quad \forall \, k,$$

where q and p are some scalars.

Show that either $f(x^k) \to -\infty$ or else $\{f(x^k)\}$ converges to a finite value and $\nabla f(x^k) \to 0$. Furthermore, every limit point of $\{x^k\}$ is a stationary point of f. *Hint:* Show that for sufficiently large k, we have

$$f(x^{k+1}) \leq f(x^k) - \alpha^k b_1 \|\nabla f(x^k)\|^2 + (\alpha^k)^2 b_2$$

for some constants b_1 and b_2. Use the line of argument of Prop. 1.2.3.

1.3 GRADIENT METHODS – RATE OF CONVERGENCE

The second major issue regarding gradient methods relates to the rate (or speed) of convergence of the generated sequences $\{x^k\}$. The mere fact that $\{x^k\}$ converges to a stationary point x^* will be of little practical value unless the points x^k are reasonably close to x^* after relatively few iterations. Thus, the study of the rate of convergence provides what are often the dominant criteria for selecting one algorithm in favor of others for solving a particular problem.

Approaches for Rate of Convergence Analysis

There are several approaches towards quantifying the rate of convergence of nonlinear programming algorithms. We will discuss briefly three possibilities and then concentrate on the third.

(a) *Computational complexity approach*: Here we try to estimate the number of elementary operations needed by a given method to find an optimal solution exactly or within an ϵ-tolerance. Usually, this approach provides worst-case estimates, that is, upper bounds on the number of required operations over a class of problems of given dimension and type (e.g. linear, convex, etc.). These estimates may also involve parameters such as the distance of the starting point from the optimal solution set, etc.

(b) *Informational complexity approach*: One difficulty with the computational complexity approach is that for a diverse class of problems, it is often difficult or meaningless to quantify the amount of computation needed for a single function or gradient evaluation. For example, in estimating the computational complexity of the gradient method applied to the entire class of differentiable convex functions, how are we to compare the overhead for finding the stepsize and for updating the x vector with the work needed to compute the cost function value and its gradient? The informational complexity approach, which is discussed in detail in the book [NeY83] (see also [TrW80]), bypasses this difficulty by estimating the number of function (and possibly gradient) evaluations needed to find an exact or approximately optimal solution (as opposed to the number of necessary computational operations). In other respects, the informational and computational complexity approaches are similar.

(c) *Local analysis*: In this approach we focus on the local behavior of the method in a neighborhood of an optimal solution. Local analysis can describe quite accurately the behavior of a method near the solution by using series approximations, but ignores entirely the behavior of the method when far from the solution.

The main potential advantage of the computational and informational complexity approaches is that they provide information about the method's progress when far from the eventual limit. Unfortunately, however, this information is usually pessimistic as it accounts for the worst possible problem instance within the class considered. This has resulted in some striking discrepancies between the theoretical model predictions and practical real-world observations. For example, the most widely used linear programming method, the simplex method, is categorized as a "bad" method by worst-case complexity analysis, because it performs very poorly on some specially constructed examples, which, however, are highly unlikely in practice. On the other hand, the ellipsoid method of Khachiyan [Kha79]† is categorized as much better than the simplex method by worst-case complexity analysis, even though it performs very poorly on most practical linear programs.

The computational complexity approach has received considerable attention in the context of interior point methods. These methods, discussed in Section 5.1, were primarily motivated by Karmarkar's development of a linear programming algorithm with a polynomial complexity bound that was more favorable than the one of the ellipsoid method [Kar84]. It turned out, however, that the worst-case predictions for the required number of iterations of these methods were off by many orders of magnitude from the practically observed number of iterations. Furthermore, the interior point methods that perform best in practice have poor worst-case complexity, while the ones with the best complexity bounds are very slow in practice.

The local analysis approach, which will be adopted almost exclusively in this text, has enjoyed considerable success in predicting the behavior of various methods near nonsingular local minima where the cost function can be well approximated by a quadratic. Moreover it is often more intuitive than the computational complexity approach, and lends itself better to geometrical interpretations. However, the local analysis approach also has some drawbacks, the most important of which is that it does not account for the rate of progress in the initial iterations. Nonetheless, in many practical situations this is not a serious omission because progress is fast in the initial iterations and slows down only in the limit (the reasons for this seem hard to understand; they are problem-dependent). Furthermore, often in practice, starting points that are near a solution are easily obtainable by a combination of heuristics and experience from problems with similar data, in which case local analysis becomes more meaningful.

Local analysis has not been very successful for problems which either involve singular local minima or which are difficult in the sense that the principal methods take many iterations to get near their solution where local analysis applies. Theoretical guidance to help a practitioner who is

---

† The ellipsoid method was chronologically the first linear programming algorithm with a polynomial complexity bound; see [BGT81] or [BeT97] for a survey and discussion of this method.

Sec. 1.3 Gradient Methods – Rate of Convergence

faced with such problems is an important subject, which is still under active development.

1.3.1 The Local Analysis Approach

We now formalize the basic ingredients of our local rate of convergence analysis approach. These are:

(a) We restrict attention to sequences $\{x^k\}$ that converge to a unique limit point x^*.

(b) Rate of convergence is evaluated using an *error function* $e : \Re^n \mapsto \Re$ satisfying $e(x) \geq 0$ for all $x \in \Re^n$ and $e(x^*) = 0$. Typical choices are the Euclidean distance

$$e(x) = \|x - x^*\|$$

and the cost difference

$$e(x) = |f(x) - f(x^*)|.$$

(c) Our analysis is asymptotic; that is, we look at the rate of convergence of the tail of the error sequence $\{e(x^k)\}$.

(d) The generated error sequence $\{e(x^k)\}$ is compared with some "standard" sequences. In our case, we compare $\{e(x^k)\}$ with the geometric progression

$$\beta^k, \qquad k = 0, 1, \ldots,$$

where $\beta \in (0,1)$ is some scalar. In particular, we say that $\{e(x^k)\}$ *converges linearly or geometrically*, if there exist $q > 0$ and $\beta \in (0,1)$ such that for all k

$$e(x^k) \leq q\beta^k.$$

It is possible to show that linear convergence is obtained if for some $\beta \in (0,1)$ we have

$$\limsup_{k \to \infty} \frac{e(x^{k+1})}{e(x^k)} \leq \beta;$$

that is, asymptotically, the error is decreasing by a factor of at least β at each iteration (see Exercise 1.3.6, which gives several additional convergence rate characterizations). If for every $\beta \in (0,1)$, there exists q such that the condition $e(x^k) \leq q\beta^k$ holds for all k, we say that $\{e(x^k)\}$ converges *superlinearly*. This is true in particular, if

$$\lim_{k \to \infty} \frac{e(x^{k+1})}{e(x^k)} = 0.$$

To quantify further the notion of superlinear convergence, we may compare $\{e(x^k)\}$ with the sequence

$$(\beta)^{p^k}, \quad k = 0, 1, \ldots,$$

(this is β raised to the power p raised to the power k) where $\beta \in (0, 1)$, and $p > 1$ are some scalars. This sequence converges much faster than a geometric progression. We say that $\{e(x^k)\}$ *converges at least superlinearly with order p*, if there exist $q > 0$, $\beta \in (0, 1)$, and $p > 1$ such that for all k

$$e(x^k) \leq q\,(\beta)^{p^k}.$$

The case where $p = 2$ is referred to as *quadratic convergence*. It is possible to show that superlinear convergence with order p is obtained if

$$\limsup_{k \to \infty} \frac{e(x^{k+1})}{e(x^k)^p} < \infty,$$

or equivalently, $e(x^{k+1}) = O\bigl(e(x^k)^p\bigr)$; see Exercise 1.3.7.

Most optimization algorithms that are of interest in practice produce sequences converging either linearly or superlinearly, at least when they converge to nonsingular local minima. Linear convergence is a fairly satisfactory rate of convergence for nonlinear programming algorithms, provided the factor β of the associated geometric progression is not too close to unity. Several nonlinear programming algorithms converge superlinearly for particular classes of problems. Newton's method is an important example, as we will see in the present section and also in Section 1.4. For convergence to singular local minima, slower than linear convergence rate is expected for most cases. One may then compare $\{e(x^k)\}$ with some standard sequences that converge sublinearly, such as $\{qk^{-p}\}$, where $q > 0$ and $p \geq 1$.

1.3.2 The Role of the Condition Number

Many of the important convergence rate characteristics of gradient methods reveal themselves when the cost function is quadratic. To see why, assume that a gradient method is applied to minimization of a twice continuously differentiable function $f : \Re^n \mapsto \Re$, and it generates a sequence $\{x^k\}$ converging to a nonsingular local minimum x^*. We have

$$f(x) = f(x^*) + \tfrac{1}{2}(x - x^*)'\nabla^2 f(x^*)(x - x^*) + o(\|x - x^*\|^2).$$

Therefore, since $\nabla^2 f(x^*)$ is positive definite, f can be accurately approximated near x^* by the quadratic function

$$f(x^*) + \tfrac{1}{2}(x - x^*)'\nabla^2 f(x^*)(x - x^*).$$

Sec. 1.3   Gradient Methods – Rate of Convergence

We thus expect that asymptotic convergence rate results obtained for the quadratic cost case have direct analogs for the general case. This conjecture has been substantiated by extensive numerical experimentation, and is one of the most reliable analytical guidelines in nonlinear programming.

For this reason, we take the positive definite quadratic case as our point of departure. In Exercise 1.3.3 we extend our analysis for the case where f has Lipschitz continuous gradient and is strongly convex (and hence also has positive definite Hessian when it is twice differentiable). We also discuss later in this section what happens when $\nabla^2 f(x^*)$ is not positive definite, in which case an analysis based on a quadratic model is inadequate.

Convergence Rate of Steepest Descent for Quadratic Functions

Suppose that the cost function f is quadratic with positive definite Hessian Q. We may assume without loss of generality that f is minimized at $x^* = 0$ and that $f(x^*) = 0$ [otherwise we can use the change of variables $y = x - x^*$ and subtract the constant $f(x^*)$ from $f(x)$]. Thus we have

$$f(x) = \tfrac{1}{2} x'Qx, \qquad \nabla f(x) = Qx, \qquad \nabla^2 f(x) = Q.$$

The steepest descent method takes the form

$$x^{k+1} = x^k - \alpha^k \nabla f(x^k) = (I - \alpha^k Q)x^k.$$

Therefore, we have

$$\|x^{k+1}\|^2 = x^{k'}(I - \alpha^k Q)^2 x^k.$$

Since by Prop. A.18(b) of Appendix A, we have for all $x \in \Re^n$

$$x'(I - \alpha^k Q)^2 x \leq \big(\text{maximum eigenvalue of } (I - \alpha^k Q)^2\big) \|x\|^2,$$

we obtain

$$\|x^{k+1}\|^2 \leq \big(\text{maximum eigenvalue of } (I - \alpha^k Q)^2\big) \|x^k\|^2.$$

Using Prop. A.13 of Appendix A, it can be seen that the eigenvalues of $(I - \alpha^k Q)^2$ are equal to $(1 - \alpha^k \lambda_i)^2$, where λ_i are the eigenvalues of Q. Therefore, we have

$$\text{maximum eigenvalue of } (I - \alpha^k Q)^2 = \max\big\{(1 - \alpha^k m)^2, (1 - \alpha^k M)^2\big\},$$

where

$$m : \text{ smallest eigenvalue of } Q,$$
$$M : \text{ largest eigenvalue of } Q.$$

It follows that for $x^k \neq 0$, we have

$$\frac{\|x^{k+1}\|}{\|x^k\|} \leq \max\{|1 - \alpha^k m|, |1 - \alpha^k M|\}. \tag{1.38}$$

It can be seen that this inequality holds as an equation if x^k is proportional to an eigenvector corresponding to m and if $|1 - \alpha^k m| \geq |1 - \alpha^k M|$. Otherwise, if $|1 - \alpha^k m| < |1 - \alpha^k M|$, the inequality holds as an equation if x^k is proportional to an eigenvector corresponding to M.

The relation (1.38) is a fundamental convergence rate bound for the steepest descent method with a constant stepsize, which admits an extension to the case of a strongly convex cost function with Lipschitz continuous gradient (see Exercise 1.3.3). Figure 1.3.1 illustrates the bound of Eq. (1.38) as a function of the stepsize α^k. It can be seen that the value of α^k that minimizes the bound is

$$\alpha^* = \frac{2}{M + m},$$

in which case

$$\frac{\|x^{k+1}\|}{\|x^k\|} \leq \frac{M - m}{M + m}. \tag{1.39}$$

This is the best convergence rate bound for steepest descent with constant stepsize.

There is another interesting convergence rate result, which holds when α^k is chosen by the line minimization rule. This result quantifies the rate at which the cost decreases and has the form

$$\frac{f(x^{k+1})}{f(x^k)} \leq \left(\frac{M - m}{M + m}\right)^2. \tag{1.40}$$

The above inequality is verified in Prop. 1.3.1, given in the next subsection, where we collect and prove the more formal results of this section. It can be shown that the inequality is sharp in the sense that given any Q, there is a starting point x^0 such that this inequality holds as an equation for all k (see Fig. 1.3.2).

The ratio M/m is called the *condition number* of Q, and problems where M/m is large are referred to as *ill-conditioned*. Such problems are characterized by very elongated elliptical level sets. The steepest descent method converges slowly for these problems as indicated by the convergence rate bounds of Eqs. (1.38) and (1.40), and as illustrated in Fig. 1.3.2.

Scaling and Steepest Descent

Consider now the more general method

$$x^{k+1} = x^k - \alpha^k D^k \nabla f(x^k), \tag{1.41}$$

Sec. 1.3   Gradient Methods – Rate of Convergence   73

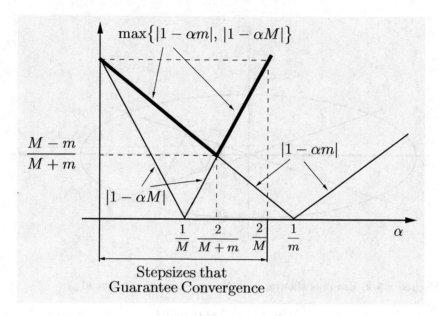

Figure 1.3.1. Illustration of the convergence rate bound

$$\frac{\|x^{k+1}\|}{\|x^k\|} \leq \max\{|1-\alpha m|, |1-\alpha M|\}$$

for steepest descent. The bound is minimized when α is such that $1 - \alpha m = \alpha M - 1$, i.e., for $\alpha = 2/(M+m)$.

where D^k is positive definite and symmetric; most of the gradient methods of interest have this form as discussed in Section 1.2. It turns out that we may view this iteration as a *scaled version of steepest descent*. In particular, this iteration is just steepest descent applied in a different coordinate system, which depends on D^k.

Indeed, let

$$S = (D^k)^{1/2}$$

denote the positive definite square root of D^k (cf. Prop. A.21 in Appendix A), and consider a transformation of variables defined by

$$x = Sy.$$

Then, in the space of y, the problem is written as

$$\begin{aligned} \text{minimize} \quad & h(y) \equiv f(Sy) \\ \text{subject to} \quad & y \in \Re^n. \end{aligned} \qquad (1.42)$$

The steepest descent method for this problem takes the form

$$y^{k+1} = y^k - \alpha^k \nabla h(y^k). \qquad (1.43)$$

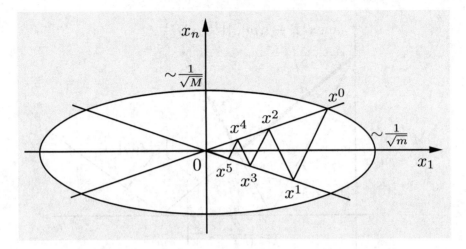

Figure 1.3.2. Example showing that the convergence rate bound

$$\frac{f(x^{k+1})}{f(x^k)} \leq \left(\frac{M-m}{M+m}\right)^2$$

is sharp for the steepest descent method with the line minimization rule. Consider the quadratic function

$$f(x) = \tfrac{1}{2}\sum_{i=1}^n \lambda_i x_i^2,$$

where $0 < m = \lambda_1 \leq \lambda_2 \leq \cdots \leq \lambda_n = M$. Any positive definite quadratic function can be put into this form by transformation of variables. Consider the starting point

$$x^0 = \left(m^{-1}, 0, \ldots, 0, M^{-1}\right)'$$

and apply the steepest descent method $x^{k+1} = x^k - \alpha^k \nabla f(x^k)$ with α^k chosen by the line minimization rule. We have $\nabla f(x^0) = (1, 0, \ldots, 0, 1)'$ and it can be verified that the minimizing stepsize is $\alpha^0 = 2/(M+m)$. Thus we obtain $x_1^1 = 1/m - 2/(M+m)$, $x_n^1 = 1/M - 2/(M+m)$, $x_i^1 = 0$ for $i = 2, \ldots, n-1$. Therefore,

$$x^1 = \left(\frac{M-m}{M+m}\right)\left(m^{-1}, 0, \ldots, 0, -M^{-1}\right)'$$

and, we can verify by induction that for all k,

$$x^{2k} = \left(\frac{M-m}{M+m}\right)^{2k} x^0, \qquad x^{2k+1} = \left(\frac{M-m}{M+m}\right)^{2k} x^1.$$

Thus, there exist starting points on the plane of points x of the form $x = (\xi_1, 0, \ldots, 0, \xi_n)'$, $\xi_1 \in \Re$, $\xi_n \in \Re$, in fact two lines shown in the figure, for which steepest descent converges in a way that the inequality

$$\frac{f(x^{k+1})}{f(x^k)} \leq \left(\frac{M-m}{M+m}\right)^2$$

is satisfied as an equation at each iteration.

Sec. 1.3 Gradient Methods – Rate of Convergence

Multiplying with S, we obtain

$$Sy^{k+1} = Sy^k - \alpha^k S \nabla h(y^k).$$

By passing back to the space of x, using the relations

$$x^k = Sy^k, \qquad \nabla h(y^k) = S\nabla f(x^k), \qquad S^2 = D^k, \qquad (1.44)$$

we obtain

$$x^{k+1} = x^k - \alpha^k D^k \nabla f(x^k).$$

Thus the above gradient iteration is nothing but the steepest descent method (1.43) in the space of y.

We now apply the convergence rate results for steepest descent to the scaled iteration $y^{k+1} = y^k - \alpha^k \nabla h(y^k)$, obtaining

$$\frac{\|y^{k+1}\|}{\|y^k\|} \leq \max\{|1 - \alpha^k m^k|, |1 - \alpha^k M^k|\}$$

and

$$\frac{h(y^{k+1})}{h(y^k)} \leq \left(\frac{M^k - m^k}{M^k + m^k}\right)^2,$$

[cf. the convergence rate bounds (1.38) and (1.40), respectively], where m^k and M^k are the smallest and largest eigenvalues of the Hessian $\nabla^2 h(y)$, which is equal to $S\nabla^2 f(x)S = (D^k)^{1/2}Q(D^k)^{1/2}$. Using the equations

$$y^k = (D^k)^{-1/2}x^k, \qquad y^{k+1} = (D^k)^{-1/2}x^{k+1}$$

to pass back to the space of x, we obtain the convergence rate bounds

$$\frac{x^{k+1'}(D^k)^{-1}x^{k+1}}{x^{k'}(D^k)^{-1}x^k} \leq \max\{(1 - \alpha^k m^k)^2, (1 - \alpha^k M^k)^2\} \qquad (1.45)$$

and

$$\frac{f(x^{k+1})}{f(x^k)} \leq \left(\frac{M^k - m^k}{M^k + m^k}\right)^2, \qquad (1.46)$$

where

m^k : smallest eigenvalue of $(D^k)^{1/2}Q(D^k)^{1/2}$,

M^k : largest eigenvalue of $(D^k)^{1/2}Q(D^k)^{1/2}$.

The stepsize that minimizes the right-hand side bound of Eq. (1.45) is

$$\frac{2}{M^k + m^k}. \qquad (1.47)$$

The important point is that if M^k/m^k is much larger than unity, the convergence rate can be very slow, even if an optimal stepsize is used. Furthermore, we see that it is desirable to choose D^k as close as possible to Q^{-1}, so that $(D^k)^{1/2}$ is close to $Q^{-1/2}$ (cf. Prop. A.21 in Appendix A) and $M^k \approx m^k \approx 1$. Note that if D^k is so chosen, Eq. (1.47) shows that the stepsize $\alpha = 1$ is near optimal.

Diagonal Scaling

Many practical problems are ill-conditioned because of poor relative scaling of the optimization variables. By this we mean that the units in which the variables are expressed are incongruent in the sense that single unit changes of different variables have disproportionate effects on the cost.

As an example, consider a financial problem with two variables, *investment* denoted x_1 and expressed in dollars, and *interest rate* denoted x_2 and expressed in percentage points. If the effect on the cost function f due to a million dollar increment of investment is comparable to the effect due to a percentage point increment of interest rate, then the condition number will be of the order of 10^{12}!! [This rough calculation is based on estimating the condition number by the ratio

$$\frac{\partial^2 f(x_1, x_2)}{(\partial x_2^2)} \bigg/ \frac{\partial^2 f(x_1, x_2)}{(\partial x_1)^2},$$

approximating the second partial derivatives by the finite difference formulas

$$\frac{\partial^2 f(x_1, x_2)}{(\partial x_1)^2} \approx \frac{f(x_1 + h_1, x_2) + f(x_1 - h_1, x_2) - 2f(x_1, x_2)}{h_1^2},$$

$$\frac{\partial^2 f(x_1, x_2)}{(\partial x_2)^2} \approx \frac{f(x_1, x_2 + h_2) + f(x_1, x_2 - h_2) - 2f(x_1, x_2)}{h_2^2},$$

and using the relations $f(x_1 + h_1, x_2) \approx f(x_1, x_2 + h_2)$, $f(x_1 - h_1, x_2) \approx f(x_1, x_2 - h_2)$, and $h_1 = 10^6$, $h_2 = 1$, which express the comparability of the effects of a million dollar investment increment and an interest rate percentage point increment.]

The ill-conditioning in such problems can be significantly alleviated by changing the units in which the optimization variables are expressed, which amounts to diagonal scaling of the variables. By this, we mean working in a new coordinate system of a vector y related to x by a transformation,

$$x = Sy,$$

where S is a diagonal matrix. In the absence of further information, a reasonable choice of S is one that makes all the diagonal elements of the Hessian of the cost

$$S\nabla^2 f(x) S$$

in the y-coordinate system approximately equal to unity. For this, we must have

$$s_i \approx \left(\frac{\partial^2 f(x)}{(\partial x_i)^2}\right)^{-1/2},$$

where s_i is the ith diagonal element of S. As discussed earlier, we may express any gradient algorithm in the space of variables y as a gradient algorithm in the space of variables x. In particular, steepest descent in the y-coordinate system, when translated in the x-coordinate system, yields the *diagonally scaled steepest descent method*

$$x^{k+1} = x^k - \alpha^k D^k \nabla f(x^k),$$

where

$$D^k = \begin{pmatrix} d_1^k & 0 & 0 & \cdots & 0 & 0 & 0 \\ 0 & d_2^k & 0 & \cdots & 0 & 0 & 0 \\ \vdots & \vdots & \vdots & \ddots & \vdots & \vdots & \vdots \\ 0 & 0 & 0 & \cdots & 0 & d_{n-1}^k & 0 \\ 0 & 0 & 0 & \cdots & 0 & 0 & d_n^k \end{pmatrix},$$

and

$$d_i^k \approx \left(\frac{\partial^2 f(x^k)}{(\partial x_i)^2} \right)^{-1}.$$

This method is also valid for nonquadratic problems as long as d_i^k are chosen to be positive. It is not guaranteed to improve the convergence rate of steepest descent, but it is simple and often surprisingly effective in practice. In particular, it tends to automatically correct mismatches of the units in which the various optimization variables are expressed.

Nonquadratic Cost Functions

It is possible to show that our main conclusions on rate of convergence carry over to the nonquadratic case for sequences converging to nonsingular local minima. Conceptually, this makes sense because in the neighborhood of a nonsingular local minimum a twice continuously differentiable cost function is very close to a positive definite quadratic (up to second order). The technical details of the proofs of such results are straightforward, but tend to be uninsightful and tedious, and for the most part will be omitted.

More specifically, let f be twice continuously differentiable and consider the gradient method

$$x^{k+1} = x^k - \alpha^k D^k \nabla f(x^k), \tag{1.48}$$

where D^k is positive definite and symmetric. Consider a generated sequence $\{x^k\}$, and assume that

$$x^k \to x^*, \qquad \nabla f(x^*) = 0, \qquad \nabla^2 f(x^*) : \text{ positive definite}, \tag{1.49}$$

and that $x^k \neq x^*$ for all k. Then, denoting

$$m^k : \text{ smallest eigenvalue of } (D^k)^{1/2} \nabla^2 f(x^k)(D^k)^{1/2},$$

$$M^k : \text{ largest eigenvalue of } (D^k)^{1/2} \nabla^2 f(x^k)(D^k)^{1/2},$$

it is possible to show the following:

(a) There holds

$$\limsup_{k\to\infty} \frac{(x^{k+1}-x^*)'(D^k)^{-1}(x^{k+1}-x^*)}{(x^k-x^*)'(D^k)^{-1}(x^k-x^*)}$$
$$= \limsup_{k\to\infty} \max\{|1-\alpha^k m^k|^2, |1-\alpha^k M^k|^2\}.$$

(b) If α^k is chosen by the line minimization rule, there holds

$$\limsup_{k\to\infty} \frac{f(x^{k+1})-f(x^*)}{f(x^k)-f(x^*)} \leq \limsup_{k\to\infty} \left(\frac{M^k-m^k}{M^k+m^k}\right)^2. \quad (1.50)$$

An alternative result for the case of the Armijo rule is given in Exercise 1.3.10 (with solution posted online).

From Eq. (1.50), we see that if D^k converges to some positive definite matrix as $x^k \to x^*$, the sequence $\{f(x^k)\}$ converges to $f(x^*)$ linearly. When

$$D^k \to \nabla^2 f(x^*)^{-1},$$

we have $\lim_{k\to\infty} M^k = \lim_{k\to\infty} m^k = 1$ and Eq. (1.50) shows that the convergence rate of $\{f(x^k)\}$ is superlinear. A somewhat more general version of this result for the case of the Armijo rule is given in Prop. 1.3.2 in the next subsection. In particular, it is shown that if the direction

$$d^k = -D^k \nabla f(x^k)$$

approaches asymptotically the Newton direction $-\left(\nabla^2 f(x^k)\right)^{-1} \nabla f(x^k)$ and the Armijo rule is used with initial stepsize equal to one, the rate of convergence is superlinear.

There is a consistent theme that emerges from our analysis, namely that to achieve asymptotically fast convergence of the gradient method

$$x^{k+1} = x^k - \alpha^k D^k \nabla f(x^k),$$

one should try to choose the matrices D^k as close as possible to $\left(\nabla^2 f(x^*)\right)^{-1}$ so that the maximum and minimum eigenvalues of $(D^k)^{1/2} \nabla^2 f(x^*)(D^k)^{1/2}$ satisfy $M^k \approx 1$ and $m^k \approx 1$. Furthermore, when D^k is so chosen, the initial stepsize $s = 1$ is a good choice for the Armijo rule and other related rules, or as a starting point for one-dimensional minimization procedures used in minimization stepsize rules. This finding has been supported by extensive numerical experience and is one of the most reliable guidelines for selecting and designing optimization algorithms for unconstrained problems. Note, however, that this guideline is valid only for problems where the cost function is twice differentiable and has positive definite Hessian near the points of interest. We discuss next problems where this condition is not satisfied.

Singular and Difficult Problems

Let us consider problems where the Hessian matrix either does not exist or is not positive definite at or near local minima of interest. Expressed mathematically, there are local minima x^* and directions d such that the slope of f along d, which is $\nabla f(x^* + \alpha d)'d$, changes very slowly or very rapidly with α, i.e., either

$$\lim_{\alpha \to 0} \frac{\nabla f(x^* + \alpha d)'d - \nabla f(x^*)'d}{\alpha} = 0, \qquad (1.51)$$

or

$$\lim_{\alpha \to 0} \frac{\nabla f(x^* + \alpha d)'d - \nabla f(x^*)'d}{\alpha} = \infty. \qquad (1.52)$$

The case of Eq. (1.51) is characterized by flatness of the cost along the direction d; large excursions from x^* along d produce small changes in cost. In the case of Eq. (1.52) the reverse is true; the cost rises steeply along d. An example is the function

$$f(x_1, x_2) = |x_1|^4 + |x_2|^{3/2},$$

where for the minimum $x^* = (0,0)$, Eq. (1.51) holds along the direction $d = (1,0)$ and Eq. (1.52) holds along the direction $d = (0,1)$. Gradient methods that use directions that are comparable in size to the gradient may require very large stepsizes in the case of Eq. (1.51) and very small stepsizes in the case of Eq. (1.52). This suggests potential difficulties in the implementation of a good stepsize rule; certainly a constant stepsize does not look like an attractive possibility. Furthermore, in the Armijo rule, the initial stepsize should not be taken constant; it should be adjusted according to a suitable scheme, although designing such a scheme may not be easy.

One may view the cases of Eqs. (1.51) and (1.52) as corresponding to an "infinite condition number," thereby suggesting slower than linear convergence rate for the method of steepest descent. Proposition 1.3.3 of the next subsection quantifies the rate of convergence of gradient methods for the case of a convex function whose gradient satisfies the Lipschitz condition

$$\|\nabla f(x) - \nabla f(y)\| \leq L\|x - y\|, \qquad (1.53)$$

for some L, and all x and y in a neighborhood of x^* [this assumption is consistent with the "flat" cost case of Eq. (1.51), but not with the "steep" cost case of Eq. (1.52)]. It is shown in particular that for a gradient method with several types of stepsize rules, we have

$$f(x^k) - f(x^*) = o(1/k).$$

This type of estimate suggests that for many practical singular problems one may be unable to obtain a highly accurate approximation of an optimal solution. In the "steep" cost case where Eq. (1.52) holds for some directions d, computational examples suggest that the rate of convergence can be slower than linear for the method of steepest descent, although a formal analysis of this conjecture does not seem to have been published.

It should be noted that problems with singular local minima are not the only ones for which gradient methods may converge slowly. There are problems where a given method may have excellent asymptotic rate of convergence, but its progress when far from the eventual limit can be very slow. A prominent example is when the cost function is continuously differentiable but its Hessian matrix is discontinuous and possibly singular in some regions that are outside a small neighborhood of the solution; such functions arise for example in augmented Lagrangian methods for inequality constrained problems (see Section 5.2). Then the powerful Newton-like methods may require a very large number of iterations to get to the small neighborhood of the eventual limit where their convergence rate is favorable. What happens here is that these methods use second derivative information in sophisticated ways, but this information may be misleading due to the Hessian discontinuities.

Generally, there is a tendency to think that difficult problems should be addressed with sophisticated methods, such as Newton-like methods. This is often true, particularly for problems with nonsingular local minima that are poorly conditioned. However, it is important to realize that *often the reverse is true*, namely that for problems with "difficult" cost functions and singular local minima, it is best to use simple methods such as (perhaps diagonally scaled) steepest descent with simple stepsize rules such as a constant or a diminishing stepsize. The reason is that methods that use sophisticated descent directions and stepsize rules often rely on assumptions that are likely to be violated in difficult problems. We also note that for difficult problems, it may be helpful to supplement the steepest descent method with features that allow it to deal better with multiple local minima and peculiarities of the cost function. An often useful modification is to introduce extrapolation based on the preceding two iterates, which we discuss next.

Steepest Descent with Extrapolation

A variant of the steepest descent method, known as *gradient method with momentum*, involves extrapolation along the direction of the difference of the preceding two iterates:

$$x^{k+1} = x^k - \alpha^k \nabla f(x^k) + \beta^k (x^k - x^{k-1}), \qquad (1.54)$$

where β^k is a scalar in $[0, 1)$, and we define $x_{-1} = x_0$. When α^k and β^k are chosen to be constant scalars α and β, respectively, the method is known as the *heavy ball method* [Pol64]; see Fig. 1.3.3. This is a sound method with

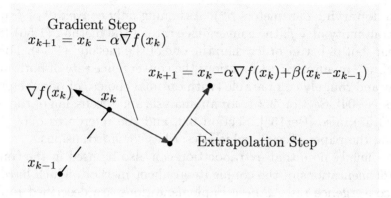

Figure 1.3.3. Illustration of the heavy ball method (1.54), where $\alpha^k \equiv \alpha$ and $\beta^k \equiv \beta$.

guaranteed convergence under a Lipschitz continuity assumption on ∇f. It can be shown to have faster convergence rate than the corresponding gradient method where $\alpha^k \equiv \alpha$ and $\beta^k \equiv 0$. In particular, for a positive definite quadratic problem with minimum at x^*, and with optimal choices of the constants α and β, the convergence rate of the heavy ball method is linear, and is governed by the formula

$$\frac{\|x^{k+1} - x^*\|}{\|x^k - x^*\|} \le \frac{\sqrt{M} - \sqrt{m}}{\sqrt{M} + \sqrt{m}}; \tag{1.55}$$

(see Exercise 1.3.8, whose solution is posted online; also see [GFJ15] for a related convergence analysis). This formula has the same form as the one for the steepest descent method, but with M/m replaced by $\sqrt{M/m}$, which is a substantial improvement. Simple examples also suggest that with extrapolation, the steepest descent method is less prone to getting trapped at "shallow" local minima, and deals better with cost functions that are alternately very flat and very steep along the path of the algorithm.

A method with similar structure as (1.54), proposed in [Nes83] and often called *Nesterov's method*, has received a lot of attention because it has theoretically interesting complexity properties. The iteration of this method is commonly described in two steps: first an extrapolation step, to compute

$$y^k = x^k + \beta^k(x^k - x^{k-1}) \tag{1.56}$$

with β^k chosen in a special way so that $\beta^k \to 1$, and then a gradient step with constant stepsize α, and gradient calculated at y^k,

$$x^{k+1} = y^k - \alpha \nabla f(y^k). \tag{1.57}$$

Compared to the method (1.54), it reverses the order of gradient calculation and extrapolation, and uses $\nabla f(y^k)$ in place of $\nabla f(x^k)$ (in addition to using time-varying parameters β^k). Assuming only convexity of f and Lipschitz continuity of ∇f, the convergence rate of the method (1.56)-(1.57) is sublinear, but of better order than the one of the method (1.54). For a positive definite quadratic cost function, the convergence rate of both methods is linear and roughly comparable (with optimal choices of parameters). We refer to [Nes04], Section 2.2.1, for an analysis and discussion of the method (1.56)-(1.57); also [Ber15a], Section 6.2, and the references quoted there, including the paper [Tse08], which describes some extensions.

We finally note that extrapolation can also be used in the context of two implementations of the conjugate gradient method, which have superlinear convergence rate. These implementations are described in Section 2.1; see Exercises 2.1.5 and 2.1.6 (with solutions posted online).

1.3.3 Convergence Rate Results

We first derive the convergence rate of steepest descent with the minimization stepsize rule when the cost is quadratic.

Proposition 1.3.1: Consider the quadratic function

$$f(x) = \tfrac{1}{2} x' Q x, \tag{1.58}$$

where Q is positive definite and symmetric, and the method of steepest descent

$$x^{k+1} = x^k - \alpha^k \nabla f(x^k), \tag{1.59}$$

where the stepsize α^k is chosen according to the minimization rule

$$f\bigl(x^k - \alpha^k \nabla f(x^k)\bigr) = \min_{\alpha \geq 0} f\bigl(x^k - \alpha \nabla f(x^k)\bigr).$$

Then, for all k,

$$f(x^{k+1}) \leq \left(\frac{M - m}{M + m}\right)^2 f(x^k),$$

where M and m are the largest and smallest eigenvalues of Q, respectively.

Proof: Let us denote

$$g^k = \nabla f(x^k) = Q x^k. \tag{1.60}$$

The result clearly holds if $g^k = 0$, so we assume $g^k \neq 0$. We first compute the minimizing stepsize α^k. We have

$$\frac{d}{d\alpha} f(x^k - \alpha g^k) = -g^{k'} Q(x^k - \alpha g^k) = -g^{k'} g^k + \alpha g^{k'} Q g^k.$$

Sec. 1.3 Gradient Methods – Rate of Convergence   83

By setting this derivative equal to zero, we obtain

$$\alpha^k = \frac{g^{k\prime} g^k}{g^{k\prime} Q g^k}. \tag{1.61}$$

We have, using Eqs. (1.58)-(1.60),

$$\begin{aligned} f(x^{k+1}) &= \tfrac{1}{2}(x^k - \alpha^k g^k)' Q (x^k - \alpha^k g^k) \\ &= \tfrac{1}{2}\left(x^{k\prime} Q x^k - 2\alpha^k g^{k\prime} Q x^k + (\alpha^k)^2 g^{k\prime} Q g^k\right) \\ &= \tfrac{1}{2}\left(x^{k\prime} Q x^k - 2\alpha^k g^{k\prime} g^k + (\alpha^k)^2 g^{k\prime} Q g^k\right) \end{aligned}$$

and using Eq. (1.61),

$$f(x^{k+1}) = \tfrac{1}{2}\left(x^{k\prime} Q x^k - \frac{(g^{k\prime} g^k)^2}{g^{k\prime} Q g^k}\right).$$

Thus, using the fact $f(x^k) = \tfrac{1}{2} x^{k\prime} Q x^k = \tfrac{1}{2} g^{k\prime} Q^{-1} g^k$, we obtain

$$f(x^{k+1}) = \left(1 - \frac{(g^{k\prime} g^k)^2}{(g^{k\prime} Q g^k)(g^{k\prime} Q^{-1} g^k)}\right) f(x^k). \tag{1.62}$$

At this point we need the following lemma.

Lemma 3.1: (Kantorovich Inequality) Let Q be a positive definite and symmetric $n \times n$ matrix. Then for any vector $y \in \Re^n, y \neq 0$, there holds

$$\frac{(y'y)^2}{(y'Qy)(y'Q^{-1}y)} \geq \frac{4Mm}{(M+m)^2},$$

where M and m are the largest and smallest eigenvalues of Q, respectively.

Proof: Let $\lambda_1, \ldots, \lambda_n$ denote the eigenvalues of Q and assume that

$$0 < m = \lambda_1 \leq \lambda_2 \leq \cdots \leq \lambda_n = M.$$

Let S be the matrix consisting of the n orthogonal eigenvectors of Q, normalized so that they have unit norm (cf. Prop. A.17 in Appendix A). Then, it can be seen that $S'QS$ is diagonal with diagonal elements $\lambda_1, \ldots, \lambda_n$. By using if necessary a transformation of the coordinate system that replaces y by Sx, we may assume that Q is diagonal and that its diagonal elements are $\lambda_1, \ldots, \lambda_n$. We have for $y = (y_1, \ldots, y_n)' \neq 0$

$$\frac{(y'y)^2}{(y'Qy)(y'Q^{-1}y)} = \frac{\left(\sum_{i=1}^n y_i^2\right)^2}{\left(\sum_{i=1}^n \lambda_i y_i^2\right)\left(\sum_{i=1}^n \frac{y_i^2}{\lambda_i}\right)}.$$

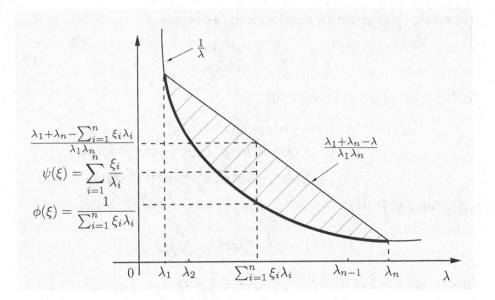

Figure 1.3.4. Proof of the Kantorovich inequality. Consider the function $1/\lambda$. The scalar $\sum_{i=1}^{n} \xi_i \lambda_i$ represents, for any $\xi = (\xi_1, \ldots, \xi_n)$ with $\xi_i \geq 0$, $\sum_{i=1}^{n} \xi_i = 1$, a point in the line segment $[\lambda_1, \lambda_n]$. Thus, the values $\phi(\xi) = 1/\sum_{i=1}^{n} \xi_i \lambda_i$ correspond to the thick part of the curve $1/\lambda$. On the other hand, the value $\psi(\xi) = \sum_{i=1}^{n} (\xi_i/\lambda_i)$ is a convex combination of $1/\lambda_1, \ldots, 1/\lambda_n$ and hence corresponds to a point in the shaded area in the figure. For the same vector ξ, both $\phi(\xi)$ and $\psi(\xi)$ are represented by points on the same vertical line. Hence,

$$\frac{\phi(\xi)}{\psi(\xi)} \geq \min_{\lambda_1 \leq \lambda \leq \lambda_n} \frac{\frac{1}{\lambda}}{\frac{\lambda_1 + \lambda_n - \lambda}{\lambda_1 \lambda_n}}.$$

The minimum is attained for $\lambda = (\lambda_1 + \lambda_n)/2$ and we obtain

$$\frac{\phi(\xi)}{\psi(\xi)} \geq \frac{4\lambda_1 \lambda_n}{(\lambda_1 + \lambda_n)^2},$$

which is used to show the result.

By letting

$$\xi_j = \frac{y_j^2}{\sum_{i=1}^{n} y_i^2}$$

and by defining

$$\phi(\xi) = \frac{1}{\sum_{i=1}^{n} \xi_i \lambda_i}, \qquad \psi(\xi) = \sum_{i=1}^{n} \frac{\xi_i}{\lambda_i},$$

we obtain

$$\frac{(y'y)^2}{(y'Qy)(y'Q^{-1}y)} = \frac{\phi(\xi)}{\psi(\xi)}.$$

Figure 1.3.4 shows that we have

$$\frac{\phi(\xi)}{\psi(\xi)} \geq \frac{4\lambda_1 \lambda_n}{(\lambda_1 + \lambda_n)^2},$$

which proves the desired inequality. **Q.E.D.**

Returning to the proof of Prop. 1.3.1, we have by using the Kantorovich inequality in Eq. (1.62)

$$f(x^{k+1}) \leq \left(1 - \frac{4Mm}{(M+m)^2}\right) f(x^k) = \left(\frac{M-m}{M+m}\right)^2 f(x^k).$$

Q.E.D.

The following proposition shows superlinear convergence for methods where d^k approaches the Newton direction $-(\nabla^2 f(x^*))^{-1} \nabla f(x^k)$ and the Armijo rule is used.

Proposition 1.3.2: (Superlinear Convergence of Newton-Like Methods) Let f be twice continuously differentiable. Consider a sequence $\{x^k\}$ generated by the gradient method $x^{k+1} = x^k + \alpha^k d^k$ and suppose that

$$x^k \to x^*, \qquad \nabla f(x^*) = 0, \qquad \nabla^2 f(x^*) : \text{ positive definite.} \quad (1.63)$$

Assume further that $\nabla f(x^k) \neq 0$ for all k and

$$\lim_{k \to \infty} \frac{\|d^k + (\nabla^2 f(x^*))^{-1} \nabla f(x^k)\|}{\|\nabla f(x^k)\|} = 0. \quad (1.64)$$

Then, if α^k is chosen by means of the Armijo rule with initial stepsize $s = 1$ and $\sigma < 1/2$, we have

$$\lim_{k \to \infty} \frac{\|x^{k+1} - x^*\|}{\|x^k - x^*\|} = 0. \quad (1.65)$$

Furthermore, there exists an integer $\bar{k} \geq 0$ such that $\alpha^k = 1$ for all $k \geq \bar{k}$ (i.e., eventually no reduction of the initial stepsize will be taking place).

Proof: We first prove that there exists a $\bar{k} \geq 0$ such that for all $k \geq \bar{k}$,

$$f(x^k + d^k) - f(x^k) \leq \sigma \nabla f(x^k)' d^k,$$

i.e., the unity initial stepsize passes the test of the Armijo rule. By the mean value theorem, we have

$$f(x^k + d^k) - f(x^k) = \nabla f(x^k)'d^k + \tfrac{1}{2}d^{k'}\nabla^2 f(\bar{x}^k)d^k,$$

where $\bar{x}^k$ is a point on the line segment joining x^k and $x^k + d^k$. Thus, it will be sufficient to show that for k sufficiently large, we have

$$\nabla f(x^k)'d^k + \tfrac{1}{2}d^k\nabla^2 f(\bar{x}^k)d^k \leq \sigma \nabla f(x^k)'d^k,$$

or equivalently,

$$(1-\sigma)p^{k'}q^k + \tfrac{1}{2}q^{k'}\nabla^2 f(\bar{x}^k)q^k \leq 0, \qquad (1.66)$$

where

$$p^k = \frac{\nabla f(x^k)}{\|\nabla f(x^k)\|}, \qquad q^k = \frac{d^k}{\|\nabla f(x^k)\|}.$$

Indeed, from the assumption (1.64), we have

$$q^k + \left(\nabla^2 f(x^*)\right)^{-1} p^k \to 0.$$

Since $\nabla^2 f(x^*)$ is positive definite and $\|p^k\| = 1$, it follows that $\{q^k\}$ is a bounded sequence, and in view of $q^k = d^k/\|\nabla f(x^k)\|$ and $\nabla f(x^k) \to 0$, we obtain $d^k \to 0$. Hence, $x^k + d^k \to x^*$, and it follows that $\bar{x}^k \to x^*$ and $\nabla^2 f(\bar{x}^k) \to \nabla^2 f(x^*)$. We now write Eq. (1.64) as

$$q^k = -\left(\nabla^2 f(x^*)\right)^{-1} p^k + \beta^k,$$

where $\{\beta^k\}$ denotes a vector sequence with $\beta^k \to 0$. By using the above relation and the fact $\nabla^2 f(\bar{x}^k) \to \nabla^2 f(x^*)$, we may write Eq. (1.66) as

$$(1-\sigma)p^{k'}\left(\nabla^2 f(x^*)\right)^{-1}p^k - \tfrac{1}{2}p^{k'}\left(\nabla^2 f(x^*)\right)^{-1}p^k \geq \gamma^k,$$

where $\{\gamma^k\}$ is some scalar sequence with $\gamma^k \to 0$. Thus Eq. (1.66) is equivalent to

$$\left(\tfrac{1}{2} - \sigma\right) p^{k'}\left(\nabla^2 f(x^*)\right)^{-1} p^k \geq \gamma^k.$$

Since $1/2 > \sigma$, $\|p^k\| = 1$, and $\nabla^2 f(x^*)$ is positive definite, the above relation holds for sufficiently large k. Thus, Eq. (1.66) holds, and it follows that the unity initial stepsize is acceptable for sufficiently large k.

To complete the proof, we note that from Eq. (1.64), we have

$$d^k + \left(\nabla^2 f(x^*)\right)^{-1}\nabla f(x^k) = \|\nabla f(x^k)\|\delta^k, \qquad (1.67)$$

where δ^k is some vector sequence with $\delta^k \to 0$. We have

$$\nabla f(x^k) = \nabla^2 f(x^*)(x^k - x^*) + o(\|x^k - x^*\|),$$

from which
$$\left(\nabla^2 f(x^*)\right)^{-1}\nabla f(x^k) = x^k - x^* + o(\|x^k - x^*\|),$$
$$\|\nabla f(x^k)\| = O(\|x^k - x^*\|).$$

Using the above two relations in Eq. (1.67), we obtain
$$d^k + x^k - x^* = o(\|x^k - x^*\|). \tag{1.68}$$

Since for sufficiently large k we have $d^k + x^k = x^{k+1}$, Eq. (1.68) yields
$$x^{k+1} - x^* = o(\|x^k - x^*\|),$$

from which
$$\lim_{k\to\infty} \frac{\|x^{k+1} - x^*\|}{\|x^k - x^*\|} = \lim_{k\to\infty} \frac{o(\|x^k - x^*\|)}{\|x^k - x^*\|} = 0.$$

Q.E.D.

Note that the equation
$$\lim_{k\to\infty} \frac{\|x^{k+1} - x^*\|}{\|x^k - x^*\|} = 0$$

[cf. Eq. (1.65)] implies that $\{\|x^k - x^*\|\}$ converges superlinearly (see Exercise 1.3.6). In particular, we see that Newton's method, combined with the Armijo rule with unity initial stepsize, has the property that when it converges to a local minimum x^* such that $\nabla^2 f(x^*)$ is positive definite, its rate of convergence is superlinear. The capture theorem (Prop. 1.2.3) together with the preceding proposition suggest that Newton-like methods with the Armijo rule and a unity initial stepsize converge to a local minimum x^* such that $\nabla^2 f(x^*)$ is positive definite, whenever they are started sufficiently close to such a local minimum. The proof of this is left as Exercise 1.3.4 for the reader (solution posted online).

We finally consider the convergence rate of gradient methods for singular problems with a convex cost function, and a stepsize that is either constant within an appropriate range (cf. Prop. 1.2.2), or is obtained by line minimization.

Proposition 1.3.3: (Convergence Rate of Gradient Methods for Singular Problems) Suppose that the cost function f is convex and its gradient satisfies for some L the Lipschitz condition

$$\|\nabla f(x) - \nabla f(y)\| \leq L\|x - y\|, \quad \forall\ x, y \in \Re^n. \tag{1.69}$$

Assume further that the set of global minima of f is nonempty and bounded. Consider a gradient method $x^{k+1} = x^k + \alpha^k d^k$ where for some $c > 0$ and all k we have

$$\nabla f(x^k)' d^k \leq -c \|\nabla f(x^k)\|^2, \qquad d^k \neq 0, \tag{1.70}$$

while α^k either satisfies for some $\epsilon \in (0, 1]$ and all k

$$\epsilon \leq \alpha^k \leq (2 - \epsilon)\bar{\alpha}^k, \tag{1.71}$$

where

$$\bar{\alpha}^k = \frac{|\nabla f(x^k)' d^k|}{L \|d^k\|^2},$$

or else satisfies $f(x^k + \alpha^k d^k) \leq f(x^k + \bar{\alpha}^k d^k)$. Then all limit points of $\{x^k\}$ are optimal and there exists at least one limit point. Moreover

$$f(x^k) - f^* = o(1/k),$$

where $f^* = \min_{x \in \Re^n} f(x)$ is the optimal value.

Proof: We assume that $\nabla f(x^k) \neq 0$; otherwise the method terminates finitely at a global minimum and the result holds trivially. Assume first that α^k is chosen by the rule (1.71). Then from the proof of Prop. 1.2.2, we have

$$f(x^k + \alpha^k d^k) - f(x^k) \leq -\tfrac{1}{2}\epsilon^2 |\nabla f(x^k)' d^k|.$$

Combining this relation with Eq. (1.70), we obtain

$$f(x^k + \alpha^k d^k) \leq f(x^k) - \frac{c\epsilon^2}{2} \|\nabla f(x^k)\|^2. \tag{1.72}$$

The above relation holds also for $\alpha^k = \bar{\alpha}^k$, so that

$$f(x^k + \bar{\alpha}^k d^k) \leq f(x^k) - \frac{c\epsilon^2}{2} \|\nabla f(x^k)\|^2.$$

Thus for any of the possible stepsizes chosen by the algorithm, we have

$$f(x^{k+1}) \leq f(x^k) - \frac{c\epsilon^2}{2} \|\nabla f(x^k)\|^2. \tag{1.73}$$

Let X^* be the set of global minima of f. Since X^* is nonempty and compact, all the level sets of f are compact (Prop. B.10 in Appendix B). This together with the monotone decrease of $\{f(x^k)\}$, shows that $\{x^k\}$ is

Sec. 1.3 Gradient Methods – Rate of Convergence

bounded. Hence, by Prop. 1.2.1, all limit points of $\{x^k\}$ belong to X^*, and the distance of x^k from X^*, defined by

$$d(x^k, X^*) = \min_{x^* \in X^*} \|x^k - x^*\|,$$

converges to 0 and $e^k \to 0$. Using the convexity of f, we also have for every global minimum x^*

$$f(x^k) - f(x^*) \leq \nabla f(x^k)'(x^k - x^*) \leq \|\nabla f(x^k)\| \cdot \|x^k - x^*\|,$$

from which, by minimizing over $x^* \in X^*$,

$$f(x^k) - f^* \leq \|\nabla f(x^k)\| d(x^k, X^*). \tag{1.74}$$

Let us denote for all k

$$e^k = f(x^k) - f^*.$$

Combining Eqs. (1.73) and (1.74), we obtain

$$e^{k+1} \leq e^k - \frac{c\epsilon^2 (e^k)^2}{2d(x^k, X^*)^2}, \qquad \forall\, k, \tag{1.75}$$

where we assume without loss of generality that $d(x^k, X^*) \neq 0$.

We will show that Eq. (1.75) implies that $e^k = o(1/k)$. Indeed we have

$$0 < e^{k+1} \leq e^k \left(1 - \frac{c\epsilon^2 e^k}{2d(x^k, X^*)^2}\right),$$

$$0 < 1 - \frac{c\epsilon^2 e^k}{2d(x^k, X^*)^2},$$

from which

$$(e^{k+1})^{-1} \geq (e^k)^{-1}\left(1 - \frac{c\epsilon^2 e^k}{2d(x^k, X^*)^2}\right)^{-1} \geq (e^k)^{-1}\left(1 + \frac{c\epsilon^2 e^k}{2d(x^k, X^*)^2}\right)$$

$$= (e^k)^{-1} + \frac{c\epsilon^2}{2d(x^k, X^*)^2}.$$

Summing this inequality over all k, we obtain

$$e^k \leq \left((e^0)^{-1} + \frac{c\epsilon^2}{2}\sum_{i=0}^{k-1} d(x^i, X^*)^{-2}\right)^{-1},$$

or

$$ke^k \leq \left(\frac{1}{ke^0} + \frac{c\epsilon^2}{2k}\sum_{i=0}^{k-1} d(x^i, X^*)^{-2}\right)^{-1}. \tag{1.76}$$

Since $d(x^i, X^*) \to 0$, we have $d(x^i, X^*)^{-2} \to \infty$ and

$$\frac{c\epsilon^2}{2k} \sum_{i=0}^{k-1} d(x^i, X^*)^{-2} \to \infty.$$

Therefore the right-hand side of Eq. (1.76) tends to 0, implying that $e^k = o(1/k)$. **Q.E.D.**

The key step in the preceding proof is that the stepsize rule is such that Eq. (1.73) holds. Indeed the proof goes through for any stepsize rule for which we have

$$f(x^{k+1}) \le f(x^k) - \beta \|\nabla f(x^k)\|^2$$

for some constant $\beta > 0$ and all k. The proof can also be modified for the case where the Lipschitz condition (1.69) holds within the level set $\{x \mid f(x) \le f(x^0)\}$. Moreover, with additional assumptions on the structure of f, some more precise convergence rate results can be obtained. In particular, if f is convex, has a unique minimum x^*, and satisfies the following growth condition

$$f(x) - f(x^*) \ge q\|x - x^*\|^\beta, \qquad \forall \, x \text{ such that } f(x) \le f(x^0),$$

for some scalars $q > 0$ and $\beta > 2$, it can be shown (see [Dun81]) that for the method of steepest descent with the Armijo rule we have

$$f(x^k) - f(x^*) = O\left(\frac{1}{k^{\frac{\beta}{\beta-2}}}\right).$$

EXERCISES

1.3.1

Estimate the rate of convergence of steepest descent with the line minimization rule when applied to the function of two variables $f(x,y) = x^2 + 1.999xy + y^2$. Find a starting point for which this estimate is sharp (cf. Fig. 1.3.2).

1.3.2

Consider a positive definite quadratic problem with Hessian matrix Q. Suppose we use scaling with the diagonal matrix whose ith diagonal element is q_{ii}^{-1}, where q_{ii} is the ith diagonal element of Q. Show that if Q is 2×2, this diagonal scaling improves the condition number of the problem and the convergence rate of steepest descent. (*Note*: This need not be true for dimensions higher than 2.)

1.3.3 (Linear Convergence under Strong Convexity)

Let f be differentiable and satisfy the Lipschitz condition

$$\|\nabla f(x) - \nabla f(y)\| \leq L\|x - y\|, \qquad \forall\, x, y \in \Re^n.$$

Assume further that f is strongly convex, i.e., for some $\sigma > 0$, we have

$$\bigl(\nabla f(x) - \nabla f(y)\bigr)'(x - y) \geq \sigma\|x - y\|^2, \qquad \forall\, x, y \in \Re^n,$$

and let x^* be the unique minimum of f.

(a) Show that the mapping $G_\alpha(x) = x - \alpha \nabla f(x)$ of the steepest descent iteration with constant stepsize α satisfies

$$\|G_\alpha(x) - G_\alpha(y)\| \leq \max\bigl\{|1 - \alpha L|, |1 - \alpha \sigma|\bigr\} \|x - y\|, \qquad \forall\, x, y, \in \Re^n,$$

and is a contraction for all $\alpha \in (0, 2/L)$. *Abbreviated Proof:* For all $x, y \in \Re^n$, we have

$$\|G_\alpha(x) - G_\alpha(y)\|^2 = \bigl\|\bigl(x - \alpha\nabla f(x)\bigr) - \bigl(y - \alpha\nabla f(y)\bigr)\bigr\|^2.$$

Expanding the quadratic on the right-hand side, and using Prop. B.5(a), the Lipschitz condition, and the strong convexity condition, we obtain

$$\begin{aligned}
\|G_\alpha(x) - G_\alpha(y)\|^2 &\leq \|x-y\|^2 - 2\alpha\bigl(\nabla f(x) - \nabla f(y)\bigr)'(x-y) \\
&\quad + \alpha^2 \|\nabla f(x) - \nabla f(y)\|^2 \\
&\leq \|x - y\|^2 - \frac{2\alpha\sigma L}{\sigma + L}\|x-y\|^2 - \frac{2\alpha}{\sigma + L}\|\nabla f(x) - \nabla f(y)\|^2 \\
&\quad + \alpha^2 \|\nabla f(x) - \nabla f(y)\|^2 \\
&= \left(1 - \frac{2\alpha\sigma L}{\sigma + L}\right)\|x - y\|^2 \\
&\quad + \alpha\left(\alpha - \frac{2}{\sigma + L}\right)\|\nabla f(x) - \nabla f(y)\|^2 \\
&\leq \left(1 - \frac{2\alpha\sigma L}{\sigma + L}\right)\|x-y\|^2 \\
&\quad + \alpha \max\left\{L^2\left(\alpha - \frac{2}{\sigma + L}\right), \sigma^2\left(\alpha - \frac{2}{\sigma + L}\right)\right\}\|x-y\|^2 \\
&= \max\bigl\{(1-\alpha L)^2, (1-\alpha\sigma)^2\bigr\}\|x-y\|^2,
\end{aligned}$$

from which the desired inequality follows.

(b) Use part (a) to show that for the steepest descent method $x^{k+1} = G_\alpha(x^k)$ we have

$$\|x^{k+1} - x^*\| \leq \max\bigl\{|1 - \alpha\sigma|, |1 - \alpha L|\bigr\}\|x^k - x^*\|,$$

and that this relation generalizes the estimate of Eq. (1.38). From this relation, argue that the ratio L/σ plays the role of the condition number, which we have defined for twice differentiable f.

1.3.4 (Superlinear Convergence) (www)

Let f be twice continuously differentiable. Consider a sequence $\{x^k\}$ generated by the gradient method $x^{k+1} = x^k + \alpha^k d^k$ and suppose that x^* is a nonsingular local minimum. Assume that, for all k, $\nabla f(x^k) \neq 0$ and $d^k = d(x^k)$, where $d(\cdot)$ is a continuous function of x with

$$\lim_{x \to x^*, \nabla f(x) \neq 0} \frac{\left\| d(x) + \left(\nabla^2 f(x)\right)^{-1} \nabla f(x) \right\|}{\|\nabla f(x)\|} = 0.$$

Furthermore, α^k is chosen by means of the Armijo rule with initial stepsize $s = 1$ and $\sigma < 1/2$. Show that there exists an $\epsilon > 0$ such that if $\|x^0 - x^*\| < \epsilon$, then:

(a) $\{x^k\}$ converges to x^*.

(b) $\alpha^k = 1$ for all k.

(c) $\lim_{k \to \infty} \left(\|x^{k+1} - x^*\| / \|x^k - x^*\| \right) = 0$.

Hint: Use the line of argument of Prop. 1.3.2 together with the capture theorem (Prop. 1.2.3). Alternatively, instead of using the capture theorem, consult the proof of the subsequent Prop. 1.4.1.

1.3.5 (Steepest Descent with Errors)

Consider the steepest descent method

$$x^{k+1} = x^k - \alpha \left(\nabla f(x^k) + e^k \right),$$

where α is a constant stepsize, e^k is an error satisfying $\|e^k\| \leq \delta$ for all k, and f is the positive definite quadratic function

$$f(x) = \tfrac{1}{2}(x - x^*)'Q(x - x^*).$$

Let

$$q = \max\{|1 - \alpha m|, |1 - \alpha M|\},$$

where

$m :$ smallest eigenvalue of Q, $M :$ largest eigenvalue of Q,

and assume that $q < 1$. Show that for all k, we have

$$\|x^k - x^*\| \leq \frac{\alpha \delta}{1 - q} + q^k \|x^0 - x^*\|.$$

1.3.6 (Convergence Rate Characterizations [Ber82a], p. 14)

Consider a scalar sequence $\{e^k\}$ with $e^k \geq 0$ for all k, and $e^k \to 0$. We say that $\{e^k\}$ converges *faster than linearly with convergence ratio* β, where $0 < \beta < 1$, if for every $\bar{\beta} \in (\beta, 1)$ and $q > 0$, there exists $\bar{k}$ such that

$$e^k \leq q\bar{\beta}^k, \qquad \forall\, k \geq \bar{k}.$$

We say that $\{e^k\}$ converges *slower than linearly with convergence ratio* β, where $0 < \beta < 1$, if for every $\bar{\beta} \in (\beta, 1)$ and $q > 0$, there exists $\bar{k}$ such that

$$q\bar{\beta}^k \leq e^k, \qquad \forall\, k \geq \bar{k}.$$

We say that $\{e^k\}$ converges *linearly with convergence ratio* β if it converges both faster and slower than linearly with convergence ratio β. Show that:

(a) $\{e^k\}$ converges faster than linearly with convergence ratio β if and only if

$$\limsup_{k \to \infty} (e^k)^{1/k} \leq \beta.$$

$\{e^k\}$ converges slower than linearly with convergence ratio β if and only if

$$\liminf_{k \to \infty} (e^k)^{1/k} \geq \beta.$$

$\{e^k\}$ converges linearly with convergence ratio β if and only if

$$\lim_{k \to \infty} (e^k)^{1/k} = \beta.$$

(b) Assume that $e^k \neq 0$ for all k, and denote

$$\beta_1 = \liminf_{k \to \infty} \frac{e^{k+1}}{e^k}, \qquad \beta_2 = \limsup_{k \to \infty} \frac{e^{k+1}}{e^k}.$$

Show that if $0 < \beta_1 \leq \beta_2 < 1$, then $\{e^k\}$ converges faster than linearly with convergence ratio β_2 and slower than linearly with convergence ratio β_1. Furthermore, if $\beta_1 = \beta_2 = 0$, then $\{e^k\}$ converges superlinearly.

1.3.7

Consider a scalar sequence $\{e^k\}$ with $e^k > 0$ for all k, and $e^k \to 0$. Show that $\{e^k\}$ converges superlinearly with order p if

$$\limsup_{k \to \infty} \frac{e^{k+1}}{(e^k)^p} < \infty.$$

1.3.8 (The Heavy Ball Method [Pol64]) (www)

Consider the following variant of the steepest descent method:
$$x^{k+1} = x^k - \alpha \nabla f(x^k) + \beta(x^k - x^{k-1}), \qquad k = 1, 2, \ldots,$$
where α is a constant positive stepsize and β is a scalar with $0 < \beta < 1$.

(a) Let f be the quadratic function $f(x) = (1/2)x'Qx + c'x$, where Q is positive definite and symmetric, and let m and M be the minimum and the maximum eigenvalues of Q, respectively. Show that the method converges linearly to the unique solution if $0 < \alpha < 2(1+\beta)/M$. Show that with optimal choices of α and β, the ratio of linear convergence for the sequence $\{\|x^k - x^*\|\}$ is
$$\frac{\sqrt{M} - \sqrt{m}}{\sqrt{M} + \sqrt{m}};$$
cf. Eq. (1.55). [This is faster than the corresponding ratio of the steepest descent method where $\beta = 0$ and α is chosen optimally; cf. Eq. (1.39).]
Hint: Write the iteration as
$$\begin{pmatrix} x^{k+1} \\ x^k \end{pmatrix} = \begin{pmatrix} (1+\beta)I - \alpha Q & -\beta I \\ I & 0 \end{pmatrix} \begin{pmatrix} x^k \\ x^{k-1} \end{pmatrix}$$
and show that v is an eigenvalue of the matrix in the above equation if and only if $v + \beta/v$ is equal to $1 + \beta - \alpha \lambda$ where λ is an eigenvalue of Q. (This is a challenging exercise.)

(b) It is generally conjectured that in comparison to steepest descent, the method is less prone to getting trapped at "shallow" local minima, and tends to behave better for difficult problems where the cost function is alternatively very flat and very steep. Argue for or against this conjecture.

(c) In support of your answer in (b), write a computer program to test the method with $\beta = 0$ and $\beta > 0$ with one-dimensional cost functions of the form
$$f(x) = \tfrac{1}{2} x^2 \big(1 + \gamma \cos(x)\big),$$
where $\gamma \in (0, 1)$, and
$$f(x) = \tfrac{1}{2} \sum_{i=1}^{m} |z_i - \tanh(xy_i)|^2,$$
where z_i and y_i are given scalars.

1.3.9 (www)

Suppose that a vector sequence $\{e^k\}$ satisfies
$$\|e^{k+1} - e^k\| \le \beta \|e^k - e^{k-1}\|, \qquad \forall\, k \ge \bar{k},$$
where $\bar{k}$ is a positive integer and $\beta \in (0,1)$ is a scalar. Show that $\{e^k\}$ converges to some vector e^* linearly, and in fact we have
$$\|e^k - e^*\| \le q \beta^k$$
for some scalar q and all k. Hint: Show that $\{e^k\}$ is a Cauchy sequence.

1.3.10 (Convergence Rate of Steepest Descent with the Armijo Rule) (www)

Let $f : \Re^n \mapsto \Re$ be a twice continuously differentiable function that satisfies

$$m\|y\|^2 \leq y'\nabla^2 f(x)y \leq M\|y\|^2, \qquad \forall\ x, y \in \Re^n,$$

where m and M are some positive scalars. Consider the steepest descent method $x^{k+1} = x^k - \alpha^k \nabla f(x^k)$ with α^k determined by the Armijo rule. Let x^* be the unique unconstrained minimum of f and let

$$r = 1 - \frac{4m\beta\sigma(1-\sigma)}{M}.$$

Show that for all k, we have

$$f(x^{k+1}) - f(x^*) \leq r\big(f(x^k) - f(x^*)\big),$$

and

$$\|x^k - x^*\|^2 \leq qr^k,$$

where q is some constant.

1.4 NEWTON'S METHOD AND VARIATIONS

In the last two sections we emphasized a basic tradeoff in gradient methods: implementation simplicity versus fast convergence. We have already discussed steepest descent, one of the simplest but also one of the slowest methods. We now consider its opposite extreme, Newton's method, which is arguably the most complex and also the fastest of the gradient methods (under appropriate conditions).

Newton's method consists of the iteration

$$x^{k+1} = x^k - \alpha^k \big(\nabla^2 f(x^k)\big)^{-1} \nabla f(x^k), \tag{1.77}$$

assuming that the Newton direction

$$d^k = -\big(\nabla^2 f(x^k)\big)^{-1} \nabla f(x^k) \tag{1.78}$$

is defined [i.e., $\nabla^2 f(x^k)$ is invertible] and is a direction of descent [i.e., $d^{k'}\nabla f(x^k) < 0$]. As explained in the preceding section, one may view this iteration as a scaled version of steepest descent where the "optimal" scaling matrix $\big(\nabla^2 f(x^k)\big)^{-1}$ is used. It is worth mentioning in this connection that *Newton's method is "scale-free,"* in the sense that it cannot be affected by a change in coordinate system as is true for steepest descent (see Exercise 1.4.1).

When the Armijo rule is used with initial stepsize $s = 1$, then no reduction of the stepsize will be necessary near a nonsingular minimum (positive definite Hessian), as shown in Prop. 1.3.2. Thus, near convergence the method takes the form

$$x^{k+1} = x^k - \left(\nabla^2 f(x^k)\right)^{-1} \nabla f(x^k), \qquad (1.79)$$

which will be referred to as the *pure form of Newton's method*. On the other hand, far from such a local minimum, the Hessian matrix may be singular or the Newton direction of Eq. (1.78) may not be a direction of descent because the Hessian $\nabla^2 f(x^k)$ is not positive definite. Thus the analysis of Newton's method has two principal aspects:

(a) Local convergence, dealing with the behavior of the pure form of the method near a nonsingular local minimum.

(b) Global convergence, addressing the modifications that are necessary to ensure that the method is valid and is likely to converge to a local minimum when started far from all local minima.

We consider these issues in this section and we also discuss some variations of Newton's method, which are aimed at reducing the overhead for computing the Newton direction.

Local Convergence

It can be shown that the pure form of Newton's method converges superlinearly when started close enough to a nonsingular local minimum. This is suggested by the local convergence result for gradient methods (the capture theorem of Prop. 1.2.4) together with the superlinear convergence result for Newton-like methods (Prop. 1.3.2). Results of this type hold for a more general form of Newton's method, which can be used to solve the system of n equations with n unknowns

$$g(x) = 0, \qquad (1.80)$$

where $g : \Re^n \mapsto \Re^n$ is a continuously differentiable function. This method has the form

$$x^{k+1} = x^k - \left(\nabla g(x^k)'\right)^{-1} g(x^k), \qquad (1.81)$$

and for the special case where $g(x)$ is equal to the gradient $\nabla f(x)$, it yields the pure form of Eq. (1.79). Note here that a continuously differentiable function $g : \Re^n \mapsto \Re^n$ need not be equal to the gradient of some function. In particular, $g(x) = \nabla f(x)$ for some $f : \Re^n \mapsto \Re$, if and only if the $n \times n$ matrix $\nabla g(x)$ is symmetric for all x (see [OrR70], p. 95). Thus, the equation version of Newton's method [cf. Eq. (1.81)] is more broadly applicable than the optimization version [cf. Eq. (1.79)].

Sec. 1.4 Newton's Method and Variations

Here is a simple argument that shows the fast convergence of Newton's method (1.81). Suppose that the method generates a sequence $\{x^k\}$ that converges to a vector x^* such that $g(x^*) = 0$ and $\nabla g(x^*)$ is invertible. Let us use a first order expansion around x^k to write

$$0 = g(x^*) = g(x^k) + \nabla g(x^k)'(x^* - x^k) + o(\|x^k - x^*\|).$$

By multiplying this relation with $(\nabla g(x^k)')^{-1}$ we have

$$x^k - x^* - (\nabla g(x^k)')^{-1} g(x^k) = o(\|x^k - x^*\|),$$

so for the pure Newton iteration, $x^{k+1} = x^k - (\nabla g(x^k)')^{-1} g(x^k)$, we obtain

$$x^{k+1} - x^* = o(\|x^k - x^*\|).$$

Thus, for $x^k \neq x^*$,

$$\lim_{k \to \infty} \frac{\|x^{k+1} - x^*\|}{\|x^k - x^*\|} = \lim_{k \to \infty} \frac{o(\|x^k - x^*\|)}{\|x^k - x^*\|} = 0,$$

implying superlinear convergence. This argument can also be used to show convergence to x^* if the initial vector x^0 is sufficiently close to x^*. The following proposition proves a more detailed result.

Proposition 1.4.1: Consider a function $g : \Re^n \mapsto \Re^n$, and a vector x^* such that $g(x^*) = 0$. For $\delta > 0$, let S_δ denote the sphere $\{x \mid \|x - x^*\| \leq \delta\}$. Assume that g is continuously differentiable within some sphere $S_{\bar{\delta}}$ and that $\nabla g(x^*)$ is invertible.

(a) There exists $\delta > 0$ such that if $x^0 \in S_\delta$, the sequence $\{x^k\}$ generated by the iteration

$$x^{k+1} = x^k - (\nabla g(x^k)')^{-1} g(x^k)$$

is defined, belongs to S_δ, and converges to x^*. Furthermore, $\{\|x^k - x^*\|\}$ converges superlinearly.

(b) Assume that for some $L > 0$, $M > 0$, $\delta \in (0, \bar{\delta}]$, and for all x and y in S_δ,

$$\|\nabla g(x) - \nabla g(y)\| \leq L\|x - y\|, \qquad \|(\nabla g(x)')^{-1}\| \leq M. \quad (1.82)$$

Then, if $x^0 \in S_\delta$, we have

$$\|x^{k+1} - x^*\| \le \frac{LM}{2}\|x^k - x^*\|^2, \qquad \forall\, k = 0, 1, \ldots,$$

so if $LM\delta/2 < 1$ and $x^0 \in S_\delta$, $\{\|x^k - x^*\|\}$ converges superlinearly with order at least two.

Proof: (a) Choose $\delta > 0$ so that $\bigl(\nabla g(x)'\bigr)^{-1}$ exists for $x \in S_\delta$, and let

$$M = \sup_{x \in S_\delta} \left\|\bigl(\nabla g(x)'\bigr)^{-1}\right\|.$$

Assuming that $x^0 \in S_\delta$, and using the relation

$$g(x^k) = \int_0^1 \nabla g\bigl(x^* + t(x^k - x^*)\bigr)' dt\,(x^k - x^*),$$

we estimate $\|x^{k+1} - x^*\|$ as

$$\|x^{k+1} - x^*\| = \left\|x^k - x^* - \bigl(\nabla g(x^k)'\bigr)^{-1} g(x^k)\right\|$$

$$= \left\|\bigl(\nabla g(x^k)'\bigr)^{-1}\bigl(\nabla g(x^k)'(x^k - x^*) - g(x^k)\bigr)\right\|$$

$$= \left\|\bigl(\nabla g(x^k)'\bigr)^{-1}\left(\nabla g(x^k)' - \int_0^1 \nabla g\bigl(x^* + t(x^k - x^*)\bigr)' dt\right)(x^k - x^*)\right\|$$

$$= \left\|\bigl(\nabla g(x^k)'\bigr)^{-1}\left(\int_0^1 \bigl[\nabla g(x^k)' - \nabla g\bigl(x^* + t(x^k - x^*)\bigr)'\bigr] dt\right)(x^k - x^*)\right\|$$

$$\le M\left(\int_0^1 \|\nabla g(x^k) - \nabla g\bigl(x^* + t(x^k - x^*)\bigr)\|\, dt\right)\|x^k - x^*\|.$$

By continuity of ∇g, we can take δ sufficiently small to ensure that $\{\|x^k - x^*\|\}$ is monotonically decreasing and that the term under the integral sign is arbitrarily small for all k. The convergence of x^k to x^* and the superlinear convergence of $\|x^k - x^*\|$ follow.

(b) If the condition (1.82) holds, the preceding relation yields

$$\|x^{k+1} - x^*\| \le M\left(\int_0^1 Lt\|x^k - x^*\| dt\right)\|x^k - x^*\| = \frac{LM}{2}\|x^k - x^*\|^2.$$

Q.E.D.

A related result is the following. Its proof requires a simple modification of the proof of Prop. 1.4.1(a), and is left as Exercise 1.4.2 for the reader.

> **Proposition 1.4.2:** Under the assumptions of Prop. 1.4.1(a), given any $r > 0$, there exists a $\delta > 0$ such that if $\|x^k - x^*\| < \delta$, then
>
> $$\|x^{k+1} - x^*\| \leq r\|x^k - x^*\|, \qquad \|g(x^{k+1})\| \leq r\|g(x^k)\|.$$

Thus, the pure form of Newton's method converges extremely fast once it gets "near" a solution x^* where $\nabla g(x^*)$ is invertible, typically taking a handful of iterations to achieve very high solution accuracy; see Fig. 1.4.1. Unfortunately, it is usually difficult to predict whether a given starting point is sufficiently near to a solution for the fast convergence rate of Newton's method to take hold right away. Thus, in practice one can only expect that *eventually* the fast convergence rate of Newton's method will take hold. Figure 1.4.2 illustrates how the method can fail to converge when started far from a solution.

Global Convergence

Newton's method in its pure form for unconstrained minimization of f has several serious drawbacks.

(a) The inverse $\left(\nabla^2 f(x^k)\right)^{-1}$ may fail to exist, in which case the method breaks down. This will happen, for example, in regions where f is linear ($\nabla^2 f = 0$).

(b) The pure form is not a descent method; i.e., possibly $f(x^{k+1}) > f(x^k)$.

(c) The pure form tends to be attracted by local maxima just as much as it is attracted by local minima. It just tries to solve the system of equations $\nabla f(x) = 0$.

For these reasons, it is necessary to modify the pure form of Newton's method to turn it into a reliable minimization algorithm. There are several schemes that accomplish this by converting the pure form into a gradient method with a gradient related direction sequence. Simultaneously the modifications are such that, near a nonsingular local minimum, the algorithm assumes the pure form of Newton's method (1.79) and achieves the attendant fast convergence rate.

A simple possibility is to replace the Newton direction by the steepest descent direction (possibly after diagonal scaling), whenever the Newton direction is either not defined or is not a descent direction.† With proper

---

† Interestingly, this motivated the development of steepest descent by M. Augustin Cauchy. In his original paper [Cau47], Cauchy states as motivation for the steepest descent method its capability to obtain a close approximation to the

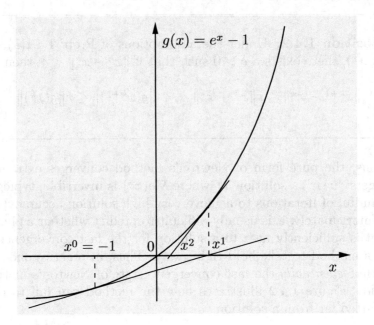

Figure 1.4.1. Fast convergence of Newton's method for solving the equation $e^x - 1 = 0$.

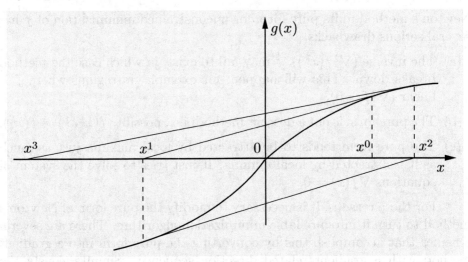

Figure 1.4.2. Divergence of Newton's method for solving an equation $g(x) = 0$ of a single variable x, when the starting point is far from the solution. This phenomenon typically occurs when $\left\|\nabla g(x)\right\|$ tends to decrease as $\|x\| \to \infty$.

safeguards, such a method has appropriate convergence and asymptotic

---

solution, in which case "... one can obtain new approximations very rapidly with the aid of the linear or Newton's method ..." (Note the attribution to Newton by Cauchy.)

Sec. 1.4   Newton's Method and Variations

rate of convergence properties (see Exercise 1.4.3, and for a related method, see Exercise 1.4.4). However, its performance at the early iterations may be quite slow, whether the Newton direction or the steepest descent direction is used in these iterations.

Generally, no modified version of Newton's method can be guaranteed to converge fast in the early iterations, but there are schemes that can use second derivative information effectively, even when the Hessian is not positive definite. These schemes are based on making diagonal modifications to the Hessian; that is, they obtain the direction d^k by solving a system of the form

$$(\nabla^2 f(x^k) + \Delta^k)d^k = -\nabla f(x^k),$$

whenever the Newton direction does not exist or is not a descent direction. Here Δ^k is a diagonal matrix such that

$$\nabla^2 f(x^k) + \Delta^k : \text{ positive definite}.$$

We outline some possibilities in the next two subsections.

1.4.1 Modified Cholesky Factorization

It can be shown that every positive definite matrix Q has a unique factorization of the form

$$Q = LL',$$

where L is a lower triangular matrix; this is known as the *Cholesky factorization of Q* (see Appendix D). Systems of equations of the form $Qx = b$ can be solved by first solving for y the triangular system $Ly = b$, and then by solving for x the triangular system $L'x = y$. These triangular systems can be solved easily [in $O(n^2)$ operations, as opposed to general systems, which require $O(n^3)$ operations; see Appendix D]. Since calculation of the Newton direction involves solution of the system

$$\nabla^2 f(x^k)d^k = -\nabla f(x^k),$$

it is natural to compute d^k by attempting to form the Cholesky factorization of $\nabla^2 f(x^k)$. During this process, one can detect whether $\nabla^2 f(x^k)$ is either nonpositive definite or nearly singular, in which case some of the diagonal elements of $\nabla^2 f(x^k)$ are suitably increased to ensure that the resulting matrix is positive definite. This is done sequentially during the factorization process, so in the end we obtain

$$L^k L^{k'} = \nabla^2 f(x^k) + \Delta^k,$$

where L^k is lower triangular and nonsingular, and Δ^k is diagonal.

As an illustration, consider the 2-dimensional case (for the general case, see Appendix D). Let

$$\nabla^2 f(x^k) = \begin{pmatrix} h_{11} & h_{12} \\ h_{21} & h_{22} \end{pmatrix}$$

and let the desired factorization be of the form

$$LL' = \begin{pmatrix} \alpha & 0 \\ \gamma & \beta \end{pmatrix} \cdot \begin{pmatrix} \alpha & \gamma \\ 0 & \beta \end{pmatrix}.$$

We choose α, β, and γ, so that $\nabla^2 f(x^k) = LL'$ if $\nabla^2 f(x^k)$ is positive definite, and we appropriately modify h_{11} and h_{22} otherwise. This determines the first diagonal element α according to the relation

$$\alpha = \begin{cases} \sqrt{h_{11}} & \text{if } h_{11} > 0 \\ \sqrt{h_{11} + \delta_1} & \text{otherwise} \end{cases}$$

where δ_1 is such that $h_{11} + \delta_1 > 0$. Given α, we can calculate γ by equating the corresponding elements of $\nabla^2 f(x^k)$ and LL'. We obtain $\gamma\alpha = h_{12}$ or

$$\gamma = \frac{h_{12}}{\alpha}.$$

We can now calculate the second diagonal element β by equating the corresponding elements of $\nabla^2 f(x^k)$ and LL', after appropriately modifying h_{22} if necessary,

$$\beta = \begin{cases} \sqrt{h_{22} - \gamma^2} & \text{if } h_{22} > \gamma^2, \\ \sqrt{h_{22} - \gamma^2 + \delta_2} & \text{otherwise,} \end{cases}$$

where δ_2 is such that $h_{22} - \gamma^2 + \delta_2 > 0$. The method for choosing the increments δ_1 and δ_2 is largely heuristic. One possibility is discussed in Appendix D, which also describes more sophisticated versions of the above procedure where a positive increment is added to the diagonal elements of the Hessian even when the corresponding diagonal elements of the factorization are positive but very close to zero.

Given the $L^k L^{k'}$ factorization, the direction d^k is obtained by solving the system

$$L^k L^{k'} d^k = -\nabla f(x^k).$$

The next iterate is

$$x^{k+1} = x^k + \alpha^k d^k,$$

where α^k is chosen according to the Armijo rule or one of the other stepsize rules we have discussed.

To guarantee convergence, the increments added to the diagonal elements of the Hessian can be chosen so that $\{d^k\}$ is gradient related (cf. Prop. 1.2.1). Also, these increments can be chosen to be zero near a nonsingular local minimum. In particular, with proper safeguards, near such a point, the method becomes identical to the pure form of Newton's method and achieves the corresponding superlinear convergence rate (see Appendix D).

1.4.2 Trust Region Methods

As explained in Section 1.2, the pure Newton step is obtained by minimizing over d the second order approximation of f around x^k, given by

$$f^k(d) = f(x^k) + \nabla f(x^k)'d + \tfrac{1}{2} d' \nabla^2 f(x^k) d.$$

We know that $f^k(d)$ is a good approximation of $f(x^k + d)$ when d is in a small neighborhood of zero, but the difficulty is that with unconstrained minimization of $f^k(d)$ one may obtain a step that lies outside this neighborhood. It therefore makes sense to consider a *restricted Newton step* d^k, which is obtained by minimizing $f^k(d)$ over a suitably small neighborhood of zero, called the *trust region*:

$$d^k \in \arg \min_{\|d\| \leq \gamma^k} f^k(d), \tag{1.83}$$

where γ^k is some positive scalar.† An approximate solution of the constrained minimization problem of Eq. (1.83) can be obtained quickly using the fact that it has only one constraint. We refer to the specialized literature, including [MoS83] and the book [CGT00], for an account of approximate solution methods.

An important observation here is that even if $\nabla^2 f(x^k)$ is not positive definite or, more generally, even if the pure Newton direction is not a descent direction, the restricted Newton step d^k improves the cost, provided $\nabla f(x^k) \neq 0$ and γ^k is sufficiently small. The reason is that, in view of Eq. (1.83), $f^k(d^k)$ is smaller than $f(x^k)$ [which is equal to $f^k(0)$], and $f(x^k + d^k)$ is very close to its second order expansion $f^k(d^k)$ when $\|d^k\|$ is small.

More specifically, we have for all d with $\|d\| \leq \gamma^k$

$$f(x^k + d) = f^k(d) + o\big((\gamma^k)^2\big),$$

so that

$$f(x^k + d^k) = f^k(d^k) + o\big((\gamma^k)^2\big)$$
$$= f(x^k) + \min_{\|d\| \leq \gamma^k} \big\{ \nabla f(x^k)'d + \tfrac{1}{2} d' \nabla^2 f(x^k) d \big\} + o\big((\gamma^k)^2\big).$$

Therefore, denoting

$$\tilde{d}^k = -\frac{\nabla f(x^k)}{\|\nabla f(x^k)\|} \gamma^k,$$

---

† It can be shown that the restricted Newton step d^k also solves a system of the form $\big(\nabla^2 f(x^k) + \delta^k I\big)d = -\nabla f(x^k)$, where I is the identity matrix and δ^k is a nonnegative scalar (a Lagrange multiplier in the terminology of Chapter 4), so the preceding method of determining d^k fits the general framework of using a correction of the Hessian matrix by a positive semidefinite matrix.

we have

$$f(x^k + d^k) \leq f(x^k) + \nabla f(x^k)'\tilde{d}^k + \tfrac{1}{2}\tilde{d}^{k'}\nabla^2 f(x^k)\tilde{d}^k + o\big((\gamma^k)^2\big)$$

$$= f(x^k) - \gamma^k \|\nabla f(x^k)\| + \frac{(\gamma^k)^2}{2\|\nabla f(x^k)\|^2} \nabla f(x^k)'\nabla^2 f(x^k)\nabla f(x^k)$$

$$+ o\big((\gamma^k)^2\big).$$

For γ^k sufficiently small, the negative term $-\gamma^k\|\nabla f(x^k)\|$ dominates the last two terms on the right-hand side above, showing that

$$f(x^k + d^k) < f(x^k).$$

It can be seen in fact from the preceding relations that a cost improvement is possible even when $\nabla f(x^k) = 0$, provided γ^k is sufficiently small and f has a direction of negative curvature at x^k, i.e., $\nabla^2 f(x^k)$ is not positive semidefinite. Thus the preceding procedure will fail to improve the cost only if $\nabla f(x^k) = 0$ and $\nabla^2 f(x^k)$ is positive semidefinite, i.e., x^k satisfies the first *and* the second order necessary conditions. In particular, one can make progress even if x^k is a stationary point that is not a local minimum.

We are thus motivated to consider a method of the form

$$x^{k+1} = x^k + d^k,$$

where d^k is the restricted Newton step corresponding to a suitably chosen scalar γ^k as per Eq. (1.83). Here, for a given x^k, γ^k should be small enough so that there is cost improvement; one possibility is to start from an initial trial γ^k and successively reduce γ^k by a certain factor as many times as necessary until a cost reduction occurs $[f(x^{k+1}) < f(x^k)]$. The choice of the initial trial value for γ^k is crucial here; if it is chosen too large, a large number of reductions may be necessary before a cost improvement occurs; if it is chosen too small the convergence rate may be poor. In particular, to maintain the superlinear convergence rate of Newton's method, as x^k approaches a nonsingular local minimum, one should select the initial trial value of γ^k sufficiently large so that the restricted Newton step and the pure Newton step coincide.

A reasonable way to adjust the initial trial value for γ^k is to increase this value when the method appears to be progressing well and to decrease this value otherwise. One can measure progress by using the ratio of actual over predicted cost improvement [based on the approximation $f^k(d)$]

$$r^k = \frac{f(x^k) - f(x^{k+1})}{f(x^k) - f^k(d^k)}.$$

In particular, it makes sense to increase the initial trial value for γ ($\gamma^{k+1} > \gamma^k$) if this ratio is close to or above unity, and decrease γ otherwise. The following algorithm is a typical example of such a method. Given x^k and an initial trial value γ^k, it determines x^{k+1} and an initial trial value γ^{k+1} by using two thresholds σ_1, σ_2 with $0 < \sigma_1 \leq \sigma_2 \leq 1$ and two factors β_1, β_2 with $0 < \beta_1 < 1 < \beta_2$ (typical values are $\sigma_1 = 0.2$, $\sigma_2 = 0.8$, $\beta_1 = 0.25$, $\beta_2 = 2$).

Step 1: Find
$$d^k \in \arg\min_{||d|| \leq \gamma^k} f^k(d), \tag{1.84}$$

If $f^k(d^k) = f(x^k)$ stop (x^k satisfies the first and second order necessary conditions for a local minimum); else go to Step 2.

Step 2: If $f(x^k + d^k) < f(x^k)$ set
$$x^{k+1} = x^k + d^k \tag{1.85}$$

calculate
$$r^k = \frac{f(x^k) - f(x^{k+1})}{f(x^k) - f^k(d^k)} \tag{1.86}$$

and go to Step 3; else set $\gamma^k := \beta_1 ||d^k||$ and go to Step 1.

Step 3: Set
$$\gamma^{k+1} = \begin{cases} \beta_1 ||d^k|| & \text{if } r^k < \sigma_1, \\ \beta_2 \gamma^k & \text{if } \sigma_2 \leq r^k \text{ and } ||d^k|| = \gamma^k, \\ \gamma^k & \text{otherwise.} \end{cases} \tag{1.87}$$

Go to the next iteration.

Assuming that f is twice continuously differentiable, it is possible to show that the above algorithm is convergent in the sense that if $\{x^k\}$ is a bounded sequence, there exists a limit point of $\{x^k\}$ that satisfies the first and the second order necessary conditions for optimality. Furthermore, if $\{x^k\}$ converges to a nonsingular local minimum x^*, then asymptotically, the method is identical to the pure form of Newton's method, thereby attaining a superlinear convergence rate; see the references given at the end of the chapter for proofs of these and other related results for trust region methods.

1.4.3 Variants of Newton's Method

We will now briefly consider approximate implementations of Newton's method. The idea is to calculate the Newton direction approximately, with the aim of economizing on computational overhead with relatively small degradation of the convergence rate.

Newton's Method with Periodic Reevaluation of the Hessian

A variation of Newton's method is obtained if the Hessian matrix $\nabla^2 f$ is recomputed every $p > 1$ iterations rather than at every iteration. In particular, this method, in unmodified form, is given by

$$x^{k+1} = x^k - \alpha^k D^k \nabla f(x^k),$$

where

$$D^{ip+j} = \left(\nabla^2 f(x^{ip})\right)^{-1}, \qquad j = 0, 1, ..., p-1, \ i = 0, 1, ...$$

The idea here is to save the computation and the inversion (or factorization) of the Hessian for the iterations where $j \neq 0$. This reduction in overhead is achieved at the expense of what is usually a small degradation in speed of convergence.

Truncated Newton Methods

We have so far assumed that the Newton system $\nabla^2 f(x^k) d^k = -\nabla f(x^k)$ will be solved exactly for the direction d^k by Cholesky factorization or Gaussian elimination, which require a finite number of arithmetic operations $[O(n^3)]$. When the dimension n is large, the calculation required for exact solution may be prohibitive. An alternative is to use an approximate solution, which may be obtained with an iterative method. This approach is often useful for solving very large linear systems of equations, arising in the solution of partial differential equations, where an adequate approximation to the solution can often be obtained by iterative methods quite fast, while the computation to find the exact solution can be overwhelming.

Generally, solving for d any system of the form $H^k d = -\nabla f(x^k)$, where H^k is a positive definite symmetric $n \times n$ matrix, can be done by solving the quadratic optimization problem

$$\begin{aligned} \text{minimize} \quad & \tfrac{1}{2} d' H^k d + \nabla f(x^k)' d \\ \text{subject to} \quad & d \in \Re^n, \end{aligned} \qquad (1.88)$$

whose cost function gradient is zero at d if and only if $H^k d = -\nabla f(x^k)$. Suppose that an iterative descent method is used for solution and the starting point is $d^0 = 0$. Since the quadratic cost is reduced at each iteration and its value at the starting point is zero, we obtain after each iteration a vector d^k satisfying $\tfrac{1}{2} d^{k'} H^k d^k + \nabla f(x^k)' d^k < 0$, from which, using the positive definiteness of H^k,

$$\nabla f(x^k)' d^k < 0.$$

Thus the approximate solution d^k of the system $H^k d = -\nabla f(x^k)$, obtained after any positive number of iterations, is a descent direction.

Possible iterative methods for solving the direction finding problem (1.88) include the conjugate gradient method to be presented in Section 2.1 and the coordinate descent method to be discussed in Section 2.3.1. For the conjugate gradient method to be practical, the calculation of matrix-vector products of the form $H^k d$ must be convenient, and for this, the presence of special structure of H^k may be important. An example of this type is network optimization problems, to be discussed in Section 3.8. Also, the idea of implementing Newton's method by using an iterative method applies more generally to constrained forms of the method, to be discussed in Chapter 3.

Conditions on the accuracy of the approximate solution d^k that ensure linear or superlinear rate of convergence are given in Exercise 1.4.5. Generally, the superlinear convergence rate property of the method to a nonsingular local minimum is maintained if the approximate Newton directions d^k satisfy

$$\lim_{k \to \infty} \frac{\left\|\nabla^2 f(x^k) d^k + \nabla f(x^k)\right\|}{\left\|\nabla f(x^k)\right\|} = 0,$$

(cf. Prop. 1.3.2). Thus, for superlinear convergence rate, the norm of the error in solving the Newton system must become negligible relative to the gradient norm in the limit.

1.4.4 Least Squares and the Gauss-Newton Method

We will now consider a specialized Newton-like method for solving least squares problems of the form

$$\begin{aligned}\text{minimize} \quad & f(x) = \tfrac{1}{2}\|g(x)\|^2 = \tfrac{1}{2}\sum_{i=1}^m \|g_i(x)\|^2 \\ \text{subject to} \quad & x \in \Re^n,\end{aligned} \qquad (1.89)$$

where g is a continuously differentiable function with component functions $g_1, \ldots, g_m$, where $g_i : \Re^n \to \Re^{r_i}$. Usually $r_i = 1$, but it is sometimes convenient to consider the more general case.

Least squares problems are common in many practical contexts. An important case arises when g consists of n scalar-valued functions and we want to solve the system of n equations with n unknowns $g(x) = 0$. We can formulate this as the least squares optimization problem (1.89) [x^* solves the system $g(x) = 0$ if and only if it minimizes $\tfrac{1}{2}\|g(x)\|^2$ and the optimal value is zero]. Here are some other examples:

Example 1.4.1 (Model Construction – Curve Fitting)

Suppose that we want to estimate n parameters of a mathematical model so that it fits well a physical system, based on a set of input-output data. In particular, we hypothesize an approximate relation of the form

$$z = h(x, y),$$

where h is a known function representing the model and

$x \in \Re^n$ is a vector of unknown parameters,
$z \in \Re^r$ is the model's output,
$y \in \Re^p$ is the model's input.

Given a set of m input-output data pairs $(y_1, z_1), \ldots, (y_m, z_m)$ from measurements of the physical system that we try to model, we want to find the vector of parameters x that matches best the data in the sense that it minimizes the sum of squared errors

$$\tfrac{1}{2} \sum_{i=1}^{m} \left\| z_i - h(x, y_i) \right\|^2.$$

For example, to fit the data pairs by a cubic polynomial approximation, we would choose

$$h(x, y) = x_3 y^3 + x_2 y^2 + x_1 y + x_0,$$

where $x = (x_0, x_1, x_2, x_3)$ is the vector of unknown coefficients of the cubic polynomial.

The next two examples are really special cases of the preceding one.

Example 1.4.2 (Dynamic System Identification)

A common model for a single input-single output dynamic system is to relate the input sequence $\{y_k\}$ to the output sequence $\{z_k\}$ by a linear equation of the form

$$\sum_{j=0}^{n} \alpha_j z_{k-j} = \sum_{j=0}^{n} \beta_j y_{k-j}.$$

Given a record of inputs and outputs $y_1, z_1, \ldots, y_m, z_m$ from the true system, we would like to find a set of parameters $\{\alpha_j, \beta_j \mid j = 0, 1, \ldots, n\}$ that matches this record best in the sense that it minimizes

$$\sum_{k=n}^{m} \left(\sum_{j=0}^{n} \alpha_j z_{k-j} - \sum_{j=0}^{n} \beta_j y_{k-j} \right)^2.$$

This is a least-squares problem.

Example 1.4.3 (Neural Networks)

A least squares modeling approach that has received a lot of attention is provided by *neural networks*. Here the model is specified by a multistage system, also called a *multilayer perceptron*. The kth stage consists of n_k *activation units*, each being a single input-single output mapping of a given form $\phi : \Re \mapsto \Re$ (examples will be given shortly). The output of the jth

Sec. 1.4   Newton's Method and Variations

activation unit of the $(k+1)$st stage is denoted by x_{k+1}^j and the input is a linear function of the output vector $x_k = (x_k^1, \ldots, x_k^{n_k})$ of the kth stage. Thus

$$x_{k+1}^j = \phi\left(u_k^{0j} + \sum_{s=1}^{n_k} x_k^s u_k^{sj}\right), \qquad j = 1, \ldots, n_{k+1}, \tag{1.90}$$

where the coefficients u_k^{sj} (also called *weights*) are to be determined. In variants of this approach there may be some constraints on the weights in order to induce desired connectivity structures between the stages, which are designed to produce a particular effect and/or exploit some known structure of the input. However, for simplicity we will not consider this possibility; the algorithmic ideas to be described generalize.

Suppose that the multilayer perceptron has N stages, and let u denote the vector of the weights of all the stages:

$$u = \{u_k^{sj} \mid k = 0, \ldots, N-1, \, s = 0, \ldots, n_k, \, j = 1, \ldots, n_{k+1}\}.$$

Then, for a given vector u of weights, an input vector x_0 to the first stage produces a unique output vector x_N from the Nth stage via Eq. (1.90). Thus, we may view the multilayer perceptron as a mapping h that is parameterized by u and transforms the input vector x_0 into an output vector of the form $x_N = h(u, x_0)$. Suppose that we have m sample input-output pairs $(y_1, z_1), \ldots, (y_m, z_m)$ from a physical system that we are trying to model. Then, by selecting u appropriately, we can try to match the mapping of the multilayer perceptron with the mapping of the physical system. A common way to do this is to minimize over u the sum of squared errors

$$\tfrac{1}{2} \sum_{i=1}^m \left\| z_i - h(u, y_i) \right\|^2.$$

In neural network terminology, finding the optimal weights u is referred to as *training the network*. Incremental gradient methods, to be discussed in Section 2.4.1, are often used for this purpose (see e.g., [BeT96], [Hay11]).

Examples of activation units are functions such as

$$\phi(\xi) = \frac{1}{1 + e^{-\xi}}, \qquad \text{(sigmoidal function),}$$

$$\phi(\xi) = \frac{e^\xi - e^{-\xi}}{e^\xi + e^{-\xi}}, \qquad \text{(hyperbolic tangent function),}$$

whose gradients are zero as the argument ξ approaches $-\infty$ and ∞. For these functions, it is possible to show that with a sufficient number of activation units and a number of stages $N \geq 2$, a multilayer perceptron can approximate arbitrarily closely very complex input-output maps; see [Cyb89]. In practice, a number N that is considerably larger than 2 is often considered, in combination with functions ϕ specially tailored to particular types of problems, giving rise to so called *deep neural networks*, which have attained considerable success in a variety of applications; see e.g., [HDY12], [SHM16].

Neural network training problems can be quite challenging. Their cost function is typically nonconvex and involves multiple local minima. For large

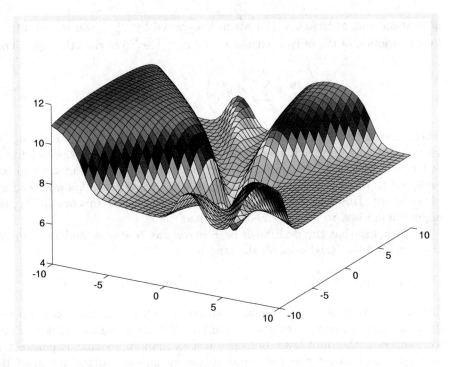

Figure 1.4.3. Three-dimensional plot of a least squares cost function

$$\tfrac{1}{2}\sum_{i=1}^{5}\big(z_i - \phi(u_1 y_i + u_0)\big)^2,$$

for a neural network training problem where there are only two weights u_0 and u_1, five data pairs, and ϕ is the hyperbolic tangent function. The data of the problem are given in Exercise 2.4.3. The cost function tends to a constant as u is changed along rays of the form $r\bar{u}$, where $r > 0$ and $\bar{u}$ is a fixed vector.

values of the weights u_k^{ij}, the cost becomes "flat." In fact, the cost function tends to a constant as u is changed along rays of the form $r\bar{u}$, where $r > 0$ and $\bar{u}$ is a fixed vector; see Fig. 1.4.3. For u near the origin, the cost function can be quite complicated, alternately involving flat and steep regions.

The next example deals with an important context where neural networks are often used.

Example 1.4.4 (Classification - Hypothesis Testing)

Let us consider a problem of classifying objects based on the values of their characteristics. Here we use the term "object" generically. In some contexts, the classification may relate to persons or situations. In other cases, an object may represent a hypothesis, and the problem is to decide which of the hypotheses is true, based on some data.

We assume that each object is presented to us with a vector y, and we wish to classify it in one of s categories $1, \ldots, s$. For example, the vector y may represent data, such as the results of a collection of tests on a medical patient, and we may wish to classify the patient as being healthy or as having one of several types of illnesses.

A classical classification approach is to assume that for each category $j = 1, \ldots, s$, we know the conditional probability $p(j \mid y)$ that an object is of category j given data y. Then, given data y, we decide on the category $j^*(y)$ having maximum posterior probability, i.e.,

$$j^*(y) \in \arg\max_{j=1,\ldots,s} p(j \mid y). \qquad (1.91)$$

This is called the Maximum a Posteriori rule (or MAP rule for short; see for example the book [BeT08] for a discussion).

Suppose now that the probabilities $p(j \mid y)$, viewed as functions of y, are unknown, but instead we have a sample consisting of data for m object-category pairs. Then we may try to estimate $p(j \mid y)$ based on the following simple fact: out of all functions $f_j(y)$ of y, $p(j \mid y)$ is the one that minimizes the expected value of $(z_j - f_j(y))^2$, where

$$z_j = \begin{cases} 1 & \text{if } y \text{ is of category } j, \\ 0 & \text{otherwise.} \end{cases}$$

To this end, we adopt a parametric approach. For each category $j = 1, \ldots, s$, we estimate the probability $p(j \mid y)$ with a function $h_j(x_j, y)$ that is parameterized by a vector x_j. The function h_j may be provided for example by a neural network (cf. Example 1.4.3). Then, denoting y_i the data vector of the ith object, we obtain x_j by minimizing the least squares function

$$\tfrac{1}{2} \sum_{i=1}^{m} \left(z_j^i - h_j(x_j, y_i)\right)^2,$$

where

$$z_j^i = \begin{cases} 1 & \text{if } y_i \text{ is of category } j, \\ 0 & \text{otherwise.} \end{cases}$$

This minimization approximates the minimization of the expected value of $\left(z_j - f_j(y)\right)^2$. Once the optimal parameter vectors x_j^*, $j = 1, \ldots, s$, have been obtained, we may use them to classify a new object with data vector y according to the rule

$$\text{Estimated Object Category} \in \arg\max_{j=1,\ldots,s} h_j(x_j^*, y),$$

which approximates the MAP rule (1.91).

For the simpler case where there are just two categories, say A and B, a similar formulation is to hypothesize a relation of the following form between data vector y and category of an object:

$$\text{Object Category} = \begin{cases} A & \text{if } h(x, y) = 1, \\ B & \text{if } h(x, y) = -1, \end{cases}$$

where h is a given function and x is an unknown vector of parameters. Given a set of m data pairs $(z_1, y_1), \ldots, (z_m, y_m)$ of representative objects of known category, where y_i is the data vector of the ith object, and

$$z_i = \begin{cases} 1 & \text{if } y \text{ is of category } A, \\ -1 & \text{if } y \text{ is of category } B, \end{cases}$$

we obtain x by minimizing the least squares function

$$\tfrac{1}{2} \sum_{i=1}^{m} \bigl(z_i - h(x, y_i)\bigr)^2.$$

The optimal parameter vector x^* is used to classify a new object with data vector y according to the rule

$$\text{Estimated Object Category} = \begin{cases} A & \text{if } h(x^*, y) > 0, \\ B & \text{if } h(x^*, y) < 0. \end{cases}$$

There are several variations on the above theme, for which we refer to the specialized literature. Furthermore, there are several alternative optimization-based methods for classification (see Example 2.4.1, and the book [Ber15a], which also gives many references to the extensive literature on this subject).

The Gauss-Newton Method

We now consider the *Gauss-Newton method*, which is a specialized method for minimizing the least squares cost $(1/2)\|g(x)\|^2$. Given a point x^k, the pure form of the Gauss-Newton iteration is based on linearizing g to obtain

$$\tilde{g}(x, x^k) = g(x^k) + \nabla g(x^k)'(x - x^k)$$

and then minimizing the norm of the linearized function $\tilde{g}$:

$$\begin{aligned}
x^{k+1} &\in \arg\min_{x \in \Re^n} \tfrac{1}{2} \|\tilde{g}(x, x^k)\|^2 \\
&= \arg\min_{x \in \Re^n} \tfrac{1}{2} \bigl\{ \|g(x^k)\|^2 + 2(x - x^k)' \nabla g(x^k) g(x^k) \\
&\qquad + (x - x^k)' \nabla g(x^k) \nabla g(x^k)'(x - x^k) \bigr\}.
\end{aligned}$$

Assuming that the $n \times n$ matrix $\nabla g(x^k) \nabla g(x^k)'$ is invertible, the above quadratic minimization yields

$$x^{k+1} = x^k - \bigl(\nabla g(x^k) \nabla g(x^k)'\bigr)^{-1} \nabla g(x^k) g(x^k). \qquad (1.92)$$

Note that if g is already a linear function, we have $\|g(x)\|^2 = \|\tilde{g}(x, x^k)\|^2$, and the method converges in a single iteration. Note also that the direction

$$-\bigl(\nabla g(x^k) \nabla g(x^k)'\bigr)^{-1} \nabla g(x^k) g(x^k)$$

used in the above iteration is a descent direction since $\nabla g(x^k) g(x^k)$ is the gradient at x^k of the least squares cost function $(1/2)\|g(x)\|^2$ and $\bigl(\nabla g(x^k) \nabla g(x^k)'\bigr)^{-1}$ is a positive definite matrix.

Sec. 1.4   Newton's Method and Variations

To deal with the case where the matrix $\nabla g(x^k)\nabla g(x^k)'$ is singular (as well as enhance convergence when this matrix is nearly singular), the method is often implemented in the modified form

$$x^{k+1} = x^k - \alpha^k \left(\nabla g(x^k)\nabla g(x^k)' + \Delta^k\right)^{-1}\nabla g(x^k)g(x^k), \qquad (1.93)$$

where α^k is a stepsize chosen by one of the stepsize rules that we have discussed, and Δ^k is a diagonal matrix such that

$$\nabla g(x^k)\nabla g(x^k)' + \Delta^k : \text{positive definite}.$$

For example, Δ^k may be chosen in accordance with the Cholesky factorization scheme outlined in Section 1.4.1. An early proposal, known as the *Levenberg-Marquardt method*, is to choose Δ^k to be a positive multiple of the identity matrix. With these choices of Δ^k, it can be seen that the directions used by the method are gradient related, and the convergence results of Section 1.2.2 apply.

Relation to Newton's Method

The Gauss-Newton method bears a close relation to Newton's method. In particular, assuming each g_i is a scalar function, the Hessian of the cost function $(1/2)\|g(x)\|^2$ is

$$\nabla g(x^k)\nabla g(x^k)' + \sum_{i=1}^{m}\nabla^2 g_i(x^k)g_i(x^k), \qquad (1.94)$$

so it is seen that the Gauss-Newton iterations (1.92) and (1.93) are approximate versions of their Newton counterparts, where the second order term

$$\sum_{i=1}^{m}\nabla^2 g_i(x^k)g_i(x^k) \qquad (1.95)$$

is neglected. Thus, in the Gauss-Newton method, we save the computation of this term at the expense of some deterioration in the convergence rate. If, however, the neglected term (1.95) is relatively small near a solution, the convergence rate of the Gauss-Newton method is satisfactory. This is often true in many applications such as for example when g is nearly linear, and also when the components $g_i(x)$ are small near the solution.

A case in point is when $m = n$ and the problem is to solve the system $g(x) = 0$. Then the neglected term is zero at a solution, and assuming $\nabla g(x^k)$ is invertible, we have

$$\left(\nabla g(x^k)\nabla g(x^k)'\right)^{-1}\nabla g(x^k)g(x^k) = \left(\nabla g(x^k)'\right)^{-1}g(x^k).$$

Thus the pure form of the Gauss-Newton method (1.92) takes the form

$$x^{k+1} = x^k - \left(\nabla g(x^k)'\right)^{-1}g(x^k),$$

and is identical to Newton's method for solving the system $g(x) = 0$ [rather than Newton's method for minimizing $\|g(x)\|^2$]. The convergence rate is typically superlinear in this case (cf. Prop. 1.4.1).

EXERCISES

1.4.1 (Scale-Free Character of Newton's Method)

The purpose of this exercise is to show that Newton's method is unaffected by linear scaling of the variables. Consider a linear invertible transformation of variables $x = Sy$. Write the pure form of Newton's method in the space of the variables y and show that it generates the sequence $y^k = S^{-1}x^k$, where $\{x^k\}$ is the sequence generated by Newton's method in the space of the variables x.

1.4.2 (www)

Show Prop. 1.4.2. *Hint*: For the second relation, let

$$M(x) = \int_0^1 \nabla g\big(x^* + t(x - x^*)\big)' dt,$$

so that $g(x) = M(x)(x - x^*)$. Argue that for some $\delta > 0$ the eigenvalues of $M(x)'M(x)$ lie between some positive scalars γ and Γ for all x with $\|x - x^*\| \leq \delta$. Show that

$$\gamma \|x - x^*\|^2 \leq \|g(x)\|^2 \leq \Gamma \|x - x^*\|^2, \qquad \forall\; x \text{ with } \|x - x^*\| \leq \delta.$$

1.4.3 (Combination of Newton and Steepest Descent Methods)

Consider the iteration $x^{k+1} = x^k + \alpha^k d^k$ where α^k is chosen by the Armijo rule with initial stepsize $s = 1$, $\sigma \in (0, 1/2)$, and d^k is equal to

$$d_N^k = -\big(\nabla^2 f(x^k)\big)^{-1} \nabla f(x^k)$$

if $\nabla^2 f(x^k)$ is nonsingular and the following two inequalities hold:

$$c_1 \|\nabla f(x^k)\|^{p_1} \leq -\nabla f(x^k)' d_N^k,$$

$$\|d_N^k\|^{p_2} \leq c_2 \|\nabla f(x^k)\|;$$

otherwise

$$d^k = -D \nabla f(x^k),$$

where D is a fixed positive definite symmetric matrix. The scalars c_1, c_2, p_1, and p_2 satisfy $c_1 > 0$, $c_2 > 0$, $p_1 > 2$, $p_2 > 1$. Show that the sequence $\{d^k\}$ is gradient related. Furthermore, every limit point of $\{x^k\}$ is stationary, and if $\{x^k\}$ converges to a nonsingular local minimum x^*, the rate of convergence of $\big\{\|x^k - x^*\|\big\}$ is superlinear.

1.4.4 (Armijo Rule Along a Curved Path)

This exercise provides a globally convergent variant of Newton's method, which combines the Newton and steepest descent directions along a curved path. Let $f : \Re^n \mapsto \Re$ be twice continuously differentiable. At a point x^k, let $d_S^k = -D\nabla f(x^k)$ be a scaled steepest descent direction, where D is a fixed positive definite symmetric matrix. Let also d_N^k be the Newton direction $-\left(\nabla^2 f(x^k)\right)^{-1}\nabla f(x^k)$ if $\nabla^2 f(x^k)$ is nonsingular, and be equal to d_S^k otherwise. Consider the method

$$x^{k+1} = x^k + \alpha^k\left((1-\alpha^k)d_S^k + \alpha^k d_N^k\right),$$

where $\alpha^k = \beta^{m^k}$ and m^k is the first nonnegative integer m such that the following three inequalities hold:

$$f(x^k) - f\left(x^k + \beta^m\left((1-\beta^m)d_S^k + \beta^m d_N\right)\right) \geq -\sigma\beta^m\nabla f(x^k)'\left((1-\beta^m)d_S^k + \beta^m d_N\right),$$

$$c_1 \min\left\{\nabla f(x^k)'D\nabla f(x^k), \|\nabla f(x^k)\|^3\right\} \leq -\nabla f(x^k)'\left((1-\beta^m)d_S^k + \beta^m d_N\right),$$

$$\left\|(1-\beta^m)d_S^k + \beta^m d_N\right\| \leq c_2 \max\left\{\|D\nabla f(x^k)\|, \|\nabla f(x^k)\|^{1/2}\right\},$$

where β and σ are scalars satisfying $0 < \beta < 1$ and $0 < \sigma < 1/2$, and c_1 and c_2 are scalars satisfying $c_1 < 1$ and $c_2 > 1$. Show that the method is well defined in the sense that the stepsize α^k will be obtained after a finite number of trials. Furthermore, every limit point of $\{x^k\}$ is stationary and if $\{x^k\}$ converges to a nonsingular local minimum x^*, the rate of convergence of $\{\|x^k - x^*\|\}$ is superlinear. *Hint*: For a given x^k, each of the three inequalities is satisfied for m sufficiently large, so α^k is obtained after a finite number of trials. The directions $d^k = (1-\alpha^k)d_S^k + \alpha^k d_N^k$ are gradient related by construction (cf. the last two inequalities). Use the line of proof of Prop. 1.2.1 to show stationarity of the limit points of $\{x^k\}$. Use the line of proof of Prop. 1.3.2 to show that if $\{x^k\}$ converges to a nonsingular local minimum x^*, the convergence is superlinear (including the fact that $\alpha^k = 1$ for all sufficiently large k).

1.4.5 (www)

Consider a truncated Newton method with the stepsize chosen by the Armijo rule with initial stepsize $s = 1$ and $\sigma < 1/2$, and assume that $\{x^k\}$ converges to a nonsingular local minimum x^*. Assume that the matrices H^k and the directions d^k satisfy

$$\lim_{k\to\infty}\left\|H^k - \nabla^2 f(x^k)\right\| = 0, \qquad \lim_{k\to\infty}\frac{\|H^k d^k + \nabla f(x^k)\|}{\|\nabla f(x^k)\|} = 0.$$

Show that $\{\|x^k - x^*\|\}$ converges superlinearly.

1.4.6 (www)

Apply Newton's method with a constant stepsize to minimization of the function $f(x) = \|x\|^3$. Identify the range of stepsizes for which convergence is obtained, and show that it includes the unit stepsize. Show that for any stepsize within this range, the method converges linearly to $x^* = 0$. Explain this fact in light of Prop. 1.4.1.

1.4.7

Consider Newton's method with the given trust region implementation for the case of a positive definite quadratic cost function. Show that the method terminates in a finite number of iterations.

1.4.8

(a) Consider the pure form of Newton's method for the case of the cost function $f(x) = \|x\|^\beta$, where $\beta > 1$. For what starting points and values of β does the method converge to the optimal solution? What happens when $\beta \leq 1$?

(b) Repeat part (a) for the case where Newton's method with the Armijo rule is used.

1.4.9 (Necessary and Sufficient Conditions for Convergence of Iterative Methods for Linear Equations)

This exercise deals with the convergence of iterative algorithms for the system of linear equations $Ax = b$, where A is a given (possibly singular) $n \times n$ matrix and b is a vector in $\Re^n$. We assume that b lies in the range of A, so that the system has at least one solution. For a given $n \times n$ matrix D, we say that the iteration

$$x^{k+1} = x^k - \alpha D(Ax^k - b) \qquad (1.96)$$

is *convergent* if there exists $\bar{\alpha} > $ such that for all $\alpha \in (0, \bar{\alpha}]$ and $x^0 \in \Re^n$ the sequence $\{x^k\}$ produced by the iteration converges to some solution of $Ax = b$. Show that the iteration is convergent if and only if the following conditions hold.

(i) Each eigenvalue of DA either has a positive real part or is equal to 0.

(ii) The dimension of the nullspace of DA is equal to the multiplicity of the 0 eigenvalue of DA.

(iii) The nullspace of A is equal to the nullspace of DA.

Note: The case where A is invertible is straightforward [then conditions (i)-(iii) are reduced to the condition that each eigenvalue of DA has positive real part]. The case where A is singular is challenging. To show that condition (ii) is necessary for a convergent iteration, use the fact that if it does not hold then there exists a vector v such that $DAv \neq 0$ and $(DA)^2 v = 0$. See [WaB13b] or [Ber12], Section 7.3.8, for a complete proof and related analysis.

1.4.10 (Iterative Solution of Nonlinear Equations)

This exercise deals with the iterative solution of the system of equations $g(x) = 0$, where $g : \Re^n \mapsto \Re^n$ is a continuously differentiable function. Consider the sequence $\{x^k\}$ generated by the iteration

$$x^{k+1} = x^k - D^k g(x^k), \qquad k = 0, 1, \ldots,$$

where $\{D^k\}$ is a sequence of $n \times n$ matrices.

(a) (*Linear Convergence*) Suppose that $\{x^k\}$ converges to some x^*. Denote

$$\bar{L} = \limsup_{k \to \infty} \left\| I - D^k \nabla g(x^*)' \right\|,$$

where $\|\cdot\|$ is the matrix norm induced by the standard Euclidean norm, and assume that $\bar{L} < 1$. Show that $g(x^*) = 0$. Moreover for every $L \in (\bar{L}, 1)$, there is an integer m such that

$$\|x^{k+1} - x^*\| \le L \|x^k - x^*\|, \qquad \forall\, k \ge m.$$

(b) (*Local Convergence*) Let x^* be such that $g(x^*) = 0$ and $\nabla g(x^*)$ is nonsingular. Let also $D(x)$ be a matrix function such that

$$\limsup_{x \to x^*} \left\| I - D(x) \nabla g(x^*)' \right\| < 1.$$

Show that there is a neighborhood N of x^* such that if $x^0 \in N$, the sequence $\{x^k\}$ generated by the algorithm with $D^k = D(x^k)$ remains in N and converges to x^*.

Note: This is a challenging exercise; see [Hes80], Section 1.4, for a complete proof and related analysis.

1.5 NOTES AND SOURCES

Section 1.2: The steepest descent method dates to Cauchy [Cau47], who attributes to Newton the unity stepsize version of what we call Newton's method. The modern theory of gradient methods evolved in the 60s, when several practical convergent stepsize rules were proposed starting with the work of Goldstein [Gol62], [Gol64], Armijo [Arm66], and others. Shortly afterwards, general methods of convergence analysis were formulated starting with the work of Zangwill [Zan69], which was followed by the works of Ortega and Rheinboldt [OrR70], Daniel [Dan71], and Polak [Pol71]. The Armijo stepsize rule is the most popular of a broad variety of rules that enforce descent and provide convergence guarantees without requiring a full line minimization.

The capture theorem (Prop. 1.2.3) was formulated and proved by the author for the case of a nonsingular local minimum in [Ber82a], Prop. 1.12. It was extended to the form given here by Dunn [Dun93c].

Gradient methods with errors are discussed in Poljak [Pol87], and Bertsekas and Tsitsiklis [BeT96], [BeT00]. Parallel and asynchronous stochastic gradient methods converge under very weak conditions; see Tsitsiklis, Bertsekas, and Athans [TBA86], Bertsekas and Tsitsiklis [BeT89], Section 7.8. The convergence rate of these methods is discussed by Duchi, Chaturapruek, and Re [DCR15].

Section 1.3: For further discussion of various measures of rate of convergence, see Ortega and Rheinboldt [OrR70], Bertsekas [Ber82a], and Barzilai and Dempster [BaD93]. The convergence rate of steepest descent with line minimization was analyzed by Kantorovich [Kan45]. The case of a constant stepsize was analyzed by Goldstein [Gol64], and Levitin and Poljak [LeP65]. For analysis of the convergence rate of steepest descent for singular problems, see Dunn [Dun81], [Dun87], and Poljak [Pol87].

Section 1.4: The modern analysis of Newton's method is generally attributed to Kantorovich [Kan39], [Kan49], although the method has a long history, reviewed among others by Deuflhard [Deu12] and Ypma [Ypm95]. For an analysis of the case where the method converges to a singular point, see Decker and Kelley [DeK80], Decker, Keller, and Kelley [DKK83], and Hughes and Dunn [HuD84]. An alternative analysis, based on the notion of self-concordance, which is related to the interior point algorithms described in Chapter 5, is given by Nesterov and Nemirovskii [NeN94] (for a more accessible account, see Boyd and Vandenbergue [BoV04]).

The modification to extend the region of convergence of Newton's method by modifying the Cholesky factorization was given by Gill and Murray [GiM74]; see also Gill, Murray, and Wright [GMW81]. The use of a trust region has been discussed in the paper by Moré and Sorensen [MoS83], and in the book by Conn, Gould, and Toint [CGT00]. Extensive accounts of various aspects of Newton-like methods are given in the books by Goldstein [Gol67], Hestenes [Hes80], Gill, Murray, and Wright [GMW81], Dennis and Schnabel [DeS83], Luenberger [Lue84], Nazareth [Naz94], Kelley [Kel99], Fletcher [Fle00], and Nocedal and Wright [NoW06]. The truncated Newton method is discussed in Dembo, Eisenstadt, and Steihaug [DES82], Nash [Nas85], and Nash and Sofer [NaS89].

There is a vast literature on least squares problems. They arise in many practical contexts, including statistical data analysis, where they are often referred to as *regression problems*. In addition to the Gauss-Newton method, they are also often solved with the incremental methods to be discussed in Section 2.4, particularly when m, the number of terms in the least squares sum, is very large.

2
Unconstrained Optimization: Additional Methods

Contents

2.1. Conjugate Direction Methods . . . . p. 120
   2.1.1. The Conjugate Gradient Method . . . . p. 125
   2.1.2. Convergence Rate of Conjugate Gradient Method . p. 132
2.2. Quasi-Newton Methods . . . . p. 138
2.3. Nonderivative Methods . . . . p. 148
   2.3.1. Coordinate Descent . . . . p. 149
   2.3.2. Direct Search Methods . . . . p. 154
2.4. Incremental Methods . . . . p. 158
   2.4.1. Incremental Gradient Methods . . . . p. 161
   2.4.2. Incremental Aggregated Gradient Methods . . . p. 172
   2.4.3. Incremental Gauss-Newton Methods . . . . . . p. 178
   2.4.3. Incremental Newton Methods . . . . p. 185
2.5. Distributed Asynchronous Algorithms . . . . p. 194
   2.5.1. Totally and Partially Asynchronous Algorithms . . p. 197
   2.5.2. Totally Asynchronous Convergence . . . . p. 198
   2.5.3. Partially Asynchronous Gradient-Like Algorithms . p. 203
   2.5.4. Convergence Rate of Asynchronous Algorithms . . p. 204
2.6. Discrete-Time Optimal Control Problems . . . . p. 210
   2.6.1. Gradient and Conjugate Gradient Methods for
       Optimal Control . . . . p. 221
   2.6.2. Newton's Method for Optimal Control . . . . . . p. 222
2.7. Solving Nonlinear Programming Problems - Some
   Practical Guidelines . . . . p. 227
2.8. Notes and Sources . . . . p. 232

In this chapter we will continue to focus on algorithms for unconstrained differentiable problems, building upon the discussion of gradient and Newton methods of Chapter 1. In Sections 2.1 and 2.2 we consider conjugate direction and quasi-Newton methods, which use first derivatives of the cost function, while emulating the ideas underlying Newton's method. We continue in Section 2.3 with methods that are based on the use of cost function values, and do not use derivatives at all. Finally in Sections 2.4-2.7 we discuss special problem structures, as well as a variety of issues involving distributed computation and practical implementation.

2.1 CONJUGATE DIRECTION METHODS

Conjugate direction methods are motivated by a desire to accelerate the convergence rate of steepest descent, while avoiding the overhead associated with Newton's method. They were originally developed for solving the quadratic problem

$$\begin{aligned} \text{minimize} \quad & f(x) = \tfrac{1}{2}x'Qx - b'x \\ \text{subject to} \quad & x \in \Re^n, \end{aligned} \quad (2.1)$$

where Q is positive definite symmetric, or equivalently, for solving the linear system

$$Qx = b. \quad (2.2)$$

They can also be used for solution of the more general system $Ax = b$, where A is invertible but not symmetric or positive definite, after conversion to the positive definite system $A'Ax = A'b$.

Conjugate direction methods can solve these problems after at most n iterations but they are best viewed as iterative methods, since usually fewer than n iterations are required to attain a sufficiently accurate solution, particularly when n is large (these methods are most often applied to high-dimensional problems for reasons that will be explained later). Conjugate direction methods can also be used to solve nonquadratic optimization problems. For such problems, they will typically not terminate after a finite number of iterations, but still, when properly implemented, they have attractive convergence and rate of convergence properties. We will first develop the methods for quadratic problems and then discuss their application to more general problems.

Given a positive definite $n \times n$ matrix Q, we say that a set of nonzero vectors $d^1, \ldots, d^k$ are *Q-conjugate*, if

$$d^{i'}Qd^j = 0, \quad \text{for all } i \text{ and } j \text{ such that } i \neq j. \quad (2.3)$$

If $d^1, \ldots, d^k$ are Q-conjugate, then they are *linearly independent*, since if one of these vectors, say d^k, were expressed as a linear combination of the others,

$$d^k = \alpha^1 d^1 + \cdots + \alpha^{k-1} d^{k-1},$$

Sec. 2.1   Conjugate Direction Methods

then by multiplication with $d^{k'}Q$ we would obtain using the Q-conjugacy of d^k and d^j, $j = 1, \ldots, k-1$,

$$d^{k'}Qd^k = \alpha^1 d^{k'}Qd^1 + \cdots + \alpha^{k-1}d^{k'}Qd^{k-1} = 0,$$

which is impossible since $d^k \neq 0$ and Q is positive definite.

For a given set of n Q-conjugate directions $d^0, \ldots, d^{n-1}$, the corresponding *conjugate direction method* for unconstrained minimization of the quadratic function

$$f(x) = \tfrac{1}{2}x'Qx - b'x, \tag{2.4}$$

is given by

$$x^{k+1} = x^k + \alpha^k d^k, \qquad k = 0, \ldots, n-1, \tag{2.5}$$

where x^0 is an arbitrary starting vector and α^k is obtained by the line minimization rule

$$f(x^k + \alpha^k d^k) = \min_{\alpha} f(x^k + \alpha d^k). \tag{2.6}$$

In particular, by setting to zero the derivative of $f(x^k + \alpha d^k)$ with respect to α, we obtain

$$0 = \frac{\partial f(x^k + \alpha^k d^k)}{\partial \alpha} = d^{k'}\nabla f(x^k + \alpha^k d^k) = d^{k'}\big(Q(x^k + \alpha^k d^k) - b\big),$$

or

$$\alpha^k = \frac{d^{k'}(b - Qx^k)}{d^{k'}Qd^k}.$$

The principal result about conjugate direction methods is that *successive iterates minimize f over a progressively expanding linear manifold that eventually includes the global minimum of f*. In particular, for each k, x^{k+1} minimizes f over the linear manifold passing through x^0 and spanned by the conjugate directions $d^0, \ldots, d^k$, i.e.,

$$x^{k+1} \in \arg\min_{x \in M^k} f(x), \tag{2.7}$$

where

$$M^k = \{x \mid x = x^0 + v,\ v \in (\text{subspace spanned by } d^0, \ldots, d^k)\}$$
$$= x^0 + (\text{subspace spanned by } d^0, \ldots, d^k).$$

In particular, x^n minimizes f over M^{n-1}, which is equal to $\Re^n$.

To show this, note that by Eq. (2.6), we have for all i

$$\left.\frac{\partial f(x^i + \alpha d^i)}{\partial \alpha}\right|_{\alpha = \alpha^i} = \nabla f(x^{i+1})' d^i = 0$$

and, for $i = 0, \ldots, k-1$,

$$\nabla f(x^{k+1})'d^i = (Qx^{k+1} - b)'d^i$$

$$= \left(x^{i+1} + \sum_{j=i+1}^{k} \alpha^j d^j\right)' Qd^i - b'd^i$$

$$= x^{i+1\prime}Qd^i - b'd^i$$

$$= \nabla f(x^{i+1})'d^i,$$

where we have used the conjugacy of d^i and d^j, $j = i+1, \ldots, k$. Combining the preceding two equations we obtain

$$\nabla f(x^{k+1})'d^i = 0, \qquad i = 0, \ldots, k, \tag{2.8}$$

so that

$$\left.\frac{\partial f(x^0 + \gamma^0 d^0 + \cdots + \gamma^k d^k)}{\partial \gamma^i}\right|_{\substack{\gamma^j = \alpha^j \\ j=0,\ldots,k}} = 0, \qquad i = 0, \ldots, k,$$

which verifies Eq. (2.7).

It is easy to visualize the expanding manifold minimization property of Eq. (2.7) when $b = 0$ and $Q = I$ (the identity matrix). In this case, the equal cost surfaces of f are concentric spheres, and the notion of Q-conjugacy reduces to usual orthogonality. By a simple algebraic argument, we see that minimization along n orthogonal directions yields the global minimum of f, i.e., the center of the spheres. (This becomes evident once we rotate the coordinate system so that the given n orthogonal directions coincide with the coordinate directions.)

The case of a general positive definite Q can be reduced to the case where $Q = I$ by means of a scaling transformation. By setting $y = Q^{1/2}x$, minimizing $\frac{1}{2}x'Qx$ is equivalent to minimizing $\frac{1}{2}\|y\|^2$. If $w^0, \ldots, w^{n-1}$ are any set of orthogonal nonzero vectors in $\Re^n$, the algorithm

$$y^{k+1} = y^k + \alpha^k w^k, \qquad k = 0, \ldots, n-1, \tag{2.9}$$

where

$$\alpha^k \in \arg\min_{\alpha} \tfrac{1}{2}\|y^k + \alpha w^k\|^2,$$

terminates in at most n steps with $y^n = 0$. To pass back to the x-coordinate system, we multiply Eq. (2.9) by $Q^{-1/2}$ and obtain

$$x^{k+1} = x^k + \alpha^k d^k, \qquad k = 0, \ldots, n-1,$$

where $d^k = Q^{-1/2}w^k$. The orthogonality of $w^0, \ldots, w^{n-1}$ ($w^{i\prime}w^j = 0$ for $i \neq j$), is equivalent to Q-conjugacy of the directions $d^0, \ldots, d^{n-1}$ ($d^{i\prime}Qd^j = 0$ for $i \neq j$); see Fig. 2.1.1.

Thus, using the transformation $y = Q^{1/2}x$, we can think of a conjugate direction method for minimizing $\frac{1}{2}x'Qx$ as a method that minimizes $\frac{1}{2}\|y\|^2$ by successive minimization along n orthogonal directions.

Sec. 2.1   Conjugate Direction Methods   123

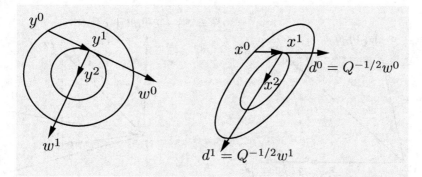

Figure 2.1.1. Geometric interpretation of conjugate direction methods in terms of successive minimization along n orthogonal directions. In the left-hand side figure the function $\|y\|^2$ is minimized successively along the directions $w^0, \ldots, w^{n-1}$, which are orthogonal in the usual sense ($w^{i'}w^j = 0$ for $i \neq j$). When this process is viewed in the coordinate system of variables $x = Q^{-1/2}y$, it yields the conjugate direction method that uses the Q-conjugate directions $d^0, \ldots, d^{n-1}$ with $d^i = Q^{-1/2}w^i$, as shown in the right-hand side figure.

Generating Q-Conjugate Directions

Given any set of linearly independent vectors $\xi^0, \ldots, \xi^k$, we can construct a set of mutually Q-conjugate directions $d^0, \ldots, d^k$ such that for all $i = 0, \ldots, k$, we have

(subspace spanned by $d^0, \ldots, d^i$) = (subspace spanned by $\xi^0, \ldots, \xi^i$). \hfill (2.10)

This can be done using the so called *Gram-Schmidt procedure*. Indeed, let us do this recursively, starting with

$$d^0 = \xi^0. \tag{2.11}$$

Suppose that, for some $i < k$, we have selected Q-conjugate $d^0, \ldots, d^i$ so that the above property holds. We then take d^{i+1} to be of the form

$$d^{i+1} = \xi^{i+1} + \sum_{m=0}^{i} c^{(i+1)m} d^m \tag{2.12}$$

and choose the coefficients $c^{(i+1)m}$ so that d^{i+1} is Q-conjugate to $d^0, \ldots, d^i$. This will be so if for each $j = 0, \ldots, i$,

$$d^{i+1'}Qd^j = \xi^{i+1'}Qd^j + \left(\sum_{m=0}^{i} c^{(i+1)m} d^m\right)' Qd^j = 0. \tag{2.13}$$

Since $d^0, \ldots, d^i$ are Q-conjugate, we have $d^{m'}Qd^j = 0$ if $m \neq j$, and Eq. (2.13) yields

$$c^{(i+1)j} = -\frac{\xi^{i+1'}Qd^j}{d^{j'}Qd^j}, \qquad j = 0, \ldots, i. \tag{2.14}$$

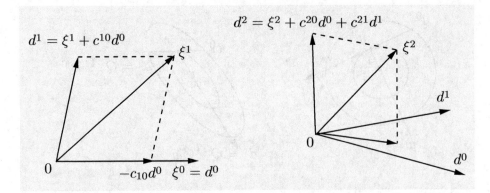

Figure 2.1.2. Illustration of the Gram-Schmidt procedure for generating Q-conjugate directions $d^0, \ldots, d^k$ from a set of linearly independent vectors $\xi^0, \ldots, \xi^k$, so that

(subspace spanned by $d^0, \ldots, d^k$) = (subspace spanned by $\xi^0, \ldots, \xi^k$).

Given $d^0, \ldots, d^{i-1}$, the ith direction is obtained as $d^i = \xi^i - \hat{\xi}^i$, where $\hat{\xi}^i$ is a vector on the subspace spanned by $d^0, \ldots, d^{i-1}$ (or $\xi^0, \ldots, \xi^{i-1}$) chosen so that d^i is Q-conjugate to $d^0, \ldots, d^{i-1}$. (It can be shown that the vector $\hat{\xi}^i$ is the projection of ξ^i on this subspace with respect to the norm $\|x\|_Q = \sqrt{x'Qx}$, i.e., minimizes $\|\xi^i - x\|_Q$ over all x in this subspace; see Exercise 2.1.1.)

Note that the denominator $d^{j'}Qd^j$ in the above equation is nonzero, since $d^0, \ldots, d^i$ are assumed Q-conjugate and are therefore nonzero. Note also that $d^{i+1} \neq 0$, since otherwise from Eqs. (2.10) and (2.12), ξ^{i+1} would be a linear combination of $\xi^0, \ldots, \xi^i$, contradicting the linear independence of $\xi^0, \ldots, \xi^k$. Finally, note from Eq. (2.12) that ξ^{i+1} lies in the subspace spanned by $d^0, \ldots, d^{i+1}$, while d^{i+1} lies in the subspace spanned by $\xi^0, \ldots, \xi^{i+1}$, since $d^0, \ldots, d^i$ and $\xi^0, \ldots, \xi^i$ span the same space [cf. Eq. (2.10)]. Thus, Eq. (2.10) is satisfied when i is increased to $i+1$ and the Gram-Schmidt procedure defined by Eqs. (2.11), (2.12), and (2.14), has the property claimed. Figure 2.1.2 illustrates the procedure.

It is also worth noting what will happen if the vectors $\xi^0, \ldots, \xi^i$ are linearly independent, but the next vector ξ^{i+1} is linearly dependent on these vectors. In this case it can be seen (compare also with Fig. 2.1.2) that the equations (2.12) and (2.14) are still valid, but the new vector d^{i+1} as given by Eq. (2.12) will be zero.† We can use this property to construct a set of Q-conjugate directions that span the same space as a set of vectors $\xi^0, \ldots, \xi^k$ that are not a priori known to be linearly independent. This

---

† From Eq. (2.12) and the linear independence of $\xi^0, \ldots, \xi^i$, it is seen that d^{i+1} can be uniquely expressed as $d^{i+1} = \sum_{m=0}^{i} \gamma^m d^m$, where γ^m are some scalars. By multiplying this equation with $d^{j'}Q$ and by using the Q-conjugacy of the directions $d^0, \ldots, d^i$ and Eq. (2.13), we see that $\gamma^m = 0$ for all $m = 0, \ldots, i$.

Sec. 2.1 Conjugate Direction Methods

construction can be accomplished with an extended version of the Gram-Schmidt procedure that generates directions via Eqs. (2.12) and (2.14), but each time the new direction d^{i+1} as given by Eq. (2.12) turns out to be zero, it is simply discarded rather than added to the set of preceding directions.

2.1.1 The Conjugate Gradient Method

The most important conjugate direction method, the conjugate gradient method, is obtained by applying the Gram-Schmidt procedure to the gradient vectors $\xi^0 = -g^0, \ldots, \xi^{n-1} = -g^{n-1}$, where we use the notation

$$g^k = \nabla f(x^k) = Qx^k - b. \tag{2.15}$$

Thus the conjugate gradient method is defined by

$$x^{k+1} = x^k + \alpha^k d^k, \tag{2.16}$$

where the stepsize α^k is obtained by line minimization, and the direction d^k is obtained by applying the kth step of the Gram-Schmidt procedure to the vector $-g^k$ and the preceding negative gradients, or equivalently, the preceding directions $d^0, \ldots, d^{k-1}$. In particular, from the Gram-Schmidt equations (2.12) and (2.14), we have

$$d^k = -g^k + \sum_{j=0}^{k-1} \frac{g^{k\prime} Q d^j}{d^{j\prime} Q d^j} d^j. \tag{2.17}$$

Note here that

$$d^0 = -g^0$$

and that the method terminates with an optimal solution if $g^k = 0$. The method also effectively stops if $d^k = 0$, but we will show that this can only happen if $g^k = 0$.

The key property of the conjugate gradient method is that the direction formula (2.17) can be greatly simplified. In particular, all but one of the coefficients in the sum of Eq. (2.17) turn out to be zero because, in view of the expanding manifold minimization property, the gradient g^k is orthogonal to the subspace spanned by $d^0, \ldots, d^{k-1}$ [cf. Eq. (2.8)]. This simplification of the direction formula will be shown as part of the following proposition.

Proposition 2.1.1: The directions of the conjugate gradient method are generated by

$$d^0 = -g^0,$$

$$d^k = -g^k + \beta^k d^{k-1}, \qquad k = 1, \ldots, n-1,$$

where β^k is given by

$$\beta^k = \frac{g^{k'} g^k}{g^{k-1'} g^{k-1}}.$$

Furthermore, the method terminates with an optimal solution after at most n steps.

Proof: We first use induction to show that all the gradients g^k generated up to termination are linearly independent. We have that g^0 by itself is linearly independent, unless $g^0 = 0$, in which case the method terminates. Suppose that the method has not terminated after k steps, and that $g^0, \ldots, g^{k-1}$ are linearly independent. Then, since the method is by definition a conjugate direction method, we have

(subspace spanned by $d^0, \ldots, d^{k-1}$) = (subspace spanned by $g^0, \ldots, g^{k-1}$) \hfill (2.18)

[cf. Eq. (2.10)]. There are two possibilities:

(a) $g^k = 0$, in which case the method terminates.

(b) $g^k \neq 0$, in which case the expanding manifold minimization property of the conjugate direction method implies that

$$g^k \text{ is orthogonal to } d^0, \ldots, d^{k-1} \qquad (2.19)$$

[cf. Eq. (2.8)]. Since the subspaces spanned by $(d^0, \ldots, d^{k-1})$ and by $(g^0, \ldots, g^{k-1})$ are the same [cf. Eq. (2.18)], we see that

$$g^k \text{ is orthogonal to } g^0, \ldots, g^{k-1}. \qquad (2.20)$$

Therefore, g^k is linearly independent of $g^0, \ldots, g^{k-1}$, thus completing the induction.

Since at most n linearly independent gradients can be generated, it follows that the gradient will be zero after at most n iterations and the method will terminate with the minimum of f.

To conclude the proof, we use the orthogonality properties (2.19) and (2.20) to verify that the calculation of the coefficients multiplying d^j in the Gram-Schmidt formula (2.17) can be simplified as stated in the proposition. We have for all j such that $g^j \neq 0$,

$$g^{j+1} - g^j = Q(x^{j+1} - x^j) = \alpha^j Q d^j, \qquad (2.21)$$

[cf. Eqs. (2.15) and (2.16)]. We note that $\alpha^j \neq 0$, since if $\alpha^j = 0$ we would have $g^{j+1} = g^j$ implying, in view of Eq. (2.20), that $g^j = 0$. Therefore, we have using Eqs. (2.20) and (2.21)

$$g^{i'} Q d^j = \frac{1}{\alpha^j} g^{i'}(g^{j+1} - g^j) = \begin{cases} 0 & \text{if } j = 0, \ldots, i-2, \\ \frac{1}{\alpha^j} g^{i'} g^i & \text{if } j = i-1, \end{cases}$$

and also that
$$d^{j'}Qd^j = \frac{1}{\alpha^j}d^{j'}(g^{j+1} - g^j).$$

Substituting the last two relations in the Gram-Schmidt formula (2.17), we obtain
$$d^k = -g^k + \beta^k d^{k-1}, \tag{2.22}$$
where
$$\beta^k = \frac{g^{k'}g^k}{d^{k-1'}(g^k - g^{k-1})}. \tag{2.23}$$

From Eq. (2.22) we have $d^{k-1} = -g^{k-1} + \beta^{k-1}d^{k-2}$. Using this equation, and the orthogonality of g^k and g^{k-1}, and of d^{k-2} and $g^k - g^{k-1}$ [cf. Eqs. (2.19) and (2.20)], the denominator in Eq. (2.23) is written as $g^{k-1'}g^{k-1}$, and the desired formula for β^k follows. **Q.E.D.**

Note that by using the orthogonality of g^k and g^{k-1} the formula
$$\beta^k = \frac{g^{k'}g^k}{g^{k-1'}g^{k-1}} \tag{2.24}$$
of Prop. 2.1.1 can also be written as
$$\beta^k = \frac{g^{k'}(g^k - g^{k-1})}{g^{k-1'}g^{k-1}}. \tag{2.25}$$

While the alternative formulas (2.24) and (2.25) produce the same results for quadratic problems, their differences become significant when the conjugate gradient method is extended to nonquadratic problems, as we will discuss shortly.

Preconditioned Conjugate Gradient Method

This method is really the conjugate gradient method implemented in a new coordinate system. Suppose we make a change of variables, $x = Sy$, where S is an invertible symmetric $n \times n$ matrix, and we apply the conjugate gradient method to the equivalent problem

$$\text{minimize} \quad h(y) = f(Sy) = \tfrac{1}{2}y'SQSy - b'Sy$$
$$\text{subject to} \quad y \in \Re^n.$$

The method is described by
$$y^{k+1} = y^k + \alpha^k \tilde{d}^k, \tag{2.26}$$

where α^k is obtained by line minimization and $\tilde{d}^k$ is generated by [cf. Eqs. (2.22) and (2.24)]
$$\tilde{d}^0 = -\nabla h(y^0), \qquad \tilde{d}^k = -\nabla h(y^k) + \beta^k \tilde{d}^{k-1}, \qquad k = 1,\ldots,n-1, \tag{2.27}$$

where
$$\beta^k = \frac{\nabla h(y^k)'\nabla h(y^k)}{\nabla h(y^{k-1})'\nabla h(y^{k-1})}. \tag{2.28}$$

Setting $x^k = Sy^k$, $\nabla h(y^k) = Sg^k$, $d^k = S\tilde{d}^k$, and $H = S^2$, we obtain from Eqs. (2.26)-(2.28) the equivalent method

$$x^{k+1} = x^k + \alpha^k d^k, \tag{2.29}$$

$$d^0 = -Hg^0, \qquad d^k = -Hg^k + \beta^k d^{k-1}, \qquad k = 1, \ldots, n-1, \tag{2.30}$$

where
$$\beta^k = \frac{g^{k'}Hg^k}{g^{k-1'}Hg^{k-1}}, \tag{2.31}$$

and α^k is obtained by line minimization.

The method described by the above equations is called the *preconditioned conjugate gradient method with scaling matrix H*. To the author's knowledge, it was first suggested in [Ber74], and used to accelerate the standard conjugate gradient method for some optimal control problems with special structure (see Exercise 2.1.10). To see that this method is a conjugate direction method, note that since $\nabla^2 h(y) = SQS$, the vectors $\tilde{d}^0, \ldots, \tilde{d}^{n-1}$ are (SQS)-conjugate. Since $d^k = S\tilde{d}^k$, we obtain that $d^0, \ldots, d^{n-1}$ are Q-conjugate. Therefore, the preconditioned method terminates with the minimum of f after at most n iterations, just as the ordinary conjugate gradient method. The motivation for preconditioning is to improve the rate of convergence within an n-iteration cycle (see the following analysis). This is important for a nonquadratic problem, but it may be important even for a quadratic problem if n is large and we want to obtain an approximate solution without waiting for the method to terminate after n steps.

Application to Nonquadratic Problems

The conjugate gradient method can be applied to the nonquadratic problem

$$\text{minimize} \quad f(x)$$
$$\text{subject to} \quad x \in \Re^n.$$

It takes the form
$$x^{k+1} = x^k + \alpha^k d^k, \tag{2.32}$$

where α^k is obtained by line minimization

$$f(x^k + \alpha^k d^k) = \min_\alpha f(x^k + \alpha d^k), \tag{2.33}$$

and d^k is generated by

$$d^k = -\nabla f(x^k) + \beta^k d^{k-1}. \tag{2.34}$$

The most common way to compute β^k is [cf. Eq. (2.25)]

$$\beta^k = \frac{\nabla f(x^k)'\big(\nabla f(x^k) - \nabla f(x^{k-1})\big)}{\nabla f(x^{k-1})'\nabla f(x^{k-1})}. \qquad (2.35)$$

The direction d^k generated by the formula $d^k = -\nabla f(x^k) + \beta^k d^{k-1}$ is a direction of descent, since from Eq. (2.33) we obtain $\nabla f(x^k)'d^{k-1} = 0$, so that

$$\nabla f(x^k)'d^k = -\|\nabla f(x^k)\|^2 + \beta^k \nabla f(x^k)'d^{k-1} = -\|\nabla f(x^k)\|^2.$$

For nonquadratic problems, the formula (2.35) is typically superior to alternative formulas such as

$$\beta^k = \frac{\nabla f(x^k)'\nabla f(x^k)}{\nabla f(x^{k-1})'\nabla f(x^{k-1})}, \qquad (2.36)$$

[cf. Eq. (2.24)]. A heuristic explanation is that due to nonquadratic terms in the objective function and possibly inaccurate line searches, conjugacy of the generated directions is progressively lost and a situation may arise where the method "jams" in the sense that the generated direction d^k is nearly orthogonal to the gradient $\nabla f(x^k)$. When this occurs, we have $\nabla f(x^{k+1}) \simeq \nabla f(x^k)$. In that case, the scalar β^{k+1}, generated by

$$\beta^{k+1} = \frac{\nabla f(x^{k+1})'\big(\nabla f(x^{k+1}) - \nabla f(x^k)\big)}{\nabla f(x^k)'\nabla f(x^k)},$$

will be nearly zero and the next direction $d^{k+1} = -\nabla f(x^{k+1}) + \beta^{k+1}d^k$ will be close to $-\nabla f(x^{k+1})$, thereby breaking the jam by restarting the conjugate gradient cycle. By contrast, when Eq. (2.36) is used, under the same circumstances the method typically continues to jam.

Regardless of the direction update formula used, one must deal with the loss of conjugacy that results from nonquadratic terms in the cost function. The conjugate gradient method is often employed in problems where the number of variables n is large, and it is not unusual for the method to start generating nonsensical and inefficient directions of search after a few iterations. For this reason it is important to operate the method in cycles of conjugate direction steps, with the first step in the cycle being a steepest descent step. Some possible restarting policies are:

(a) Restart with a steepest descent step n iterations after the preceding restart.

(b) Restart with a steepest descent step k iterations after the preceding restart with $k < n$. This is recommended when the problem has special structure so that the resulting method has good convergence rate (see Prop. 2.1.2, and Exercises 2.1.2, 2.1.3, and 2.1.10).

(c) Restart with a steepest descent step if either n iterations have taken place since the preceding restart or if

$$\left|\nabla f(x^k)'\nabla f(x^{k-1})\right| > \gamma \left\|\nabla f(x^{k-1})\right\|^2,$$

where γ is a fixed scalar with $0 < \gamma < 1$. The above relation is a test on loss of conjugacy, for if the generated directions were conjugate, then we would have $\nabla f(x^k)'\nabla f(x^{k-1}) = 0$.

Note that in all these restart procedures the steepest descent iteration serves as a spacer step and guarantees global convergence (Prop. 1.2.5 in Section 1.2). If the preconditioned conjugate gradient method is used, then a scaled steepest descent iteration is used to restart a cycle. The scaling matrix may change at the beginning of a cycle but should remain unchanged during the cycle.

An important practical issue relates to the line search accuracy that is necessary for efficient computation. On one hand, an accurate line search is needed to limit the loss of direction conjugacy and the attendant deterioration of convergence rate. On the other hand, insisting on a very accurate line search can be computationally expensive. Some trial and error may therefore be required in practice. For a discussion of implementations that are tolerant of line search inaccuracies see the papers [Per78] and [Sha78] (also Exercise 2.2.6). Note also that as mentioned in Section 1.2, the overhead for line search is significantly reduced in the presence of special structure. A prominent example is when f has the form

$$f(x) = h(Ax),$$

where A is a matrix such that the calculation of the vector $y = Ax$ for a given x is far more expensive than the calculation of $h(y)$ and its gradient.

Conjugate Gradient-Like Methods for Linear Systems

The conjugate gradient method can be used to solve the linear system of equations

$$Ax = b,$$

where A is an invertible $n \times n$ matrix and b is a given vector in $\Re^n$. One way to do this is to apply the conjugate gradient method to the positive definite quadratic optimization problem

$$\text{minimize } \tfrac{1}{2}x'A'Ax - b'Ax$$
$$\text{subject to } x \in \Re^n,$$

which corresponds to the equivalent linear system $A'Ax = A'b$. This, however, has several disadvantages, including the need to form the matrix $A'A$, which may have a much less favorable sparsity structure than A.

Sec. 2.1 Conjugate Direction Methods

An alternative possibility is to introduce the vector z defined by

$$x = A'z$$

and to solve the system $AA'z = b$ or equivalently the positive definite quadratic problem

$$\text{minimize } \tfrac{1}{2}z'AA'z - b'z$$
$$\text{subject to } z \in \Re^n,$$

whose cost function gradient is zero at z if and only if $AA'z = b$. By streamlining the computations, it is possible to write the conjugate gradient method for the preceding problem directly in terms of the vector x, and without explicitly forming the product AA'. The resulting method is known as *Craig's method*; it is given by the following iteration where H is a positive definite symmetric preconditioning matrix:

$$x^{k+1} = x^k + \alpha^k d^k, \qquad \alpha^k = \frac{r^{k'} r^k}{d^{k'} d^k},$$

where the vectors r^k and d^k are generated by the recursions

$$r^{k+1} = r^k + \alpha^k H A d^k, \qquad d^{k+1} = -A'Hr^{k+1} + \frac{r^{k+1'} r^{k+1}}{r^{k'} r^k} d^k$$

with the initial conditions

$$r^0 = H(Ax^0 - b), \qquad d^0 = -A'Hr^0.$$

The verification of these equations is left for the reader.

There are other conjugate gradient-like methods for the system $Ax = b$, which are not really equivalent to the conjugate gradient method for any quadratic optimization problem. One possibility, suggested in [SaS86], and known as the *Generalized Minimum Residual* method (GMRES), is to start with a vector x^0 and obtain x^k as the vector that minimizes $\|Ax - b\|^2$ over the linear manifold $x^0 + S^k$, where

$$S^k = (\text{subspace spanned by the vectors } r, Ar, A^2r, \ldots, A^{k-1}r),$$

and r is the initial residual

$$r = Ax^0 - b.$$

This successive subspace minimization process can be efficiently implemented, but we will not get into the details further (see [SaS86]). It can be shown that x^k is a solution of the system $Ax = b$ if and only if $A^k r$ belongs to the subspace S^k (write the minimization of $\|Ax-b\|^2$ over $x^0 + S^k$ as the equivalent minimization of $\|\xi - r\|^2$ over all ξ in the subspace AS^k). Thus if none of the vectors $x^0, \ldots, x^{n-2}$ is a solution, the subspace S^{n-1} is equal to $\Re^n$, implying that x^{n-1} is an unconstrained minimum of $\|Ax - b\|^2$, and therefore solves the system $Ax = b$. It follows that the method will terminate after at most n iterations.

GMRES can be viewed as a conjugate gradient method only in the special case where A *is positive definite and symmetric*. In that case it can be shown that the method is equivalent to a preconditioned conjugate gradient method applied to the quadratic cost $\|Ax - b\|^2$. This is based on the expanding subspace minimization property of the conjugate gradient method (see also Exercise 2.1.4). Note, however, that GMRES can be used for any matrix A that is invertible. For recent related work, see [CPS11], [FoS11], [FoS12], and the references quoted there.

2.1.2 Convergence Rate of Conjugate Gradient Method

There are a number of convergence rate results for the conjugate gradient method. Since the method terminates in at most n steps for a quadratic cost, one would expect that when viewed in cycles of n steps, its rate of convergence for a nonquadratic cost would be comparable to the rate of Newton's method. Indeed there are results which roughly state that if the method is restarted every n iterations and $\{x^k\}$ converges to a nonsingular local minimum x^*, then the error $e^k = \|x^{nk} - x^*\|$ converges superlinearly. (Note that here the error is considered at the end of cycles of n iterations rather than at the end of each iteration.)

Such superlinear convergence results are reassuring but not terribly interesting because the conjugate gradient method is most useful in problems where n is large (see the discussion at the end of Section 2.2), and for such problems, one often hopes that practical convergence will occur after fewer than n iterations. Therefore, the single-step rate of convergence of the method is typically more interesting than its rate of convergence in terms of n-step cycles. The following analysis gives a result of this type, based on an interpretation of the conjugate gradient method as an optimal process.

Assume that the cost is positive definite quadratic of the form

$$f(x) = \tfrac{1}{2} x' Q x.$$

(To simplify the following exposition, we have assumed that the linear term $b'x$ is zero, but with minor modifications, the following analysis holds also when $b \neq 0$.) Let g^i denote as usual the gradient $\nabla f(x^i)$ and consider an algorithm of the form

$$x^1 = x^0 + \gamma^{00} g^0,$$
$$x^2 = x^0 + \gamma^{10} g^0 + \gamma^{11} g^1,$$
$$\ldots \quad \ldots$$

Sec. 2.1   Conjugate Direction Methods   133

$$x^{k+1} = x^0 + \gamma^{k0}g^0 + \cdots + \gamma^{kk}g^k, \qquad (2.37)$$

where γ^{ij} are arbitrary scalars. Since $g^i = Qx^i$, we see that for suitable scalars c^{ki}, the above algorithm can be written for all k as

$$x^{k+1} = x^0 + c^{k0}Qx^0 + c^{k1}Q^2x^0 + \cdots + c^{kk}Q^{k+1}x^0 = \bigl(I + QP^k(Q)\bigr)x^0,$$

where P^k is a polynomial of degree k.

Among algorithms of the form (2.37), the conjugate gradient method is optimal in the sense that for every k, it minimizes $f(x^{k+1})$ over all sets of coefficients $\gamma^{k0}, \ldots, \gamma^{kk}$. It follows from the preceding equation that in the conjugate gradient method we have, for every k,

$$f(x^{k+1}) = \min_{P^k} \tfrac{1}{2} x^{0\prime} Q\bigl(I + QP^k(Q)\bigr)^2 x^0. \qquad (2.38)$$

Let $\lambda_1, \ldots, \lambda_n$ be the eigenvalues of Q, and let $e_1, \ldots, e_n$ be corresponding orthogonal eigenvectors, normalized so that $\|e_i\| = 1$. Since $e_1, \ldots, e_n$ form a basis, any vector $x^0 \in \Re^n$ can be written as

$$x^0 = \sum_{i=1}^{n} \xi_i e_i$$

for some scalars ξ_i. Since

$$Qx^0 = \sum_{i=1}^{n} \xi_i Q e_i = \sum_{i=1}^{n} \xi_i \lambda_i e_i,$$

we have, using the orthogonality of $e_1, \ldots, e_n$ and the fact $\|e_i\| = 1$,

$$f(x^0) = \tfrac{1}{2} x^{0\prime} Q x^0 = \tfrac{1}{2} \left(\sum_{i=1}^{n} \xi_i e_i\right)' \left(\sum_{i=1}^{n} \xi_i \lambda_i e_i\right) = \tfrac{1}{2} \sum_{i=1}^{n} \lambda_i \xi_i^2.$$

Applying the same process to Eq. (2.38), we obtain for *any* polynomial P^k of degree k

$$f(x^{k+1}) \leq \tfrac{1}{2} \sum_{i=1}^{n} \bigl(1 + \lambda_i P^k(\lambda_i)\bigr)^2 \lambda_i \xi_i^2,$$

and it follows that

$$f(x^{k+1}) \leq \max_{i} \bigl(1 + \lambda_i P^k(\lambda_i)\bigr)^2 f(x^0), \qquad \forall\ P^k,\ k. \qquad (2.39)$$

One can use this relationship for different choices of polynomials P^k to obtain a number of convergence rate results. We provide one such result, which explains a behavior often observed in practice: after a few relatively ineffective iterations, the conjugate gradient method produces substantial and often spectacular cost improvement. In particular, the following proposition shows that the first k conjugate gradient iterations in an n-iteration cycle eliminate the effect of the k largest eigenvalues of Q.

> **Proposition 2.1.2:** Assume that Q has $n-k$ eigenvalues in an interval $[a,b]$ with $a > 0$, and the remaining k eigenvalues are greater than b. Then for every x^0, the vector x^{k+1} generated after $k+1$ steps of the conjugate gradient method satisfies
>
> $$f(x^{k+1}) \leq \left(\frac{b-a}{b+a}\right)^2 f(x^0).$$
>
> This relation also holds for the preconditioned conjugate gradient method (2.29)-(2.31) if the eigenvalues of Q are replaced by those of $H^{1/2}QH^{1/2}$.

Proof: Let $\lambda_1, \ldots, \lambda_k$ be the eigenvalues of Q that are greater than b and consider the polynomial P^k defined by

$$1 + \lambda P^k(\lambda) = \frac{2}{(a+b)\lambda_1 \cdots \lambda_k}\left(\frac{a+b}{2} - \lambda\right)(\lambda_1 - \lambda)\cdots(\lambda_k - \lambda). \quad (2.40)$$

Since $1 + \lambda_i P^k(\lambda_i) = 0$, we have, using Eqs. (2.39), (2.40), and a simple calculation,

$$f(x^{k+1}) \leq \max_{a \leq \lambda \leq b} \frac{\left(\lambda - \frac{1}{2}(a+b)\right)^2}{\left(\frac{1}{2}(a+b)\right)^2} f(x^0) = \left(\frac{b-a}{b+a}\right)^2 f(x^0).$$

(See [LuY16], Section 7.5, for a more detailed argument.) **Q.E.D.**

One consequence of the above proposition is that if the conjugate gradient method is restarted every $k+1$ steps, its convergence rate is identical to steepest descent, but with a "reduced condition number," which is determined by the smallest $n-k$ eigenvalues of Q. Another consequence is that if the eigenvalues of Q take only k distinct values, then the conjugate gradient method will find the minimum of the quadratic function f in at most k iterations (take $a = b$). Some other possibilities are explored in Exercises 2.1.2, 2.1.3, and 2.1.10.

It is worth mentioning some additional rate of convergence results regarding the conjugate gradient method, as applied to the positive definite quadratic function

$$f(x) = \tfrac{1}{2}(x - x^*)'Q(x - x^*).$$

Let M and m be the largest and smallest eigenvalues of Q, respectively. Then, for any starting x^0 and any iteration index k, it can be shown that

$$\|x^k - x^*\| \leq 2\left(\frac{M}{m}\right)^{1/2}\left(\frac{\sqrt{M} - \sqrt{m}}{\sqrt{M} + \sqrt{m}}\right)^k \|x^0 - x^*\|,$$

Sec. 2.1   Conjugate Direction Methods

a result that relates to the view of the conjugate gradient method as a two-step iterative process (see Exercise 2.1.5 and compare also with the heavy ball method of Exercise 1.3.8). This estimate suggests a more favorable convergence rate than the one of steepest descent; compare with the results of Section 1.3.

EXERCISES

2.1.1

Show that the Gram-Schmidt procedure has the projection property stated in Fig. 2.1.2.

2.1.2 [Lue84]

Assume that Q has all its eigenvalues concentrated in two intervals of the form
$$[a, b], \qquad [a + \delta, b + \delta],$$
where a, b, δ are some positive scalars. Show that for every x^0, the vector x^2 generated after two steps of the conjugate gradient method satisfies
$$f(x^2) \leq \left(\frac{b-a}{b+a}\right)^2 f(x^0).$$

Hint: Proceed as in the proof of Prop. 2.1.2, using an appropriate polynomial.

2.1.3 (Hessian with Clustered Eigenvalues [Ber82a], p. 56) (www)

Assume that Q has all its eigenvalues concentrated at k intervals of the form
$$[z_i - \delta_i, z_i + \delta_i], \qquad i = 1, \ldots, k,$$
where we assume that $\delta_i \geq 0$, $i = 1, \ldots, k$, $0 < z_1 - \delta_1$, and
$$0 < z_1 < z_2 < \cdots < z_k, \qquad z_i + \delta_i \leq z_{i+1} - \delta_{i+1}, \qquad i = 1, \ldots, k-1.$$
Show that the vector x^{k+1} generated after $k+1$ steps of the conjugate gradient method satisfies
$$f(x^{k+1}) \leq R f(x^0),$$
where
$$R = \max \left\{ \frac{\delta_1^2}{z_1^2}, \frac{\delta_2^2 (z_2 + \delta_2 - z_1)^2}{z_1^2 z_2^2}, \ldots, \right.$$
$$\left. \frac{\delta_k^2 (z_k + \delta_k - z_1)^2 (z_k + \delta_k - z_2)^2 \cdots (z_k + \delta_k - z_{k-1})^2}{z_1^2 z_2^2 \cdots z_k^2} \right\}.$$

2.1.4 (www)

Consider the conjugate gradient method applied to the minimization of $f(x) = \frac{1}{2}x'Qx - b'x$, where Q is positive definite and symmetric. Show that the iterate x^k minimizes f over the linear manifold

$$x^0 + (\text{subspace spanned by } g^0, Qg^0, \ldots, Q^{k-1}g^0),$$

where $g^0 = \nabla f(x^0)$.

2.1.5 (Conjugate Gradients Obtained by Extrapolation) (www)

Consider a method whereby the first iteration is a steepest descent iteration with the stepsize determined by line minimization, and subsequent iterations have the form

$$x^{k+1} = x^k - \alpha^k g^k + \beta^k(x^k - x^{k-1}), \qquad k = 1, 2, \ldots,$$

where α^k and β^k are chosen to minimize f (compare with the heavy ball method of Exercise 1.3.8). Thus, for $k \geq 1$, the $(k+1)$st iteration finds x^{k+1} that minimizes f over the two-dimensional linear manifold

$$x^k + (\text{subspace spanned by } g^k \text{ and } x^k - x^{k-1}).$$

Show that if $f(x) = \frac{1}{2}x'Qx - b'x$, where Q is positive definite and symmetric, this method is equivalent to the conjugate gradient method, in the sense that the two methods generate identical iterates if they start at the same point.

2.1.6 (PARTAN) (www)

Let the cost function be $f(x) = \frac{1}{2}x'Qx - b'x$, where Q is positive definite and symmetric. Show that in the conjugate gradient method implementation of Exercise 2.1.5 the minimization over the two-dimensional manifold

$$x^k + (\text{subspace spanned by } g^k \text{ and } x^k - x^{k-1}),$$

to obtain x^{k+1}, can be done using just two line searches:

(1) Find the vector y^k that minimizes f over the line $y_\alpha = x^k - \alpha g^k$, $\alpha \geq 0$.

(2) Find x^{k+1} as the vector that minimizes f over the line passing through x^{k-1} and y^k.

(This is known as the method of *parallel tangents* or PARTAN, first proposed in the paper [SBK64].) *Hint*: Show that g^k is orthogonal to both g^{k-1} and $\nabla f(y^k)$. Conclude that g^k is orthogonal to the gradient $\nabla f(y)$ of any vector y on the line passing through x^{k-1} and y^k. *Note*: When applied to nonquadratic functions, PARTAN seems to be quite resilient to line minimization errors, relative to other conjugate gradient implementations.

Sec. 2.1   Conjugate Direction Methods   137

2.1.7 (www)

Let the cost function be $f(x) = \frac{1}{2}x'Qx - b'x$, where Q is positive semidefinite and symmetric. Apply the conjugate gradient method to this function. Show that if an optimal solution exists, the method will find one such solution in at most m steps, where m is the rank of Q. What happens if the problem has no optimal solution?

2.1.8 (Conjugate Gradient Implementation without Gradient Evaluation [Pow64], [Zan67a]) (www)

Let the cost function be $f(x) = \frac{1}{2}x'Qx - b'x$, where Q is positive definite and symmetric. Suppose that x_1 and x_2 minimize f over linear manifolds that are parallel to subspaces S_1 and S_2, respectively. Show that if $x_1 \neq x_2$, then $x_1 - x_2$ is Q-conjugate to all vectors in the intersection of S_1 and S_2. Use this property to suggest a conjugate direction method that does not evaluate gradients and uses only line minimizations.

2.1.9

Suppose that $d^0, \ldots, d^k$ are Q-conjugate directions, let $x^1, \ldots, x^{k+1}$ be the vectors generated by the corresponding conjugate direction method, and assume that $x^{i+1} \neq x^i$ for all $i = 0, \ldots, k$. Show that a vector d^{k+1} is Q-conjugate to $d^0, \ldots, d^k$ if and only if $d^{k+1} \neq 0$ and d^{k+1} is orthogonal to the gradient differences $g^{i+1} - g^i$, $i = 0, \ldots, k$.

2.1.10 (Preconditioning for Special Structures [Ber74])

Let the cost function be $f(x) = \frac{1}{2}x'Qx - b'x$, where Q has the form

$$Q = M + \sum_{i=1}^{k} v_i v_i',$$

where M is positive definite and symmetric, and v_i are some vectors in $\Re^n$. Show that the vector x^{k+1} generated after $k+1$ steps of the conjugate gradient method satisfies

$$f(x^{k+1}) \leq \left(\frac{b-a}{b+a}\right)^2 f(x^0),$$

where a and b are the smallest and largest eigenvalues of M, respectively. Show also that the vector x^{k+1} generated by the preconditioned conjugate gradient method with $H = M^{-1}$ minimizes f. *Hint:* Use the interlocking eigenvalues lemma [Prop. A.18(d) in Appendix A].

2.2 QUASI-NEWTON METHODS

Quasi-Newton methods are gradient methods of the form

$$x^{k+1} = x^k + \alpha^k d^k, \qquad (2.41)$$

$$d^k = -D^k \nabla f(x^k), \qquad (2.42)$$

where D^k is a positive definite matrix, which is adjusted from one iteration to the next so that the direction d^k tends to approximate the Newton direction. Some of these methods are quite popular because they typically converge fast, while avoiding the second derivative calculations associated with Newton's method. Their main drawback relative to the conjugate gradient method is that they require storage of the matrix D^k as well as the matrix-vector multiplication overhead associated with the calculation of the direction d^k (see the subsequent discussion).

An important idea for many quasi-Newton methods is that two successive iterates x^k, x^{k+1} together with the corresponding gradients $\nabla f(x^k)$, $\nabla f(x^{k+1})$, yield curvature information by means of the approximate relation

$$q^k \approx \nabla^2 f(x^{k+1}) p^k, \qquad (2.43)$$

where

$$p^k = x^{k+1} - x^k, \qquad (2.44)$$

$$q^k = \nabla f(x^{k+1}) - \nabla f(x^k). \qquad (2.45)$$

In particular, given n linearly independent iteration increments $p^0, \ldots, p^{n-1}$ together with the corresponding gradient increments $q^0, \ldots, q^{n-1}$, we can obtain approximately the Hessian as

$$\nabla^2 f(x^n) \approx \begin{bmatrix} q^0 & \cdots & q^{n-1} \end{bmatrix} \begin{bmatrix} p^0 & \cdots & p^{n-1} \end{bmatrix}^{-1},$$

and the inverse Hessian as

$$\nabla^2 f(x^n)^{-1} \approx \begin{bmatrix} p^0 & \cdots & p^{n-1} \end{bmatrix} \begin{bmatrix} q^0 & \cdots & q^{n-1} \end{bmatrix}^{-1}.$$

When the cost is quadratic, this relation is exact. Many interesting quasi-Newton methods use similar but more sophisticated ways to build curvature information into the matrix D^k so that it progressively approaches the inverse Hessian. This is true in particular for the quasi-Newton methods we will focus on in this section, and which we now introduce.

In the most popular class of quasi-Newton methods, the matrix D^{k+1} is obtained from D^k, and the vectors p^k and q^k by means of the equation

$$D^{k+1} = D^k + \frac{p^k p^{k'}}{p^{k'} q^k} - \frac{D^k q^k q^{k'} D^k}{q^{k'} D^k q^k} + \xi^k \tau^k v^k v^{k'}, \qquad (2.46)$$

Sec. 2.2 Quasi-Newton Methods

where
$$v^k = \frac{p^k}{p^{k'}q^k} - \frac{D^k q^k}{\tau^k}, \qquad (2.47)$$

$$\tau^k = q^{k'} D^k q^k, \qquad (2.48)$$

the scalars ξ^k satisfy, for all k,

$$0 \leq \xi^k \leq 1, \qquad (2.49)$$

and D^0 is an arbitrary positive definite matrix. The scalars ξ^k parameterize the method. If $\xi^k = 0$ for all k, we obtain the *Davidon-Fletcher-Powell (DFP) method*, which is historically the first quasi-Newton method. If $\xi^k = 1$ for all k, we obtain the *Broyden-Fletcher-Goldfarb-Shanno (BFGS) method*, for which there is substantial evidence that it is an excellent, perhaps the best, general purpose quasi-Newton method currently known.

We will show that the matrices D^k generated in the above manner indeed encode curvature information as indicated in Eqs. (2.43)-(2.45). In particular, we will see in the proof of the subsequent Prop. 2.2.2 that for a quadratic problem and all k we have

$$D^{k+1} q^i = p^i, \qquad i = 0, \ldots, k,$$

from which after $n - 1$ iterations, we obtain

$$D^n = \begin{bmatrix} p^0 & \cdots & p^{n-1} \end{bmatrix} \begin{bmatrix} q^0 & \cdots & q^{n-1} \end{bmatrix}^{-1} = \nabla^2 f(x^n)^{-1}.$$

We first show that under a mild assumption, the matrices D^k generated by Eq. (2.46) are positive definite. This is a very important property, since it guarantees that d^k is a descent direction.

Proposition 2.2.1: If D^k is positive definite and the stepsize α^k is chosen so that x^{k+1} satisfies

$$\nabla f(x^k)' d^k < \nabla f(x^{k+1})' d^k, \qquad (2.50)$$

then D^{k+1} as given by Eq. (2.46) is positive definite.

Note: If x^k is not a stationary point, we have $\nabla f(x^k)' d^k < 0$, so in order to satisfy condition (2.50), it is sufficient to carry out the line search to a point where
$$|\nabla f(x^{k+1})' d^k| < |\nabla f(x^k)' d^k|.$$

In particular, if α^k is determined by the line minimization rule, then we have $\nabla f(x^{k+1})' d^k = 0$ and Eq. (2.50) is satisfied.

Proof: We first note that Eq. (2.50) implies that $\alpha^k \neq 0$, $q^k \neq 0$, and
$$p^{k'}q^k = \alpha^k d^{k'}(\nabla f(x^{k+1}) - \nabla f(x^k)) > 0. \qquad (2.51)$$
Thus all denominator terms in Eqs. (2.46) and (2.47) are nonzero, and D^{k+1} is well defined.

Now for any $z \neq 0$ we have
$$z'D^{k+1}z = z'D^k z + \frac{(z'p^k)^2}{p^{k'}q^k} - \frac{(q^{k'}D^k z)^2}{q^{k'}D^k q^k} + \xi^k \tau^k (v^{k'}z)^2.$$
Using the notation $a = (D^k)^{1/2}z$, $b = (D^k)^{1/2}q^k$, this equation is written as
$$z'D^{k+1}z = \frac{\|a\|^2\|b\|^2 - (a'b)^2}{\|b\|^2} + \frac{(z'p^k)^2}{p^{k'}q^k} + \xi^k \tau^k (v^{k'}z)^2. \qquad (2.52)$$

From Eqs. (2.48) and (2.51), and the Schwarz inequality [Eq. (A.2) in Appendix A], we have that all terms on the right-hand side of Eq. (2.52) are nonnegative. In order that $z'D^{k+1}z > 0$, it will suffice to show that we cannot have simultaneously
$$\|a\|^2\|b\|^2 = (a'b)^2 \qquad \text{and} \qquad z'p^k = 0.$$
Indeed if $\|a\|^2\|b\|^2 = (a'b)^2$, we must have $a = \lambda b$ or equivalently, $z = \lambda q^k$. Since $z \neq 0$, it follows that $\lambda \neq 0$, so if $z'p^k = 0$, we must have $q^{k'}p^k = 0$, which is impossible by Eq. (2.51). **Q.E.D.**

An important property of the algorithm is that when applied to the positive definite quadratic function $f(x) = \frac{1}{2}x'Qx - b'x$, with the stepsize α^k determined by line minimization, it generates a Q-conjugate direction sequence, while simultaneously constructing the inverse Hessian Q^{-1} after n iterations. This is the subject of the next proposition.

Proposition 2.2.2: Let $\{x^k\}$, $\{d^k\}$, and $\{D^k\}$ be sequences generated by the quasi-Newton algorithm (2.41)-(2.42), (2.46)-(2.49), applied to minimization of the positive definite quadratic function
$$f(x) = \tfrac{1}{2}x'Qx - b'x,$$
with α^k chosen by the line minimization rule,
$$f(x^k + \alpha^k d^k) = \min_\alpha f(x^k + \alpha d^k).$$
Assume that none of the vectors $x^0, \ldots, x^{n-1}$ is optimal. Then:

(a) The vectors $d^0, \ldots, d^{n-1}$ are Q-conjugate.

(b) There holds
$$D^n = Q^{-1}.$$

Proof: We will show that for all k

$$d^{i'}Qd^j = 0, \quad 0 \le i < j \le k, \tag{2.53}$$

$$D^{k+1}Qp^i = p^i, \quad 0 \le i \le k. \tag{2.54}$$

Equation (2.53) proves part (a) and it can be shown that Eq. (2.54) proves part (b). Indeed, since for $i < n$, none of the vectors x^i is optimal and d^i is a descent direction [cf. Eq. (2.42) and Prop. 2.2.1], we have that $p^i \ne 0$. Since $p^i = \alpha^i d^i$ and $d^0, \ldots, d^{n-1}$ are Q-conjugate, it follows that $p^0, \ldots, p^{n-1}$ are linearly independent and therefore, Eq. (2.54) implies that $D^n Q$ is equal to the identity matrix.

We first verify that

$$D^{k+1}Qp^k = p^k, \quad \forall\, k. \tag{2.55}$$

From the equation $Qp^k = q^k$ and the updating formula (2.46), we have

$$D^{k+1}Qp^k = D^{k+1}q^k$$
$$= D^k q^k + \frac{p^k p^{k'} q^k}{p^{k'} q^k} - \frac{D^k q^k q^{k'} D^k q^k}{q^{k'} D^k q^k} + \xi^k \tau^k v^k v^{k'} q^k$$
$$= D^k q^k + p^k - D^k q^k + \xi^k \tau^k v^k v^{k'} q^k$$
$$= p^k + \xi^k \tau^k v^k v_k{'} q^k.$$

From Eqs. (2.47) and (2.48), we have that $v^{k'} q^k = 0$ and Eq. (2.55) follows.

We now show Eqs. (2.53) and (2.54) simultaneously by induction. For $k = 0$ there is nothing to show for Eq. (2.53), while Eq. (2.54) holds in view of Eq. (2.55). Assuming that Eqs. (2.53) and (2.54) hold for k, we prove them for $k + 1$. We have, for $i < k$,

$$\nabla f(x^{k+1}) = \nabla f(x^{i+1}) + Q(p^{i+1} + \cdots + p^k). \tag{2.56}$$

The vector p^i is orthogonal to each vector in the right-hand side of this equation; it is orthogonal to $Qp^{i+1}, \ldots, Qp^k$ because $p^0, \ldots, p^k$ are Q-conjugate (since $p^i = \alpha^i d^i$) and it is orthogonal to $\nabla f(x^{i+1})$ because of the line minimization property of the stepsize. Therefore from Eq. (2.56) we obtain

$$p^{i'} \nabla f(x^{k+1}) = 0, \quad 0 \le i < k.$$

From this equation and Eq. (2.54),

$$p^{i'} Q D^{k+1} \nabla f(x^{k+1}) = 0, \quad 0 \le i \le k,$$

and since $p^i = \alpha^i d^i$ and $d^{k+1} = -D^{k+1} \nabla f(x^{k+1})$, we obtain

$$d^{i'} Q d^{k+1} = 0, \quad 0 \le i \le k.$$

This proves Eq. (2.53) for $k+1$.

From the induction hypothesis (2.54) and Eq. (2.55), we have for all i with $0 \leq i \leq k$,

$$q^{k+1'}D^{k+1}Qp^i = q^{k+1'}p^i = p^{k+1'}Qp^i = \alpha^{k+1}\alpha^i d^{k+1'}Qd^i = 0.$$

From Eq. (2.46), we have, for $0 \leq i \leq k$,

$$D^{k+2}q^i = D^{k+1}q^i + \frac{p^{k+1}p^{k+1'}q^i}{p^{k+1'}q^{k+1}} - \frac{D^{k+1}q^{k+1}q^{k+1'}D^{k+1}q^i}{q^{k+1'}D^{k+1}q^{k+1}} \quad (2.57)$$
$$+ \xi^{k+1}\tau^{k+1}v^{k+1}v^{k+1'}q^i.$$

Since $p^{k+1'}q^i = p^{k+1'}Qp^i = \alpha^{k+1}\alpha^i d^{k+1'}Qd^i = 0$, we see that the second term in the right-hand side of Eq. (2.57) is zero. Similarly, Eq. (2.54) implies that

$$q^{k+1'}D^{k+1}q^i = q^{k+1'}D^{k+1}Qp^i = q^{k+1'}p^i = p^{k+1'}Qp^i = 0,$$

and we see that the third term in the right-hand side of Eq. (2.57) is zero. Finally, a similar argument using the definition (2.47) of v^{k+1}, shows that the fourth term in the right-hand side of Eq. (2.57) is zero as well. Therefore, Eqs. (2.57) and (2.54) yield

$$D^{k+2}Qp^i = D^{k+2}q^i = D^{k+1}q^i = D^{k+1}Qp^i = p^i, \quad 0 \leq i \leq k.$$

Taking into account also Eq. (2.55), this proves Eq. (2.54) for $k+1$. **Q.E.D.**

The following proposition sharpens the preceding one and clarifies further the relation between quasi-Newton and conjugate gradient methods.

Proposition 2.2.3: The sequence of iterates $\{x^k\}$ generated by the quasi-Newton algorithm applied to minimization of a positive definite quadratic function as in Prop. 2.2.2 is identical to the one that would be generated by the preconditioned conjugate gradient method with scaling matrix $H = D^0$.

Proof: It is sufficient to show that for $k = 0, 1, \ldots, n-1$, the vector x^{k+1} minimizes f over the linear manifold

$$M^k = \{z \mid z = x^0 + \gamma^0 D^0 \nabla f(x^0) + \cdots + \gamma^k D^0 \nabla f(x^k), \; \gamma^0, \ldots, \gamma^k \in \Re\}.$$

This can be proved for the case where $D^0 = I$ by verifying through induction that for all k there exist scalars β_{ij}^k such that

$$D^k = I + \sum_{i=0}^{k} \sum_{j=0}^{k} \beta_{ij}^k \nabla f(x^i) \nabla f(x^j)'.$$

Sec. 2.2   Quasi-Newton Methods   143

Therefore, for some scalars b_i^k and all k, we have

$$d^k = -D^k \nabla f(x^k) = \sum_{i=0}^{k} b_i^k \nabla f(x^i).$$

Hence, for all i, x^{i+1} lies on the manifold

$$M^i = \{z \mid z = x^0 + \gamma^0 \nabla f(x^0) + \cdots + \gamma^i \nabla f(x^i), \ \gamma^0, \ldots, \gamma^i \in \Re\},$$

and since, by Prop. 2.2.2, the algorithm is a conjugate direction method, x^{i+1} minimizes f over M^i based on the results of the preceding section. Thus, when $D^0 = I$, the algorithm satisfies the defining property of the conjugate gradient method (for all i, x^{i+1} is the unique minimum of f over M^i). For the case where $D^0 \neq I$, the proof follows by making a transformation of variables so that in the transformed space the initial matrix is the identity. **Q.E.D.**

A consequence of the above proposition is that if the minimization stepsize rule is used and the cost is quadratic, the generated iterates by the quasi-Newton algorithm do not depend on the values of the scalar ξ^k. Thus the DFP and the BFGS methods perform identically on a quadratic function if the line minimization is perfect. It turns out that this can be shown even when the cost is nonquadratic ([Dix72a], [Dix72b]), which is a rather surprising result. Thus the choice of ξ^k makes a difference only if the line minimization is inaccurate.

We finally note that multiplying the initial matrix D^0 by a positive scaling factor can have a significant beneficial effect on the behavior of the algorithm in the initial iterations of an n-iteration cycle, and also more generally in the case of a nonquadratic problem. A popular choice is to compute

$$\tilde{D}^0 = \frac{p^{0\prime} q^0}{q^{0\prime} D^0 q^0} D^0, \qquad (2.58)$$

once the vector x^1 (and hence also p^0 and q^0) has been obtained, and use $\tilde{D}^0$ in place of D^0 in computing D^1. Sometimes it is beneficial to scale D^k even after the first iteration by multiplication with $p^{k\prime} q^k / q^{k\prime} D^k q^k$. This is the idea behind the *self-scaling algorithms* introduced in the papers [OrL74] and [Ore73]. The rationale for these algorithms is that if the scaling (2.58) is used, then the condition number M^k/m^k, where

$$M^k = \text{max eigenvalue of } (D^k)^{1/2} Q (D^k)^{1/2},$$

$$m^k = \text{min eigenvalue of } (D^k)^{1/2} Q (D^k)^{1/2},$$

is not increased (and is usually decreased) at each iteration (see Exercise 2.2.5).

Comparison of Quasi-Newton Methods with Other Methods

Let us now consider a nonquadratic problem, and compare the quasi-Newton method of Eqs. (2.41)-(2.42), (2.46)-(2.49) with the conjugate gradient method. One advantage of the quasi-Newton method is that when the line search is accurate, the algorithm not only tends to generate conjugate directions but also constructs an approximation to the inverse Hessian matrix. As a result, near convergence to a local minimum with positive definite Hessian, it tends to approximate Newton's method thereby attaining a fast convergence rate. It is significant that this property does not depend on the starting matrix D^0, and as a result it is not usually necessary to periodically restart the method with a steepest descent-type step. By contrast, this is essential for the conjugate gradient method.

A second advantage is that the quasi-Newton method is not as sensitive to accuracy in the line search as the conjugate gradient method. This has been verified by extensive computational experience and can be substantiated to some extent by analysis. A partial explanation is that, under essentially no restriction on the line search accuracy, the method generates positive definite matrices D^k and hence directions of descent (Prop. 2.2.1).

To compare further the conjugate gradient method and the quasi-Newton method, we consider their computational requirements per iteration when n is large. The kth iteration of the conjugate gradient method requires computation of the cost function and its gradient (perhaps several times in the course of the line minimization) together with $O(n)$ operations to compute the conjugate direction d^k and the next point x^{k+1}. The quasi-Newton method requires roughly the same amount of computation for function and gradient evaluations together with $O(n^2)$ operations to compute the matrix D^k and the next point x^{k+1}. If the computation needed for a function and gradient evaluation is larger or comparable to $O(n^2)$ operations, the quasi-Newton method requires only slightly more computation per iteration than the conjugate gradient method and holds the edge in view of its other advantages mentioned earlier. In problems where a function and gradient evaluation requires computation time much less than $O(n^2)$ operations, the conjugate gradient method is typically preferable. As an example, we will see in Section 2.6, that in optimal control problems where typically n is very large, a function and a gradient evaluation typically requires $O(n)$ operations. For this reason the conjugate gradient method is typically preferable for these problems.

In general, both the conjugate gradient method and the quasi-Newton algorithm require less computation per iteration than Newton's method, which requires a function, gradient, and Hessian evaluation, as well as $O(n^3)$ operations at each step for computing the Newton direction. This is counterbalanced by the faster speed of convergence of Newton's method. Furthermore, in some cases, special structure can be exploited to compute the Newton direction efficiently. For example in optimal control problems,

Newton's method typically requires $O(n)$ operations per iteration versus $O(n^2)$ operations for the quasi-Newton method (see Section 2.6).

EXERCISES

2.2.1 (Rank One Quasi-Newton Methods) (www)

Suppose that D^0 is symmetric and that D^k is updated according to the formula

$$D^{k+1} = D^k + \frac{y^k y^{k'}}{q^{k'} y^k},$$

where $y^k = p^k - D^k q^k$. Show that we have

$$D^{k+1} q^i = p^i, \quad \text{for all } k \text{ and } i \leq k.$$

Conclude that for a positive definite quadratic problem, after n steps for which n linearly independent increments $q^0, \ldots, q^{n-1}$ are obtained, D^n is equal to the inverse Hessian of the cost function.

2.2.2 (BFGS Update) (www)

Verify the following formula for the BFGS update:

$$D^{k+1} = D^k + \left(1 + \frac{q^{k'} D^k q^k}{p^{k'} q^k}\right) \frac{p^k p^{k'}}{p^{k'} p^k} - \frac{D^k q^k p^{k'} + p^k q^{k'} D^k}{p^{k'} q^k}.$$

2.2.3 (Limited Memory BFGS Method [Noc80]) (www)

A major drawback of quasi-Newton methods for large problems is the large storage requirement. This motivates methods that construct the quasi-Newton direction $d^k = -D^k \nabla f(x^k)$ using only a limited number of the vectors p^i and q^i (for example, the last m). This exercise shows one way to do this.

(a) Use Exercise 2.2.2 to show that the BFGS updating formula can be written as

$$D^{k+1} = V^{k'} D^k V^k + \rho^k p^k p^{k'},$$

where

$$\rho^k = \frac{1}{q^{k'} p^k}, \quad V^k = I - \rho^k q^k p^{k'}.$$

(b) Show how to calculate the direction $d^k = -D^k \nabla f(x^k)$ using D^0 and the past vectors $p^i, q^i, i = 0, 1, \ldots, k-1$.

2.2.4 (Hessian Approximation Updating Formula) (www)

The matrices D^k generated by the updating formula of this section can be viewed as approximations of the inverse Hessian. An alternative is to generate and update approximations of the Hessian. It turns out that the inverses $H^k = (D^k)^{-1}$ can be updated by formulas similar to the ones for D^k, except that q^k is replaced by p^k and reversely. This can be traced to the fact that the relations

$$D^{k+1}q^i = D^{k+1}Qp^i = p^i, \qquad 0 \le i \le k,$$

shown in Prop. 2.2.1 can also be written as

$$H^{k+1}p^i = H^{k+1}Q^{-1}q^i = q^i, \qquad 0 \le i \le k.$$

In particular, use the BFGS formula given in Exercise 2.2.2 to show that if D^k (or H^k) is updated by the DFP formula, then H^k (or D^k, respectively) is updated by the BFGS formula.

2.2.5 (Condition Number of a Quasi-Newton Iteration [Ore73], [OrL74]) (www)

Consider a positive definite quadratic problem. If we view the quasi-Newton iteration $x^{k+1} = x^k - \alpha^k D^k \nabla f(x^k)$ as a scaled gradient method, its convergence rate (per step) is governed by the condition number of the matrix $(D^k)^{1/2} Q (D^k)^{1/2}$, or equivalently the condition number of the matrix

$$R^k = Q^{1/2} D^k Q^{1/2}.$$

(a) Suppose that D^k is updated using the DFP formula. Let $\lambda_1 \le \cdots \le \lambda_n$ be the eigenvalues of R^k and suppose that $1 \in [\lambda_1, \lambda_n]$. Show that the eigenvalues of R^{k+1} are contained in the range $[\lambda_1, \lambda_n]$, so that the condition number of R^{k+1} is no worse than the condition number of R^k. Furthermore, 1 is an eigenvalue of R^{k+1}. *Hint*: Let

$$r^k = Q^{1/2} p^k.$$

Show that R^k is generated by

$$R^{k+1} = R^k + \frac{r^k r^{k'}}{r^{k'} r^k} - \frac{R^k r^k r^{k'} R^k}{r^{k'} R^k r^k}.$$

Use the interlocking eigenvalues lemma [Prop. A.18(d) in Appendix A] to show that the eigenvalues $\mu_1, \ldots, \mu_n$ of the matrix

$$R^k - \frac{R^k r^k r^{k'} R^k}{r^{k'} R^k r^k}$$

satisfy

$$0 = \mu_1 \le \lambda_1 \le \mu_2 \le \cdots \le \mu_n \le \lambda_n.$$

Furthermore, r^k is an eigenvector corresponding to the eigenvalue $\mu_1 = 0$. Use the interlocking eigenvalues lemma again to show that the eigenvalues of R^{k+1} are $\mu_2, \ldots, \mu_n$ and 1.

(b) Show that the property $1 \in [\lambda_1, \lambda_n]$ of part (a) holds if the scaling (2.58) is used.

(c) Show the result of part (a) for the case of the BFGS update. *Hint*: Work with an update formula for $(R^k)^{-1}$; see Exercise 2.2.4.

2.2.6 (Memoryless Quasi-Newton Method [Per78], [Sha78]) (www)

This exercise provides the basis for an alternative implementation of the conjugate gradient method, which incorporates some of the advantages of quasi-Newton methods. Denote for all k, $g^k = \nabla f(x^k)$, and consider an algorithm of the form

$$x^{k+1} = x^k + \alpha^k d^k,$$

where

$$d^0 = -g^0, \qquad d^{k+1} = -D^{k+1} g^{k+1}, \qquad k = 0, 1, \ldots.$$

The matrix D^{k+1} is obtained via the BFGS update of Exercise 2.2.2 starting with the identity matrix:

$$D^{k+1} = I + \left(1 + \frac{q^{k'} q^k}{p^{k'} q^k}\right) \frac{p^k p^{k'}}{p^{k'} p^k} - \frac{q^k p^{k'} + p^k q^{k'}}{p^{k'} q^k}.$$

Furthermore, the line search is exact, so that $p^{k'} g^{k+1} = 0$. Assume that the cost function is positive definite quadratic.

(a) Show that

$$d^{k+1} = -g^{k+1} + \frac{(g^{k+1} - g^k)' g^{k+1}}{\|g^k\|^2} d^k,$$

so that the method coincides with the conjugate gradient method.

(b) Let D be a positive definite symmetric matrix and assume that

$$d^0 = -Dg^0, \qquad d^{k+1} = -D^{k+1} g^{k+1}, \qquad k = 0, 1, \ldots,$$

where

$$D^{k+1} = D + \left(1 + \frac{q^{k'} Dq^k}{p^{k'} q^k}\right) \frac{p^k p^{k'}}{p^{k'} p^k} - \frac{Dq^k p^{k'} + p^k q^{k'} D}{p^{k'} q^k}.$$

Show that the method coincides with a preconditioned version of the conjugate gradient method.

2.3 NONDERIVATIVE METHODS

All the gradient methods examined so far require calculation of at least the gradient $\nabla f(x^k)$ and possibly the Hessian matrix $\nabla^2 f(x^k)$ at each generated point x^k. In many problems, however, either these derivatives are not available in explicit form or they are given by very complicated expressions. In such cases, it may be preferable to use variants of the algorithms discussed so far, where all unavailable derivatives are approximated by finite differences.

First derivatives may be approximated by the *forward difference formula*

$$\frac{\partial f(x^k)}{\partial x_i} \approx \frac{1}{h}\big(f(x^k + he_i) - f(x^k)\big) \qquad (2.59)$$

or by the *central difference formula*

$$\frac{\partial f(x^k)}{\partial x_i} \approx \frac{1}{2h}\big(f(x^k + he_i) - f(x^k - he_i)\big). \qquad (2.60)$$

In these relations, h is a small positive scalar and e_i is the ith unit vector (ith column of the identity matrix). In some cases the same value of h can be used for all partial derivatives, but in other cases, particularly when the problem is poorly scaled, it is essential to use a different value of h for each partial derivative. This is a tricky process that often requires trial and error, as we will discuss shortly.

The central difference formula requires twice as much computation as the forward difference formula. However, it is much more accurate. This can be seen by forming the corresponding first and second order expansions, and by verifying that the absolute value of the error between the approximation and the actual derivatives is $O(h)$ for the forward difference formula, while it is $O(h^2)$ for the central difference formula. Note that if the central difference formula is used, one obtains at essentially no extra cost an approximation of each diagonal element of the Hessian using the formula

$$\frac{\partial^2 f(x^k)}{(\partial x_i)^2} \approx \frac{1}{h^2}\big(f(x^k + he_i) + f(x^k - he_i) - 2f(x^k)\big).$$

These approximations can be used in schemes based on diagonal scaling.

To reduce the approximation error, we would like to choose the finite difference interval h as small as possible. Unfortunately, there is a limit on how much h can be reduced due to the roundoff error that occurs when quantities of similar magnitude are subtracted by the computer. In particular, an error δ due to finite precision arithmetic in evaluating the numerator in Eq. (2.59) [or Eq. (2.60)], results in an error of δ/h (or $\delta/2h$, respectively) in the first derivative evaluation. Roundoff error is particularly evident in the approximate formulas (2.59) and (2.60) near a stationary point where ∇f is nearly zero, and the relative error size in the gradient approximation becomes very large.

Practical experience suggests that a good policy is to keep the scalar h for each derivative at a fixed value, which roughly balances the approximation error against the roundoff error. Based on the preceding calculations, this leads to the guideline

$$\frac{\delta}{h} = O(h) \quad \text{or} \quad h = O(\delta^{1/2}), \quad \text{for the forward difference formula (2.59)},$$

$$\frac{\delta}{2h} = O(h^2) \quad \text{or} \quad h = O(\delta^{1/3}), \quad \text{for the central difference formula (2.60)},$$

where δ is the error due to finite precision arithmetic in evaluating the numerator in Eq. (2.59) [or Eq. (2.60)]. Thus, a much larger value of h can be used in conjunction with the central difference formula. A good practical rule is to use the forward formula (2.59) until the absolute value of the corresponding approximate derivative becomes less than a certain tolerance; i.e.,

$$\left| \frac{f(x^k + he_i) - f(x^k)}{h} \right| \leq \epsilon,$$

where $\epsilon > 0$ is some prespecified scalar. At that point, a switch to the central difference formula should be made.

Second derivatives may be approximated by the forward difference formula

$$\frac{\partial^2 f(x^k)}{\partial x_i \partial x_j} \approx \frac{1}{h} \left(\frac{\partial f(x^k + he_j)}{\partial x^i} - \frac{\partial f(x^k)}{\partial x_i} \right) \qquad (2.61)$$

or the central difference formula

$$\frac{\partial^2 f(x^k)}{\partial x_i \partial x_j} \approx \frac{1}{2h} \left(\frac{\partial f(x^k + he_j)}{\partial x_i} - \frac{\partial f(x^k - he_j)}{\partial x_i} \right). \qquad (2.62)$$

Practical experience suggests that in discretized forms of Newton's method, extreme accuracy in approximating second derivatives is not very important in terms of rate of convergence. For this reason, exclusive use of the forward difference formula (2.61) is adequate in most cases. However, one should certainly check for positive definiteness of the discretized Hessian approximation and introduce corresponding modifications if necessary, as discussed in Section 1.4.

2.3.1 Coordinate Descent

There are several other nonderivative approaches for minimizing differentiable functions. A particularly important one is the *coordinate descent method*. Here the cost is minimized along one coordinate direction at each

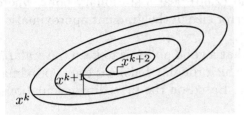

Figure 2.3.1. Illustration of the coordinate descent method.

iteration. This not only simplifies the calculation of the search direction, but often also facilitates the stepsize selection, since a line minimization along the coordinate directions may be possible analytically. The order in which coordinates are chosen may vary in the course of the algorithm. In the case where this order is *cyclic*, given x^k, the ith coordinate of x^{k+1} is determined by

$$x_i^{k+1} \in \arg\min_{\xi \in \Re} f(x_1^{k+1}, \ldots, x_{i-1}^{k+1}, \xi, x_{i+1}^k, \ldots, x_n^k); \qquad (2.63)$$

see Fig. 2.3.1. The method can also be used for minimization of f subject to upper and lower bounds on the variables x^i (the minimization over $\xi \in \Re$ in the preceding equation is replaced by minimization over the appropriate interval). We will analyze the method within this more general context in the next chapter.

An important advantage of the coordinate descent method is that it is well suited for *parallel and distributed computation*. In particular, suppose that there is a subset of coordinates $x_{i_1}, x_{i_2}, \ldots, x_{i_m}$, which are not coupled through the cost function, i.e., $f(x)$ can be written as $\sum_{r=1}^m f_{i_r}(x)$, where for each r, $f_{i_r}(x)$ does not depend on the coordinates x_{i_s} for all $s \neq r$. Then one can perform the m coordinate descent iterations

$$x_{i_r}^{k+1} \in \arg\min_\xi f_{i_r}(x^k + \xi e_{i_r}), \qquad r = 1, \ldots, m,$$

independently and in parallel. Thus, in problems with special structure where the set of coordinates can be partitioned into p subsets with the independence property just described, one can perform a full cycle of coordinate descent iterations in p (as opposed to n) parallel steps (assuming of course that a sufficient number of parallel processors is available); see Section 2.5 for a discussion of parallel and distributed computation issues.

The coordinate descent method generally yields sequences whose limit points are stationary, although the proof of this is more complicated than for gradient methods, and requires some additional assumptions. In particular, the following proposition requires that the minimum of the cost function along each coordinate is uniquely attained. This is a special case of a more general result to be shown in Section 3.7 (Prop. 3.7.1), which deals with a constrained version of coordinate descent.

> **Proposition 2.3.1: (Convergence of Coordinate Descent)** Suppose that f is continuously differentiable, and that for each $x = (x_1, \ldots, x_n) \in \Re^n$ and i,
>
> $$f(x_1, \ldots, x_{i-1}, \xi, x_{i+1}, \ldots, x_m)$$
>
> viewed as a function of ξ, attains a unique minimum $\bar{\xi}$ over $\Re$, and is monotonically nonincreasing in the interval from x_i to $\bar{\xi}$. Let $\{x^k\}$ be the sequence generated by the coordinate descent method (2.63). Then, every limit point of $\{x^k\}$ is a stationary point.

It may be said that the convergence behavior of coordinate descent for differentiable cost functions is fairly similar to the one of steepest descent. Both methods are convergent but can be very slow. Still, for many practical contexts, they can be quite effective. Coordinate descent also works in the presence of separable nondifferentiable terms in the cost function, such as the ℓ_1 norm $\sum_{i=1}^{n} |x_i|$, as we will see in Section 3.7; this is useful in some important machine learning contexts (see e.g., the subsequent Example 2.4.1, and [Ber15a], Section 1.3). In practice the choice between the two methods is typically dictated by the structure of the cost function.

Convergence and Convergence Rate for a Constant Stepsize

We will now discuss the convergence properties of some variants of the coordinate descent method for strongly convex functions. Instead of minimization along each coordinate, we will assume a constant but sufficiently small stepsize. The results to be obtained hold qualitatively for the case of the minimization rule as well, but the analysis is more complicated. We will also consider different orders of coordinate selection as alternatives to the cyclic order.

In the subsequent analysis, we assume that the function $f : \Re^n \mapsto \Re$ has Lipschitz continuous gradient along the ith coordinate, i.e., for some $L > 0$, we have

$$\left|\nabla_i f(x+\alpha e_i) - \nabla_i f(x)\right| \leq L|\alpha|, \qquad \forall\, x \in \Re^n,\, \alpha \in \Re,\, i = 1, \ldots, n, \quad (2.64)$$

where e_i is the ith coordinate direction, $e_i = (0, \ldots, 0, 1, 0, \ldots, 0)$ with the 1 in the ith position, and $\nabla_i f$ is the ith component of the gradient. We also assume that f is strongly convex in the sense that for some $\sigma > 0$,

$$f(y) \geq f(x) + \nabla f(x)'(y-x) + \frac{\sigma}{2}\|x - y\|^2, \qquad \forall\, x, y \in \Re^n, \quad (2.65)$$

and we denote by x^* the minimum of f (which is unique by Prop. 1.1.2).

Consider the coordinate descent method with constant stepsize $\alpha = \frac{1}{L}$

$$x^{k+1} = x^k - \frac{1}{L} \nabla_{i_k} f(x^k) e_{i_k}, \qquad (2.66)$$

where i_k is the index selected at iteration k. [The following analysis also applies for any $\alpha < 1/L$, since the constant L in Eq. (2.64) can be increased as much as desired.] By using the descent lemma (Prop. A.24 in Appendix A) and Eq. (2.64), we have

$$\begin{aligned} f(x^{k+1}) &\leq f(x^k) + \nabla_{i_k} f(x^k)(x^{k+1}_{i_k} - x^k_{i_k}) + \frac{L}{2}(x^{k+1}_{i_k} - x^k_{i_k})^2 \\ &= f(x^k) - \frac{1}{2L}\left|\nabla_{i_k} f(x^k)\right|^2. \end{aligned} \qquad (2.67)$$

Thus the cost function is reduced at each iteration. Consider now any order of component selection such that there is some integer M such that every component is iterated at least once within M successive iterations (an example with $M = n$ is the cyclic order). Then from Eq. (2.67) it can be seen that the sequence $\{x^k\}$ converges to the optimal solution x^*.

Linear convergence rate results can also be shown for different implementations of the method. As an example, a short proof can be obtained for the case of a positive definite quadratic function and a cyclic component selection rule. Then the iterates that mark the beginning of each cycle are generated by a stationary linear iteration, which must be a contraction since it converges from every starting point. A more complicated analysis also shows that for the case of a general strongly convex twice differentiable function f, the iteration with the cyclic rule is a contraction within a neighborhood of the minimum x^* and hence has a linear convergence rate.

We next discuss the convergence rate of the coordinate descent iteration (2.66) with various noncyclic component selection rules. We first consider the case where the index i_k is selected by uniform randomization over the index set $\{1, \ldots, n\}$, independently of previous selections. Following [LeL10], [Nes12], we will show that for all k, we have

$$E\{f(x^{k+1})\} - f(x^*) \leq \left(1 - \frac{\sigma}{Ln}\right)(f(x^k) - f(x^*)), \qquad (2.68)$$

where $E\{\cdot\}$ denotes expected value. Indeed, by minimizing over y both sides of Eq. (2.65) with $x = x^k$, we have

$$f(x^*) \geq f(x^k) - \frac{1}{2\sigma}\|\nabla f(x^k)\|^2. \qquad (2.69)$$

Taking conditional expected value, given x^k, in Eq. (2.67), we obtain

$$E\{f(x^{k+1})\} \le E\left\{f(x^k) - \frac{1}{2L}|\nabla_{i_k}f(x^k)|^2\right\}$$

$$= f(x^k) - \frac{1}{2L}\sum_{i=1}^n \frac{1}{n}|\nabla_{i_k}f(x^k)|^2$$

$$= f(x^k) - \frac{1}{2Ln}\|\nabla f(x^k)\|^2.$$

Subtracting $f(x^*)$ from both sides and using Eq. (2.69), we obtain Eq. (2.68).

Let us now consider the possibility of using a different stepsize along each coordinate. In particular,

$$x^{k+1} = x^k - \frac{1}{L_{i_k}}\nabla_{i_k}f(x^k)\,e_{i_k},$$

where L_i is a Lipschitz constant for the gradient along the ith coordinate, i.e.,

$$|\nabla_i f(x + \alpha e_i) - \nabla_i f(x)| \le L_i|\alpha|, \qquad \forall\, x \in \Re^n,\ \alpha \in \Re,\ i = 1,\ldots,n.$$

Here i_k is again selected by uniform randomization over the index set $\{1,\ldots,n\}$, independently of previous selections. Then with a similar argument it can be shown that for all k,

$$E\{f(x^{k+1})\} - f(x^*) \le \left(1 - \frac{\sigma}{\bar{L}}\right)\left(f(x^k) - f(x^*)\right),$$

where $\bar{L} = \sum_{i=1}^n L_i$. Note that this is a stronger convergence rate estimate than the one of Eq. (2.68), which applies with $L = \min\{L_1,\ldots,L_n\}$. The use of different stepsizes for different coordinates may be viewed as a form of diagonal scaling. In practice, L_{i_k} may be approximated by a finite difference approximation of the second derivative of f along e_{i_k} or some other crude line search scheme.

Let us finally consider a different order selection method, which aims to identify promising coordinates to iterate on. Here i_k is selected according to the so called Gauss-Southwell scheme

$$i_k \in \arg\max_{i=1,\ldots,n}|\nabla_i f(x^k)|.$$

We assume that the following strong convexity assumption holds for some $\sigma_1 > 0$,

$$f(y) \ge f(x) + \nabla f(x)'(y-x) + \frac{\sigma_1}{2}\|x-y\|_1^2, \qquad \forall\, x,y \in \Re^n, \qquad (2.70)$$

where $\|\cdot\|_1$ denotes the ℓ_1 norm. We will show that for all k,

$$f(x^{k+1}) - f(x^*) \leq \left(1 - \frac{\sigma_1}{L}\right)\left(f(x^k) - f(x^*)\right). \qquad (2.71)$$

Indeed, by minimizing with respect to y both sides of Eq. (2.70) with $x = x^k$, we obtain

$$f(x^*) \geq f(x^k) - \frac{1}{2\sigma_1}\|\nabla f(x^k)\|_\infty^2,$$

where $\|\cdot\|_\infty$ denotes the sup-norm. Combining this with Eq. (2.67) and using the definition of i_k, which implies that

$$\left|\nabla_{i_k} f(x^k)\right|^2 = \left\|\nabla f(x^k)\right\|_\infty^2,$$

we obtain Eq. (2.71). An important fact is that the convergence rate estimate (2.71) is more favorable than the estimate (2.68), because it can be shown that $\frac{\sigma}{n} \leq \sigma_1 \leq \sigma$; see [ScF14]. For further analysis along this line we refer to [ScF14] and [NSL15], which discuss additional variants of coordinate descent.

2.3.2 Direct Search Methods

In the coordinate descent method we search along the fixed set of coordinate directions and we are guaranteed a cost improvement at a nonstationary point because these directions are linearly independent. This idea can be generalized by using a different set of directions and by occasionally changing this set of directions with the aim of accelerating convergence. There are a number of methods of this type: Rosenbrock's method [Ros60a], the pattern search algorithm of Hooke and Jeeves [HoJ61], and the simplex algorithms of Spendley, Hext, and Himsworth [SHH62], and Nelder and Mead [NeM65]. Unfortunately, the rationale of these methods often borders on the heuristic, and their theoretical convergence properties are often not reassuring. However, these methods are often fairly simple to implement and like the coordinate descent method, they do not require gradient calculations. We describe the *Nelder-Mead simplex method* [NeM65] (not to be confused with the simplex method of linear programming), which has enjoyed considerable popularity.

At the typical iteration of this method, we start with a *simplex*, i.e., the convex hull of $n+1$ points, $x^0, x^1, \ldots, x^n$, and we end up with another simplex. Let x_{min} and x_{max} denote the "best" and "worst" vertices of the simplex, that is the vertices satisfying

$$f(x_{min}) = \min_{i=0,1,\ldots,n} f(x^i), \qquad (2.72)$$

$$f(x_{max}) = \max_{i=0,1,\ldots,n} f(x^i). \qquad (2.73)$$

Let also $\hat{x}$ denote the centroid of the face of the simplex formed by the vertices other than x_{max}

$$\hat{x} = \frac{1}{n}\left(-x_{max} + \sum_{i=0}^{n} x^i\right). \qquad (2.74)$$

The iteration replaces the worst vertex x_{max} by a "better" one. In particular, the *reflection point* $x_{ref} = 2\hat{x} - x_{max}$ is computed, which lies on the line passing through x_{max} and $\hat{x}$, and is symmetric to x_{max} with respect to $\hat{x}$. Depending on the cost value of x_{ref} relative to the points of the simplex other than x_{max}, a new vertex x_{new} is computed, and a new simplex is formed from the old by replacing the vertex x_{max} by x_{new}, while keeping the other n vertices.

Typical Iteration of the Nelder-Mead Simplex Method

Step 1: (Reflection Step) Compute

$$x_{ref} = 2\hat{x} - x_{max}. \qquad (2.75)$$

Then compute x_{new} according to the following three cases:

(1) (x_{ref} **has min cost**) If $f(x_{min}) > f(x_{ref})$, go to Step 2.

(2) (x_{ref} **has intermediate cost**) If $\max\{f(x^i) \mid x^i \neq x_{max}\} > f(x_{ref}) \geq f(x_{min})$, go to Step 3.

(3) (x_{ref} **has max cost**) If $f(x_{ref}) \geq \max\{f(x^i) \mid x^i \neq x_{max}\}$, go to Step 4.

Step 2: (Attempt Expansion) Compute

$$x_{exp} = 2x_{ref} - \hat{x}. \qquad (2.76)$$

Define

$$x_{new} = \begin{cases} x_{exp} & \text{if } f(x_{exp}) < f(x_{ref}), \\ x_{ref} & \text{otherwise,} \end{cases}$$

and form the new simplex by replacing the vertex x_{max} with x_{new}.

Step 3: (Use Reflection) Define $x_{new} = x_{ref}$, and form the new simplex by replacing the vertex x_{max} with x_{new}.

Step 4: (Perform Contraction) Define

$$x_{new} = \begin{cases} \frac{1}{2}(x_{max} + \hat{x}) & \text{if } f(x_{max}) \leq f(x_{ref}), \\ \frac{1}{2}(x_{ref} + \hat{x}) & \text{otherwise,} \end{cases} \qquad (2.77)$$

and form the new simplex by replacing the vertex x_{max} with x_{new}.

The reflection step and the subsequent possible steps of the iteration are illustrated in Fig. 2.3.2(a)-(d). The entire method is illustrated in Fig. 2.3.3. Exercise 2.3.3 shows a cost improvement property of the method in the case where f is strictly convex. However, there are no known convergence results for the method. Furthermore, when the cost function is not convex, it is possible that the new simplex vertex x_{new} has larger cost value than the old vertex x_{max}. In this case a modification that has been suggested is to "shrink" the old simplex towards the best vertex x_{min}, i.e., form a new simplex by replacing all the vertices x^i, $i = 0, 1, \ldots, n$, by

$$\bar{x}^i = \tfrac{1}{2}(x^i + x_{min}), \qquad i = 0, 1, \ldots, n.$$

The method as just described seems to work reasonably well in practice, particularly for problems of relatively small dimension (say up to 10). However, its convergence is not guaranteed, and in fact a convergence counterexample is given in [McK98]. The paper [Tse95a] provides a relatively simple modification with satisfactory convergence properties. Another convergence analysis is given in the paper [LRW98]. There are also a number of related methods, some of which have demonstrable convergence properties; see [DeT91] and [Tor91]. Note that the constants used in Eqs. (2.75), (2.76), and (2.77) are somewhat arbitrary, as suggested by the interpretation of the method given in Figs. 2.3.2 and 2.3.3. More general forms of these equations are

$$x_{ref} = \hat{x} + \beta(\hat{x} - x_{max}),$$
$$x_{exp} = x_{ref} + \gamma(x_{ref} - \hat{x}),$$
$$x_{con} = \begin{cases} \theta x_{max} + (1-\theta)\hat{x} & \text{if } f(x_{max}) \leq f(x_{ref}), \\ \theta x_{ref} + (1-\theta)\hat{x} & \text{otherwise,} \end{cases}$$

where $\beta > 0$, $\gamma > 0$, and $\theta \in (0, 1)$ are scalars known as the *reflection coefficient*, the *expansion coefficient*, and the *contraction coefficient*, respectively. The formulas of Eqs. (2.75)-(2.77) correspond to $\beta = 1$, $\gamma = 1$, and $\theta = 1/2$, respectively.

EXERCISES

2.3.1

Let $f : \Re^n \mapsto \Re$ be continuously differentiable, let $p_1, \ldots, p_n$ be linearly independent vectors, and suppose that for some x^*, $\alpha = 0$ is a stationary point of each of the one-dimensional functions $g_i(\alpha) = f(x^* + \alpha p_i)$, $i = 1, \ldots, n$. Show that x^* is a stationary point of f. Suggest a minimization method that does not use derivatives based on this property.

Sec. 2.3   Nonderivative Methods   157

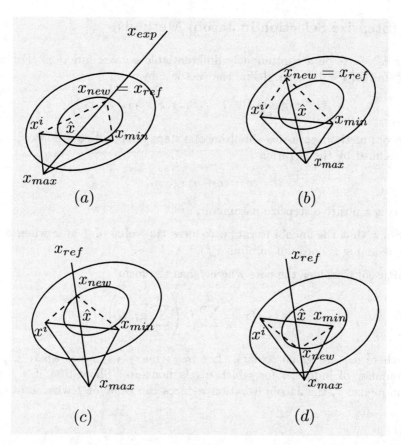

Figure 2.3.2. Illustration of the reflection step and the possible subsequent steps of an iteration of the simplex method. In (a), the new vertex x_{new} is determined via the expansion Step 2. In (b), x_{new} is determined via Step 3, and the reflection step is accepted. In (c) and (d), x_{new} is determined via the contraction Step 4.

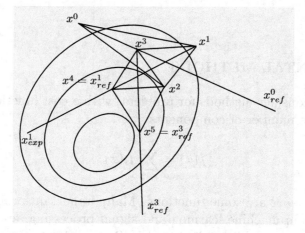

Figure 2.3.3. Illustration of three iterations of the simplex method, which generate the points x^3, x^4, and x^5, starting from the simplex x^0, x^1, x^2. The simplex obtained after these three iterations consists of x^2, x^4, and x^5.

2.3.2 (Stepsize Selection in Jacobi Methods)

Let $f: \Re^n \mapsto \Re$ be a continuously differentiable convex function. For a given $x \in \Re^n$ and all $i = 1, \ldots, n$, define the vector $\bar{x}$ by

$$\bar{x}_i \in \arg\min_{\xi \in \Re} f(x_1, \ldots, x_{i-1}, \xi, x_{i+1}, \ldots, x_n).$$

The Jacobi method performs simultaneous steps along all coordinate directions, and is defined by the iteration

$$x := x + \alpha(\bar{x} - x),$$

where α is a positive stepsize parameter.

(a) Show that the Jacobi iteration reduces the value of f at x when $\alpha = 1/n$ (assuming x does not minimize f).

(b) [Rus95] Consider the case where f has the form

$$f(x) = \sum_{j=1}^{J} f_j \left(\sum_{i=1}^{n} a_{ji} x_i \right),$$

where a_{ji} are given scalars. Let $m = \max_{j=1,\ldots,J} m_j$, where m_j is the number of indices i for which a_{ji} is nonzero. Show that if x does not minimize f, the Jacobi iteration reduces the value of f when $\alpha = 1/m$.

2.3.3

Consider the Nelder-Mead simplex method applied to a strictly convex function f. Show that at each iteration, either $f(x_{max})$ decreases strictly, or else the number of vertices x^i of the simplex such that $f(x^i) = f(x_{max})$ decreases by at least one.

2.4 INCREMENTAL METHODS

We will now consider methods for problems with a cost function that is the sum of a large number of components:

$$f(x) = \sum_{i=1}^{m} f_i(x),$$

where $f_i : \Re^n \mapsto \Re$ are some functions. Many important problems, arising among others in machine learning and signal processing, are of this type. The least squares problems discussed in Section 1.4.4 are a special case. The author's convex optimization algorithms textbook [Ber15a] describes a broad variety of other problems, which are not of the least squares type. The following is a representative example.

Sec. 2.4   Incremental Methods   159

Example 2.4.1: (Binary Classification)

Here we aim to construct a parametric model for predicting whether an object with certain characteristics (also called features) belongs to a given category or not. We assume that each object is characterized by a *feature vector c* that belongs to $\Re^n$ and a *label b* that takes the values $+1$ or -1, if the object belongs to the category or not, respectively. As illustration consider a credit card company that wishes to classify applicants as "low risk" ($+1$) or "high risk" (-1), with each customer characterized by n scalar features of financial and personal type.

We are given data, which is a set of feature-label pairs (c_i, b_i), $i = 1, \ldots, m$. Based on this data, we want to construct a model that allows us to generically assign labels to objects based on their feature vectors. In particular, we want to find a parameter vector $x \in \Re^n$ and a scalar $y \in \Re$ such that the sign of $c'x + y$ is a good predictor of the label of an object with feature vector c. Thus, loosely speaking, x and y should be such that for "most" of the given feature-label data (c_i, b_i) we have

$$c_i'x + y > 0, \qquad \text{if } b_i = +1,$$

$$c_i'x + y < 0, \qquad \text{if } b_i = -1.$$

Since a classification error for object i is made when $b_i(c_i'x + y) < 0$, it makes sense to formulate classification as an optimization problem where negative values of $b_i(c_i'x + y)$ are penalized. This leads to the problem

$$\text{minimize} \quad \sum_{i=1}^{m} h\big(b_i(c_i'x + y)\big)$$

$$\text{subject to} \quad x \in \Re^n, \ y \in \Re,$$

where $h : \Re \mapsto \Re$ is a convex function that penalizes negative values of its argument. It would make sense to use a penalty of one unit for misclassification, i.e.,

$$h(z) = \begin{cases} 0 & \text{if } z \geq 0, \\ 1 & \text{if } z < 0, \end{cases}$$

but such a penalty function is discontinuous. To obtain a continuous cost function, we allow a continuous transition of h from negative to positive values, leading to a variety of nonincreasing functions h. The choice of h depends on the given application and other theoretical considerations for which we refer to the literature. Some common examples are

$$h(z) = e^{-z}, \qquad \text{(exponential loss)},$$

$$h(z) = \log\left(1 + e^{-z}\right), \qquad \text{(logistic loss)},$$

$$h(z) = \max\{0, 1 - z\}, \qquad \text{(hinge loss)}.$$

For the case of logistic loss the method comes under the methodology of *logistic regression*, and for the case of hinge loss the method comes under the methodology of *support vector machines*. Both of these methodologies have extensive applications for which we refer to the literature.

Often the cost function is augmented with a regularization function $R : \Re^n \mapsto \Re$, and takes the form

$$R(x) + \sum_{i=1}^{m} h\big(b_i(c_i'x + y)\big).$$

Common choices for R are various quadratic penalties, such as

$$R(x) = \|x - \bar{x}\|^2,$$

where $\bar{x}$ is some known vector, the ℓ_1 norm

$$R(x) = \|x\|_1 = \sum_{i=1}^{n} |x_i|,$$

or some scaled version or combination thereof.

The ℓ_1 norm is popular, because it tends to produce optimal solutions where a greater number of coordinates x_i are zero, relative to the case of quadratic regularization. This is considered desirable in many statistical applications, where the number of parameters to include in a model may not be known a priori. There is extensive literature on the use of regularization functions, their properties, and their applications, to which we refer for further discussion (see [Ber15a], for some orientation, references, and related applications). Note that the functions R or h in the cost function may be nondifferentiable, in which case the methodology of this section is not applicable. However, in Chapter 7 we will consider related incremental methods that can be used with convex nondifferentiable cost functions.

Incremental methods are interesting when m is very large, so calculating the full gradient is very costly. For such problems one hopes to make progress with approximate but much cheaper incremental steps that use *only one component function f_i*. Incremental methods are also well-suited for problems where m is large and the component functions f_i become known sequentially. Then one may be able to operate on each component as it reveals itself, without waiting for the other components to become known, i.e., in an *on-line* fashion.

In this section we will explain the ideas underlying incremental methods, and support our arguments with some convergence analysis for the case where the component functions f_i are differentiable. We will consider four specific types of incremental methods in the next four subsections, respectively: *gradient, aggregated gradient, Gauss-Newton, and Newton methods*. In Chapter 3, we will consider the related class of *incremental proximal methods*. In Chapter 7, we will discuss how the ideas of the present section can be applied to constrained minimization, in the context

Sec. 2.4   Incremental Methods   **161**

of *incremental augmented Lagrangian methods*. In Chapter 7, we will also discuss *incremental subgradient methods*, where the component functions f_i may be nondifferentiable. We finally note that incremental algorithms are well-suited for *distributed asynchronous implementation*, and we will explain the underlying ideas in Section 2.5.

2.4.1 Incremental Gradient Methods

Consider minimization of the sum $f(x) = \sum_{i=1}^{m} f_i(x)$, where $f_i : \Re^n \mapsto \Re$ are continuously differentiable functions. Incremental gradient methods have the general form

$$x^{k+1} = x^k - \alpha^k \nabla f_{i_k}(x^k), \qquad (2.78)$$

where α^k is a positive stepsize, and i_k is some index from the set $\{1, \ldots, m\}$, chosen by some deterministic or randomized rule. Three common rules are:

(1) A *cyclic order*, whereby the indexes are taken up in the fixed deterministic order $1, \ldots, m$, so that i_k is equal to (k modulo m) plus 1. A contiguous block of iterations involving $f_1, \ldots, f_m$ in this order and exactly once is called a *cycle*.

(2) A *uniform random order*, whereby the index i_k chosen randomly by sampling over all indexes with a uniform distribution, independently of the past history of the algorithm.

(3) A *cyclic order with random reshuffling*, whereby the indexes are taken up one by one within each cycle, but their order after each cycle is reshuffled randomly (and independently of the past).

Note that it is essential to include all components in a cycle in the cyclic case, and to sample according to the uniform distribution in the random order case, for otherwise some components will be sampled more often than others, leading to a bias in the convergence process.

Focusing for the moment on the cyclic rule, we note that the motivation for the incremental gradient method is faster convergence: we hope that far from the solution, a single cycle of the method will be as effective as several (as many as m) iterations of the ordinary gradient method (think of the case where the components f_i are similar in structure). Near a solution, however, the incremental method may not be as effective.

To be more specific, we note that there are two complementary performance issues to consider in comparing incremental and nonincremental methods:

(a) *Progress when far from convergence.* Here the incremental method can be much faster. For an extreme case take m large and all components f_i identical to each other. Then an incremental iteration requires m times less computation than a classical gradient iteration, but gives exactly the same result, when the stepsize is scaled to be m times larger. While this is an extreme example, it reflects the essential mechanism by which incremental methods can be much superior: far from the minimum a single component gradient will point to "more or less" the right direction, at least most of the time.

(b) *Progress when close to convergence.* Here the incremental method can be inferior. As a case in point, assume that all components f_i are differentiable functions. Then the nonincremental gradient method can be shown to converge with a constant stepsize under reasonable assumptions, as we have seen in Section 1.2.2. However, the incremental method requires a diminishing stepsize, and its ultimate rate of convergence can be much slower.

This type of behavior is illustrated in the following example.

Example 2.4.2

Assume that x is a scalar, and that the problem is

$$\text{minimize} \quad f(x) = \tfrac{1}{2} \sum_{i=1}^{m} (c_i x - b_i)^2$$

$$\text{subject to} \quad x \in \Re,$$

where c_i and b_i are given scalars with $c_i \neq 0$ for all i. The minimum of each of the components $f_i(x) = \tfrac{1}{2}(c_i x - b_i)^2$ is

$$x_i^* = \frac{b_i}{c_i},$$

while the minimum of the least squares cost function f is

$$x^* = \frac{\sum_{i=1}^{m} c_i b_i}{\sum_{i=1}^{m} c_i^2}.$$

It can be seen that x^* lies within the range of the component minima

$$R = \left[\min_i x_i^*, \ \max_i x_i^* \right],$$

and that for all x *outside* the range R, the gradient

$$\nabla f_i(x) = c_i(c_i x - b_i)$$

has the same sign as $\nabla f(x)$ (see Fig. 2.4.1). As a result, when outside the region R, the incremental gradient method

$$x^{k+1} = x^k - \alpha^k c_{i_k}(c_{i_k} x^k - b_{i_k})$$

Sec. 2.4   Incremental Methods   **163**

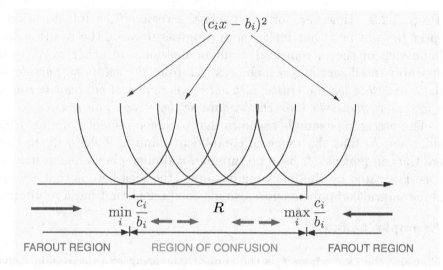

Figure 2.4.1. Illustrating the advantage of incrementalism when far from the optimal solution. The region of component minima

$$R = \left[\min_i x_i^*, \max_i x_i^*\right],$$

is labeled as the "region of confusion." It is the region where the method does not have a clear direction towards the optimum. The ith step in an incremental gradient cycle is a gradient step for minimizing $(c_i x - b_i)^2$, so if x lies outside the region of component minima $R = \left[\min_i x_i^*, \max_i x_i^*\right]$, (labeled as the "farout region") and the stepsize is small enough, progress towards the solution x^* is made.

approaches x^* at each step, provided the stepsize α^k is small enough. In fact it can be verified that it is sufficient that

$$\alpha^k \leq \min_i \frac{1}{c_i^2}.$$

However, for x *inside* the region R, the ith step of a cycle of the incremental gradient method need not make progress. It will approach x^* (for small enough stepsize α^k) only if the current point x^k does not lie in the interval connecting x_i^* and x^*. This induces an oscillatory behavior within the region R, and as a result, the incremental gradient method will typically not converge to x^* unless $\alpha^k \to 0$. By contrast, the steepest descent method, which takes the form

$$x^{k+1} = x^k - \alpha^k \sum_{i=1}^m c_i(c_i x^k - b_i),$$

converges to x^* for any constant stepsize satisfying

$$\alpha^k \leq \frac{1}{\sum_{i=1}^m c_i^2};$$

cf. Prop. 1.2.2. However, for x outside the region R, a full iteration of steepest descent need not make more progress towards the solution than a single step of the incremental gradient method. In other words, with comparably intelligent stepsize choices, *far from the solution (outside R), a single pass through the entire data set by incremental gradient is roughly as effective as m passes through the data set by steepest descent.*

The preceding example assumes that each component function f_i has a minimum, so that the range of component minima is defined. In cases where the components f_i have no minima, a similar phenomenon may occur, as illustrated by the following example (the idea here is that we may combine several components into a single component that has a minimum).

Example 2.4.3:

Consider the case where f is the sum of increasing and decreasing convex exponentials, i.e.,

$$f_i(x) = a_i e^{b_i x}, \qquad x \in \Re,$$

where a_i and b_i are scalars with $a_i > 0$ and $b_i \neq 0$. Let

$$I^+ = \{i \mid b_i > 0\}, \qquad I^- = \{i \mid b_i < 0\},$$

and assume that I^+ and I^- have roughly equal numbers of components. Let also x^* be the minimum of $\sum_{i=1}^{m} f_i$.

Consider the incremental gradient method that given the current point, call it x^k, chooses some component f_{i_k} and iterates according to the incremental gradient iteration

$$x^{k+1} = x^k - \alpha^k \nabla f_{i_k}(x^k).$$

Then it can be seen that if $x^k >> x^*$, x^{k+1} will be substantially closer to x^* if $i \in I^+$, and negligibly further away than x^* if $i \in I^-$. The net effect, averaged over many incremental iterations, is that if $x^k >> x^*$, an incremental gradient iteration makes roughly one half the progress of a full gradient iteration, with m times less overhead for calculating gradients. The same is true if $x^k << x^*$. On the other hand as x^k gets closer to x^* the advantage of incrementalism is reduced, similar to the preceding example. In fact in order for the incremental method to converge, a diminishing stepsize is necessary, which will ultimately make the convergence slower than the one of the nonincremental gradient method with a constant stepsize.

The discussion of the preceding examples relies on x being one-dimensional, but in many multidimensional problems the same qualitative behavior can be observed. In particular, a pass through the ith component f_i by the incremental gradient method can make progress towards the solution in the region where the component gradient $\nabla f_{i_k}(x^k)$ makes an angle less than 90 degrees with the cost function gradient $\nabla f(x^k)$. If the components f_i are not "too dissimilar", this is likely to happen in a region of points

Sec. 2.4   Incremental Methods   165

that are not too close to the optimal solution set; see also Exercise 2.4.7. This behavior has been verified in many practical contexts, including the training of neural networks (cf. Section 1.4.4), where incremental gradient methods have been used extensively, frequently under the name *backpropagation methods*. There is an extensive literature on this subject, for which we refer to the books [Bis95], [BeT96], and [Hay11].

Stepsize Choice

The choice of the stepsize α^k plays an important role in the performance of incremental gradient methods. On close examination, it turns out that the direction used by the method differs from the gradient direction by an error that is proportional to the stepsize, and for this reason a diminishing stepsize is essential for convergence to a stationary point of f, as discussed in Section 1.2.

However, it turns out that a peculiar form of convergence also typically occurs for a constant but sufficiently small stepsize. In this case, the iterates converge to a "limit cycle", whereby the ith iterates ψ_i within the cycles converge to a different limit than the jth iterates ψ_j for $i \neq j$. The sequence $\{x^k\}$ of the iterates obtained at the end of cycles converges, except that the limit obtained *need not* be a stationary point of f (see Exercise 2.4.2 and the following analysis). The limit tends to be close to a stationary point when the constant stepsize is small (see the following Prop. 2.4.1). In practice, it is common to use a constant stepsize for a (possibly prespecified) number of iterations, then decrease the stepsize by a certain factor, and repeat, up to the point where the stepsize reaches a prespecified minimum.

An alternative possibility is to use a diminishing stepsize rule of the form

$$\alpha^k = \min\left\{\gamma, \frac{\gamma_1}{k + \gamma_2}\right\},$$

where γ, γ_1, and γ_2 are some positive scalars. There are also variants of the incremental gradient method that use a constant stepsize throughout, and generically converge to a stationary point of f at a linear rate. In one type of such method the degree of incrementalism gradually diminishes as the method progresses (see Exercise 2.4.4 and [Ber97]). Another incremental approach with similar aims, is the aggregated incremental gradient method, which will be discussed in the next subsection.

Regardless of whether a constant or a diminishing stepsize is ultimately used, to maintain the advantage of faster convergence when far from the solution, the incremental method must use a much larger stepsize than the corresponding nonincremental gradient method (as much as m times larger so that the size of the incremental gradient step is comparable to the size of the nonincremental gradient step).

One possibility is to use an adaptive stepsize rule, whereby the stepsize is reduced (or increased) when the progress of the method indicates that the algorithm is oscillating because it operates within (or outside, respectively) the region of confusion. There are formal ways to implement such stepsize rules with sound convergence properties (see [Tse98], [MYF03], [GOP15a]). The idea is to look at a batch of incremental updates $\psi_i, \ldots, \psi_{i+M}$, for some relatively large $M \leq m$, and consider the ratio

$$\frac{\|\psi_i - \psi_{i+M}\|}{\sum_{\ell=1}^{M} \|\psi_{1+\ell-1} - \psi_{i+\ell}\|}.$$

If this ratio is "small" this suggests that the method is oscillating.

The difficulty with stepsize selection may also be addressed with diagonal scaling, similar to the case of the nonincremental steepest descent method (cf. the discussions of Chapter 1). Second derivatives, or approximations thereof, can be very effective for this purpose; see also the discussion on incremental Newton methods in Section 2.4.4.

Convergence Analysis of Incremental Gradient Methods

We will now analyze the convergence of the incremental gradient method. The main idea is that the method can be viewed as a steepest descent iteration where the gradient is perturbed by an error term that is proportional to the stepsize. It follows that the qualitative behavior of the method is similar to the one of gradient methods with errors, which was described in Section 1.2.1. We will first consider a quadratic cost function, which arises among others in linear least squares. We provide an analysis of the nonquadratic case in the exercises.

Proposition 2.4.1: Consider the case of quadratic component functions,

$$f_i(x) = \tfrac{1}{2} x' Q_i x - b_i' x, \qquad i = 1, \ldots, m, \tag{2.79}$$

where Q_i are positive semidefinite matrices, and the incremental gradient method

$$x^{k+1} = x^k + \alpha^k \sum_{i=1}^{m} (b_i - Q_i \psi_{i-1}), \tag{2.80}$$

where $\psi_0 = x^k$ and

$$\psi_i = \psi_{i-1} + \alpha^k (b_i - Q_i \psi_{i-1}), \qquad i = 1, \ldots, m. \tag{2.81}$$

Assume that $\sum_{i=1}^{m} Q_i$ is positive definite and let x^* be the minimum of $f = \sum_{i=1}^{m} f_i$. Then:

Sec. 2.4   Incremental Methods   167

> (a) There exists $\bar{\alpha} > 0$ such that if α^k is equal to some constant $\alpha \in (0, \bar{\alpha}]$ for all k, $\{x^k\}$ converges to some vector $x(\alpha)$. Furthermore, the error $\|x^k - x(\alpha)\|$ converges to 0 linearly. In addition, we have $\|x(\alpha) - x^*\| = O(\alpha)$, and in particular $\lim_{\alpha \to 0} x(\alpha) = x^*$.
>
> (b) If $\alpha^k > 0$ for all k, and
>
> $$\alpha^k \to 0, \qquad \sum_{k=0}^{\infty} \alpha^k = \infty,$$
>
> then $\{x^k\}$ converges to x^*.

Proof: (a) We first show by induction that for all $i = 1, \ldots, m$, the vectors ψ_i of Eq. (2.81) have the form

$$\psi_i = x^k + \alpha \sum_{j=1}^{i} (b_j - Q_j x^k) + \sum_{j=1}^{i-1} \alpha^{j+1} (\Phi_{ij} x^k + \phi_{ij}), \qquad (2.82)$$

where Φ_{ij} and ϕ_{ij} are some matrices and vectors, respectively, which are independent of k. Indeed, this relation holds by definition for $i = 1$ [the last term in Eq. (2.82) is zero when $i = 1$]. Suppose that it holds for $i = p$. We have, using the induction hypothesis,

$$\psi_{p+1} = \psi_p + \alpha(b_{p+1} - Q_{p+1} \psi_p)$$
$$= x^k + \alpha \sum_{j=1}^{p} (b_j - Q_j x^k) + \sum_{j=1}^{p-1} \alpha^{j+1} (\Phi_{pj} x^k + \phi_{pj})$$
$$+ \alpha(b_{p+1} - Q_{p+1} x^k) - \alpha Q_{p+1} (\psi_p - x^k)$$
$$= x^k + \alpha \sum_{j=1}^{p+1} (b_j - Q_j x^k) + \sum_{j=1}^{p-1} \alpha^{j+1} (\Phi_{pj} x^k + \phi_{pj})$$
$$- \alpha Q_{p+1} \left(\alpha \sum_{j=1}^{p} (b_j - Q_j x^k) + \sum_{j=1}^{p-1} \alpha^{j+1} (\Phi_{pj} x^k + \phi_{pj}) \right),$$

which is of the form (2.82), thereby completing the induction.

For $i = m$, Eq. (2.82) yields

$$x^{k+1} = A(\alpha) x^k + b(\alpha), \qquad (2.83)$$

where

$$A(\alpha) = I - \alpha \sum_{j=1}^{m} Q_j + \sum_{j=1}^{m-1} \alpha^{j+1} \Phi_{mj}, \qquad (2.84)$$

$$b(\alpha) = \alpha \sum_{j=1}^{m} b_j + \sum_{j=1}^{m-1} \alpha^{j+1} \phi_{mj}. \tag{2.85}$$

Let us choose α small enough so that the eigenvalues of $A(\alpha)$ are all strictly within the unit circle; this is possible since $\sum_{j=1}^{m} Q_j$ is assumed positive definite and the last term in Eq. (2.84) involves powers of α that are greater or equal to 2. Define

$$x(\alpha) = (I - A(\alpha))^{-1} b(\alpha). \tag{2.86}$$

Then $b(\alpha) = (I - A(\alpha))x(\alpha)$, and by substituting this expression in Eq. (2.83), it can be seen that

$$x^{k+1} - x(\alpha) = A(\alpha)(x^k - x(\alpha)),$$

from which

$$x^k - x(\alpha) = A(\alpha)^k (x^0 - x(\alpha)), \quad \forall\, k. \tag{2.87}$$

Since all the eigenvalues of $A(\alpha)$ are strictly within the unit circle, we have $A(\alpha)^k \to 0$ (Prop. A.16 in Appendix A), so $\|x^k - x(\alpha)\| \to 0$ linearly.

To prove that $\lim_{\alpha \to 0} x(\alpha) = x^*$, we first calculate x^*. We set the gradient of $\|g(x)\|^2$ to zero, to obtain

$$\sum_{i=1}^{m} (Q_i x^* - b_i) = 0,$$

so that

$$x^* = \left(\sum_{i=1}^{m} Q_i \right)^{-1} \sum_{i=1}^{m} b_i. \tag{2.88}$$

Then, we use Eqs. (2.84)-(2.86) to write

$$x(\alpha) = \left(\sum_{i=1}^{m} Q_i + O(\alpha) \right)^{-1} \left(\sum_{i=1}^{m} b_i + O(\alpha) \right) = x^* + O(\alpha).$$

so from the preceding two relations we have $\|x(\alpha) - x^*\| = O(\alpha)$.

(b) For $i = m$ and $\alpha = \alpha^k$, Eq. (2.82) yields

$$x^{k+1} = x^k + \alpha^k \sum_{j=1}^{m} (b_j - Q_j x^k) + (\alpha^k)^2 H^k (x^k - x^*) + (\alpha^k)^2 h^k, \tag{2.89}$$

where

$$H^k = \sum_{j=1}^{m-1} (\alpha^k)^{j-1} \Phi_{mj}, \tag{2.90}$$

Sec. 2.4   Incremental Methods   169

$$h^k = \sum_{j=1}^{m-1} (\alpha^k)^{j-1}(\Phi_{mj}x^* + \phi_{mj}). \qquad (2.91)$$

Using also the expression (2.88) for x^*, we can write Eq. (2.89) as

$$x^{k+1} - x^* = \left(I - \alpha^k \sum_{j=1}^{m} Q_j + (\alpha^k)^2 H^k\right)(x^k - x^*) + (\alpha^k)^2 h^k. \qquad (2.92)$$

Since $\alpha^k \to 0$, for large enough k, the eigenvalues of $\alpha^k \sum_{j=1}^{m} Q_j$ are bounded from above by 1, and hence the matrix $I - \alpha^k \sum_{j=1}^{m} Q_j$ is positive definite. Without loss of generality, we assume that this is so for all k. Then we have

$$\left\|\left(I - \alpha^k \sum_{j=1}^{m} Q_j\right)(x^k - x^*)\right\| \leq (1 - \alpha^k \lambda)\|x^k - x^*\|, \qquad (2.93)$$

where λ is the largest eigenvalue of $\sum_{j=1}^{m} Q_j$. Let also γ and δ be positive scalars such that for all k we have

$$\|H^k(x^k - x^*)\| \leq \gamma\|x^k - x^*\|, \qquad \|h^k\| \leq \delta. \qquad (2.94)$$

Combining Eqs. (2.92)-(2.94), we have

$$\begin{aligned}
\|x^{k+1} - x^*\| &\leq \left\|\left(I - \alpha^k \sum_{j=1}^{m} Q_j\right)(x^k - x^*)\right\| \\
&\quad + (\alpha^k)^2 \|H^k(x^k - x^*)\| + (\alpha^k)^2 \|h^k\| \\
&\leq \left(1 - \alpha^k \lambda + (\alpha^k)^2 \gamma\right)\|x^k - x^*\| + (\alpha^k)^2 \delta.
\end{aligned} \qquad (2.95)$$

Let $\bar{k}$ be such that $\alpha^k \gamma \leq \lambda/2$ for all $k \geq \bar{k}$. Then from Eq. (2.95) we obtain

$$\|x^{k+1} - x^*\| \leq (1 - \alpha^k \lambda/2)\|x^k - x^*\| + (\alpha^k)^2 \delta, \qquad \forall\ k \geq \bar{k}.$$

From Prop. A.30 of Appendix A it follows that $\|x^k - x^*\| \to 0$. **Q.E.D.**

The preceding proof also quantifies the linear convergence rate of the method (within the region of confusion) for a constant stepsize α. It is governed by the spectral radius of the matrix

$$A(\alpha) = I - \alpha \sum_{j=1}^{m} Q_j + \sum_{j=1}^{m-1} \alpha^{j+1} \Phi_{mj}$$

of Eq. (2.84). For small values of α, this matrix is approximately equal to $I - \alpha \sum_{j=1}^{m} Q_j$, which is the matrix whose spectral radius governs the convergence rate of the *nonincremental* gradient method with stepsize α (see the analysis of Section 1.2.2). Thus *the incremental and nonincremental gradient methods have asymptotically (for small values of constant stepsize α) the same convergence rate*. The difference is that the incremental method converges to $x(\alpha)$, which differs from x^*, with the error $\|x(\alpha) - x^*\|$ being of order $O(\alpha)$. Thus at the expense of an asymptotic error $\|x(\alpha) - x^*\|$, we can improve the convergence rate of the incremental gradient method within the region of confusion to make it comparable to the one of the nonincremental gradient method, when using the same stepsize. We reiterate, however, that outside the region of confusion, a larger stepsize is appropriate for the incremental method.

In the case of nonquadratic cost components, stationarity of the limit points of sequences $\{x^k\}$ generated by the incremental gradient method has been shown under certain assumptions (including Lipschitz continuity of the component gradients) for the case of the stepsize $\alpha^k = \gamma/(k+1)$, where γ is a positive scalar [see Exercises 2.4.5, 2.4.6, and [BeT96], Prop. 3.8, [BeT00], and [MaS94]]. The convergence proof is similar to the one of the preceding proposition, but it is technically more involved. In the case of a constant stepsize and nonquadratic cost components, it is also possible to show a result analogous to Prop. 2.4.1(a), but again the proof is technically complex.

Stochastic Gradient Methods

Incremental gradient methods are related to methods that aim to minimize an expected value
$$f(x) = E\{F(x, w)\},$$
where w is a random variable, and $F(\cdot, w) : \Re^n \mapsto \Re$ is a differentiable function for each possible value of w. The stochastic gradient method for minimizing f is given by
$$x^{k+1} = x^k - \alpha^k \nabla_x F(x^k, w^k), \tag{2.96}$$
where w^k is a sample of w and $\nabla_x F$ denotes gradient of F with respect to x. This method has a rich theory and a long history, and it is strongly related to the classical algorithmic field of *stochastic approximation*; see the books [KuC78], [BeT96], [KuY03], [Spa03], [Mey07], [Bor08], [BPP13]. The method is also often referred to as *stochastic gradient descent*, particularly in the context of machine learning applications.

If we view the expected value cost $E\{F(x, w)\}$ as a weighted sum of cost function components, we see that the stochastic gradient method (2.96) is related to the incremental gradient method
$$x^{k+1} = x^k - \alpha^k \nabla f_{i_k}(x^k) \tag{2.97}$$

Sec. 2.4   Incremental Methods   171

for minimizing a finite sum $\sum_{i=1}^{m} f_i$, when randomization is used for component selection. An important difference is that the former method involves stochastic sampling of cost components $F(x,w)$ from a possibly infinite population, under some probabilistic assumptions, while in the latter the set of cost components f_i is predetermined and finite. However, it is possible to view the incremental gradient method (2.97), with uniform randomized selection of the component function f_i (i.e., with i_k chosen to be any one of the indexes $1, \ldots, m$, with equal probability $1/m$, and independently of preceding choices), as a stochastic gradient method.

Despite the apparent similarity of the incremental and the stochastic gradient methods, the view that the problem

$$\text{minimize} \quad f(x) = \sum_{i=1}^{m} f_i(x) \qquad (2.98)$$
$$\text{subject to} \quad x \in \Re^n,$$

can simply be treated as a special case of the problem

$$\text{minimize} \quad f(x) = E\{F(x,w)\}$$
$$\text{subject to} \quad x \in \Re^n,$$

is questionable.

One reason is that once we convert the finite sum problem to a stochastic problem, we preclude the use of methods that exploit the finite sum structure, such as the incremental aggregated gradient methods to be discussed in the next subsection. Another reason is that the finite-component problem (2.98) is often genuinely deterministic, and to view it as a stochastic problem at the outset may mask some of its important characteristics, such as the number m of cost components, or the sequence in which the components are ordered and processed. These characteristics may potentially be algorithmically exploited. For example, with insight into the problem's structure, one may be able to discover a special deterministic or partially randomized order of processing the component functions that is superior to a uniform randomized order. On the other hand analysis indicates that in the absence of problem-specific knowledge that can be exploited to select a favorable deterministic order, a uniform randomized order (each component f_i chosen with equal probability $1/m$ at each iteration, independently of preceding choices) has superior worst-case complexity; see [NeB01], [BNO03], and [Ber15a].

Finally, let us note a popular hybrid technique, which reshuffles randomly the order of the components after each cycle through the data set. Practical experience and recent analysis [GOP15c] indicates that it has somewhat better performance to the uniform randomized order when m is large. One possible reason is that random reshuffling allocates exactly one computation slot to each component in an m-slot cycle, while uniform sampling allocates one computation slot to each component *on the average*. A nonzero variance in the number of slots that any fixed component gets within a cycle, may be detrimental to performance. While it seems difficult to establish this fact analytically, a justification is suggested by the view of the incremental gradient method as a gradient method with error in the computation of the gradient. The error has apparently greater variance in the uniform sampling method than in the random reshuffling method, and heuristically, if the variance of the error is larger, the direction of descent deteriorates, suggesting slower convergence.

2.4.2 Incremental Aggregated Gradient Methods

Another incremental algorithm is the *incremental aggregated gradient method*, which has the form

$$x^{k+1} = x^k - \alpha^k \sum_{\ell=0}^{m-1} \nabla f_{i_{k-\ell}}(x^{k-\ell}), \qquad (2.99)$$

where f_{i_k} is the new component function selected for iteration k.† In the most common version of the method the component indexes i_k are selected in a cyclic order $[i_k = (k \text{ modulo } m) + 1]$. Random selection of the index i_k has also been suggested.

From Eq. (2.99) it can be seen that the method computes the gradient incrementally, one component per iteration. However, in place of the single component gradient $\nabla f_{i_k}(x^k)$, used in the incremental gradient method (2.78), it uses the sum of the component gradients computed in the past m iterations, which is an approximation to the total cost gradient $\nabla f(x^k)$.

The idea of the method is that by aggregating the component gradients one may be able to reduce the error between the true gradient $\nabla f(x^k)$ and the incrementally computed approximation used in Eq. (2.99), and thus attain a faster asymptotic convergence rate. Indeed, it turns out that for a strongly convex cost function and with a nondiminishing stepsize the method exhibits a linear convergence rate, just like in the nonincremental gradient method, without incurring the cost of a full gradient evaluation at each iteration. This is in contrast with the incremental gradient method (2.78), for which a linear convergence rate can be achieved only at the expense of asymptotic error, as discussed earlier. We will demonstrate this convergence rate advantage of the aggregated method later in this section, by viewing it as a gradient method with delays, i.e., a method that uses the "delayed" iterates $x^{k-\ell}$ rather than the current iterate x^k, cf. Eq. (2.99).

---

† In the case where $k < m$, the summation in Eq. (2.99) should go up to $\ell = k$, and the stepsize should be replaced by a corresponding larger value.

Sec. 2.4   Incremental Methods

A disadvantage of the aggregated gradient method (2.99) is that it requires that the most recent component gradients be kept in memory, so that when a component gradient is reevaluated at a new point, the preceding gradient of the same component is discarded from the sum of gradients of Eq. (2.99). There have been alternative implementations of the incremental aggregated gradient method idea that ameliorate this memory issue, by recalculating the full gradient periodically and replacing an old component gradient by a new one. More specifically, instead of the gradient sum

$$s^k = \sum_{\ell=0}^{m-1} \nabla f_{i_{k-\ell}}(x^{k-\ell}),$$

in Eq. (2.99), these methods use

$$\tilde{s}^k = \nabla f_{i_k}(x^k) - \nabla f_{i_k}(\tilde{x}^k) + \sum_{\ell=0}^{m-1} \nabla f_{i_{k-\ell}}(\tilde{x}^k),$$

where $\tilde{x}^k$ is the most recent point where the full gradient has been calculated. To calculate $\tilde{s}_k$ one only needs to compute the difference of the two gradients

$$\nabla f_{i_k}(x^k) - \nabla f_{i_k}(\tilde{x}^k)$$

and add it to the full gradient $\sum_{\ell=0}^{m-1} \nabla f_{i_{k-\ell}}(\tilde{x}^k)$. This bypasses the need for extensive memory storage, and with proper implementation, typically leads to small degradation in performance. However, periodically calculating the full gradient when m is very large can be a drawback.

Convergence Analysis

We will discuss the convergence of the incremental aggregated method (2.99) with a constant stepsize by embedding it within a general class of gradient methods involving iterates that are "delayed" by some maximum amount of time b. In particular, we will consider the method

$$x^{k+1} = x^k - \alpha \sum_{i=1}^{m} \nabla f_i(x^{\ell_i}), \tag{2.100}$$

where the index ℓ_i satisfies

$$\max\{0, k-b\} \le \ell_i \le k,$$

for all i and k, where b is a positive integer. Clearly, this algorithm contains as a special case the incremental aggregated method (2.99) and its variants, which we discussed earlier. For example it contains as a special case (with

$b = 2k$) the variant which involves random reshuffling of the indexes after each cycle.

We will assume that the component gradients satisfy the Lipschitz condition

$$\|\nabla f_i(x) - \nabla f_i(y)\| \le L_i \|x - y\|, \qquad \forall\, x, y \in \Re^n, \tag{2.101}$$

for some constants $L_i > 0$. Moreover, we will assume that the cost function f is strongly convex, so that there is a unique global minimum, denoted x^*. We will then show that the distance $\|x^k - x^*\|$ converges to zero linearly.

In particular, we have the following proposition.

Proposition 2.4.2: Assume that the functions f_i are convex and differentiable, and satisfy the Lipschitz condition (2.101) for some constants L_i. Assume further that the function $\sum_{i=1}^m f_i$ is strongly convex with parameter σ. Then there exists $\bar\alpha > 0$ such that for all $\alpha \in (0, \bar\alpha]$, the sequence $\{x^k\}$ generated by the incremental aggregated iteration (2.100) with constant stepsize α converges to x^* linearly, in the sense that

$$\|x^k - x^*\| \le \gamma \rho^k, \qquad \forall\, k = 0, 1, \ldots,$$

for some scalars $\gamma > 0$ and $\rho \in (0, 1)$.

Proof: In the course of the proof we take the stepsize α as small as is needed for the various calculations to be valid. Also for convenience in expressing various formulas involving delays we consider the algorithm for large enough iteration indexes, so that all the delayed iteration indexes in the following calculations are larger than 0 (for this it will be sufficient to consider the algorithm as starting at an iteration $k \ge 2b$). For parts of the proof, we will use only the Lipschitz assumption on ∇f_i. This assumption, also implies a Lipschitz condition and a bound on ∇f. More precisely, by denoting

$$L = \sum_{i=1}^m L_i,$$

we have for all $x, z \in \Re^n$,

$$\|\nabla f(x) - \nabla f(z)\| = \left\| \sum_{i=1}^m \nabla f_i(x) - \sum_{i=1}^m \nabla f_i(z) \right\|$$

$$\le \sum_{i=1}^m \|\nabla f_i(x) - \nabla f_i(z)\| \tag{2.102}$$

$$\le \sum_{i=1}^m L_i \|x - z\| = L\|x - z\|,$$

Sec. 2.4   Incremental Methods   175

so that in particular, we have

$$\|\nabla f(x^\ell)\| = \|\nabla f(x^\ell) - \nabla f(x^*)\| \le L\|x^\ell - x^*\|, \qquad \forall\, \ell \ge 0. \qquad (2.103)$$

The proof is long so we break it down in steps:

(a) We write the iteration (2.100) as a gradient method with errors

$$x^{k+1} = x^k - \alpha\bigl(\nabla f(x^k) + e^k\bigr), \qquad (2.104)$$

where the error term e_k is given by

$$e^k = \sum_{i=1}^{m} \bigl(\nabla f_i(x^{\ell_i}) - \nabla f_i(x^k)\bigr). \qquad (2.105)$$

(b) We relate the distance $\|x^k - x^*\|$ to the gradient error e^k by verifying the relation

$$\|x^{k+1} - x^*\|^2 = \|x^k - x^*\|^2 - 2\alpha \nabla f(x^k)'(x^k - x^*) + \alpha^2 \|\nabla f(x^k)\|^2 + E^k, \qquad (2.106)$$

where

$$E^k = \alpha^2 \|e^k\|^2 - 2\alpha\bigl(x^k - \alpha \nabla f(x^k) - x^*\bigr)' e^k. \qquad (2.107)$$

This is done by subtracting x^* from both sides of Eq. (2.104), norm-squaring both sides, and carrying out the straightforward calculation.

(c) We use Eq. (2.107) to bound $|E^k|$ according to

$$|E^k| \le \alpha^2 \|e^k\|^2 + 2\alpha \|e^k\|\,\|x^k - x^*\|, \qquad (2.108)$$

for all sufficiently small α. In particular, from Eq. (2.107), we have

$$|E^k| \le \alpha^2 \|e^k\|^2 + 2\alpha \|e^k\|\,\|x^k - x^* - \alpha \nabla f(x^k)\|,$$

and Eq. (2.108) is obtained from the preceding relation by using the inequality

$$\|x^k - x^* - \alpha \nabla f(x^k)\| \le \|x_k - x^*\|.$$

which holds for α sufficiently small; this is a consequence of the fact that under our assumptions, a gradient iteration (with no error) reduces the distance to x^* for $\alpha \in (0, 2/L)$ (see Prop. 1.2.2).

(d) We use the strong convexity assumption

$$\bigl(\nabla f(x) - \nabla f(y)\bigr)'(x - y) \ge \sigma\|x - y\|^2, \qquad \forall\, x,y \in \Re^n, \qquad (2.109)$$

where σ is the coefficient of strong convexity and the Lipschitz condition (2.102), to invoke the relation

$$\nabla f(x^k)'(x^k - x^*) \ge \frac{\sigma L}{\sigma + L}\|x^k - x^*\|^2 + \frac{1}{\sigma + L}\|\nabla f(x^k)\|^2; \qquad (2.110)$$

see Prop. B.5(a) in Appendix B. This will be used to bound the term $\nabla f(x^k)'(x^k - x^*)$ of Eq. (2.106).

(e) We show that for $\alpha \leq \frac{2}{\sigma+L}$, we have

$$\|x^{k+1} - x^*\|^2 \leq \left(1 - 2\alpha\frac{\sigma L}{\sigma + L}\right)\|x^k - x^*\|^2 + |E^k|. \qquad (2.111)$$

In particular, using the relations (2.106) and (2.110), we obtain

$$\|x^{k+1} - x^*\|^2 \leq \|x^k - x^*\|^2 - 2\alpha\left(\frac{\sigma L}{\sigma + L}\|x^k - x^*\|^2 + \frac{1}{\sigma + L}\|\nabla f(x^k)\|^2\right)$$
$$+ \alpha^2\|\nabla f(x^k)\|^2 + |E^k|$$
$$\leq \left(1 - 2\alpha\frac{\sigma L}{\sigma + L}\right)\|x^k - x^*\|^2 + \alpha\left(\alpha - \frac{2}{\sigma + L}\right)\|\nabla f(x^k)\|^2 + |E^k|, \qquad (2.112)$$

from which Eq. (2.112) follows.

(f) We prove that the error e_k is proportional to the stepsize α, and to the sum of the distances of a finite number of delayed iterates from the optimum x^*:

$$\|e^k\| \leq O(\alpha) \sum_{\ell=k-2b}^{k} \|x^\ell - x^*\|. \qquad (2.113)$$

This is straightforward, using the Lipschitz assumption on ∇f_i and the bound (2.103) on ∇f.

In particular, from Eq. (2.105), we have

$$\|e^k\| \leq \sum_{i=1}^{m} \|\nabla f_i(x^{\ell_i}) - \nabla f_i(x^k)\|$$
$$\leq \sum_{i=1}^{m} L_i \|x^k - x^{\ell_i}\| \qquad (2.114)$$
$$\leq \sum_{i=1}^{m} L_i \left(\|x^k - x^{k-1}\| + \cdots + \|x^{\ell_i+1} - x^{\ell_i}\|\right).$$

Moreover from Eqs. (2.103) and (2.104),

$$\|x^{\ell+1} - x^\ell\| = \alpha\|\nabla f(x^\ell)\| + \alpha\|e^\ell\| \leq \alpha L\|x^\ell - x^*\| + \alpha\|e^\ell\|, \qquad \forall\, \ell \geq 0.$$

Using this relation for ℓ in the range $[k-b, k]$ in Eq. (2.114), we obtain

$$\|e^k\| \leq O(\alpha)\left(\sum_{\ell=k-b}^{k} \|x^\ell - x^*\| + \sum_{\ell=k-b}^{k-1} \|e^\ell\|\right). \qquad (2.115)$$

From Eq. (2.105), we also have

$$\|e^\ell\| \le \sum_{i=1}^m L_i \|x^\ell - x^{\ell_i}\| \le \sum_{i=1}^m L_i\big(\|x^\ell - x^*\| + \|x^{\ell_i} - x^*\|\big).$$

Since for ℓ in the range $[k-b, k-1]$, ℓ_i lies in the range $[k-2b, k-1]$, it follows that $\|e^\ell\|$ is bounded by

$$c \sum_{\ell=k-2b}^{k} \|x^\ell - x^*\|, \qquad \forall\, \ell \in [k-b, k-1],$$

where c is some constant. Combining this with Eq. (2.115), we obtain Eq. (2.113).

(g) We use Eqs. (2.113), (2.112), and (2.108) to obtain

$$\|x^{k+1}-x^*\|^2 \le \left(1 - 2\alpha \frac{\sigma L}{\sigma + L}\right)\|x^k - x^*\|^2 + o(\alpha^2) \max_{\max\{0, k-2b\} \le \ell \le k} \|x_\ell - x^*\|^2. \tag{2.116}$$

In particular, the two terms bounding $|E^k|$ in Eq. (2.108) are $\alpha^2 \|e^k\|^2$ and $\alpha \|e^k\|\,\|x^k - x^*\|$, which in view of Eq. (2.113) are bounded by terms that are $O(\alpha^4)$ and $O(\alpha^2)$ times $\max_{\max\{0,k-2b\} \le \ell \le k} \|x_\ell - x^*\|^2$, respectively.

(h) We use Eq. (2.116) and Prop. A.29 of Appendix A, with $\beta^k = \|x^k - x^*\|^2$, $p = 1 - 2\alpha \frac{\sigma L}{\sigma + L}$, and $q = o(\alpha^2)$, so that $p + q < 1$ for sufficiently small α. This shows that β^k converges linearly to 0, and the same is true for its square root, $\|x^k - x^*\|$. This completes the proof. **Q.E.D.**

Note from Eq. (2.116) that the ratio of linear convergence, asymptotically, for small α is the same as the one that would be obtained for the standard (nonincremental) gradient method (cf. Exercise 1.3.3). However, there is an important difference. The cost for calculation of the aggregated gradient direction is far smaller than the calculation of the full gradient.

The ratio of linear convergence, asymptotically, for small α, is also the same as the one for the incremental gradient method (cf. the discussion following Prop. 2.4.1). However, again there is an important difference: the nonaggregated incremental method converges to $x(\alpha) \ne x^*$, and thus incurs an error $\|x(\alpha) - x^*\|$, which is proportional to α [cf. Prop. 2.4.1(a)]. By contrast the aggregated method converges to x^*. Roughly speaking, the convergence rate of the two incremental methods is comparable outside the region of confusion, and also during the first cycle of subiterations (when the aggregated method has not yet processed a full set of component gradients). Within the region of confusion and after the first cycle, the aggregated method is far superior.

2.4.3 Incremental Gauss-Newton Methods

We now consider incremental versions of the Gauss-Newton method for the nonlinear least squares problem

$$\begin{aligned}\text{minimize} \quad & f(x) = \tfrac{1}{2}\|g(x)\|^2 = \tfrac{1}{2}\sum_{i=1}^{m}\|g_i(x)\|^2 \\ \text{subject to} \quad & x \in \Re^n,\end{aligned} \quad (2.117)$$

where g is a continuously differentiable function with component functions $g_i : \Re^n \to \Re^{r_i}$, $i = 1, \ldots, m$.

An example of such a method starts with some x^0, then updates x via a Gauss-Newton-like iteration aimed at minimizing

$$\|g_1(x)\|^2,$$

then updates x via a Gauss-Newton-like iteration aimed at minimizing

$$\lambda\|g_1(x)\|^2 + \|g_2(x)\|^2,$$

where λ is a scalar with

$$0 \leq \lambda \leq 1,$$

and similarly continues, with the ith step consisting of a Gauss-Newton-like iteration aimed at minimizing the weighted partial sum

$$\sum_{j=1}^{i}\lambda^{i-j}\|g_j(x)\|^2. \quad (2.118)$$

Once the entire data set is processed, the cycle is restarted.

The parameter λ determines the influence of old cost components on new estimates. Generally, with smaller values of λ, the effect of old cost components is discounted faster, and successive estimates produced by the method tend to change more rapidly. Thus one may obtain a faster rate of progress of the method when $\lambda < 1$, and this is often desirable, particularly when the cost components are obtained in real time from a model whose parameters are slowly changing.

The Kalman Filter for Linear Least Squares

When the functions g_i are linear functions, it takes a single pure Gauss-Newton iteration to find the least squares estimate, and it turns out that this iteration can be implemented with an incremental algorithm known as the *Kalman filter*. This algorithm has many important applications in control and communication theory, and has been studied extensively. We

Sec. 2.4   Incremental Methods   179

first develop this algorithm for linear g_i, and we then extend it to the nonlinear case.

Suppose that
$$g_i(x) = z_i - C_i x, \tag{2.119}$$
where $z_i \in \Re^{r_i}$ are given vectors and C_i are given $r_i \times n$ matrices. In other words, we are trying to fit a linear model to the set of measurements $z_1, \ldots, z_m$. Let us consider the incremental method that sequentially generates the vectors
$$\psi_i \in \arg\min_{x \in \Re^n} \sum_{j=1}^{i} \lambda^{i-j} \|z_j - C_j x\|^2, \qquad i = 1, \ldots, m. \tag{2.120}$$

Then, for $\lambda = 1$, the least squares solution is obtained at the last step as
$$x^* = \psi_m.$$

Furthermore, the method can be conveniently implemented as shown by the following proposition.

Proposition 2.4.3: (Kalman Filter) Assume that the matrix $C_1' C_1$ is positive definite. Then the least squares estimates ψ_i, where
$$\psi_i \in \arg\min_{x \in \Re^n} \sum_{j=1}^{i} \lambda^{i-j} \|z_j - C_j x\|^2, \qquad i = 1, \ldots, m,$$
can be generated by the algorithm
$$\psi_i = \psi_{i-1} + H_i^{-1} C_i' (z_i - C_i \psi_{i-1}), \qquad i = 1, \ldots, m, \tag{2.121}$$
where ψ_0 is an arbitrary vector, and the positive definite matrices H_i are generated by
$$H_i = \lambda H_{i-1} + C_i' C_i, \qquad i = 1, \ldots, m, \tag{2.122}$$
with $H_0 = 0$. More generally, for all integers i and $\bar{i}$ with $1 \leq \bar{i} < i \leq m$ we have
$$\psi_i = \psi_{\bar{i}} + H_i^{-1} \sum_{j=\bar{i}+1}^{i} \lambda^{i-j} C_j' (z_j - C_j \psi_{\bar{i}}). \tag{2.123}$$

Proof: We first establish the result for the case of two cost components in the following proposition:

Proposition 2.4.4: Let ζ_1, ζ_2 be given vectors, and Γ_1, Γ_2 be given matrices such that $\Gamma_1'\Gamma_1$ is positive definite. Then the vectors ψ_1, ψ_2, where

$$\psi_1 \in \arg\min_{x\in\Re^n} \|\zeta_1 - \Gamma_1 x\|^2, \qquad (2.124)$$

and

$$\psi_2 \in \arg\min_{x\in\Re^n} \left\{ \|\zeta_1 - \Gamma_1 x\|^2 + \|\zeta_2 - \Gamma_2 x\|^2 \right\}, \qquad (2.125)$$

are also given by

$$\psi_1 = \psi_0 + (\Gamma_1'\Gamma_1)^{-1}\Gamma_1'(\zeta_1 - \Gamma_1\psi_0), \qquad (2.126)$$

and

$$\psi_2 = \psi_1 + (\Gamma_1'\Gamma_1 + \Gamma_2'\Gamma_2)^{-1}\Gamma_2'(\zeta_2 - \Gamma_2\psi_1), \qquad (2.127)$$

where ψ_0 is an arbitrary vector.

Proof: By carrying out the minimization in Eq. (2.124), we obtain

$$\psi_1 = (\Gamma_1'\Gamma_1)^{-1}\Gamma_1'\zeta_1, \qquad (2.128)$$

yielding for any ψ_0,

$$\psi_1 = \psi_0 - (\Gamma_1'\Gamma_1)^{-1}\Gamma_1'\Gamma_1\psi_0 + (\Gamma_1'\Gamma_1)^{-1}\Gamma_1'\zeta_1,$$

from which the desired Eq. (2.126) follows.

Also, by carrying out the minimization in Eq. (2.125), we obtain

$$\psi_2 = (\Gamma_1'\Gamma_1 + \Gamma_2'\Gamma_2)^{-1}(\Gamma_1'\zeta_1 + \Gamma_2'\zeta_2),$$

or equivalently, using also Eq. (2.128),

$$(\Gamma_1'\Gamma_1 + \Gamma_2'\Gamma_2)\psi_2 = \Gamma_1'\zeta_1 + \Gamma_2'\zeta_2$$
$$= \Gamma_1'\Gamma_1\psi_1 + \Gamma_2'\zeta_2$$
$$= (\Gamma_1'\Gamma_1 + \Gamma_2'\Gamma_2)\psi_1 - \Gamma_2'\Gamma_2\psi_1 + \Gamma_2'\zeta_2,$$

from which, by multiplying both sides with $(\Gamma_1'\Gamma_1 + \Gamma_2'\Gamma_2)^{-1}$, the desired Eq. (2.127) follows. **Q.E.D.**

Proof of Prop. 2.4.3: Equation (2.123) follows by applying Prop. 2.4.4 with the notational identifications $\psi_0 \sim \psi_0$, $\psi_1 \sim \psi_{\bar{i}}$, $\psi_2 \sim \psi_i$, and

$$\zeta_1 \sim \begin{pmatrix} \sqrt{\lambda^{i-1}}z_1 \\ \vdots \\ \sqrt{\lambda^{i-\bar{i}}}z_{\bar{i}} \end{pmatrix}, \qquad \Gamma_1 \sim \begin{pmatrix} \sqrt{\lambda^{i-1}}C_1 \\ \vdots \\ \sqrt{\lambda^{i-\bar{i}}}C_{\bar{i}} \end{pmatrix},$$

Sec. 2.4   Incremental Methods

$$\zeta_2 \sim \begin{pmatrix} \sqrt{\lambda^{i-\bar{i}-1}} z_{\bar{i}+1} \\ \vdots \\ z_i \end{pmatrix}, \quad \Gamma_2 \sim \begin{pmatrix} \sqrt{\lambda^{i-\bar{i}-1}} C_{\bar{i}+1} \\ \vdots \\ C_i \end{pmatrix},$$

and by carrying out the straightforward algebra. Equation (2.121) is the special case of Eq. (2.123) corresponding to $\bar{i} = i - 1$. **Q.E.D.**

Note that the positive definiteness assumption on $C_1'C_1$ in Prop. 2.4.3 is needed to guarantee that the first matrix H_1 is positive definite and hence invertible; then the positive definiteness of the subsequent matrices $H_2, \ldots, H_m$ follows from Eq. (2.122). As a practical matter, it is possible to guarantee the positive definiteness of $C_1'C_1$ by lumping a sufficient number of measurements into the first cost component (C_1 should contain n linearly independent columns). An alternative is to redefine ψ_i as

$$\psi_i \in \arg\min_{x \in \Re^n} \left\{ \delta \lambda^i \|x - \psi_0\|^2 + \sum_{j=1}^{i} \lambda^{i-j} \|z_j - C_j x\|^2 \right\}, \quad i = 1, \ldots, m,$$

where δ is a small positive scalar. Then it can be seen from the proof of Prop. 2.4.3 that ψ_i is generated by the same equations (2.121) and (2.122), except that the initial condition $H_0 = 0$ is replaced by

$$H_0 = \delta I,$$

so that $H_1 = \lambda \delta I + C_1'C_1$ is positive definite even if $C_1'C_1$ is not. Note, however, that in this case, the last estimate ψ_m is only approximately equal to the least squares estimate x^*, even if $\lambda = 1$ (the approximation error depends on the size of δ).

The Extended Kalman Filter

Consider now the general case where the functions g_i are nonlinear. Then a generalization of the Kalman filter, known as the *extended Kalman filter* (EKF for short), can be used. Like the Gauss-Newton method, it involves linearization of the functions g_i and solution of linear least squares problems. However, these problems are solved incrementally using the Kalman filtering algorithm, and *the linearization of each g_i is done at the latest iterate that is available when g_i is processed*. In particular, a cycle through the data set of the algorithm sequentially generates the vectors

$$\psi_i \in \arg\min_{x \in \Re^n} \sum_{j=1}^{i} \lambda^{i-j} \|\tilde{g}_j(x, \psi_{j-1})\|^2, \quad i = 1, \ldots, m, \quad (2.129)$$

where $\tilde{g}_j(x, \psi_{j-1})$ are the linearized functions

$$\tilde{g}_j(x, \psi_{j-1}) = g_j(\psi_{j-1}) + \nabla g_j(\psi_{j-1})'(x - \psi_{j-1}), \quad (2.130)$$

and ψ_0 is an initial estimate of x. Using the formulas (2.121) and (2.122) of Prop. 2.4.3 with the identifications

$$z_i = g_i(\psi_{i-1}) - \nabla g_i(\psi_{i-1})'\psi_{i-1}, \qquad C_i = -\nabla g_i(\psi_{i-1})',$$

this algorithm can be written in the incremental form

$$\psi_i = \psi_{i-1} - H_i^{-1}\nabla g_i(\psi_{i-1})g_i(\psi_{i-1}), \qquad i = 1, \ldots, m, \qquad (2.131)$$

where the matrices H_i are generated by

$$H_i = \lambda H_{i-1} + \nabla g_i(\psi_{i-1})\nabla g_i(\psi_{i-1})', \qquad i = 1, \ldots, m, \qquad (2.132)$$

with

$$H_0 = 0. \qquad (2.133)$$

To contrast the EKF with the pure form of the Gauss-Newton method (unit stepsize), note that a single iteration of the latter can be written as

$$x^{k+1} \in \arg\min_{x \in \Re^n} \sum_{i=1}^m \left\|\tilde{g}_i(x, x^k)\right\|^2. \qquad (2.134)$$

Using the formulas of Prop. 2.4.3 with the identifications

$$z_i = g_i(x^k) - \nabla g_i(x^k)'x^k, \qquad C_i = -\nabla g_i(x^k)',$$

we can generate x^{k+1} by an incremental algorithm as

$$x^{k+1} = \bar{\psi}_m,$$

where

$$\bar{\psi}_i = \bar{\psi}_{i-1} - \bar{H}_i^{-1}\nabla g_i(x^k)\big(g_i(x^k) + \nabla g_i(x^k)'(\bar{\psi}_{i-1} - x^k)\big), \qquad i = 1, \ldots, m, \qquad (2.135)$$

$\bar{\psi}_0 = x^k$, and the matrices $\bar{H}_i$ are generated by

$$\bar{H}_i = \bar{H}_{i-1} + \nabla g_i(x^k)\nabla g_i(x^k)', \qquad i = 1, \ldots, m, \qquad (2.136)$$

with

$$\bar{H}_0 = 0. \qquad (2.137)$$

Thus, by comparing Eqs. (2.131)-(2.133) with Eqs. (2.135)-(2.137), we see that, if $\lambda = 1$, a cycle of the EKF through the data set differs from a pure Gauss-Newton iteration only in that the linearization of the functions g_i is done at the corresponding current estimates ψ_{i-1} rather than at the estimate x^k available at the start of the cycle.

Convergence Issues for the Extended Kalman Filter

We have considered so far a single cycle of the EKF. To obtain an algorithm that cycles through the data set multiple times, we can simply create a larger data set by concatenating multiple copies of the original data set, i.e., by forming what we refer to as *the extended data set*

$$(g_1, g_2, \ldots, g_m, g_1, g_2, \ldots, g_m, g_1, g_2, \ldots). \tag{2.138}$$

The basic reason why this algorithm works is that asymptotically it resembles a gradient method with diminishing stepsize of the type described in Section 1.2. To get a sense of this, assume that the EKF is applied to the extended data set (2.138) with $\lambda = 1$. Let us denote by x^k the iterate at the end of the kth cycle through the data set, i.e.,

$$x^k = \psi_{km}, \qquad k = 1, 2, \ldots$$

Then by using Eq. (2.133) with $i = (k+1)m$ and $\bar{i} = km$, we obtain

$$x^{k+1} = x^k - H_{(k+1)m}^{-1}\left(\sum_{i=1}^{m}\nabla g_i(\psi_{km+i-1})g_i(\psi_{km+i-1})\right). \tag{2.139}$$

Now $H_{(k+1)m}$ grows roughly in proportion to $k+1$ because, by Eq. (2.122), we have

$$H_{(k+1)m} = \sum_{j=0}^{k}\sum_{i=1}^{m}\nabla g_i(\psi_{jm+i-1})\nabla g_i(\psi_{jm+i-1})'. \tag{2.140}$$

It is therefore reasonable to expect that the method tends to make slow progress when k is large, which means that the vectors ψ_{km+i-1} in Eq. (2.139) are roughly equal to x^k. Thus for large k, the sum in the right-hand side of Eq. (2.139) is roughly equal to the gradient $\nabla g(x^k)g(x^k)$, while from Eq. (2.140), $H_{(k+1)m}$ is roughly equal to $(k+1)\nabla g(x^k)\nabla g(x^k)'$, where $g = (g_1, g_2, \ldots, g_m)$ is the original data set. It follows that for large k, the EKF iteration (2.139) can be written approximately as

$$x^{k+1} \approx x^k - \frac{1}{k+1}\left(\nabla g(x^k)\nabla g(x^k)'\right)^{-1}\nabla g(x^k)g(x^k), \tag{2.141}$$

i.e., as an approximate Gauss-Newton iteration with diminishing stepsize.

When $\lambda < 1$, the matrix H_i^{-1} generated by the EKF recursion (2.132) will typically not diminish to zero, and $\{x^k\}$ may not converge to a stationary point of $\sum_{i=1}^{m}\lambda^{m-i}\|g_i(x)\|^2$. Furthermore, as the following example shows, the sequences $\{\psi_{km+i}\}$ produced by the EKF using Eq. (2.131), may converge to different limits for different i:

Example 2.4.4

Consider the case where there are two functions, $g_1(x) = x - c_1$ and $g_2(x) = x - c_2$, where c_1 and c_2 are given scalars. Each cycle of the EKF consists of two steps. At the second step of the kth cycle, we minimize

$$\sum_{i=1}^{k} \left(\lambda^{2i-1}(x-c_1)^2 + \lambda^{2i-2}(x-c_2)^2 \right),$$

which is equal to the following scalar multiple of $\lambda(x-c_1)^2 + (x-c_2)^2$,

$$(1+\lambda^2+\cdots+\lambda^{2k-2})\left(\lambda(x-c_1)^2 + (x-c_2)^2\right).$$

Thus at the second step, we obtain the minimizer of $\lambda(x-c_1)^2 + (x-c_2)^2$,

$$\psi_{2k} = \frac{\lambda c_1 + c_2}{\lambda + 1}.$$

At the first step of the kth cycle, we minimize

$$(x-c_1)^2 + \lambda \sum_{i=1}^{k-1} \left(\lambda^{2i-1}(x-c_1)^2 + \lambda^{2i-2}(x-c_2)^2 \right),$$

which is equal to the following scalar multiple of $(x-c_1)^2 + \lambda(x-c_2)^2$

$$(1+\lambda^2+\cdots+\lambda^{2k-4})\left((x-c_1)^2 + \lambda(x-c_2)^2\right),$$

plus the diminishing term $\lambda^{2k-2}(x-c_1)^2$. Thus at the first step, we obtain approximately (for large k) the minimizer of $(x-c_1)^2 + \lambda(x-c_2)^2$,

$$\psi_{2k-1} \approx \frac{c_1 + \lambda c_2}{1+\lambda}.$$

We see therefore that within each cycle, there is an oscillation around the minimizer $(c_1+c_2)/2$ of $(x-c_1)^2 + (x-c_2)^2$. The size of the oscillation diminishes as λ approaches 1.

Generally, for a nonlinear least squares problem, the convergence rate tends to be faster when $\lambda < 1$ than when $\lambda = 1$, essentially because the implicit stepsize does not diminish as in the case $\lambda = 1$. For this reason, a hybrid method that uses a different value of λ within each cycle may work best in practice. One may start with a relatively small λ to attain a fast initial rate of convergence, and then progressively increase λ towards 1 in order to attain high solution accuracy. The following proposition shows convergence for the case where λ tends to 1 at a sufficiently fast rate.

Sec. 2.4   Incremental Methods   185

> **Proposition 2.4.5:** Assume that $\nabla g_i(x)$ has full rank for all x and $i = 1, \ldots, m$, and that for some $L > 0$, we have
>
> $$\|\nabla g_i(x)g_i(x) - \nabla g_i(y)g_i(y)\| \leq L\|x-y\|, \qquad \forall\, x, y \in \Re^n,\ i = 1, \ldots, m.$$
>
> Assume also that there is a constant $c > 0$ such that the scalar λ used in the updating formula (2.132) within the kth cycle, call it $\lambda(k)$, satisfies
>
> $$0 \leq 1 - \bigl(\lambda(k)\bigr)^m \leq \frac{c}{k}, \qquad \forall\, k = 1, 2, \ldots.$$
>
> Then if the EKF applied to the extended data set (2.138) generates a bounded sequence of vectors ψ_i, each of the limit points of $\{x^k\}$ is a stationary point of the least squares problem.

Proof: *(Abbreviated; for a complete proof see [Ber96b])* We have using the Kalman filter recursion (2.123) that x^k satisfies

$$x^{k+1} = x^k - H_{(k+1)m}^{-1} \left(\sum_{i=1}^{m} \bigl(\lambda(k)\bigr)^{m-i} \nabla g_i(\psi_{km+i-1}) g_i(\psi_{km+i-1}) \right).$$

It can be shown using the rank assumption on $\nabla g_i(x)$, the growth assumption on λ^k, the boundedness of $\{x^k\}$, and the preceding analysis that the eigenvalues of the matrices H_{km} are within an interval $[c_1 k, c_2 k]$, where c_1 and c_2 are some positive constants (see Exercise 2.4.11). The proof then follows the line of argument of the convergence proof of gradient methods with diminishing stepsize (Prop. 1.2.3 in Section 1.2). **Q.E.D.**

2.4.4   Incremental Newton Methods

We will now consider an incremental version of Newton's method for unconstrained minimization of an additive cost function of the form

$$f(x) = \sum_{i=1}^{m} f_i(x),$$

where the functions $f_i : \Re^n \mapsto \Re$ are convex and twice continuously differentiable.† Consider the quadratic approximation $\tilde{f}_i$ of a function f_i at a vector $\psi \in \Re^n$, i.e., the second order expansion of f_i at ψ:

$$\tilde{f}_i(x;\psi) = \nabla f_i(\psi)'(x-\psi) + \tfrac{1}{2}(x-\psi)'\nabla^2 f_i(\psi)(x-\psi), \quad \forall\, x, \psi \in \Re^n.$$

---

† A beneficial consequence of assuming convexity of f_i is that the Hessian matrices $\nabla^2 f_i(x)$ are positive semidefinite, which facilitates the implementation of the algorithms to be described. On the other hand, the algorithmic ideas of this section may also be adapted for the case where f_i are nonconvex.

Similar to Newton's method, which minimizes a quadratic approximation at the current point of the cost function, the incremental form of Newton's method minimizes a sum of quadratic approximations of components. As in the case of the incremental gradient and Gauss-Newton methods, we view an iteration as a cycle of m subiterations, each involving a single additional component f_i, and its gradient and Hessian at the current point within the cycle. In particular, if x^k is the vector obtained after k cycles, the vector x^{k+1} obtained after one more cycle is

$$x^{k+1} = \psi_{m,k},$$

where starting with $\psi_{0,k} = x^k$, we obtain $\psi_{m,k}$ after the m steps

$$\psi_{i,k} \in \arg\min_{x \in \Re^n} \sum_{\ell=1}^{i} \tilde{f}_\ell(x; \psi_{\ell-1,k}), \qquad i = 1, \ldots, m. \qquad (2.142)$$

If all the functions f_i are quadratic, it can be seen that the method finds the solution in a single cycle.† The reason is that when f_i is quadratic, each $f_i(x)$ differs from $\tilde{f}_i(x; \psi)$ by a constant, which does not depend on x. Thus the difference

$$\sum_{i=1}^{m} f_i(x) - \sum_{i=1}^{m} \tilde{f}_i(x; \psi_{i-1,k})$$

is a constant that is independent of x, and minimization of either sum in the above expression gives the same result.

It is important to note that the computations of Eq. (2.142) can be carried out efficiently. For simplicity, let as assume that $\tilde{f}_1(x; \psi)$ is a positive definite quadratic, so that for all i, $\psi_{i,k}$ is well defined as the unique solution of the minimization problem in Eq. (2.142). We will show that the incremental Newton method (2.142) can be implemented in terms of the incremental update formula

$$\psi_{i,k} = \psi_{i-1,k} - D_{i,k} \nabla f_i(\psi_{i-1,k}), \qquad (2.143)$$

---

† Here we assume that the m quadratic minimizations (2.142) to generate $\psi_{m,k}$ have a solution. For this it is sufficient that the first Hessian matrix $\nabla^2 f_1(x^0)$ be positive definite, in which case there is a unique solution at every iteration. A simple possibility to deal with this requirement is to add to f_1 a small positive definite quadratic term, such as $\frac{\epsilon}{2}\|x - x^0\|^2$. A more sound possibility is to lump together several of the component functions (enough to ensure that the sum of their quadratic approximations at x^0 is positive definite), and to use them in place of f_1. This is generally a good idea and leads to smoother initialization, as it ensures a relatively stable behavior of the algorithm for the initial iterations.

Sec. 2.4   Incremental Methods   187

where $D_{i,k}$ is given by

$$D_{i,k} = \left(\sum_{\ell=1}^{i} \nabla^2 f_\ell(\psi_{\ell-1,k}) \right)^{-1}, \qquad (2.144)$$

and is generated iteratively as

$$D_{i,k} = \left(D_{i-1,k}^{-1} + \nabla^2 f_i(\psi_{i,k}) \right)^{-1}. \qquad (2.145)$$

Indeed, from the definition of the method (2.142), the quadratic function $\sum_{\ell=1}^{i-1} \tilde{f}_\ell(x; \psi_{\ell-1,k})$ is minimized by $\psi_{i-1,k}$ and its Hessian matrix is $D_{i-1,k}^{-1}$, so we have

$$\sum_{\ell=1}^{i-1} \tilde{f}_\ell(x; \psi_{\ell-1,k}) = \tfrac{1}{2}(x - \psi_{\ell-1,k})' D_{i-1,k}^{-1}(x - \psi_{\ell-1,k}) + \text{constant}.$$

Thus, by adding $\tilde{f}_i(x; \psi_{i-1,k})$ to both sides of this expression, we obtain

$$\sum_{\ell=1}^{i} \tilde{f}_\ell(x; \psi_{\ell-1,k}) = \tfrac{1}{2}(x - \psi_{\ell-1,k})' D_{i-1,k}^{-1}(x - \psi_{\ell-1,k}) + \text{constant}$$
$$+ \tfrac{1}{2}(x - \psi_{i-1,k})' \nabla^2 f_i(\psi_{i-1,k})(x - \psi_{i-1,k}) + \nabla f_i(\psi_{i-1,k})'(x - \psi_{i-1,k}).$$

Since by definition $\psi_{i,k}$ minimizes this function, we obtain Eqs. (2.143)-(2.145).

The recursion (2.145) for the matrix $D_{i,k}$ can often be efficiently implemented by using convenient formulas for the inverse of the sum of two matrices. In particular, if f_i is given by

$$f_i(x) = h_i(a_i' x - b_i),$$

for some twice differentiable convex function $h_i : \Re \mapsto \Re$, vector a_i, and scalar b_i, we have

$$\nabla^2 f_i(\psi_{i-1,k}) = \nabla^2 h_i(\psi_{i-1,k}) \, a_i a_i',$$

and the recursion (2.145) can be written as

$$D_{i,k} = D_{i-1,k} - \frac{D_{i-1,k} a_i a_i' D_{i-1,k}}{\nabla^2 h_i(\psi_{i-1,k})^{-1} + a_i' D_{i-1,k} a_i};$$

this is the well-known Sherman-Morrison formula for the inverse of the sum of an invertible matrix and a rank-one matrix (see the matrix inversion formula in Section A.1 of Appendix A).

We have considered so far a single cycle of the incremental Newton method. Similar to the case of the extended Kalman filter, we may cycle through the component functions multiple times. In particular, we may apply the incremental Newton method to the extended set of components

$$f_1, f_2, \ldots, f_m, f_1, f_2, \ldots, f_m, f_1, f_2, \ldots.$$

The resulting method asymptotically resembles a scaled incremental gradient method with diminishing stepsize of the type described earlier. Indeed, from Eq. (2.144)], the matrix $D_{i,k}$ diminishes roughly in proportion to $1/k$. From this it follows that the asymptotic convergence properties of the incremental Newton method are similar to those of an incremental gradient method with diminishing stepsize of order $O(1/k)$. Thus its convergence rate is slower than linear.

To accelerate the convergence of the method one may employ a form of restart, so that $D_{i,k}$ does not converge to 0. For example $D_{i,k}$ may be reinitialized and increased in size at the beginning of each cycle. For problems where f has a unique nonsingular minimum x^* [one for which $\nabla^2 f(x^*)$ is nonsingular], one may design incremental Newton schemes with restart that converge linearly to within a neighborhood of x^* (and even superlinearly if x^* is also a minimum of all the functions f_i, so there is no region of confusion). Alternatively, the update formula (2.145) may be modified to

$$D_{i,k} = \left(\lambda_k D_{i-1,k}^{-1} + \nabla^2 f_\ell(\psi_{i,k})\right)^{-1}, \qquad (2.146)$$

by introducing a parameter $\lambda_k \in (0,1)$, which can be used to accelerate the practical convergence rate of the method; cf. the discussion of the incremental Gauss-Newton methods.

Incremental Newton Method with Diagonal Approximation

Generally, with proper implementation, the incremental Newton method is often substantially faster than the incremental gradient method, in terms of numbers of iterations (there are theoretical results suggesting this property for stochastic versions of the two methods; see the end-of-chapter references). However, in addition to computation of second derivatives, the incremental Newton method involves greater overhead per iteration due to matrix-vector calculations in Eqs. (2.143), (2.145), and (2.146), so it is suitable only for problems where n, the dimension of x, is relatively small.

A way to remedy in part this difficulty is to approximate $\nabla^2 f_i(\psi_{i,k})$ by a diagonal matrix, and recursively update a diagonal approximation of $D_{i,k}$ using Eqs. (2.145) or (2.146). In particular, one may set to 0 the off-diagonal components of $\nabla^2 f_i(\psi_{i,k})$. In this case, the iteration (2.143) becomes a diagonally scaled version of the incremental gradient method, and involves comparable overhead per iteration (assuming the required diagonal second derivatives are easily computed). As an additional scaling

EXERCISES

2.4.1

Consider the least squares problem (2.117) for the case where $m < n$ and $r_i = 1$ for all i.

(a) Show that the Hessian matrix is singular at any optimal solution x^* for which $g(x^*) = 0$.

(b) Consider the case where g is linear and of the form $g(x) = z - Ax$, where A is an $m \times n$ matrix. Show that there are infinitely many optimal solutions. Show also that if A has linearly independent rows, $x^* = A'(AA')^{-1}z$ is one of these solutions.

2.4.2 [Luo91]

Consider the least squares problem

$$\text{minimize} \quad \tfrac{1}{2}\{(z_1 - x)^2 + (z_2 - x)^2\}$$
$$\text{subject to} \quad x \in \Re,$$

and the incremental gradient method that generates x^{k+1} from x^k according to

$$x^{k+1} = y^k - \alpha(y^k - z_2),$$

where

$$y^k = x^k - \alpha(x^k - z_1),$$

and α is positive stepsize. Assuming that $\alpha < 1$, show that $\{x^k\}$ and $\{y^k\}$ converge to limits $x(\alpha)$ and $y(\alpha)$, respectively. However, unless $z_1 = z_2$, $x(\alpha)$ and $y(\alpha)$ are neither equal to each other, nor equal to the least squares solution $x^* = (z_1 + z_2)/2$. Consistently with Prop. 2.4.1, verify that

$$\lim_{\alpha \to 0} x(\alpha) = \lim_{\alpha \to 0} y(\alpha) = x^*.$$

2.4.3 (Computational Problem)

Consider the least squares cost function of Fig. 1.4.3 for the case where the five data pairs (y_s, z_s) are $(1.165, 1)$, $(0.626, -1)$, $(0.075, -1)$, $(0.351, 1)$, $(-0.696, 1)$ (these correspond to the three-dimensional plot of Fig. 1.4.3). Minimize this cost function using appropriate versions of the Gauss-Newton method, the Extended Kalman Filter, the incremental gradient method, and the incremental aggregated gradient method.

2.4.4 (A Generalized Incremental Gradient Method [Ber97])

The purpose of this exercise is to embed the incremental gradient method and the method of steepest descent within a one-parameter family of methods for the least squares problem. In contrast with the incremental gradient method, some of the methods in this family have a generically linear convergence rate. For a fixed $\mu \geq 0$, define

$$\xi_i(\mu) = \frac{1}{1 + \mu + \cdots + \mu^{m-i}}, \qquad i = 1, \ldots, m.$$

Consider a method which given x^k, generates x^{k+1} according to $x^{k+1} = \psi_m$, where

$$\psi_i = x^k - \alpha^k h_i,$$

$$h_i = \mu h_{i-1} + \sum_{j=1}^{i} \xi_j(\mu) \nabla g_j(\psi_{j-1}) g_j(\psi_{j-1}),$$

with the initial condition $h_0 = 0$.

(a) Show that when $\mu = 0$, the method coincides with the incremental gradient method, and when $\mu \to \infty$, the method approaches the steepest descent method.

(b) Show that parts (a) and (b) of Prop. 2.4.1, which were proved for the case $\mu = 0$, hold as stated when $\mu > 0$ as well.

(c) Suppose that in the kth iteration, a k-dependent value of μ, say $\mu(k)$, and a constant stepsize $\alpha^k = \alpha$ are used. Under the assumptions of Prop. 2.4.1, show that if for some $q > 1$ and all k greater than some index $\bar{k}$, we have $\mu(k) \geq q^k$, then there exists $\bar{\alpha} > 0$ such that for all $\alpha \in (0, \bar{\alpha}]$, the iteration converges linearly to the optimal solution x^*.

2.4.5 (Convergence of Gradient Method with Errors [BeT00])

Consider the problem of unconstrained minimization of a continuously differentiable function $f : \Re^n \mapsto \Re$. Let $\{x^k\}$ be a sequence generated by the method

$$x^{k+1} = x^k - \alpha^k \left(\nabla f(x^k) + w^k \right),$$

Sec. 2.4 Incremental Methods

where α^k is a positive stepsize, and w^k is an error vector satisfying for some positive scalars p and q,

$$\|w^k\| \leq \alpha^k \big(q + p\|\nabla f(x^k)\|\big), \qquad k = 0, 1, \ldots. \tag{2.147}$$

Assume that for some constant $L > 0$, we have

$$\|\nabla f(x) - \nabla f(y)\| \leq L\|x - y\|, \qquad \forall\, x, y \in \Re^n,$$

and that

$$\sum_{k=0}^{\infty} \alpha^k = \infty, \qquad \sum_{k=0}^{\infty} (\alpha^k)^2 < \infty. \tag{2.148}$$

Show that either $f(x^k) \to -\infty$ or else $f(x^k)$ converges to a finite value and $\lim_{k\to\infty} \nabla f(x^k) = 0$. Furthermore, every limit point $\bar{x}$ of $\{x^k\}$ satisfies $\nabla f(\bar{x}) = 0$. *Abbreviated Proof*: The descent inequality of Prop. A.24 of Appendix A yields

$$f(x^{k+1}) \leq f(x^k) - \alpha^k \nabla f(x^k)'\big(\nabla f(x^k) + w^k\big) + \frac{(\alpha^k)^2 L}{2} \big\|\nabla f(x^k) + w^k\big\|^2.$$

Using Eq. (2.147), we have

$$-\nabla f(x^k)'\big(\nabla f(x^k) + w^k\big) \leq -\big\|\nabla f(x^k)\big\|^2 + \big\|\nabla f(x^k)\big\|\,\|w^k\|$$
$$\leq -\big\|\nabla f(x^k)\big\|^2 + \alpha^k q \big\|\nabla f(x^k)\big\| + \alpha^k p \big\|\nabla f(x^k)\big\|^2,$$

and

$$\tfrac{1}{2}\big\|\nabla f(x^k) + w^k\big\|^2 \leq \big\|\nabla f(x^k)\big\|^2 + \|w^k\|^2$$
$$\leq \big\|\nabla f(x^k)\big\|^2 + (\alpha^k)^2 \big(q^2 + 2pq\big\|\nabla f(x^k)\big\| + p^2\big\|\nabla f(x^k)\big\|^2\big).$$

Combining the preceding three relations and collecting terms, it follows that

$$f(x^{k+1}) \leq f(x^k) - \alpha^k\big(1 - \alpha^k L - \alpha^k p - (\alpha^k)^3 p^2 L\big)\big\|\nabla f(x^k)\big\|^2$$
$$+ (\alpha^k)^2\big(q + 2(\alpha^k)^2 pqL\big)\big\|\nabla f(x^k)\big\| + (\alpha^k)^4 q^2 L.$$

Since $\alpha^k \to 0$, we have for some positive constants c and d, and all k sufficiently large

$$f(x^{k+1}) \leq f(x^k) - \alpha^k c\big\|\nabla f(x^k)\big\|^2 + (\alpha^k)^2 d\big\|\nabla f(x^k)\big\| + (\alpha^k)^4 q^2 L.$$

Using the inequality $\big\|\nabla f(x^k)\big\| \leq 1 + \big\|\nabla f(x^k)\big\|^2$, the above relation yields for all k sufficiently large

$$f(x^{k+1}) \leq f(x^k) - \alpha^k(c - \alpha^k d)\big\|\nabla f(x^k)\big\|^2 + (\alpha^k)^2 d + (\alpha^k)^4 q^2 L.$$

By applying the Supermartingale Convergence Theorem (Prop. A.4.4 in Appendix A), using also the assumption (2.148), it follows that either $f(x^k) \to -\infty$ or else $f(x^k)$ converges to a finite value and $\sum_{k=0}^{\infty} \alpha^k \big\|\nabla f(x^k)\big\|^2 < \infty$. In the latter case, in view of the assumption $\sum_{k=0}^{\infty} \alpha^k = \infty$, we must have $\liminf_{k\to\infty} \|\nabla f(x^k)\| = 0$. This implies that $\nabla f(x^k) \to 0$; for a detailed proof of this last step see [BeT00]. This reference also provides a stochastic version of the result of this exercise, which, however, requires a proof that does not rely on supermartingale convergence arguments.

2.4.6 (Convergence of the Incremental Gradient Method)

Consider the minimization of a cost function

$$f(x) = \sum_{i=1}^{m} f_i(x),$$

where $f_i : \Re^n \mapsto \Re$ are continuously differentiable, and let $\{x^k\}$ be a sequence generated by the incremental gradient method. Assume that for some constants L, C, D, and all $i = 1, \ldots, m$, we have

$$\left\|\nabla f_i(x) - \nabla f_i(y)\right\| \leq L\|x - y\|, \qquad \forall\, x, y \in \Re^n,$$

and

$$\left\|\nabla f_i(x)\right\| \leq C + D\left\|\nabla f(x)\right\|, \qquad \forall\, x \in \Re^n.$$

Assume also that

$$\sum_{k=0}^{\infty} \alpha^k = \infty, \qquad \sum_{k=0}^{\infty} (\alpha^k)^2 < \infty.$$

Then either $f(x^k)$ converges to a finite value and $\lim_{k\to\infty} \nabla f(x^k) = 0$, or else $f(x^k) \to -\infty$. Furthermore, every limit point $\bar{x}$ of $\{x^k\}$ satisfies $\nabla f(\bar{x}) = 0$. *Abbreviated Solution*: The idea is to view the incremental gradient method as a gradient method with errors, so that the result of Exercise 2.4.5 can be used. For simplicity we assume that $m = 2$. The proof is similar when $m > 2$. We have

$$\psi_1 = x^k - \alpha^k \nabla f_1(x^k), \qquad x^{k+1} = \psi_1 - \alpha^k \nabla f_2(\psi_1).$$

By adding these two relations, we obtain

$$x^{k+1} = x^k + \alpha^k \left(-\nabla f(x^k) + w^k\right),$$

where

$$w^k = \nabla f_2(x^k) - \nabla f_2(\psi_1).$$

We have

$$\|w^k\| \leq L\|x^k - \psi_1\| = \alpha^k L \left\|\nabla f_1(x^k)\right\| \leq \alpha^k \left(LC + LD\left\|\nabla f(x^k)\right\|\right).$$

Thus Exercise 2.4.5 applies and the result follows.

2.4.7 (Incremental Gradient Method for Infinitely Many Cost Components)

Consider the gradient method
$$x^{k+1} = x^k - \alpha \nabla f_k(x^k) \qquad k = 0, 1, \ldots,$$
where $f_0, f_1, \ldots$, are quadratic functions with eigenvalues lying within some interval $[\gamma, \Gamma]$, where $\gamma > 0$. Suppose that for a given $\epsilon > 0$, there is a vector x^* such that
$$\|\nabla f_k(x^*)\| \leq \epsilon, \qquad \forall\, k = 0, 1, \ldots.$$
Show that for all α with $0 < \alpha \leq 2/(\gamma + \Gamma)$, we have
$$\limsup_{k \to \infty} \|x^k - x^*\| \leq \frac{2\epsilon}{\gamma}.$$
Hint: Let Q_k be the positive definite symmetric matrix corresponding to f_k, and write
$$x^{k+1} - x^* = (I - \alpha Q_k)(x^k - x^*) - \alpha \nabla f_k(x^*).$$
Use this relation to show that
$$\|x^k - x^*\| > \frac{2\epsilon}{\gamma} \quad \Rightarrow \quad \|x^{k+1} - x^*\| < \left(1 - \frac{\alpha\gamma}{2}\right)\|x^k - x^*\|,$$
while
$$\|x^k - x^*\| \leq \frac{2\epsilon}{\gamma} \quad \Rightarrow \quad \|x^{k+1} - x^*\| \leq \frac{2\epsilon}{\gamma}.$$

2.4.8

Show that if the sequence $\{x^k\}$ produced by the EKF with $\lambda = 1$ converges to some $\bar{x}$, then $\bar{x}$ must be a stationary point of $\sum_{i=1}^m \|g_i(x)\|^2$.

2.4.9

Consider the EKF of Eqs. (2.131)-(2.133) for the data set $g_1(x) = x^2 - 2$ and $g_2(x) = x^2$, where $x \in \Re$.

(a) Verify that each cycle of the EKF consists of executing sequentially the iterations
$$H_y := \lambda H_x + (2x)^2,$$
$$y := x - H_y^{-1}(2x)(x^2 - 2),$$
$$H_x := \lambda H_y + (2y)^2,$$
$$x := y - H_x^{-1}(2y)(y^2).$$

(b) Write a computer program that implements the EKF, and compare, for different values of λ, the limit of $(x + y)/2$ with the least squares solutions, which are 1 and -1. Try also a version of the method where λ is progressively increased towards 1.

2.4.10 (Incremental Gauss-Newton Method with Restart)

Consider a version of the EKF where the matrix H is reset to 0 at the end of each cycle and a stepsize α^k is used. In particular, this method, given x^k, generates x^{k+1} according to $x^{k+1} = \psi_m$, where

$$\psi_i = \psi_{i-1} - \alpha^k H_i^{-1} \nabla g_i(\psi_{i-1}) g_i(\psi_{i-1}), \qquad i = 1, \ldots, m,$$

$$H_i = H_{i-1} + \nabla g_i(\psi_{i-1}) \nabla g_i(\psi_{i-1})', \qquad i = 1, \ldots, m,$$

with the initial conditions

$$\psi_0 = x^k, \qquad H_0 = 0.$$

(a) Show that if the functions g_i are linear and α^k is for all k equal to a constant α with $0 < \alpha < 2$, then $\{x^k\}$ converges to the optimal solution.

(b) Show that if $\sum_{k=0}^{\infty} \alpha^k = \infty$, $\sum_{k=0}^{\infty} (\alpha^k)^2 < \infty$, and the generated sequence $\{x^k\}$ is bounded, then every limit point of $\{x^k\}$ is a stationary point of the least squares problem.

2.4.11

Verify a key part of the proof of Prop. 2.4.5, which is to show that the eigenvalues of the matrices H_{km} lie within an interval $[c_1 k, c_2 k]$ for some positive constants c_1 and c_2. *Hint*: Let X be a compact set containing all vectors ψ_i generated by the algorithm, and let B and b be an upper bound and a lower bound, respectively, on the eigenvalues of $\nabla g_i(x) \nabla g_i(x)'$ as x ranges over X. Show that all eigenvalues of H_{km} are less or equal to kmB. If v_k is the smallest eigenvalue of H_{km}, show that

$$v_{k+1} \geq (1 - c/k) v_k + m(1 - c/k) b,$$

and use this relation to show that $v_k \geq k\gamma$ for a sufficiently small but positive value of γ.

2.5 DISTRIBUTED ASYNCHRONOUS ALGORITHMS

We will now consider distributed asynchronous counterparts of some of the algorithms discussed earlier. Here an iterative algorithm, such as a gradient method or a coordinate descent method, is parallelized by separating it into several local algorithms operating concurrently at different processors. However, the local algorithms do not have to wait at predetermined points to synchronize the information that they use. Thus, some processors execute more iterations than others, some processors communicate more frequently than others, and the interprocessor communication is subject to substantial and unpredictable delays. In particular, at iteration k, a given processor, instead of using the vector x^k, uses a different vector, the components of which have been computed at some earlier iterations, possibly at some other processors. This models adequately a very broad variety of parallel and distributed computing systems.

Examples of Asynchronous Optimization Algorithms

In this section, we focus on the unconstrained minimization of a differentiable function $f : \Re^n \mapsto \Re$. The ideas that we discuss generally also apply to a constrained context, based on some of the algorithms of the next chapter. Out of the unconstrained algorithms developed so far, there are four types that are suitable for asynchronous distributed computation. Their asynchronous versions are as follows:

(a) *Gradient methods*, where we assume that each coordinate x_i is updated at a subset of times $\mathcal{R}_i \subset \{0, 1, \ldots\}$, according to

$$x_i^{k+1} = \begin{cases} x_i^k & \text{if } k \notin \mathcal{R}_i, \\ x_i^k - \alpha^k \dfrac{\partial f\left(x_1^{\tau_{i1}(k)}, \ldots, x_n^{\tau_{in}(k)}\right)}{\partial x_i} & \text{if } k \in \mathcal{R}_i, \end{cases} \quad i = 1, \ldots, n,$$
(2.149)

where α^k is a positive stepsize. Here $\tau_{ij}(k)$ is the time at which the jth coordinate used in this update was computed, and the difference $k - \tau_{ij}(k)$ is commonly called the *communication delay* from j to i at time k. Note that each coordinate x_i (or block of coordinates) may be updated by a separate processor, on the basis of values of coordinates made available by other processors, with some delay. Of course, several coordinates may be updated at the same time k, i.e., that k is contained in the sets $\mathcal{R}_i$ of multiple processors i.

(b) *Coordinate descent methods*, where for simplicity we consider a block size of one; cf. Eq. (2.43). We assume that the ith scalar coordinate is updated at a subset of times $\mathcal{R}_i \subset \{0, 1, \ldots\}$, according to

$$x_i^{k+1} \in \arg\min_{\xi \in \Re} f\left(x_1^{\tau_{i1}(k)}, \ldots, x_{i-1}^{\tau_{i,i-1}(k)}, \xi, x_{i+1}^{\tau_{i,i+1}(k)}, \ldots, x_n^{\tau_{in}(k)}\right),$$
(2.150)

and is left unchanged ($x_i^{k+1} = x_i^\ell$) if $k \notin \mathcal{R}_i$. There are also similar asynchronous versions of variants of the above coordinate descent method (for example involving a stepsize), which we have discussed in Section 2.3.1. The meanings of the subsets of updating times $\mathcal{R}_i$ and indexes $\tau_{ij}(k)$ are the same as in the case of gradient methods. Also the distributed environment where the method can be applied is similar to the case of the gradient method. Another practical setting that may be modeled well by this iteration is when all computation takes place at a single computer, but any number of coordinates may be simultaneously updated at a time, with the order of coordinate selection being irregular and possibly random.

(c) *Incremental gradient methods* for the case where $f(x) = \sum_{i=1}^m f_i(x)$. Here the ith component is used to update x at a subset of times $\mathcal{R}_i$:

$$x^{k+1} = x^k - \alpha^k \nabla f_i\left(x_1^{\tau_{i1}(k)}, \ldots, x_n^{\tau_{in}(k)}\right), \quad k \in \mathcal{R}_i, \quad (2.151)$$

where we assume that a single component gradient ∇f_i is used at each time (i.e., $\mathcal{R}_i \cap \mathcal{R}_j = \emptyset$ for $i \neq j$). The meaning of $\tau_{ij}(k)$ is the same as in the preceding cases. Here the entire vector x is updated at a central computer, based on component gradients ∇f_i that are computed at other computers and are communicated with some delay to the central computer. The central computer also communicates the computed values x^k to the other computers, possibly with some delay. For validity of these methods, it is essential that all the components f_i are used in the iteration with the same asymptotic frequency, $1/m$. Moreover, for this type of asynchronous implementation to make sense, the computation of ∇f_i must be substantially more time-consuming than the update of x^k using the incremental iteration (2.151). The method is a generalization of the incremental gradient method of Section 2.4.1. Let us also note that in the case in the case of nondifferentiable f_i, the gradient ∇f_i in Eq. (2.151) can be replaced by a subgradient (see [NBB01], and the corresponding discussion in Chapter 7).

(d) *Incremental aggregated gradient methods* for the case where $f(x) = \sum_{i=1}^{m} f_i(x)$. Here all of the m components of f are used to update x, at every time k:

$$x^{k+1} = x^k - \alpha^k \sum_{i=1}^{m} \nabla f_i \left(x_1^{\tau_{i1}(k)}, \ldots, x_n^{\tau_{in}(k)} \right), \qquad (2.152)$$

but the gradients ∇f_i of the different components are calculated with different delays $k - \tau_{ij}(k)$. As in the preceding case, the entire vector x is updated at a central computer, based on component gradients ∇f_i that are computed at other computers and are communicated with some delay to the central computer. This method generalizes the incremental aggregated gradient method of Section 2.4.2, which is the special case where $\tau_{i1}(k) = \cdots = \tau_{in}(k) = \ell_i$ for each i and k, where ℓ_i is the index notation of Section 2.4.2 [cf. Eq. (2.100)]. Moreover, in the case of nondifferentiable f_i, the gradients ∇f_i in Eq. (2.152) can be replaced by subgradients (see the original proposal of the paper [NBB01], and the discussion in Section 7.5.2).

For our discussion in this section we assume that the sets $\mathcal{R}_i$ and the delay sequences $\{k - \tau_{ij}(k)\}$ are deterministically given. It follows therefore that for any initial condition x^0, the generated sequence $\{x^k\}$ by an asynchronous algorithm is also deterministic. The objective of the analysis is to show that given an initial condition x^0 (perhaps satisfying some assumptions), the generated sequence $\{x^k\}$ has limit points that are optimal solutions of the problem at hand.

In natural stochastic versions of asynchronous algorithms, the sets $\mathcal{R}_i$ and the delay sequences $\{\tau_{ij}(k)\}$ may be random. In this case, the convergence results are somewhat different in character (since $\{x^k\}$ will be a random sequence, its convergence should be considered within a stochastic framework, e.g., convergence with probability one). However, the deterministic results can be used as a conceptual starting point for a stochastic analysis.

2.5.1 Totally and Partially Asynchronous Algorithms

An interesting fact is that some asynchronous algorithms, called *totally asynchronous*, can tolerate arbitrarily large delays $k - \tau_{ij}(k)$, while other algorithms, called *partially asynchronous*, are not guaranteed to work unless there is an upper bound on these delays. As we will explain shortly, *the convergence mechanisms at work in each of these two cases are fundamentally different*. In what follows in this section, we will provide a broad overview of the convergence analysis of these two types of algorithms. For a detailed discussion we refer to the book [BeT89], where totally and partially asynchronous algorithms, and various special cases including gradient and coordinate descent methods, are treated in Chapter 6, Section 6.3, and Chapter 7, Section 7.5, respectively, while asynchronous stochastic gradient methods are discussed in Section 7.8.

Essential assumptions, shared by all asynchronous algorithms, are:

(a) No processor stops computing (i.e., the set of computing times $\mathcal{R}_i$ is an infinite set for all i).

(b) We have
$$\lim_{k \to \infty} \tau_{ij}(k) = \infty, \qquad \forall \ i,j = 1, \ldots, n. \tag{2.153}$$

In words, no processor stops using more recent information than the information it used in the past (i.e., outdated information is eventually purged from the system, in the sense that an iterate produced at some time k will not be used in the future after a certain time).

In totally asynchronous algorithms, we make no assumption other than the preceding two. In particular, the analysis allows for the delays $k - \tau_{ij}(k)$ to become progressively larger and unbounded, i.e., we may have
$$\sup_{k \geq 0} \left(k - \tau_{ij}(k) \right) = \infty$$

for some i and j. In partially asynchronous algorithms, in addition to Eq. (2.153), we require that
$$\sup_{k \geq 0} \left(k - \tau_{ij}(k) \right) < \infty, \tag{2.154}$$

for all i and j.

There are variations and refinements of the partially asynchronous assumption (2.154), which further differentiate the convergence behavior, depending on various algorithmic parameters, such as stepsize. An important example is the distributed gradient method (2.149), implemented with a constant stepsize ($\alpha_k \equiv \alpha$). Its convergence generally depends on the relation between α and the size of delays $k - \tau_{ij}(k)$. In particular, even when it is not convergent under the totally asynchronous condition (2.152), the method is typically convergent assuming that

$$\sup_{k \geq 0} \left(k - \tau_{ij}(k) \right) < b(\alpha), \qquad \forall\, i, j,$$

where $b(\alpha)$ is an α-dependent bound. Moreover, under reasonable conditions, it may be shown that

$$\lim_{\alpha \downarrow 0} b(\alpha) = \infty; \qquad (2.155)$$

see [BeT89], Section 7.5. In words, any finite size of delays may be tolerated provided the stepsize is sufficiently small.

The idea here is that the asynchronous gradient method (2.149) with constant stepsize α may be viewed as a *synchronous gradient method with errors* in the calculation of the gradient; cf. the analysis of incremental aggregated gradient methods in Section 2.4.2. The errors are proportional to α, as well as the size of delays $k - \tau_{ij}(k)$. Therefore, as α becomes smaller we can tolerate larger delays, which is consistent with Eq. (2.155). A detailed example illustrating this phenomenon is given in Section 7.1 of [BeT89] (Example 1.3).

There are also interesting situations where the distributed gradient method (2.149) with constant stepsize α is convergent with

$$b(\alpha) = \infty, \qquad \forall\, \alpha \in (0, \bar{\alpha}],$$

where $\bar{\alpha}$ is some scalar. In words, there is a range of values of α for which any size of finite delay is tolerable, but still for convergence, the partially asynchronous condition (2.154) must hold. Exercise 2.5.1 provides examples illustrating the asynchronous behavior of the method with a constant stepsize. The next two subsections describe more formally the corresponding lines of analysis.

2.5.2 Totally Asynchronous Convergence

Totally asynchronous algorithms are valid only under special conditions, which guarantee that any progress in the computation of the individual processors, is consistent with progress in the collective computation. We will focus on a convergence theorem that applies to asynchronous iterations for solution of stationary fixed point equations of the form $x = F(x)$, where F is a given function; for example a gradient method with a constant stepsize [cf. Eq. (2.149)] and coordinate descent [cf. Eq. (2.150)].

We represent x as $x = (x_1, \ldots, x_m)$, where $x_i \in R^{n_i}$ with n_i being some positive integer. Thus $x \in \Re^n$, where $n = n_1 + \cdots + n_m$, and F maps $\Re^n$ to $\Re^n$. We denote by $F_i : \Re^n \mapsto \Re^{n_i}$ the ith component of F, so $F(x) = \big(F_1(x), \ldots, F_m(x)\big)$. Our computation framework involves m interconnected processors, the ith of which updates the ith component x^i by applying the corresponding mapping F_i. Thus, in a (synchronous) distributed fixed point algorithm, processor i iterates at time k according to
$$x_i^{k+1} = F_i(x_1^k, \ldots, x_m^k), \qquad \forall\ i = 1, \ldots, m.$$

In an asynchronous version of the algorithm, processor i updates x_i only for k in a selected subset $\mathcal{R}_i$ of iterations, and with components x_j, $j \neq i$, supplied by other processors with delays $k - \tau_{ij}(k)$,
$$x_i^{k+1} = \begin{cases} F_i\left(x_1^{\tau_{i1}(k)}, \ldots, x_m^{\tau_{im}(k)}\right) & \text{if } k \in \mathcal{R}_i, \\ x_i^k & \text{if } k \notin \mathcal{R}_i. \end{cases} \qquad (2.156)$$

Here $\tau_{ij}(k)$ is the time at which the jth coordinate used in the kth update was computed, and the difference $k - \tau_{ij}(k)$ is the communication delay from j to i at time k; cf. the gradient method (2.149) with a constant stepsize $\alpha^k \equiv \alpha$, and the coordinate descent method (2.150).

To discuss the convergence of the asynchronous fixed point algorithm (2.156), we introduce the following assumption.

Assumption 2.5.1: (Continuous Updating and Information Renewal)

(1) The set of times $\mathcal{R}_i$ at which processor i updates x_i is infinite, for each $i = 1, \ldots, m$.

(2) $\lim_{k \to \infty} \tau_{ij}(k) = \infty$ for all $i, j = 1, \ldots, m$.

Assumption 2.5.1 is natural, and is essential for any kind of convergence result about the algorithm. In particular, the condition $\tau_{ij}(k) \to \infty$ guarantees that outdated information about the processor updates will eventually be purged from the computation. It is also natural to assume that $\tau_{ij}(k)$ is monotonically increasing with k, but this assumption is not necessary for the subsequent analysis.

We have the following convergence theorem for totally asynchronous iterations from [Ber83]. The theorem has served as the basis for the treatment of totally asynchronous iterations in the book [BeT89] (Chapter 6), including gradient-based and coordinate descent methods. However, the applications of the theorem extend to other optimization domains, including dynamic programming, where the asynchronous iteration (2.156) corresponds to the classical value iteration algorithm; see [Ber82d], [Ber12], [Ber13].

Proposition 2.5.1: (Asynchronous Convergence Theorem) Let F have a unique fixed point x^*, let Assumption 2.5.1 hold, and assume that there is a sequence of nonempty subsets $\{S(k)\} \subset \Re^n$ with

$$S(k+1) \subset S(k), \quad k = 0, 1, \ldots,$$

such that if $\{y^k\}$ is any sequence with $y^k \in S(k)$, for all $k \geq 0$, then $\{y^k\}$ converges to x^*. Assume further the following:

(1) *Synchronous Convergence Condition:* We have

$$F(x) \in S(k+1), \quad \forall\, x \in S(k),\ k = 0, 1, \ldots.$$

(2) *Box Condition:* For all k, $S(k)$ is a Cartesian product of the form

$$S(k) = S_1(k) \times \cdots \times S_m(k),$$

where $S_i(k)$ is a set of real-valued functions on X_i, $i = 1, \ldots, m$.

Then for every $x^0 \in S(0)$, the sequence $\{x^k\}$ generated by the asynchronous algorithm (2.156) converges to x^*.

Proof: To explain the idea of the proof, let us note that our assumptions imply that updating any component x_i, by applying F to $x \in S(k)$, while leaving all other components unchanged, yields a function in $S(k)$. Thus, once enough time passes so that the delays become "irrelevant," then after x enters $S(k)$, it stays within $S(k)$. Moreover, once a component x_i enters the subset $S_i(k)$ and the delays become "irrelevant," x_i gets permanently within the smaller subset $S_i(k+1)$ at the first time that x_i is iterated on with $x \in S(k)$. Once every component x_i, $i = 1, \ldots, m$, gets within $S_i(k+1)$, the entire vector x is within $S(k+1)$ by the box condition. Thus the iterates from $S(k)$ eventually get into $S(k+1)$ and so on, and converge pointwise to x^* in view of the assumed properties of $\{S(k)\}$.

With this idea in mind, we show by induction that for each $k \geq 0$, there is a time t_k such that:

(1) $x^t \in S(k)$ for all $t \geq t_k$.

(2) For all i and $t \in \mathcal{R}_i$ with $t \geq t_k$, we have

$$\left(x_1^{\tau_{i1}(t)}, \ldots, x_m^{\tau_{im}(t)}\right) \in S(k).$$

[In words, after some time, all fixed point estimates will be in $S(k)$ and all estimates used in iteration (2.156) will come from $S(k)$.]

The induction hypothesis is true for $k = 0$ since $x^0 \in S(0)$. Assuming it is true for a given k, we will show that there exists a time t_{k+1} with the required properties. For each $i = 1, \ldots, m$, let $t(i)$ be the first element of $\mathcal{R}_i$ such that $t(i) \geq t_k$. Then by the synchronous convergence condition, we have $F(x^{t(i)}) \in S(k+1)$, implying (in view of the box condition) that

$$x_i^{t(i)+1} \in S_i(k+1).$$

Similarly, for every $t \in \mathcal{R}_i$, $t \geq t(i)$, we have $x_i^{t+1} \in S_i(k+1)$. Between elements of $\mathcal{R}_i$, x_i^t does not change. Thus,

$$x_i^t \in S_i(k+1), \qquad \forall\, t \geq t(i) + 1.$$

Let $t'_k = \max_i \{t(i)\} + 1$. Then, using the box condition we have

$$x^t \in S(k+1), \qquad \forall\, t \geq t'_k.$$

Finally, since by Assumption 2.5.1, we have $\tau_{ij}(t) \to \infty$ as $t \to \infty$, $t \in \mathcal{R}_i$, we can choose a time $t_{k+1} \geq t'_k$ that is sufficiently large so that $\tau_{ij}(t) \geq t'_k$ for all i, j, and $t \in \mathcal{R}_i$ with $t \geq t_{k+1}$. We then have, for all $t \in \mathcal{R}_i$ with $t \geq t_{k+1}$ and $j = 1, \ldots, m$, $x_j^{\tau_{ji}(t)} \in S_j(k+1)$, which (by the box condition) implies that

$$\left(x_1^{\tau_{i1}(t)}, \ldots, x_m^{\tau_{im}(t)}\right) \in S(k+1).$$

The induction is complete. **Q.E.D.**

The main issue in applying the theorem is to identify the set sequence $\{S(k)\}$ and to verify the assumptions of Prop. 2.5.1. These assumptions hold in two primary contexts of interest. The first is based on monotonicity conditions, and is particularly useful in dynamic programming algorithms, for which we refer to the papers [Ber82d], [BeY10], and the books [Ber12], [Ber13], [Ber15a]; see also Exercises 2.5.3 and 2.5.4. The second context is when $S(k)$ are spheres centered at x^* with respect to the weighted sup-norm

$$\|x\|_\xi = \max_{i=1,\ldots,n} \frac{|x_i|}{\xi_i},$$

where $\xi = (\xi_1, \ldots, \xi_n)$ is a vector of positive weights, as shown in the following proposition.

Proposition 2.5.2: Let F be a contraction with respect to a weighted sup-norm $\|\cdot\|_\xi$ and modulus $\rho < 1$, and let Assumption 2.5.1 hold. Then a sequence $\{x^k\}$ generated by the asynchronous algorithm (2.156) converges to x^*.

Proof: We apply Prop. 2.5.1 with
$$S(k) = \{x \in \Re^n \mid \|x^k - x^*\|_\xi \le \rho^k \|x^0 - x^*\|_\xi\}, \qquad k = 0, 1, \ldots.$$
Since F is a contraction with modulus ρ, the synchronous convergence condition is satisfied. Since F is a weighted sup-norm contraction, the box condition is also satisfied, and the result follows. **Q.E.D.**

The contraction property of the preceding proposition can be verified in a few interesting special cases. In particular, let F be linear of the form
$$F(x) = Ax + b,$$
where A and b are given $n \times n$ matrix and vector in $\Re^n$. Then F is a contraction with respect to the sup-norm $\|x\|_\xi = \max_{i=1,\ldots,n} \frac{|x_i|}{\xi_i}$, with modulus $\rho \in (0, 1)$ if and only if
$$\frac{\sum_{j=1}^n |a_{ij}|\xi_j}{\xi_i} \le \rho, \qquad \forall \, i = 1, \ldots, n, \qquad (2.157)$$
[cf. Prop. A.27(a) of Appendix A].

Example 2.5.1 (Totally Asynchronous Convergence of Gradient Method)

Consider the positive definite quadratic cost function
$$f(x) = \tfrac{1}{2} x' Q x + b' x.$$
and the gradient method
$$x^{k+1} = x^k - \alpha \nabla f(x^k) = (I - \alpha Q) x^k - \alpha b.$$
Then the mapping $F(x) = (I - \alpha Q)x - \alpha b$ is a contraction with respect to the weighted sup-norm $\|\cdot\|_\xi$ if and only if α and ξ satisfy
$$\xi_i |1 - \alpha q_{ii}| + \alpha \sum_{j \ne i}^n \xi_j |q_{ij}| < \xi_i, \qquad \forall \, i = 1, \ldots, n, \qquad (2.158)$$
[cf. Eq. (2.157)]. In particular, if Q is weighted-diagonally dominant, in the sense that
$$\xi_i q_{ii} > \sum_{j \ne i}^n \xi_j |q_{ij}|, \qquad i = 1, \ldots, n,$$
then F is a contraction with respect to $\|\cdot\|_\xi$ when α satisfies Eq. (2.158), which can in turn be shown to be equivalent to
$$0 < \alpha < \min_{i=1,\ldots,n} \frac{2\xi_i}{\xi_i q_{ii} + \sum_{j \ne i}^n \xi_j |q_{ij}|}.$$
In this case the gradient method $x^{k+1} = x^k - \alpha \nabla f(x^k)$ converges to the unique solution $x^* = -Q^{-1} b$ under totally asynchronous conditions.

2.5.3 Partially Asynchronous Gradient-Like Algorithms

Partially asynchronous gradient-like algorithms are characterized by the condition

$$\sup_{k \geq 0} \bigl(k - \tau_{ij}(k)\bigr) < \infty,$$

[cf. Eq. (2.154) and the associated discussion in Section 2.5.1]. They typically require either a diminishing stepsize, or a constant stepsize that is small and is inversely proportional to the size of the delays. The idea is that when the delays are bounded and the stepsize is small enough, the asynchronous algorithm resembles its synchronous counterpart sufficiently closely, so that the convergence properties of the latter are maintained. This mechanism for convergence is similar to the one for incremental methods, discussed in Section 2.4. For this reason, incremental gradient and coordinate descent methods are natural candidates for partially asynchronous implementation; see [BeT89], Section 7.5, for gradient and coordinate descent methods, and [NBB01] for the incremental methods (2.151) and (2.152).

For an extensive discussion of the implementation and convergence of partially asynchronous algorithms, we refer to the paper [TBA86], to the books [BeT89] (Chapter 7) and [Bor08] (for the case of deterministic and stochastic gradient and coordinate descent methods), and to the paper [NBB01] (for the case of incremental gradient and subgradient methods). In what follows in this section we will focus on gradient and coordinate descent type methods for minimization, and we will describe a convergence result from [BeT89], Section 7.5.

Consider the unconstrained minimization of a (not necessarily convex) continuously differentiable function $f : \Re^n \mapsto \Re$ that is bounded below [$\inf_{x \in \Re^n} f(x) > -\infty$] and satisfies the gradient Lipschitz condition

$$\bigl\|\nabla f(x) - \nabla f(y)\bigr\| \leq L\|x - y\|, \qquad \forall\, x, y \in \Re^n.$$

We consider the partially asynchronous version of the iteration

$$x^{k+1} = x^k + \alpha d^k, \tag{2.159}$$

where $\alpha > 0$ is a constant stepsize and $d^k = (d_1^k, \ldots, d_n^k)$ is a descent direction (cf. Section 1.2), for which we additionally assume that

$$d_i^k \frac{\partial f(x^k)}{\partial x^i} \leq 0, \qquad \forall\, k \geq 0,\ i = 1, \ldots. \tag{2.160}$$

This condition specifies that d^k and $-\nabla f(x^k)$ are in the same quadrant of $\Re^n$. It is satisfied by several methods of interest, including the diagonally scaled steepest descent method [$d^k = -D^k \nabla f(x^k)$ where D^k is a diagonal positive definite matrix] and the coordinate descent method. Moreover, to satisfy the gradient related condition of Section 1.2, we assume that for some positive scalars $\underline{\beta}$ and $\bar{\beta}$, and all i and k, we have

$$\underline{\beta}\left|\frac{\partial f(x^k)}{\partial x^i}\right| \leq |d_i^k| \leq \bar{\beta}\left|\frac{\partial f(x^k)}{\partial x^i}\right|.$$

We consider the asynchronous algorithm

$$x_i^{k+1} = \begin{cases} x_i^k & \text{if } k \notin \mathcal{R}_i, \\ x_i^k + \alpha d_i^{\tau_i(k)} & \text{if } k \in \mathcal{R}_i, \end{cases} \quad i = 1, \ldots, n,$$

where $\mathcal{R}_i \subset \{0, 1, \ldots\}$ is the subset of times where the coordinate x_i is updated, and $k - \tau_i(k)$ is the communication delay. The result shown in [BeT89], Section 7.5, is that under the preceding assumptions, and the partially asynchronous condition

$$\sup_{k \geq 0}\big(k - \tau_i(k)\big) < b < \infty,$$

for some $b > 0$, there exists a range of stepsizes

$$0 < \alpha \leq \bar{\alpha}(b)$$

such that $\lim_{k \to \infty} \nabla f(x^k) = 0$. The threshold $\bar{\alpha}(b)$ depends on b, and the analysis of [BeT89], Section 7.5, shows that $\bar{\alpha}(b)$ is roughly inversely proportional to b. In particular, as b increases to ∞, $\bar{\alpha}(b)$ decreases to 0. This analysis also discusses the dependence of the threshold $\bar{\alpha}(b)$ on the scalars $\underline{\beta}$, $\bar{\beta}$, L, and the dimension n. Note that convergence is attained for any value of the bound b on the communication delays, provided the stepsize α is sufficiently small.

2.5.4 Convergence Rate of Asynchronous Algorithms

We will now summarize the convergence rate characteristics of asynchronous optimization algorithms. We draw on the analysis of the book [BeT89], the survey [BeT91], and the paper [NeB00], which collectively contain a far more extensive discussion. The recent doctoral thesis [Fey16] contains additional results.

Generally, for the partially asynchronous algorithms discussed in this book, we cannot expect a faster convergence rate than the one of their synchronous counterparts (unless we take into account speedup due to concurrency of computations when a distributed computing system is used). The question is really how much the convergence rate is degraded due to the presence and the size of the communication delays; clearly when larger delays are allowed, the convergence rate deteriorates. However, it is generally true that for the optimization algorithms of interest to us, if the synchronous version of the algorithm converges linearly or sublinearly, the same is true for the asynchronous version. The various constants in the convergence rate estimates are affected, however, by the size of the communication delays. In particular, let us consider the four types of algorithms that we mentioned earlier in this section:

(a) *Gradient methods*, where each coordinate x_i is updated at a subset of times $\mathcal{R}_i \subset \{0, 1, \ldots\}$, according to

$$x_i^{k+1} = \begin{cases} x_i^k & \text{if } k \notin \mathcal{R}_i, \\ x_i^k - \alpha^k \dfrac{\partial f\left(x_1^{\tau_{i1}(k)}, \ldots, x_n^{\tau_{in}(k)}\right)}{\partial x_i} & \text{if } k \in \mathcal{R}_i, \end{cases} \quad i = 1, \ldots, n,$$
(2.161)

where α^k is a positive stepsize [cf. Eq. (2.149)]. We discussed this and related methods in the preceding section [cf. Eqs. (2.159) and (2.160)]. In particular, under the partially asynchronous condition

$$\sup_{k \geq 0} \left(k - \tau_i(k)\right) < b < \infty,$$

for some $b > 0$, there exists a range of stepsizes $0 < \alpha \leq \bar{\alpha}(b)$ such that $\lim_{k \to \infty} \nabla f(x^k) = 0$. If in addition, f is strongly convex, a result similar to the one of Prop. 2.4.2 can be proved, namely that there exists $\bar{\alpha} > 0$ such that for all $\alpha \in (0, \bar{\alpha}]$, the sequence $\{x^k\}$ generated by the asynchronous gradient iteration (2.161), with constant stepsize α within a sufficiently small range, converges to x^* linearly, in the sense that $\|x^k - x^*\| \leq \gamma \rho^k$ for some scalars $\gamma > 0$ and $\rho \in (0, 1)$, and all k. This result is also proved in [BeT89] for various special cases where f is convex quadratic (see Exercise 1.2 of Section 7.1, and Section 7.2.2 of [BeT89]).

(b) *Coordinate descent methods*, where the ith scalar coordinate is updated at a subset of times $\mathcal{R}_i \subset \{0, 1, \ldots\}$, according to

$$x_i^{k+1} \in \arg\min_{\xi \in \Re} f\left(x_1^{\tau_{i1}(k)}, \ldots, x_{i-1}^{\tau_{i,i-1}(k)}, \xi, x_{i+1}^{\tau_{i,i+1}(k)}, \ldots, x_n^{\tau_{in}(k)}\right),$$

and is left unchanged ($x_i^{k+1} = x_i^\ell$) if $k \notin \mathcal{R}_i$ [cf. Eq. (2.150)], as well as variants involving a stepsize. This case is also covered by the analysis of Section 7.5 of [BeT89], which was described in the preceding section. When f is a sup-norm contraction, an additional result may be proved, which shows that the convergence rate is linear, and under certain conditions, it is superior to the convergence rate of the corresponding synchronous iteration (even without the benefit of speedup due to parallelization). This result was shown for the case of linear fixed point iterations in Section 6.3.5 of [BeT89], and its proof is outlined in Exercise 2.5.5.

(c) *Incremental gradient methods* for $f(x) = \sum_{i=1}^{m} f_i(x)$, where the ith component is used to update x at a subset of times $\mathcal{R}_i$:

$$x^{k+1} = x^k - \alpha^k \nabla f_i\left(x_1^{\tau_{i1}(k)}, \ldots, x_n^{\tau_{in}(k)}\right), \qquad k \in \mathcal{R}_i,$$

where we assume that a single component gradient ∇f_i is used at each time (i.e., $\mathcal{R}_i \cap \mathcal{R}_j = \emptyset$ for $i \neq j$) [cf. Eq. (2.151)]. In this case, when a diminishing stepsize α^k is used, the synchronous version of the method (no delays) has sublinear convergence rate, and the same is true for the asynchronous version. The paper [NeB00] discusses extensively results of this type, which quantify further the convergence rate under various assumptions. The same paper discusses the case of a strongly convex function f, and the use of a constant but sufficiently small stepsize α. In particular, it is proved that the method converges linearly to a neighborhood of the unique solution. The size of the neighborhood depends, among others, on the size of the delays (exact convergence in the case of a constant stepsize is not generally expected for this type of incremental method, as we have noted earlier).

(d) *Incremental aggregated gradient methods* for $f(x) = \sum_{i=1}^{m} f_i(x)$, where all of the m components of f are used to update x, at every time k,

$$x^{k+1} = x^k - \alpha^k \sum_{i=1}^{m} \nabla f_i\left(x_1^{\tau_{i1}(k)}, \ldots, x_n^{\tau_{in}(k)}\right), \qquad (2.162)$$

but the gradients ∇f_i of the different components are calculated with different delays $k - \tau_{ij}(k)$ [cf. Eq. (2.152)]. We have discussed the convergence rate of this method in Section 2.4.2 (cf. Prop. 2.4.2). In particular, the convergence rate of the distributed asynchronous algorithm (2.162) is linear or sublinear under the same conditions as for the synchronous version.

EXERCISES

2.5.1 (Examples of Asynchronous Convergence Behavior)

Consider the two-dimensional quadratic optimization problem

$$\text{minimize} \quad f(x) = \tfrac{1}{2} x' Q x$$
$$\text{subject to} \quad x \in \Re^2,$$

where
$$Q = \begin{pmatrix} 1 & -b \\ -b & c \end{pmatrix}.$$

(a) Assume that $b^2 < c$, so that Q is positive definite. Show that the totally asynchronous version of the gradient method converges to $x^* = 0$ for all stepsizes α such that
$$0 < \alpha < \min\left\{\frac{2\xi_1}{\xi_1 + \xi_2|b|}, \frac{2\xi_2}{\xi_1|b| + \xi_2 c}\right\},$$
where ξ_1 and ξ_2 are any positive numbers such that
$$|b| < \frac{\xi_1}{\xi_2} < \frac{c}{|b|}.$$

Hint: Use the analysis of Example 2.5.1. *Note*: Assuming a two-dimensional problem is critical for this part. For an example of a three-dimensional positive definite quadratic problem, where the totally asynchronous version of the gradient method does not converge to x^*, see the following exercise.

(b) Assume that $c = b = 1$, so that Q is positive semidefinite/singular and there are multiple minima, the set of x^* where $x_1^* = x_2^*$. Let the stepsize in the asynchronous gradient method be $\alpha = \frac{1}{2}$, and consider a scenario where the two processors 1 and 2 update the values of x_1^k and x_2^k, respectively, and exchange the computed values of x_1^k and x_2^k every $T_\ell > 0$ iterations, where $\{T_\ell\}$ is an increasing sequence with $T_\ell \to \infty$. Show that $\{x^k\}$ need not converge to the optimal solution set. *Hint*: Defining $y^0 = x^0$ and $y^\ell = x^{T_0 + \cdots + T_{\ell-1}}$ for $\ell \geq 1$, we have
$$y_1^{\ell+1} = (1/2)^{T_\ell} y_1^\ell + \left(1 - (1/2)^{T_\ell}\right) y_2^\ell, \qquad \forall\, \ell = 0, 1, \ldots,$$
$$y_2^{\ell+1} = \left(1 - (1/2)^{T_\ell}\right) y_1^\ell + (1/2)^{T_\ell} y_2^\ell, \qquad \forall\, \ell = 0, 1, \ldots,$$
and it can be verified that if the sequence $\{T_\ell\}$ is increasing sufficiently fast, the sequence $\{y^\ell\}$ need not converge to the optimal solution set (this counterexample to totally asynchronous convergence is developed in much greater detail in [BeT89], Section 7.1, Example 1.2).

2.5.2 (Counterexample to Totally Asynchronous Convergence of Gradient Method; [BeT89], Section 6.3.2, Example 3.1)

Consider the positive definite quadratic cost function
$$f(x) = \tfrac{1}{2} x' Q x + b' x,$$
and the asynchronous gradient method of Example 2.5.1. Consider the case where
$$Q = \begin{pmatrix} 1+\epsilon & 1 & 1 \\ 1 & 1+\epsilon & 1 \\ 1 & 1 & 1+\epsilon \end{pmatrix},$$
where $0 < \epsilon < 1$, so the diagonal dominance condition of Example 2.5.1 is violated. For any constant stepsize $\alpha > 0$, provide a sequence of iterates $\{x^k\}$ generated by the algorithm for which $\|x^k\| \to \infty$.

2.5.3 (Totally Asynchronous Convergence of Monotone Fixed Point Iterations [Ber83])

Let U be a finite set and consider mappings $H_i : \Re^n \times U \mapsto \Re$ that are monotone in the sense

$$H_i(x, u) \le H_i(\tilde{x}, u), \qquad \text{for all } i = 1, \ldots, n, \ u \in U, \text{ and } x, \tilde{x} \in \Re^n \text{ with } x \le \tilde{x}.$$

Let $F_i : \Re^n \mapsto \Re$, $i = 1, \ldots, n$, be the mappings defined by

$$F_i(x) = \min_{u \in U} H_i(x, u), \qquad x \in \Re^n.$$

Assume that the mapping $F = (F_1, \ldots, F_n)$ has a unique fixed point within $\Re^n$, denoted x^*, and that there exist vectors $\underline{x}$ and $\bar{x}$ such that

$$\underline{x} \le F(\underline{x}) \le F(\bar{x}) \le \bar{x}. \qquad (2.163)$$

Let $\{x^k\}$ be a sequence generated by the totally asynchronous fixed point iteration (2.156). Use Prop. 2.5.1 to show that $\{x^k\}$ converges to x^*.

2.5.4 (Totally Asynchronous Convergence of Nonexpansive Monotone Fixed Point Iterations)

Let U, H_i, and F_i, $i = 1, \ldots, n$, be as in Exercise 2.5.3. Instead of Eq. (2.163), assume that each mapping $H_i(\cdot, u)$, $u \in U$, is nonexpansive with respect to the weighted sup-norm $\|\cdot\|_\xi$, i.e., for all $x, \bar{x} \in \Re^n$, H_i satisfies

$$\frac{\left| H_i(x, u) - H_i(\bar{x}, u) \right|}{\xi_i} \le \|x - \bar{x}\|_\xi, \qquad \forall\, i = 1, \ldots, n,\ u \in U,$$

and the mapping $F = (F_1, \ldots, F_n)$ has a unique fixed point within $\Re^n$, denoted x^*. Let $\{x^k\}$ be a sequence generated by the totally asynchronous fixed point iteration (2.156). Use Prop. 2.5.1 to show that $\{x^k\}$ converges to x^*. (Unpublished joint research of the author with Huizhen Yu.)

2.5.5 (Convergence Rate of Asynchronous Sup-Norm Contractive Fixed Point Iterations [BeT89], Section 6.3.5)

Consider the asynchronous fixed point algorithm

$$x_i^{k+1} = f_i\left(x_1^{\tau_{i1}(k)}, \ldots, x_n^{\tau_{ni}(k)} \right), \qquad i = 1, \ldots, n,$$

assuming that the function $f : \Re^n \mapsto \Re^n$ is a sup-norm contraction with the unique fixed point denoted by x^*. This exercise quantifies the effect of different variables being incorporated in the computations with different communication delays. In particular, we assume that there is a positive integer B such that

$$k - \tau_{ij}(k) \le B, \qquad \forall\, k,\ i, j = 1, \ldots, n.$$

Moreover for some nonnegative integer b with $0 \leq b < B$, there exists for each index i, a nonempty subset of coordinates $F(i)$ such that

$$k - \tau_{ij}(k) \leq b, \qquad \forall \ k, \ i = 1, \ldots, n, \text{ and } j \in F(i).$$

The interpretation here is that the variables x_j, $j \in F(i)$, are "special" in that they are communicated "fast" to the processor updating x_i (within $b < B$ time units). Consider the following two conditions:

(1) There exist scalars $\alpha \in [0, 1)$ and $A \in [0, 1)$ such that for all i and x,

$$\left| f_i(x) - x_i^* \right| \leq \max \left\{ \alpha \max_{j \in F(i)} |x_j - x_j^*|, \ A \max_{j \notin F(i)} |x_j - x_j^*| \right\}.$$

(2) There exist scalars $\alpha \geq 0$ and $A \geq 0$ with $\alpha + A < 1$, and such that for all i and x,

$$\left| f_i(x) - x_i^* \right| \leq \alpha \max_{j \in F(i)} |x_j - x_j^*| + A \max_{j \notin F(i)} |x_j - x_j^*|.$$

Show that the sequence $\{x^k\}$ satisfies

$$\|x^k - x^*\| \leq \bar{\rho}^k \|x^0 - x^*\|,$$

where, under condition (1), $\bar{\rho}$ is the unique nonnegative solution of the equation

$$\rho = \max \left\{ \alpha \rho^{-b}, \ A \rho^{-B} \right\},$$

while, under condition (2), $\bar{\rho}$ is the unique nonnegative solution of the equation

$$\rho = \alpha \rho^{-b} + A \rho^{-B}.$$

Verify that for a fixed B, this convergence rate is faster when $b < B$ than when $b = B$. *Note:* This result can be extended to the case where for each i, there is a partition $F_1(i) \cup \cdots \cup F_m(i)$ of the index set $\{1, \ldots, n\}$, where $m \geq 2$ is some integer, and for some nonnegative integers $b_1, \ldots, b_m$, we have

$$k - \tau_{ij}(k) \leq b_\ell, \qquad \forall \ k, \ i = 1, \ldots, n, \ \ell = 1, \ldots, m, \text{ and } j \in F_\ell(i),$$

under the assumption

$$\left| f_i(x) - x_i^* \right| \leq \max_{\ell = 1, \ldots, m} \left\{ \alpha_\ell \max_{j \in F_\ell(i)} |x_j - x_j^*| \right\}, \qquad i = 1, \ldots, n,$$

where $\alpha_\ell \in [0, 1)$, and under the assumption

$$\left| f_i(x) - x_i^* \right| \leq \sum_{\ell=1}^m \alpha_\ell \max_{j \in F_\ell(i)} |x_j - x_j^*|,$$

where $\sum_{\ell=1}^m \alpha_\ell < 1$, $\alpha_\ell \geq 0$ (see [BeT89], Section 6.3.5, Exercise 3.2).

2.6 DISCRETE-TIME OPTIMAL CONTROL PROBLEMS

In this section, we consider a class of optimization problems involving a discrete-time dynamic system, i.e., a vector difference equation. Such problems arise often in applications and are frequently challenging because of their large dimension. Continuous-time versions of these problems are the focus of the modern theory of the calculus of variations and the Pontryagin maximum principle, and give rise to some of the most fascinating mathematical problems of optimization. We will focus on some structural aspects of optimal control problems, which can be effectively exploited both for analysis and computation.

Let us consider the problem of finding sequences $u = (u_0, u_1, \ldots, u_{N-1})$ and $x = (x_1, x_2, \ldots, x_N)$, which minimize

$$g_N(x_N) + \sum_{i=0}^{N-1} g_i(x_i, u_i), \tag{2.164}$$

subject to the constraints

$$x_{i+1} = f_i(x_i, u_i), \qquad i = 0, \ldots, N-1, \tag{2.165}$$

$$x_0 : \text{given},$$

$$x_i \in X_i \subset \Re^n, \qquad i = 1, \ldots, N, \tag{2.166}$$

$$u_i \in U_i \subset \Re^m, \qquad i = 0, \ldots, N-1. \tag{2.167}$$

We refer to u_i as the *control vectors*, and to $u = (u_0, u_1, \ldots, u_{N-1})$ as a *control trajectory*. We refer to x_i as the *state vectors*, and to the sequence $x = (x_0, x_1, \ldots, x_N)$ as a *state trajectory*. The equation $x_{i+1} = f_i(x_i, u_i)$ is called the *system equation*, and for a given initial state x_0, specifies uniquely the state trajectory, which corresponds to a given control trajectory. The functions $f_i : \Re^n \times \Re^m \mapsto \Re^n$ are given and they will be assumed once or twice differentiable for the most part of the following. The sets X_i and U_i specify constraints on the state and control vectors, respectively. They are usually represented by equations or inequalities. However, for the time being we will not specify them further. A more general constraint, that we will discuss on occasion has the form

$$(x_i, u_i) \in \Omega_i \subset \Re^n \times \Re^m, \qquad i = 0, \ldots, N-1.$$

The functions $g_N : \Re^n \mapsto \Re$ and $g_i : \Re^n \times \Re^m \mapsto \Re$, $i = 0, \ldots, N-1$, are given and for the most part, they will be assumed to be once or twice differentiable.

Example 2.6.1 (Reservoir Regulation)

Let x_i denote the volume of water held in a reservoir at the ith of N time periods. The volume x_i evolves according to

$$x_{i+1} = x_i - u_i, \qquad \forall\ i = 0, \ldots, N-1,$$

where u_i is water used for some productive purpose in period i. This is the system equation, with the volume x_i viewed as the state and the outflow u_i viewed as the control. There is a cost $G(x_N)$ for the terminal state being x_N and there is a cost $g_i(u_i)$ for outflow u_i at period i. For example, when u_i is used for electric power generation, $g_i(u_i)$ may be equal to minus the value of power produced from u_i. We want to choose the outflows $u_0, u_1, \ldots, u_{N-1}$ so as to minimize

$$G(x_N) + \sum_{i=0}^{N-1} g_i(u_i),$$

while observing some constraints on the volume x_i (e.g. x_i should lie between some upper and lower bounds) and some constraints on the outflow u_i (e.g. $u_i \geq 0$).

There are also multidimensional versions of the problem, involving several reservoirs that are interconnected in the sense that the outflow u_i from one becomes inflow to another (in addition to serving some other productive purpose). This leads to an optimal control formulation where the state and control vectors are multidimensional with dimension equal to the number of reservoirs.

Let us consider the case where there are no state constraints, i.e.,

$$X_i = \Re^n, \qquad i = 1, \ldots, N.$$

Then, one may reduce problem (2.164)-(2.167) (which is a constrained problem in x and u) to a problem which involves only the control variables $u_0, \ldots, u_{N-1}$. To see this, note that to any given control trajectory

$$u = (u_0, u_1, \ldots, u_{N-1}),$$

there corresponds a unique state trajectory via the system equation $x_{i+1} = f_i(x_i, u_i)$. We may write this correspondence abstractly as

$$x_i = \phi_i(u) = \phi_i(u_0, \ldots, u_{N-1}), \qquad i = 1, \ldots, N, \qquad (2.168)$$

where ϕ_i are appropriate functions, determined by the functions f_i. (Actually in each of the functions ϕ_i, only the variables $u_0, \ldots, u_{i-1}$ enter explicitly. However, our notation is technically correct and will prove convenient.)

We may substitute x_i, using the functions ϕ_i of Eq. (2.168), into the cost function (2.164) and write the problem as

$$\text{minimize} \quad J(u) = g_N\big(\phi_N(u)\big) + \sum_{i=0}^{N-1} g_i\big(\phi_i(u), u_i\big) \tag{2.169}$$
$$\text{subject to} \quad u_i \in U_i, \quad i = 0, \ldots, N-1.$$

If the controls are also unconstrained, i.e.,

$$U_i = \Re^m, \quad i = 0, \ldots, N-1,$$

problem (2.169) is an unconstrained minimization problem of the type that we have examined in this chapter. Thus, once we calculate the gradient and Hessian matrix of J, we can write explicitly necessary conditions and sufficient conditions for optimality.

We will calculate the derivatives of J in two different ways, both of which are valuable, since they provide complementary insights.

Calculation of $\nabla J(u)$ (1st Method)

Let us first assume that the cost function has the form

$$J(u) = g_N(x_N) = g_N\big(\phi_N(u)\big)$$

i.e., there is only a terminal state cost. As we shall see shortly, the general case can be reduced to this case.

We have for every $i = 0, \ldots, N-1$, using the chain rule,

$$\nabla_{u_i} J(u) = \nabla_{u_i} \phi_N \cdot \nabla g_N, \tag{2.170}$$

where the derivatives are calculated along the current control trajectory u and the corresponding state trajectory x, and where x and u are related by

$$x_i = \phi_i(u), \quad i = 1, \ldots, N,$$

[cf. Eq. (2.168)]. From Eq. (2.170), we obtain, using again the chain rule,

$$\nabla_{u_i} J(u) = \nabla_{u_i} x_{i+1} \cdot \nabla_{x_{i+1}} x_{i+1} \cdots \nabla_{x_{N-1}} x_N \cdot \nabla g_N$$
$$= \nabla_{u_i} f_i \cdot \nabla_{x_{i+1}} f_{i+1} \cdots \nabla_{x_{N-1}} f_{N-1} \cdot \nabla g_N.$$

By defining the, so called, *costate* vectors $p_i \in \Re^n$, $i = 1, \ldots, N$, via the equations

$$p_i = \nabla_{x_i} f_i \cdots \nabla_{x_{N-1}} f_{N-1} \cdot \nabla g_N, \quad i = 1, \ldots, N-1, \tag{2.171}$$

$$p_N = \nabla g_N, \tag{2.172}$$

Sec. 2.6 Discrete-Time Optimal Control Problems

we obtain
$$\nabla_{u_i} J(u) = \nabla_{u_i} f_i \cdot p_{i+1}, \qquad i = 0, \ldots, N-1. \tag{2.173}$$

Note also that $p_1, \ldots, p_N$ are generated (backwards in time) by means of the so called *adjoint equation*:
$$p_i = \nabla_{x_i} f_i \cdot p_{i+1}, \qquad i = 1, \ldots, N-1, \tag{2.174}$$

starting from
$$p_N = \nabla g_N. \tag{2.175}$$

Consider now the general case of the cost function
$$J(u) = g_N(x_N) + \sum_{i=0}^{N-1} g_i(x_i, u_i), \tag{2.176}$$

which involves the intermediate cost terms $g_i(x_i, u_i)$. We may reduce the corresponding problem to one which involves a terminal state cost only, by introducing an additional state variable y_i and a corresponding equation as follows:
$$y_{i+1} = y_i + g_i(x_i, u_i),$$
$$y_0 = 0.$$

Defining a new state vector
$$\tilde{x}_i = \begin{pmatrix} x_i \\ y_i \end{pmatrix} \tag{2.177}$$

and a new system equation
$$\tilde{x}_{i+1} = \begin{pmatrix} f_i(x_i, u_i) \\ y_i + g_i(x_i, u_i) \end{pmatrix} \equiv \tilde{f}_i(\tilde{x}_i, u_i), \tag{2.178}$$

the cost function (2.176) takes the form
$$J(u) = g_N(x_N) + y_N \equiv \tilde{g}_N(\tilde{x}_N). \tag{2.179}$$

This cost function involves only a terminal state cost and is of the type considered previously. We have from Eqs. (2.173)-(2.175),
$$\nabla_{u_i} J(u) = \nabla_{u_i} \tilde{f}_i \cdot \tilde{p}_{i+1}, \qquad i = 0, \ldots, N-1, \tag{2.180}$$

where
$$\tilde{p}_i = \nabla_{\tilde{x}_i} \tilde{f}_i \cdot \tilde{p}_{i+1}, \qquad i = 1, \ldots, N-1,$$
$$\tilde{p}_N = \nabla \tilde{g}_N.$$

From Eqs. (2.177)-(2.179), we have by writing

$$\tilde{p}_i = \begin{pmatrix} p_i \\ z_i \end{pmatrix}, \qquad p_i \in \Re^n, \; z_i \in \Re,$$

the following form of the adjoint equation

$$\tilde{p}_N = \begin{pmatrix} p_N \\ z_N \end{pmatrix} = \begin{pmatrix} \nabla g_N \\ 1 \end{pmatrix},$$

$$\tilde{p}_i = \begin{pmatrix} p_i \\ z_i \end{pmatrix} = \begin{pmatrix} \nabla_{x_i} f_i & \nabla_{x_i} g_i \\ 0 & 1 \end{pmatrix} \begin{pmatrix} p_{i+1} \\ z_{i+1} \end{pmatrix}.$$

As a result, we have

$$z_i = 1, \qquad i = 1, \ldots, N,$$

and we obtain the final form of the adjoint equation

$$p_i = \nabla_{x_i} f_i \cdot p_{i+1} + \nabla_{x_i} g_i, \tag{2.181}$$

$$p_N = \nabla g_N. \tag{2.182}$$

Furthermore, from Eq. (2.180),

$$\nabla_{u_i} J(u) = \begin{pmatrix} \nabla_{u_i} f_i & \nabla_{u_i} g_i \end{pmatrix} \begin{pmatrix} p_{i+1} \\ z_{i+1} \end{pmatrix}$$

and finally

$$\nabla_{u_i} J(u) = \nabla_{u_i} f_i \cdot p_{i+1} + \nabla_{u_i} g_i. \tag{2.183}$$

Equation (2.183) yields the cost gradient in terms of the costate vectors obtained from the adjoint equation (2.181) and (2.182), with all derivatives evaluated along the control trajectory u and the corresponding state trajectory under consideration. Thus to compute the gradient $\nabla J(u)$ for a given u:

(1) We calculate the state trajectory corresponding to u by forwards propagation of the system equation $x_{i+1} = f_i(x_i, u_i)$, starting from the given initial condition x_0.

(2) We generate the costate vectors by backwards propagation of the adjoint equation Eqs. (2.181), starting from the terminal condition $p_N = \nabla g_N$. (All partial derivatives are evaluated at the current state and control trajectories.)

(3) We use Eq. (2.183) to compute the components $\nabla_{u_i} J(u)$ of the gradient $\nabla J(u)$.

Calculation of $\nabla J(u)$ and $\nabla^2 J(u)$ (2nd Method)

The alternative method for calculating the derivatives of J is based on writing all the system equations $x_{i+1} = f_i(x_i, u_i)$ compactly as

$$h(\phi(u), u) = 0. \tag{2.184}$$

Here, $u = (u_0, u_1, \ldots, u_{N-1})$ is the control trajectory and $\phi(u)$ is the corresponding state trajectory

$$x = \phi(u),$$

where the function ϕ maps $\Re^{Nm}$ into $\Re^{Nn}$ and is defined in terms of the functions ϕ_i of Eq. (2.168) as

$$\phi(u) = \begin{pmatrix} \phi_1(u) \\ \vdots \\ \phi_N(u) \end{pmatrix} = \begin{pmatrix} x_1 \\ \vdots \\ x_N \end{pmatrix}. \tag{2.185}$$

The equation $h(\phi(u), u) = 0$ is a compact representation of the equations

$$\begin{aligned} f_0(x_0, u_0) - x_1 &= 0, \\ f_1(x_1, u_1) - x_2 &= 0, \\ &\vdots \\ f_{N-1}(x_{N-1}, u_{N-1}) - x_N &= 0, \end{aligned} \tag{2.186}$$

and is satisfied by every control trajectory u.

Similarly the cost function may be written abstractly as

$$F(\phi(u), u) = g_N(\phi_N(u)) + \sum_{i=0}^{N-1} g_i(\phi_i(u), u_i), \tag{2.187}$$

and the optimal control problem becomes

$$\text{minimize } J(u) = F(\phi(u), u)$$
$$\text{subject to } u \in \Re^{Nm}.$$

To obtain the derivatives of $J(u)$, we use the trick of writing for any vector $p \in \Re^{Nn}$,

$$J(u) = F(\phi(u), u) + h(\phi(u), u)' p,$$

[cf. Eq. (2.184)] and we select cleverly p so that the derivatives of $J(u)$ are easily calculated. We have

$$\begin{aligned} \nabla J(u) =& \nabla \phi(u) \big(\nabla_x F(\phi(u), u) + \nabla_x h(\phi(u), u) \cdot p \big) \\ &+ \nabla_u F(\phi(u), u) + \nabla_u h(\phi(u), u) \cdot p, \end{aligned} \tag{2.188}$$

where in the preceding equation, ∇_x denotes gradient with respect to the first argument $[x = \phi(u)]$, ∇_u denotes gradient with respect to the second argument u, and $\nabla\phi(u)$ is the $Nm \times Nn$ gradient matrix of ϕ evaluated at u, i.e.,

$$\nabla\phi(u) = \begin{bmatrix} \nabla_u\phi_1(u) \vdots \nabla_u\phi_2(u) \vdots \cdots \vdots \nabla_u\phi_N(u) \end{bmatrix}.$$

Equation (2.188) holds for every $p \in \Re^{Nn}$. Suppose that p is selected so that it satisfies the equation

$$\nabla_x F\big(\phi(u), u\big) + \nabla_x h\big(\phi(u), u\big) \cdot p(u) = 0, \qquad (2.189)$$

where we denote by $p(u)$ the particular value of p satisfying the equation above. We will see shortly that Eq. (2.189) has a unique solution with respect to p, so our notation is justified. We have then from Eq. (2.187),

$$\nabla J(u) = \nabla_u F\big(\phi(u), u\big) + \nabla_u h\big(\phi(u), u\big) \cdot p(u). \qquad (2.190)$$

For notational convenience, we introduce the function $L : \Re^{Nn} \times \Re^{Nm} \times \Re^{Nn} \mapsto \Re$ given by

$$L(x, u, p) = F(x, u) + h(x, u)'p. \qquad (2.191)$$

We then have
$$\nabla J(u) = \nabla_u L\big(\phi(u), u, p(u)\big). \qquad (2.192)$$

To compute the Hessian matrix of the cost, we write Eq. (2.188) as

$$\nabla J(u) = \nabla\phi(u)\nabla_x L\big(\phi(u), u, p\big) + \nabla_u L\big(\phi(u), u, p\big)$$

and we differentiate this equation with respect to u. By using also the equation

$$\nabla_x L\big(\phi(u), u, p(u)\big) = 0$$

[cf. Eq. (2.189)] we obtain for $p = p(u)$

$$\nabla^2 J(u) = \nabla\phi(u)\nabla^2_{xx} L\big(\phi(u), u, p(u)\big)\nabla\phi(u)' + \nabla\phi(u)\nabla^2_{xu} L\big(\phi(u), u, p(u)\big)$$
$$+ \nabla^2_{ux} L\big(\phi(u), u, p(u)\big)\nabla\phi(u)' + \nabla^2_{uu} L\big(\phi(u), u, p(u)\big)'. \qquad (2.193)$$

In order to calculate the derivatives of J, we must calculate the solution $p(u)$ of Eq. (2.189) and then evaluate the derivatives of L. We have:

$$L(x, u, p) = g_N(x_N) + \sum_{i=0}^{N-1} g_i(x_i, u_i) + \sum_{i=1}^{N} \big(f_{i-1}(x_{i-1}, u_{i-1}) - x_i\big)' p_i. \qquad (2.194)$$

Equation (2.189) can be written as $\nabla_x L(\phi(u), u, p(u)) = 0$, or by using Eq. (2.194),

$$\begin{pmatrix} \nabla g_N \\ \nabla_{x_{N-1}} g_{N-1} \\ \nabla_{x_{N-2}} g_{N-2} \\ \vdots \\ \nabla_{x_1} g_1 \end{pmatrix} + \begin{pmatrix} -I & 0 & 0 & \cdots & 0 \\ \nabla_{x_{N-1}} f_{N-1} & -I & 0 & \cdots & 0 \\ 0 & \nabla_{x_{N-2}} f_{N-2} & -I & \cdots & 0 \\ \vdots & \vdots & \vdots & \cdots & \vdots \\ 0 & 0 & 0 & \cdots & -I \end{pmatrix} \begin{pmatrix} p_N \\ p_{N-1} \\ p_{N-2} \\ \vdots \\ p_1 \end{pmatrix} = 0.$$

By rewriting this equation, we have

$$p_i = \nabla_{x_i} f_i \cdot p_{i+1} + \nabla_{x_i} g_i, \qquad i = 1, \ldots, N-1, \qquad (2.195)$$

$$p_N = \nabla g_N, \qquad (2.196)$$

which is the same as the adjoint equation derived earlier [Eqs. (2.181) and (2.182)]. Thus, the adjoint equation specifies uniquely $p(u)$.

Also, from Eqs. (2.191), (2.192), and (2.194),

$$\nabla_{u_i} J(u) = \nabla_{u_i} f_i \cdot p_{i+1} + \nabla_{u_i} g_i, \qquad i = 0, \ldots, N-1, \qquad (2.197)$$

which is the same equation for the gradient ∇J as the one obtained earlier.

Concerning the Hessian matrix $\nabla^2 J$, it is given by Eqs. (2.193) and (2.194). We only give here the form of the gradient matrix $\nabla \phi(u)$

$$\nabla \phi(u) = \begin{pmatrix} \nabla_{u_0} f_0 & \nabla_{x_1} f_1 \cdot \nabla_{u_0} f_0 & \cdots & \nabla_{x_{N-1}} f_{N-1} \cdots \nabla_{x_1} f_1 \cdot \nabla_{u_0} f_0 \\ 0 & \nabla_{u_1} f_1 & \cdots & \nabla_{x_{N-1}} f_{N-1} \cdots \nabla_{x_2} f_2 \cdot \nabla_{u_1} f_1 \\ \vdots & \vdots & \cdots & \vdots \\ 0 & 0 & \cdots & \nabla_{u_{N-1}} f_{N-1} \end{pmatrix}$$

The expression (2.193) for $\nabla^2 J$ is quite complex, but will be useful when we will discuss Newton's method later in this section.

Example 2.6.2 (Neural Networks)

Recall Example 1.4.3 in Section 1.4.4, which describes how the neural network training problem can be posed as a least squares problem. It can be seen that this problem can also be viewed as an optimal control problem, where the weight vector at each stage is the control vector for that stage. Mathematically, the problem is to find the control trajectory u that minimizes

$$J(u) = \sum_{j=1}^{m} J_j(u),$$

where

$$J_j(u) = \tfrac{1}{2} \big\| z_j - h(u, y_j) \big\|^2,$$

(z_j, y_j) is the jth input-output pair in the training set, and h is an appropriate function defined by the functional relationship between input and output of the neural network. Each gradient $\nabla J_j(u)$ can be calculated as described earlier via the corresponding adjoint equation, and the cost gradient is obtained as

$$\nabla J(u) = \sum_{j=1}^{m} \nabla J_j(u).$$

The process of calculating the gradient by means of the adjoint equation is sometimes called *backpropagation* and corresponding gradient-like methods are sometimes called *backpropagation methods*; see the books [BeT96], [Bis95], [Hay11] for more details.

First Order Necessary Condition

The preceding expressions for the cost gradient can be used to write the first order necessary optimality condition

$$\nabla J(u^*) = 0,$$

for

$$u^* = (u_0^*, u_1^*, \ldots, u_{N-1}^*)$$

to be a local minimum of J. It is customary to write this condition in terms of the *Hamiltonian function*, defined for each i by

$$H_i(x_i, u_i, p_{i+1}) = g_i(x_i, u_i) + p_{i+1}' f_i(x_i, u_i). \tag{2.198}$$

By using the expression of $\nabla J(u^*)$ [cf. Eq. (2.195)-(2.197)] we obtain:

Proposition 2.6.1: Let $u^* = (u_0^*, u_1^*, \ldots, u_{N-1}^*)$ be a local minimum control trajectory and let $x^* = (x_1^*, x_2^*, \ldots, x_N^*)$ be the corresponding state trajectory. Then we have

$$\nabla_{u_i} H_i\big(x_i^*, u_i^*, p_{i+1}^*\big) = 0, \qquad i = 0, \ldots, N-1, \tag{2.199}$$

where the costate vectors $p_1^*, \ldots, p_N^*$ are obtained from the adjoint equation

---

$$p_i^* = \nabla_{x_i} H_i(x_i^*, u_i^*, p_{i+1}^*), \qquad i = 1, \ldots, N-1, \qquad (2.200)$$

with the terminal condition

$$p_N^* = \nabla g_N(x_N^*). \qquad (2.201)$$

Example 2.6.3 (Linear System and Quadratic Cost)

Consider the case where the system is linear

$$x_{i+1} = A_i x_i + B_i u_i, \qquad i = 0, \ldots, N-1 \qquad (2.202)$$

and the cost function is quadratic of the form

$$J(u) = \tfrac{1}{2} \left\{ x_N' Q_N x_N + \sum_{i=0}^{N-1} (x_i' Q_i x_i + u_i' R_i u_i) \right\}. \qquad (2.203)$$

The $n \times n$ matrices A_i and the $n \times m$ matrices B_i are given. The matrices Q_i are assumed symmetric and positive semidefinite, and the matrices R_i are assumed symmetric and positive definite. There are no constraints on the state or control vectors.

Let $u^* = (u_0^*, u_1^*, \ldots, u_{N-1}^*)$ be an optimal control trajectory and let $x^* = (x_1^*, x_2^*, \ldots, x_N^*)$ be the corresponding state trajectory. The adjoint equation is given by

$$p_i^* = A_i' p_{i+1}^* + Q_i x_i^*, \qquad i = 1, \ldots, N-1, \qquad (2.204)$$

$$p_N^* = Q_N x_N^*. \qquad (2.205)$$

It can be seen from Eq. (2.202), that the state trajectory x is linearly related to the control trajectory u, so the cost function $J(u)$ of Eq. (2.203) is a quadratic function of u. Because the matrices Q_i and R_i are assumed positive semidefinite and positive definite, respectively, $J(u)$ is a convex, positive definite, quadratic function. Therefore, the necessary optimality condition $\nabla J(u^*) = 0$ of Prop. 2.6.1 is also sufficient. It can be written as

$$\nabla_{u_i} H_i(x_i^*, u_i^*, p_{i+1}^*) = \nabla_{u_i} \left\{ p_{i+1}^{*'} B_i u_i^* + \tfrac{1}{2} u_i^{*'} R_i u_i^* + \tfrac{1}{2} x_i^{*'} Q_i x_i^* \right\} = 0$$

or equivalently

$$R_i u_i^* + B_i' p_{i+1}^* = 0, \qquad i = 0, \ldots, N-1,$$

from which

$$u_i^* = -R_i^{-1} B_i' p_{i+1}^*, \qquad i = 0, \ldots, N-1. \qquad (2.206)$$

We will obtain a more convenient expression for u_i^* by verifying a relation of the form

$$p_{i+1}^* = K_{i+1} x_{i+1}^*, \qquad (2.207)$$

where K_{i+1} is a positive semidefinite symmetric matrix, which can be explicitly calculated. We show this relation by induction, first noting that it holds for $i = N - 1$ with

$$K_N = Q_N,$$

[cf. Eq. (2.205)]. Assume that it holds for some $i \leq N-1$. Then, by combining Eqs. (2.206) and (2.207), we have

$$u_i^* = -R_1^{-1} B_i' p_{i+1}^* = -R_i^{-1} B_i' K_{i+1} x_{i+1}^*. \qquad (2.208)$$

With the substitution $x_{i+1}^* = A_i x_i^* + B_i u_i^*$, this equation yields

$$u_i^* = -R_i^{-1} B_i' K_{i+1} \big(A_i x_i^* + B_i u_i^*\big)$$

or equivalently, by solving for u_i^*,

$$u_i^* = -(R_i + B_i' K_{i+1} B_i)^{-1} B_i' K_{i+1} A_i x_i^*, \qquad i = 0, \ldots, N-1. \qquad (2.209)$$

We thus obtain

$$x_{i+1}^* = A_i x_i^* + B_i u_i^* = A_i x_i^* - B_i (R_i + B_i' K_{i+1} B_i)^{-1} B_i' K_{i+1} A_i x_i^*.$$

Multiplying both sides with $A_i' K_{i+1}$ and using Eq. (2.207), we see that

$$A_i' p_{i+1}^* = A_i' K_{i+1} A_i x_i^* - A_i' K_{i+1} B_i (R_i + B_i' K_{i+1})^{-1} B_i' K_{i+1} B_i A_i x_i^*.$$

Adding $Q_i x_i^*$ to both sides, we obtain

$$p_i^* = A_i' p_{i+1}^* + Q_i x_i^* = K_i x_i^*,$$

where

$$K_i = A_i' \Big(K_{i+1} - K_{i+1} B_i (R_i + B_i' K_{i+1} B_i)^{-1} B_i' K_{i+1}\Big) A_i + Q_i, \qquad (2.210)$$

thus completing the induction proof of Eq. (2.207).

Equation (2.210) is known as the *Riccati equation*. It generates the matrices $K_1, K_2, \ldots K_{N-1}$ from the terminal condition

$$K_N = Q_N.$$

Given K_1, we may obtain u_0^* using Eq. (2.209) and the given initial state x_0. Then, we may calculate x_1^* from the system equation (2.202) and the vector u_1^* from Eq. (2.209) using K_2. Similarly we may calculate $x_1^*, u_2^*, \ldots, u_{N-1}^*, x_N^*$.

2.6.1 Gradient and Conjugate Gradient Methods for Optimal Control

The application of gradient methods to unconstrained optimal control problems is straightforward in principle. For example the steepest descent method takes the form

$$u_i^{k+1} = u_i^k - \alpha^k \nabla_{u_i} H_i\left(x_i^k, u_i^k, p_{i+1}^k\right), \qquad i = 0, \ldots, N-1, \qquad (2.211)$$

where H_i denotes the Hamiltonian function

$$H_i(x_i, u_i, p_{i+1}) = g_i(x_i, u_i) + p_{i+1}' f_i(x_i, u_i),$$

$u^k = \{u_0^k, u_1^k, \ldots, u_{N-1}^k\}$ is the kth control trajectory, $x^k = (x_0^k, x_1^k, \ldots, x_N^k)$ denotes the kth state trajectory, and $p^k = (p_1,^k, \ldots, p_N^k)$ denotes the kth costate trajectory

$$p_i^k = \nabla_{x_i} H_i\left(x_i^k, u_i^k, p_{i+1}^k\right), \qquad i = 1, \ldots, N-1,$$

$$p_N^k = \nabla g_N(x_N^k).$$

Thus, given u^k, one computes x^k by forward propagation of the system equation, and then p^k by backward propagation of the adjoint equation. Subsequently, the steepest descent iteration (2.211) is performed with the stepsize α^k chosen, for example, by the Armijo rule or some line minimization rule. An important point that guides the selection of the stepsize rule is that a gradient evaluation in optimal control problems with many control variables is usually not much more expensive than a function evaluation. Therefore, it is worth considering stepsize rules that aim at efficiency by using intelligently gradient calculations to save substantially on the number of function evaluations.

While steepest descent is simple, its convergence rate in optimal control problems is often very poor. This is particularly so when the underlying dynamic system tends to be unstable (see Exercise 2.6.1). For this reason scaling of the control variables is recommended if suitable scaling factors can be determined with reasonable effort. Otherwise, one should use either the conjugate gradient method or Newton's method.

Conjugate Gradient Method

The application of the conjugate gradient method to unconstrained optimal control problems is also straightforward. The search direction is a linear combination of the current gradient and the preceding search direction, as discussed in Section 2.1. The performance of the method can often be improved through the use of preconditioning. One possibility is to use a diagonal approximation to the Hessian matrix as preconditioning matrix, but in some cases, far more effective preconditioning schemes are possible; see Exercise 2.6.2 and [Ber74].

2.6.2 Newton's Method for Optimal Control

To derive an efficient form of Newton's method, it is important to recognize that it is impractical to evaluate explicitly the Hessian matrix $\nabla^2 J(u)$ and find the control trajectory u^{k+1} from

$$u^{k+1} = u^k - \left(\nabla^2 J(u^k)\right)^{-1} \nabla J(u^k).$$

Instead we will use a different approach, which exploits the optimal control problem structure.

To this end, we note that the next iterate of the pure form of Newton's method

$$x^{k+1} = x^k - \left(\nabla^2 f(x^k)\right)^{-1} \nabla f(x^k) \qquad (2.212)$$

can be obtained by minimizing the second order approximation of f around x^k given by

$$f(x^k) + \nabla f(x^k)'(x - x^k) + \tfrac{1}{2}(x - x^k)'\nabla^2 f(x^k)(x - x^k).$$

Carrying the argument one step further, assume that $\tilde{f}^k : \Re^n \mapsto \Re$ is a *quadratic* function such that

$$\nabla \tilde{f}^k(x^k) = \nabla f(x^k), \qquad \nabla^2 \tilde{f}^k(x^k) = \nabla^2 f(x^k).$$

Then $\tilde{f}^k$ has the form

$$\tilde{f}^k(x) = c + \nabla f(x^k)'(x - x^k) + \tfrac{1}{2}(x - x^k)'\nabla^2 f(x^k)(x - x^k),$$

where c is some constant, and hence the next iterate x^{k+1} of Eq. (2.212) is obtained by minimizing the quadratic function $\tilde{f}^k$.

Motivated by this alternative possibility, we formulate, for given u^k, a *linear-quadratic* optimal control problem with cost function $\tilde{J}_k(u)$ such that

$$\nabla \tilde{J}_k(u^k) = \nabla J(u^k), \qquad \nabla^2 \tilde{J}_k(u^k) = \nabla^2 J(u^k). \qquad (2.213)$$

The optimal solution u^{k+1} of the linear quadratic problem can be obtained conveniently via a Riccati equation (cf. Example 2.6.3) and according to the preceding discussion, it represents the next iterate of the pure form of Newton's method as applied to minimization of $J(u)$.

The linear quadratic problem that corresponds to a control trajectory $u^k = (u_0^k, u_1^k, \ldots, u_{N-1}^k)$ and the associated states $x_0, x_1^k, \ldots, x_N^k$, and costates $p_1^k, p_2^k, \ldots, p_N^k$, involves a linearized version of the original system. It has the form

$$\text{minimize } \tilde{J}_k(\delta u) = \tfrac{1}{2}\delta x_N' Q_N \delta x_N + a_N' \delta x_N + \sum_{i=0}^{N-1} \tfrac{1}{2}\delta x_i' Q_i \delta x_i + a_i' \delta x_i$$

$$+ \sum_{i=0}^{N-1} \tfrac{1}{2}\delta u_i' R_i \delta u_i + b_i' \delta u_i + \sum_{i=1}^{N-1} \delta u_i' M_i \delta x_i$$

$$\text{subject to } \delta x_0 = 0, \qquad \delta x_{i+1} = A_i \delta x_i + B_i \delta u_i, \quad i = 1, \ldots, N-1,$$

$$\qquad (2.214)$$

Sec. 2.6   Discrete-Time Optimal Control Problems   223

where
$$A_i = \nabla_{x_i} f'_i, \qquad B_i = \nabla_{u_i} f'_i, \qquad i = 0, \ldots, N-1,$$
$$Q_N = \nabla^2 g_N, \qquad a_N = \nabla g_N,$$
$$Q_i = \nabla^2_{x_i x_i} H_i, \qquad a_i = \nabla_{x_i} g_i, \qquad i = 1, \ldots, N-1,$$
$$R_i = \nabla^2_{u_i u_i} H_i, \qquad b_i = \nabla_{u_i} g_i, \qquad i = 0, \ldots, N-1,$$
$$M_i = \nabla^2_{u_i x_i} H_i, \qquad i = 0, \ldots, N-1,$$
$$H_i(x_i, u_i, p^k_{i+1}) = g_i(x_i, u_i) + p^{k\,\prime}_{i+1} f_i(x_i, u_i). \qquad (2.215)$$

The partial derivatives appearing above are evaluated along the current control and state trajectories

$$u^k = (u^k_0, u^k_1, \ldots, u^k_{N-1}), \qquad x^k = (x_0, x^k_1, \ldots, x^k_N), \qquad (2.216)$$

and the corresponding costate vectors $p^k_1, p^k_2, \ldots, p^k_N$. The problem (2.214) involves the variations

$$\delta x_i = x_i - \tilde{x}^k_i$$
$$\delta u_i = u_i - u^k_i$$

from the nominal trajectories $u^k, \tilde{x}^k$, where $\tilde{x}^k$ is the trajectory corresponding to the linearized system

$$\tilde{x}^k_{i+1} = A_i \tilde{x}^k_i + B_i u^k_i, \qquad i = 0, \ldots, N-1 \qquad (2.217)$$

$$\tilde{x}^k_0 = x_0 : \text{given initial state}.$$

It is straightforward to verify that the cost function $J(u)$ for our original problem has a gradient $\nabla J(u^k)$ and Hessian matrix $\nabla^2 J(u^k)$ evaluated at u^k, which are equal to the corresponding gradient and Hessian matrix $\nabla \tilde{J}_k(\delta u^k)$, $\nabla^2 \tilde{J}_k(\delta u^k)$ of the problem (2.214)-(2.215). This can be shown by comparing the expressions for the gradient and Hessian of the two problems given earlier [cf. Eqs. (2.192)-(2.194)]. Hence, according to our earlier discussion, the solution of problem (2.214)-(2.215) yields the next iterate of Newton's method.

We now turn to the linear-quadratic problem (2.214)-(2.215). It is possible to show using the necessary conditions for optimality, similar to Example 2.6.3, that the solution of this problem (provided it exists and is unique) may be obtained in feedback form as follows:

$$\delta u^k_i = -(R_i + B'_i K_{i+1} B_i)^{-1}\big((M_i + B'_i K_{i+1} A_i)\delta x^k_i + b_i + B'_i \lambda_{i+1}\big),$$
$$i = 0, \ldots, N-1,$$
$$(2.218)$$

where
$$\delta x^k_0 = 0,$$

$$\delta x_{i+1}^k = A_i \delta x_i^k + B_i \delta u_i^k, \qquad i = 0, \ldots, N-1, \qquad (2.219)$$

the matrices $K_1, K_2, \ldots, K_N$ are given recursively by the equation

$$K_N = Q_N$$

$$K_i = A_i' K_{i+1} A_i + Q_i - (M_i + B_i' K_{i+1} A_i)'(R_i + B_i' K_{i+1} B_i)^{-1} \\ (M_i + B_i' K_{i+1} A_i), \qquad i = 1, \ldots, N-1, \qquad (2.220)$$

and the vectors $\lambda_1, \lambda_2, \ldots, \lambda_N$ are given recursively by

$$\lambda_N = a_N,$$

$$\lambda_i = a_i + A_i' \lambda_{i+1} - (M_i + B_i' K_{i+1} A_i)'(R_i + B_i' K_{i+1} B_i)^{-1}(b_i + B_i' \lambda_{i+1}). \qquad (2.221)$$

The next iterate of the pure form of Newton's method is obtained from Eq. (2.218) and if a stepsize α^k is introduced, then we obtain the iteration

$$u_i^{k+1} = u_i^k + \alpha^k \delta u_i^k, \qquad i = 0, \ldots, N-1, \qquad (2.222)$$

with

$$\delta u_i^k = -(R_i + B_i' K_{i+1} B_i)^{-1}\big((M_i + B_i' K_{i+1} A_i)\delta x_i^k + b_i + B_i' \lambda_{i+1}\big). \qquad (2.223)$$

The computations of the typical iteration of Newton's method are carried out in the following sequence.

---

kth Iteration of Newton's Method

Step 1: Given the current control trajectory u^k the corresponding state trajectory x^k is calculated.

Step 2: The solution $K_1, \ldots, K_N$ of Eq. (2.220) and the solution $\lambda_1, \ldots, \lambda_N$ of Eq. (2.221) are calculated (backwards, from the terminal conditions) together with all the necessary data for calculation of the optimal control in equation (2.218). This is done by calculating simultaneously the costate vectors $p_1^k, \ldots, p_N^k$, which are needed to evaluate the second derivatives of the Hamiltonian (i.e., Q_i, R_i, M_i).

Step 3: The Newton direction δu^k is generated as follows:

1) Calculate δu_0^k using Eq. (2.223),

$$\delta u_0^k = -(R_0 + B_0' K_1 B_0)^{-1}(b_0 + B_0' \lambda_1).$$

2) Calculate δx_1^k using Eq. (2.219) and the fact $\delta x_0^k = 0$.

Sec. 2.6   Discrete-Time Optimal Control Problems

3) Calculate δu_1^k using δx_1^k and Eq. (2.223).

4) Continue similarly to compute δu_i^k and δx_i^k for all i.

Step 4: A stepsize α^k is obtained using some stepsize rule; the next control trajectory is

$$u_i^{k+1} = u_i^k + \alpha^k \delta u_i^k, \qquad i = 0, \ldots, N-1.$$

In the computation of the solution of the Riccati equation it is necessary that $(R_i + B_i' K_{i+1} B_i)^{-1}$ exist and be positive definite. If any of these inverses is not positive definite, it can be seen that $\nabla^2 J(u^k)$ is not positive definite and hence the Newton iteration must be modified. Under these circumstances one should replace R_i by $(R_i + E_i)$ where E_i is a positive definite diagonal matrix such that $(R_i + E_i + B_i' K_{i+1} B_i)$ is positive definite, as discussed in Section 1.4. In a neighborhood of a local minimum u^* with $\nabla^2 J(u^*) > 0$, the solution of the Riccati equation exists and the Newton iteration can be carried out as described.

It is interesting to note that our analysis yields as a byproduct a sufficient condition for optimality of a control trajectory u^*. Clearly u^* will be optimal if the corresponding linear-quadratic problem (2.214) with all derivatives evaluated at u^* has $\delta u = 0$ as an optimal solution. This can be guaranteed when

$$\nabla_{u_i} H_i = 0, \qquad i = 0, \ldots, N-1, \quad \text{(1st order condition)},$$

$$R^* + B_i^{*'} K_{i+1}^* B_i^* > 0, \qquad i = 0, \ldots, N-1, \quad \text{(2nd order condition)},$$

where

$$R_i^* = \nabla_{u_i u_i}^2 H_i, \qquad B_i^* = \nabla_{u_i} f_i', \qquad i = 0, \ldots, N-1,$$

K_{i+1}^* is given by the Riccati equation (2.220) corresponding to u^*, and all derivatives above are evaluated along u^* and the corresponding state and costate trajectories.

Finally, we mention a variant of Newton's method that offers some computational advantages. It uses in place of Eqs. (2.222) and (2.223) the following two equations:

$$u_i^{k+1} = u_i^k - \alpha^k (R_i + B_i' K_{i+1} B_i)^{-1} \big((M_i + B_i' K_{i+1} A_i)(x_i^{k+1} - x_i^k) \\ + b_i + B_i' \lambda_{i+1} \big),$$

(2.224)

where

$$x_i^{k+1} - x_i^k = f_{i-1}(x_{i-1}^{k+1}, u_{i-1}^{k+1}) - f_{i-1}(x_{i-1}^k, u_{i-1}^k). \qquad (2.225)$$

In other words, the first order variation δx_i^k, which is obtained from the linearized system equation (2.219), is replaced in Eq. (2.223) by the actual variation $(x_i^{k+1} - x_i^k)$ of Eq. (2.225). Since $(x_i^{k+1} - x_i^k)$ and δx_i^k differ by terms which are of second or higher order, the two methods are asymptotically equivalent. The advantage of using Eqs. (2.224) and (2.225) is that the matrices A_i, B_i need not be stored (or recalculated) for use in Eq. (2.219). In addition, the trajectory x^{k+1} corresponding to u^{k+1} may be used in the next iteration, if the pure form of Newton's method is used or the stepsize α^k turns out to be equal to one.

EXERCISES

2.6.1

Calculate the condition number of the optimal control problem involving the quadratic cost function $x_N^2 + \sum_{i=0}^{N-1} u_i^2$ and the scalar system $x_{i+1} = ax_i + u_i$, where $a > 1$. Show that it tends to ∞ as $N \to \infty$. (This example illustrates a general phenomenon: a problem involving a system that is unstable in the absence of control tends to be ill-conditioned.)

2.6.2 (Preconditioning for Special Structures [Ber74])

Consider the special case of the linear quadratic problem of Eqs. (2.202) and (2.203) where $Q_i = 0$ for all $i = 0, \ldots, N-1$. Derive a preconditioned conjugate gradient method that converges in at most $n+1$ steps. *Hint*: Consider the structure of the Hessian matrix and use the result of Exercise 2.1.10.

2.6.3

Consider a discrete-time optimal control problem, which is similar to the one of this section, except that the system equation has the form

$$x_{i+1} = f_i(x_{i+1}, x_i, u_i),$$

and the continuously differentiable function f_i is such that the above equation can be solved uniquely for x_{i+1} in terms of x_i and u_i. Assume also that when linearized along any control trajectory and corresponding state trajectory, the above equation can also be solved uniquely for δx_{i+1} in terms of δx_i and δu_i. Derive analogs of the steepest descent and the conjugate gradient methods.

2.7 SOLVING NONLINEAR PROGRAMMING PROBLEMS: SOME PRACTICAL GUIDELINES

The practical solution of a nonlinear programming problem can be very challenging. Even if the problem is well-formulated mathematically, slow convergence and the existence of local minima that are not global are serious potential difficulties. It is thus useful to follow a few basic practical guidelines. Keep in mind, however, that in nonlinear programming, just as there is no foolproof method, there is also no foolproof advice.

Problem Formulation

In the real world, optimization problems seldom come neatly formulated as mathematical problems. Thus, usually the first step is to formalize mathematically the optimization model by selecting the cost function and delineating the constraints. One potential difficulty here is that there may be multiple and possibly competing objectives; for example in an engineering design problem one may simultaneously want to minimize cost and maximize efficiency. In this case one usually encodes all objectives except one in the constraints and incorporates the remaining objective in the cost function. However, the division between objective function and constraints may be arbitrary and one may have to reconsider the formulation after evaluating the results of the optimization. It is important here to properly correlate the problem formulation with the objectives of the investigation. For example if only an approximate answer is required, there is no point in constructing a highly detailed model.

The most critical task in problem formulation is to capture the realism of the practical context while obtaining an analytically or computationally tractable model. This is where experience and insight into both the application and the methodology are very important. In particular, it is important to be able to recognize the models that are easily solvable and those that are essentially impossible to solve. Two rules of thumb here are:

(a) Convex cost and constraint sets are preferable to nonconvex ones.

(b) Linear programming problems are easier than nonlinear programming problems, which are in turn easier than integer/discrete combinatorial problems.

While (a) is generally true because of the lack of local minima and other spurious stationary points in convex problems, there are many exceptions to (b). For example there is a widespread practice of replacing nonlinear functions by piecewise linear ones that lead to linear programming formulations, so that linear programming codes can be used. This may work out fine, but unfortunately it may also lead to vastly increased dimensionality, and loss of insight (solutions to continuous problems often have elegant features that are lost in discrete approximations). Another example along similar lines arises when essentially combinatorial problems are formulated as highly nonconvex nonlinear programming problems whose local minima correspond to the feasible solutions of the original problem. For a somewhat absurd example, think of discarding an integrality constraint on a variable x_i and adding to the cost function the (differentiable) term $c(x_i - n(x_i))^2$ where $n(x_i)$ is the nearest integer to x_i and c is a very large penalty parameter. Generally, when a convex problem formulation seems impossible and the presence of many local minima is an important concern, one may wish to bring to bear the methods of global optimization, for which we refer to the specialized literature (see e.g., the books [PaR87], [FlP95], [Flo95], [HPT00], [RuK04]).

We note, however, that depending on the practical situation, there is often merit in replacing integer variables with continuous variables and using a nonlinear programming formulation to obtain a noninteger solution that can then be rounded to integer. For example, consider a variable such as number of telephone lines to establish between two points in a communication network. This number is naturally discrete but if its range is in the hundreds, little will be lost by replacing it by a continuous variable. On the other hand, other variables may naturally be discrete-valued, such as a $\{0,1\}$-valued variable modeling whether route A or route B is used to send a message in a communication network. A continuous approximation of such a variable may be meaningless.

An interesting case where approximating discrete $\{0,1\}$-variables by continuous ones may be useful is when through analysis or experience one can establish that the overwhelming majority of the $\{0,1\}$-variables take an integer (0 or 1) value at an optimal solution of the continuous approximating problem (the one where all variables are treated as continuous and constrained to take values in $[0,1]$). Once such an optimal solution is obtained, one may try to round up or down its few noninteger values to obtain a feasible solution. This rounded solution is often near-optimal. An example of such a case arises in large-scale separable integer programming problems with few separable constraints, which can be solved by Lagrangian relaxation as (continuous) dual problems with small duality gap; see Section 6.1.5. For another typical example, which arises in integer-constrained multicommodity flow problems, see [OzB03].

Another important consideration in formulating optimization problems is whether the cost function is or is not differentiable. The most powerful methods in nonlinear programming require a once or twice differentiable cost function. This motivates the smoothing of whatever natural nondifferentiabilities may exist in the cost. An important special case is when nondifferentiable terms of the form $\max\{f_1(x), f_2(x)\}$ are replaced in the cost function and/or the constraints by a smooth approximation (see Fig. 2.7.1). On the other hand there are situations, particularly arising in the context of duality where nondifferentiable cost functions are unavoidable, and the use of a nondifferentiable optimization method is naturally

Sec. 2.7   Some Practical Guidelines   229

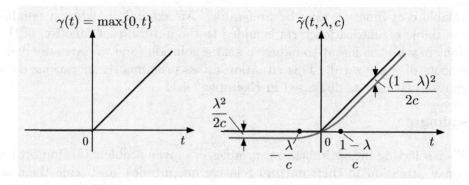

Figure 2.7.1. Smoothing of nondifferentiable terms. The function $f : \Re^n \to \Re$ given by

$$f(x) = \max\big\{f_1(x), f_2(x)\big\} = f_1(x) + \max\big\{0, f_2(x) - f_1(x)\big\}$$

is replaced by

$$f_1(x) + \tilde{\gamma}\big(f_2(x) - f_1(x), \lambda, c\big),$$

where $\tilde{\gamma}(t, \lambda, c)$ is a differentiable function that approximates $\gamma(t) = \max\{0, t\}$. This function depends on a parameter $\lambda \in [0, 1]$ and a parameter $c > 0$. It has the form

$$\tilde{\gamma}(t, \lambda, c) = \begin{cases} t - \dfrac{(1-\lambda)^2}{2c} & \text{if } \dfrac{1-\lambda}{c} \leq t \\ \lambda t + \dfrac{c}{2} t^2 & \text{if } -\dfrac{\lambda}{c} \leq t \leq \dfrac{1-\lambda}{c} \\ -\dfrac{\lambda^2}{2c} & \text{if } t \leq -\dfrac{\lambda}{c} \end{cases}$$

shown in the figure. The accuracy of the approximation increases as c increases. The scalar λ plays the role of a Lagrange multiplier; see Exercise 6.4.3 in Section 6.4. This smoothing method can be generalized for the case

$$f(x) = \max\big\{f_1(x), \ldots, f_m(x)\big\};$$

see [Ber75c], [Ber77], [Geo77], [Pol79], [Pap81], [Ber82a], [PiZ94] for further analysis. An alternative smooth approximation of f is the function

$$\frac{1}{c} \ln\left\{\sum_{i=1}^{m} \lambda_i e^{cf_i(x)}\right\}.$$

where $c > 0$ and λ_i satisfy $\lambda_i > 0$ and $\sum_{i=1}^{m} \lambda_i = 1$ (see [Ber82a], Section 5.1.3).

suited to the character of the problem (see Chapter 7). Note that gradient methods can benefit from the use of automatic differentiation programs that compute first and second derivatives from user-supplied programs that compute only function values; see the discussions in the books [NoW06], [GrW08], and the references quoted there.

Let us finally mention that for some types of problems a nondifferentiable cost function may be preferable. An example is when a nondifferentiable regularization term is added to the natural cost function of the problem with the intent to induce a sparse solution (one where many components of x are zero). This situation arises prominently in various data analysis settings, as discussed in Example 2.4.1.

Scaling

When selecting the optimization variables in a given problem it is important to pay attention to their natural relative magnitudes, and scale them so that their values are neither too large nor too small. This is useful to control roundoff error and it is also often helpful for improving the condition number of the problem, and the natural convergence rate of steepest descent and other algorithms. However, one should also pay attention to the range of the variables as well, and use a translation to somewhere in the middle of that range. In particular, suppose that the optimization variable x_i is expected to take values in the range $[\alpha_i, \beta_i]$. Then it may be helpful to use the transformation of variables

$$y_i = \frac{x_i - (\alpha_i + \beta_i)/2}{(\beta_i - \alpha_i)/2},$$

so that the new variable y_i takes values in the range $[-1, 1]$. Note that this preliminary scaling of the variables complements but is not a substitute for the iteration-dependent scaling that we discussed in Section 1.3 and which is inherent in Newton-like methods.

Some least squares problems are special because the variables can be scaled automatically by scaling some of the coefficients in the data blocks. As an example consider the linear least squares problem

$$\text{minimize} \sum_{i=1}^{m} (y_i - a_i'x)^2$$
$$\text{subject to } x \in \Re^n,$$

where y_i are given scalars and a_i are given vectors in $\Re^n$ with components a_{ij}, $j = 1, \ldots, n$. The jth diagonal element of the Hessian matrix is $\sum_{i=1}^{m} a_{ij}^2$. Thus reasonable scaling of the variables is often obtained by multiplying for each j, all components a_{ij} with a common scalar s_j so that they all lie in the range $[-1, 1]$. Then the diagonal elements of the Hessian are roughly equalized, while the optimization variables x_j are divided by s_j.

Note that for problems involving sums of functions, including least squares, the incremental Gauss-Newton and Newton methods of Sections 2.4.3 and 2.4.4 provide a natural form of scaling. Their diagonally scaled versions are well suited for problems where matrix-vector multiplications are expensive because of large dimension of the optimization vector x.

Method Selection

An important consideration here is whether a one-time solution of a given problem is sought or whether the problem (with data variations) will be solved many times on a production basis. In the former case any method that will do the job is sufficient, so one should aim for speed of code development, perhaps through the use of an existing optimization package. In the latter case, efficiency and accuracy of solution become important, so careful experimentation with a variety of methods may be appropriate in order to obtain a method that solves the problem fast and exploits its special structure.

Throughout this chapter we emphasized the inherent scaling properties and the fast convergence rate of Newton-like methods when derivatives are available and the problem is nonsingular. Indeed, Newton-like methods should be seriously considered in most practical optimization settings where the requisite differentiability conditions are satisfied. It is worth repeating, however, that for singular and difficult problems, simpler steepest descent-like methods with simple scaling may be more effective. Generally, in lack of a clear choice it is best to start with a simple method that can be easily coded and understood, and be prepared to proceed with more sophisticated choices.

Once a type of method is selected, it must be tuned to the problem at hand. In particular, stepsize rule parameters and termination criteria must be chosen. This often requires trial and error, particularly for first order methods, for a diminishing or constant stepsize, and for difficult (e.g., singular) problems. Termination criteria should be scale-free to the extent possible. For example it is preferable to use as termination tests such as

$$\|x^k - x^{k-1}\| \leq \epsilon \|x^k\| \quad \text{or} \quad \|\nabla f(x^k)\| \leq \epsilon \|\nabla f(x^0)\|$$

rather than $\|x^k - x^{k-1}\| \leq \epsilon$ or $\|\nabla f(x^k)\| \leq \epsilon$, respectively.

Validation

Finally one must strive to be convinced that the answers obtained from the computation are reasonable approximations to a (global) minimum. There are no systematic methods for this, but a few heuristics tailored to the problem at hand are adequate in many cases. In particular, suppose that some special case or variation of the given problem has a known optimal solution. Then it is reassuring if the coded method succeeds in finding this solution. Similarly, if a lower bound to the optimal value is known, it is reassuring when the algorithm comes close to achieving this lower bound. For example, in least squares problems, a small cost function value is a strong indication of success.

Another frequently used technique is to start the algorithm from widely varying starting points and see if it will give a better answer. This is particularly recommended in the case of multiple local minima. Finally, one may vary the parameters of the algorithm and also try alternative algorithms to see if substantial improvements can be obtained.

2.8 NOTES AND SOURCES

Section 2.1: Conjugate direction methods were introduced by Hestenes and Stiefel [HeS52]. The convergence rate of the conjugate gradient method is discussed in Fadeev and Fadeeva [FaF63], and also in Luenberger [Lue84], which we follow in our discussion. The conjugate gradient method with the formula (2.35) is known as the Poljak-Polak-Ribiere method and its convergence was analyzed by Poljak [Pol69a], and Polak and Ribiere [PoR69]. An extensive discussion of conjugate direction methods can be found in the monograph by Hestenes [Hes80].

Section 2.2: The DFP method is due to Davidon [Dav59]. It was popularized primarily through the work of Fletcher and Powell [FlP63]. The BFGS method was independently derived by Broyden [Bro70], Fletcher [Fle70a], Goldfarb [Gol70], and Shanno [Sha70]. For surveys and extensive discussions of quasi-Newton methods, including convergence analysis for a nonquadratic cost function, see Dennis and Moré [DeM77], Dennis and Schnabel [DeS83], Nazareth [Naz94], and Polak [Pol97].

Section 2.3: Coordinate descent has a long history. For a discussion of recent work in this area, see Section 3.7, where constrained versions of the coordinate descent approach are considered. Direct search methods also have a long history. For further discussion and additional methods, see Gill and Murray [GiM74], Avriel [Avr76], Kelley [Kel99], Lucidi, Sciandrone, and Tseng [LST01], Nazareth and Tseng [NaT02], and Kolda, Lewis, and Torczon [KLT03].

Section 2.4: Incremental gradient methods for linear least squares problems attracted attention with the influential paper by Widrow and Hoff [WiH60]. For more recent analyses, see Gaivoronski [Gai94], Grippo [Gri94], Luo and Tseng [LuT94a], Mangasarian and Solodov [MaS94], Tseng [Tse98], Solodov [Sol98], and Bertsekas and Tsitsiklis [BeT96], [BeT00]. Proposition 2.4.1 is due to Luo [Luo91] and stems from an earlier result of Kohonen [Koh74]. For a description of methods for neural network training problems, see Bishop [Bis95], Bertsekas and Tsitsiklis [BeT96], and Haykin [Hay11]. The ill-conditioned character of these problems is discussed in Saarinen, Bramley, and Cybenko [SBC93].

Incremental aggregated gradient (and subgradient) methods were first proposed, to our knowledge, by Nedić, Bertsekas, and Borkar [NBB01]. They were motivated primarily by distributed asynchronous solution of dual separable problems, similar to the ones that will be discussed in Section 6.1.5 and in Chapter 7. A convergence result was shown in [NBB01] assuming that the delays are bounded, and that the stepsize sequence $\{a^k\}$ is diminishing and satisfies the conditions

$$\sum_{k=0}^{\infty} \alpha^k = \infty, \qquad \sum_{k=0}^{\infty} (\alpha^k)^2 < \infty.$$

The constant stepsize version of the method attracted considerable attention through the paper by Blatt, Hero, and Gauchman [BHG08], who first proved the important linear convergence result of Prop. 2.4.2 for the case where the cost components f_i are quadratic and the delayed indexes ℓ_i satisfy certain restrictions that are consistent with a cyclic selection of components for iteration (see also [AFB06]). This result was subsequently extended for nonquadratic problems and for variants of the method by several other authors, including Schmidt, Le Roux, and Bach [SLB13], Mairal [Mai13], [Mai14], Defazio, Caetano, and Domke [DCD14], and Reddi et al. [RSP16]. The use of arbitrary indexes $\ell_i \in [k-b, k]$ in the aggregated gradient method was introduced in the paper by Gurbuzbalaban, Ozdaglar, and Parillo [GOP15a], which proved Prop. 2.4.2 in the form given here.

The incremental Gauss-Newton method was proposed by Davidon [Dav76] on an empirical basis. The development given here in terms of the extended Kalman filter and the corresponding convergence analysis is due to Bertsekas [Ber96b]; see also Moriyama, Yamashita, and Fukushima [MYF03]. In the case where $\lambda < 1$, and for some x^* we have $g(x^*) = 0$ and $\nabla g(x^*)$: nonsingular, the method has been shown to converge to x^* linearly with convergence ratio λ when started sufficiently close to x^* (Pappas [Pap82]). There are also many stochastic analyses and applications of the extended Kalman filter in the literature of estimation and control of dynamic systems.

The incremental Newton and related methods have been considered by several authors in a stochastic approximation framework; e.g., by Sakrison [Sak66], Venter [Ven67], Fabian [Fab73], Poljak and Tsypkin [PoT80b], [PoT81], and more recently by Bottou and LeCun [BoL05], and Bhatnagar, Prasad, and Prashanth [BPP13]. Among others, these references quantify the convergence rate advantage that stochastic Newton methods have over stochastic gradient methods. Deterministic incremental Newton methods have received little attention (for a recent work see Gurbuzbalaban, Ozdaglar, and Parrilo [GOP15b]). They admit an analysis that is similar to a deterministic analysis of the extended Kalman filter (see [Ber96b], [MYF03]). The incremental Newton method with diagonal Hessian approximation (Section 2.4.4) seems to be new.

Section 2.5: The totally asynchronous convergence theorem was first given in the author's paper [Ber83], where it was applied to a variety of algorithms, including gradient methods for unconstrained optimization (see also Bertsekas and Tsitsiklis [BeT89], where a textbook account is presented with many references, as well as the survey paper [BeT91]). Partially asynchronous gradient algorithms were developed in the paper by Tsitsiklis, Bertsekas, and Athans [TBA85], and more extensively in [BeT89] (Chapter 7), including the convergence and rate of convergence aspects of distributed asynchronous gradient and coordinate descent methods (Section 7.5); see also the papers by Tseng, Bertsekas, and Tsitsiklis [TBT90], Bertsekas and Tsitsiklis [BeT00], and the rate of convergence analysis of stochastic partially asynchronous algorithms by Duchi, Chaturapruek, and Re [DCR15]. The doctoral thesis by Feyzmahdavian [Fey16] provides an extensive discussion of the convergence rate of deterministic asynchronous algorithms.

Section 2.6: An extensive reference on optimality conditions for discrete-time optimal control is Canon, Cullum, and Polak [CCP70]. For computational methods see Polak [Pol71], [Pol73], and [Pol97]. The implementation of Newton's method given here was derived in an early draft of this book [Ber82e], and was further formalized in Dunn and Bertsekas [DuB89]. It was also independently derived by Pantoja and Mayne [PaM89] (see Mitter [Mit66] for a continuous-time version of this algorithm). An alternative approach to discrete-time optimal control is based on dynamic programming; see e.g., Bertsekas [Ber05a]. Discrete-time optimal control problems often arise from discretization of their continuous-time counterparts. The associated issues are discussed by Cullum [Cul71] and Polak [Pol97].

The discrete-time optimal control algorithms that we have discussed find considerable application in *stochastic* optimal feedback control, within the context of a variety of suboptimal control methods known as *certainty equivalent control*, *model predictive control*, and *receding horizon control*; see Keerthi and Gilbert [KeG88], the surveys by Wright [Wri97b], Morari and Lee [MoL99], and Mayne et. al. [MRR00], and the references quoted there. Here, at each time step i, and given the state x_i, the uncertain quantities in the system equation are fixed at some typical or estimated values, the corresponding optimal control problem is solved (over the time periods from i to N, and starting at x_i), and the control u_i applied at time i is the first one in the optimizing sequence $\{u_i, u_{i+1}, \ldots, u_{N-1}\}$. The process is then repeated at the next time step. For an extensive discussion of this and other related suboptimal control methods, such as rollout algorithms, see the author's dynamic programming texts [Ber05a], [Ber12], and survey paper [Ber05b].

Section 2.7: A great deal of material on the implementation of nonlinear programming methods is given by Gill, Murray, and Wright [GMW81]. A description of available software packages is provided by Moré and Wright [MoW93], and Schittkowski [Sch12].

3
Optimization over a Convex Set

Contents

3.1. Constrained Optimization Problems	p. 236
3.1.1. Necessary and Sufficient Conditions for Optimality	p. 236
3.1.2. Existence of Optimal Solutions	p. 246
3.2. Feasible Directions - Conditional Gradient Method	p. 257
3.2.1. Descent Directions and Stepsize Rules	p. 257
3.2.2. The Conditional Gradient Method	p. 262
3.3. Gradient Projection Methods	p. 272
3.3.1. Feasible Directions and Stepsize Rules Based on Projection	p. 272
3.3.2. Convergence Analysis	p. 283
3.4. Two-Metric Projection Methods	p. 292
3.5. Manifold Suboptimization Methods	p. 298
3.6. Proximal Algorithms	p. 307
3.6.1. Rate of Convergence	p. 312
3.6.2. Variants of the Proximal Algorithm	p. 318
3.7. Block Coordinate Descent Methods	p. 323
3.7.1. Variants of Coordinate Descent	p. 327
3.8. Network Optimization Algorithms	p. 331
3.9. Notes and Sources	p. 338

In this chapter we consider the constrained optimization problem

$$\text{minimize } f(x)$$
$$\text{subject to } x \in X,$$

where, in the absence of an explicit statement to the contrary, we assume throughout that:

(a) X is a nonempty and convex subset of $\Re^n$. When dealing with algorithms, we assume in addition that X is closed.

(b) The function $f : \Re^n \mapsto \Re$ is continuously differentiable over an open set that contains X.

This problem generalizes the unconstrained optimization problem of the preceding chapters, where $X = \Re^n$. We will see that the main algorithmic ideas for solving the unconstrained and the constrained problems are quite similar.

Usually the set X has structure specified by equations and inequalities. If we take into account this structure, some new algorithmic ideas, based on Lagrange multipliers and duality theory, come into play. These ideas will not be discussed in the present chapter, but they will be the focus of subsequent chapters.

Similar to the unconstrained case, the methods of this chapter are based on iterative descent along suitably obtained directions. However, these directions must have the additional property that they maintain feasibility of the iterates. Such directions are called *feasible*, and as we will see later, they are usually obtained by solving certain optimization subproblems. We will consider various ways to construct feasible descent directions following the discussion of optimality conditions in the next section.

3.1 CONSTRAINED OPTIMIZATION PROBLEMS

In this section we consider the main analytical techniques for our problem, and we provide some examples of their application.

3.1.1 Necessary and Sufficient Conditions for Optimality

We first expand the unconstrained optimality conditions of Section 1.1 for the problem of minimizing the continuously differentiable function f over the convex set X. Recalling the definitions of Section 1.1, a vector $x \in X$ is referred to as a *feasible* vector, and a vector $x^* \in X$ is a *local minimum* of f over X if it is no worse than its feasible neighbors; that is, if there exists an $\epsilon > 0$ such that

$$f(x^*) \leq f(x), \qquad \forall \, x \in X \text{ with } \|x - x^*\| < \epsilon.$$

Sec. 3.1   Constrained Optimization Problems   237

A vector $x^* \in X$ is a *global minimum* of f over X if it is no worse than all other feasible vectors, i.e.,

$$f(x^*) \leq f(x), \qquad \forall\ x \in X.$$

The local or global minimum x^* is said to be *strict* if the corresponding inequality above is strict for $x \neq x^*$. Local and global maxima of f over X are similarly defined; they are the local and global minima of the function $-f$ over X. If in addition to X, the cost function f is also convex, a local minimum is also global, as shown in Prop. 1.1.2.

As in unconstrained optimization, we expect that at a local minimum x^*, the first order variation, due to a small feasible variation Δx is nonnegative:

$$\nabla f(x^*)' \Delta x \geq 0.$$

Because X is convex, a feasible variation has the form $\Delta x = x - x^*$, where $x \in X$. Thus the preceding optimality condition can be written as in the following proposition, and is also a sufficient condition for optimality when f is convex.

Proposition 3.1.1: (Optimality Condition)

(a) If x^* is a local minimum of f over X, then

$$\nabla f(x^*)'(x - x^*) \geq 0, \qquad \forall\ x \in X. \tag{3.1}$$

(b) If f is convex over X, then the condition of part (a) is also sufficient for x^* to minimize f over X.

Proof: (a) Fix $x \in X$ and define the function $g(\alpha) = f\big(x^* + \alpha(x - x^*)\big)$, which is continuously differentiable in an open interval containing $[0, 1]$. By using the chain rule to differentiate g, we have

$$0 \leq \lim_{\alpha \downarrow 0} \frac{f\big(x^* + \alpha(x - x^*)\big) - f(x^*)}{\alpha} = \frac{dg(0)}{d\alpha} = \nabla f(x^*)'(x - x^*),$$

where the inequality follows from the assumption that x^* is a local minimum. This proves Eq. (3.1).

(b) See Prop. 1.1.3(a). **Q.E.D.**

Figure 3.1.1 interprets the optimality condition (3.1) geometrically. Figure 3.1.2 shows how the condition can fail in the absence of convexity of the constraint set X.

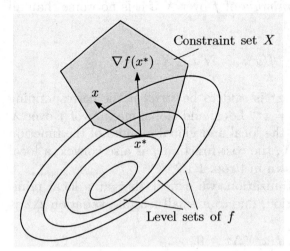

Figure 3.1.1. Geometric interpretation of the optimality condition of Prop. 3.1.1(a). At a local minimum x^*, the gradient $\nabla f(x^*)$ makes an angle less than or equal to 90 degrees with all feasible variations $x - x^*$, $x \in X$.

A vector x^* satisfying the optimality condition (3.1) is referred to as a *stationary point*. In the absence of convexity of f, this condition may also be satisfied by local maxima and other points. To see this, note that if $X = \Re^n$ or if x^* is an interior point of X, the condition (3.1) reduces to the stationarity condition $\nabla f(x^*) = 0$ of unconstrained optimization. We now illustrate the condition (3.1) with some examples.

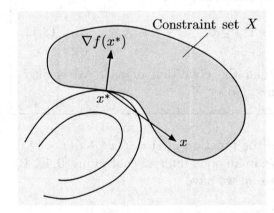

Figure 3.1.2. Illustration of how the necessary optimality condition may fail when X is not convex. Here x^* is a local minimum but we have

$$\nabla f(x^*)'(x - x^*) < 0$$

for the feasible vector x shown.

Example 3.1.1 (Optimization Subject to Bounds)

Consider a positive orthant constraint
$$X = \{x \mid x \geq 0\}.$$
Then the necessary condition (3.1) for $x^* = (x_1^*, \ldots, x_n^*)$ to be a local minimum is

$$\sum_{i=1}^{n} \frac{\partial f(x^*)}{\partial x_i}(x_i - x_i^*) \geq 0, \qquad \forall\, x_i \geq 0,\ i = 1, \ldots, n. \tag{3.2}$$

Sec. 3.1 Constrained Optimization Problems 239

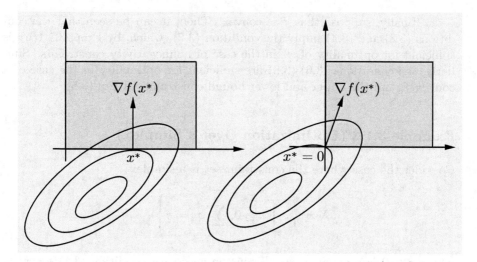

Figure 3.1.3. Illustration of the optimality condition for an orthant constraint. At a minimum x^*, all the partial derivatives $\partial f(x^*)/\partial x_i$ are nonnegative, and they are zero for the inactive constraint indices, i.e. the indices i with $x_i^* > 0$. If all constraints are inactive, the optimality condition reduces to the unconstrained optimization condition $\nabla f(x^*) = 0$.

Let us fix i. By letting $x_j = x_j^*$ for $j \neq i$ and $x_i = x_i^* + 1$ in Eq. (3.2), we obtain

$$\frac{\partial f(x^*)}{\partial x_i} \geq 0, \qquad \forall\, i. \tag{3.3}$$

Furthermore, if we have $x_i^* > 0$, by letting $x_j = x_j^*$ for $j \neq i$ and $x_i = \tfrac{1}{2} x_i^*$ in Eq. (3.2), we obtain $\partial f(x^*)/\partial x_i \leq 0$, which when combined with Eq. (3.3) yields

$$\frac{\partial f(x^*)}{\partial x_i} = 0, \qquad \text{if } x_i^* > 0. \tag{3.4}$$

The preceding two conditions are equivalent to the necessary condition (3.1) for x^* to be a local minimum, and are illustrated in Fig. 3.1.3.

Consider also the case where the constraints are upper and lower bounds on the variables, i.e.,

$$X = \{x \mid \xi_i \leq x_i \leq \zeta_i, i = 1, \ldots, n\}, \tag{3.5}$$

where ξ_i and ζ_i are given scalars. A variant of the preceding argument shows that if x^* is a local minimum, then

$$\frac{\partial f(x^*)}{\partial x_i} \geq 0, \qquad \text{if } x_i^* = \xi_i, \tag{3.6}$$

$$\frac{\partial f(x^*)}{\partial x_i} \leq 0, \qquad \text{if } x_i^* = \zeta_i, \tag{3.7}$$

$$\frac{\partial f(x^*)}{\partial x_i} = 0, \qquad \text{if } \xi_i < x_i^* < \zeta_i. \tag{3.8}$$

Finally, suppose that f is convex. Then, it can be seen that the conditions (3.3) and (3.4) imply the condition (3.2), which, by Prop. 3.1.1(b), is sufficient for optimality of x^* in the case of nonnegativity constraints. Similarly, the conditions (3.6)-(3.8) are sufficient for optimality in the case of a convex f, and the upper and lower bound constraints of Eq. (3.5).

Example 3.1.2 (Optimization Over a Simplex)

Consider the case where the constraint set is a simplex

$$X = \left\{ x \mid x \geq 0, \sum_{i=1}^{n} x_i = r \right\},$$

where $r > 0$ is a given scalar. Then the necessary condition (3.1) for $x^* = (x_1^*, \ldots, x_n^*)$ to be a local minimum is

$$\sum_{i=1}^{n} \frac{\partial f(x^*)}{\partial x_i}(x_i - x_i^*) \geq 0, \qquad \forall \ x \geq 0 \text{ with } \sum_{i=1}^{n} x_i = r. \qquad (3.9)$$

Fix an index i for which $x_i^* > 0$ and let j be any other index. By using the feasible vector $x = (x_1, \ldots, x_n)$ with $x_i = 0$, $x_j = x_j^* + x_i^*$, and $x_m = x_m^*$ for all $m \neq i, j$ in Eq. (3.9), we obtain

$$\left(\frac{\partial f(x^*)}{\partial x_j} - \frac{\partial f(x^*)}{\partial x_i} \right) x_i^* \geq 0,$$

or equivalently

$$x_i^* > 0 \quad \Longrightarrow \quad \frac{\partial f(x^*)}{\partial x_i} \leq \frac{\partial f(x^*)}{\partial x_j}, \qquad \forall \ j. \qquad (3.10)$$

Thus, all coordinates which are positive at the optimum must have minimal (and equal) partial cost derivatives.

Assuming that f is convex, we can show that Eq. (3.10) is also sufficient for global optimality of x^*. Indeed, suppose that x^* is feasible and satisfies Eq. (3.10), and let

$$M = \min_{i=1,\ldots,n} \frac{\partial f(x^*)}{\partial x_i}.$$

For every $x \in X$, we have $\sum_{i=1}^{n}(x_i - x_i^*) = 0$, so that

$$0 = \sum_{i=1}^{n} M(x_i - x_i^*) \leq \sum_{\{i \mid x_i > x_i^*\}} \frac{\partial f(x^*)}{\partial x_i}(x_i - x_i^*) + \sum_{\{i \mid x_i < x_i^*\}} M(x_i - x_i^*).$$

If i is such that $x_i < x_i^*$, we must have $x_i^* > 0$ and, by condition (3.10), $M = \partial f(x^*)/\partial x_i$. Thus M can be replaced by $\partial f(x^*)/\partial x_i$ in the right-hand side of the preceding inequality, thereby yielding Eq. (3.9), which by Prop. 3.1.1(b), implies that x^* is optimal. Thus, when f is convex, condition (3.10) is necessary and sufficient for optimality of x^*.

The preceding argument can be straightforwardly generalized to the case where X is a Cartesian product of several simplices. Then, there is a separate condition of the form (3.10) for each simplex; that is, the condition

$$\frac{\partial f(x^*)}{\partial x_i} \leq \frac{\partial f(x^*)}{\partial x_j}$$

holds for all i with $x_i^* > 0$ and all j for which x_j is constrained by the same simplex as x_i.

Example 3.1.3 (Optimal Routing in a Communication Network)

We are given a directed graph with a set of directed arcs $\mathcal{A}$, which is viewed as a model of a data communication network. We are also given a set W of ordered node pairs $w = (i, j)$. The nodes i and j are referred to as the *origin* and the *destination* of w, respectively, and w is referred to as an OD pair. For each w, we are given a scalar r_w referred to as the *input traffic* of w. In the context of routing of data in a communication network, r_w (measured in data units/second) is the arrival rate of traffic entering and exiting the network at the origin and the destination of w, respectively. The routing objective is to divide each r_w among the many paths from origin to destination in a way that the resulting total arc flow pattern minimizes a suitable cost function. We denote:

P_w: A given set of paths that start at the origin and end at the destination of w. All arcs on each of these paths are oriented in the direction from the origin to the destination.

x_p: The portion of r_w assigned to path p, also called the *flow of path p*.

The collection of all path flows $\{x_p \mid p \in P_w, w \in W\}$ must satisfy the constraints

$$\sum_{p \in P_w} x_p = r_w, \qquad \forall\ w \in W, \tag{3.11}$$

$$x_p \geq 0, \qquad \forall\ p \in P_w,\ w \in W, \tag{3.12}$$

as shown in Fig. 3.1.4. The total flow f_a of arc a is the sum of all path flows traversing the arc:

$$f_a = \sum_{\substack{\text{all paths } p \\ \text{containing } a}} x_p. \tag{3.13}$$

Consider a cost function of the form

$$\sum_{a \in \mathcal{A}} D_a(f_a). \tag{3.14}$$

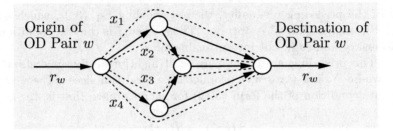

Figure 3.1.4. Constraints for the path flows of an OD pair w. The path flows x_p of the paths $p \in P_w$, should be nonnegative and add up to the given traffic input r_w of the OD pair.

The problem is to find a set of path flows $\{x_p\}$ that minimize this cost function subject to the constraints of Eqs. (3.11)-(3.13). We assume that D_a is a convex and continuously differentiable function of f_a with first derivative denoted by D'_a. In data routing applications, the form of D_a is often based on a queueing model of average delay (see [BeG92]).

The preceding problem is known as a *multicommodity network flow problem*. The terminology reflects the fact that the arc flows consist of several different commodities; in the present example, the different commodities are the data of the distinct OD pairs.

By expressing the total flows f_a in terms of the path flows in the cost function (3.14) [using Eq. (3.13)], the problem can be formulated in terms of the path flow variables $\{x_p \mid p \in P_w, w \in W\}$ as

$$\text{minimize } D(x)$$
$$\text{subject to } \sum_{p \in P_w} x_p = r_w, \quad \forall \, w \in W,$$
$$x_p \geq 0, \quad \forall \, p \in P_w, \, w \in W,$$

where

$$D(x) = \sum_a D_a \left(\sum_{\substack{\text{all paths } p \\ \text{containing } a}} x_p \right)$$

and x is the vector of path flows x_p.

This problem, viewed as a problem in the variables $\{x_p\}$, has a convex and differentiable cost function, and a constraint set that is a Cartesian product of simplices. Therefore, we can apply the condition (3.10) developed in the preceding example, separately, for the path flows of each OD pair. In particular, we will show that optimal routing directs traffic exclusively along paths that are shortest with respect to arc lengths that depend on the flows carried by the arcs. To this end, we first derive the partial derivative of D with respect to x_p. It is given by

$$\frac{\partial D(x)}{\partial x_p} = \sum_{\substack{\text{all arcs } a \\ \text{on path } p}} D'_a(f_a), \qquad (3.15)$$

Sec. 3.1   Constrained Optimization Problems   243

where the first derivatives D'_a are evaluated at the arc flows f_a corresponding to x. From this equation, it is seen that $\partial D / \partial x_p$ is *the length of path p when the length of each arc a is taken to be the first derivative D'_a evaluated at f_a*. Consequently, in what follows, $\partial D / \partial x_p$ is called the *first derivative length of path p*.

By using the necessary and sufficient optimality condition (3.10) of the preceding example, we obtain for all $w \in W$ and $p \in P_w$

$$x_p^* > 0 \implies \frac{\partial D(x^*)}{\partial x_p} \leq \frac{\partial D(x^*)}{\partial x_{p'}}, \qquad \forall \, p' \in P_w. \tag{3.16}$$

In words, the above condition says that *a set of path flows is optimal if and only if path flow is positive only on paths with a minimum first derivative length*. The condition also implies that at an optimum, the paths along which the input flow r_w of OD pair w is split must have *equal* length (and less than or equal length to that of all other paths of w).

Example 3.1.4 (Traffic Assignment)

We are given a directed graph, which is viewed as a model of a transportation network. The arcs of the graph represent transportation links such as highways, rail lines, etc. The nodes of the graph represent junction points where traffic can exit from one transportation link and enter another. Similar to the preceding example, we are given a set W of OD pairs. For OD pair $w = (i, j)$, there is a known input $r_w > 0$ representing traffic entering the network at the origin node i of w and exiting the network at the destination node j of w. The input r_w is to be divided among a set P_w of paths starting at the origin node of w and ending at the destination node of w. Let x_p denote the portion of r_w carried by path p and let x be the vector having as coordinates all the path flows x_p, $p \in P_w$, $w \in W$. Thus, the constraint set is the Cartesian product of simplices defined by Eqs. (3.11) and (3.12) of the preceding example.

For each arc a, we are given a function $T_a(f_a)$ of the total traffic f_a carried by arc a [cf. Eq. (3.13)]. This function models the time required for traffic to travel from the start node to the end node of the arc a. An important problem is to find a path flow vector $x^* \in X$ that consists of path flows that are positive only on paths of minimum travel time. By this we mean that for all $w \in W$ and paths $p \in P_w$, we require that

$$x_p^* > 0 \implies t_p(x^*) \leq t_{p'}(x^*), \qquad \forall \, p' \in P_w, \, \forall \, w \in W, \tag{3.17}$$

where $t_p(x)$, the travel time of path p, is defined as the sum of the travel times of the arcs of the path,

$$t_p(x) = \sum_{\substack{\text{all arcs } a \\ \text{on path } p}} T_a(f_a), \qquad \forall \, p \in P_w, \, \forall \, w \in W.$$

The preceding problem draws its validity from a hypothesis, called the *user-optimization principle*, which asserts that traffic equilibrium is established when each user of the network chooses, among all available paths, a path requiring minimum travel time. Thus, assuming that the user-optimization principle holds, a path flow vector x^* that solves the problem also models accurately the distribution of traffic through the network, and can be used for traffic projections when planning modifications to the network.

We now observe that the minimum travel time hypothesis (3.17) is identical with the routing optimality condition (3.16) if we identify the arc travel time $T_a(f_a)$ with the cost derivative $D'_a(f_a)$ [cf. Eq. (3.15)]. It follows that we can solve the transportation problem by converting it to the optimal routing problem of the preceding example with the identification

$$D_a(f_a) = \int_0^{f_a} T_a(\xi) d\xi. \tag{3.18}$$

If we assume that T_a is continuous and monotonically nondecreasing, as is natural in a transportation context, it is straightforward to show that the function D_a as defined by Eq. (3.18) is convex with derivative D'_a equal to T_a, so a minimum first derivative length path is a path of minimum travel time.

Example 3.1.5 (Equality-Constrained Quadratic Programming)

Consider the quadratic programming problem

$$\begin{aligned} \text{minimize} \quad & \tfrac{1}{2}\|x\|^2 + c'x \\ \text{subject to} \quad & Ax = 0, \end{aligned} \tag{3.19}$$

where c is a given vector in $\Re^n$ and A is an $m \times n$ matrix of rank m.

By adding the constant term $\tfrac{1}{2}\|c\|^2$ to the cost function, we can equivalently write this problem as

$$\begin{aligned} \text{minimize} \quad & \tfrac{1}{2}\|c + x\|^2 \\ \text{subject to} \quad & Ax = 0, \end{aligned}$$

which is the problem of projecting the vector $-c$ on the subspace $X = \{x \mid Ax = 0\}$. By Eq. (3.1), as specialized in the projection theorem (Prop. 1.1.4), a vector x^* such that $Ax^* = 0$ is the unique projection if and only if

$$(c + x^*)'x = 0, \quad \forall \, x \text{ with } Ax = 0.$$

It can be seen that the vector

$$x^* = -\bigl(I - A'(AA')^{-1}A\bigr)c \tag{3.20}$$

satisfies this condition and is thus the unique solution of the quadratic programming problem (3.19). [The matrix AA' is invertible because A has rank m (Prop. A.20 in Appendix A).]

Sec. 3.1 Constrained Optimization Problems

Note that x^* is zero if and only if c is orthogonal to the subspace $X = \{x \mid Ax = 0\}$, or equivalently from Eq. (3.20),

$$c = A'(AA')^{-1}Ac.$$

Thus, in this case, c can be expressed in the form $A'\mu$, for some $\mu \in \Re^m$, i.e., as a linear combination of the rows of A. Conversely, it is seen that every vector of the form $A'\mu$, with $\mu \in \Re^m$, is orthogonal to the subspace X.

The conclusion is that *the orthogonal complement $X^\perp$ of X (the set of all vectors orthogonal to all x with $Ax = 0$), is the subspace $\{x \mid x = A'\mu, \mu \in \Re^m\}$*. Our proof of this fact assumed that A has rank m but can be easily modified for the case where the rows of A are linearly dependent by eliminating a sufficient number of dependent rows of A.

Consider now the more general quadratic program

$$\begin{aligned}&\text{minimize} \quad \tfrac{1}{2}(x-\bar{x})'H(x-\bar{x}) + c'(x-\bar{x})\\ &\text{subject to} \quad Ax = b,\end{aligned} \quad (3.21)$$

where c and A are as before, H is a positive definite symmetric matrix, b is a given vector in $\Re^m$, and $\bar{x}$ is a given vector in $\Re^n$, which is feasible, i.e., satisfies $A\bar{x} = b$. By introducing the vector $y = H^{1/2}(x - \bar{x})$, we can write this problem as

$$\begin{aligned}&\text{minimize} \quad \tfrac{1}{2}\|y\|^2 + \left(H^{-1/2}c\right)' y\\ &\text{subject to} \quad AH^{-1/2}y = 0.\end{aligned}$$

Using Eq. (3.20) we see that the solution of this problem is

$$y^* = -\left(I - H^{-1/2}A'\left(AH^{-1}A'\right)^{-1}AH^{-1/2}\right)H^{-1/2}c,$$

and by passing to the x-coordinate system through the transformation $x^* - \bar{x} = H^{-1/2}y^*$, we obtain the optimal solution

$$x^* = \bar{x} - H^{-1}(c - A'\lambda), \quad (3.22)$$

where the vector λ is given by

$$\lambda = \left(AH^{-1}A'\right)^{-1} AH^{-1}c. \quad (3.23)$$

The quadratic program (3.21) contains as a special case the program

$$\begin{aligned}&\text{minimize} \quad \tfrac{1}{2}x'Hx + c'x\\ &\text{subject to} \quad Ax = b.\end{aligned} \quad (3.24)$$

This special case is obtained when $\bar{x}$ is given by

$$\bar{x} = H^{-1}A'(AH^{-1}A')^{-1}b. \quad (3.25)$$

Indeed $\bar{x}$ as given above satisfies $A\bar{x} = b$ as required, and for all x with $Ax = b$, we have
$$x'H\bar{x} = x'A'(AH^{-1}A')^{-1}b = b'(AH^{-1}A')^{-1}b,$$
which implies that for all x with $Ax = b$

$$\tfrac{1}{2}(x-\bar{x})'H(x-\bar{x}) + c'(x-\bar{x}) = \tfrac{1}{2}x'Hx + c'x + \left(\tfrac{1}{2}\bar{x}'H\bar{x} - c'\bar{x} - b'(AH^{-1}A')^{-1}b\right).$$

The last term in parentheses on the right-hand side above is constant, thus establishing that the programs (3.21) and (3.24) have the same optimal solution when $\bar{x}$ is given by Eq. (3.25). By combining Eqs. (3.22) and (3.25), we obtain the optimal solution of program (3.24):

$$x^* = -H^{-1}\left(c - A'\lambda - A'(AH^{-1}A')^{-1}b\right),$$

where λ is given by Eq. (3.23).

3.1.2 Existence of Optimal Solutions

As discussed in Section 1.1, the principal method for asserting the existence of a minimum of a continuous function $f : \Re^n \mapsto \Re$ over a subset X of $\Re^n$ is Weierstrass' theorem (Prop. A.8 in Appendix A), which applies when X is compact, or when X is closed and f is coercive over X. However, there are many types of problems, including linear and quadratic programming, where the assumptions of Weierstrass' theorem are often not satisfied. In this section, we will develop some tools to deal with such cases. †

Our main approach will be to *view the problem of existence of an optimal solution as a question of nonemptiness of the intersection of a nested sequence of sets*. Indeed, the set of minima X^* of f over X is equal to the intersection of the nonempty level sets of f when its domain is restricted to X, i.e.,

$$X^* = \cap_{k=0}^{\infty} \{x \in X \mid f(x) \leq \gamma^k\},$$

where $\{\gamma^k\}$ is any decreasing sequence such that

$$\gamma^k \downarrow f^* = \inf_{x \in X} f(x).$$

With this in mind, we introduce some definitions.

---

† The analysis in this section is theoretically important, but will not be used later, except for the existence of optimal solutions of linear and quadratic programming problems that have finite optimal value (Prop. 3.1.3). It may be skipped at first reading.

Sec. 3.1   *Constrained Optimization Problems*   **247**

Let $\{S^k\}$ be a sequence of nonempty closed sets such that $S^{k+1} \subset S^k$ for all k. We say that a vector $d \neq 0$ is an *asymptotic direction of* $\{S^k\}$ if there exists a sequence $\{x^k\}$ such that
$$x^k \in S^k, \qquad x^k \neq 0, \qquad k = 0, 1, \ldots$$
and
$$\|x^k\| \to \infty, \qquad \frac{x^k}{\|x^k\|} \to \frac{d}{\|d\|}.$$

A sequence $\{x^k\}$ associated with an asymptotic direction d as above is called an *asymptotic sequence corresponding to* d. Such a sequence is called *retractive* if there exists a bounded sequence of positive numbers $\{\alpha^k\}$ and an index $\bar{k}$ such that
$$x^k - \alpha^k d \in S^k, \qquad \forall \, k \geq \bar{k}.$$

Roughly speaking, an asymptotic direction is a direction along which we can escape towards ∞ through each of the sets S^k. A retractive sequence is one that goes to infinity along an asymptotic direction d, but still belongs to the corresponding sets S^k when shifted in the opposite direction $-d$ by some bounded positive amounts α^k. The importance of the notion of a retractive sequence may not be apparent at first sight, but is motivated by the following proposition. The idea of the proof is that $\cap_{k=0}^{\infty} S^k$ is empty if and only if there is an unbounded asymptotic sequence consisting of minimum norm vectors from the sets S^k. It will be shown that such a sequence cannot be retractive. This line of analysis is developed in several sources, including [AuT03], [BeT07], [Ber09], to which we refer for further discussion.

Proposition 3.1.2: Let $\{S^k\}$ be a sequence of nonempty closed sets such that $S^{k+1} \subset S^k$ for all k, and assume that all asymptotic sequences corresponding to asymptotic directions of $\{S^k\}$ are retractive. Then $\cap_{k=0}^{\infty} S^k$ is nonempty.

Proof: For each k, let x^k be a vector of minimum norm on the closed set S^k (such a vector exists by Weierstrass' theorem, since it can be obtained by minimizing $\|x\|$ over all x in the compact set $S^k \cap \{x \mid \|x\| \leq \|\bar{x}^k\|\}$, where $\bar{x}^k$ is any vector in S^k). It will be sufficient to show that a subsequence $\{x^k\}_{k \in \mathcal{K}}$ is bounded. Then, since for each m, we have $x^k \in S^m$ for all $k \in \mathcal{K}$, $k \geq m$, and S^m is closed, each of the limit points of $\{x^k\}_{k \in \mathcal{K}}$ will belong to $\cap_{m=0}^{\infty} S^m$, thereby proving the proposition. Thus, we will prove the proposition by showing that it is impossible to have $\|x^k\| \to \infty$.

Indeed, assume the contrary, let d be the limit of a subsequence $\{x^k / \|x^k\|\}_{k \in \mathcal{K}}$, and for each k, define $y^k = x^m$, where m is the smallest index $m \in \mathcal{K}$ with $k \leq m$. Then d is an asymptotic direction of $\{S^k\}$

and $\{y^k\}$ is an asymptotic sequence corresponding to d, so $\{y^k\}$ is retractive. Let $\{\alpha^k\}$ be a bounded sequence of positive numbers and $\bar{k}$ be such that $y^k - \alpha^k d \in S^k$ for all $k \geq \bar{k}$. We have $d'y^k \to \infty$, since $\|d\| = 1$ and $d'y^k/\|y^k\| \to 1$, so for all $k \geq \bar{k}$ with $2d'y^k > \alpha^k$, we obtain

$$\|y^k - \alpha^k d\|^2 = \|y^k\|^2 - \alpha^k(2d'y^k - \alpha^k) < \|y^k\|^2.$$

This is a contradiction, since for infinitely many k, y^k is the vector of minimum norm on S^k. **Q.E.D.**

For an example where the above proposition applies, consider the subsets of $\Re^2$

$$S^k = \left\{ (x_1, x_2) \ \Big| \ x_1 \geq -\frac{1}{k}, \ |x_2| \leq \frac{1}{k} \right\}.$$

It can be verified that every asymptotic sequence is retractive and indeed the intersection $\cap_{k=0}^\infty S^k$ is nonempty. On the other hand, the condition for nonemptiness of $\cap_{k=0}^\infty S^k$ of the proposition is far from necessary. For example, the intersection of the subsets $S^k = \{(x_1, x_2) \mid x_2 \geq kx_1^2\}$ of $\Re^2$ is nonempty, but the asymptotic sequence $\{(k, k^3)\}$ is not retractive. However, the proposition is useful in asserting existence of optimal solutions for important classes of problems. The next proposition considers the most prominent such class, and the exercises deal with several other types of problems.

Proposition 3.1.3: (Existence of Solutions of Linear and Quadratic Programming Problems) Let Q be a positive semidefinite symmetric $n \times n$ matrix, let c and $a_1, \ldots, a_r$ be vectors in $\Re^n$, and let $b_1, \ldots, b_r$ be scalars. Assume that the optimal value of the problem

$$\begin{aligned}
\text{minimize} \quad & x'Qx + c'x \\
\text{subject to} \quad & a_j'x + b_j \leq 0, \quad j = 1, \ldots, r,
\end{aligned} \quad (3.26)$$

is finite. Then the problem has at least one optimal solution.

Proof: Let f^* be the optimal value and let F be the feasible region

$$F = \{ x \mid a_j'x + b_j \leq 0, \ j = 1, \ldots, r \}.$$

Let $\{\gamma^k\}$ be a decreasing sequence with $\gamma^k \downarrow f^*$, and denote

$$S^k = \{ x \in F \mid x'Qx + c'x \leq \gamma^k \}.$$

Then the set of optimal solutions of the problem is $\cap_{k=0}^\infty S^k$, so by Prop. 3.1.2, it will suffice to show that for each asymptotic direction of $\{S^k\}$, all

Sec. 3.1 Constrained Optimization Problems

corresponding asymptotic sequences are retractive. Let d be an asymptotic direction and let $\{x^k\}$ be a corresponding asymptotic sequence.

We claim that
$$Qd = 0, \qquad c'd \leq 0, \tag{3.27}$$
$$a_j'd \leq 0, \qquad j = 1, \ldots, r. \tag{3.28}$$

Indeed, since $x^k \in S^k$, by denoting $d^k = x^k/\|x^k\|$, we have
$$d^{k\prime}Qd^k + \frac{c'd^k}{\|x^k\|} \leq \frac{\gamma^k}{\|x^k\|^2}.$$

Taking the limit as $k \to \infty$, and using the fact $d^k \to d$ and $\|x^k\| \to \infty$, we see that $d'Qd \leq 0$. Since Q is positive semidefinite, it follows that $Qd = 0$. We also have, using the positive semidefiniteness of Q,
$$c'd^k \leq \|x^k\| d^{k\prime}Qd^k + c'd^k \leq \frac{\gamma^k}{\|x^k\|},$$

so taking the limit as $k \to \infty$, we obtain $c'd \leq 0$. Also, for all j, we have $a_j'x^k + b_j \leq 0$, so that $a_j'd^k \leq -b_j/\|x^k\|$, and by taking the limit as $k \to \infty$, we obtain $a_j'd \leq 0$, for all j.

Next, using the finiteness of f^*, we will show that in fact $c'd = 0$. Indeed, let $\bar{x}$ be a feasible vector, and consider the vectors $\tilde{x}^k = \bar{x} + kd$. From the fact $a_j'd \leq 0$ [cf. Eq. (3.28)], it can be seen that $\tilde{x}^k$ is feasible. Thus, using the fact $Qd = 0$ [cf. Eq. (3.27)], we see that the cost function value corresponding to $\tilde{x}^k$ satisfies
$$f^* \leq (\bar{x} + kd)'Q(\bar{x} + kd) + c'(\bar{x} + kd) = \bar{x}'Q\bar{x} + c'\bar{x} + kc'd.$$

From the finiteness of f^*, it follows that $c'd \geq 0$. Since we have also shown that $c'd \leq 0$ [cf. Eq. (3.27)], it follows that $c'd = 0$. Thus, for all $\alpha > 0$ and k, the cost function value of the vector $x^k - \alpha d$ satisfies
$$(x^k - \alpha d)'Q(x^k - \alpha d) + c'(x^k - \alpha d) = x^{k\prime}Qx^k + c'x^k \leq \gamma^k.$$

It follows that to show that $\{x^k\}$ is retractive, it will suffice to find $\alpha > 0$ such that $x^k - \alpha d$ is feasible, i.e.,
$$a_j'(x^k - \alpha d) + b_j \leq 0, \tag{3.29}$$

for all $j = 1, \ldots, r$, and all sufficiently large k.

Indeed, Eq. (3.29) holds for all $\alpha > 0$ and all j such that $a_j'd = 0$. For any j such that $a_j'd \neq 0$, in view of Eq. (3.28), we must have for some $\epsilon > 0$, $a_j'd < -\epsilon$, so for sufficiently large k, we have $a_j'd^k < -\epsilon$, or
$$a_j'x^k < -\epsilon \|x^k\|.$$

Using this relation and the unboundedness of $\{x^k\}$, it follows that for any $\alpha > 0$, we have for sufficiently large k,

$$a'_j(x^k - \alpha d) + b_j \leq -\epsilon \|x^k\| - \alpha a'_j d + b_j < 0.$$

Q.E.D.

Note that the preceding proposition applies to the linear programming problem, where $Q = 0$. The proposition is generalized in Exercises 3.1.19 and 3.1.20. In particular, it is shown in Exercise 3.1.19 that *every (possibly nonconvex) quadratic programming problem with finite optimal value has an optimal solution*. This is a classical result, known as the *Frank-Wolfe theorem* [FrW56].

EXERCISES

3.1.1 (Heron's Problem)

In three-dimensional space, consider a two-dimensional plane and two points z_1 and z_2 lying outside the plane. Use the optimality condition of Prop. 3.1.1 to characterize the vector x^*, which minimizes $\|z_1 - x\| + \|z_2 - x\|$ over all x in the plane.

3.1.2 (Euclid's Problem)

Among all parallelograms that are contained in a given triangle, find the ones that have maximal area. Show that the maximal area is equal to half the area of the triangle. *Hint*: First argue that an optimal solution can be found among the parallelograms that share one angle with the triangle. Then, reduce the problem to maximizing xy subject to $x + y = \alpha$, $x \geq 0$, $y \geq 0$, where α is a constant.

3.1.3 (Kepler's Planimetric Problem)

Among all rectangles contained in a given circle show that the one that has maximal area is a square.

3.1.4 (Tartaglia's Problem)

Divide the number 8 into two nonnegative parts x and y so as to maximize $xy(x - y)$. Answer: $x = 4 + 4/\sqrt{3}$, $y = 4 - 4/\sqrt{3}$.

3.1.5

Among all pyramids whose base is a given triangle and whose height is given, show that the one that has minimal surface area (sum of areas of the triangular sides) has the following property: its vertex that lies outside the triangle projects on the point in the triangle that is at equal distance from all the sides. *Hint*: First argue that the projection point must lie within the triangle. Let x_1, x_2, x_3 be the distances to the sides of the triangle and let a_1, a_2, a_3 be the lengths of the corresponding sides. Express the surface of the pyramid in terms of x_1, x_2, x_3, and show that $\sum_{i=1}^{3} a_i x_i$ is constant. Formulate the problem as a problem of minimization over a simplex.

3.1.6

Consider the problem

$$\text{maximize} \quad x_1^{a_1} x_2^{a_2} \cdots x_n^{a_n}$$
$$\text{subject to} \quad \sum_{i=1}^{n} x_i = 1, \quad x_i \geq 0, \quad i = 1, \ldots, n,$$

where a_i are given positive scalars. Find a global maximum and show that it is unique.

3.1.7

Let Q be a positive definite symmetric matrix. State and prove a generalization of the projection theorem that involves the cost function $(z-x)'Q(z-x)$ in place of $\|z-x\|^2$. *Hint*: Use a transformation of variables.

3.1.8

Consider a closed convex set X on a plane and a point z outside the plane whose projection on the plane is $\hat{z}$. Characterize the projection of z on X in terms of the position of $\hat{z}$ relative to X. Work out the details of the case where X is a triangle.

3.1.9 (www)

Verify the necessary optimality conditions of Eqs. (3.6)-(3.8). Show that, assuming f is convex, they are also sufficient for optimality of x^*.

3.1.10 (Second Order Necessary Optimality Condition) (www)

Show that if x^* is a local minimum of the twice continuously differentiable function $f : \Re^n \mapsto \Re$ over the convex set X, then

$$(x - x^*)'\nabla^2 f(x^*)(x - x^*) \geq 0$$

for all $x \in X$ such that $\nabla f(x^*)'(x - x^*) = 0$.

3.1.11 (Second Order Sufficiency Conditions) (www)

Show that a vector $x^* \in X$ is a local minimum of a twice continuously differentiable function $f : \Re^n \mapsto \Re$ over the convex set X if

$$\nabla f(x^*)'(x - x^*) \geq 0, \qquad \forall\, x \in X,$$

and one of the following three conditions holds:

(1) X is polyhedral and we have

$$(x - x^*)'\nabla^2 f(x^*)(x - x^*) > 0$$

for all $x \in X$ satisfying $x \neq x^*$ and $\nabla f(x^*)'(x - x^*) = 0$.

(2) We have $p'\nabla^2 f(x^*)p > 0$ for all nonzero p that are in the closure of the set $\{d \mid d = \alpha(x - x^*),\ x \in X,\ \alpha \geq 0\}$ and satisfy $\nabla f(x^*)'p = 0$.

(3) For some $\gamma > 0$, we have

$$(x - x^*)'\nabla^2 f(x^*)(x - x^*) \geq \gamma\|x - x^*\|^2, \qquad \forall\, x \in X.$$

Give an example showing that the polyhedral assumption is needed in (1) above.

3.1.12 (Projection on a Simplex)

(a) Develop an algorithm to find the projection of a vector z on a simplex. This algorithm should be almost as simple as a closed form solution.

(b) Modify the algorithm of part (a) so that it finds the minimum of the cost function

$$f(x) = \sum_{i=1}^n \left(\alpha_i x_i + \tfrac{1}{2}\beta_i x_i^2\right)$$

over a simplex, where α_i and β_i are given scalars with $\beta_i > 0$. Consider also the case where β_i may be zero for some indices i.

3.1.13

Let $a_1, \ldots, a_m$ be given vectors in $\Re^n$, and consider the problem of minimizing $\sum_{j=1}^m \|x - a_j\|^2$ over a convex set X. Show that this problem is equivalent to the problem of projecting on X the center of gravity $\frac{1}{m} \sum_{j=1}^m a_j$.

3.1.14 (Optimal Control – Minimum Principle)

Consider the problem of finding sequences $u = (u_0, u_1, \ldots, u_{N-1})$ and $x = (x_1, x_2, \ldots, x_N)$ that minimize

$$g_N(x_N) + \sum_{i=0}^{N-1} g_i(x_i, u_i),$$

subject to the constraints

$$x_{i+1} = f_i(x_i, u_i), \quad i = 0, \ldots, N-1, \quad x_0 : \text{given},$$

$$u_i \in U_i \subset \Re^m, \quad i = 0, \ldots, N-1.$$

All functions g_i and f_i are assumed continuously differentiable, and the sets U_i are assumed convex. Show that if u^* and x^* are optimal, then

$$\nabla_{u_i} H_i\big(x_i^*, u_i^*, p_{i+1}^*\big)'(u_i - u_i^*) \geq 0, \quad \forall\, u_i \in U_i,\ i = 0, \ldots, N-1,$$

where

$$H_i(x_i, u_i, p_{i+1}) = g_i(x_i, u_i) + p_{i+1}' f_i(x_i, u_i)$$

is the Hamiltonian function, and the vectors $p_1^*, \ldots, p_N^*$ are obtained from the adjoint equation

$$p_i^* = \nabla_{x_i} H_i\big(x_i^*, u_i^*, p_{i+1}^*\big), \quad i = 1, \ldots, N-1,$$

with the terminal condition

$$p_N^* = \nabla g_N(x_N^*).$$

If, in addition, the Hamiltonian H_i is a convex function of u_i for any fixed x_i and p_{i+1}, we have

$$u_i^* \in \arg\min_{u_i \in U_i} H_i\big(x_i^*, u_i, p_{i+1}^*\big), \quad \forall\, i = 0, \ldots, N-1.$$

(The last condition is known as the *minimum principle*.) *Hint*: Apply Prop. 3.1.1 and use the results of Section 2.6.

3.1.15

A farmer annually producing x_i units of a certain crop stores $(1 - u_i)x_i$ units of his production, where $0 \leq u_i \leq 1$, and invests the remaining $u_i x_i$ units, thus increasing the next year's production to a level x_{i+1} given by

$$x_{i+1} = x_i + w u_i x_i, \quad i = 0, 1, \ldots, N-1,$$

where w is a given positive scalar. The problem is to find the optimal investment sequence $u_0, \ldots, u_{N-1}$ that maximizes the total product stored over N years

$$x_N + \sum_{i=0}^{N-1} (1 - u_i) x_i.$$

Show that one optimal sequence is given by:

(1) If $w > 1$, $u_0^* = \cdots = u_{N-1}^* = 1$.

(2) If $0 < w < 1/N$, $u_0^* = \cdots = u_{N-1}^* = 0$.

(3) If $1/N \leq w \leq 1$,
$$u_0^* = \cdots = u_{N-\bar{i}-1}^* = 1,$$
$$u_{N-\bar{i}}^* = \cdots = u_{N-1}^* = 0,$$

where $\bar{i}$ is such that $1/(\bar{i}+1) < w \leq 1/\bar{i}$.

3.1.16

Let x_i denote the number of educators in a certain country at time i and let y_i denote the number of research scientists at time i. New scientists (potential educators or research scientists) are produced during the ith period by educators at a rate γ per educator, while educators and research scientists leave the field due to death, retirement, and transfer at a rate δ, where γ and δ are given scalars with $0 < \gamma$ and $0 < \delta < 1$. By means of incentives a science policy maker can determine the proportion u_i of new scientists produced at time i who become educators. Thus the number of research scientists and educators evolves according to the equations

$$x_{i+1} = (1 - \delta) x_i + u_i \gamma x_i,$$

$$y_{i+1} = (1 - \delta) y_i + (1 - u_i) \gamma x_i.$$

The initial numbers x_0, y_0 are known, and the u_i are constrained by

$$0 < \alpha \leq u_i \leq \beta < 1, \quad i = 0, 1, \ldots, N-1,$$

where the scalars α and β are given. Find a sequence $\{u_0^*, \ldots, u_{N-1}^*\}$ that maximizes the final number y_N of research scientists.

3.1.17 (Fractional Programming)

Consider the problem
$$\text{minimize } \frac{f(x)}{g(x)}$$
$$\text{subject to } x \in X,$$
where $f : \Re^n \mapsto \Re$ and $g : \Re^n \mapsto \Re$ are given functions and X is a given subset such that $g(x) > 0$ for all $x \in X$. For $\lambda \in \Re$, define
$$Q(\lambda) = \inf_{x \in X} \{f(x) - \lambda g(x)\},$$
and suppose that a scalar λ^* and a vector $x^* \in X$ satisfy $Q(\lambda^*) = 0$ and
$$x^* \in \arg\min_{x \in X} \{f(x) - \lambda^* g(x)\}.$$
Show that x^* is an optimal solution of the original problem. Use this observation to suggest a solution method that does not require dealing with fractions of functions.

3.1.18 (www)

Let $f : \Re^n \mapsto \Re$ be a twice continuously differentiable function that satisfies
$$m\|y\|^2 \le y' \nabla^2 f(x) y \le M\|y\|^2, \quad \forall \ x, y \in \Re^n,$$
where m and M are some positive scalars. Let also X be a closed convex set. Show that f has a unique global minimum x^* over X, which satisfies
$$\theta_M(x) \le f(x) - f(x^*) \le \theta_m(x), \quad \forall \ x \in \Re^n,$$
where for all $\delta > 0$, we denote
$$\theta_\delta(x) = -\inf_{y \in X} \left\{ \nabla f(x)'(y - x) + \frac{\delta}{2}\|y - x\|^2 \right\}.$$

3.1.19 (Existence of Solutions of Nonconvex Quadratic Programming Problems) (www)

Consider the problem of minimizing over a set X the quadratic function
$$f(x) = x'Qx + c'x,$$
where Q is assumed to be symmetric, but not necessarily positive semidefinite. Show that if the optimal value of the problem is finite, there exists an optimal solution in each of the following two cases:

(i) X is specified by linear inequalities as in Prop. 3.1.3.

(ii) X is the vector sum of a compact set and a closed cone.

Hint: In the proof of Prop. 3.1.3, show that $d'Qd \le 0$, instead of Eq. (3.27). Then show that the finiteness of f^* implies that $(c + 2Qx)'d \ge 0$ for every $x \in X$. Use these facts to show that for every $\alpha > 0$ and k, we have $x^k - \alpha d \in S^k$.

3.1.20 (www)

Show a generalization of Prop. 3.1.3, whereby the quadratic cost function $x'Qx + c'x$ is replaced by any function of the form $f(Ax) + c'x$, where A is a matrix, and f is a convex function satisfying $\liminf_{\|y\|\to\infty}\{f(y)/\|y\|\} = \infty$. Show that the problem has at least one optimal solution, assuming that its optimal value is finite.

3.1.21 (www)

Consider the problem of minimizing f over X, where $f: \Re^n \mapsto \Re$ is continuous, and X is the vector sum of a compact set and a closed cone N. Let f^* be the optimal value of the problem, and assume that for some sequence $\{\gamma^k\}$ with $\gamma^k \downarrow f^*$, the asymptotic directions of $\{x \mid f(x) \leq \gamma^k\}$ that belong to N have asymptotic sequences that are retractive. Then the problem has at least one optimal solution. *Abbreviated Proof*: Let $S^k = \{x \in X \mid f(x) \leq \gamma^k\}$, and let d be an asymptotic direction of $\{S^k\}$, and let $\{x^k\}$ be a corresponding asymptotic sequence. Then d is an asymptotic direction of $\{x \mid f(x) \leq \gamma^k\}$, and by hypothesis for some bounded positive sequence $\{\alpha^k\}$ and some positive integer $\bar{k}$, we have $f(x^k - \alpha^k d) \leq \gamma^k$ for all $k \geq \bar{k}$. For sufficiently large k, because $d \in N$ and $\{x^k\}$ is unbounded, we also have $x^k - \alpha^k d \in X$, so $x^k - \alpha^k d \in S^k$. Thus $\{x^k\}$ is retractive, and by applying Prop. 3.1.2, the result follows.

3.1.22 (Linear Transformations of Closed Sets) (www)

The closedness of images of closed sets under linear transformations is a fundamental issue in convexity and duality theory; see the books [Roc70], [BNO03], and [Ber09]. This exercise shows how to cast this issue in terms of the nonemptiness of a closed set intersection, and address it by using Prop. 3.1.2. Let S be a nonempty closed subset of $\Re^n$. We say that a vector $d \neq 0$ is an asymptotic direction of S if there exists a sequence $\{x^k\}$ such that $x^k \in S$, $x^k \neq 0$, for all k, $\|x^k\| \to \infty$, and $x^k/\|x^k\| \to d/\|d\|$, i.e., if it is an asymptotic direction of the sequence $\{S^k\}$, where $S^k = S$ for all k. Let A be an $m \times n$ matrix, and assume that for every asymptotic direction of S which is in the nullspace of A, every corresponding asymptotic sequence is retractive. Show that AS is a closed set. *Hint*: Let $\{y^k\}$ be a sequence in AS, which converges to some $\bar{y}$. To show that $\bar{y} \in AS$, consider the sets

$$W^k = \{z \mid \|z - \bar{y}\| \leq \|y^k - \bar{y}\|\}, \qquad S^k = \{x \in S \mid Ax \in W^k\}.$$

Use Prop. 3.1.2 to show that the intersection $\cap_{k=0}^\infty S^k$ is nonempty.

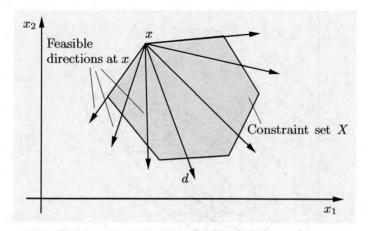

Figure 3.2.1. Feasible directions d at a feasible point x. By definition, d is a feasible direction if changing x by a small amount in the direction d maintains feasibility.

3.2 FEASIBLE DIRECTIONS – CONDITIONAL GRADIENT METHOD

We now turn to computational methods for solving the constrained problem of the preceding section:

$$\text{minimize } f(x)$$
$$\text{subject to } x \in X,$$

where f is continuously differentiable, and X is a nonempty, closed, and convex set. We will see that there is a great variety of algorithms for this problem. However, in this chapter we will restrict ourselves to a limited class of methods that have the following characteristics:

(a) They generate sequences of feasible points $\{x^k\}$ by searching along descent directions. In this sense, they can be viewed as constrained versions of the unconstrained descent algorithms of the previous chapter.

(b) For the most part, they do not rely on any structure of the constraint set other than its convexity.

Most of the algorithms in this chapter belong to the class of the, so called, feasible direction methods, which we proceed to discuss.

3.2.1 Descent Directions and Stepsize Rules

Given a feasible vector x, a *feasible direction* at x is a vector d such that $x + \alpha d$ is feasible for all $\alpha > 0$ that are sufficiently small. Figure 3.2.1 illustrates the set of feasible directions at a point.

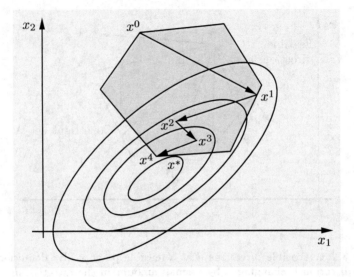

Figure 3.2.2. Sample path of a feasible direction method. At the current point x^k, we choose a feasible direction d^k that is also a direction of descent $[\nabla f(x^k)'d^k < 0]$, and we obtain a new feasible point $x^{k+1} = x^k + \alpha^k d^k$ along d^k.

A *feasible direction method* starts with a feasible vector x^0 and generates a sequence of feasible vectors $\{x^k\}$ according to

$$x^{k+1} = x^k + \alpha^k d^k,$$

where, if x^k is not stationary, d^k is a feasible direction at x^k, which is also a descent direction, i.e.,

$$\nabla f(x^k)'d^k < 0,$$

and the stepsize α^k is chosen to be positive and such that

$$x^k + \alpha^k d^k \in X.$$

If x^k is stationary, the method stops, i.e., $x^{k+1} = x^k$ (equivalently, we choose $d^k = 0$). We will primarily concentrate on feasible direction methods that are also descent algorithms; that is, the stepsize α^k is selected so that

$$f(x^k + \alpha^k d^k) < f(x^k), \qquad \forall\ k.$$

Figure 3.2.2 illustrates a feasible direction method.

In our case where X is convex, there is an alternative and equivalent characterization of feasible direction methods. In particular, it can be seen that the feasible directions at x^k are the vectors of the form

$$d^k = \gamma(\bar{x}^k - x^k), \qquad \gamma > 0,$$

where $\bar{x}^k$ is some feasible vector. Thus, taking into account that $x^k + \alpha^k d^k \in X$, a feasible direction method, if x^k is nonstationary, can be written in the form

$$x^{k+1} = x^k + \alpha^k(\bar{x}^k - x^k),$$

where

$$\alpha^k \in (0,1], \qquad \bar{x}^k \in X,$$

and we have

$$\nabla f(x^k)'(\bar{x}^k - x^k) < 0. \tag{3.30}$$

Because the constraint set X is convex, we have $x^k + \alpha^k(\bar{x}^k - x^k) \in X$ for all $\alpha^k \in [0,1]$ when $x^k \in X$, so that the generated sequence of iterates $\{x^k\}$ is feasible. Furthermore, if x^k is nonstationary, there always exists a feasible direction $\bar{x}^k - x^k$ with the descent property (3.30), since otherwise we would have $\nabla f(x^k)'(x - x^k) \geq 0$ for all $x \in X$, implying that x^k is stationary.

We will henceforth assume that the feasible direction d^k in any feasible direction method under consideration is of the form

$$d^k = \bar{x}^k - x^k,$$

where $\bar{x}^k$ belongs to X and has the descent property (3.30).

Stepsize Selection in Feasible Direction Methods

Most of the rules for choosing the stepsize α^k in gradient methods apply also to feasible direction methods. Here are some of the most popular ones, where we assume that the direction d^k is scaled so that the vector $x^k + d^k$ is feasible and $\alpha^k \in (0,1]$.

Limited Minimization Rule

Here, α^k is chosen so that

$$f(x^k + \alpha^k d^k) = \min_{\alpha \in [0,1]} f(x^k + \alpha d^k).$$

Note that there is no loss of generality in limiting the stepsize to the interval $[0, 1]$, since different stepsize ranges can in effect be used by redefining the direction d^k. Let us also recall that, as mentioned in Section 1.2, the overhead for line search is significantly reduced in the presence of special structure. A prominent example is when f has the form

$$f(x) = h(Ax),$$

where A is a matrix such that the calculation of the vector $y = Ax$ for a given x is far more expensive than the calculation of $h(y)$ and its gradient. This structure arises for example in the context of multicommodity flow problems (cf. Examples 3.1.3 and 3.1.4).

Armijo Rule

Here, fixed scalars β, and $\sigma > 0$, with $\beta \in (0,1)$, and $\sigma \in (0,1)$ are chosen, and we set $\alpha^k = \beta^{m_k}$, where m_k is the first nonnegative integer m for which

$$f(x^k) - f(x^k + \beta^m d^k) \geq -\sigma \beta^m \nabla f(x^k)'d^k.$$

In other words, the stepsizes $1, \beta, \beta^2, \ldots$, are tried successively until the above inequality is satisfied for $m = m_k$.

Constant Stepsize

Here, a fixed stepsize

$$\alpha^k = 1, \qquad k = 0, 1, \ldots$$

is used. This choice is not as simple or as restrictive as it may seem. In fact, in any feasible direction method, we can use a constant unity stepsize if we rescale or redefine appropriately the direction d^k. In this case, the burden is placed in effect on the direction finding procedure to yield a direction that guarantees descent and convergence.

Convergence Analysis of Feasible Direction Methods

The convergence analysis of feasible direction methods for the case of the minimization rule and the Armijo rule is very similar to the one for gradient methods. As in Section 1.2, we say that the direction sequence $\{d^k\}$ is *gradient related* to $\{x^k\}$ if the following property can be shown:

For any subsequence $\{x^k\}_{k \in \mathcal{K}}$ that converges to a nonstationary point, the corresponding subsequence $\{d^k\}_{k \in \mathcal{K}}$ is bounded and satisfies

$$\limsup_{k \to \infty,\, k \in \mathcal{K}} \nabla f(x^k)'d^k < 0.$$

A verbatim repetition of the proof of Prop. 1.2.1 of Section 1.2 shows the following result:

Proposition 3.2.1: (Stationarity of Limit Points for Feasible Direction Methods) Let $\{x^k\}$ be a sequence generated by the feasible direction method $x^{k+1} = x^k + \alpha^k d^k$. Assume that $\{d^k\}$ is gradient related and that α^k is chosen by the limited minimization rule or the Armijo rule. Then every limit point of $\{x^k\}$ is a stationary point.

There are also convergence results for feasible direction methods that use a constant stepsize, but these results are best considered in the context of individual methods.

Finding an Initial Feasible Point

To apply a feasible direction method, it is necessary to have an initial feasible point. Finding such a point may be difficult if the constraint set is specified by nonlinear inequality constraints. If, however, the constraint set is polyhedral (i.e., is specified by linear equality and inequality constraints), one can find an initial feasible point by solving the linear programming problem involving the same constraints and an arbitrary linear cost function.

An alternative method is based on introducing an additional artificial variable y, which is constrained to be nonnegative but is heavily penalized in the cost so that it is zero at the optimum. In particular, consider the problem

$$\text{minimize } f(x)$$
$$\text{subject to } a_i'x = b_i, \quad i = 1, \ldots, m, \quad x \geq 0,$$

where a_i and b_i are given vectors in $\Re^n$ and scalars, respectively. This problem can be replaced by the problem

$$\text{minimize } f(x) + cy$$
$$\text{subject to } a_i'x + \left(b_i - \sum_{j=1}^{n} a_{ij}\right) y = b_i, \quad i = 1, \ldots, m, \quad x \geq 0, \ y \geq 0,$$

where c is a very large cost coefficient, a_{ij} is the jth coordinate of the vector a_i. The vector $(\bar{x}, \bar{y})$ with $\bar{x}_j = 1$, $j = 1, \ldots, n$, $\bar{y} = 1$ is feasible for the modified problem. Note that this vector is interior to the bound constraints $x \geq 0$, $y \geq 0$; this is significant for interior point methods, which will be discussed in Section 5.1.

Similarly, the problem

$$\text{minimize } f(x)$$
$$\text{subject to } a_j'x \leq b_j, \quad j = 1, \ldots, r,$$

can be converted to the problem

$$\text{minimize } f(x) + cy$$
$$\text{subject to } a_j'x - y \leq b_j, \quad j = 1, \ldots, r, \quad y \geq 0,$$

where c is a very large cost coefficient. For any vector $\bar{x}$ that is infeasible for the original problem, the vector $(\bar{x}, \bar{y})$ where

$$\bar{y} = \max_{j=1,\ldots,r} \{a_j'\bar{x} - b_j\}$$

is feasible for the modified problem, while the vector $(\bar{x}, \bar{y} + 1)$ is feasible, as well as interior to the inequality constraints of the modified problem.

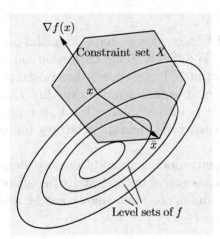

Figure 3.2.3. Finding the feasible direction $\bar{x} - x$ at a point x in the conditional gradient method. $\bar{x}$ is a point of X that lies furthest along the negative gradient direction $-\nabla f(x)$.

3.2.2 The Conditional Gradient Method

The most straightforward way to generate a feasible direction $\bar{x}^k - x^k$ that satisfies the descent condition $\nabla f(x^k)'(\bar{x}^k - x^k) < 0$, and can be used in the method

$$x^{k+1} = x^k + \alpha^k(\bar{x}^k - x^k),$$

is to solve the optimization problem

$$\begin{aligned} \text{minimize} \quad & \nabla f(x^k)'(x - x^k) \\ \text{subject to} \quad & x \in X, \end{aligned} \qquad (3.31)$$

and obtain $\bar{x}^k$ as the solution, i.e.,

$$\bar{x}^k \in \arg\min_{x \in X} \nabla f(x^k)'(x - x^k).$$

Here we assume that X is compact, so that the direction finding subproblem (3.31) has a solution. The corresponding feasible direction method is the, so called, *conditional gradient method*, also known as the *Frank-Wolfe method*. The process to obtain $\bar{x}^k$ is illustrated in Fig. 3.2.3. In particular, $\bar{x}^k$ is the remotest point of X along the negative gradient direction.

Naturally, in order for the method to make practical sense, the direction finding subproblem (3.31) must be much simpler than the original. This is typically the case when f is nonlinear, and X is specified by linear constraints. Then, the subproblem (3.31) is a linear program, which in many applications is very easy to solve. A principal example is when X is a simplex or a Cartesian product of simplices, as in the communication and transportation examples of the previous section.

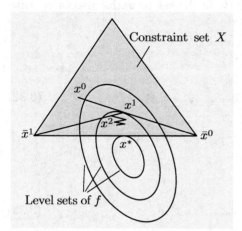

Figure 3.2.4. Successive iterates of the conditional gradient method with the limited minimization stepsize rule, for the case of a simplex constraint.

Example 3.2.1 (Simplex Constraint)

Let
$$X = \left\{ x \mid x \geq 0, \sum_{i=1}^{n} x_i = r \right\},$$
where $r > 0$ is a given scalar (cf. Example 3.1.2). Then, the direction finding subproblem (3.31) takes the form

$$\text{minimize} \sum_{i=1}^{n} \frac{\partial f(x^k)}{\partial x_i}(x_i - x_i^k)$$

$$\text{subject to} \sum_{i=1}^{n} x_i = r, \qquad x_i \geq 0$$

and a solution $\bar{x}^k$ has all coordinates equal to zero except for a single coordinate, say j, which is equal to r; the jth coordinate corresponds to a minimal partial derivative, i.e.,

$$j \in \arg \min_{i=1,\ldots,n} \frac{\partial f(x^k)}{\partial x_i}.$$

This choice, together with the path followed by the conditional gradient method is illustrated in Fig. 3.2.4. At each iteration, the component of x^k that has minimal partial derivative is increased, while the remaining components are decreased by a factor equal to the stepsize α^k.

Example 3.2.2 (Optimal Routing and Multicommodity Flows)

Consider the routing problem that we discussed in the context of communication and transportation networks (Examples 3.1.3 and 3.1.4, respectively). The optimization is with respect to the path flow vector

$$x = \{x_\pi \mid p \in P_w, w \in W\},$$

where W is the set of OD pairs, and P_w is the set of paths associated with OD pair w. The problem has the form

$$\text{minimize} \quad \sum_{a \in \mathcal{A}} D_a(f_a)$$
$$\text{subject to} \quad \sum_{p \in P_w} x_p = r_w, \quad \forall\, w \in W, \tag{3.32}$$
$$x_p \geq 0, \quad \forall\, p \in P_w,\ w \in W,$$

where f_a is the total flow on arc a,

$$f_a = \sum_{\substack{\text{all paths } p \\ \text{containing } a}} x_p, \quad a \in \mathcal{A}, \tag{3.33}$$

We may reformulate the problem over the lower dimensional space of the arc flow vector

$$f = \{f_a \mid a \in \mathcal{A}\} \tag{3.34}$$

as

$$\begin{aligned}
\text{minimize} \quad & G(f) \\
\text{subject to} \quad & f \in \mathcal{F},
\end{aligned} \tag{3.35}$$

where

$$G(f) = \sum_{a \in \mathcal{A}} D_a(f_a), \tag{3.36}$$

and $\mathcal{F}$ is the set of all possible arc flow vectors, i.e., the set of all f given by Eqs. (3.33)-(3.34) for some feasible path flow vector x.

We note that $\mathcal{F}$ is a polyhedral set since it is obtained by a linear transformation of the set of feasible path flow vectors, which is a polyhedral set. It turns out that the conditional gradient method is well-suited for solution of the problem in the form (3.35). Given the current flow vector f^k, we find $\bar{f}^k$ as

$$\bar{f}^k \in \arg\min_{f \in \mathcal{F}} \nabla G(f^k)'(f - f^k), \tag{3.37}$$

and we obtain f^{k+1} as

$$f^{k+1} = f^k + \alpha^k(\bar{f}^k - f^k),$$

where α^k is obtained by the line minimization

$$\alpha^k \in \arg\min_{\alpha \in (0,1]} G\bigl(f^k + \alpha(\bar{f}^k - f^k)\bigr),$$

or some other stepsize rule, such as the Armijo rule.

A key fact here is that the direction-finding subproblem (3.37) can be solved as a shortest path problem, where the length of each arc a is the

Sec. 3.2   Feasible Directions – Conditional Gradient Method   **265**

partial derivative $\partial G(f^k)/\partial f_a$, as we have seen in Example 3.1.3.† Thus the major portion of the work of each iteration is the solution of this shortest path problem which has relatively small dimension (the number of arcs of the graph), a much smaller number than the number of paths, which is the dimension of the original problem (3.32).

Convergence of the Conditional Gradient Method

We will show that the direction sequence of the conditional gradient method is gradient related, so Prop. 3.2.1 applies. Indeed, suppose that $\{x^k\}_{k \in K}$ converges to a nonstationary point $\tilde{x}$. We must prove that

$$\limsup_{k \to \infty, \, k \in K} \|\bar{x}^k - x^k\| < \infty, \tag{3.38}$$

$$\limsup_{k \to \infty, \, k \in K} \nabla f(x^k)'(\bar{x}^k - x^k) < 0. \tag{3.39}$$

Equation (3.38) holds because $\bar{x}^k \in X$, $x^k \in X$, and the set X is assumed compact. To show Eq. (3.39), we note that, by the definition of $\bar{x}^k$,

$$\nabla f(x^k)'(\bar{x}^k - x^k) \leq \nabla f(x^k)'(x - x^k), \qquad \forall \, x \in X,$$

and by taking limit we obtain

$$\limsup_{k \to \infty, \, k \in K} \nabla f(x^k)'(\bar{x}^k - x^k) \leq \nabla f(\tilde{x})'(x - \tilde{x}), \qquad \forall \, x \in X.$$

By taking the minimum over $x \in X$ and by using the nonstationarity of $\tilde{x}$, we have

$$\limsup_{k \to \infty, \, k \in K} \nabla f(x^k)'(\bar{x}^k - x^k) \leq \min_{x \in X} \nabla f(\tilde{x})'(x - \tilde{x}) < 0,$$

thereby proving Eq. (3.39).

We have thus shown that every limit point of the conditional gradient method with the limited minimization rule or the Armijo rule is stationary. An alternative line of convergence analysis is given in Exercise 3.2.4.

---

† It is worth noting that the suitability of the conditional gradient method for solution of the problem (3.35) does not depend on the separability of the cost function (3.36). In most practical cases, however, the cost function G is separable.

Rate of Convergence of the Conditional Gradient Method

Unfortunately, the asymptotic rate of convergence of the conditional gradient method is not very fast when the set X is a polyhedral set, i.e., it is specified by linear equality and inequality constraints. A partial explanation is that the vectors $\bar{x}^k$ used in the algorithm are typically extreme points (vertices) of X. Thus, the feasible direction used may tend to be orthogonal to the direction leading to the minimum (see Fig. 3.2.4). In fact one may construct simple examples where the error sequences $\{f(x^k) - f(x^*)\}$ and $\{\|x^k - x^*\|\}$ do not converge linearly (see Exercise 3.2.3). Thus, when X is a polyhedral set, the method is recommended only for problems where solution accuracy is not very important.

We note that there have been proposals for modifications of the conditional gradient method, which are aimed at improving its convergence rate (see the end-of-chapter references). However, such modifications may detract from the basic simplicity of the method.

Somewhat peculiarly, the practical performance of the conditional gradient method tends to improve as the number of constraints becomes very large, even if these constraints are linear. An explanation for this is given in the papers [Dun79], [DuS83], where it is shown that the convergence rate of the method is linear when the constraint set is not polyhedral but rather has a "positive curvature" property (for example it is a sphere). When there are many linear constraints, the constraint set tends to have very many closely spaced extreme points, and has this "positive curvature" property in an approximate sense.

EXERCISES

3.2.1 (Computational Problem)

Consider the three-dimensional problem

$$\text{minimize} \quad f(x) = \tfrac{1}{2}\left(x_1^2 + x_2^2 + 0.1x_3^2\right) + 0.55x_3$$
$$\text{subject to} \quad x_1 + x_2 + x_3 = 1, \quad 0 \leq x_1,\ 0 \leq x_2,\ 0 \leq x_3.$$

Show that the global minimum is $x^* = (1/2, 1/2, 0)$. Write a computer program implementing the conditional gradient method with the line minimization stepsize rule. (Here, there is a closed form expression for the minimizing stepsize.) Verify computationally that for a starting point (ξ_1, ξ_2, ξ_3) with $\xi_i > 0$ for all i, and $\xi_1 \neq \xi_2$, the rate of convergence is not linear in the sense that

$$\frac{f(x^{k+1}) - f(x^*)}{f(x^k) - f(x^*)} \to 1.$$

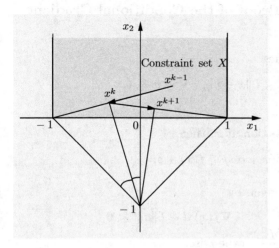

Figure 3.2.5. Successive iterates of the conditional gradient method for the problem of Exercise 3.2.3. The directions used (asymptotically) become horizontal and the convergence rate is sublinear.

3.2.2 (Termination Criterion for the Conditional Gradient Method)

Show that if f is convex, then in the conditional gradient method we have the easily computable bounds

$$f(x^k) \geq \min_{x \in X} f(x) \geq f(x^k) + \nabla f(x^k)'(\bar{x}^k - x^k),$$

and the upper and lower bound difference, $\nabla f(x^k)'(\bar{x}^k - x^k)$, tends to zero as $k \to \infty$.

3.2.3 (Sublinear Rate of Convergence of the Conditional Gradient Method)

Consider the two-dimensional problem

$$\text{minimize } f(x) = x_1^2 + (x_2 + 1)^2$$
$$\text{subject to } -1 \leq x_1 \leq 1, \quad 0 \leq x_2,$$

with global minimum $x^* = (0,0)$. Successive iterates of the conditional gradient method are shown in Fig. 3.2.5. Show that

$$\lim_{k \to \infty} \frac{f(x^{k+1}) - f(x^*)}{f(x^k) - f(x^*)} = 1,$$

and therefore the rate of convergence of $\{f(x^k) - f(x^*)\}$ is not linear. *Hint:* This is a tricky problem. Show that for all $k \geq 1$, we have

$$f(x^k) - f(x^*) = 2(\cos \beta^k)^2 - 1 = \sin(2\alpha^k),$$

where α^k and β^k are the angles shown in Fig. 3.2.5.

3.2.4 (Another Convergence Proof of the Conditional Gradient Method)

Assume that the gradient of f satisfies
$$\|\nabla f(x) - \nabla f(y)\| \leq L\|x - y\|, \quad \forall\, x, y \in X,$$
where L is a positive constant.

(a) Show that if d is a descent direction at x, then
$$\min_{\alpha \in [0,1]} f(x + \alpha d) \leq f(x) + \delta,$$
where δ is the negative scalar given by
$$\delta = \begin{cases} \frac{1}{2}\nabla f(x)'d & \text{if } \nabla f(x)'d + L\|d\|^2 < 0, \\ -\dfrac{|\nabla f(x)'d|^2}{2LR^2} & \text{otherwise,} \end{cases}$$
where R is the diameter of X, i.e.,
$$R = \max_{x,y \in X} \|x - y\|.$$

Hint: Use the descent lemma (Prop. A.24 of Appendix A), which shows that
$$f(x + \alpha d) \leq f(x) + \alpha \nabla f(x)'d + \frac{\alpha^2 L}{2}\|d\|^2.$$
Minimize over $\alpha \in [0, 1]$ both sides of this inequality.

(b) Consider the conditional gradient method
$$x^{k+1} = x^k + \alpha^k d^k$$
with the limited minimization stepsize rule. Show that every limit point of $\{x^k\}$ is stationary. *Hint*: Argue that, if δ^k corresponds to x^k as in part (a), then $\delta^k \to 0$, and therefore we have $\nabla f(x^k)'d^k \to 0$. Take the limit in the relation
$$\nabla f(x^k)'d^k \leq \nabla f(x^k)'(x - x^k), \quad \forall\, x \in X.$$

3.2.5 (Conditional Gradient Method Without a Line Search)

Consider the conditional gradient method, assuming that the gradient of f satisfies
$$\|\nabla f(x) - \nabla f(y)\| \leq L\|x - y\|, \quad \forall\, x, y \in X,$$
where L is a positive constant. Show that if the stepsize α^k is given by
$$\alpha^k = \min\left\{1, \frac{\nabla f(x^k)'(x^k - \bar{x}^k)}{L\|x^k - \bar{x}^k\|^2}\right\},$$
then every limit point of $\{x^k\}$ is stationary. *Hint*: Use the line of analysis of Exercise 3.2.4.

3.2.6 (Zoutendijk's Method of Feasible Directions [Zou60])

This exercise describes one of the main algorithms from the first book to formalize feasible direction methods. Consider the problem

$$\text{minimize } f(x)$$
$$\text{subject to } g_j(x) \leq 0, \quad j = 1, \ldots, r,$$

where $f : \Re^n \mapsto \Re$ is continuously differentiable, and $g_j : \Re^n \mapsto \Re$ are convex and continuously differentiable functions. For any $\epsilon \geq 0$ and feasible vector x, define by $A(x; \epsilon)$ the set of ϵ-active constraints, i.e.,

$$A(x; \epsilon) = \{j \mid -\epsilon \leq g_j(x) \leq 0\}$$

and let $\phi(x; \epsilon)$ be the optimal value of the following linear programming problem in the vector $(d, z) \in \Re^{n+1}$

$$\text{minimize } z$$
$$\text{subject to } \|d\|_\infty \leq 1, \quad \nabla f(x)'d \leq z \qquad (P_{x,\epsilon})$$
$$\nabla g_j(x)'d \leq z, \quad \forall \, j \in A(x; \epsilon)$$

where $\|\cdot\|_\infty$ denotes the maximum norm.

Zoutendijk's method uses two scalars $\bar{\epsilon} > 0$ and $\gamma \in (0, 1)$, and determines a feasible descent direction d^k at a nonstationary vector x^k so that $\bigl(d^k, \phi(x^k; \epsilon^k)\bigr)$ is a solution of problem (P_{x^k, ϵ^k}), where

$$\epsilon^k = \gamma^{m_k}\bar{\epsilon}$$

and m_k is the first nonnegative integer m for which

$$\phi\bigl(x^k, \gamma^m \bar{\epsilon}\bigr) \leq -\gamma^m \bar{\epsilon}.$$

(a) Show that d^k is obtained after solving a finite number of problems (P_{x^k, ϵ^k}). Furthermore, d^k is a feasible descent direction at x^k.

(b) Prove that $\{d^k\}$ is gradient related, thus establishing stationarity of the limit points of $\{x^k\}$.

3.2.7 (Simplicial Decomposition Method [Hol74], [Hoh77], [HLV87], [VeH93], [BeY11], [Ber15a]) (www)

Assume that X is a bounded polyhedral set. Let $\bar{x}^k$ be defined as an extreme point of X satisfying

$$\bar{x}^k \in \arg\min_{x \in X} \nabla f(x^k)'(x - x^k),$$

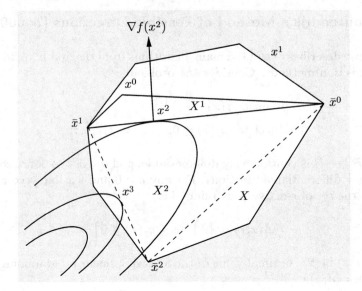

Figure 3.2.6. Successive iterates of the simplicial decomposition method; cf. Exercise 3.2.7. The figure shows how given the initial point x^0, and the calculated extreme points $\bar{x}^0$, $\bar{x}^1$, we determine the next iterate x^2 as a stationary point of f over the convex hull X^1 of x^0, $\bar{x}^0$, and $\bar{x}^1$. Then the new extreme point $\bar{x}^2$ is determined from x^2 as in the conditional gradient method, and the process is repeated to obtain x^3.

and suppose that x^{k+1} is a stationary point of the optimization problem

$$\text{minimize } f(x)$$
$$\text{subject to } x \in X^k,$$

where X^k is the convex hull of x^0 and the extreme points $\bar{x}^0, \ldots, \bar{x}^k$,

$$X^k = \left\{ x \mid x = \beta x^0 + \sum_{i=0}^{k} \gamma_i \bar{x}^i, \beta \geq 0, \gamma_i \geq 0, \beta + \sum_{i=0}^{k} \gamma_i = 1 \right\};$$

see Fig. 3.2.6. Using the fact that the number of extreme points of X is finite, show that the method finds a stationary point of f over X in a finite number of iterations.

Notes: Since the linear program to find $\bar{x}^k$ is the same in the conditional gradient and simplicial decomposition methods, the two methods can be conveniently applied for the same types of problems. Generally, the simplicial decomposition method requires fewer iterations in practice, at the expense of solving a more complicated problem at each iteration (minimizing f over X^k in place of the line search of the conditional gradient method). Both methods are well suited for problems where f has the form $f(x) = h(Ax)$, where A is a matrix such that the calculation of the vector $y = Ax$ for a given x is far more expensive than the calculation of $h(y)$ and its gradient; see e,g., the multicommodity flow problem of Example 3.2.2. For an extensive discussion of the simplicial decomposition

3.2.8 (A Variation of the Simplicial Decomposition Method)

Consider the method of Exercise 3.2.7 with the difference that the set X^k is any closed subset of the set

$$\left\{ x \mid x = \beta x_0 + \sum_{i=0}^{k} \gamma_i \bar{x}^i, \ \beta \geq 0, \ \gamma_i \geq 0, \ \beta + \sum_{i=0}^{k} \gamma_i = 1 \right\}$$

that contains the segment connecting x^k and $\bar{x}^k$. Prove that every limit point of a sequence $\{x^k\}$ generated by this method is a stationary point. *Hint*: Use the fact that for every k, the cost reduction $f(x^k) - f(x^{k+1})$ is larger than the cost reduction that would be obtained by using an Armijo rule along the segment connecting x^k and $\bar{x}^k$ (cf. Props. 1.2.1 and 3.2.1).

3.2.9 (Min-H Method for Optimal Control)

Consider the problem of finding sequences $u = (u_0, u_1, \ldots, u_{N-1})$ and $x = (x_1, x_2, \ldots, x_N)$ that minimize

$$g_N(x_N) + \sum_{i=0}^{N-1} g_i(x_i, u_i),$$

subject to the constraints

$$u_i \in U_i, \qquad i = 0, \ldots, N-1,$$

$$x_{i+1} = \phi_i(x_i) + B_i u_i, \quad i = 0, \ldots, N-1, \qquad x_0 : \text{given},$$

where B_i are given matrices of appropriate dimension. We assume that the functions g_i and ϕ_i are continuously differentiable, and the sets U_i are convex. We further assume that $g_i(x_i, \cdot)$ is convex as a function of u_i for all x_i. Consider the algorithm that, given $u^k = (u_0^k, \ldots, u_{N-1}^k)$, and the corresponding state and costate trajectories $x^k = (x_1^k, \ldots, x_N^k)$ and $p^k = (p_1^k, \ldots, p_N^k)$, respectively, generates u^{k+1} by

$$u_i^{k+1} = u_i^k + \alpha^k (\bar{u}_i^k - u_i^k), \qquad i = 0, \ldots, N-1,$$

where α^k is generated by the Armijo rule or the limited minimization rule, and $\bar{u}_i^k$ is given by

$$\bar{u}_i^k \in \arg\min_{u_i \in U_i} H_i(x_i^k, u_i, p_{i+1}^k), \qquad i = 0, \ldots, N-1,$$

with

$$H_i(x_i, u_i, p_{i+1}) = g_i(x_i, u_i) + p_{i+1}' f_i(x_i, u_i)$$

denoting the Hamiltonian function. Show that this is a feasible direction method.

3.3 GRADIENT PROJECTION METHODS

The conditional gradient method uses a feasible direction obtained by solving a subproblem with linear cost. Gradient projection methods use instead a subproblem with quadratic cost. While this subproblem may be more complex, the resulting convergence rate is typically better. There are many variations of gradient projection methods, which may be viewed collectively as constrained analogs of the gradient methods of Section 1.2. The next section discusses some of the most common variations and provides an overview of their properties. Section 3.3.2 provides a convergence analysis.

3.3.1 Feasible Directions and Stepsize Rules Based on Projection

We first develop informally the main ideas regarding projection methods, leaving the detailed results and proofs for the next subsection.

The simplest gradient projection method is a feasible direction method of the form
$$x^{k+1} = x^k + \alpha^k(\bar{x}^k - x^k), \tag{3.40}$$
where
$$\bar{x}^k = \left[x^k - s^k \nabla f(x^k)\right]^+. \tag{3.41}$$

Here, $[\cdot]^+$ denotes projection on the set X, $\alpha^k \in (0,1]$ is a stepsize, and s^k is a positive scalar. Thus, to obtain the vector $\bar{x}^k$:

(1) We take a step $-s^k \nabla f(x^k)$ along the negative gradient, as in steepest descent.

(2) We project the result $x^k - s^k \nabla f(x^k)$ on X, thereby obtaining the feasible vector $\bar{x}^k$.

(3) We take a step along the feasible direction $\bar{x}^k - x^k$ using the stepsize α^k.

Note that the projection on X in iteration (3.41) can be implemented by minimizing a quadratic cost over X, so equivalently $\bar{x}^k$ can be defined as the vector that minimizes the expression
$$\left\|x - x^k + s^k \nabla f(y^k)\right\|^2 = (s^k)^2 \left\|\nabla f(x^k)\right\|^2 + 2s^k \nabla f(x^k)'(x - x^k) + \left\|x - x^k\right\|^2,$$
over $x \in X$. By neglecting the constant term $(s^k)^2 \|\nabla f(x^k)\|^2$ and by dividing by $2s^k$ in the right-hand side of the above expression, we see that
$$\bar{x}^k = \arg\min_{x \in X} \left\{ \nabla f(y^k)'(x - x^k) + \frac{1}{2s^k} \|x - x^k\|^2 \right\}. \tag{3.42}$$

The scalar s^k determines the feasible direction $\bar{x}^k - x^k$, but it may also be viewed as a stepsize. This becomes evident when we select $\alpha^k = 1$ for all k, in which case $x^{k+1} = \bar{x}^k$ and the method becomes
$$x^{k+1} = \left[x^k - s^k \nabla f(x^k)\right]^+ \tag{3.43}$$

Sec. 3.3 Gradient Projection Methods

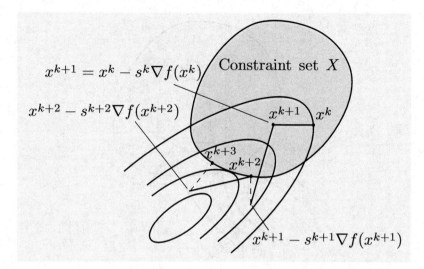

Figure 3.3.1. Illustration of a few iterations of the gradient projection method for the case where $\alpha^k = 1$ for all k. Note that when $x^k - s^k \nabla f(x^k)$ belongs to X, the iteration reduces to an unconstrained steepest descent iteration.

(see Fig. 3.3.1). If $x^k - s^k \nabla f(x^k)$ is feasible, the gradient projection iteration becomes an unconstrained steepest descent iteration, in either the form (3.40) or the form (3.43). In this case, Eq. (3.40) takes the form $x^{k+1} = x^k - \alpha^k \nabla f(x^k)$, while Eq. (3.43) takes the form $x^{k+1} = x^k - s^k \nabla f(x^k)$.

Note that we have $x^* = \left[x^* - s \nabla f(x^*)\right]^+$ for all $s > 0$ if and only if x^* is stationary (see Fig. 3.3.2). Thus, the method stops if and only if it encounters a stationary point.

In order for the method to make practical sense, it is necessary that the projection operation is fairly simple. This will be so if X has a relatively simple structure. For example, when the constraints are bounds on the variables,

$$X = \{x \mid \xi_i \leq x_i \leq \zeta_i, \ i = 1, \ldots, n\},$$

the ith coordinate of the projection of a vector x is given by

$$[x]_i^+ = \begin{cases} \xi_i & \text{if } x_i \leq \xi_i, \\ \zeta_i & \text{if } x_i \geq \zeta_i, \\ x_i & \text{otherwise.} \end{cases}$$

Stepsize Selection and Convergence

There are several stepsize selection approaches in the gradient projection context. The following two are straightforward, although we will later argue that they are often not the best options.

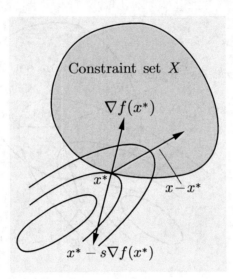

Figure 3.3.2. Illustrating why the gradient projection method stops if and only if it finds a stationary point x^*. By definition, x^* is a stationary point if

$$\nabla f(x^*)'(x - x^*) \geq 0, \qquad \forall\, x \in X,$$

which is equivalent to

$$\left(\left(x^* - s\nabla f(x^*)\right) - x^*\right)'(x - x^*) \leq 0, \qquad \forall\, x \in X,\ s > 0.$$

This holds if and only if x^* is the projection of $x^* - s\nabla f(x^*)$ on X (cf. Prop. 1.1.4).

Limited Minimization Rule

Here
$$s^k = s: \text{constant}, \qquad k = 0, 1, \ldots$$
and α^k is chosen by minimization over $[0, 1]$; i.e.,

$$f\left(x^k + \alpha^k(\bar{x}^k - x^k)\right) = \min_{\alpha \in [0,1]} f\left(x^k + \alpha(\bar{x}^k - x^k)\right).$$

Armijo Rule Along the Feasible Direction

Here
$$s^k = s: \text{constant}, \qquad k = 0, 1, \ldots$$
and α^k is chosen by the Armijo rule over the interval $[0, 1]$. In particular, fixed scalars β and $\sigma > 0$, with $\beta \in (0, 1)$ and $\sigma \in (0, 1)$ are chosen, and we set $\alpha^k = \beta^{m_k}$, where m_k is the first nonnegative integer m for which

$$f(x^k) - f\left(x^k + \beta^m(\bar{x}^k - x^k)\right) \geq -\sigma\beta^m \nabla f(x^k)'(\bar{x}^k - x^k).$$

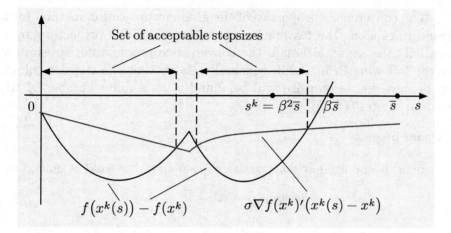

Figure 3.3.3. Illustration of the successive tests of the Armijo inequality (3.45).

Our general convergence result for feasible direction methods (Prop. 3.2.1) applies to the gradient projection method with the above two stepsize rules, provided we can show that the direction sequence is gradient related. Indeed this can be proved (see the following Prop. 3.3.1), so every limit point of a sequence generated by the gradient projection method with the line minimization rule or the Armijo rule is stationary.

Armijo Rule Along the Projection Arc

Here the stepsize α^k is fixed at unity,

$$\alpha^k = 1, \quad k = 0, 1, \ldots$$

and the stepsize s^k is determined by successive reduction until an Armijo-like inequality is satisfied. This means that x^{k+1} is determined by an Armijo-like search on the "projection arc"

$$\{x^k(s) \mid s > 0\},$$

where, for all $s > 0$, $x^k(s)$ is defined by

$$x^k(s) = \left[x^k - s\nabla f(x^k)\right]^+. \tag{3.44}$$

In particular, fixed scalars $\bar{s}$, β, and σ, with $\bar{s} > 0$, $\beta \in (0, 1)$, and $\sigma \in (0, 1)$ are chosen, and we set $s^k = \beta^{m_k}\bar{s}$, where m_k is the first nonnegative integer m for which

$$f(x^k) - f\bigl(x^k(\beta^m \bar{s})\bigr) \geq \sigma \nabla f(x^k)'\bigl(x^k - x^k(\beta^m \bar{s})\bigr). \tag{3.45}$$

Figure 3.3.3 illustrates the Armijo rule on a projection arc. A useful variant of this rule is to allow the initial stepsize $\bar{s}$ to vary at each iteration and to be chosen by some heuristic procedure with the safeguard that it is bounded away from 0.

The convergence properties of the gradient projection method for the Armijo rules along the feasible direction and along the projection arc are essentially the same, although the convergence proofs differ substantially [see the following Prop. 3.3.3, which also shows that the stepsize rules are well defined, i.e., a stepsize will be found after a finite number of trials based on the test (3.45)].

Constant Stepsize

Here, s^k is also fixed at some constant $s > 0$ and α^k is fixed at unity, i.e.,

$$s^k = s : \text{constant}, \qquad \alpha^k = 1, \qquad k = 0, 1, \ldots$$

Similar to unconstrained gradient methods, it is possible to show that the limit points of a sequence generated by the gradient projection method with a constant stepsize are stationary, provided s is sufficiently small and the gradient satisfies a Lipschitz continuity condition (see the following Prop. 3.3.2).

Diminishing Stepsize

Here, α^k is fixed at unity and

$$s^k \to 0, \qquad \sum_{k=0}^{\infty} s^k = \infty.$$

The convergence properties of the diminishing stepsize rule are similar to the ones of their unconstrained counterpart. Descent is not guaranteed at each iteration, but descent is more likely as the stepsize diminishes. The convergence rate tends to be slow, so this stepsize rule is used primarily for singular problems or when there are errors in the gradient calculation.

Identification of the Set of Active Constraints

The main difference between the Armijo rules along the feasible direction and along the projection arc, is that the iterates produced by the latter are more likely to be at the boundary of the constraint set X than the iterates produced by the former. To formalize this idea, let us assume that X is specified by linear inequality constraints

$$X = \{x \mid a_j'x \leq b_j\}.$$

The *set of active constraints at a point* $x \in X$ is the set $A(x)$ of indices of the constraints that are satisfied as equations at x,

$$A(x) = \{j \mid a_j'x = b_j\}.$$

We say that at the kth iteration the jth constraint is *active*, if $a_j' x^k = b_j$.

It can be shown under fairly mild assumptions, that the *set of active constraints at x^* are finitely identified* using the Armijo rule along the projection arc (this was first shown in [Ber76c] for the case of bound constraints; for analysis of more general cases, see [GaB84], [Dun87], [BuM88], [Bon89a], [Fla92], [DeT93], [Wri93a]). By this we mean that the set of active constraints at x^k is the same as the set of active constraints at x^* for all sufficiently large k. By contrast an analogous result cannot be shown for the gradient projection method using the Armijo rule along the feasible direction. Generally, the property of finite identification of the active constraints is significant, because it leads to a sharper rate of convergence analysis and because it facilitates the combination of the gradient projection method with other iterations, which are applicable when the set of active constraints stays fixed.

Rate of Convergence

The convergence rate properties of the gradient projection method are essentially the same as those of the unconstrained steepest descent method. As an example, assume that f is quadratic of the form

$$f(x) = \tfrac{1}{2} x' Q x - b' x,$$

where Q is positive definite, and let x^* denote the unique minimum of f over X. Consider the case of a constant stepsize ($a^k = 1$ and $s^k = s$ for all k). Using the nonexpansive property of the projection [Prop. 1.1.4(c)] and the gradient formula $\nabla f(x) = Qx - b$, we have

$$\begin{aligned}
\|x^{k+1} - x^*\| &= \left\| \left[x^k - s\nabla f(x^k) \right]^+ - \left[x^* - s\nabla f(x^*) \right]^+ \right\| \\
&\leq \left\| \left(x^k - s\nabla f(x^k) \right) - \left(x^* - s\nabla f(x^*) \right) \right\| \\
&= \left\| (I - sQ)(x^k - x^*) \right\| \\
&\leq \max\{|1 - sm|, |1 - sM|\} \|x^k - x^*\|,
\end{aligned} \qquad (3.46)$$

where m and M are the minimum and maximum eigenvalues of Q. This is precisely the rate of convergence estimate obtained for the unconstrained steepest descent method with constant stepsize [cf. Eq. (3.40) and Fig. 1.3.1 in Section 1.3]. This estimate may be generalized for the case where f is strongly convex with Lipschitz continuous gradient, using an analysis similar to the one for the unconstrained steepest descent method; see Exercise 1.3.3.

We conclude that the gradient projection method suffers from the same type of slow convergence as steepest descent. This motivates us to consider scaling, i.e., application of the gradient projection method in a different coordinate system.

Scaled Gradient Projection

Scaling for gradient projection is derived similar to steepest descent. In particular, at the kth iteration, let H^k be a positive definite matrix and consider a transformation of variables defined by

$$x = (H^k)^{-1/2} y.$$

Then, in the space of y, the problem is written as

$$\text{minimize } h^k(y) \equiv f\left((H^k)^{-1/2} y\right)$$
$$\text{subject to } y \in Y^k,$$

where Y^k is the set

$$Y^k = \{y \mid (H^k)^{-1/2} y \in X\}. \tag{3.47}$$

The gradient projection iteration for this problem takes the form

$$y^{k+1} = y^k + \alpha^k (\bar{y}^k - y^k), \tag{3.48}$$

where

$$\bar{y}^k = \left[y^k - s^k \nabla h^k(y^k)\right]^+,$$

or equivalently,

$$\bar{y}^k = \arg\min_{y \in Y^k} \left\{ \nabla h^k(y^k)'(y - y^k) + \frac{1}{2s^k} \|y - y^k\|^2 \right\}; \tag{3.49}$$

cf. Eq. (3.42).

By using the identifications

$$x = (H^k)^{-1/2} y, \qquad x^k = (H^k)^{-1/2} y^k, \qquad \bar{x}^k = (H^k)^{-1/2} \bar{y}^k,$$

$$\nabla h^k(y^k) = (H^k)^{-1/2} \nabla f(x^k)$$

and the definition (3.47) of the set Y^k, it follows that the iteration (3.48) can be written as

$$x^{k+1} = x^k + \alpha^k (\bar{x}^k - x^k), \tag{3.50}$$

where $\bar{x}^k$ is given by [cf. Eq. (3.49)]

$$\bar{x}^k = \arg\min_{x \in X} \left\{ \nabla f(x^k)'(x - x^k) + \frac{1}{2s^k} (x - x^k)' H^k (x - x^k) \right\}. \tag{3.51}$$

We refer to the iteration defined by the two preceding equations as the *scaled gradient projection method*.

Note that we can view the quadratic problem in Eq. (3.51) as a generalized projection problem. In particular, we can verify that $\bar{x}^k$ is the vector of X, which is at minimum distance from the vector $x^k - s^k(H^k)^{-1}\nabla f(x^k)$, but with distance measured in terms of the norm $\|z\|_{H^k} = \sqrt{z'H^k z}$ (see Exercise 3.3.1).

Note also that positive definiteness of the scaling matrix H^k is not strictly necessary for the validity of the method. What is needed is that H^k satisfies

$$(x - x^k)'H^k(x - x^k) > 0, \quad \forall\, x \in X \text{ with } x \neq x^k.$$

In particular, if the set X is contained in some linear manifold of $\Re^n$, H^k need only be positive definite over the subspace that is parallel to this manifold.

The convergence properties of the scaled gradient projection method are the same as the ones of the unscaled version, provided that the sequence $\{H^k\}$ satisfies a condition guaranteeing that the generated direction sequence is gradient related (see the following Prop. 3.3.4).

The convergence rate of the scaled gradient projection method is governed by m^k and M^k, the smallest and largest eigenvalues of the Hessian $\nabla^2 H^k(y^k)$, which is equal to $(H^k)^{-1/2}\nabla^2 f(x^k)(H^k)^{-1/2}$. As in unconstrained optimization, this suggests that one should try to choose the scaling matrix H^k as close as possible to the Hessian matrix $\nabla^2 f(x^k)$. Using a diagonal approximation to the Hessian is a particularly useful choice because it maintains the simplicity of the quadratic subproblem of Eq. (3.51), when the constraint set X is simple (see Exercise 3.3.3). If $H^k = \nabla^2 f(x^k)$, we obtain a constrained version of Newton's method, which we proceed to describe.

Constrained Newton's Method

Let us assume that f is twice continuously differentiable and that the Hessian matrix $\nabla^2 f(x^k)$ is positive definite for all $x \in X$. Consider the scaled gradient projection method with scaling matrix $H^k = \nabla^2 f(x^k)$. It is given by

$$x^{k+1} = x^k + \alpha^k(\bar{x}^k - x^k), \tag{3.52}$$

where

$$\bar{x}^k = \arg\min_{x \in X}\left\{\nabla f(x^k)'(x - x^k) + \frac{1}{2s^k}(x - x^k)'\nabla^2 f(x^k)(x - x^k)\right\}. \tag{3.53}$$

Note that if $s^k = 1$, the quadratic cost above is the second order expansion of f around x^k (except for a constant term). In particular, when

$$\alpha^k = 1, \qquad s^k = 1,$$

x^{k+1} is the vector that minimizes the second order expansion, just as in the case of unconstrained optimization. We expect, therefore, that for a starting point x^0 sufficiently close to a local minimum x^*, the method with unity stepsizes α^k and s^k converges to x^* superlinearly. Indeed this can be shown using a nearly identical proof to the one for the corresponding unconstrained case (see the following Prop. 3.3.5).

When a good starting point is unknown, one of the Armijo rules with unity initial stepsize may be used to improve the convergence properties of the method. The convergence and rate of convergence results that can be proved for such methods hold no surprises and are very similar to the corresponding unconstrained optimization results of Sections 1.3 and 1.4 (see Exercise 3.3.2).

The main difficulty with Newton's method is that the quadratic direction finding subproblem (3.53) may not be simple, even when the constraint set X has a simple structure. Thus the method typically makes practical sense only for problems of small dimension. This motivates the development of methods that attain the fast convergence of Newton's method, while maintaining the simplicity of the direction finding subproblem when X is a simple set. We will discuss methods of this type in Section 3.4.

Variations of the Scaled Gradient Projection Method

There are several variations of the scaled gradient projection method. We describe some of the simpler possibilities here and we discuss two major variations in Sections 3.4 and 3.5. In all these variations, X is specified by linear inequality constraints

$$X = \{x \mid a_j'x \leq b_j, j = 1, \ldots, r\}.$$

Projecting on an Expanded Constraint Set

In this variation, we modify the direction finding quadratic subproblem of Eq. (3.51) by considering only the active or nearly active constraints at the current point x^k. In particular, we fix a positive scalar ϵ and obtain a vector $\tilde{x}^k$ as

$$\tilde{x}^k = \arg\min_{x \in X^k} \left\{ \nabla f(x^k)'(x - x^k) + \frac{1}{2s^k}(x - x^k)'H^k(x - x^k) \right\}, \quad (3.54)$$

where

$$X^k = \left\{ x \mid a_j'x \leq b_j, \text{ for all } j \text{ with } b_j - \epsilon \leq a_j'x^k \leq b_j \right\}. \quad (3.55)$$

Note that X^k potentially involves a much smaller number of constraints than X, so there may be an advantage in solving the quadratic subproblem of Eq. (3.54) rather than the subproblem of Eq. (3.51). Note also that $\tilde{x}^k$ need not be a feasible point because X^k may contain X strictly. A feasible descent direction can be obtained, however, as $\bar{x}^k - x^k$, where

$$\bar{x}^k = \bar{\gamma}\tilde{x}^k + (1 - \bar{\gamma})x^k \tag{3.56}$$

and $\bar{\gamma}$ is the largest $\gamma \in [0, 1]$ such that $\gamma\tilde{x}^k + (1-\gamma)x^k \in X$. The next iterate x^{k+1} is obtained by the usual formula

$$x^{k+1} = x^k + \alpha^k(\bar{x}^k - x^k), \tag{3.57}$$

where α^k is calculated using either the limited minimization or the Armijo rule. The convergence and rate of convergence analysis of this section apply with minor modifications to this method; see Exercise 3.3.4.

Projecting on a Restricted Constraint Set

This variation is similar to the simplicial decomposition method described in Exercise 3.2.7. At the kth iteration, let $\tilde{x}^k$ be defined as an extreme point of X satisfying

$$\tilde{x}^k \in \arg\min_{x \in X} \nabla f(x^k)'(x - x^k) \tag{3.58}$$

and define $\bar{x}^k$ as

$$\bar{x}^k = \arg\min_{x \in X^k} \left\{ \nabla f(x^k)'(x - x^k) + \frac{1}{2s^k}(x - x^k)'H^k(x - x^k) \right\}, \tag{3.59}$$

where X^k is the convex hull of x^0 and the extreme points $\tilde{x}^0, \ldots, \tilde{x}^k$,

$$X^k = \left\{ x \; \middle| \; x = \beta x^0 + \sum_{i=0}^{k} \gamma_i \tilde{x}^i, \; \beta \geq 0, \; \gamma_i \geq 0, \; \beta + \sum_{i=0}^{k} \gamma_i = 1 \right\}. \tag{3.60}$$

Note that X^k is the convex hull of the preceding subset X^{k-1} and the extreme point $\tilde{x}^k$ and that it is possible that $\tilde{x}^k \in X^{k-1}$, in which case $X^k = X^{k-1}$. Using the fact that the number of extreme points of X is finite, we can show that the method

$$x^{k+1} = x^k + \alpha^k(\bar{x}^k - x^k), \tag{3.61}$$

with α^k obtained by either the limited minimization or the Armijo rule has the same convergence properties as the regular scaled gradient projection method (Exercise 3.3.5). Note that by using the subset X^k in place of X, the direction finding subproblem may be greatly simplified. For an example where this simplification is important, see Exercise 3.3.6.

Combinations with Unconstrained Optimization Methods

We mentioned earlier that when the Armijo rule along the projection arc is used, the gradient projection method tends to identify the active constraints at a minimum in a finite number of iterations. Thus, for sufficiently large k the method becomes equivalent to steepest descent over a linear manifold

$$\{x \mid a_j'x = b_j, j \in A(x^*)\},$$

where $A(x^*)$ is the set of active constraints at the eventual limit x^*. Once this occurs, one may introduce iterations that implement a Newton-like method along this manifold. There is an endless range of possibilities here; any kind of unconstrained minimization method can be adapted fruitfully in this context. The stepsize rule for these methods should be modified so that the generated iterates are feasible, i.e. the stepsize α^k is small enough so that the vector $x^k + \alpha^k(\bar{x}^k - x^k)$ does not violate any of the inactive constraints $a_j'x \leq b_j$, $j \notin A(x^k)$.

A difficulty with the type of combined method just described is that one can never really be sure that the active constraints have been identified. Thus a trial and error approach is usually followed, whereby one switches to the unconstrained method after one or more consecutive gradient projection iterations do not change the active set, and one switches to the gradient projection method following a change in the set of active constraints. There are many possibilities along these lines, some of which may be found in [Ber76c], [DeT83], [MoT89], [CGT91].

The Main Limitation of the Gradient Projection Method – Alternatives

The principal drawback of the gradient projection method is the substantial overhead for computing the projection at each iteration. To obtain a good convergence rate, nondiagonal (Newton-like) scaling must often be used, as for example in the iteration (3.52)-(3.53). Unfortunately, in this case, even if the constraints are simple, such as bounds on the variables, the corresponding quadratic program can be very time-consuming.

The next two sections discuss different ways to overcome the difficulty for the case where the constraint set is polyhedral. This is done by modifying the basic method to solve *equality constrained quadratic programs* in place of inequality constrained quadratic programs. Solving an equality constrained quadratic program is much simpler, because it amounts to projection on a subspace and involves just the solution of an associated system of linear equations (see Example 3.1.3).

In Section 3.4, we discuss the *two-metric projection method*, given by

$$x^{k+1} = \left[x^k - \alpha^k D^k \nabla f(x^k)\right]^+,$$

for the case of bound constraints. This is a natural and simple adaptation of unconstrained Newton-like methods. The main difficulty here is that an

Sec. 3.3   Gradient Projection Methods   283

arbitrary positive definite matrix D^k will not necessarily yield a descent direction. However, it turns out that if a suitable subset of the off-diagonal terms of D^k are set to zero, one can obtain descent. Furthermore, one can select D^k as the inverse of a partially diagonalized version of the Hessian matrix $\nabla^2 f(x^k)$ and attain the superlinear rate of Newton's method.

In Section 3.5, we discuss *manifold suboptimization methods*, which are based on scaled gradient projection on the manifold of active constraints. The motivation for these methods is similar to the one for the previously discussed combinations of gradient projection iterations (aimed at identifying the active constraints) with unconstrained iterations on the manifold of active constraints. The main difference is that at the typical iteration of a manifold suboptimization method, at most one constraint can be added or subtracted from the active set.

3.3.2 Convergence Analysis

Generally, the convergence results for (unscaled) gradient projection are similar to those for steepest descent. The details, however, are somewhat more complicated, particularly for the Armijo rule along the projection arc. The following three propositions, each relating to different stepsize rules, are the main results.

Proposition 3.3.1: (Armijo Rule and Limited Minimization Rule Along the Descent Direction) Let $\{x^k\}$ be a sequence generated by the gradient projection method with α^k chosen by the limited minimization rule or by the Armijo rule along the feasible direction. Then every limit point of $\{x^k\}$ is stationary.

Proof: We will show that the direction sequence $\{\bar{x}^k - x^k\}$ is gradient related. Then, application of Prop. 3.2.1 proves the result. Indeed, suppose that $\{x^k\}_{k \in K}$ converges to a nonstationary point $\tilde{x}$. We must prove that

$$\limsup_{k \to \infty,\, k \in K} \|\bar{x}^k - x^k\| < \infty, \tag{3.62}$$

$$\limsup_{k \to \infty,\, k \in K} \nabla f(x^k)'(\bar{x}^k - x^k) < 0. \tag{3.63}$$

By the continuity of the projection [Prop. 1.1.4(c)], we have

$$\lim_{k \to \infty,\, k \in K} \bar{x}^k = [\tilde{x} - s\nabla f(\tilde{x})]^+, \tag{3.64}$$

so Eq. (3.62) holds because $\{\|\bar{x}^k - x^k\|\}_{k \in K}$ converges to $\|[\tilde{x} - s\nabla f(\tilde{x})]^+ - \tilde{x}\|$. To show Eq. (3.63), we note that, by the characteristic property of the projection [Prop. 1.1.4(b)], we have

$$\bigl(x^k - s\nabla f(x^k) - \bar{x}^k\bigr)'(x - \bar{x}^k) \leq 0, \qquad \forall\, x \in X.$$

Applying this relation with $x = x^k$, we obtain

$$\nabla f(x^k)'(\bar{x}^k - x^k) \leq -\frac{1}{s}\|x^k - \bar{x}^k\|^2. \qquad (3.65)$$

By taking limit in the above relation, we obtain

$$\limsup_{k \to \infty,\, k \in K} \nabla f(x^k)'(\bar{x}^k - x^k) \leq -\frac{1}{s}\left\|\tilde{x} - [\tilde{x} - s\nabla f(\tilde{x})]^+\right\|^2.$$

Since $\tilde{x}$ is nonstationary, the right-hand side of the above inequality is negative (cf. Fig. 3.3.2), proving Eq. (3.63). **Q.E.D.**

Proposition 3.3.2: (Constant Stepsize) Let $\{x^k\}$ be a sequence generated by the gradient projection method with $\alpha^k = 1$ and $s^k = s$ for all k. Assume that for some constant $L > 0$, we have

$$\|\nabla f(x) - \nabla f(y)\| \leq L\|x - y\|, \qquad \forall\ x, y \in X. \qquad (3.66)$$

Then, if $0 < s < 2/L$, every limit point of $\{x^k\}$ is stationary.

Proof: By using the descent lemma (Prop. A.24 in Appendix A), we have

$$f(x^{k+1}) - f(x^k) = f(\bar{x}^k) - f(x^k) \leq \nabla f(x^k)'(\bar{x}^k - x^k) + \frac{L}{2}\|\bar{x}^k - x^k\|^2.$$

From this equation and Eq. (3.65), we obtain

$$f(x^{k+1}) - f(x^k) \leq \left(\frac{L}{2} - \frac{1}{s}\right)\|\bar{x}^k - x^k\|^2.$$

If $s < 2/L$, the right-hand side of the above relation is nonpositive, so if $\{x^k\}$ has a limit point, the left-hand side tends to 0. Therefore, $\|\bar{x}^k - x^k\| \to 0$, which implies that for every limit point $\tilde{x}$ of $\{x^k\}$ we have $[\tilde{x} - s\nabla f(\tilde{x})]^+ = \tilde{x}$ [cf. Eq. (3.64))], so $\tilde{x}$ is stationary (cf. Fig. 3.3.2). **Q.E.D.**

An extension of the above proposition that substantially weakens the Lipschitz assumption (3.66), is given in Exercise 3.3.7. The next proposition, originally shown in the paper [GaB82], requires a fairly sophisticated proof.

Sec. 3.3   Gradient Projection Methods   285

Proposition 3.3.3: (Armijo Rule Along the Projection Arc)

(a) For every $x \in X$ there exists a scalar $s_x > 0$ such that

$$f(x) - f(x(s)) \geq \sigma \nabla f(x)'(x - x(s)), \qquad \forall \, s \in [0, s_x], \quad (3.67)$$

where $x(s) = [x - s\nabla f(x)]^+$ [cf. Eq. (3.44)].

(b) Let $\{x^k\}$ be a sequence generated by the gradient projection method with $\alpha^k = 1$ for all k and with the stepsize s^k chosen by the Armijo rule along the projection arc. Then every limit point of $\{x^k\}$ is stationary.

Proof: We first show the following lemma.

Lemma 2.3.1: For every $x \in X$ and $z \in \Re^n$, the function $g : [0, \infty) \mapsto \Re$ defined by

$$g(s) = \frac{\|[x + sz]^+ - x\|}{s}, \qquad \forall \, s > 0,$$

is monotonically nonincreasing.

Proof: We take two scalars s_1 and s_2 with $s_1 > 0$ and $s_2 > s_1$, and we show that

$$\frac{\|[x + s_2 z]^+ - x\|}{s_2} \leq \frac{\|[x + s_1 z]^+ - x\|}{s_1}.$$

Defining

$$y = s_1 z, \qquad \gamma = \frac{s_2}{s_1},$$

$$a = x + y, \qquad b = x + \gamma y,$$

this inequality is written as

$$\|\bar{b} - x\| \leq \gamma \|\bar{a} - x\|, \quad (3.68)$$

where $\bar{a}$ and $\bar{b}$ are the projections on X of a and b, respectively.

If $\bar{a} = x$ then clearly $\bar{b} = x$, so Eq. (3.68) holds. Also if $a \in X$, then $\bar{a} = a = x + y$, so Eq. (3.68) becomes $\|\bar{b} - x\| \leq \gamma \|y\| = \|b - x\|$, which again holds since the projection is a nonexpansive mapping [cf. Prop. 1.1.4(c)]. Finally, if $\bar{a} = \bar{b}$, then Eq. (3.68) also holds. Therefore, it will suffice to

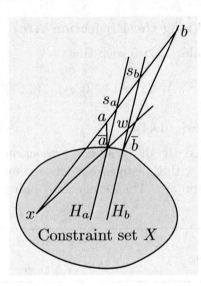

Figure 3.3.4. Construction used in the proof of Lemma 2.3.1.

show Eq. (3.68) in the case where $\bar{a} \neq \bar{b}$, $\bar{a} \neq x$, $\bar{b} \neq x$, $a \notin X$, shown in Fig. 3.3.4.

Let H_a and H_b be the two hyperplanes that are orthogonal to $\bar{b} - \bar{a}$ and pass through $\bar{a}$ and $\bar{b}$, respectively. Since $(\bar{b} - \bar{a})'(b - \bar{b}) \geq 0$ and $(\bar{b} - \bar{a})'(a - \bar{a}) \leq 0$, we have that neither a nor b lie strictly between the two hyperplanes H_a and H_b. Furthermore, x lies on the same side of H_a as a, so $x \notin H_a$. Denote the intersections of the line $\{x + \alpha(b-x) \mid \alpha \in \Re\}$ with H_a and H_b by s_a and s_b, respectively. Denote the intersection of the line $\{x + \alpha(\bar{a} - x) \mid \alpha \in \Re\}$ with H_b by w. We have

$$\begin{aligned}
\gamma &= \frac{\|b - x\|}{\|a - x\|} \geq \frac{\|s_b - x\|}{\|s_a - x\|} \\
&= \frac{\|w - x\|}{\|\bar{a} - x\|} = \frac{\|w - \bar{a}\| + \|\bar{a} - x\|}{\|\bar{a} - x\|} \\
&\geq \frac{\|\bar{b} - \bar{a}\| + \|\bar{a} - x\|}{\|\bar{a} - x\|} \geq \frac{\|\bar{b} - x\|}{\|\bar{a} - x\|},
\end{aligned} \qquad (3.69)$$

where the second equality is by similarity of triangles, the next to last inequality follows from the orthogonality relation $(w - \bar{b})'(\bar{b} - \bar{a}) = 0$, and the last inequality is obtained from the triangle inequality. From Eq. (3.69), we obtain Eq. (3.68), which was to be proved. **Q.E.D.**

We now return to the proof of Prop. 2.3.3:

(a) By the projection theorem, we have

$$\bigl(x - x(s)\bigr)'\bigl(x - s\nabla f(x) - x(s)\bigr) \leq 0, \qquad \forall\, x \in X,\ s > 0.$$

Hence

$$\nabla f(x)'\bigl(x - x(s)\bigr) \geq \frac{\|x - x(s)\|^2}{s} \qquad \forall\, x \in X,\ s > 0. \qquad (3.70)$$

Sec. 3.3   Gradient Projection Methods

If x is stationary, the conclusion holds with s_x being any positive scalar, so assume that x is nonstationary and therefore $\|x - x(s)\| \neq 0$ for all $s > 0$. By the mean value theorem, we have for all $x \in X$ and $s \geq 0$,

$$f(x) - f(x(s)) = \nabla f(x)'(x - x(s)) + (\nabla f(\xi_s) - \nabla f(x))'(x - x(s)),$$

where ξ_s lies on the line segment joining x and $x(s)$. Therefore, Eq. (3.67) can be written as

$$(1 - \sigma)\nabla f(x)'(x - x(s)) \geq (\nabla f(x) - \nabla f(\xi_s))'(x - x(s)). \qquad (3.71)$$

From Eq. (3.70) and Lemma 2.3.1, we have for all $s \in (0, 1]$,

$$\nabla f(x)'(x - x(s)) \geq \frac{\|x - x(s)\|^2}{s} \geq \|x - x(1)\| \cdot \|x - x(s)\|.$$

Therefore, Eq. (3.71) is satisfied for all $s \in (0, 1]$ such that

$$(1 - \sigma)\|x - x(1)\| \geq (\nabla f(x) - \nabla f(\xi_s))' \frac{x - x(s)}{\|x - x(s)\|}.$$

It is seen that there exists $s_x > 0$ such that the above relation, and therefore also Eqs. (3.71) and (3.67), are satisfied for $s \in (0, s_x]$.

(b) Part (a), together with Eq. (3.70) and the definition of the Armijo rule along the projection arc [cf. Eqs. (3.44) and (3.45)], show that s^k is well defined as a positive number for all k. Let $\bar{x}$ be a limit point of $\{x^k\}$ and let $\{x^k\}_K$ be a subsequence converging to $\bar{x}$. Since $\{f(x^k)\}$ is monotonically nonincreasing, we have $f(x^k) \to f(\bar{x})$. Consider two cases:

Case 1: $\liminf_{k \to \infty, k \in K} s^k > \hat{s}$ for some $\hat{s} > 0$. Then, from Eq. (3.70) and Lemma 2.3.1, we have for all $k \in K$ that are sufficiently large

$$\begin{aligned}
f(x^k) - f(x^{k+1}) &\geq \sigma \nabla f(x^k)'(x^k - x^{k+1}) \\
&\geq \sigma \frac{\|x^k - x^{k+1}\|^2}{s^k} \\
&= \frac{\sigma s^k \|x^k - x^{k+1}\|^2}{(s^k)^2} \\
&\geq \frac{\sigma \hat{s} \|x^k - x^k(\bar{s})\|^2}{\bar{s}^2},
\end{aligned}$$

where $\bar{s}$ is the initial stepsize of the Armijo rule. Taking limit as $k \to \infty$, $k \in K$, we obtain

$$0 \geq \frac{\sigma \hat{s} \|\bar{x} - \bar{x}(\bar{s})\|^2}{\bar{s}^2}.$$

Hence $\bar{x} = \bar{x}(\bar{s})$ and it follows that $\bar{x}$ is stationary (cf. Fig. 3.3.2).

Case 2: $\liminf_{k\to\infty, k\in K} s^k = 0$. Then there exists a subsequence $\{s^k\}_{\bar{K}}$, $\bar{K} \subset K$, converging to zero. It follows that for all $k \in \bar{K}$, which are sufficiently large, the Armijo test (3.45) will be failed at least once (i.e., $m^k \geq 1$) and therefore

$$f(x^k) - f\big(x^k(\beta^{-1}s^k)\big) < \sigma \nabla f(x^k)'\big(x^k - x^k(\beta^{-1}s^k)\big). \qquad (3.72)$$

Furthermore, for all such $k \in \bar{K}$, x^k cannot be stationary, since for stationary x^k, we have $s^k = \bar{s}$. Therefore,

$$\big\| x^k - x^k(\beta^{-1}s^k) \big\| > 0. \qquad (3.73)$$

By the mean value theorem, we have

$$f(x^k) - f\big(x^k(\beta^{-1}s^k)\big) = \nabla f(x^k)'\big(x^k - x^k(\beta^{-1}s^k)\big) \\ + \big(\nabla f(\xi^k) - \nabla f(x^k)\big)'\big(x^k - x^k(\beta^{-1}s^k)\big), \qquad (3.74)$$

where ξ^k lies in the line segment joining x^k and $x^k(\beta^{-1}s^k)$. Combining Eqs. (3.72) and (3.74), we obtain for all $k \in \bar{K}$ that are sufficiently large

$$(1-\sigma)\nabla f(x^k)'\big(x^k - x^k(\beta^{-1}s^k)\big) < \big(\nabla f(x^k) - \nabla f(\xi^k)\big)'\big(x^k - x^k(\beta^{-1}s^k)\big).$$

Using Eq. (3.70) and Lemma 2.3.1, we obtain

$$\nabla f(x^k)'\big(x^k - x^k(\beta^{-1}s^k)\big) \geq \frac{\big\|x^k - x^k(\beta^{-1}s^k)\big\|^2}{\beta^{-1}s^k}$$

$$\geq \frac{1}{\bar{s}}\big\|x^k - x^k(\bar{s})\big\| \cdot \big\|x^k - x^k(\beta^{-1}s^k)\big\|.$$

Combining the preceding two relations, and using the Schwarz inequality, we obtain for all $k \in \bar{K}$ that are sufficiently large

$$\frac{1-\sigma}{\bar{s}}\big\|x^k - x^k(\bar{s})\big\| \cdot \big\|x^k - x^k(\beta^{-1}s^k)\big\| \\ < \big(\nabla f(x^k) - \nabla f(\xi^k)\big)'\big(x^k - x^k(\beta^{-1}s^k)\big) \qquad (3.75) \\ \leq \big\|\nabla f(x^k) - \nabla f(\xi^k)\big\| \cdot \big\|x^k - x^k(\beta^{-1}s^k)\big\|.$$

Using Eqs. (3.73) and (3.75), we obtain

$$\frac{1-\sigma}{\bar{s}}\big\|x^k - x^k(\bar{s})\big\| < \big\|\nabla f(x^k) - \nabla f(\xi^k)\big\|.$$

Since $s^k \to 0$ and $x^k \to \bar{x}$ as $k \to \infty$, $k \in \bar{K}$, it follows that $\xi^k \to \bar{x}$, as $k \to \infty$, $k \in \bar{K}$. Taking the limit in the above relation as $k \to \infty$, $k \in \bar{K}$, we obtain

$$\big\|\bar{x} - \bar{x}(\bar{s})\big\| \leq 0.$$

Sec. 3.3 Gradient Projection Methods 289

Hence $\bar{x} = \bar{x}(\bar{s})$ and it follows that $\bar{x}$ is stationary. **Q.E.D.**

Proposition 3.3.4: (Convergence of Scaled Gradient Projection) Let $\{x^k\}$ be a sequence generated by the scaled gradient projection method with α^k chosen by the limited minimization rule or by the Armijo rule along the feasible direction. Assume that, for some positive scalars c_1 and c_2, the scaling matrices H^k satisfy

$$c_1 \|z\|^2 \leq z'H^k z \leq c_2 \|z\|^2, \qquad \forall \, z \in \Re^n, \ k = 0, 1, \ldots$$

Then every limit point of $\{x^k\}$ is stationary.

Proof: Almost identical to the proof of Prop. 3.3.1; left for the reader. **Q.E.D.**

Proposition 3.3.5: (Convergence of Newton's Method) Let f be twice continuously differentiable with positive definite Hessian for all $x \in X$ and let x^* be a local minimum of f over X. There exists a $\delta > 0$ such that if $\|x^0 - x^*\| < \delta$, then the sequence $\{x^k\}$ generated by the constrained version of Newton's method [Eqs. (3.52) and (3.53)] with $\alpha^k = 1$ and $s^k = 1$ for all k, satisfies $\|x^k - x^*\| < \delta$ for all k and $x^k \to x^*$. Furthermore, $\|x^k - x^*\|$ converges to zero superlinearly.

Proof: (*Abbreviated*) Denote $H^k = \nabla^2 f(x^k)$ and consider the vector norms defined by $\|z\|_{H^k} = \sqrt{z'H^k z}$ for $z \in \Re^n$, and the corresponding norms $\|A\|_{H^k} = \sup_{z \neq 0}(\|Az\|_{H^k}/\|z\|_{H^k})$ for $n \times n$ matrices A. Let also $[z]^+$ denote the projection of z on X with respect to the norm $\|\cdot\|_{H^k}$ (cf. Exercise 3.3.1). We have, using the result of Exercise 3.3.1 and the nonexpansive property of projections [Prop. 1.1.4(c) in Section 1.1, which can be generalized for the case of projections with respect to the norm $\|\cdot\|_{H^k}$],

$$\|x^{k+1} - x^*\|_{H^k} = \left\|\left[x^k - (H^k)^{-1}\nabla f(x^k)\right]^+ - x^*\right\|_{H^k}$$
$$\leq \left\|x^k - (H^k)^{-1}\nabla f(x^k) - x^*\right\|_{H^k}.$$

Using this relation, we obtain as in the proof of Prop. 1.4.1

$$\|x^{k+1} - x^*\|_{H^k} \leq M \left(\int_0^1 \left\|\nabla^2 f(x^k) - \nabla^2 f\left(x^* + t(x^k - x^*)\right)\right\|_{H^k} dt\right)$$
$$\cdot \|x^k - x^*\|_{H^k},$$

for some $M > 0$. By continuity of $\nabla^2 f$, we can take δ sufficiently small to ensure that for $\|x^k - x^*\| < \delta$, the term under the integral sign is arbitrarily small. From this, the superlinear convergence of $\|x^k - x^*\|$ to zero follows. **Q.E.D.**

EXERCISES

3.3.1 (Scaled Gradient Projection Method)

Let H^k be a positive definite matrix and define $\bar{x}^k$ by

$$\bar{x}^k = \arg\min_{x \in X} \left\{ \nabla f(x^k)'(x - x^k) + \frac{1}{2s^k}(x - x^k)' H^k (x - x^k) \right\}.$$

Show that $\bar{x}^k$ solves the problem

$$\text{minimize } \left\| x - \left(x^k - s^k (H^k)^{-1} \nabla f(x^k) \right) \right\|_{H^k}^2$$
$$\text{subject to } x \in X,$$

where $\|\cdot\|_{H^k}$ is the norm defined by $\|z\|_{H^k} = \sqrt{z' H^k z}$.

3.3.2

Let f be twice continuously differentiable with positive definite Hessian for all $x \in X$. Consider a sequence $\{x^k\}$ generated by the constrained Newton method of Eqs. (3.52) and (3.53), where for each k, either $s^k = 1$ and α^k is chosen by means of the Armijo rule along the feasible direction, or $\alpha^k = 1$ and s^k is chosen by the Armijo rule along the projection arc with initial stepsize $s = 1$. Assume that x^k converges to a local minimum x^* with positive definite Hessian. Show that $\{\|x^k - x^*\|\}$ converges superlinearly and that there exists an integer $\bar{k} \geq 0$ such that $s^k = 1$ for all $k \geq \bar{k}$.

3.3.3 (Gradient Projection with Diagonal Scaling)

Consider the gradient projection method with a diagonal scaling matrix having the terms H_i^k along the diagonal.

(a) Show that for the case of bound constraints, $X = \{x \mid \xi_i \leq x_i \leq \zeta_i, i = 1, \ldots, n\}$, the ith coordinate of the vector $\bar{x}^k$ is given by

$$\bar{x}_i^k = \begin{cases} \xi_i & \text{if } x_i^k - \frac{s^k}{H_i^k} \frac{\partial f(x^k)}{\partial x_i} \leq \xi_i \\ \zeta_i & \text{if } x_i^k - \frac{s^k}{H_i^k} \frac{\partial f(x^k)}{\partial x_i} \geq \zeta_i, \\ x_i^k - \frac{s^k}{H_i^k} \frac{\partial f(x^k)}{\partial x_i} & \text{otherwise.} \end{cases}$$

Sec. 3.3 Gradient Projection Methods

(b) Derive a simple algorithm for calculating $\bar{x}^k$ in the case of a simplex constraint.

3.3.4 (www)

Consider the unscaled gradient projection method with projection on an expanded constraint set [Eqs. (3.54)-(3.57) with $s^k = s$]. Show that:

(a) $\bar{x}^k - x^k$ is a feasible descent direction when x^k is nonstationary.

(b) Every limit point of a sequence $\{x^k\}$ generated by the method is stationary.

(c) Consider a variant of the method involving a constant stepsize and show that for this variant, the convergence rate estimate of Eq. (3.46) holds for x^k sufficiently close to x^*.

3.3.5

Consider the unscaled method with projection on a restricted constraint set [Eqs. (3.58)-(3.61)]. Show that the conclusions of parts (a)-(c) of Exercise 3.3.4 hold.

3.3.6 (Gradient Projection for Optimal Routing [BeT89], [BeG92], [LuT94b])

Describe the application of the diagonally scaled gradient projection method with projection on a restricted constraint set to the optimal routing problem of Example 3.1.3. Why is this method preferable to the ordinary diagonally scaled gradient projection method? *Hint:* The number of all possible paths may be astronomical, but typically only a few of these paths carry positive flow.

3.3.7 (www)

Suppose that the Lipschitz condition

$$\|\nabla f(x) - \nabla f(y)\| \leq L\|x - y\|, \quad \forall\, x, y \in X,$$

[cf. Eq. (3.66)] is replaced by the following two conditions:

(i) For every bounded set $A \subset X$, there exists some constant L such that

$$\|\nabla f(x) - \nabla f(y)\| \leq L\|x - y\|, \quad \forall\, x, y \in A.$$

(ii) The set $\{x \in X \mid f(x) \leq c\}$ is bounded for every $c \in \Re$.

Show that the convergence result of Prop. 3.3.2 remains valid provided that the constant stepsize s is allowed to depend on the choice of the initial vector x^0. *Hint:* Choose a stepsize that guarantees that x^k stays within the level set $\{x \in X \mid f(x) \leq f(x^0)\}$.

3.3.8 (Projected Contractive Iterations)

This exercise shows that projection preserves the contractive properties of algorithmic mappings. Let H be a positive definite $n \times n$ matrix and let $\|\cdot\|_H$ denote the norm defined by $\|z\|_H = \sqrt{z'Hz}$. Let $g : \Re^n \mapsto \Re^n$ be a contraction mapping of modulus $\beta \in (0,1)$ with respect to this norm, and consider the iteration defined by

$$x^{k+1} = \left[g(x^k)\right]^+, \qquad k = 0, 1, \ldots$$

where $[\cdot]^+$ denotes projection on a closed convex set X with respect to $\|\cdot\|_H$.

(a) Show that the mapping $\left[g(\cdot)\right]^+$ is a contraction mapping of modulus β.

(b) Let $\hat{x}$ and x^* be the fixed points of the mappings $\left[g(\cdot)\right]^+$ and $g(\cdot)$, respectively. Show that

$$\|\hat{x} - x^*\|_H \leq \frac{1}{1-\beta} \left\|[x^*]^+ - x^*\right\|_H.$$

3.4 TWO-METRIC PROJECTION METHODS

We mentioned in the preceding section that gradient projection methods make practical sense only when the projection can be carried out fairly easily. A typical example is the case of an unscaled or diagonally scaled gradient projection method and an orthant constraint

$$X = \{x \mid x \geq 0\}. \tag{3.76}$$

On the other hand we have seen that the convergence rate of the diagonally scaled gradient projection method is often unacceptably slow. This motivated the use of more general, nondiagonal scaling. Then, however, the scaled projection becomes a fairly complex quadratic programming problem, even for the case of the simple orthant constraint (3.76).

To resolve the slow convergence/complex implementation dilemma one may think of the method

$$x^{k+1} = \left[x^k - \alpha^k D^k \nabla f(x^k)\right]^+. \tag{3.77}$$

Here, D^k is a positive definite, not necessarily diagonal, matrix, and $[\cdot]^+$ denotes the usual easy projection on the orthant with respect to the Euclidean norm. Thus, except for the projection operation, this method is identical with the usual scaled gradient iteration for unconstrained minimization. We call this method a *two-metric projection method* because, in contrast with the methods of the preceding section, it embodies two different scaling matrices; one is the matrix D^k, which scales the gradient, and the other is the identity matrix, which is used in the (Euclidean) projection

Sec. 3.4 Two-Metric Projection Methods 293

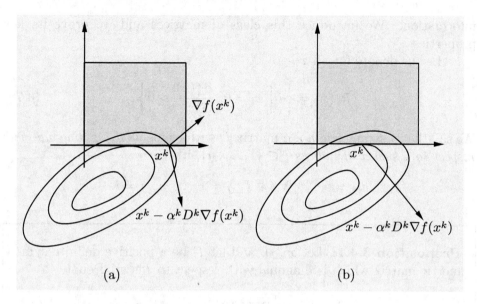

Figure 3.4.1. (a) An example where the iteration

$$x^{k+1} = \left[x^k - \alpha^k D^k \nabla f(x^k)\right]^+$$

fails to make progress at a nonstationary point with a poor choice of the scaling matrix D^k. (b) An example where the iteration fails to stop at a local minimum with a poor choice of the scaling matrix D^k.

norm. This allows the possibility of incorporating second derivative information within D^k [e.g. $D^k = \left(\nabla^2 f(x^k)\right)^{-1}$, as in unconstrained Newton's method], while maintaining the simplicity of the Euclidean norm projection on the orthant.

In this section we discuss the convergence of the two-metric projection method (3.77) for minimizing a continuously differentiable function $f : \Re^n \mapsto \Re$ subject to the orthant constraint (3.76). A similar analysis applies to the case of a box constraint, i.e., upper and lower bounds on the coordinates of x (see Exercise 3.4.1), the case of a Cartesian product of simplices (see Exercise 3.4.2), and more general cases (see the references cited at the end of the chapter).

There is a fundamental difficulty with the two-metric gradient projection method (3.77): it is not in general a descent iteration. In particular we may have $f(x^{k+1}) > f(x^k)$ for all positive stepsizes α^k. This situation is illustrated in Fig. 3.4.1, where it can be seen that with an unfavorable choice of D^k, the method does not even recognize (i.e., stop at) a local minimum.

It turns out, however, that there is a class of nondiagonal matrices D^k for which descent is guaranteed. This class is sufficiently wide to allow superlinear convergence when D^k properly embodies second derivative information. We introduce this class of matrices and we prove its key properties.

Let us denote for all $x \geq 0$

$$I^+(x) = \left\{ i \;\middle|\; x_i = 0, \; \frac{\partial f(x)}{\partial x_i} > 0 \right\}. \tag{3.78}$$

We say that a symmetric $n \times n$ matrix D with elements d_{ij} is *diagonal with respect to a subset of indices* $I \subset \{1, \ldots, n\}$, if

$$d_{ij} = 0, \qquad \forall \; i \in I, \; j = 1, \ldots, n, \; j \neq i.$$

Proposition 3.4.1: Let $x \geq 0$ and let D be a positive definite symmetric matrix which is diagonal with respect to $I^+(x)$. Denote

$$x(\alpha) = \big[x - \alpha D \nabla f(x)\big]^+, \qquad \forall \; \alpha \geq 0.$$

(a) The vector x is stationary if and only if

$$x = x(\alpha), \qquad \forall \; \alpha \geq 0.$$

(b) If x is not stationary, there exists a scalar $\bar{\alpha} > 0$ such that

$$f\big(x(\alpha)\big) < f(x), \qquad \forall \; \alpha \in (0, \bar{\alpha}]. \tag{3.79}$$

Proof: (a) By relabeling the coordinates of x if necessary, we will assume that for some integer r we have

$$I^+(x) = \{r+1, \ldots, n\}.$$

The following proof also applies with minor modifications when $I^+(x)$ is empty. Since D is diagonal with respect to $I^+(x)$, it has the form

$$D = \begin{pmatrix} \bar{D} & 0 & 0 & \cdots & 0 \\ 0 & d_{r+1} & 0 & \cdots & 0 \\ 0 & 0 & d_{r+2} & \cdots & 0 \\ \vdots & \vdots & \vdots & \vdots & \vdots \\ 0 & 0 & 0 & \cdots & d_n \end{pmatrix}, \tag{3.80}$$

where $\bar{D}$ is positive definite and $d_i > 0$, $i = r+1, \ldots, n$. Denote

$$p = D \nabla f(x). \tag{3.81}$$

Sec. 3.4 Two-Metric Projection Methods 295

Assume that x is a stationary point. Then, from the necessary optimality conditions [Eqs. (3.3) and (3.4) in Section 3.1] and the definition (3.78) of $I^+(x)$, we have

$$\frac{\partial f(x)}{\partial x_i} = 0, \quad \forall\, i = 1, \ldots, r,$$

$$\frac{\partial f(x)}{\partial x_i} > 0, \quad \forall\, i = r+1, \ldots, n.$$

These relations and the positivity of d_i imply that

$$p_i = 0, \quad \forall\, i = 1, \ldots, r,$$

$$p_i > 0, \quad \forall\, i = r+1, \ldots, n.$$

Since $x_i(\alpha) = [x_i - \alpha p_i]^+$ and $x_i = 0$ for $i = r+1, \ldots, n$, it follows that $x_i(\alpha) = x_i$ for all i, and $\alpha \geq 0$.

Conversely assume that $x = x(\alpha)$ for all $\alpha \geq 0$. Then, we must have

$$p_i = 0, \quad \forall\, i = 1, \ldots, n \text{ with } x_i > 0,$$

$$p_i \geq 0, \quad \forall\, i = 1, \ldots, n \text{ with } x_i = 0.$$

Now by the definition of $I^+(x)$, we have that if $x_i = 0$ and $i \notin I^+(x)$, then $\partial f(x)/\partial x_i \leq 0$. This, together with the preceding relations, imply that

$$\sum_{i=1}^{r} p_i \frac{\partial f(x)}{\partial x_i} \leq 0.$$

Since by Eqs. (3.80) and (3.81),

$$\begin{bmatrix} p_1 \\ \vdots \\ p_r \end{bmatrix} = \bar{D} \begin{bmatrix} \frac{\partial f(x)}{\partial x_1} \\ \vdots \\ \frac{\partial f(x)}{\partial x_r} \end{bmatrix},$$

while $\bar{D}$ is positive definite, we also have

$$\sum_{i=1}^{r} p_i \frac{\partial f(x)}{\partial x_i} \geq 0,$$

and it follows that

$$p_i = \frac{\partial f(x)}{\partial x_i} = 0, \quad \forall\, i = 1, \ldots, r.$$

Since for $i = r+1, \ldots, n$, we have $\partial f(x)/\partial x_i > 0$ and $x_i = 0$, we see that x is a stationary point.

(b) Let r be as in the proof of part (a). For $i = r+1, \ldots, n$, we have $\partial f(x)/\partial x_i > 0$, $x_i = 0$, and from Eqs. (3.80) and (3.81), we see that

$$x_i = x_i(\alpha) = 0, \quad \forall\, \alpha \geq 0, \quad i = r+1, \ldots, n. \tag{3.82}$$

Consider the set of indices

$$I_1 = \{i \mid x_i > 0 \text{ or } (x_i = 0 \text{ and } p_i < 0), i = 1, \ldots, r\}, \tag{3.83}$$

$$I_2 = \{i \mid x_i = 0 \text{ and } p_i \geq 0, i = 1, \ldots, r\}. \tag{3.84}$$

Let

$$\alpha_1 = \sup\{\alpha \mid x_i - \alpha p_i \geq 0, \forall\, i \in I_1\}. \tag{3.85}$$

Note that, in view of the definition of I_1, α_1 is either positive or $+\infty$. Define the vector $\bar{p}$ with coordinates

$$\bar{p}_i = \begin{cases} p_i & \text{if } i \in I_1, \\ 0 & \text{if } i \in I_2 \text{ or } i \in I^+(x). \end{cases} \tag{3.86}$$

Using Eqs. (3.82)-(3.86), we have

$$x(\alpha) = x - \alpha \bar{p}, \quad \forall\, \alpha \in (0, \alpha_1). \tag{3.87}$$

In view of the definition of I_2 and $I^+(x)$, we obtain

$$\frac{\partial f(x)}{\partial x_i} \leq 0, \quad \forall\, i \in I_2$$

and hence

$$\sum_{i \in I_2} \frac{\partial f(x)}{\partial x_i} p_i \leq 0. \tag{3.88}$$

Now using Eqs. (3.86) and (3.88), we have

$$\nabla f(x)' \bar{p} = \sum_{i \in I_1} \frac{\partial f(x)}{\partial x_i} p_i \geq \sum_{i=1}^{r} \frac{\partial f(x)}{\partial x_i} p_i. \tag{3.89}$$

Since x is not a stationary point, by part (a) and Eq. (3.87), we must have $x \neq x(\alpha)$ for some $\alpha > 0$ and hence also, in view of Eq. (3.82), $p_i \neq 0$ for some $i \in \{1, \ldots, r\}$. In view of the positive definiteness of $\bar{D}$, and Eqs. (3.80) and (3.81),

$$\sum_{i=1}^{r} \frac{\partial f(x)}{\partial x_i} p_i > 0.$$

It follows from Eq. (3.89) that

$$\nabla f(x)' \bar{p} > 0.$$

Combining this relation with Eq. (3.87) and the fact $\alpha_1 > 0$, yields that $\bar{p}$ is a feasible descent direction at x and there exists a scalar $\bar{\alpha} > 0$ for which the desired relation (3.79) is satisfied. **Q.E.D.**

Based on Prop. 3.4.1 we conclude that to guarantee descent, the matrix D^k in the iteration

$$x^{k+1} = \left[x^k - \alpha^k D^k \nabla f(x^k)\right]^+$$

should be chosen diagonal with respect to a subset of indices that contains

$$I^+(x^k) = \left\{ i \;\middle|\; x_i^k = 0, \frac{\partial f(x^k)}{\partial x_i} > 0 \right\}.$$

However, it turns out that to guarantee convergence one should implement the iteration more carefully. The reason is that the set $I^+(x^k)$ exhibits an undesirable discontinuity at the boundary of the constraint set, whereby given a sequence $\{x^k\}$ of interior points that converges to a boundary point $\bar{x}$ the set $I^+(x^k)$ may be strictly smaller than the set $I^+(\bar{x})$. This causes difficulties in proving convergence of the algorithm and may have an adverse effect on its rate of convergence. To bypass these difficulties one may add to the set $I^+(x^k)$ the indices of those variables x_i^k that satisfy $\partial f(x^k)/\partial x_i > 0$ and are "near" zero (i.e., $0 \leq x_i^k \leq \epsilon$, where ϵ is a small fixed scalar). With such a modification and with a variation of the Armijo rule on the projection arc given in Section 3.3, one can prove a satisfactory convergence result. It is also possible to construct Newton-like algorithms, where the nondiagonal portion of the matrix D^k consists of the inverse of the Hessian submatrix corresponding to the indices not in $I^+(x^k)$, and to show a quadratic rate of convergence result under the appropriate assumptions. This analysis can be found in [Ber82a] (Section 1.5), and [Ber82b].

EXERCISES

3.4.1

Consider the problem
$$\text{minimize} \;\; f(x)$$
$$\text{subject to} \;\; \alpha \leq x \leq \beta,$$

where α and β are given vectors. Formulate and prove a result that parallels Prop. 2.4.1.

3.4.2 (Simplex Constraints [Ber82b])

Consider the problem

$$\text{minimize } f(x)$$
$$\text{subject to } x \geq 0, \quad \sum_{i=1}^{n} x_i = 1.$$

For a feasible vector x, let $\bar{i}$ be an index such that $x_{\bar{i}} > 0$, and to simplify notation, assume without loss of generality that $\bar{i} = n$. Let $y = (x_1, \ldots, x_{n-1})$, let

$$h(y) = f\left(x_1, \ldots, x_{n-1}, 1 - \sum_{i=1}^{n-1} x_i\right),$$

and let

$$I^+(y) = \left\{i \;\Big|\; y_i = 0, \; \frac{\partial h(y)}{\partial y_i} > 0\right\}.$$

Suppose that D is an $(n-1) \times (n-1)$ positive definite matrix, which is diagonal with respect to $I^+(y)$, and define

$$y(\alpha) = [y - \alpha D \nabla h(y)]^+, \qquad \alpha \geq 0,$$

$$x(\alpha) = \left(1 - \sum_{i=1}^{n-1} y_i(\alpha)\right)$$

Show that:

(a) x is a stationary point if and only if $x = x(\alpha)$ for all $\alpha \geq 0$.

(b) If x is not stationary, there exists a scalar $\bar{\alpha} > 0$ such that

$$f(x(\alpha)) < f(x), \qquad \forall \, \alpha \in (0, \bar{\alpha}].$$

(c) State a descent method of the Newton type in terms of the first and second derivatives of f.

3.5 MANIFOLD SUBOPTIMIZATION METHODS

The methods of this section are feasible direction methods for the linearly constrained problem

$$\text{minimize } f(x)$$
$$\text{subject to } a_j' x \leq b_j, \qquad j = 1, \ldots, r.$$

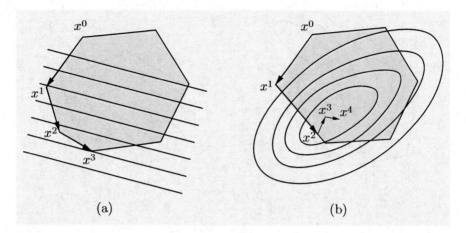

Figure 3.5.1. Illustration of the manifold suboptimization method. The method searches through a sequence of manifolds. Two successive manifolds typically differ by at most one constraint. In (a), the optimal solution is at a vertex, as in the case of linear programs. In (b), the optimal solution is in the interior of the constraint set.

They may be viewed as variants of gradient projection methods, the main difference being that, to obtain a feasible descent direction, the gradient is projected on a linear manifold of active constraints rather than on the entire constraint set (see Fig. 3.5.1). This greatly simplifies the projection and constitutes the main advantage of the methods of this section. Furthermore, once the set of active constraints at a solution is identified, the methods behave identically with unconstrained (scaled) gradient methods. The early portion of the computation may be viewed as a systematic effort to identify the set of active constraints at a solution by searching through a sequence of successive manifolds typically differing by a single constraint. Thus the number of iterations required to identify the set of active constraints is typically at least as large as the number of constraints whose active/inactive status is different at the starting point than at the solution. It follows that the methods of this section are well suited only for problems with a relatively small number of constraints.

Throughout this section, we assume that at every feasible point x, the set of vectors

$$\{a_j \mid j \in A(x)\}$$

is linearly independent, where $A(x)$ is the set of indices of active constraints at x

$$A(x) = \{j \mid a_j'x = b_j, \ j = 1, \ldots, r\}.$$

This assumption can be relaxed at the expense of some technical complications [basically enough indices have to be dropped from $A(x)$ so that the remaining vectors a_j are linearly independent, but the indices dropped must be chosen carefully so that they define redundant constraints].

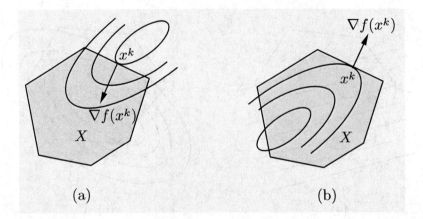

Figure 3.5.2. Illustration of the manifold suboptimization method when x^k is stationary over the current manifold of active constraints. In (a), x^k is stationary over the entire constraint set and the algorithm stops. In (b), one of the active constraints is relaxed and the search continues over a larger manifold.

The iteration at a vector x^k proceeds roughly as follows. We try to find a feasible descent direction from the subspace

$$S(x^k) = \{d \mid a_j'd = 0,\, j \in A(x^k)\}. \tag{3.90}$$

This is the subspace which is parallel to the manifold of active constraints, so a small movement along any $d \in S(x^k)$ does not change the set of active constraints.

There are two possibilities:

(a) A feasible descent direction $d^k \in S(x^k)$ is found, in which case the next point x^{k+1} is given by

$$x^{k+1} = x^k + \alpha^k d^k,$$

where α^k is obtained by some stepsize rule within the interval of stepsizes

$$\{\alpha > 0 \mid x^k + \alpha d^k \text{ is feasible}\}.$$

(b) No feasible descent direction $d \in S(x^k)$ can be found because x^k is stationary over the manifold $x^k + S(x^k)$ of active constraints.

In case (b), there are two possibilities: either x^k is stationary over the entire constraint set

$$\{x \mid a_j'x \leq b_j,\, j = 1, ..., r\},$$

in which case the algorithm stops, or else one of the active constraints, say $\bar{j}$, is relaxed and a feasible descent direction belonging to the subspace

$$\bar{S}(x^k) = \{d \mid a_j'd = 0,\, j \in A(x^k),\, j \neq \bar{j}\}$$

is obtained (see Fig. 3.5.2). We will describe shortly how $\bar{j}$ is selected and how the corresponding direction is computed.

Central to manifold optimization methods are quadratic programming problems of the form

$$\text{minimize } \nabla f(x^k)'d + \tfrac{1}{2}d'H^k d$$
$$\text{subject to } d \in S(x^k), \tag{3.91}$$

where H^k is a symmetric positive definite matrix and $S(x^k)$ is the subspace given by Eq. (3.90). Note that the solution of this problem can be viewed as a scaled projection of the gradient $\nabla f(x^k)$ on the subspace $S(x^k)$. The unique optimal solution is (see Example 3.1.5)

$$d^k = -(H^k)^{-1}\bigl(\nabla f(x^k) + A^{k'}\mu\bigr), \tag{3.92}$$

$$\mu = -\bigl(A^k(H^k)^{-1}A^{k'}\bigr)^{-1} A^k(H^k)^{-1}\nabla f(x^k), \tag{3.93}$$

where A^k is the matrix that has as rows the vectors a_j, $j \in A(x^k)$. [It is actually sufficient that H^k be positive definite on just the subspace $S(x^k)$; then there will again be a unique solution d^k, but Eqs. (3.92) and (3.93) must be appropriately modified.] Let us describe the typical iteration of the method:

We first note that since d^k is the optimal solution of the quadratic programming problem (3.91), and since the vector $d = 0$ is feasible for this problem, we must have

$$\nabla f(x^k)'d^k + \tfrac{1}{2}d'^k H^k d^k \leq 0.$$

If $d^k \neq 0$, we see that

$$\nabla f(x^k)'d^k \leq -\tfrac{1}{2}d'^k H^k d^k < 0, \tag{3.94}$$

and d^k is a feasible descent direction at x^k. The iteration is then given by

$$x^{k+1} = x^k + \alpha^k d^k, \tag{3.95}$$

where α^k is obtained by some stepsize rule (e.g., Armijo, limited minimization, etc.) over the interval of stepsizes α for which $x^k + \alpha d^k$ is feasible, i.e., the set

$$\{\alpha > 0 \mid a_j'(x^k + \alpha d^k) \leq b_j,\ j \notin A(x^k)\}.$$

[Note that, by construction, we have $a_j'd^k = 0$ for all $j \in A(x^k)$, so by moving along the direction d^k, none of the active constraints will be violated or become inactive.]

On the other hand, if $d^k = 0$, it follows from Eq. (3.92) that $\nabla f(x^k) + A^{k'}\mu = 0$, or equivalently, that for some scalars $\mu_j, j \in A(x^k)$, we have

$$\nabla f(x^k) + \sum_{j \in A(x^k)} \mu_j a_j = 0. \tag{3.96}$$

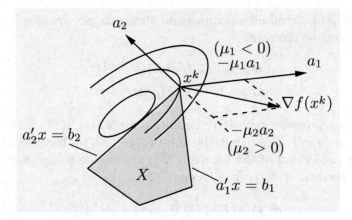

Figure 3.5.3. Dropping a constraint when x^k is a stationary point on the manifold of active constraints. In this example, x^k is a vertex and coincides with the manifold of active constraints. Therefore, x^k is by default stationary on the manifold of active constraints, and one of the two constraints must be dropped. This is the constraint $a_1'x = b_1$ for which the corresponding scalar μ_1 is negative (notice that dropping the other constraint does not allow further descent).

Now note that the feasible directions at x^k are the vectors d with $a_j'd \leq 0$, for all $j \in A(x^k)$, so if $\mu_j \geq 0$ for all $j \in A(x^k)$, then we have

$$\nabla f(x^k)'d = -\sum_{j \in A(x^k)} \mu_j a_j' d \geq 0$$

for all feasible directions d, so that x^k is stationary. Hence if $d^k = 0$ and x^k is nonstationary, there must exist some index $\bar{j} \in A(x^k)$ such that $\mu_{\bar{j}} < 0$ (see Fig. 3.5.3 for an interpretation). Let $\bar{d}^k$ be the unique solution of the problem

$$\begin{aligned}\text{minimize} \quad & \nabla f(x^k)'d + \tfrac{1}{2} d' \bar{H}^k d \\ \text{subject to} \quad & d \in \bar{S}(x^k) = \{d \mid a_j'd = 0,\, j \in A(x^k),\, j \neq \bar{j}\}\end{aligned} \quad (3.97)$$

where $\bar{H}^k$ is a positive definite symmetric matrix. We claim that $\bar{d}^k$ is a feasible descent direction, so that the iteration takes the form

$$x^{k+1} = x^k + \alpha^k \bar{d}^k, \quad (3.98)$$

where α^k is a stepsize chosen as earlier to maintain feasibility of x^{k+1}.

To verify this, we first show that $\bar{d}^k \neq 0$. Indeed, if $\bar{d}^k = 0$, then

$$\nabla f(x^k) + \sum_{j \in A(x^k),\, j \neq \bar{j}} \bar{\mu}_j a_j = 0, \quad (3.99)$$

Sec. 3.5   Manifold Suboptimization Methods   303

for some $\bar{\mu}_j$, $j \in A(x^k)$, $j \neq \bar{j}$. By subtracting Eqs. (3.96) and (3.99), we see that

$$\mu_{\bar{j}} a_{\bar{j}} = \sum_{j \in A(x^k), j \neq \bar{j}} (\bar{\mu}_j - \mu_j) a_j,$$

which, in view of the choice $\mu_{\bar{j}} < 0$, contradicts the linear independence of $\{a_j \mid j \in A(x^k)\}$. Therefore, $\bar{d}^k \neq 0$ and as earlier [cf. Eq. (3.94)], we obtain

$$\nabla f(x^k)' \bar{d}^k \leq -\tfrac{1}{2} \bar{d}'^k \bar{H}^k \bar{d}^k < 0,$$

implying that $\bar{d}^k$ is a descent direction. To show that $\bar{d}^k$ is also a feasible direction, we must show that

$$a'_j \bar{d}^k \leq 0, \qquad \forall\, j \in A(x^k).$$

We know that $a'_j \bar{d}^k = 0$ for all $j \in A(x^k)$ except $j = \bar{j}$, and we will show that $a'_{\bar{j}} \bar{d}^k < 0$. Indeed, by taking inner product of Eq. (3.96) with $\bar{d}^k$, we obtain

$$\nabla f(x^k)' \bar{d}^k + \sum_{j \in A(x^k)} \mu_j a'_j \bar{d}^k = 0$$

and since by construction [cf. Eq. (3.97)], we have $a'_j \bar{d}^k = 0$ for all $j \in A(x^k)$, $j \neq \bar{j}$, we obtain

$$a'_{\bar{j}} \bar{d}^k = -\frac{\nabla f(x^k)' \bar{d}^k}{\mu_{\bar{j}}}.$$

Since $\mu_{\bar{j}} < 0$ and $\nabla f(x^k)' \bar{d}^k < 0$, we see that $a'_{\bar{j}} \bar{d}^k < 0$. Thus, $\bar{d}^k$ is a feasible descent direction and a small movement along $\bar{d}^k$ makes the $\bar{j}$th constraint inactive, while it maintains the active status of all other constraints that are active at x^k.

We have thus completed the description of the manifold suboptimization iteration, together with a method for modifying the manifold of active constraints when no more progress is possible on the current active constraint manifold. It is given by iterations (3.95) or (3.98), depending on whether the scaled projection is done on the manifold of active constraints, or on the manifold obtained after one of the constraints is dropped, respectively.

Positive Definite Quadratic Programming

Suppose that f is the positive definite quadratic function

$$f(x) = \tfrac{1}{2} x' Q x + c' x.$$

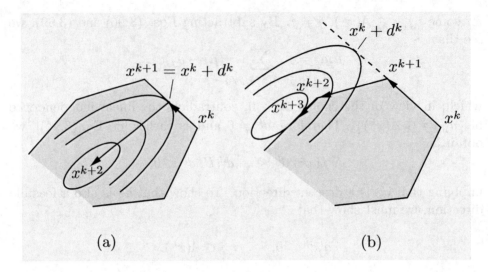

Figure 3.5.4. Illustration of the quadratic programming algorithm. In (a) the vector $x^k + d^k$ is feasible and one of the active constraints must be dropped at the next iteration; in (b) the vector $x^k + d^k$ is not feasible, in which case a new constraint is added to the active set.

Consider the algorithm of this section with the matrices H^k and $\bar{H}^k$ in the quadratic subproblems (3.91) and (3.97) chosen to be equal to Q. In this case, the quadratic subproblem (3.91) takes the form

$$\begin{aligned}\text{minimize} \quad & (Qx^k + c)'d + \tfrac{1}{2}d'Qd \\ \text{subject to} \quad & a_j'd = 0, \quad j \in A(x^k).\end{aligned} \qquad (3.100)$$

This problem can be viewed as a restricted version of the original quadratic program. Indeed, since $a_j'x^k = b_j$ for $j \in A(x^k)$, by letting $x = x^k + d$ and by adding the constant term $\tfrac{1}{2}x^{k'}Qx^k + c'x^k$ to its cost function, this problem can be written equivalently as

$$\begin{aligned}\text{minimize} \quad & \tfrac{1}{2}x'Qx + c'x \\ \text{subject to} \quad & a_j'x = b_j, \quad j \in A(x^k).\end{aligned} \qquad (3.101)$$

This is the same problem as the original except that the cost function $f(x)$ is minimized over the currently active manifold

$$\{x \mid a_j'x = b_j, \, j \in A(x^k)\}$$

rather than over the full constraint set

$$\{x \mid a_j'x \leq b_j, \, j = 1, \ldots, r\}.$$

Let now d^k be the unique optimal solution of problem (3.100). There are two cases (see Fig. 3.5.4):

Sec. 3.5 Manifold Suboptimization Methods

(1) $d^k = 0$. Then, one of the active constraints $\bar{j} \in A(x^k)$ involving a negative coefficient $\mu_{\bar{j}}$, must be dropped to form the enlarged manifold of active constraints. Once this is done, we obtain d^k as the unique solution of the problem

$$\text{minimize} \quad (Qx^k + c)'d + \tfrac{1}{2}d'Qd$$
$$\text{subject to} \quad a_j'd = 0, \quad j \in A(x^k), \ j \neq \bar{j}$$

and proceed as in the following case (2).

(2) $d^k \neq 0$. Then, there are two possibilities:

(a) $x^k + d^k$ is feasible, in which case we set

$$x^{k+1} = x^k + d^k.$$

Then, since problems (3.100) and (3.101) are equivalent, x^{k+1} minimizes $f(x)$ over the current manifold

$$\{x \mid a_j'x = b_j, \ j \in A(x^k)\},$$

and in the next iteration, we will drop a constraint as in case (1) above.

(b) $x^k + d^k$ is not feasible, in which case we set

$$x^{k+1} = x^k + \alpha^k d^k,$$

where

$$\alpha^k = \max\{\alpha > 0 \mid x^k + \alpha d^k : \text{feasible}\}.$$

Then, at least one (and typically just one) new constraint will be added to the active set for the next iteration.

In a practical algorithm, the operations of adding and dropping constraints from the active set, together with the solution of the corresponding quadratic subproblems (3.101) can be done using efficient linear algebra operations (see the books [GiM74], [GMW81], [GMW91] for a more detailed discussion).

We finally show that the quadratic programming algorithm terminates with the unique optimal solution in a finite number of iterations. To this end, let us classify the manifold of active constraints $x^k + S(x^k)$ as one of two types: if the unique minimizer of f over $x^k + S(x^k)$ is feasible, we say that the manifold is of type F and otherwise we say that the manifold is of type N. We note that a manifold of type F can be visited at most once because if x^k belongs to such a manifold, then x^{k+1} is the unique minimizer of f over the manifold and a descent algorithm cannot generate the same point twice. On the other hand, each time the algorithm visits a manifold of type N, at the next iteration it visits a manifold involving at least one more constraint. Therefore, there can be at most n successive iterations in manifolds of type N. Thus, if the number of manifolds of type F is m, the number of iterations is bounded by mn.

Manifold Suboptimization and Linear Programming

Manifold suboptimization methods relate to one of the oldest and most popular optimization algorithms, the *simplex method* for linear programming. This method applies to problems of the form

$$\text{minimize } c'x$$
$$\text{subject to } x \in X,$$

where X is a polyhedral set in $\Re^n$ that has at least one vertex (extreme point) and c is a given vector. The idea of the simplex method is to start at some vertex of X and then to generate a sequence of vertices of X. These vertices satisfy two properties:

(a) Two successive vertices are connected by an edge; that is, they both lie on some one-dimensional active constraint manifold.

(b) Each vertex has a lower cost than the preceding vertices.

Assuming the linear cost function attains a minimum over X, the method is guaranteed to terminate in a finite number of iterations at an optimal vertex. The reason is that X has a finite number of vertices (as shown in Appendix B), a vertex cannot be repeated because of property (b) above, and according to the fundamental theorem of linear programming [Prop. B.20(c) in Appendix B], if the linear cost function attains a minimum over X, then this minimum is attained at some vertex of X.

The selection of the next vertex starting from a given vertex in the simplex method is guided by iterative descent; that is, out of all the edges that lead to neighboring vertices, one that corresponds to a descent direction is chosen. If we view a vertex as an active constraint manifold of dimension 0 and an edge as an active constraint manifold of dimension 1, it is seen that the movement to a neighboring vertex of lower cost encapsulates the basic operations of the manifold suboptimization method: first dropping a constraint at a vertex, then moving along a descent direction on the corresponding one-dimensional manifold (edge), until a new constraint is encountered and the active manifold becomes a vertex again. The choice of a constraint to drop is identical with the one of the manifold suboptimization method (except for streamlining the linear algebra). Thus the behavior of the manifold suboptimization method, when it is applied to linear programming problems and it is started at a vertex of the constraint polyhedron, is identical to the one of the simplex method. The versions of the simplex method used in practice involve a lot of important implementation technology, which is beyond the scope of this book. This technology is needed to enhance efficiency and also to deal with degeneracy (more than n linear constraints active at a given vertex). Manifold suboptimization methods can benefit from the use of some of this technology. We refer to the literature (particularly, [GiM74], [GMW81], [GMW91]) for more details.

EXERCISES

3.5.1

Use the method of this section to solve the three-dimensional quadratic problem

$$\text{minimize } f(x) = x_1^2 + 2x_2^2 + 3x_3^2$$
$$\text{subject to } x_1 + x_2 + x_3 \geq 1, \quad 0 \leq x_1,\ 0 \leq x_2,\ 0 \leq x_3,$$

starting from the point $x^0 = (0, 0, 1)$.

3.5.2

Show by example that in the method of this section, if several constraints $a_j'x \leq b_j$ with negative values μ_j are simultaneously dropped, then the corresponding vector $\bar{d}^k$ need not be a feasible direction.

3.6 PROXIMAL ALGORITHMS

In this section, we discuss the proximal algorithm for minimizing a convex function $f : \Re^n \mapsto \Re$ over a closed convex set X. It is given by

$$x^{k+1} = \arg\min_{x \in X} \left\{ f(x) + \frac{1}{2c^k} \|x - x^k\|^2 \right\}, \tag{3.102}$$

where x^0 is an arbitrary starting point and c^k is a positive scalar parameter; see Fig. 3.6.1. This algorithm is somewhat different from the feasible direction methods of the preceding sections, because it does not require that f has a gradient. In particular, the entire cost function $f(x)$ is used in the proximal iteration (3.102) rather than the linear approximation $\nabla f(x^k)'(x - x^k)$ that is used in the gradient projection method [cf. Eq. (3.42)]. Note that this makes the algorithm applicable to nondifferentiable convex cost functions f.†

---

† The analysis of this section also holds for the case of a convex function $f : X \mapsto \Re$, where X is a convex set that need not be closed; instead the epigraph set $\{(x, w) \mid x \in X, f(x) \leq w\}$ is assumed to be closed (see e.g., [Ber15a], Section 5.1). The proximal algorithm can also be extended to the far more general problem of finding a zero of a monotone multivalued operator (see the end-of-chapter references). The present section follows closely Section 5.1 of [Ber15a], where the analysis is given in a form that applies more generally.

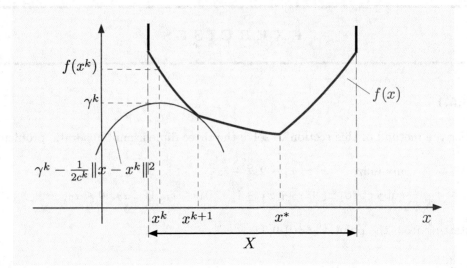

Figure 3.6.1. Geometric view of the proximal algorithm (3.102). The minimum of

$$f(x) + \frac{1}{2c^k}\|x - x^k\|^2$$

is attained at a unique point x^{k+1} as shown. In this figure, γ_k is the scalar by which the graph of the quadratic

$$-\frac{1}{2c^k}\|x - x^k\|^2$$

must be raised so that it just touches the graph of f.

The main motivation of the algorithm is regularization: the quadratic term $\|x - x^k\|^2$ makes the function that is minimized in iteration (3.102) strictly convex with compact level sets, thereby guaranteeing, among others, that x^{k+1} is well-defined as the unique minimum in Eq. (3.102) [cf. Props. 1.1.2 and B.10(b)]. Thus the algorithm is useful only for problems that can benefit from this regularization. We will see, however, that many interesting problems fall in this category, and often in unexpected and diverse ways. In particular, the creative application of the proximal algorithm can allow the elimination of constraints and nondifferentiabilities, the effective exploitation of special problem structures, and the stabilization of other algorithms, such as the cutting plane methods of Section 7.5.3.

We note that the proximal algorithm, when applied to a differentiable (but not necessarily convex) function f, is related to some of the algorithms we discussed earlier, including the gradient projection and the block coordinate descent methods (see Exercises 3.6.1 and 3.7.2). However, the algorithm has excellent convergence properties, even when f is nondifferentiable, provided f is convex. We will first develop these properties, and to this end we derive some preliminary results in the next two

Sec. 3.6   Proximal Algorithms   309

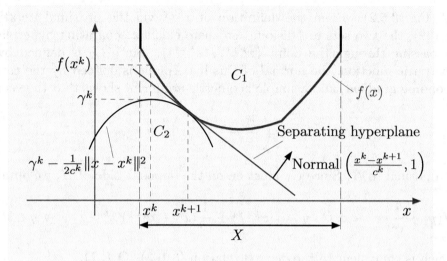

Figure 3.6.2. Separating hyperplane argument for the proof of Prop. 3.6.1. The normal of the hyperplane is defined by

$$\frac{x^k - x^{k+1}}{c^k},$$

the gradient of $-(1/2c^k)\|x - x^k\|^2$ at x^{k+1} [which is also equal to $\nabla f(x^{k+1})$ if f is differentiable at x^{k+1}].

propositions.

The essence of the following proposition may be seen from Fig. 3.6.2. It admits a simple calculus-based proof when f is differentiable. To deal with the case where f is convex but nondifferentiable, we provide a proof that is based on a hyperplane separation argument.

Proposition 3.6.1: For the proximal algorithm, we have for all k

$$f(y) \geq f(x^{k+1}) + \frac{1}{c^k}(x^k - x^{k+1})'(y - x^{k+1}), \qquad \forall\, y \in X. \quad (3.103)$$

Proof: Consider the following two convex subsets of $\Re^{n+1}$:

$$C_1 = \{(x, w) \mid f(x) < w,\ x \in X\},$$

$$C_2 = \left\{(x, w) \mid w \leq \gamma^k - \frac{1}{2c^k}\|x - x^k\|^2,\ x \in \Re^n\right\},$$

where γ^k is given by

$$\gamma^k = f(x^{k+1}) - \frac{1}{2c^k}\|x^{k+1} - x^k\|;$$

(cf. Fig. 3.6.2). From the definition of x^{k+1} via the proximal iteration (3.102), the two sets are disjoint, so there exists a separating hyperplane H passing through the point $(x^{k+1}, f(x^{k+1}))$. Since C_2 is defined by a quadratic function, the normal of this hyperplane is defined by the corresponding gradient, and a simple geometric argument shows that the vector

$$\left(\frac{x^k - x^{k+1}}{c^k}, 1\right)$$

is a normal to H. Since C_1 must lie on the opposite side of H, we obtain

$$f(y) + \frac{1}{c^k}(x^k - x^{k+1})'y \geq f(x^{k+1}) + \frac{1}{c^k}(x^k - x^{k+1})'x^{k+1}, \qquad \forall\, y \in \Re^n,$$

which is equivalent to the desired relation (3.103). **Q.E.D.**

Generally, starting from any nonoptimal point x^k, the cost function value is reduced at each iteration, since from Eq. (3.102), by setting $x = x^k$, we have

$$f(x^{k+1}) + \frac{1}{2c^k}\|x^{k+1} - x^k\|^2 \leq f(x^k).$$

The following proposition provides an inequality, which among others shows that the iterate distance to every optimal solution is also reduced.

Proposition 3.6.2: For the proximal algorithm, we have for all k,

$$\|x^{k+1} - y\|^2 \leq \|x^k - y\|^2 - 2c^k\big(f(x^{k+1}) - f(y)\big) - \|x^k - x^{k+1}\|^2, \quad \forall\, y \in X.$$

Proof: We have

$$\|x^k - y\|^2 = \|x^k - x^{k+1} + x^{k+1} - y\|^2$$
$$= \|x^k - x^{k+1}\|^2 + 2(x^k - x^{k+1})'(x^{k+1} - y) + \|x^{k+1} - y\|^2.$$

By combining this relation with Eq. (3.103), the result follows. **Q.E.D.**

Let us denote by f^* the optimal value

$$f^* = \inf_{x \in X} f(x),$$

(which may be $-\infty$) and by X^* the set of minima of f (which may be empty),

$$X^* = \arg\min_{x \in X} f(x).$$

Sec. 3.6   *Proximal Algorithms*   **311**

The following is the basic convergence result for the proximal algorithm.

Proposition 3.6.3: (Convergence) Let $\{x^k\}$ be a sequence generated by the proximal algorithm (3.102). Then, if $\sum_{k=0}^{\infty} c^k = \infty$, we have
$$f(x^k) \downarrow f^*,$$
and if X^* is nonempty, $\{x^k\}$ converges to some point in X^*.

Proof: We first note that since x^{k+1} minimizes $f(x) + \frac{1}{2c^k}\|x - x^k\|^2$, we have by setting $x = x^k$,
$$f(x^{k+1}) + \frac{1}{2c^k}\|x^{k+1} - x^k\|^2 \leq f(x^k), \quad \forall\, k.$$

It follows that $\{f(x^k)\}$ is monotonically nonincreasing. Hence $f(x^k) \downarrow f_\infty$, where f_∞ is either a scalar or $-\infty$. Moreover, we have $f_\infty \geq f^*$.

From Prop. 3.6.2, we have for all $y \in X$,
$$\|x^{k+1} - y\|^2 \leq \|x^k - y\|^2 - 2c^k\big(f(x^{k+1}) - f(y)\big). \tag{3.104}$$

By adding this inequality over $k = 0, \ldots, N$, we obtain
$$\|x^{N+1} - y\|^2 + 2\sum_{k=0}^{N} c^k\big(f(x^{k+1}) - f(y)\big) \leq \|x^0 - y\|^2, \quad \forall\, y \in X,\, N \geq 0,$$

so that
$$2\sum_{k=0}^{N} c^k\big(f(x^{k+1}) - f(y)\big) \leq \|x^0 - y\|^2, \quad \forall\, y \in X,\, N \geq 0.$$

Taking the limit as $N \to \infty$, we have
$$2\sum_{k=0}^{\infty} c^k\big(f(x^{k+1}) - f(y)\big) \leq \|x^0 - y\|^2, \quad \forall\, y \in X. \tag{3.105}$$

Assume to arrive at a contradiction that $f_\infty > f^*$, and let $\hat{y} \in X$ be such that
$$f_\infty > f(\hat{y}) \geq f^*.$$

Since $\{f(x^k)\}$ is monotonically nonincreasing, we have
$$f(x^{k+1}) - f(\hat{y}) \geq f_\infty - f(\hat{y}) > 0.$$

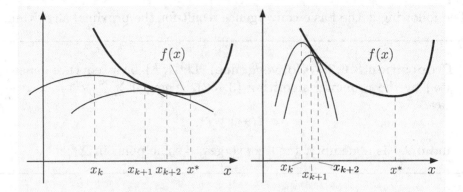

Figure 3.6.3. Illustration of the role of the parameter c^k in the convergence process of the proximal algorithm. In the figure on the left, c^k is large, the graph of the quadratic term is "blunt," and the method makes fast progress toward the optimal solution. In the figure on the right, c^k is small, the graph of the quadratic term is "pointed," and the method makes slow progress.

Then in view of the assumption $\sum_{k=0}^{\infty} c^k = \infty$, Eq. (3.105), with $y = \hat{y}$, leads to a contradiction. Thus $f_\infty = f^*$.

Consider now the case where X^* is nonempty, and let x^* be any point in X^*. Applying Eq. (3.104) with $y = x^*$, we have

$$\|x^{k+1} - x^*\|^2 \leq \|x^k - x^*\|^2 - 2c^k \big(f(x^{k+1}) - f(x^*)\big), \qquad k = 0, 1, \ldots. \tag{3.106}$$

From this relation it follows that $\|x^k - x^*\|^2$ is monotonically nonincreasing, so $\{x^k\}$ is bounded. If $\bar{x}$ is a limit point of $\{x^k\}$, we have $\bar{x} \in X$, since X is closed and

$$f(\bar{x}) = \lim_{k \to \infty,\, k \in \mathcal{K}} f(x^k) = f_\infty = f^*$$

for any subsequence $\{x^k\}_{\mathcal{K}} \to \bar{x}$, since $\{f(x^k)\}$ monotonically decreases to f_∞ (which is equal to f^* as shown earlier) and f is continuous being real-valued (Prop. B.9 of Appendix B). Hence $\bar{x}$ minimizes f over X and must belong to X^*. Finally, by Eq. (3.106), the distance of x^k to every $x^* \in X^*$ is monotonically nonincreasing, so $\{x^k\}$ must converge to a unique point in X^*. **Q.E.D.**

Note a remarkable fact from the preceding proposition: convergence to the optimal value is obtained even if X^* is empty or $f^* = -\infty$. Moreover, when X^* is nonempty, convergence to a single point of X^* occurs.

3.6.1 Rate of Convergence

We will now discuss the convergence rate of the proximal algorithm. We first note that the algorithm converges faster when c^k is increased; see Fig. 3.6.3. The following proposition shows that the convergence rate of the

Sec. 3.6   Proximal Algorithms   313

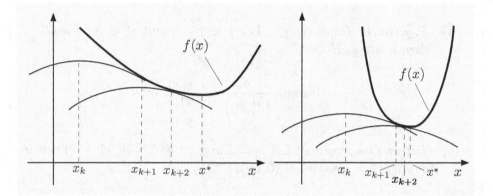

Figure 3.6.4. Illustration of the convergence rate of the proximal algorithm and the effect of the growth properties of f near the optimal solution set. In the figure on the left, f grows slowly and the convergence is slow. In the figure on the right, f grows fast and the convergence is fast.

proximal algorithm depends not only on c^k but also on the order of growth of f near the optimal solution set; see also Fig. 3.6.4.

Proposition 3.6.4: (Rate of Convergence) Assume that X^* is nonempty and that for some scalars $\beta > 0$, $\delta > 0$, and $\gamma \geq 1$, we have

$$f^* + \beta\big(d(x)\big)^\gamma \leq f(x), \qquad \forall\, x \in X \text{ with } d(x) \leq \delta, \tag{3.107}$$

where

$$d(x) = \inf_{x^* \in X^*} \|x - x^*\|.$$

Let also

$$\sum_{k=0}^{\infty} c^k = \infty,$$

so that the sequence $\{x^k\}$ generated by the proximal algorithm (3.102) converges to some point in X^* by Prop. 3.6.3. Then:

(a) For all k sufficiently large, we have

$$d(x^{k+1}) + \beta c^k \big(d(x^{k+1})\big)^{\gamma-1} \leq d(x^k), \tag{3.108}$$

if $\gamma > 1$, and

$$d(x^{k+1}) + \beta c^k \leq d(x^k), \tag{3.109}$$

if $\gamma = 1$ and $x^{k+1} \notin X^*$.

(b) (*Superlinear Convergence*) Let $1 < \gamma < 2$ and $x^k \notin X^*$ for all k. Then if $\inf_{k \geq 0} c^k > 0$,

$$\limsup_{k \to \infty} \frac{d(x^{k+1})}{\left(d(x^k)\right)^{1/(\gamma-1)}} < \infty.$$

(c) (*Linear Convergence*) Let $\gamma = 2$ and $x^k \notin X^*$ for all k. Then if $\lim_{k \to \infty} c^k = \bar{c}$ with $\bar{c} \in (0, \infty)$,

$$\limsup_{k \to \infty} \frac{d(x^{k+1})}{d(x^k)} \leq \frac{1}{1 + \beta \bar{c}},$$

while if $\lim_{k \to \infty} c^k = \infty$,

$$\lim_{k \to \infty} \frac{d(x^{k+1})}{d(x^k)} = 0.$$

(d) (*Sublinear Convergence*) Let $\gamma > 2$. Then

$$\limsup_{k \to \infty} \frac{d(x^{k+1})}{d(x^k)^{2/\gamma}} < \infty.$$

Proof: (a) Since the conclusion clearly holds when $x^{k+1} \in X^*$, we assume that $x^{k+1} \notin X^*$ and we denote by $\hat{x}^{k+1}$ the projection of x^{k+1} on X^*. From Eq. (3.103), we have

$$f(x^{k+1}) + \frac{1}{c^k}(x^k - x^{k+1})'(\hat{x}^{k+1} - x^{k+1}) \leq f(\hat{x}^{k+1}) = f^*.$$

Using the hypothesis, $\{x^k\}$ converges to some point in X^*, so it follows from Eq. (3.107) that

$$f^* + \beta\big(d(x^{k+1})\big)^\gamma \leq f(x^{k+1}),$$

for k sufficiently large. Combining the preceding two relations, we obtain

$$\beta c^k \big(d(x^{k+1})\big)^\gamma \leq (x^{k+1} - \hat{x}^{k+1})'(x^k - x^{k+1}),$$

for k sufficiently large. From the projection theorem (cf. Prop. 1.1.4), since $\hat{x}^{k+1}$ is the projection of x^{k+1} on X^* and $\hat{x}^k \in X^*$,

$$\|x^{k+1} - \hat{x}^{k+1}\|^2 - (x^{k+1} - \hat{x}^{k+1})'(x^{k+1} - \hat{x}^k) = (x^{k+1} - \hat{x}^{k+1})'(\hat{x}^k - \hat{x}^{k+1}) \leq 0.$$

Combining the last two relations and using the Schwarz inequality, we obtain

$$\|x^{k+1} - \hat{x}^{k+1}\|^2 + \beta c^k \big(d(x^{k+1})\big)^\gamma \le (x^{k+1} - \hat{x}^{k+1})'(x^k - \hat{x}^k)$$
$$\le \|x^{k+1} - \hat{x}^{k+1}\|\,\|x^k - \hat{x}^k\|.$$

Dividing with $\|x^{k+1} - \hat{x}^{k+1}\|$ (which is nonzero since we assumed that $x^{k+1} \notin X^*$), Eqs. (3.108)-(3.109) follow.

(b) From Eq. (3.108) and the fact $\gamma < 2$, we obtain the desired relation.

(c) For $\gamma = 2$, Eq. (3.108) becomes

$$(1 + \beta c^k) d(x^{k+1}) \le d(x^k),$$

from which the result follows.

(d) We have for all sufficiently large k,

$$\beta \big(d(x^{k+1})\big)^\gamma \le f(x^{k+1}) - f^* \le \frac{d(x^k)^2}{2c^k},$$

where the first inequality follows from the hypothesis (3.107) and the second inequality follows from Prop. 3.6.2, with y equal to the projection of x^k onto X^*. **Q.E.D.**

Proposition 3.6.4 shows that as the growth order γ in Eq. (3.107) increases, the rate of convergence becomes slower. An important threshold value is $\gamma = 2$; in this case the distance of the iterates to X^* decreases at least linearly if c^k remains bounded, and decreases even faster (superlinearly) if $c^k \to \infty$. Generally, the convergence is accelerated if c^k is increased with k, rather than kept constant; this is illustrated most clearly when $\gamma = 2$ [cf. Prop. 3.6.4(c)]. When $1 < \gamma < 2$, the convergence rate is superlinear [cf. Prop. 3.6.4(b)]. When $\gamma > 2$, the convergence rate is generally slower than when $\gamma = 2$, and examples show that $d(x^k)$ may converge to 0 slower than any geometric progression [cf. Prop. 3.6.4(d)].

The threshold value of $\gamma = 2$ for linear convergence is related to the quadratic growth property of the regularization term. A generalized version of the proposition, with similar proof, is possible for proximal algorithms that use nonquadratic regularization functions (see [KoB76], [Ber82a], Chapter 5, and also [Ber15a], Example 6.6.5). In this context, the threshold value for linear convergence is related to the order of growth of the regularization function.

When $\gamma = 1$, f is said to have *sharp minima*. Then the proximal algorithm converges finitely (in fact in a single iteration for c^0 sufficiently large). This is shown in the following proposition (see also Fig. 3.6.5).

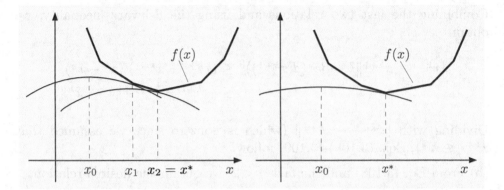

Figure 3.6.5. Finite convergence of the proximal algorithm when $f(x)$ grows at a linear rate near the optimal solution set (e.g., when f is polyhedral). In the figure on the right, convergence occurs in a single iteration for sufficiently large c^0.

Proposition 3.6.5: (Finite Convergence) Assume that X^* is nonempty and that there exists a scalar $\beta > 0$ such that

$$f^* + \beta d(x) \leq f(x), \qquad \forall\, x \in \Re^n, \tag{3.110}$$

where $d(x) = \min_{x^* \in X^*} \|x - x^*\|$. Then if $\sum_{k=0}^{\infty} c^k = \infty$, the proximal algorithm (3.102) converges to X^* finitely (i.e., there exists $\bar{k} > 0$ such that $x^k \in X^*$ for all $k \geq \bar{k}$). Furthermore, if $c^0 \geq d(x^0)/\beta$, the algorithm converges in a single iteration (i.e., $x^1 \in X^*$).

Proof: The assumption (3.107) of Prop. 3.6.4 holds with $\gamma = 1$ and all $\delta > 0$, so Eq. (3.109) yields

$$d(x^{k+1}) + \beta c^k \leq d(x^k), \qquad \text{if } x^{k+1} \notin X^*. \tag{3.111}$$

If $\sum_{k=0}^{\infty} c^k = \infty$ and $x^k \notin X^*$ for all k, by adding Eq. (3.111) over all k, we obtain a contradiction. Hence we must have $x^k \in X^*$ for k sufficiently large. Also if $c^0 \geq d(x^0)/\beta$, Eq. (3.111) cannot hold with $k = 0$, so we must have $x^1 \in X^*$. **Q.E.D.**

The growth condition (3.110) is illustrated in Fig. 3.6.6. It can be shown that this condition holds when f is a polyhedral function and X^* is nonempty (see [Ber15a], Prop. 5.1.6).

Gradient Interpretation

An interesting interpretation of the proximal iteration is obtained by con-

Sec. 3.6   Proximal Algorithms

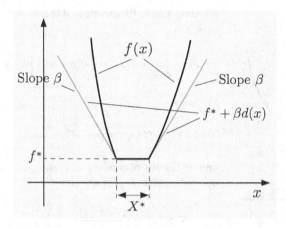

Figure 3.6.6. Illustration of the condition

$$f^* + \beta d(x) \leq f(x), \quad \forall\, x \in \Re^n,$$

for a sharp minimum [cf. Eq. (3.110)].

sidering the function

$$\phi_c(z) = \inf_{x \in X}\left\{f(x) + \frac{1}{2c}\|x - z\|^2\right\} \tag{3.112}$$

for a fixed positive value of c (cf. Fig. 3.6.7). It can be seen that

$$\inf_{x \in \Re^n} f(x) \leq \phi_c(z) \leq f(z), \quad \forall\, z \in \Re^n,$$

from which it follows that the set of minima of f and ϕ_c coincide (this is also evident from the geometric view of the proximal minimization given in Fig. 3.6.1). Moreover it turns out that ϕ_c is a differentiable convex function. The following proposition shows this fact and derives the form of the gradient $\nabla\phi_c$. For the proof, we refer to [Ber15a], Prop. 5.1.7.

Proposition 3.6.6: If $f : \Re^n \mapsto \Re$ is a convex function and X is a closed and convex set, the function ϕ_c of Eq. (3.112) is convex and differentiable, and we have

$$\nabla\phi_c(z) = \frac{z - x_c(z)}{c}, \quad \forall\, z \in \Re^n, \tag{3.113}$$

where $x_c(z)$ is the unique minimizer in Eq. (3.112). Moreover, if f is differentiable, we have

$$\nabla\phi_c(z) = \nabla f\bigl(x_c(z)\bigr), \quad \forall\, z \in \Re^n.$$

Using the gradient formula (3.113), we see that the proximal iteration can be written as

$$x^{k+1} = x^k - c^k \nabla\phi_{c^k}(x^k), \tag{3.114}$$

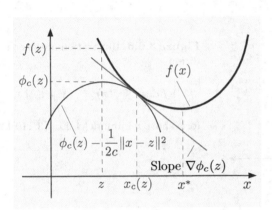

Figure 3.6.7. Illustration of the function

$$\phi_c(z) = \inf_{x \in X} \left\{ f(x) + \frac{1}{2c} \|x - z\|^2 \right\}.$$

We have

$$\phi_c(z) \le f(z), \quad \forall\, z \in \Re^n,$$

and at the set of minima of f, ϕ_c coincides with f. We also have

$$\nabla \phi_c(z) = \frac{z - x_c(z)}{c};$$

cf. Prop. 3.6.6.

so *it is a gradient iteration for minimizing ϕ_{c^k} with stepsize equal to c^k*. This interpretation provides insight into the working mechanism of the algorithm and has formed the basis for various acceleration schemes, based on gradient ideas, as well as Newton-like schemes that have superlinear convergence rate.

3.6.2 Variants of the Proximal Algorithm

There are several variations of the proximal algorithm, which involve modification of the proximal minimization problem of Eq. (3.102), and are motivated by the need for a convenient solution of this problem. We briefly describe some of the major variants, and we refer to [Ber15a] and the literature cited there for their analysis:

(a) *The use of a general positive definite quadratic proximal term* in Eq. (3.102), in place of $(1/2c^k)\|x - x^k\|^2$, i.e., the iteration

$$x^{k+1} = \arg\min_{x \in X} \left\{ f(x) + \frac{1}{2c^k}(x - x^k)' M(x - x^k) \right\},$$

where M is a positive definite symmetric matrix. This approach may be useful when a suitable matrix M is known that results in better performance characteristics, and is well suited to some special purpose. The method has the same convergence properties as in the case where $M = I$. Moreover, it achieves superlinear, sublinear, or finite convergence under the same circumstances as when $M = I$. The linear convergence rate is, however, affected by the choice of M (see Exercise 3.6.1).

(b) *The use of a nonquadratic proximal term $D_k(x;x^k)$ in Eq. (3.102)*, in place of $(1/2c^k)\|x-x^k\|^2$, i.e., the iteration

$$x^{k+1} \in \arg\min_{x \in X} \{f(x) + D_k(x;x^k)\}. \qquad (3.115)$$

This approach may be useful in specialized contexts; see the *exponential method of multipliers* in Sections 5.2.5 and 7.3.2.

(c) *Linear approximation of f using its gradient at x^k*

$$f(x) \approx f(x^k) + \nabla f(x^k)'(x-x^k),$$

assuming that f is differentiable. Then, in place of Eq. (3.115), we obtain the iteration

$$x^{k+1} \in \arg\min_{x \in X} \{f(x^k) + \nabla f(x^k)'(x-x^k) + D_k(x;x^k)\}. \qquad (3.116)$$

When the proximal term $D_k(x;x^k)$ is the quadratic $(1/2c^k)\|x-x^k\|^2$, this iteration can be seen to be equivalent to the gradient projection iteration (see also Exercise 3.6.1):

$$x^{k+1} = P_X(x^k - c^k \nabla f(x^k)),$$

but there are other choices of D_k that lead to interesting methods, such as the *mirror descent algorithm* (see Exercise 3.6.3). We refer to [Ber15a], Section 6.6, and the references cited there for further discussion.

(d) The *proximal gradient algorithm*, which applies to the problem

$$\text{minimize} \quad f(x) + h(x)$$
$$\text{subject to } x \in X,$$

where $f : \Re^n \mapsto \Re$ is a differentiable convex function, $h : \Re^n \mapsto \Re$ is a convex function, and X is a closed convex set. This algorithm combines ideas from the gradient projection method and the proximal method. It replaces f with a linear approximation in the proximal minimization, i.e.,

$$x^{k+1} \in \arg\min_{x \in X} \left\{\nabla f(x^k)'(x-x^k) + h(x) + \frac{1}{2\alpha^k}\|x-x^k\|^2\right\}, \qquad (3.117)$$

where $\alpha^k > 0$ is a parameter. Thus when f is a linear function, we obtain the proximal algorithm for minimizing $f+h$. When h is identically zero, we obtain the gradient projection method. Note that there is an alternative/equivalent way to write the algorithm (3.117):

$$z^k = x^k - \alpha^k \nabla f(x^k), \quad x^{k+1} \in \arg\min_{x \in X}\left\{h(x) + \frac{1}{2\alpha^k}\|x-z^k\|^2\right\}, \qquad (3.118)$$

as can be verified by expanding the quadratic

$$\|x - z^k\|^2 = \|x - x^k + \alpha^k \nabla f(x^k)\|^2.$$

Thus the method alternates gradient steps on f with proximal steps on h. The advantage that this method may have over the proximal algorithm is that the proximal step in Eq. (3.118) is executed with h rather than with $f + h$, and this may be significant if h has simple/favorable structure (e.g., h is the ℓ_1 norm or a distance function to a simple constraint set), while f has unfavorable structure. Under relatively mild assumptions, it can be shown that the method has a cost function descent property, provided the stepsize α is sufficiently small; see [Ber15a], Section 6.3, and the references cited there.

(e) *Incremental versions of the proximal algorithm* for minimizing a sum $\sum_{i=1}^m f_i$ of differentiable cost components, in analogy with the incremental gradient methods of Sections 2.4.1 and 2.4.2; see [Ber10b], [Ber11a], [Ber12], [Ber15a] for an analysis and extensions to the case where the components are convex but nondifferentiable. These methods will be revisited in a dual setting in Section 7.3.3.
The *incremental proximal algorithm* takes the form

$$x^{k+1} = \arg\min_{x \in X} \left\{ f_{i_k}(x) + \frac{1}{2\alpha^k} \|x - x^k\|^2 \right\}, \qquad (3.119)$$

where f_{i_k} is a single cost component selected at iteration k. The choice of stepsize α^k and the convergence properties are similar to the ones of the incremental gradient method of Section 2.4.1. In fact incremental gradient and incremental proximal methods can be combined in the same algorithm to provide flexibility in cases where some of the cost components f_i are well suited for the proximal minimization of Eq. (3.119), while others are not.
The *incremental aggregated proximal algorithm* takes the form

$$x^{k+1} = \arg\min_{x \in X} \left\{ f_{i_k}(x) + \sum_{i \neq i_k} \nabla f_i(x^{\ell_i})'(x - x^k) + \frac{1}{2\alpha^k} \|x - x^k\|^2 \right\},$$
(3.120)

where each $\nabla f_i(x^{\ell_i})$, $i \neq i_k$, is a "delayed" gradient of f_i at some earlier iterate x^{ℓ_i}. It can equivalently be written in the form

$$x^{k+1} = \arg\min_{x \in X} \left\{ f_{i_k}(x) + \frac{1}{2\alpha^k} \|x - z^k\|^2 \right\},$$

where

$$z^k = x^k - \sum_{i \neq i_k} \nabla f_i(x^{\ell_i}).$$

As in Section 2.4.2, the "delays" are bounded, in the sense that the indexes ℓ_i satisfy
$$k - b \le \ell_i \le k, \qquad \forall\, i, k,$$
where b is a fixed nonnegative integer. Thus the algorithm uses outdated gradients from previous iterations for the components f_i, $i \ne i_k$, and need not compute the gradient of these components at iteration k. Intuitively, the idea is that the term
$$\sum_{i \ne i_k} \nabla f_i(x^{\ell_i})'(x - x^k)$$
in the proximal minimization (3.120) is a linear approximation to the term
$$\sum_{i \ne i_k} f_i(x)$$
[minus the constant $\sum_{i \ne i_k} f_i(x^k)$], which would be used in the standard proximal algorithm, which is
$$x^{k+1} = \arg\min_{x \in X} \left\{ \sum_{i=1}^m f_i(x) + \frac{1}{2\alpha^k} \|x - x^k\|^2 \right\}.$$

The choice of stepsize α^k and the convergence properties are similar to the ones of the incremental aggregated gradient method of Section 2.4.2. In particular, for $X = \Re^n$, a linear convergence result with a constant but sufficiently small stepsize can be proved for the case where f is strongly convex, similar to the one of Section 2.4.2 (cf. Prop. 2.4.2); see Exercise 3.6.2 and [Ber15b].

(f) *Local versions of the proximal algorithm* for minimizing a function $f: \Re^n \mapsto \Re$ subject to equality constraints or other constraints, which do not require that f is convex. These algorithms apply to the problem
$$\begin{aligned} \text{minimize} \quad & f(x) \\ \text{subject to} \quad & h(x) = 0, \end{aligned} \tag{3.121}$$
where $f: \Re^n \mapsto \Re$ and $h: \Re^n \mapsto \Re^r$ are twice continuously differentiable functions, such that f is "locally convex" over the set $\{x \mid h(x) = 0\}$ (this is defined in terms of assumptions that relate to second order sufficiency conditions of Sections 4.2 and 4.3; see [Ber78], [Ber79a]). The local proximal algorithm has the form
$$x^{k+1} = \arg\min_{h(x)=0} \left\{ f(x) + \frac{1}{2c^k} \|x - x^k\|^2 \right\}. \tag{3.122}$$

A Newton-like version of this algorithm is also given in [Ber79a].

EXERCISES

3.6.1 (Relation of the Proximal and Gradient Projection Algorithms)

Consider the proximal algorithm

$$x^{k+1} = \arg\min_{x \in X} \left\{ f(x) + \frac{1}{2c^k} \|x - x^k\|^2 \right\}$$

for the case of the quadratic function $f(x) = (1/2)x'Qx + b'x$, and for a sequence of positive scalars $\{c^k\}$ such that $Q + (1/c^k)I$ is positive definite for all k (Q is symmetric but not necessarily positive definite; possibly $Q = 0$).

(a) Show that this algorithm is equivalent to the scaled gradient projection method (3.50)-(3.51) for the special choice of the scaling matrix $H^k = Q + (1/c^k)I$ and for stepsizes $s^k = 1$, $\alpha^k = 1$. *Hint*: Write f as

$$f(x) = \nabla f(x^k)'(x - x^k) + \tfrac{1}{2}(x - x^k)'Q(x - x^k) + \nabla f(x^k)'x^k - \tfrac{1}{2}x^{k'}Qx^k.$$

(b) Consider the following generalized version of the algorithm:

$$x^{k+1} = x^k + \alpha^k(\bar{x}^k - x^k),$$

where $\alpha^k \in (0, 1]$,

$$\bar{x}^k = \arg\min_{x \in X} \left\{ f(x) + \tfrac{1}{2}(x - x^k)'M^k(x - x^k) \right\},$$

and M^k is such that $Q + M^k$ is positive definite. Show that this is a scaled gradient projection method with scaling matrix $Q + M^k$.

(c) Assume that $X = \Re^n$, $M^k = Q$, and that $\alpha^k = 2$. Assuming Q is positive definite, show that the algorithm of part (b) converges in a single step.

3.6.2 (Linear Convergence of Incremental Aggregated Proximal Algorithm [Ber15b])

Consider the algorithm (3.120), and assume that $X = \Re^n$, and that the functions f_i are convex and differentiable and satisfy

$$\|\nabla f_i(x) - \nabla f_i(z)\| \le L_i \|x - z\|, \qquad \forall\, x, z \in \Re^n,$$

for some constants L_i. Assume further that the cost function $\sum_{i=1}^m f_i$ is strongly convex. Show that there exists $\bar{\alpha} > 0$ such that for all $\alpha \in (0, \bar{\alpha}]$, the sequence $\{x^k\}$ generated by the iteration (3.120) with constant stepsize $\alpha^k \equiv \alpha$ converges to x^* linearly, in the sense that $\|x^k - x^*\| \le \gamma \rho^k$ for some scalars $\gamma > 0$ and $\rho \in (0, 1)$, and all k. *Hint*: Show that the iteration can be written as

$$x^{k+1} = x^k - \alpha^k \left(\nabla f_{i_k}(x^{k+1}) + \sum_{i \ne i_k} \nabla f_i(x^{\ell_i}) \right),$$

and follow the line of proof of Prop. 2.4.2; see [Ber15b] for a complete proof.

3.6.3 (Mirror Descent and Entropic Descent Algorithm)

Consider the mirror descent algorithm (3.116) for the case where the regularization term is the *entropy function*,

$$D_k(x;y) = \frac{1}{c^k} \sum_{i=1}^{n} x_i \left(\ln\left(\frac{x_i}{y_i}\right) - 1 \right),$$

and the constraint set X is the unit simplex

$$X = \left\{ x \geq 0 \,\Big|\, \sum_{i=1}^{n} x_i = 1 \right\}.$$

Show that the algorithm can be written in closed form as follows:

$$x_i^{k+1} = \frac{x_i^k e^{-c^k \nabla_i f(x^k)}}{\sum_{j=1}^{n} x_j^k e^{-c^k \nabla_j f(x^k)}}, \qquad i = 1, \ldots, n,$$

where $\nabla_j f(x^k) = \partial f(x^k)/\partial x_j$. Note that this algorithm involves less overhead per iteration than the corresponding gradient projection iteration, which requires projection on the unit simplex.

3.7 BLOCK COORDINATE DESCENT METHODS

We discussed unconstrained coordinate descent methods in Section 2.3.1. We will now generalize these methods to solve the constrained problem

$$\begin{aligned} \text{minimize} \quad & f(x) \\ \text{subject to} \quad & x \in X, \end{aligned} \qquad \text{(CP)}$$

where X is a Cartesian product of closed convex sets $X_1, \ldots, X_m$:

$$X = X_1 \times X_2 \times \cdots \times X_m. \qquad (3.123)$$

We assume that X_i is a closed convex subset of $\Re^{n_i}$ and $n = n_1 + \cdots + n_m$. The vector x is partitioned as

$$x = (x_1, x_2, \ldots, x_m),$$

where each x_i belongs to $\Re^{n_i}$, so the constraint $x \in X$ is equivalent to

$$x_i \in X_i, \qquad i = 1, \ldots, m.$$

Let us assume that for every $x \in X$ and every $i = 1, \ldots, m$, the optimization problem

$$\text{minimize } f(x_1, \ldots, x_{i-1}, \xi, x_{i+1}, \ldots, x_m)$$
$$\text{subject to } \xi \in X_i,$$

has at least one solution. The following algorithm, known as *block coordinate descent* or *Gauss-Seidel* method, generates the next iterate $x^{k+1} = (x_1^{k+1}, \ldots, x_m^{k+1})$, given the current iterate $x^k = (x_1^k, \ldots, x_m^k)$, according to the iteration

$$x_i^{k+1} \in \arg\min_{\xi \in X_i} f(x_1^{k+1}, \ldots, x_{i-1}^{k+1}, \xi, x_{i+1}^k, \ldots, x_m^k), \qquad i = 1, \ldots, m. \tag{3.124}$$

Thus, at each iteration, the cost is minimized with respect to each of the "block coordinate" vectors x_i^k, taken in cyclic order. Naturally, the method makes practical sense if the minimization in Eq. (3.124) is fairly easy. This is frequently so when each x_i is a scalar, but there are also other cases of interest, where x_i is a multidimensional vector.

The following proposition gives the basic convergence result for the method. It turns out that it is necessary to make an assumption implying that the minimum in Eq. (3.124) is uniquely attained, as first suggested in [Zan69]. The need for this assumption is not obvious but has been demonstrated by an example given in [Pow73]. Exercise 3.7.2 provides a modified version of the algorithm, which does not require this assumption.

Proposition 3.7.1: (Convergence of Block Coordinate Descent) Suppose that f is continuously differentiable over the set X of Eq. (3.123). Furthermore, suppose that for each $x = (x_1, \ldots, x_m) \in X$ and i,

$$f(x_1, \ldots, x_{i-1}, \xi, x_{i+1}, \ldots, x_m)$$

viewed as a function of ξ, attains a unique minimum $\bar{\xi}$ over X_i, and is monotonically nonincreasing in the interval from x_i to $\bar{\xi}$. Let $\{x^k\}$ be the sequence generated by the block coordinate descent method (3.124). Then, every limit point of $\{x^k\}$ is a stationary point.

Proof: Denote

$$z_i^k = \left(x_1^{k+1}, \ldots, x_i^{k+1}, x_{i+1}^k, \ldots, x_m^k\right).$$

Using the definition (3.124) of the method, we obtain

$$f(x^k) \geq f(z_1^k) \geq f(z_2^k) \geq \cdots \geq f(z_{m-1}^k) \geq f(x^{k+1}), \qquad \forall\, k. \tag{3.125}$$

Let $\bar{x} = (\bar{x}_1, \ldots, \bar{x}_m)$ be a limit point of the sequence $\{x^k\}$, and note that $\bar{x} \in X$ since X is closed. Equation (3.125) implies that the sequence $\{f(x^k)\}$ converges to $f(\bar{x})$. We will show that $\bar{x}$ is a stationary point.

Let $\{x^{k_j} \mid j = 0, 1, \ldots\}$ be a subsequence of $\{x^k\}$ that converges to $\bar{x}$. From the definition (3.124) of the algorithm and Eq. (3.125), we have

$$f(x^{k_j+1}) \le f(z_1^{k_j}) \le f(x_1, x_2^{k_j}, \ldots, x_m^{k_j}), \quad \forall\ x_1 \in X_1.$$

Taking the limit as j tends to infinity, we obtain

$$f(\bar{x}) \le f(x_1, \bar{x}_2, \ldots, \bar{x}_m), \quad \forall\ x_1 \in X_1. \qquad (3.126)$$

Using the optimality conditions for minimization over a convex set (Prop. 3.1.1), we conclude that

$$\nabla_1 f(\bar{x})'(x_1 - \bar{x}_1) \ge 0, \quad \forall\ x_1 \in X_1,$$

where $\nabla_i f$ denotes the gradient of f with respect to the component x_i.

The idea of the proof is now to show that $\{z_1^{k_j}\}$ converges to $\bar{x}$ as $j \to \infty$, so that by repeating the preceding argument with $\{z_1^{k_j}\}$ in place of $\{x^{k_j}\}$, we will have

$$\nabla_2 f(\bar{x})'(x_2 - \bar{x}_2) \ge 0, \quad \forall\ x_2 \in X_2.$$

We can then continue similarly to obtain

$$\nabla_i f(\bar{x})'(x_i - \bar{x}_i) \ge 0, \quad \forall\ x_i \in X_i,$$

for all $i = 1, \ldots, m$. By adding these inequalities, and using the Cartesian product structure of the set X, it follows that $\nabla f(\bar{x})'(x - \bar{x}) \ge 0$ for all $x \in X$, i.e., $\bar{x}$ is stationary, thereby completing the proof.

To show that $\{z_1^{k_j}\}$ converges to $\bar{x}$ as $j \to \infty$, we assume the contrary, or equivalently that $\{z_1^{k_j} - x^{k_j}\}$ does not converge to zero. Let $\gamma^{k_j} = \|z_1^{k_j} - x^{k_j}\|$. By possibly restricting to a subsequence of $\{k_j\}$, we may assume that there exists some $\bar{\gamma} > 0$ such that $\gamma^{k_j} \ge \bar{\gamma}$ for all j. Let $s_1^{k_j} = (z_1^{k_j} - x^{k_j})/\gamma^{k_j}$. Thus, $z_1^{k_j} = x^{k_j} + \gamma^{k_j} s_1^{k_j}$, $\|s_1^{k_j}\| = 1$, and $s_1^{k_j}$ differs from zero only along the first block-component. Notice that $s_1^{k_j}$ belongs to a compact set and therefore has a limit point $\bar{s}_1$. By restricting to a further subsequence of $\{k_j\}$, we assume that $s_1^{k_j}$ converges to $\bar{s}_1$.

Let us fix some $\epsilon \in [0, 1]$. Since $0 \le \epsilon\bar{\gamma} \le \gamma^{k_j}$, the vector $x^{k_j} + \epsilon\bar{\gamma} s_1^{k_j}$ lies on the segment joining x^{k_j} and $x^{k_j} + \gamma^{k_j} s_1^{k_j} = z_1^{k_j}$, and belongs to X because X is convex. Using the fact that f is monotonically nonincreasing on the interval from x^{k_j} to $z_1^{k_j}$, we obtain

$$f(z_1^{k_j}) = f(x^{k_j} + \gamma^{k_j} s_1^{k_j}) \le f(x^{k_j} + \epsilon\bar{\gamma} s_1^{k_j}) \le f(x^{k_j}).$$

Since $f(x^k)$ converges to $f(\bar{x})$, Eq. (3.125) shows that $f(z_1^{k_j})$ also converges to $f(\bar{x})$. Taking the limit as j tends to infinity, we obtain $f(\bar{x}) \le f(\bar{x} + \epsilon\bar{\gamma}\bar{s}_1) \le f(\bar{x})$. We conclude that $f(\bar{x}) = f(\bar{x} + \epsilon\bar{\gamma}\bar{s}_1)$, for every $\epsilon \in [0,1]$. Since $\bar{\gamma}\bar{s}_1 \ne 0$ and by Eq. (3.126), $\bar{x}_1$ attains the minimum of $f(x_1, \bar{x}_2, \ldots, \bar{x}_m)$ over $x_1 \in X_1$, this contradicts the hypothesis that f is uniquely minimized when viewed as a function of the first block-component. This contradiction establishes that $z_1^{k_j}$ converges to $\bar{x}$, which as remarked earlier, shows that $\nabla_2 f(\bar{x})'(x_2 - \bar{x}_2) \ge 0$ for all $x_2 \in X_2$.

By using $\{z_1^{k_j}\}$ in place of $\{x^{k_j}\}$, and $\{z_2^{k_j}\}$ in place of $\{z_1^{k_j}\}$ in the preceding arguments, we can show that $\nabla_3 f(\bar{x})'(x_3 - \bar{x}_3) \ge 0$ for all $x_3 \in X_3$, and similarly $\nabla_i f(\bar{x})'(x_i - \bar{x}_i) \ge 0$ for all $x_i \in X_i$ and i. **Q.E.D.**

Note that the uniqueness of minimum and monotonic nonincrease assumption in Prop. 3.7.1 is satisfied if f is strictly convex in each block-component when all other block-components are held fixed. An alternative assumption under which the conclusion of Prop. 3.7.1 can be shown is that the sets X_i are compact (as well as convex), and that for each i and $x \in X$, the function of the ith block-component ξ

$$f(x_1, \ldots, x_{i-1}, \xi, x_{i+1}, \ldots, x_m)$$

attains a unique minimum over X_i, when all all other block-components are held fixed. The proof is similar (in fact simpler) to the proof of Prop. 3.7.1 [to show that $\{z_1^{k_j}\}$ converges to $\bar{x}$, we note that $\{x_1^{k_j+1}\}$ lies in the compact set X_1, so a limit point, call it $\bar{\xi}$, is a minimizer of $f(x_1, \bar{x}_2, \ldots, \bar{x}_m)$ over $x_1 \in X_1$, and since $\bar{x}_1$ is also a minimizer, it follows that $\bar{\xi} = \bar{x}_1$].

Block coordinate descent methods are often useful in contexts where the cost function and the constraints have a partially decomposable structure with respect to the problem's optimization variables. The following example illustrates the idea.

Example 3.7.1 (Hierarchical Decomposition)

Consider an optimization problem of the form

$$\text{minimize} \sum_{i=1}^{m} f_i(x, y_i)$$

$$\text{subject to} \quad x \in X, \quad y_i \in Y_i, \quad i = 1, \ldots, m,$$

where X and Y_i, $i = 1, \ldots, m$, are closed, convex subsets of corresponding Euclidean spaces, and the functions f_i are continuously differentiable. This problem is associated with a paradigm of optimization of a system consisting of m subsystems, with the cost function f_i associated with the operations of the ith subsystem. Here y_i is viewed as vector of local decision variables that influences the cost of the ith subsystem only, and x is viewed as a vector of

Sec. 3.7   Block Coordinate Descent Methods   327

global or coordinating decision variables that affects the operation of all the subsystems.

The coordinate descent method takes advantage of the decomposable structure and has the form

$$y_i^{k+1} \in \arg\min_{y_i \in Y_i} f_i(x^k, y_i), \qquad i = 1, \ldots, m,$$

$$x^{k+1} \in \arg\min_{x \in X} \sum_{i=1}^{m} f_i(x, y_i^{k+1}).$$

The method has a natural real-life interpretation: at each iteration, each subsystem optimizes its own cost, taking the global variables as fixed at their current values, and then the coordinator optimizes the overall cost for the current values of the local variables.

3.7.1 Variants of Coordinate Descent

There are many variations of coordinated descent, which are aimed at improved efficiency, application-specific structures, and distributed computing environments. We describe some of the possibilities here and in the exercises, and we refer to the literature for a more detailed analysis.

(a) A more general and sometimes useful version of the block coordinate descent method is one where the block-components are iterated in an irregular order instead of a fixed cyclic order. The convergence result of Prop. 3.7.1 can also be shown for a method where the order of iteration may be arbitrary as long as there is an integer M such that each block-component is iterated at least once in every group of M contiguous iterations. The proof is similar to the one of Prop. 3.7.1. [Take a limit point of the sequence generated by the algorithm, and a subsequence converging to it, with the properties that 1) every two successive elements of the subsequence are separated by at least M block-component iterations, and 2) every group of M contiguous iterations that starts with an element of the subsequence corresponds to the same order of block-components. Then, proceed as in the proof of Prop. 3.7.1.]

(b) There is a *combination of coordinate descent with the proximal algorithm*, which aims to circumvent the need for the "uniqueness of minimum" assumption in Prop. 3.7.1; see [Tse91a], [Aus92], [GrS00]. This method is obtained by applying cyclically the coordinate iteration

$$x_i^{k+1} \in \arg\min_{\xi \in X_i} \left\{ f(x_1^{k+1}, \ldots, x_{i-1}^{k+1}, \xi, x_{i+1}^k, \ldots, x_m^k) + \frac{1}{2c} \|\xi - x_i^k\|^2 \right\},$$

where c is a positive scalar. Assuming that f is convex and differentiable, it can be seen that every limit point of the sequence $\{x^k\}$ is a global minimum. This is easily proved by applying the result of Prop. 3.7.1 to the cost function

$$F(x,y) = f(x) + \frac{1}{2c}\|x-y\|^2;$$

see Exercise 3.7.3.

(c) The preceding combination of coordinate descent with the proximal algorithm is an example of a broader class of methods whereby instead of carrying out exactly the coordinate minimization

$$x_i^{k+1} \in \arg\min_{\xi \in X_i} f(x_1^{k+1}, \ldots, x_{i-1}^{k+1}, \xi, x_{i+1}^k, \ldots, x_m^k),$$

[cf. Eq. (3.124)], we perform one or more iterations of a descent algorithm aimed at solving this minimization. Aside from the proximal algorithm of (b) above, there are *combinations with other descent algorithms*, including conditional gradient, gradient projection, and two-metric projection methods. A key fact here is that there is guaranteed cost function descent after cycling through all the block components. See e.g., [Lin07], [LJS12], [Spa12], [Jag13], [HDR13], [RHL13], [RiT14], [LuX15].

(d) When f is convex but nondifferentiable, the coordinate descent approach may fail because there may exist nonoptimal points starting from which cost function descent is impossible along all the scalar coordinates [as an example consider minimization of $f(x_1, x_2) = x_1 + x_2 + 2|x_1 - x_2|$ over $x_1, x_2 \geq 0$ and points (x_1, x_2) with $x_1 = x_2 > 0$]. There is an important special case, however, where the nondifferentiability of f is inconsequential, and at each nonoptimal point, it is possible to find a direction of descent among the scalar coordinate directions. This is the case where *the nondifferentiable portion of f is separable*, i.e., f has the form

$$f(x) = F(x) + \sum_{i=1}^{n} G_i(x_i), \qquad (3.127)$$

where F is convex and differentiable, and each G_i is a function of the ith (scalar) coordinate component that is convex but not necessarily differentiable.

A case of special interest, which arises in machine learning problems involving ℓ_1-regularization, is when $\sum_{i=1}^{n} G_i(x_i)$ is a positive multiple of the ℓ_1 norm. For this case the convergence result of Prop. 3.7.1 extends easily. Indeed, the minimization of

$$F(x) + \sum_{i=1}^{n} |x_i| \qquad (3.128)$$

Sec. 3.7   Block Coordinate Descent Methods   329

over $x \in X$ is equivalent to minimization of

$$F(x) + \sum_{i=1}^{n} z_i,$$

subject to

$$x_i \in X_i, \quad x_i \leq z_i, \quad -x_i \leq z_i, \quad i = 1, \ldots, n.$$

The block coordinate descent method, where the blocks are (x_i, z_i), $i = 1, \ldots, n$, applies to this problem. Each iteration [with respect to block (x_i, z_i)] can then be written as

$$x_i^{k+1} \in \arg\min_{\xi \in X_i} f(x_1^{k+1}, \ldots, x_{i-1}^{k+1}, \xi, x_{i+1}^k, \ldots, x_n^k) + |x_i^k|, \quad i = 1, \ldots, n,$$

[cf. Eq. (3.124)]. The method is convergent subject to the uniqueness restriction of Prop. 3.7.1.

We finally note that the coordinate descent method we discussed so far uses line minimization along feasible directions. Other stepsize rules are possible, such as the Armijo rule, diminishing stepsize, and constant stepsize (see also the discussion of Section 2.3.1).

EXERCISES

3.7.1 (Parallel Projections Algorithm)

We are given m closed convex sets $X_1, X_2, \ldots, X_m$ in $\Re^n$, and we want to find a point in their intersection. Consider the equivalent problem

$$\text{minimize} \quad \tfrac{1}{2} \sum_{i=1}^{m} \|y_i - x\|^2$$
$$\text{subject to} \quad x \in \Re^n, \ y_i \in X_i, \ i = 1, \ldots, m,$$

where the variables of the optimization are $x, y_1, \ldots, y_m$. Derive a block coordinate descent algorithm involving projections on each of the sets X_i that can be carried out independently of each other. State a convergence result for this algorithm.

3.7.2 (Relation of the Proximal and Coordinate Descent Algorithms)

Let $f : \Re^n \mapsto \Re$ be a continuously differentiable function, let X be a closed convex set, and let c be a positive scalar.

(a) Show that the proximal algorithm

$$x^{k+1} = \arg\min_{x \in X} \left\{ f(x) + \frac{1}{2c} \|x - x^k\|^2 \right\}$$

is a special case of the block coordinate descent method applied to the problem

$$\text{minimize } f(x) + \frac{1}{2c} \|x - y\|^2$$
$$\text{subject to } x \in X, \ y \in \Re^n,$$

which is equivalent to the problem of minimizing f over X.

(b) Derive a convergence result based on Prop. 3.7.1 for the algorithm of part (a).

3.7.3 (Combination of Coordinate Descent and Proximal Algorithm [Tse91a])

Consider the minimization of a continuously differentiable function f of the vector $x = (x_1, \ldots, x_m)$ subject to $x^i \in X_i$, where X_i are closed convex sets. Consider the following variation of the block coordinate descent method (3.124):

$$x_i^{k+1} = \arg\min_{\xi \in X_i} \left\{ f(x_1^{k+1}, \ldots, x_{i-1}^{k+1}, \xi, x_{i+1}^k, \ldots, x_m^k) + \frac{1}{2c} \|\xi - x_i^k\|^2 \right\},$$

where c is a positive scalar. Assuming that f is convex, show that every limit point of the sequence of vectors $x^k = (x_1^k, \ldots, x_m^k)$ is a global minimum. *Hint*: Apply the result of Prop. 3.7.1 to the cost function

$$g(x, y) = f(x^1, \ldots, x^m) + \frac{1}{2c} \sum_{i=1}^m \|x^i - y^i\|^2.$$

For a related analysis of this type of algorithm see [Aus92], and for a recent analysis see [BST14]. For an analysis of this algorithm without the convexity assumption on f, see [GrS00].

3.7.4 (Block Coordinate Descent with Nondifferentiable Terms)

State and show an extension of the convergence result of Prop. 3.7.1, which applies to the minimization of

$$F(x_1, \ldots, x_m) + \sum_{i=1}^m \max \left\{ a'_{i1} x_i - b_{i1}, \ldots, a'_{im_i} x_i - b_{im_i} \right\}$$

over $x_i \in X_i$, $i = 1, \ldots, m$, where X_i is a subset of R^{n_i}, and a_{ij}, b_{ij} are given vectors and scalars, respectively. *Hint*: Extend the discussion for the case of the ℓ_1 norm given in this section [cf. Eq. (3.128)].

3.8 NETWORK OPTIMIZATION ALGORITHMS

In this section we consider some algorithms for single and multiple commodity network flow problems, such as the routing and traffic assignment problems of Examples 3.1.3 and 3.1.4. We discussed the application of the conditional gradient method to such problems in Example 3.2.2. In this section, we will consider Newton-type methods, which aim at a faster convergence rate. We will focus on efficient implementation through the use of the network structure.

A common type of nonlinear network flow optimization involves a graph with a set of directed arcs $\mathcal{A}$, and a set of paths $\mathcal{P}$. Each path p consists of a set of arcs $A_p \subset \mathcal{A}$. A set of path flows $x = \{x_p \mid p \in \mathcal{P}\}$ produces a total flow f_a on each arc $a \in \mathcal{A}$, given by

$$f_a = \sum_{\{p \mid a \in A_p\}} x_p.$$

The objective is to find x that minimizes

$$F(x) = \sum_{p \in \mathcal{P}} R_p(x_p) + \sum_{a \in \mathcal{A}} D_a(f_a), \qquad (3.129)$$

where D_a and R_p are twice continuously differentiable functions. Usually, either $R_p = 0$ or else R_p encodes a penalty for a range of values of x_p (such as small nonnegative values), while D_a encodes a penalty for a range of values of f_a (such as values close to a "capacity" of arc a).

The minimization may be subject to some constraints on x, such as upper and lower bounds on each x_p, supply/demand constraints (sum of path flows with the same origin and destination must take a given value), additional constraints coupling the arc flows of different arcs, etc. Sometimes capacity constraints on the arc flows are imposed explicitly rather than through the functions D_a. We will discuss primarily the unconstrained case and refer to the literature for various ways to handle constraints, either directly through some form of projection, or indirectly through the use of penalty and augmented Lagrangian functions.

There are several feasible direction methods for solving this type of problem, and their implementation is facilitated by the network structure. For example, in Section 3.2.2 we discussed briefly the application of the conditional gradient method for the case where the sum of path flows with the same origin and destination is given (cf. Example 3.2.2, where we showed that the conditional gradient direction can be implemented through simple shortest path computations). Other possibilities include the simplicial decomposition method of Exercise 3.2.7, and the projection methods of Section 3.3 and 3.4 (see the end-of-chapter references).

Even though we have assumed that the set of paths $\mathcal{P}$ is fixed, the methods just mentioned, as well as the ones to be discussed in this section, apply to schemes where $\mathcal{P}$ is gradually enlarged in the course of the algorithm. In particular, in such a scheme, $\mathcal{P}$ initially contains just a few paths, which are augmented at each iteration with additional paths that are obtained by shortest path computations, such as the ones described in Example 3.2.2. This type of path generation approach to forming $\mathcal{P}$ arises in the context of the simplicial decomposition method of Exercise 3.2.7 (see the end-of-chapter references).

We will now discuss second order projection-type methods, based on truncated Newton-type ideas (cf. Section 1.4.3). A common concern with such methods is that they require the calculation of the gradient and Hessian of F, and the solution of a quadratic optimization problem, which may be high-dimensional, as it typically is in network applications. However, it turns out that we can exploit the network structure to approximate the Newton direction y, satisfying

$$\nabla^2 F(x)\, y = -\nabla F(x), \tag{3.130}$$

by using a limited number of steps of the conjugate gradient method (cf. Section 1.4.3). It turns out that the conjugate gradient steps can be implemented with graph operations that do not require the explicit formation and storage of $\nabla^2 F(x)$ or its inverse. In the extreme case where a single conjugate gradient step is used per iteration, we obtain a gradient-like method.

Implementation of the Approximate Newton Step

We focus on a fixed path flow vector x and the calculation of the Newton direction from Eq. (3.130). We write

$$g = \nabla F(x), \qquad H = \nabla^2 F(x),$$

and for each path p, we denote by g_p and H_{pp} the corresponding component of g and diagonal component of H (to simplify notation, we suppress the arguments x, x_p, and f_a from various function expressions).

The Newton direction is obtained by solving the quadratic problem

$$\min_y C(y) \stackrel{\text{def}}{=} g'y + \tfrac{1}{2} y' H y. \tag{3.131}$$

The idea is to solve this problem with the conjugate gradient method. We assume that H is positive definite, so that the Newton direction $-H^{-1}g$ is defined and is obtained by the method in a finite number of steps.

The quantities needed to implement the conjugate gradient method are g and Hv for various vectors v. We first show how to compute efficiently these quantities. In particular, we have

$$g_p = R'_p + \sum_{a \in A_p} D'_a, \qquad p \in \mathcal{P}, \tag{3.132}$$

where R'_p and D'_a denote the first derivatives of R_p and D_a, calculated at the current arguments x_p and f_a, respectively. Moreover

$$H = \nabla^2 R + E'\nabla^2 D E, \qquad (3.133)$$

where $\nabla^2 R$ is the diagonal matrix with the second derivatives R''_p along the diagonal, $\nabla^2 D$ is the diagonal matrix with the second derivatives D''_a along the diagonal, and E is the matrix with components

$$E_{ap} = \begin{cases} 1 & \text{if } a \in A_p, \\ 0 & \text{if } a \notin A_p. \end{cases}$$

For some implementations, we also need the diagonal elements of H, which are

$$H_{pp} = R''_p + \sum_{a \in A_p} D''_a, \qquad p \in \mathcal{P}. \qquad (3.134)$$

By exploiting the special structure of g and H, we can perform the required computations as follows:

(a) *Vector g and diagonal of H*: For each path p, we can calculate g_p by accumulating first derivatives along p [cf. Eq. (3.132)]. Moreover, we can similarly calculate the diagonal elements H_{pp} by accumulating second derivatives along p [cf. Eq. (3.134)].

(b) *Vector-matrix products of the form Hv*: For any vector $v = \{v_p \mid p \in \mathcal{P}\}$, we can calculate the matrix-vector product Hv [cf. Eq. (3.133)] as the gradient of the function

$$\tfrac{1}{2} v' H v = \tfrac{1}{2} \sum_{p \in \mathcal{P}} R''_p v_p^2 + \tfrac{1}{2} \sum_{a \in \mathcal{A}} D''_a f_{a,v}^2,$$

where

$$f_{a,v} = \sum_{\{p \mid a \in A_p\}} v_p, \qquad a \in \mathcal{A}. \qquad (3.135)$$

Thus, once we calculate $f_{a,v}$ for all $a \in \mathcal{A}$ by accumulating v_p along path p, we can obtain the pth component of Hv as

$$(Hv)_p = R''_p v_p + \sum_{a \in A_p} D''_a f_{a,v}, \qquad p \in \mathcal{P}, \qquad (3.136)$$

using a similar computation to the gradient (3.132) and the diagonal Hessian components (3.134).

Conjugate Gradient Implementation

We now describe the iterative conjugate gradient process for solving the quadratic problem (3.131). We start with $y^0 = 0$ as the initial iterate for the minimum of the quadratic cost C, and we generate a sequence $\{y^k\}$ of iterates by successively minimizing along a sequence of conjugate directions $\{p^k\}$, starting with $p^0 = -g$. These directions are obtained therough corresponding gradients, which we denote by r^k:

$$r^k = \nabla C(y^k) = g + Hy^k, \qquad k = 0, 1, \ldots.$$

Given the current iterate-gradient-conjugate direction triplet (y^k, r^k, p^k), we generate the next triplet $(y^{k+1}, r^{k+1}, p^{k+1})$ by the conjugate gradient iteration, which has the following form (cf. Section 2.1):

$$y^{k+1} = y^k + \alpha^k p^k, \qquad \text{where } \alpha^k = -\frac{p^{k'}r^k}{p^{k'}Hp^k},$$

$$r^{k+1} = g + Hy^{k+1},$$

$$p^{k+1} = -r^{k+1} + \beta^k p^k, \qquad \text{where } \beta^k = \frac{r^{k+1'}r^{k+1}}{r^{k'}r^k}.$$

Here α^k is the stepsize that minimizes $C(y)$ over $\{y^k + \alpha p^k \mid \alpha \in \Re\}$.†
A frequently used equivalent formula for α^k is

$$\alpha^k = \frac{r^{k'}r^k}{p^{k'}Hp^k}$$

(since $p^k = -r^k + \beta^k p^{k-1}$ and r^k is orthogonal to $p^{k-1}, \ldots, p^0$, a basic property of the conjugate gradient method). The matrix-vector products to be computed at each iteration are Hp^k and Hy^{k+1}, and they can be obtained by using Eqs. (3.135)-(3.136), with $v = p^k$ and $v = y^{k+1}$, respectively.

According to the theory of Section 2.1, either y^{k+1} minimizes $C(y)$ (i.e., $r^{k+1} = 0$) and hence y^{k+1} is equal to the Newton direction, or else we have
$$C(y^k) < C(y^0) = 0, \qquad \forall\ k > 0,$$
so that
$$g'y^k < -y^{k'}Hy^k < 0, \qquad \forall\ k > 0. \tag{3.137}$$

---

† To verify the formula given for α^k, note that we have

$$0 = \frac{\partial C(y^k + \alpha p^k)}{\partial \alpha}\bigg|_{\alpha = a^k} = p^{k'}\big(g + H(y^k + \alpha^k p^k)\big) = p^{k'}r^k + \alpha^k p^{k'}Hp^k.$$

We may either let the process terminate naturally (which will happen after a number of iterations no larger than the number of paths in $\mathcal{P}$), or more practically, terminate once a certain termination criterion is satisfied, in which case, by Eq. (3.137), y^k is a descent direction of the cost function F at the current iterate.

An important variant, which is often far superior in practice, is to use a preconditioning matrix S, which is a diagonal approximation to the Hessian matrix, i.e, S is diagonal with H_{pp} along the diagonal. This method starts with $y^0 = 0$, $r^0 = g$, and $p^0 = -Sg$, and has the form

$$y^{k+1} = y^k + \alpha^k p^k, \qquad \text{where } \alpha^k = \frac{r^{k'} S r^k}{p^{k'} H p^k},$$

$$r^{k+1} = g + H y^{k+1},$$

$$p^{k+1} = -S r^{k+1} + \beta^k p^k, \qquad \text{where } \beta^k = \frac{r^{k+1'} S r^{k+1}}{r^{k'} S r^k}.$$

The non-preconditioned method bears the same relation to the preconditioned version that the gradient method bears to the diagonally scaled gradient method, with second-order diagonal scaling. The use of this type of preconditioning not only improves the rate of convergence (typically), but also facilitates the choice of stepsize (a stepsize of 1 typically works, regardless of the number of conjugate gradient iterations used). These advantages of preconditioning have been confirmed by extensive computational experience, although theoretically speaking there are rare exceptions where diagonal second-order preconditioning does not improve performance.

A somewhat different type of diagonal preconditioning can be used with advantage in the case where the number of arcs is substantially smaller than the number of paths in $\mathcal{P}$. Then by using $S = \nabla^2 R(x)$ as a preconditioning matrix, it can be shown that the number of conjugate gradient iterations to find the Newton direction is at most equal to the number of arcs (see Exercise 2.1.10).

The choice of number of conjugate gradient steps to obtain an approximate Newton direction is an interesting practical implementation issue, and may ultimately be settled by experimentation (see also the discussion of Section 1.4.3). Generally, to attain a superlinear rate, the conjugate gradient process for approximating the Newton step must become asymptotically exact, with the number of conjugate gradient steps per iteration approaching the number of paths in $\mathcal{P}$. This appears highly inefficient for large network flow problems. Typically just a few conjugate gradient steps per iteration are sufficient to attain a good convergence rate, particularly when second order diagonal preconditioning is used.

Distributed Implementation of the Newton Step

The algorithm just described can also be executed in a distributed fashion, assuming that there is a processor assigned to each path and a processor assigned to each arc. To compute the Newton step, for each p, the processor assigned to path p may update, maintain, and communicate to other processors its own path variable x_p, and compute the corresponding component y_p of the Newton direction. For each arc a, the processor assigned to a may compute and communicate to the relevant path processors, its accumulated flow variable f_a. Similar computations are used to execute the intermediate conjugate gradient iterations. The path processors may also collaborate to compute a stepsize guaranteeing descent at every iteration, although in many network applications a constant stepsize can be used reliably, as experience has shown (equal or nearly equal to one when preconditioning is used).

Information exchange between the processors may be conveniently and accurately performed by using a spanning tree with a designated node serving as a synchronizer and leader for the distributed computation. Schemes of this type have been considered extensively in data networks and distributed computing systems (see [BeT89], [BeG92]). On the other hand, it should be noted that in certain application contexts, the need for synchronization to perform Newton and conjugate gradient iterations may be a major drawback over diagonally scaled gradient and gradient projection methods, which can be implemented asynchronously with satisfactory convergence properties (see [TsB86], [BeT89] for partially asynchronous methods that are specific to the network flow problems we are considering).

Variations and Constrained Extensions

There are several network flow problem formulations and corresponding algorithms, where the preceding Newton direction computation may be potentially used. An important issue is the treatment of constraints in a network context. Here are some potential approaches, leading to viable Newton-type algorithms:

(a) Constraints may be eliminated via a penalty or augmented Lagrangian approach (see Section 5.2) to yield an unconstrained problem of minimizing a function $F(x)$ of the form (3.129). The convergence analysis of such approaches requires (1) a convergence guarantee for the unconstrained optimization of the penalized or augmented Lagrangian objective, and (2) a convergence guarantee for the overall penalty or augmented Lagrangian objective. These guarantees may be obtained in straightforward fashion as applications of our standard results for gradient-related and Newton-like methods for unconstrained minimization, as well as standard analyses of penalty and augmented Lagrangian methods (see Sections 5.2 and 7.3).

(b) Nonnegativity constraints can be treated with two-metric Newton-like methods that require computation of Newton directions such as the ones given earlier in this section.

(c) Simplex constraints, relating to supply/demand specifications, can be used to eliminate some of the variables and essentially reduce the constraint set to the nonnegativity case of (b) above (see [Ber82b], [BeG83], [GaB84] for convergence and rate of convergence analysis).

Regardless of how constraints are treated, for a convergent algorithm one must address the issue of stepsize selection. Experience has shown that in network algorithms one may often use reliably a constant stepsize, particularly when the directions used embody second order information, which makes a stepsize close to one a typically good choice. An alternative to a constant stepsize is a line search rule based on line minimization or successive stepsize reduction. The use of such rules improves the reliability of algorithms but introduces additional complexity, particularly in a distributed context.

There are several problem generalizations, involving in some cases an extended network or even a non-network structure, which admit a treatment similar to the one of the preceding sections:

(a) Cost functions F that are defined over just a subset of the path flow space arise when constraints are eliminated by means of an interior point method approach. Newton directions can still be used within this context, provided a stepsize procedure is used to ensure that the successive iterates stay within the feasible region (see Section 5.1).

(b) More general linear dependence of arc flows on path flows can be treated by generalization of the terms $D_a(f_a)$ in the cost function. In particular, we may redefine f_a to be a general linear function $c'_a x$ where c_a is some vector. For example the scalar components of c_a may represent arc gains (cf. [Ber98], Section 8.5). In this case, the approach of this section generalizes straightforwardly. What is essential is that the Hessian of F should have the generic form

$$\nabla^2 R + E' \nabla^2 D E$$

of Eq. (3.133), with $\nabla^2 R$ and $\nabla^2 D$ being diagonal matrices, and E being a matrix that encodes a linear dependence between x and the arguments f_a of the cost terms $D_a(f_a)$.

(c) Linear side constraints may be treated by using a penalty or augmented Lagrangian approach, thereby reducing to case (b) above.

(d) The basic Hessian structure that is important for the convenient computation of gradients and Hessian matrix-vector products is

$$E' D E,$$

where D and E are matrices with D diagonal. Therefore the methodology of this section will also work for structures of the form

$$R + E'_1 D_1 E_1 + \cdots + E'_m D_m E_m,$$

where R, D_i, and E_i, $i = 1, \ldots, m$, are matrices with R and D_i diagonal. There are also potential extensions in cases where R and $D_1, \ldots, D_m$ are symmetric and nearly diagonal (e.g., tridiagonal).

3.9 NOTES AND SOURCES

Section 3.1: The optimality conditions of this section are classical. For more material on multicommodity flow problems and their use in optimal routing in communication and transportation networks, see the books by Bertsekas and Gallager [BeG92], Bertsekas [Ber98], and Patriksson [Pat98], and the survey by Florian and Hearn [FlH95]. For further applications of optimization models involving convex sets, we refer to Ben-Tal and Nemirovski [BeN01], and Boyd and Vandenbergue [BoV04].

The ideas underlying retractive sequences and their use in showing existence of optimal solutions are central in the work of Auslender [Aus96], [Aus97], and are also used extensively in the book by Auslender and Teboulle [AuT03]. The notion of an asymptotic direction of a nested sequence $\{S^k\}$ of closed sets, the corresponding nonemptiness of intersection result (Prop. 3.1.2), and its use in proving existence of optimal solutions, were developed in Bertsekas and Tseng [BeT07]. The set of asymptotic directions, when specialized to a nonconvex set (rather than a nested sequence of nonconvex sets), is essentially the horizon cone described by Rockafellar and Wets [RoW98], and the asymptotic cone described by Auslender and Teboulle [AuT03].

In the case where the closed sets of the nested sequence are convex, there is a strong connection between the notion of an asymptotic direction and the notion of a direction of recession. This latter notion is fundamental in the analysis of existence of optimal solutions and duality theory in convex programming; see the book by Rockafellar [Roc70], and also the books by Bertsekas, Nedić, and Ozdaglar [BNO03], and Bertsekas [Ber09], which emphasize the fundamental role of closed set intersection results. In particular, the set of asymptotic directions of a nested sequence $\{S^k\}$ of closed convex sets is in effect the intersection of the recession cones of the sets S^k. Using this fact, it can be shown that Prop. 3.1.2 contains as special cases several results on the nonemptiness of the intersection of closed convex sets given in Section 1.5.1 of [BNO03], and in Section 1.4 of [Ber09].

Section 3.2: Feasible direction methods were first systematically investigated by Zoutendijk [Zou60]. Convergence analyses and additional methods are given by Topkis and Veinott [ToV67], Polak [Pol71], Pironneau and Polak [PiP73], and Zoutendijk [Zou76].

The conditional gradient method was first proposed for convex problems with linear constraints by Frank and Wolfe [FrW56]. Extensive discussions of its application to more general problems are given in Levitin and Poljak [LeP65], and Demjanov and Rubinov [DeR70]. The convergence rate of the method was addressed by Canon and Cullum [CaC68], Dunn [Dun79], [Dun80a], Dunn and Sachs [DuS83], Guelat and Marcotte [GuM86], and more recently by Jaggi [Jag13], Lacoste-Julien and Jaggi [LaJ13], Beck and Shtern [BeS15], and Freund and Grigas [FrG16]. A linear convergence rate can be proved for modifications of the algorithm that include so-called "away-steps," which involve maximization of the linearized cost function over the constraint set [in addition to the minimization of Eq. (3.31)]. Note, however, that minimizing and maximizing the linearized cost function of a given problem may have different structure, and solution difficulty. This is true in particular for the multicommodity flow problems of Example 3.2.2, where maximization of the linearized cost function involves solving *longest* path problems, which are typically far more difficult than shortest path problems.

Another approach to improve the convergence rate of the conditional gradient method, also based on minimizations of the linearized cost function, is the simplicial decomposition method of Exercise 3.2.7. This method substitutes the line search of the conditional gradient method with a low-dimensional minimization of the cost function over the convex hull of a finite number of extreme points of the constraint set. Contrary to the conditional gradient method, it can be applied to nondifferentiable convex problems; see [BeY11]. An extensive discussion of this method may be found in [Ber15a], Ch. 4. See also the references for Section 3.8 regarding the solution of multicommodity network flow problems.

Section 3.3: Gradient projection methods with a constant stepsize were first proposed by Goldstein [Gol64], and Levitin and Poljak [LeP65]. The Armijo rule along the projection arc was proposed by Bertsekas [Ber76c]. For convergence analysis, see Dunn [Dun81], Gafni and Bertsekas [GaB82], [GaB84], Calamai and Moré [CaM87], and Dunn [Dun87], [Dun88a]. For analysis of various aspects of the convergence rate, see Bertsekas [Ber76c], Dunn [Dun81], Bertsekas and Gafni [BeG82], Dunn and Sachs [DuS83], Dunn [Dun87], Tseng [Tse91a], Luo and Tseng [LuT92b], [LuT93a], [LuT93b], [LuT94b], and Gilmore and Kelley [GiK95]. Surveys are given by Dunn [Dun88a], [Dun94]. Variations of gradient projection methods that are based on active constraint identification are given by Bertsekas [Ber76c], Dembo and Tulowitzki [DeT83], Moré and Toraldo [MoT89], Burke, Moré, and Toraldo [BMT90], and McKenna, Mesirov, and Zenios [MMZ95].

The gradient projection method is particularly well-suited for large-scale problems with relatively simple constraint structure. Examples of such problems arise in optimal control (see e.g., Polak [Pol73], Bertsekas [Ber76c], [Ber82b], Dunn [Dun88a], [Dun91a], [Dun94]), and in multicommodity flow problems from communications and transportation (see e.g., Bertsekas and Gafni [BeG83], Bertsekas, Gafni, and Gallager [BGG84], Tsitsiklis and Bertsekas [TsB86], Bertsekas and Tsitsiklis [BeT89], Bertsekas and Gallager [BeG92], Luo and Tseng [LuT94b], Florian and Hearn [FlH95]). Newton's method for constrained problems is discussed by Dunn [Dun80b], and Hughes and Dunn [HuD84].

Section 3.4: Two metric-projection methods were first proposed by Bertsekas [Ber82b] for problems with simple bounds and, more generally, linear constraints. The methods were generalized by Gafni and Bertsekas [GaB84] to the case of general convex constraint sets; see also Bertsekas and Gafni [BeG83] for the case of multicommodity flow problems. For subsequent work see Dunn [Dun88b], [Dun91a], [Dun91b], [Dun93a], [Dun93b], [Dun94], Gawande and Dunn [GaD88], Pytlak [Pyt98], Talischi and Paulino [TaP13], and Wytock, Sra, and Kolter [WSK14].

The advantage that the two-metric projection approach can offer is to identify quickly the constraints that are active at an optimal solution. After this happens, the method reduces essentially to an unconstrained scaled gradient method (possibly Newton method, if D_k is a partially diagonalized Hessian matrix), and attains a fast convergence rate. This property has also motivated variants of the two-metric projection method for problems involving the ℓ_1 norm in their cost function or constraints; see Schmidt, Fung, and Rosales [SFR09], Schmidt [Sch10], Gupta, Kumar, and Xiao [GKX10], Schmidt, Kim, and Sra [SKS12], and Landi [Lan15].

Section 3.5: Manifold suboptimization methods draw their origin from the gradient projection method of Rosen [Ros60b] (which differs substantially from the gradient projection methods of Section 3.3). Extensive related discussions can be found in the books by Gill and Murray [GiM74], and Gill, Murray, and Wright [GMW81]. An important method of this type, which has been used as the basis for general purpose nonlinear programming software, is the *reduced gradient method*; see Lasdon and Waren [LaW78]. A convergence analysis of Rosen's gradient projection method is given by Du and Zhang [DuZ89].

The application of manifold suboptimization methods to quadratic programming is discussed by Zangwill [Zan69]. Efficient implementations are given by Gill and Murray [GiM74], and Gill, Murray, and Wright [GMW81], [GMW91]. Quadratic programming is a special case of the linear complementarity problem, which is discussed extensively by Cottle, Pang, and Stone [CPS92], and by Facchinei and Pang [FaP03].

Section 3.6: The proximal algorithm was proposed in the early 70s by Martinet [Mar70], [Mar72]. It was generalized in the influential paper by Rockafellar [Roc76a] in a form that aims to find a zero of a maximal monotone operator. The literature on the algorithm and its variations is voluminous, and reflects the central importance of proximal ideas in convex optimization and other problems. For textbook discussions, see Rockafellar and Wets [RoW98], Facchinei and Pang [FaP03], Ruszczynski [Rus06], Bauschke and Combettes [BaC11], and Bertsekas [Ber15a], which include many references.

The rate of convergence analysis given here (Prop. 3.6.4) is due to Kort and Bertsekas [KoB76], and has been extensively discussed in the monograph [Ber82a] (Section 5.4) in a more general form where the regularization term may be nonquadratic; see also Luque [Luq84]. The finite convergence result of Prop. 3.6.5 was shown in a dual context that applies to augmented Lagrangian methods in Bertsekas [Ber75a], and was generalized to the maximal monotone operator setting by Luque [Luq84].

The incremental proximal algorithm, and its combinations with incremental gradient and subgradient methods were introduced in Bertsekas [Ber10b], [Ber11a], including extensions to the case where the component functions are convex but nondifferentiable. For recent work and applications, see Andersen and Hansen [AnH13], Couellan and Trafalis [CoT13], Weinmann, Demaret, and Storath [WDS13], Bacak [Bac14], [Bac16], Bergmann et al. [BSL14], Richard, Gaiffas, and Vayatis [RGV14], You, Song, and Qiu [YSQ14], and Toulis, Tran, and Airoldi [TTA15]. Extensions of incremental methods that involve incremental treatment of constraint sets consisting of the intersection of many simple sets are given by Bertsekas [Ber11a], Nedić [Ned11], Wang and Bertsekas [WaB15], [WaB16], Bianchi [Bia15], and Iusem, Jofre, and Thompson [IJT15a], [IJT15b]. The incremental aggregated proximal algorithm was studied in Bertsekas [Ber15b].

Section 3.7: The convergence to a unique limit and the rate of convergence of the coordinate descent method are analyzed by Luo and Tseng [LuT91], [LuT92a]. Coordinate descent methods are often well suited for solving dual problems, and within this specialized context there has been much convergence analysis (see the references given for Section 7.2).

Section 3.8: The multicommodity network optimization framework in this section is standard and has many applications. It is discussed for example in the author's books ([BeT89], Section 7.6, [BeG92], Section 5.7.3, [Ber98], pp. 391-398), and in many other sources, including the survey by Florian and Hearn [FlH95], which contains extensive references, and the books by Rockafellar [Roc84] and Patriksson [Pat98].

Several types of algorithms have been proposed for solution. Among first-order gradient-based methods, we note the conditional gradient method (see Fratta, Gerla, and Kleinrock [FGK73]), gradient projection methods with and without diagonal scaling (see Gallager [Gal77], Bertsekas [Ber80a], Bertsekas, Gafni, and Gallager [BGG84], Bertsekas and Gallager [BeG92]), and simplicial decomposition methods (see Cantor and Gerla [CaG74], Holloway [Hol74], Hohenbalken [Hoh77], Hearn, Lawphongpanich, and Ventura [HLV87], Patriksson [Pat98], Bertsekas and Yu [BeY11]).

Similar methods have also been proposed for traffic equilibrium analysis, where the same types of optimization problems arise (see the survey [FlH95] and the book [Pat98]). Traffic equilibrium problems also arise in a more general setting that involves solution of a monotone variational inequality, i.e., a problem of finding $x^* \in X$ satisfying

$$T(x^*)'(x - x^*) \geq 0, \qquad \forall\ x \in X, \tag{3.138}$$

where X is a closed convex set, and $T : \Re^n \mapsto \Re^n$ is a mapping satisfying the monotonicity condition

$$\bigl(T(x) - T(y)\bigr)'(x - y) \geq 0, \qquad \forall\ x, y \in X. \tag{3.139}$$

Solving the variational inequality (3.138) is a more general problem than minimizing a differentiable convex function f over X, which is the case where $T = \nabla f$ (cf. Prop. 3.1.1). While the conditional gradient method cannot be readily extended to solve the variational inequality (3.138), the gradient projection method can, and takes the form

$$x^{k+1} = \bigl[x^k - \alpha^k T(x^k)\bigr]^+,$$

where $[\cdot]^+$ denotes projection on X and α^k is a positive stepsize. However, for convergence with a sufficiently small constant stepsize, it requires (in addition to Lipschitz continuity of T) a stronger assumption than Eq. (3.138) (a so called strong monotonicity condition); for an analysis and a counterexample to convergence, see [BeT89], Section 3.5.3. Several schemes have been proposed to address this difficulty, including the so called *extragradient method* proposed by Korpelevich [Kor76]. It turns out that for broad classes of problems where X is a polyhedral set, including typical traffic equilibrium problems, this convergence difficulty does not arise, as shown by Bertsekas and Gafni [BeG82].

Gradient projection methods for multicommodity network optimization are also well-suited for distributed implementation. Several such methods have been proposed for routing and flow control, dating to the early days of data networking (see [Gal77], [Ber80], [BGG84]). The paper by Tsitsiklis and Bertsekas [TsB86] analyzes the convergence of asynchronous distributed gradient projection in the context of data network routing. The Newton-like method presented in this section, based on the conjugate gradient method, follows the paper by Bertsekas and Gafni [BeG83], and is discussed in more detail in [Ber11b].

4
Lagrange Multiplier Theory

Contents

4.1. Necessary Conditions for Equality Constraints . . . . p. 345
   4.1.1. The Penalty Approach . . . . p. 349
   4.1.2. The Elimination Approach . . . . p. 352
   4.1.3. The Lagrangian Function . . . . p. 356
4.2. Sufficient Conditions and Sensitivity Analysis . . . . p. 364
   4.2.1. The Augmented Lagrangian Approach . . . . . . p. 365
   4.2.2. The Feasible Direction Approach . . . . p. 369
   4.2.3. Sensitivity . . . . p. 370
4.3. Inequality Constraints . . . . p. 376
   4.3.1. Karush-Kuhn-Tucker Necessary Conditions . . . . p. 378
   4.3.2. Sufficient Conditions and Sensitivity . . . . p. 383
   4.3.3. Fritz John Optimality Conditions . . . . p. 386
   4.3.4. Constraint Qualifications and Pseudonormality . . p. 392
   4.3.5. Abstract Set Constraints and the Tangent Cone . . p. 399
   4.3.6. Abstract Set Constraints, Equality, and Inequality
        Constraints . . . . p. 415
4.4. Linear Constraints and Duality . . . . p. 429
   4.4.1. Convex Cost Function and Linear Constraints . . . p. 429
   4.4.2. Duality Theory: A Simple Form for Linear
        Constraints . . . . p. 432
4.5. Notes and Sources . . . . p. 441

> **The methods I set forth require neither constructions
> nor geometric or mechanical considerations.
> They require only algebraic operations ...**
>
> > Lagrange

The constraint set of an optimization problem is usually specified in terms of equality and inequality constraints. If we take into account this structure, we obtain a sophisticated collection of optimality conditions, involving some auxiliary variables called *Lagrange multipliers*. These variables facilitate the characterization of optimal solutions, but also provide valuable sensitivity information, quantifying up to first order the variation of the optimal cost caused by variations in problem data.

In this chapter, we develop Lagrange multiplier theory for differentiable cost problems, with equality and inequality constraints. Our development is layered, and involves increasingly complex problem structures. The basic theory is developed in Sections 4.1-4.3.2. The reader may wish to proceed to Chapter 5 after Section 4.3.2, and return to the remainder of Chapter 4 later, as needed. Here are the types of problems that we will consider in this chapter:

(a) *Equality-constrained problems*, and their associated necessary conditions, sufficiency conditions, and sensitivity analysis, are discussed in Sections 4.1 and 4.2. This is the simplest class of problems, and forms the foundation upon which more complicated problems can be considered.

(b) *Problems with both equality and inequality constraints* are discussed in Sections 4.3.1-4.3.4. The simplest case of the classical Karush-Kuhn-Tucker optimality conditions is developed in Sections 4.3.1-4.3.2, based on the theory of equality constraints of Sections 4.1-4.2. More general necessary conditions, involving assumptions commonly known as *constraint qualifications*, are obtained in Sections 4.3.3-4.3.4, through the use of the classical theorem of Fritz John. To this end, we use a strengthened version of the Fritz John theorem, which allows both more powerful results and a unified theory through the concept of *pseudonormality*.

(c) *Problems involving an abstract set constraint* $x \in X$, in addition to equality and inequality constraints, are discussed in Sections 4.3.5-4.3.6. To deal with the potentially complicated nonconvex structure of the set X, we introduce conical approximations, such as the *tangent cone of X*, and use them to express the associated Lagrange multiplier conditions. For some of the more sophisticated parts of this theory we restrict ourselves to the case where X is a closed convex set, and refer to the book [BNO03] for the nonconvex case.

(d) *Problems with convex differentiable cost function and linear constraints* are discussed in Section 4.4. This class of problems includes linear and convex quadratic programming. Here there is sufficiently favorable structure to allow a simple version of duality theory. The development of Section 4.4 is sufficient for the analysis of Lagrange multiplier algorithms in Chapter 5, and also provides a conceptual foundation for the more sophisticated duality theory of Chapter 6, where the cost function is assumed to be convex (but possibly nondifferentiable), and for the dual algorithms of Chapter 7.

There are several different approaches to the development of Lagrange multiplier theory. In this chapter we will follow two basic lines of analysis:

(1) The *penalty viewpoint*, whereby we disregard the constraints, while adding to the cost a high penalty for violating them. By then working with the "penalized" unconstrained problem, and by passing to the limit as the penalty increases, we obtain the desired Lagrange multiplier theorems. This approach is surprisingly simple and powerful, and forms the basis for important algorithms, which will be more fully explored in Chapter 5.

(2) The *feasible direction viewpoint*, which similar to Chapter 3, relies on the fact that at a local minimum there can be no cost improvement when traveling a small distance along a direction that leads to feasible points. This viewpoint needs to be modified somewhat when the constraint set is not convex, resulting in a fair amount of mathematical complication, but in the end works well, and yields powerful results and useful geometric insights.

We will use primarily the simpler penalty viewpoint, but we will occasionally develop the feasible direction viewpoint as well, in a parallel and self-contained manner. It is possible to study the present chapter selectively, by focusing initially on just one of these two viewpoints, and the choice may be largely based on taste and background. We note, however, that after gaining some basic familiarity with Lagrange multipliers, the reader will benefit from studying both viewpoints, because they reinforce and complement each other in terms of analysis and algorithmic insight.

4.1 NECESSARY CONDITIONS FOR EQUALITY CONSTRAINTS

In this section we consider problems with equality constraints of the form

$$\text{minimize } f(x)$$
$$\text{subject to } h_i(x) = 0, \quad i = 1, \ldots, m. \quad \text{(ECP)}$$

We assume that $f : \Re^n \mapsto \Re$, $h_i : \Re^n \mapsto \Re$, $i = 1, \ldots, m$, are continuously differentiable functions. *All the necessary and the sufficient conditions of* this chapter relating to a local minimum can also be shown to hold if f and h_i are defined and are continuously differentiable within just an open set containing the local minimum. The proofs are essentially identical to those given here.

In what follows we will often use compact vector notation. In particular, we introduce the constraint function $h : \Re^n \mapsto \Re^m$, where $h = (h_1, \ldots, h_m)$. We can then write the constraints in the more compact form

$$h(x) = 0.$$

Our basic Lagrange multiplier theorem states that for a given local minimum x^*, there exist scalars $\lambda_1, \ldots, \lambda_m$, called *Lagrange multipliers*, such that

$$\nabla f(x^*) + \sum_{i=1}^{m} \lambda_i \nabla h_i(x^*) = 0. \tag{4.1}$$

There are two ways to interpret this equation:

(a) The cost gradient $\nabla f(x^*)$ belongs to the subspace spanned by the constraint gradients at x^*. The example of Fig. 4.1.1 illustrates this interpretation.

(b) The cost gradient $\nabla f(x^*)$ is orthogonal to the subspace of *first order feasible variations*

$$V(x^*) = \{\Delta x \mid \nabla h_i(x^*)' \Delta x = 0, \; i = 1, \ldots, m\}.$$

This is the subspace of variations Δx for which the vector $x = x^* + \Delta x$ satisfies the constraint $h(x) = 0$ up to first order. Thus, according to the Lagrange multiplier condition of Eq. (4.1), at the local minimum x^*, the first order cost variation $\nabla f(x^*)' \Delta x$ is zero for all variations Δx in this subspace. This statement is analogous to the "zero gradient condition" $\nabla f(x^*) = 0$ of unconstrained optimization.

Here is a formal statement of the most common type of Lagrange multiplier theorem. For easy reference, in this theorem and later, a feasible vector x for which the constraint gradients $\nabla h_1(x), \ldots, \nabla h_m(x)$ are linearly independent will be called *regular*.

Proposition 4.1.1: (Lagrange Multiplier Theorem – Necessary Conditions for Regular Local Minima) Let x^* be a local minimum of f subject to $h(x) = 0$, and assume that x^* is regular. Then there exists a unique vector $\lambda^* = (\lambda_1^*, \ldots, \lambda_m^*)$, called a *Lagrange multiplier vector*, such that

$$\nabla f(x^*) + \sum_{i=1}^{m} \lambda_i^* \nabla h_i(x^*) = 0. \tag{4.2}$$

Sec. 4.1 Necessary Conditions for Equality Constraints

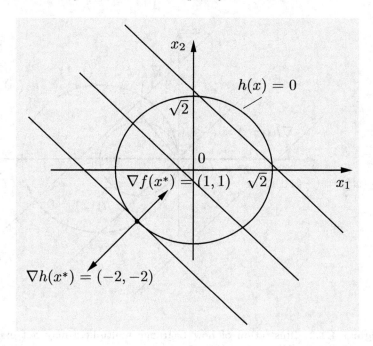

Figure 4.1.1. The Lagrange multiplier condition (4.1) for the problem

$$\text{minimize } x_1 + x_2$$
$$\text{subject to } x_1^2 + x_2^2 = 2.$$

At the local minimum $x^* = (-1, -1)$, the cost gradient

$$\nabla f(x^*) = (1, 1)$$

is normal to the constraint surface and is therefore, collinear with the constraint gradient

$$\nabla h(x^*) = (-2, -2).$$

The Lagrange multiplier is $\lambda = 1/2$.

If in addition f and h are twice continuously differentiable, we have

$$y' \left(\nabla^2 f(x^*) + \sum_{i=1}^m \lambda_i^* \nabla^2 h_i(x^*) \right) y \geq 0, \qquad \text{for all } y \in V(x^*), \quad (4.3)$$

where $V(x^*)$ is the subspace of first order feasible variations

$$V(x^*) = \{ y \mid \nabla h_i(x^*)' y = 0, \ i = 1, \ldots, m \}. \qquad (4.4)$$

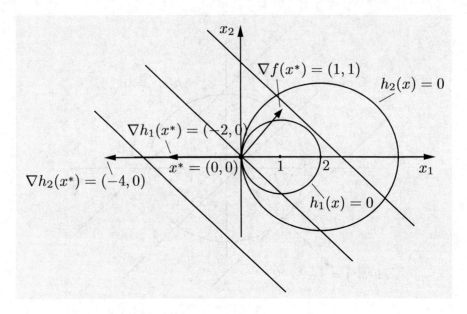

Figure 4.1.2. Illustration of how Lagrange multipliers may not exist (cf. Example 4.1.1). The problem here is

$$\text{minimize } f(x) = x_1 + x_2$$
$$\text{subject to } h_1(x) = (x_1 - 1)^2 + x_2^2 - 1 = 0,$$
$$h_2(x) = (x_1 - 2)^2 + x_2^2 - 4 = 0,$$

with a local minimum $x^* = (0,0)$. The cost gradient cannot be expressed as a linear combination of the constraint gradients, so there are no Lagrange multipliers.

Before going into the proof of the theorem, let us illustrate how there may not exist Lagrange multipliers for a local minimum that is not regular.

Example 4.1.1 (A Problem with no Lagrange Multipliers)

Consider the problem of minimizing

$$f(x) = x_1 + x_2$$

subject to the two constraints

$$h_1(x) = (x_1 - 1)^2 + x_2^2 - 1 = 0, \quad h_2(x) = (x_1 - 2)^2 + x_2^2 - 4 = 0.$$

The geometry of this problem is illustrated in Fig. 4.1.2. It can be seen that at the local minimum $x^* = (0, 0)$ (the only feasible solution), the cost gradient $\nabla f(x^*) = (1, 1)$ cannot be expressed as a linear combination of the constraint gradients $\nabla h_1(x^*) = (-2, 0)$ and $\nabla h_2(x^*) = (-4, 0)$. Thus the Lagrange multiplier condition

$$\nabla f(x^*) + \lambda_1^* \nabla h_1(x^*) + \lambda_2^* \nabla h_2(x^*) = 0$$

cannot hold for any λ_1^* and λ_2^*.

Sec. 4.1   Necessary Conditions for Equality Constraints   **349**

The difficulty here is that the subspace of first order feasible variations

$$V(x^*) = \{y \mid \nabla h_1(x^*)'y = 0, \nabla h_2(x^*)'y = 0\}$$

[cf. Eq. (4.4)], which is $\{y \mid y_1 = 0, y_2 \in \Re\}$, has larger dimension than the true set of feasible variations $\{y \mid y = 0\}$. The optimality of x^* implies that $\nabla f(x^*)$ is orthogonal to the true set of feasible variations, but for a Lagrange multiplier to exist, $\nabla f(x^*)$ must be orthogonal to the subspace of first order feasible variations. If the constraint gradients $\nabla h_1(x^*)$ and $\nabla h_2(x^*)$ were linearly independent, the set of feasible variations and the subspace of first order feasible variations would have the same dimension and this difficulty would not have occurred. It turns out that the equality of the two sets of feasible variations plays a fundamental role in the existence of Lagrange multipliers, and will be explored further in Section 4.3.6 within a broader context.

We will provide two different proofs of the Lagrange multiplier theorem, each providing important insights. These proofs are based on transforming the constrained problem to an unconstrained one, but in different ways. The constrained first and second order necessary conditions are obtained by applying the corresponding unconstrained conditions to the appropriate unconstrained problem. The approaches are:

(a) *The penalty approach.* Here we disregard the constraints, while adding to the cost a high penalty for violating them. By writing the necessary conditions for the "penalized" unconstrained problems, and by passing to the limit as the penalty increases, we obtain the Lagrange multiplier theorem. This approach is simple, and applies to inequality constraints, as we will see in Section 4.3.

(b) *The elimination approach.* Here we view the constraints as a system of m equations with n unknowns, and we express m of the variables in terms of the remaining $n - m$, thereby reducing the problem to an unconstrained problem. By then applying the corresponding first and second order necessary conditions for unconstrained minima, the Lagrange multiplier theorem follows. This approach requires the use of the implicit function theorem (Prop. A.25 in Appendix A), but is otherwise simple and insightful. Its extension, however, to inequality constraints is complicated.

4.1.1 The Penalty Approach

We will approximate the original constrained problem by an unconstrained optimization problem that involves a penalty for violation of the constraints. In particular, for $k = 1, 2, \ldots$, we introduce the cost function

$$F^k(x) = f(x) + \frac{k}{2}\|h(x)\|^2 + \frac{\alpha}{2}\|x - x^*\|^2,$$

where x^* is the local minimum of the constrained problem and α is some positive scalar. The term $(k/2)\|h(x)\|^2$ imposes a penalty for violating the constraint $h(x) = 0$, while the term $(\alpha/2)\|x - x^*\|^2$ is introduced for technical proof-related reasons [to ensure that x^* is a *strict* local minimum of the function $f(x) + (\alpha/2)\|x - x^*\|^2$ subject to $h(x) = 0$].

Since x^* is a local minimum, we can select $\epsilon > 0$ such that $f(x^*) \leq f(x)$ for all feasible x in the *closed* sphere

$$S = \{x \mid \|x - x^*\| \leq \epsilon\}.$$

Let x^k be an optimal solution of the problem

$$\begin{aligned} & \text{minimize} && F^k(x) \\ & \text{subject to} && x \in S. \end{aligned} \quad (4.5)$$

[An optimal solution exists because of Weierstrass' theorem (Prop. A.8 in Appendix A), since S is compact.] We will show that the sequence $\{x^k\}$ converges to x^*.

We have for all k

$$F^k(x^k) = f(x^k) + \frac{k}{2}\|h(x^k)\|^2 + \frac{\alpha}{2}\|x^k - x^*\|^2 \leq F^k(x^*) = f(x^*) \quad (4.6)$$

and since $f(x^k)$ is bounded over S, we obtain

$$\lim_{k \to \infty} \|h(x^k)\| = 0;$$

otherwise the left-hand side of Eq. (4.6) would become unbounded above as $k \to \infty$. Therefore, every limit point $\bar{x}$ of $\{x^k\}$ satisfies $h(\bar{x}) = 0$. Furthermore, Eq. (4.6) yields $f(x^k) + (\alpha/2)\|x^k - x^*\|^2 \leq f(x^*)$ for all k, so by taking the limit as $k \to \infty$, we obtain

$$f(\bar{x}) + \frac{\alpha}{2}\|\bar{x} - x^*\|^2 \leq f(x^*).$$

Since $\bar{x} \in S$ and $\bar{x}$ is feasible, we have $f(x^*) \leq f(\bar{x})$, which when combined with the preceding inequality yields $\|\bar{x} - x^*\| = 0$ so that $\bar{x} = x^*$. Thus the sequence $\{x^k\}$ converges to x^*, and it follows that x^k is an interior point of the closed sphere S for sufficiently large k. Therefore x^k is an *unconstrained* local minimum of $F^k(x)$ for sufficiently large k. We will now prove the Lagrange multiplier theorem by working with the corresponding unconstrained necessary optimality conditions.

Proof of the Lagrange multiplier theorem: From the first order necessary condition, we have for sufficiently large k

$$0 = \nabla F^k(x^k) = \nabla f(x^k) + k\nabla h(x^k)h(x^k) + \alpha(x^k - x^*). \quad (4.7)$$

Sec. 4.1 Necessary Conditions for Equality Constraints

Since $\nabla h(x^*)$ has rank m, the same is true for $\nabla h(x^k)$ if k is sufficiently large. For such k, $\nabla h(x^k)'\nabla h(x^k)$ is invertible (Prop. A.20 in Appendix A), and by premultiplying Eq. (4.7) with $(\nabla h(x^k)'\nabla h(x^k))^{-1}\nabla h(x^k)'$, we obtain

$$kh(x^k) = -(\nabla h(x^k)'\nabla h(x^k))^{-1}\nabla h(x^k)'(\nabla f(x^k) + \alpha(x^k - x^*)).$$

By taking the limit as $k \to \infty$ and $x^k \to x^*$, we see that $\{kh(x^k)\}$ converges to the vector

$$\lambda^* = -(\nabla h(x^*)'\nabla h(x^*))^{-1}\nabla h(x^*)'\nabla f(x^*).$$

By taking the limit as $k \to \infty$ in Eq. (4.7), we obtain

$$\nabla f(x^*) + \nabla h(x^*)\lambda^* = 0,$$

proving the first order Lagrange multiplier condition (4.2).

By using the second order unconstrained optimality condition for problem (4.5), we see that the matrix

$$\nabla^2 F^k(x^k) = \nabla^2 f(x^k) + k\nabla h(x^k)\nabla h(x^k)' + k\sum_{i=1}^{m} h_i(x^k)\nabla^2 h_i(x^k) + \alpha I$$

is positive semidefinite, for all sufficiently large k and for all $\alpha > 0$. Fix any $y \in V(x^*)$ [i.e., $\nabla h(x^*)'y = 0$], and let y^k be the projection of y on the nullspace of $\nabla h(x^k)'$, i.e.,

$$y^k = y - \nabla h(x^k)(\nabla h(x^k)'\nabla h(x^k))^{-1}\nabla h(x^k)'y, \tag{4.8}$$

(cf. Example 3.1.5 in Section 3.1). Since $\nabla h(x^k)'y^k = 0$ and $\nabla^2 F^k(x^k)$ is positive semidefinite, we have

$$0 \le y^{k'}\nabla^2 F^k(x^k)y^k = y^{k'}\left(\nabla^2 f(x^k) + k\sum_{i=1}^{m} h_i(x^k)\nabla^2 h_i(x^k)\right)y^k + \alpha\|y^k\|^2.$$

Since $kh_i(x^k) \to \lambda_i^*$, while from Eq. (4.8), together with the facts $x^k \to x^*$ and $\nabla h(x^*)'y = 0$, we have $y^k \to y$, we obtain

$$0 \le y'\left(\nabla^2 f(x^*) + \sum_{i=1}^{m} \lambda_i^*\nabla^2 h_i(x^*)\right)y + \alpha\|y\|^2, \quad \forall \, y \in V(x^*).$$

Since α can be taken arbitrarily close to 0, we obtain

$$0 \le y'\left(\nabla^2 f(x^*) + \sum_{i=1}^{m} \lambda_i^*\nabla^2 h_i(x^*)\right)y, \quad \forall \, y \in V(x^*),$$

which is the second order Lagrange multiplier condition. The proof of the Lagrange multiplier theorem is thus complete. **Q.E.D.**

4.1.2 The Elimination Approach

We introduce the elimination approach by considering first the easier case where the constraints are linear.

Example 4.1.2 (Lagrange Multipliers for Linear Constraints)

Consider the problem
$$\text{minimize } f(x)$$
$$\text{subject to } Ax = b, \tag{4.9}$$

where A is an $m \times n$ matrix with linearly independent rows and $b \in \Re^m$ is a given vector. By rearranging the components of x if necessary, we may assume that the first m columns of A are linearly independent, so that A can be partitioned as
$$A = (B \quad R),$$

where B is an invertible $m \times m$ matrix and R is an $m \times (n-m)$ matrix. We also partition x as
$$x = \begin{pmatrix} x_B \\ x_R \end{pmatrix},$$

where $x_B \in \Re^m$ and $x_R \in \Re^{n-m}$. We can then write problem (4.9) as
$$\text{minimize } f(x_B, x_R)$$
$$\text{subject to } Bx_B + Rx_R = b. \tag{4.10}$$

By using the constraint equation to express x_B in terms of x_R as
$$x_B = B^{-1}(b - Rx_R)$$

and by substitution in the cost function, we can convert problem (4.10) into the unconstrained optimization problem
$$\text{minimize } F(x_R) \equiv f\big(B^{-1}(b - Rx_R), x_R\big)$$
$$\text{subject to } x_R \in \Re^{n-m}. \tag{4.11}$$

If (x_B^*, x_R^*) is a local minimum of the constrained problem (4.9), then x_R^* is an unconstrained local minimum of the "reduced" cost function F, so we have
$$0 = \nabla F(x_R^*) = -R'(B')^{-1} \nabla_B f(x^*) + \nabla_R f(x^*), \tag{4.12}$$

where $\nabla_B f$ and $\nabla_R f$ denote the gradients of f with respect to x_B and x_R, respectively. By defining
$$\lambda^* = -(B')^{-1} \nabla_B f(x^*), \tag{4.13}$$

we write Eq. (4.12) as
$$\nabla_R f(x^*) + R'\lambda^* = 0,$$

Sec. 4.1 Necessary Conditions for Equality Constraints

and we also write Eq. (4.13) as

$$\nabla_B f(x^*) + B'\lambda^* = 0.$$

By combining the preceding two equations, we obtain

$$\nabla f(x^*) + A'\lambda^* = 0, \qquad (4.14)$$

which is the Lagrange multiplier condition of Eq. (4.2) specialized to the case of the linearly constrained problem (4.9). Note that the vector λ^* satisfying this condition is unique because of the linear independence of the columns of A'.

We next show the second order necessary condition (4.3), by showing that it is equivalent to the second order unconstrained necessary condition

$$0 \leq d'\nabla^2 F(x_R^*)d, \qquad \forall\, d \in \Re^{n-m}. \qquad (4.15)$$

We have using Eqs. (4.11) and (4.12)

$$\nabla^2 F(x_R) = \nabla\Big(-R'(B')^{-1}\nabla_B f\big(B^{-1}(b - Rx_R), x_R\big)$$
$$+ \nabla_R f\big(B^{-1}(b - Rx_R), x_R\big)\Big).$$

By evaluating this expression at $x_R = x_R^*$, and by partitioning the Hessian $\nabla^2 f(x^*)$ as

$$\nabla^2 f(x^*) = \begin{pmatrix} \nabla^2_{BB} f(x^*) & \nabla^2_{BR} f(x^*) \\ \nabla^2_{RB} f(x^*) & \nabla^2_{RR} f(x^*) \end{pmatrix},$$

we obtain

$$\nabla^2 F(x_R^*) = R'(B')^{-1}\nabla^2_{BB} f(x^*)B^{-1}R$$
$$- R'(B')^{-1}\nabla^2_{BR} f(x^*) - \nabla^2_{RB} f(x^*)B^{-1}R + \nabla^2_{RR} f(x^*).$$

From this expression together with the positive semidefiniteness of $\nabla^2 F(x_R^*)$ [cf. Eq. (4.15)] and the linearity of the constraints [implying that $\nabla^2 h_i(x^*) = 0$], it follows that for all $d \in \Re^{n-m}$,

$$0 \leq d'\nabla^2 F(x_R^*)d = y'\nabla^2 f(x^*)y = y'\left(\nabla^2 f(x^*) + \sum_{i=1}^{m} \lambda_i^* \nabla^2 h_i(x^*)\right)y, \qquad (4.16)$$

where y is the vector

$$y = \begin{pmatrix} -B^{-1}Rd \\ d \end{pmatrix}.$$

It can be seen that the subspace $V(x^*)$ of feasible variations of Eq. (4.4) is given by

$$V(x^*) = \big\{(y_B, y_R) \mid By_B + Ry_R = 0\big\}$$
$$= \big\{(y_B, y_R) \mid y_B = -B^{-1}Rd,\ y_R = d,\ d \in \Re^{n-m}\big\},$$

so Eq. (4.16) is equivalent to the second order Lagrange multiplier condition (4.3).

Note that based on the preceding analysis, the Lagrange multiplier conditions of Prop. 4.1.1 are nothing but the "zero gradient" and "positive semidefinite Hessian" conditions for the unconstrained problem (4.11), which is defined over the reduced space of the vector x_R.

Actually for the case of linear constraints, a stronger version of the Lagrange multiplier theorem can be shown. In particular, the regularity assumption is not needed for the existence of a Lagrange multiplier vector; see Exercise 4.1.4 and Section 4.3.4.

We now prove the Lagrange multiplier theorem by generalizing the analysis of the preceding example.

Proof of the Lagrange multiplier theorem: The proof assumes that $m < n$; if $m = n$, any vector, including $-\nabla f(x^*)$, can be expressed as a linear combination of the linearly independent vectors $\nabla h_1(x^*), \ldots, \nabla h_m(x^*)$, thereby proving the theorem. By reordering the components of x if necessary, we partition the vector x as $x = (x_B, x_R)$, where the square submatrix $\nabla_B h(x^*)$ (the gradient matrix of h with respect to x_B) is invertible. The constraint equation

$$h(x_B, x_R) = 0$$

has the solution (x_B^*, x_R^*), and the implicit function theorem (Prop. A.25 in Appendix A) can be used to express x_B in terms of x_R via a unique continuously differentiable function $\phi : S \mapsto \Re^m$ defined over a sphere S centered at x_R^* (ϕ is twice continuously differentiable if h is). In particular, we have $x_B^* = \phi(x_R^*)$, $h(\phi(x_R), x_R) = 0$ for all $x_R \in S$, and

$$\nabla \phi(x_R) = -\nabla_R h(\phi(x_R), x_R)\bigl(\nabla_B h(\phi(x_R), x_R)\bigr)^{-1}, \quad \forall\, x_R \in S,$$

where $\nabla_R h$ is the gradient matrix of h with respect to x_R.

We now proceed as in the earlier case of linear constraints. We observe that x_R^* is an unconstrained minimum of the "reduced" cost function

$$F(x_R) = f(\phi(x_R), x_R),$$

and we apply the corresponding unconstrained first and second order necessary conditions. The first order Lagrange multiplier condition (4.2) follows by repeating the calculation of Eqs. (4.13)-(4.14) with the definitions

$$B' = \nabla_B h(x^*), \qquad R' = \nabla_R h(x^*), \qquad A' = \nabla h(x^*),$$

$$\lambda^* = -(B')^{-1} \nabla_B f(x^*).$$

The proof of the second order necessary condition (4.3) requires a lengthy calculation, which resembles the one that yielded Eq. (4.16) in the preceding example. In particular, for $d \in \Re^{n-m}$, denote

$$y = \begin{pmatrix} -B^{-1} R d \\ d \end{pmatrix}. \tag{4.17}$$

Sec. 4.1   Necessary Conditions for Equality Constraints

Denote also
$$H_i(x_R) = h_i\big(\phi(x_R), x_R\big), \qquad i = 1, \ldots, m,$$
and let $\phi_i(x_R)$ be the ith component of ϕ, so that
$$\phi(x_R) = \begin{pmatrix} \phi_1(x_R) \\ \vdots \\ \phi_m(x_R) \end{pmatrix}.$$

Then by differentiating twice the relation
$$F(x_R) = f\big(\phi(x_R), x_R\big),$$
we obtain by a straightforward calculation [compare with the derivation of Eq. (4.16)]
$$d' \nabla^2 F(x_R) d = y' \nabla^2 f(x^*) y + d' \left(\sum_{j=1}^{m} \nabla^2 \phi_j(x_R^*) \frac{\partial f(x^*)}{\partial x_j} \right) d. \qquad (4.18)$$

Similarly by twice differentiating the equation $H_i(x_R) = 0$, we have
$$0 = d' \nabla^2 H_i(x_R) d = y' \nabla^2 h_i(x^*) y + d' \left(\sum_{j=1}^{m} \nabla^2 \phi_j(x_R^*) \frac{\partial h_i(x^*)}{\partial x_j} \right) d.$$

By multiplying the above equation with λ_i^*, and by adding over all i we obtain
$$0 = \sum_{i=1}^{m} \lambda_i^* y' \nabla^2 h_i(x^*) y + d' \left(\sum_{j=1}^{m} \nabla^2 \phi_j(x_R^*) \sum_{i=1}^{m} \lambda_i^* \frac{\partial h_i(x^*)}{\partial x_j} \right) d. \qquad (4.19)$$

Adding Eqs. (4.18) and (4.19), and using the relations $d' \nabla^2 F(x_R) d \geq 0$ and
$$\frac{\partial f(x^*)}{\partial x_j} + \sum_{i=1}^{m} \lambda_i^* \frac{\partial h_i(x^*)}{\partial x_j} = 0, \qquad j = 1, \ldots, m,$$
we obtain
$$0 \leq y' \left(\nabla^2 f(x^*) + \sum_{i=1}^{m} \lambda_i^* \nabla^2 h_i(x^*) \right) y,$$
for all y of the form defined by Eq. (4.17). As shown in Example 4.1.2, y belongs to the subspace $V(x^*)$ if and only if y has this form, thus proving the second order Lagrange multiplier condition (4.3). **Q.E.D.**

4.1.3 The Lagrangian Function

Sometimes it is convenient to write our necessary conditions in terms of the *Lagrangian function* $L : \Re^{n+m} \mapsto \Re$ defined by

$$L(x, \lambda) = f(x) + \sum_{i=1}^{m} \lambda_i h_i(x).$$

Then, if x^* is a local minimum that is regular, the Lagrange multiplier conditions of Prop. 4.1.1 together with the equation $h(x^*) = 0$ are written compactly as

$$\nabla_x L(x^*, \lambda^*) = 0, \qquad \nabla_\lambda L(x^*, \lambda^*) = 0, \qquad (4.20)$$

$$y' \nabla_{xx}^2 L(x^*, \lambda^*) y \geq 0, \qquad \text{for all } y \in V(x^*). \qquad (4.21)$$

The first order necessary conditions (4.20) represent a system of $n+m$ equations with $n+m$ unknowns – the components of x^* and λ^*. Every local minimum x^* that is regular, together with its associated Lagrange multiplier vector, will be a solution of this system. However, a solution of the system need not correspond to a local minimum, as our experience with unconstrained problems indicates; it could be for example a local maximum.

Example 4.1.3

Consider the problem

$$\begin{aligned} \text{minimize} \quad & \tfrac{1}{2}\left(x_1^2 + x_2^2 + x_3^2\right) \\ \text{subject to} \quad & x_1 + x_2 + x_3 = 3. \end{aligned} \qquad (4.22)$$

The first order necessary conditions (4.20) yield

$$x_1^* + \lambda^* = 0,$$

$$x_2^* + \lambda^* = 0,$$

$$x_3^* + \lambda^* = 0,$$

$$x_1^* + x_2^* + x_3^* = 3.$$

This is a system of four equations and four unknowns. It has the unique solution

$$x_1^* = x_2^* = x_3^* = 1, \qquad \lambda^* = -1.$$

The constraint gradient here is $(1, 1, 1)$, so all feasible vectors are regular. Therefore, $x^* = (1, 1, 1)$ is the unique candidate for a local minimum. Furthermore, since $\nabla_{xx}^2 L(x^*, \lambda^*)$ is the identity matrix for this problem, the second order necessary condition (4.21) is satisfied. We can argue that x^* is a

Sec. 4.1   Necessary Conditions for Equality Constraints

global minimum by using the convexity of the cost function and the convexity of the constraint set to verify that x^* satisfies the sufficiency condition of Prop. 3.1.1(b) (alternatively, we can use a sufficiency condition to be given in Section 4.3.2).

If instead we consider the maximization version of problem (4.22), i.e.,

$$\text{minimize } -\tfrac{1}{2}\left(x_1^2 + x_2^2 + x_3^2\right)$$
$$\text{subject to } x_1 + x_2 + x_3 = 3, \tag{4.23}$$

then the first order condition (4.20) yields $x^* = (1,1,1)$ and $\lambda^* = 1$. However, the second order condition (4.21) is not satisfied and since every feasible vector is also regular, we conclude that the problem has no solution.

Example 4.1.4 (Diffraction Law in Optics)

Consider a smooth curve on the 2-dimensional plane described by the equation

$$h(x) = 0,$$

where $h : \Re^2 \mapsto \Re$ is continuously differentiable. Imagine that the curve separates the plane into two regions and that the velocity of light is different in each region. Let y and z be two points that lie on opposite sides of the curve as shown in part (a) of Fig. 4.1.3. Suppose that light going from y to z follows a ray from y to some point x^* of the curve with velocity v_y, and then follows a ray from x^* to z with velocity v_z. Assuming that light follows a path of minimum travel time, we will show that x^* is characterized by the following diffraction law:

$$\frac{\sin \phi_y}{v_y} = \frac{\sin \phi_z}{v_z}, \tag{4.24}$$

where ϕ_y and ϕ_z are the angles shown in Fig. 4.1.3.

Indeed, the travel time can be expressed as

$$T(x^*) = \frac{\|y - x^*\|}{v_y} + \frac{\|z - x^*\|}{v_z},$$

and the Lagrange multiplier theorem states that the gradient

$$\nabla T(x^*) = \frac{x^* - y}{v_y \|x^* - y\|} + \frac{x^* - z}{v_z \|x^* - z\|} \tag{4.25}$$

is a scalar multiple of $\nabla h(x^*)$. [For this we need that x^* is regular, i.e., $\nabla h(x^*) \neq 0$, which we assume.] To interpret this condition, consider the vectors

$$\bar{y} = x^* + \frac{y - x^*}{v_y \|y - x^*\|}, \qquad \bar{z} = x^* + \frac{z - x^*}{v_z \|z - x^*\|},$$

and the parallelogram with sides $\bar{y} - x^*$ and $\bar{z} - x^*$, which by Eq. (4.25), has $-\nabla T(x^*)$ as one of its diagonals. From part (b) of Fig. 4.1.3, we see that the Lagrange multiplier condition states that the diagonal $-\nabla T(x^*)$ of the parallelogram is normal to the curve at x^*. From Euclidean geometry it follows that the distances of $\bar{y}$ and $\bar{z}$ from the vertical diagonal are equal. Since $\|\bar{y} - x^*\| = 1/v_y$ and $\|\bar{z} - x^*\| = 1/v_z$, this is equivalent to the diffraction law (4.24).

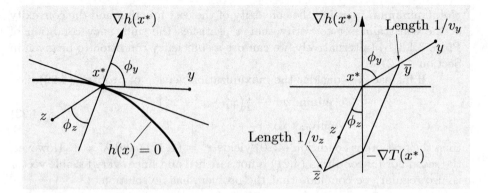

Figure 4.1.3. Diffraction law; cf. Example 4.1.4. The angles ϕ_y and ϕ_z are formed by the normal line to the curve at x^* [which is $\nabla h(x^*)$], and the vectors $y - x^*$ and $z - x^*$, respectively.

Example 4.1.5 (Optimal Portfolio Selection)

Consider an investor who wants to allocate one unit of wealth among n assets offering random rates of return $e_1, \ldots, e_n$, respectively. The means $\bar{e}_i = E\{e_i\}, i = 1, \ldots, n$, and the covariance matrix

$$Q = \begin{pmatrix} E\{(e_1 - \bar{e}_1)^2\} & \cdots & E\{(e_1 - \bar{e}_1)(e_n - \bar{e}_n)\} \\ \vdots & \cdots & \vdots \\ E\{(e_n - \bar{e}_n)(e_1 - \bar{e}_1)\} & \cdots & E\{(e_n - \bar{e}_n)^2\} \end{pmatrix}$$

are known, and we assume that Q is invertible. If x_i is the amount invested in the ith asset, the mean and the variance of the return of the investment $y = \sum_{i=1}^{n} e_i x_i$ are

$$\bar{y} = E\{y\} = \sum_{i=1}^{n} \bar{e}_i x_i,$$

and

$$\sigma^2 = E\{(y - \bar{y})^2\} = E\left\{\left(\sum_{i=1}^{n} (e_i - \bar{e}_i) x_i\right)^2\right\}$$

$$= \sum_{i=1}^{n} \sum_{j=1}^{n} E\{(e_i - \bar{e}_i)(e_j - \bar{e}_j)\} x_i x_j = x'Qx.$$

The investor's problem is to find the portfolio $x = (x_1, \ldots x_n)$ that minimizes the variance $E\{(y - \bar{y})^2\}$ to achieve a given level of mean return $E\{y\}$, say $E\{y\} = m$. Thus the problem is

$$\text{minimize } x'Qx$$

$$\text{subject to } \sum_{i=1}^{n} x_i = 1, \quad \sum_{i=1}^{n} \bar{e}_i x_i = m.$$

Sec. 4.1   Necessary Conditions for Equality Constraints

We want to see how the solution varies with m. Note that the solution of the problem "scales" with the amount of wealth available for investment. In particular, it can be seen that if w is the amount available for investment and wm is the desired mean, then the optimal portfolio is wx^*, where x^* is the optimal portfolio corresponding to one unit of wealth and a desired mean of m.

Let λ_1 and λ_2 be Lagrange multipliers for the constraints $\sum_{i=1}^{n} x_i = 1$ and $\sum_{i=1}^{n} \bar{e}_i x_i = m$, respectively. The first order optimality condition is

$$2Qx^* + \lambda_1 u + \lambda_2 \bar{e} = 0,$$

where $u = (1,\ldots,1)'$ and $\bar{e} = (\bar{e}_1,\ldots,\bar{e}_n)'$. (Strictly speaking, in order to use the Lagrange multiplier theorem of Prop. 4.1.1, we must assume that the vectors u and $\bar{e}$ are linearly independent. However, even if they are not, the existence of a Lagrange multiplier is guaranteed because the constraints are linear; see Example 4.1.2 and Exercise 4.1.4.) Equivalently,

$$x^* = -\tfrac{1}{2}Q^{-1}u\lambda_1 - \tfrac{1}{2}Q^{-1}\bar{e}\lambda_2, \tag{4.26}$$

and by substitution in the constraints $u'x^* = 1$ and $\bar{e}'x^* = m$, we obtain

$$1 = u'x^* = -\tfrac{1}{2}u'Q^{-1}u\lambda_1 - \tfrac{1}{2}u'Q^{-1}\bar{e}\lambda_2,$$

$$m = \bar{e}'x^* = -\tfrac{1}{2}\bar{e}'Q^{-1}u\lambda_1 - \tfrac{1}{2}\bar{e}'Q^{-1}\bar{e}\lambda_2.$$

By solving these equations for λ_1 and λ_2 we obtain

$$\lambda_1 = \xi_1 + \zeta_1 m,$$

$$\lambda_2 = \xi_m + \zeta_m m,$$

for some scalars $\xi_1, \zeta_1, \xi_m, \zeta_m$. Thus substitution in Eq. (4.26) yields

$$x^* = mv + w \tag{4.27}$$

for some vectors v and w that depend on Q and $\bar{e}$. The corresponding variance of return is

$$\sigma^2 = (mv + w)'Q(mv + w) = (\alpha m + \beta)^2 + \gamma, \tag{4.28}$$

where α, β, and γ are some scalars that depend on Q and $\bar{e}$.

Equation (4.28) specifies the locus of pairs (σ, m) that are achievable by optimal portfolio selection. Suppose now that one of the assets is riskless, i.e., its return is fixed at some known value $\bar{e}_f$ and the variance of its return is zero. Then γ must be zero since when $m = \bar{e}_f$, the variance is minimized (set to zero) by investing exclusively in the riskless asset. In this case, from Eq. (4.28) we obtain

$$\sigma = |\alpha m + \beta|.$$

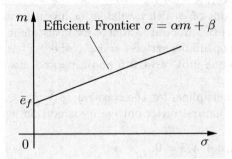

Figure 4.1.4. Efficient frontier for optimal portfolio selection. Every portfolio corresponding to a mean-variance pair on the efficient frontier can be achieved as a linear combination of two portfolios corresponding to two pairs on the efficient frontier.

Thus, assuming $\alpha > 0$, portfolios of interest correspond to mean-variance pairs of the form
$$\sigma = \alpha m + \beta, \qquad m \geq \bar{e}_f,$$
as shown in Fig. 4.1.4. This is known as the *efficient frontier*. To characterize the efficient frontier it is sufficient to determine one more point on it. Every other point is determined as a linear combination of this point and the point $(0, \bar{e}_f)$. Furthermore, by Eq. (4.27), every pair (σ, m) on the efficient frontier can be achieved by an appropriate linear combination of two portfolios, one consisting of exclusive investment on the riskless asset, and the other corresponding to some pair (σ, m) on the efficient frontier with $\sigma \neq 0$. (Economic arguments can be used to construct a second portfolio on the efficient frontier, leading to the celebrated CAPM theory of finance; see e.g. [HuL88], [Lue98].) Thus, while different investors with different preferences towards risk may prefer different points on the efficient frontier, their preferences can be obtained by appropriate mixtures of just two portfolios.

EXERCISES

4.1.1

Use the Lagrange multiplier theorem to solve the following problems:
 (a) $f(x) = ||x||^2$, $h(x) = \sum_{i=1}^{n} x_i - 1$.
 (b) $f(x) = \sum_{i=1}^{n} x_i$, $h(x) = ||x||^2 - 1$.
 (c) $f(x) = ||x||^2$, $h(x) = x'Qx - 1$, where Q is positive definite.

4.1.2

Consider all rectangular parallelepipeds with given length of diagonal; i.e., $x_1^2 + x_2^2 + x_3^2$: fixed where x_1, x_2, x_3 are the lengths of the edges. Find one that has

Sec. 4.1   Necessary Conditions for Equality Constraints   361

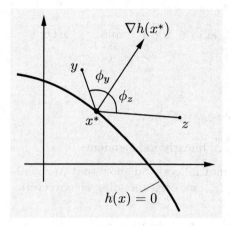

Figure 4.1.5. Fermat's principle; cf. Exercise 4.1.3. The angles ϕ_y, ϕ_z must be equal.

maximal surface area (sum of areas of the rectangular sides) and one that has maximal perimeter (sum of lengths of the edges).

4.1.3 (Fermat's Principle in Optics)

Consider a smooth curve on the plane described by the equation

$$h(x) = 0,$$

where $h : \Re^2 \mapsto \Re$ is continuously differentiable. Let y and z be two points that lie in relation to the curve as shown in Fig. 4.1.5. Show that if a point x^* minimizes over all points x on the curve, the sum of the Euclidean distances

$$\|y - x\| + \|z - x\|,$$

then the angles ϕ_y, ϕ_z shown in Fig. 4.1.5 must be equal.

4.1.4

Show that if the constraints are linear, the regularity assumption is not needed in order for the Lagrange multiplier conditions to hold, except that the Lagrange multiplier vector need not be unique. *Hint*: Discard the redundant equality constraints, and assign to them zero Lagrange multipliers. Use Prop. 4.1.1.

4.1.5

Consider a symmetric $n \times n$ matrix Q. Define

$$\lambda_1 = \min_{\|x\|^2=1} x'Qx, \qquad e_1 \in \arg\min_{\|x\|^2=1} x'Qx,$$

and for $k = 0, \ldots, n-1$,

$$\lambda_{k+1} = \min_{\substack{\|x\|^2=1 \\ e'_i x=0,\ i=1,\ldots,k}} x'Qx, \qquad e_{k+1} \in \arg\min_{\substack{\|x\|^2=1 \\ e'_i x=0,\ i=1,\ldots,k}} x'Qx.$$

(a) Show that
$$\lambda_1 \leq \lambda_2 \leq \cdots \leq \lambda_n.$$

(b) Show that the vectors $e_1, \ldots, e_n$ are linearly independent.

(c) Interpret $\lambda_1, \ldots, \lambda_n$ as Lagrange multipliers, and show that $\lambda_1, \ldots, \lambda_n$ are the eigenvalues of Q, while $e_1, \ldots, e_n$ are corresponding eigenvectors.

4.1.6 (Arithmetic-Geometric Mean Inequality)

Let $\alpha_1, \ldots, \alpha_n$ are positive scalars with $\sum_{i=1}^{n} \alpha_i = 1$. Use a Lagrange multiplier to solve the problem

$$\text{minimize } \alpha_1 x_1 + \alpha_2 x_2 + \cdots + \alpha_n x_n$$
$$\text{subject to } x_1^{\alpha_1} x_2^{\alpha_2} \cdots x_n^{\alpha_n} = 1, \qquad x_i > 0, \quad i = 1, \ldots, n.$$

Establish the arithmetic-geometric mean inequality

$$x_1^{\alpha_1} x_2^{\alpha_2} \cdots x_n^{\alpha_n} \leq \sum_{i=1}^{n} \alpha_i x_i,$$

for a set of positive numbers x_i, $i = 1, \ldots, n$. *Hint*: Use the change of variables $y_i = \ln(x_i)$.

4.1.7

Let $a_1, \ldots, a_m$ be given vectors in $\Re^n$, and consider the problem of minimizing

$$\sum_{j=1}^{m} \|x - a_j\|^2$$

subject to $\|x\|^2 = 1$. Consider the center of gravity $\hat{a} = \frac{1}{m} \sum_{j=1}^{m} a_j$ of the vectors $a_1, \ldots, a_m$. Show that if $\hat{a} \neq 0$, the problem has a unique maximum and a unique minimum. What happens if $\hat{a} = 0$?

4.1.8

Show the angles x, y, and z of a triangle maximize $\sin x \sin y \sin z$ if and only if the triangle is equilateral.

4.1.9

(a) Show that of all triangles circumscribed about a given circle, the one possessing minimal area is the equilateral. *Hint*: Let x, y, and z be the lengths of the tangent lines from the vertices of the triangle to the circle, and let ρ be the radius of the circle. The area of the triangle is $A = \rho(x+y+z)$, and it can also be expressed as $A = \sqrt{xyz(x+y+z)}$ (a theorem by Heron of Alexandria). Consider the problem of minimizing $x+y+z$ subject to $xyz = \rho^2(x+y+z)$.

(b) Consider the unit circle and two points a and b in the plane. Find a point x on the circle such that the triangle with vertices a, b, and x has maximal area, by maximizing $\|x - \hat{x}\|^2$ over $\|x\|^2 = 1$, where $\hat{x}$ is the projection of x on the line that passes through a and b. Show that the line connecting x and $\hat{x}$ passes through the center of the circle.

(c) Show that of all triangles inscribed in a given circle, the one possessing maximal area is the equilateral. *Hint*: Use part (b).

(d) Show that among all polygons in the plane with a given number of sides that are inscribed in a given circle the one that has maximal area is regular (all its sides are equal). *Hint*: Use part (b).

4.1.10

Consider the unit circle and two points a and b in the plane.

(a) Find points x on the circle such that the triangle with vertices a, b, and x has minimal and maximal perimeter. Characterize the optimal solutions by relating the problem to Fermat's principle (Exercise 4.1.3). Use your analysis to show that of all triangles inscribed in a given circle, the one possessing maximal perimeter is the equilateral.

(c) Find points x on the circle such that the inner product $(x-a)'(x-b)$ is minimum or maximum.

4.1.11

Consider the problem $\min_{h(x)=0} f(x)$, suppose that x^* is a local minimum, which is a regular point, and let $I = \{i \mid \lambda_i^* \neq 0\}$, where λ^* is the corresponding Lagrange multiplier. Extend the proof based on the penalty approach of Section 4.1.1 to show that for each neighborhood N of x^*, there exists an $x \in N$ such that
$$\lambda_i^* h_i(x) > 0, \qquad \forall\, i \in I.$$

Hint: In the proof of Section 4.1.1, use the fact that $\lambda_i^k = k h_i(x^k)$ converges to λ_i^* for each i.

4.1.12

This exercise shows that for each problem for which there are nonzero Lagrange multipliers, there are infinitely many corresponding problems with no Lagrange multipliers. Consider the problem $\min_{h(x)=0} f(x)$, and suppose that x^* is a local minimum such that $\nabla f(x^*) \neq 0$. Show that x^* is a local minimum of the equality constrained problem

$$\min_{\|h(x)\|^\rho = 0} f(x),$$

where $\rho > 1$, and that for this problem there are no Lagrange multipliers.

4.1.13

Verify the Schwarz inequality, $x'y \leq \|x\| \|y\|$ for all $x, y \in \Re^n$, by solving the problem

$$\max_{\|x\|^2=1,\, \|y\|^2=1} x'y.$$

4.2 SUFFICIENT CONDITIONS AND SENSITIVITY ANALYSIS

As illustrated by Example 4.1.3 of the preceding section, the first order necessary condition may be satisfied by both local minima and local maxima (and possibly other vectors). The second order necessary condition is useful in narrowing down the field of candidates for local minima. To guarantee that a given vector is a local minimum, we need sufficient conditions for optimality, which are given by the following proposition.

Proposition 4.2.1: (Second Order Sufficiency Conditions) Assume that f and h are twice continuously differentiable, and let $x^* \in \Re^n$ and $\lambda^* \in \Re^m$ satisfy

$$\nabla_x L(x^*, \lambda^*) = 0, \qquad \nabla_\lambda L(x^*, \lambda^*) = 0, \qquad (4.29)$$

$$y' \nabla^2_{xx} L(x^*, \lambda^*) y > 0, \qquad \text{for all } y \neq 0 \text{ with } \nabla h(x^*)'y = 0. \qquad (4.30)$$

Then x^* is a strict local minimum of f subject to $h(x) = 0$. In fact, there exist scalars $\gamma > 0$ and $\epsilon > 0$ such that

$$f(x) \geq f(x^*) + \frac{\gamma}{2}\|x - x^*\|^2, \qquad \forall\, x \text{ with } h(x) = 0 \text{ and } \|x - x^*\| < \epsilon.$$

Note that the above sufficient conditions do not include regularity of the vector x^*. We will prove Prop. 4.2.1 in two different ways, in Sections 4.2.1 and 4.2.2, respectively. We first give an example.

Example 4.2.1

Consider the problem

$$\text{minimize} \quad -(x_1x_2 + x_2x_3 + x_1x_3)$$
$$\text{subject to} \quad x_1 + x_2 + x_3 = 3.$$

(If x_1, x_2, and x_3 represent the length, width, and height of a rectangular parallelepiped P, respectively, the problem can be interpreted as maximizing the surface area of P, subject to the sum of the edge lengths of P being fixed.) The first order necessary conditions are

$$-x_2^* - x_3^* + \lambda^* = 0,$$
$$-x_1^* - x_3^* + \lambda^* = 0,$$
$$-x_1^* - x_2^* + \lambda^* = 0,$$
$$x_1^* + x_2^* + x_3^* = 3,$$

which have the unique solution $x_1^* = x_2^* = x_3^* = 1$, $\lambda^* = 2$. The Hessian of the Lagrangian is

$$\nabla_{xx}^2 L(x^*, \lambda^*) = \begin{pmatrix} 0 & -1 & -1 \\ -1 & 0 & -1 \\ -1 & -1 & 0 \end{pmatrix}.$$

We have for all $y \in V = \{y \mid \nabla h(x^*)'y = 0\} = \{y \mid y_1 + y_2 + y_3 = 0\}$ with $y \neq 0$,

$$y'\nabla_{xx}^2 L(x^*, \lambda^*)y = -y_1(y_2 + y_3) - y_2(y_1 + y_3) - y_3(y_1 + y_2)$$
$$= y_1^2 + y_2^2 + y_3^2 > 0.$$

Hence, the sufficient conditions of Prop. 4.2.1 are satisfied and x^* is a strict local minimum.

4.2.1 The Augmented Lagrangian Approach

We will first prove the sufficiency conditions, using penalty function concepts that will be used later in Section 5.2 as the basis for some important algorithms. An alternative proof, based on a feasible direction approach will be given in Section 4.2.2.

We begin with a useful lemma. Consider a symmetric matrix P and a positive semidefinite matrix Q. For each x that is not in the nullspace of Q, we have $x'Qx > 0$, so $x'(P + cQ)x > 0$ for sufficiently large scalars c. Thus, if we assume that $x'Px > 0$ for all $x \neq 0$ in the nullspace of Q, then for every $x \neq 0$, we have $x'(P + cQ)x > 0$, provided that c is sufficiently

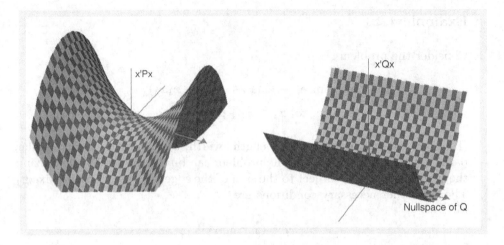

Figure 4.2.1. Illustration of Lemma 4.2.1. The possibly negative curvature of P outside the nullspace of Q is corrected by the positive curvature of Q appropriately magnified by the parameter c.

large. The essence of the lemma is that we can choose a threshold value of c that works for all x, so that $P + cQ$ is positive definite for all c greater than that threshold (cf. Fig. 4.2.1).

Lemma 4.2.1: Let P and Q be two symmetric matrices. Assume that Q is positive semidefinite and P is positive definite on the nullspace of Q, that is, $x'Px > 0$ for all $x \neq 0$ with $x'Qx = 0$. Then there exists a scalar $\bar{c}$ such that

$$P + cQ : \text{positive definite}, \qquad \forall\, c > \bar{c}.$$

Proof: Assume the contrary. Then for every integer k, there exists a vector x^k with $\|x^k\| = 1$ such that

$$x^{k\prime} P x^k + k x^{k\prime} Q x^k \leq 0.$$

Since $\{x^k\}$ is bounded, there is a subsequence $\{x^k\}_{k \in K}$ converging to some $\bar{x}$ [Prop. A.5(c) in Appendix A], and since $\|x^k\| = 1$ for all k, we have $\|\bar{x}\| = 1$. Taking upper limit in the preceding inequality, we obtain

$$\bar{x}' P \bar{x} + \limsup_{k \to \infty,\, k \in K} (k x^{k\prime} Q x^k) \leq 0. \qquad (4.31)$$

Since, by the positive semidefiniteness of Q, $x^{k\prime} Q x^k \geq 0$, we see that $\{x^{k\prime} Q x^k\}_{k \in K}$ must converge to zero, for otherwise the left-hand side of the above inequality would be $+\infty$. Therefore, $\bar{x}'Q\bar{x} = 0$ and from our hypothesis we obtain $\bar{x}'P\bar{x} > 0$. This contradicts Eq. (4.31). **Q.E.D.**

Let us introduce now the *augmented Lagrangian* function

$$L_c(x, \lambda) = f(x) + \lambda' h(x) + \frac{c}{2}\|h(x)\|^2,$$

where c is a scalar. This is the Lagrangian function for the problem

$$\text{minimize } f(x) + \frac{c}{2}\|h(x)\|^2$$

$$\text{subject to } h(x) = 0,$$

which has the same local minima as our original problem of minimizing $f(x)$ subject to $h(x) = 0$. The gradient and Hessian of L_c with respect to x are

$$\nabla_x L_c(x, \lambda) = \nabla f(x) + \nabla h(x)\bigl(\lambda + ch(x)\bigr),$$

$$\nabla^2_{xx} L_c(x, \lambda) = \nabla^2 f(x) + \sum_{i=1}^{m} \bigl(\lambda_i + ch_i(x)\bigr)\nabla^2 h_i(x) + c\nabla h(x)\nabla h(x)'.$$

In particular, if x^* and λ^* satisfy the sufficiency conditions of Prop. 4.2.1, we have

$$\nabla_x L_c(x^*, \lambda^*) = \nabla f(x^*) + \nabla h(x^*)\bigl(\lambda^* + ch(x^*)\bigr) = \nabla_x L(x^*, \lambda^*) = 0, \quad (4.32)$$

$$\nabla^2_{xx} L_c(x^*, \lambda^*) = \nabla^2 f(x^*) + \sum_{i=1}^{m} \lambda_i^* \nabla^2 h_i(x^*) + c\nabla h(x^*)\nabla h(x^*)'$$

$$= \nabla^2_{xx} L(x^*, \lambda^*) + c\nabla h(x^*)\nabla h(x^*)'.$$

By the sufficiency condition (4.30), we have that $y' \nabla^2_{xx} L(x^*, \lambda^*) y > 0$ for all $y \neq 0$ such that $y' \nabla h(x^*) \nabla h(x^*)' y = 0$, so by applying Lemma 4.2.1 with $P = \nabla^2_{xx} L(x^*, \lambda^*)$ and $Q = \nabla h(x^*) \nabla h(x^*)'$, it follows that there exists a $\bar{c}$ such that

$$\nabla^2_{xx} L_c(x^*, \lambda^*) : \text{positive definite}, \quad \forall \, c > \bar{c}. \quad (4.33)$$

Using the sufficient optimality condition for unconstrained optimization (cf. Prop. 1.1.5), we conclude from Eqs. (4.32) and (4.33), that for $c > \bar{c}$, x^* is an unconstrained local minimum of $L_c(\cdot, \lambda^*)$. In particular, there exist $\gamma > 0$ and $\epsilon > 0$ such that

$$L_c(x, \lambda^*) \geq L_c(x^*, \lambda^*) + \frac{\gamma}{2}\|x - x^*\|^2, \quad \forall \, x \text{ with } \|x - x^*\| < \epsilon.$$

Since for all x with $h(x) = 0$ we have $L_c(x, \lambda^*) = f(x)$, and by Eq. (4.29), $\nabla_\lambda L(x^*, \lambda^*) = h(x^*) = 0$, it follows that

$$f(x) \geq f(x^*) + \frac{\gamma}{2}\|x - x^*\|^2, \quad \forall \, x \text{ with } h(x) = 0, \text{ and } \|x - x^*\| < \epsilon.$$

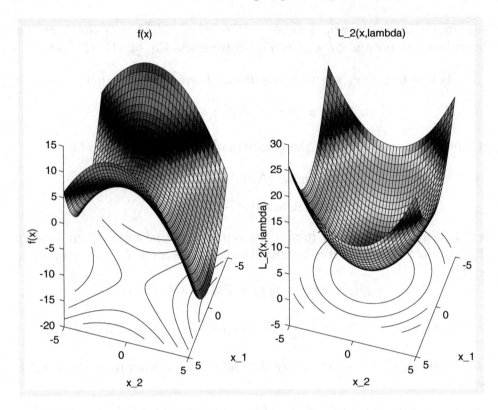

Figure 4.2.2. Illustration of how x^* is an unconstrained minimum of the augmented Lagrangian $L_c(\cdot, \lambda^*)$ for sufficiently large c. Consider the two-dimensional problem

$$\text{minimize } f(x) = \tfrac{1}{2}\left(x_1^2 - x_2^2\right) - x_2$$

$$\text{subject to } x_2 = 0.$$

Here, $x^* = (0,0)$ is the unique global minimum, and $\lambda^* = 1$ is a corresponding Lagrange multiplier. The augmented Lagrangian function is

$$L_c(x, \lambda^*) = \tfrac{1}{2}\left(x_1^2 - x_2^2\right) - x_2 + \lambda^* x_2 + \tfrac{c}{2} x_2^2 = \tfrac{1}{2}\left(x_1^2 - x_2^2\right) + \tfrac{c}{2} x_2^2$$

and has $x^* = (0,0)$ as its unique unconstrained minimum for $c > 1$. The figure shows the equal cost surfaces of f (left side) and of $L_c(\cdot, \lambda^*)$ for $c = 2$ (right side).

Thus x^* is a strict local minimum of f over $h(x) = 0$. The proof of Prop. 4.2.1 is complete.

The preceding analysis also shows that we can try to minimize f over $h(x) = 0$ by computing an unconstrained minimum of the augmented Lagrangian $L_c(\cdot, \lambda^*)$, as illustrated in Fig. 4.2.2. The difficulty here is that the Lagrange multiplier λ^* is unknown. However, it turns out that by using in place of λ^*, readily computable approximations to λ^*, we can obtain useful algorithms, the so-called *augmented Lagrangian methods*, also called *methods of multipliers*; see Section 5.2.

4.2.2 The Feasible Direction Approach

Let us now prove the sufficiency conditions of Prop. 4.2.1 using a descent/feasible direction approach. In particular, we assume that there are points of lower cost arbitrarily close to x^*, focus on the directions leading to these points, and come to a contradiction. This method of proof can also be used to obtain refined sufficiency conditions for problems with inequality constraints (see Exercise 4.3.7).

We first note that x^* is feasible, since, by Eq. (4.29), $\nabla_\lambda L(x^*, \lambda^*) = h(x^*) = 0$. Assume, to obtain a contradiction, that the conclusion of the proposition does not hold, so that there is a sequence $\{x^k\}$ such that $x^k \to x^*$, and for all k, $x^k \neq x^*$, $h(x^k) = 0$, and $f(x^k) < f(x^*) + (1/k)\|x^k - x^*\|^2$. Let us write $x^k = x^* + \delta^k y^k$, where

$$\delta^k = \|x^k - x^*\|, \qquad y^k = \frac{x^k - x^*}{\|x^k - x^*\|}.$$

The sequence $\{y^k\}$ is bounded, so it must have a subsequence converging to some y with $\|y\| = 1$. Without loss of generality, we assume that the whole sequence $\{y^k\}$ converges to y. By taking the limit as $\delta^k \to 0$ in the relation

$$0 = \frac{h_i(x^k) - h_i(x^*)}{\delta^k} = \frac{h_i(x^* + \delta^k y^k) - h_i(x^*)}{\delta^k} = \nabla h_i(x^*)' y^k + \frac{o(\delta^k)}{\delta^k},$$

we see that $\nabla h(x^*)' y = 0$.

We will now show that $y' \nabla^2_{xx} L(x^*, \lambda^*) y \leq 0$, thus coming to a contradiction [cf. Eq. (4.30)]. Since $x^k = x^* + \delta^k y^k$, by the mean value theorem [Prop. A.23(b) in Appendix A], we have

$$0 = h_i(x^k) - h_i(x^*) = \delta^k \nabla h_i(x^*)' y^k + \frac{(\delta^k)^2}{2} y^{k'} \nabla^2 h_i(\bar{\xi}_i^k) y^k, \qquad (4.34)$$

$$\frac{1}{k}\|x^k - x^*\|^2 > f(x^k) - f(x^*) = \delta^k \nabla f(x^*)' y^k + \frac{(\delta^k)^2}{2} y^{k'} \nabla^2 f(\tilde{\xi}^k) y^k, \qquad (4.35)$$

where all the vectors $\bar{\xi}_i^k$ and $\tilde{\xi}^k$ lie on the line segment joining x^* and x^k. Multiplying Eq. (4.34) by λ_i^* and adding Eq. (4.35) to it, we obtain

$$\frac{1}{k}\|x^k - x^*\|^2 > \delta^k \left(\nabla f(x^*) + \sum_{i=1}^m \lambda_i^* \nabla h_i(x^*) \right)' y^k$$

$$+ \frac{(\delta^k)^2}{2} y^{k'} \left(\nabla^2 f(\tilde{\xi}^k) + \sum_{i=1}^m \lambda_i^* \nabla^2 h_i(\bar{\xi}_i^k) \right) y^k.$$

Since $\delta^k = \|x^k - x^*\|$ and $\nabla f(x^*) + \sum_{i=1}^m \lambda_i^* \nabla h_i(x^*) = 0$, we obtain

$$\frac{2}{k} > y^{k'} \left(\nabla^2 f(\tilde{\xi}^k) + \sum_{i=1}^m \lambda_i^* \nabla^2 h_i(\bar{\xi}_i^k) \right) y^k.$$

By taking the limit as $k \to \infty$,

$$0 \geq y' \left(\nabla^2 f(x^*) + \sum_{i=1}^{m} \lambda_i^* \nabla^2 h_i(x^*) \right) y,$$

which contradicts the hypothesis (4.30). The proof of Prop. 4.2.1 is complete.

4.2.3 Sensitivity

Lagrange multipliers frequently have an interesting interpretation in specific practical contexts. In economic applications they can often be interpreted as prices, while in other problems they represent quantities with concrete physical meaning. It turns out that within our mathematical framework, they can be viewed as rates of change of the optimal cost as the level of constraint changes. This is fairly easy to show for the case of linear constraints, as indicated in Fig. 4.2.3. The following proposition provides a formal proof for the case of nonlinear constraints.

Proposition 4.2.2: (Sensitivity Theorem) Let x^* and λ^* be a local minimum and Lagrange multiplier, respectively, satisfying the second order sufficiency conditions of Prop. 4.2.1, and assume that x^* is a regular point. Consider the family of problems

$$\begin{aligned} \text{minimize} \quad & f(x) \\ \text{subject to} \quad & h(x) = u, \end{aligned} \quad (4.36)$$

parameterized by the vector $u \in \Re^m$. Then there exists an open sphere S centered at $u = 0$ such that for every $u \in S$, there is an $x(u) \in \Re^n$ and a $\lambda(u) \in \Re^m$, which are a local minimum-Lagrange multiplier pair of problem (4.36). Furthermore, $x(\cdot)$ and $\lambda(\cdot)$ are continuously differentiable functions within S and we have

$$x(0) = x^*, \qquad \lambda(0) = \lambda^*.$$

In addition, for all $u \in S$ we have

$$\nabla p(u) = -\lambda(u),$$

where $p(u)$ is the optimal cost parameterized by u, i.e.,

$$p(u) = f\big(x(u)\big).$$

Sec. 4.2 Sufficient Conditions and Sensitivity Analysis 371

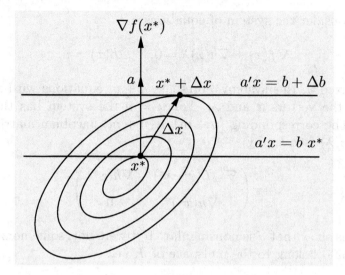

Figure 4.2.3. Illustration of the sensitivity theorem for a problem involving a single linear constraint,

$$\text{minimize } f(x)$$
$$\text{subject to } a'x = b.$$

Here, x^* is a local minimum and λ^* is a corresponding Lagrange multiplier. If the level of constraint b is changed to $b + \Delta b$, the minimum x^* will change to $x^* + \Delta x$. Since $b + \Delta b = a'(x^* + \Delta x) = a'x^* + a'\Delta x = b + a'\Delta x$, we see that the variations Δx and Δb are related by

$$a'\Delta x = \Delta b.$$

Using the Lagrange multiplier condition $\nabla f(x^*) = -\lambda^* a$, the corresponding cost change can be written as

$$\Delta\text{cost} = f(x^* + \Delta x) - f(x^*) = \nabla f(x^*)'\Delta x + o(\|\Delta x\|) = -\lambda^* a'\Delta x + o(\|\Delta x\|).$$

By combining the above two relations, we obtain $\Delta\text{cost} = -\lambda^*\Delta b + o(\|\Delta x\|)$, so up to first order we have

$$\lambda^* = -\frac{\Delta\text{cost}}{\Delta b}.$$

Thus, the Lagrange multiplier λ^* gives the rate of optimal cost decrease as the level of constraint increases.

In the case where there are multiple constraints $a_i'x = b_i$, $i = 1, \ldots, m$, the preceding argument can be appropriately modified. In particular, we have

$$\Delta\text{cost} = f(x^* + \Delta x) - f(x^*)$$
$$= \nabla f(x^*)'\Delta x + o(\|\Delta x\|)$$
$$= -\sum_{i=1}^{m} \lambda_i^* a_i'\Delta x + o(\|\Delta x\|),$$

and $a_i'\Delta x = \Delta b_i$ for all i, so we obtain $\Delta\text{cost} = -\sum_{i=1}^{m} \lambda_i^* \Delta b_i + o(\|\Delta x\|)$.

Proof: Consider the system of equations

$$\nabla f(x) + \nabla h(x)\lambda = 0, \qquad h(x) = u. \tag{4.37}$$

For each fixed u, this system represents $n + m$ equations with $n + m$ unknowns – the vectors x and λ. For $u = 0$ the system has the solution (x^*, λ^*). The corresponding $(n+m) \times (n+m)$ Jacobian matrix with respect to (x, λ) is given by

$$J = \begin{pmatrix} \nabla^2_{xx} L(x^*, \lambda^*) & \nabla h(x^*) \\ \nabla h(x^*)' & 0 \end{pmatrix}.$$

Let us show that J is nonsingular. If it were not, some nonzero vector $(y', z')'$ would belong to the nullspace of J, i.e.,

$$\nabla^2_{xx} L(x^*, \lambda^*) y + \nabla h(x^*) z = 0, \tag{4.38}$$

$$\nabla h(x^*)' y = 0. \tag{4.39}$$

Premultiplying Eq. (4.38) by y' and using Eq. (4.39), we obtain

$$y' \nabla^2_{xx} L(x^*, \lambda^*) y = 0.$$

In view of Eq. (4.39), it follows that $y = 0$, for otherwise our second order sufficiency assumption would be violated [cf. Eq. (4.30)]. Since $y = 0$, Eq. (4.38) yields $\nabla h(x^*) z = 0$, which in view of the linear independence of the columns $\nabla h_i(x^*)$, $i = 1, \ldots, m$, of $\nabla h(x^*)$, yields $z = 0$. Thus, we obtain $y = 0$, $z = 0$, which is a contradiction. Hence, J is nonsingular.

Returning now to the system (4.37), it follows from the nonsingularity of J and the implicit function theorem (Prop. A.25 in Appendix A) that for all u in some open sphere S centered at $u = 0$, there exist $x(u)$ and $\lambda(u)$ such that $x(0) = x^*$, $\lambda(0) = \lambda^*$, the functions $x(\cdot)$ and $\lambda(\cdot)$ are continuously differentiable, and

$$\nabla f(x(u)) + \nabla h(x(u)) \lambda(u) = 0, \tag{4.40}$$

$$h(x(u)) = u. \tag{4.41}$$

For u sufficiently close to 0, the vectors $x(u)$ and $\lambda(u)$ satisfy the second order sufficiency conditions for problem (4.36), since they satisfy them by assumption for $u = 0$. This is straightforward to verify by using our continuity assumptions. [If it were not true, there would exist a sequence $\{u^k\}$ with $u^k \to 0$, and a sequence $\{y^k\}$ with $\|y^k\| = 1$ and $\nabla h(x(u^k))' y^k = 0$ for all k, such that

$$y^{k'} \nabla^2_{xx} L(x(u^k), \lambda(u^k)) y^k \leq 0, \qquad \forall\ k.$$

Sec. 4.2   Sufficient Conditions and Sensitivity Analysis

By taking the limit along a convergent subsequence of $\{y^k\}$, we would obtain a contradiction of the second order sufficiency condition at (x^*, λ^*); cf. Eq. (4.30).] Hence, $x(u)$ and $\lambda(u)$ are a local minimum-Lagrange multiplier pair for problem (4.36).

There remains to show that

$$\nabla p(u) = \nabla_u \{f(x(u))\} = -\lambda(u).$$

To this end, we multiply Eq. (4.40) by $\nabla x(u)$, and obtain

$$\nabla x(u) \nabla f(x(u)) + \nabla x(u) \nabla h(x(u)) \lambda(u) = 0.$$

By differentiating the relation

$$u = h(x(u)),$$

[cf. Eq. (4.41)], it follows that

$$I = \nabla_u \{h(x(u))\} = \nabla x(u) \nabla h(x(u)), \qquad (4.42)$$

where I is the $m \times m$ identity matrix. Finally, by using the chain rule, we have

$$\nabla p(u) = \nabla_u \{f(x(u))\} = \nabla x(u) \nabla f(x(u)).$$

Combining the above three relations, we obtain

$$\nabla p(u) + \lambda(u) = 0, \qquad (4.43)$$

and the proof is complete. **Q.E.D.**

The function
$$p(u) = f(x(u))$$
in the sensitivity theorem is called the *primal function* and plays an important role in contexts other than sensitivity. When we discuss duality theory later, in Section 4.4 and in Chapter 6, we will see that it is closely related to the, so-called, *dual function*. Within that context, the relation $\nabla p(0) = -\lambda^*$, illustrated in Fig. 4.2.4, will also turn out to be significant.

Note that the regularity of x^* is essential for the primal function to be defined within a sphere centered at 0. For example, for the problem

$$\text{minimize } \tfrac{1}{2}(x_1^2 - x_2^2) - x_2$$
$$\text{subject to } (x_2)^2 = 0,$$

which is equivalent to the problem of Fig. 4.2.4, the primal function is undefined for $u < 0$.

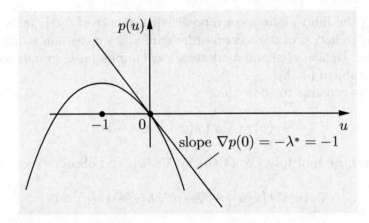

Figure 4.2.4. Illustration of the primal function $p(u) = f(x(u))$ for the two-dimensional problem of Fig. 4.2.2.

$$\text{minimize } f(x) = \tfrac{1}{2}\left(x_1^2 - x_2^2\right) - x_2$$
$$\text{subject to } h(x) = x_2 = 0.$$

Here,

$$p(u) = \min_{h(x)=u} f(x) = -\tfrac{1}{2}u^2 - u$$

and we have

$$\lambda^* = -\nabla p(0) = 1$$

consistently with the sensitivity theorem.

EXERCISES

4.2.1

Consider the problem

$$\text{maximize } x_1 + x_2$$
$$\text{subject to } x_1^2 + x_2^2 = 2.$$

(This is the problem of maximizing the perimeter of a rectangle inscribed in a given circle.)

(a) Show that it has a unique global maximum and a unique global minimum.

(b) Calculate the primal function corresponding to the maximum and the minimum and verify that its gradient is related to the Lagrange multiplier as specified by the sensitivity theorem.

4.2.2

Consider the problem of Example 4.1.3 in Section 4.1. Calculate the primal function corresponding to the minimum and verify that its gradient is related to the Lagrange multiplier as specified by the sensitivity theorem.

4.2.3

Use sufficiency conditions to verify optimality in the following problems:

(a) Among all rectangular parallelepipeds with given sum of lengths of edges, find one that has maximal volume.

(b) Among all rectangular parallelepipeds with given volume, find one that has minimal sum of lengths of edges.

(c) Among all rectangular parallelepipeds that are inscribed in an ellipsoid

$$\left\{ (x,y,z) \;\Big|\; \frac{x^2}{a^2} + \frac{y^2}{b^2} + \frac{z^2}{c^2} = 1 \right\},$$

and have sides that are parallel to the axes of the ellipsoid, find one that has maximal volume.

4.2.4

The purpose of this exercise is to show that under certain circumstances one may exchange the cost function with one of the constraints and obtain a new optimization problem whose solution is the same as the original. Consider the problem of minimizing $f(x)$ subject to $h_i(x) = 0$, $i = 1, \ldots, m$. Assume that x^* together with Lagrange multipliers λ_i^* satisfy the second order sufficiency conditions of Prop. 4.2.1. Let j be a constraint index such that $\lambda_j^* \neq 0$.

(a) Show that if $\lambda_j^* > 0$, then x^* is a local minimum of $h_j(x)$ subject to the constraints $f(x) = f(x^*)$ and $h_i(x) = 0$, $i = 1, \ldots, j-1, j+1, \ldots, m$.

(b) Show that if $\lambda_j^* < 0$, then x^* is a local maximum of $h_j(x)$ subject to the constraints $f(x) = f(x^*)$ and $h_i(x) = 0$, $i = 1, \ldots, j-1, j+1, \ldots, m$.

4.2.5 (Archimedes' Problem)

A spherical segment is the intersection of a sphere with one of the halfspaces corresponding to a 2-dimensional plane that meets the sphere. The area of a spherical segment is the area of the portion of its surface that is also part of the surface of the sphere. Note that if r is the radius of the sphere and h is the height of a spherical segment, the volume enclosed by the segment is $\pi h^2(r - h/3)$ and its spherical area is $2\pi rh$. Show that among all spherical segments with the same spherical area, the one that encloses the largest volume is a hemisphere.

4.2.6 (Alternative Statement of Sufficiency Conditions) (www)

Let x^* be a feasible point that is regular and together with some λ^* satisfies the first and second order necessary conditions of Prop. 4.1.1. Show that x^* and λ^* satisfy the sufficiency conditions of Prop. 4.2.1 if and only if the matrix

$$\begin{pmatrix} \nabla^2_{xx} L(x^*, \lambda^*) & \nabla h(x^*) \\ \nabla h(x^*)' & 0 \end{pmatrix}$$

is nonsingular. *Hint*: For the reverse assertion, see the proof of the sensitivity theorem. For the forward assertion, suppose that there exists a $\bar{y}$ such that $\nabla h(x^*)' \bar{y} = 0$ and $\bar{y}' \nabla^2_{xx} L(x^*, \lambda^*) \bar{y} = 0$. Use the fact that $\bar{y}$ minimizes $y' \nabla^2_{xx} L(x^*, \lambda^*) y$ over all y with $\nabla h(x^*)' y = 0$.

4.2.7 (Hessian of the Primal Function) (www)

Under the assumptions of the sensitivity theorem, show that for every scalar c for which the matrix

$$A_c(u) = \nabla^2_{xx} L\big(x(u), \lambda(u)\big) + c \nabla h\big(x(u)\big) \nabla h\big(x(u)\big)'$$

is invertible, we have

$$\nabla^2 p(u) = \left(\nabla h\big(x(u)\big)' A_c(u)^{-1} \nabla h\big(x(u)\big) \right)^{-1} - cI.$$

Hint: Differentiate Eqs. (4.40) and (4.43), and use Eq. (4.42).

4.3 INEQUALITY CONSTRAINTS

We now consider the following problem, which involves both equality and inequality constraints:

$$\begin{aligned} &\text{minimize} \quad f(x) \\ &\text{subject to} \quad h_1(x) = 0, \ldots, h_m(x) = 0, \\ &\qquad\qquad\quad g_1(x) \leq 0, \ldots, g_r(x) \leq 0, \end{aligned} \qquad \text{(ICP)}$$

where f, h_i, g_j are continuously differentiable functions from $\Re^n$ to $\Re$. More succinctly, we can write this problem as

$$\begin{aligned} &\text{minimize} \quad f(x) \\ &\text{subject to} \quad h(x) = 0, \quad g(x) \leq 0, \end{aligned}$$

Sec. 4.3   Inequality Constraints   377

where $h : \Re^n \mapsto \Re^m$ and $g : \Re^n \mapsto \Re^r$ are the functions

$$h = (h_1, ..., h_m), \qquad g = (g_1, ..., g_r).$$

In lack of an explicit statement to the contrary, our comments and analysis in this section will refer to problem (ICP).

We will first use a simple approach to this problem that relies on the theory for equality constraints of the preceding sections. For any feasible point x, the *set of active inequality constraints* is denoted by

$$A(x) = \{j \mid g_j(x) = 0\}. \qquad (4.44)$$

If $j \notin A(x)$, we say that the jth constraint is *inactive* at x. We note that if x^* is a local minimum of the inequality constrained problem (ICP), then x^* is also a local minimum for a problem identical to (ICP) except that the inactive constraints at x^* have been discarded. Thus, in effect, *inactive constraints at x^* don't matter*; they can be ignored in the statement of optimality conditions.

On the other hand, at a local minimum, *active inequality constraints can be treated to a large extent as equalities*. In particular, if x^* is a local minimum of the inequality constrained problem (ICP), then x^* is also a local minimum for the equality constrained problem

$$\begin{aligned}&\text{minimize} \ \ f(x)\\ &\text{subject to} \ \ h_1(x) = 0, \ldots, h_m(x) = 0, \qquad g_j(x) = 0, \quad \forall \ j \in A(x^*).\end{aligned} \qquad (4.45)$$

Thus, if x^* is regular for the latter problem, there exist Lagrange multipliers $\lambda_1^*, \ldots, \lambda_m^*$, and μ_j^*, $j \in A(x^*)$ such that

$$\nabla f(x^*) + \sum_{i=1}^m \lambda_i^* \nabla h_i(x^*) + \sum_{j \in A(x^*)} \mu_j^* \nabla g_j(x^*) = 0.$$

Assigning zero Lagrange multipliers to the inactive constraints, we obtain

$$\nabla f(x^*) + \sum_{i=1}^m \lambda_i^* \nabla h_i(x^*) + \sum_{j=1}^r \mu_j^* \nabla g_j(x^*) = 0, \qquad (4.46)$$

$$\mu_j^* = 0, \qquad \forall \ j \notin A(x^*), \qquad (4.47)$$

which may be viewed as an analog of the first order optimality condition for the equality constrained problem.

There is one more important fact about the Lagrange multipliers μ_j^*: *they are nonnegative*. For a geometric illustration, see Fig. 4.3.1. For an algebraic argument, note that if the jth constraint is relaxed to $g_j(x) \leq u_j$, where $u_j > 0$, the optimal cost will tend to decrease because the constraint

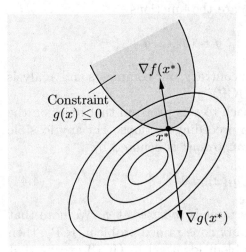

Figure 4.3.1. Illustration of the nonnegativity of the Lagrange multiplier for a problem with a single inequality constraint. If the constraint is inactive, then $\mu^* = 0$. Otherwise, $\nabla f(x^*)$ is normal to the constraint surface and points to the inside of the constraint set, while $\nabla g(x^*)$ is normal to the constraint surface and points to the outside of the constraint set. Thus $\nabla f(x^*)$ and $\nabla g(x^*)$ are collinear and have opposite signs, implying that the Lagrange multiplier is nonnegative.

set will become larger. Therefore, the sensitivity interpretation of the Lagrange multiplier $[\mu_j^* = -(\Delta\text{cost due to } u_j)/u_j$, cf. Prop. 4.2.2] suggests that $\mu_j^* \geq 0$.

The preceding development will be made rigorous in Sections 4.3.1 and 4.3.2. While straightforward, it is nonetheless limited by regularity assumptions. Such assumptions are quite general for problems with nonlinear equality constraints, but are somewhat restrictive for problems with inequalities. The reason is that in many types of important problems (linear programs is an example) there may be many (possibly more than n) inequality constraints that are satisfied as equalities at a local minimum, and the corresponding constraint gradients are often dependent. With this motivation, we will develop later, starting in Section 4.3.3, a Lagrange multiplier theory that is not based on regularity-type assumptions. Still this theory is not too complicated and relies on the penalty approach that we have used for equality constraints in Section 4.1.1.

4.3.1 Karush-Kuhn-Tucker Optimality Conditions

The preceding informal discussion will now be made rigorous. Our first line of development is based on an extended version of the notion of regularity that we used for equality constrained problems. This leads to necessary conditions and sufficient conditions that are analogous to the ones for equality constraints.

A feasible vector x is said to be *regular* if the equality constraint and the active inequality constraint gradients

$$\nabla h_i(x), \quad i = 1, \ldots, m, \qquad \nabla g_j(x), \quad j \in A(x),$$

are linearly independent. (For completeness, we also say that x is regular in the case where there are no equality constraints and all the inequality

Sec. 4.3   Inequality Constraints

constraints are inactive at x.) The following proposition generalizes the Lagrange multiplier theorem of Section 4.1 (Prop. 4.1.1). It will be stated in terms of the Lagrangian function defined by

$$L(x, \lambda, \mu) = f(x) + \sum_{i=1}^{m} \lambda_i h_i(x) + \sum_{j=1}^{r} \mu_j g_j(x). \qquad (4.48)$$

Proposition 4.3.1: (Karush-Kuhn-Tucker Necessary Conditions) Let x^* be a local minimum of problem (ICP), and assume that x^* is regular. Then there exist unique Lagrange multiplier vectors $\lambda^* = (\lambda_1^*, \ldots, \lambda_m^*)$, $\mu^* = (\mu_1^*, \ldots, \mu_r^*)$, such that

$$\nabla_x L(x^*, \lambda^*, \mu^*) = 0,$$

$$\mu_j^* \geq 0, \qquad j = 1, \ldots, r,$$

$$\mu_j^* = 0, \qquad \forall \, j \notin A(x^*),$$

where $A(x^*)$ is the set of active constraints at x^* [cf. Eq. (4.44)]. If in addition f, h, and g are twice continuously differentiable, there holds

$$y' \nabla_{xx}^2 L(x^*, \lambda^*, \mu^*) y \geq 0, \qquad (4.49)$$

for all $y \in \Re^n$ such that

$$\nabla h_i(x^*)' y = 0, \, \forall \, i = 1, \ldots, m, \qquad \nabla g_j(x^*)' y = 0, \, \forall \, j \in A(x^*).$$

Proof: All the assertions of the proposition follow from our earlier discussion [cf. Eqs. (4.46) and (4.47)], and the first and second order optimality conditions for the equality constrained problem (4.45) (cf. Prop. 4.1.1), except for the assertion $\mu_j^* \geq 0$ for $j \in A(x^*)$. We give a proof of this assertion using the penalty approach of Section 4.1.1. We introduce the functions

$$g_j^+(x) = \max\{0, g_j(x)\}, \qquad j = 1, \ldots, r,$$

and for each $k = 1, 2, \ldots$, the "penalized" problem

$$\text{minimize} \quad F^k(x) \equiv f(x) + \frac{k}{2} \|h(x)\|^2 + \frac{k}{2} \sum_{j=1}^{r} \left(g_j^+(x)\right)^2 + \frac{\alpha}{2} \|x - x^*\|^2$$

subject to $x \in S$,

where α is a fixed positive scalar, $S = \{x \mid \|x - x^*\| \leq \epsilon\}$, and $\epsilon > 0$ is such that $f(x^*) \leq f(x)$ for all feasible x with $x \in S$. Note that the function

$\left(g_j^+(x)\right)^2$ is continuously differentiable with gradient $2g_j^+(x)\nabla g_j(x)$. If x^k minimizes $F^k(x)$ over S, a similar argument to the one used for the equality constraint case in Section 4.1.1 shows that $x^k \to x^*$, and that the Lagrange multipliers λ_i^* and μ_j^* are given by

$$\lambda_i^* = \lim_{k\to\infty} kh_i(x^k), \qquad i = 1,\ldots,m,$$

$$\mu_j^* = \lim_{k\to\infty} kg_j^+(x^k), \qquad j = 1,\ldots,r.$$

Since $g_j^+(x^k) \geq 0$, we obtain $\mu_j^* \geq 0$ for all j. **Q.E.D.**

The condition $\mu_j^* = 0$ for all $j \notin A(x^*)$ can also be compactly written as

$$\mu_j^* g_j(x^*) = 0, \qquad j = 1,\ldots,r,$$

and is called *complementary slackness condition*, (CS for short). The name derives from the fact that for each j, whenever the constraint $g_j(x^*) \leq 0$ is slack [meaning that $g_j(x^*) < 0$], the constraint $\mu_j^* \geq 0$ must not be slack (meaning that $\mu_j^* = 0$), and reversely.

One approach for using necessary conditions to solve inequality constrained problems is to consider separately all the possible combinations of constraints being active or inactive. The following example illustrates the process for the case of a single constraint.

Example 4.3.1

Consider the problem

$$\text{minimize } \tfrac{1}{2}\left(x_1^2 + x_2^2 + x_3^2\right)$$
$$\text{subject to } x_1 + x_2 + x_3 \leq -3.$$

Then for a local minimum x^*, the first order necessary condition [cf. Eq. (4.48)] yields

$$x_1^* + \mu^* = 0,$$
$$x_2^* + \mu^* = 0,$$
$$x_3^* + \mu^* = 0.$$

There are two possibilities:

(a) The constraint is inactive,

$$x_1^* + x_2^* + x_3^* < -3,$$

in which case $\mu^* = 0$. Then, we obtain $x_1^* = x_2^* = x_3^* = 0$, which contradicts the inequality $x_1^* + x_2^* + x_3^* < -3$. Hence, this possibility is excluded.

Sec. 4.3   Inequality Constraints

(b) The constraint is active,

$$x_1^* + x_2^* + x_3^* = -3.$$

Then we obtain $x_1^* = x_2^* = x_3^* = -1$ and $\mu^* = 1$, which satisfy all the necessary conditions for a local minimum. Since every point is regular it follows that $x^* = (-1, -1, -1)$ is the unique candidate for a local minimum. We may proceed further and verify that x^* and μ^* satisfy the second order sufficiency conditions for optimality (to be provided in Section 4.3.2) and conclude that x^* is indeed the unique local minimum for this problem. Alternatively, we can verify the sufficiency conditions based on convexity (also to be provided in Section 4.3.2) and conclude that x^* is the unique global minimum.

Analysis by Conversion to the Equality Case

We now provide an alternative proof of the Karush-Kuhn-Tucker conditions of Prop. 4.3.1. It is based on converting the inequality constrained problem (ICP) into the equality constrained problem

$$\begin{aligned}
\text{minimize} \quad & f(x) \\
\text{subject to} \quad & h_1(x) = 0, \ldots, h_m(x) = 0, \\
& g_1(x) + z_1^2 = 0, \ldots, g_r(x) + z_r^2 = 0,
\end{aligned} \quad (4.50)$$

where we have introduced additional variables $z_1, \ldots, z_r$. By using the necessary conditions for equality constraints, developed in Section 4.1, we will obtain necessary conditions for problem (ICP). This approach is straightforward and is also useful in other contexts. For example, it will be used in Section 4.3.2 to provide an easy proof of sufficiency conditions for inequality constraints. The drawback of the approach, however, is that it requires that f, h_i, and g_j be twice continuously differentiable.

Let x^* be a local minimum for our original problem (ICP). Then (x^*, z^*) is a local minimum for problem (4.50), where $z^* = (z_1^*, \ldots, z_r^*)$,

$$z_j^* = \left(-g_j(x^*)\right)^{1/2}, \qquad j = 1, \ldots, r.$$

It is straightforward to verify that x^* is regular for problem (ICP), if and only if (x^*, z^*) is regular for problem (4.50); we leave this as an exercise for the reader. By applying the first order necessary conditions for equality constraints (Prop. 4.1.1) to problem (4.50), we see that there exist Lagrange multipliers $\lambda_1^*, \ldots, \lambda_m^*, \mu_1^*, \ldots, \mu_r^*$ such that

$$\nabla f(x^*) + \sum_{i=1}^{m} \lambda_i^* \nabla h_i(x^*) + \sum_{j=1}^{r} \mu_j^* \nabla g_j(x^*) = 0,$$

$$2\mu_j^* z_j^* = 0, \qquad j = 1, \ldots, r.$$

Since
$$z_j^* = \left(-g_j(x^*)\right)^{1/2} > 0, \qquad \forall\, j \notin A(x^*),$$
the last equation can also be written as
$$\mu_j^* = 0, \qquad \forall\, j \notin A(x^*). \tag{4.51}$$

Thus, to prove Prop. 4.3.1, there remains to show the nonnegativity condition $\mu_j^* \geq 0$ and the second order condition (4.49). To this end, we require the twice continuous differentiability of f, h_i, and g_j, in order to use the second order necessary condition for the equivalent equality constrained problem (4.50). It yields [cf. Eq. (4.21) in Section 4.1]

$$(y' \; w') \begin{pmatrix} \nabla_{xx}^2 L(x^*, \lambda^*, \mu^*) & 0 \\ & 2\mu_1^* & 0 & \cdots & 0 \\ & 0 & 2\mu_2^* & \cdots & 0 \\ 0 & \vdots & \vdots & \vdots & \vdots \\ & 0 & 0 & \cdots & 2\mu_r^* \end{pmatrix} \begin{pmatrix} y \\ w \end{pmatrix} \geq 0, \tag{4.52}$$

for all $y \in \Re^n$, $w \in \Re^r$ satisfying
$$\nabla h(x^*)' y = 0, \qquad \nabla g_j(x^*)' y + 2z_j^* w_j = 0, \qquad j = 1, \ldots, r. \tag{4.53}$$

We will now try different pairs (y, w) satisfying Eq. (4.53) and obtain the desired results from Eq. (4.52). First let us select w by
$$w_j = \begin{cases} 0 & \text{if } j \in A(x^*), \\ -\dfrac{\nabla g_j(x^*)' y}{2z_j^*} & \text{if } j \notin A(x^*). \end{cases}$$

Then
$$\nabla g_j(x^*)' y + 2z_j^* w_j = 0$$
for all $j \notin A(x^*)$, $z_j^* w_j = 0$ for all $j \in A(x^*)$, and by taking into account the fact $\mu_j^* = 0$ for $j \notin A(x^*)$ [cf. Eq. (4.51)], we also have
$$\mu_j^* w_j = 0, \qquad \forall\, j = 1, \ldots, r.$$
Thus we obtain from Eqs. (4.52) and (4.53)
$$y' \nabla_{xx}^2 L(x^*, \lambda^*, \mu^*) y \geq 0,$$
for all y such that $\nabla h(x^*)' y = 0$ and $\nabla g_j(x^*)' y = 0$ for all $j \in A(x^*)$, thereby verifying the second order condition (4.49).

Next let us select, for every $j \in A(x^*)$, a vector (y, w) with $y = 0$, $w_j \neq 0$, $w_k = 0$ for all $k \neq j$. Such a vector satisfies the condition of Eq. (4.53). By using such a vector in Eq. (4.52), we obtain $2\mu_j^* w_j^2 \geq 0$, and
$$\mu_j^* \geq 0, \qquad \forall\, j \in A(x^*).$$

Thus, the conversion to the equivalent equality constrained problem (4.50) has yielded all the Karush-Kuhn-Tucker conditions for the inequality constrained problem (ICP).

4.3.2 Sufficient Conditions and Sensitivity

Let us consider again the transformation of the inequality constrained problem (ICP) to an equality constrained problem, which we used in the preceding section to derive necessary conditions for optimality. By applying the sufficient conditions for optimality of Prop. 4.2.1, we obtain the following.

Proposition 4.3.2: (Second Order Sufficiency Conditions) Consider problem (ICP), assume that f, h, and g are twice continuously differentiable, and let $x^* \in \Re^n$, $\lambda^* \in \Re^m$, and $\mu^* \in \Re^r$ satisfy

$$\nabla_x L(x^*, \lambda^*, \mu^*) = 0, \qquad h(x^*) = 0, \qquad g(x^*) \leq 0,$$

$$\mu_j^* \geq 0, \qquad j = 1, \ldots, r,$$

$$\mu_j^* = 0, \qquad \forall\, j \notin A(x^*),$$

$$y' \nabla_{xx}^2 L(x^*, \lambda^*, \mu^*) y > 0,$$

for all $y \neq 0$ such that

$$\nabla h_i(x^*)' y = 0,\ \forall\, i = 1, \ldots, m, \qquad \nabla g_j(x^*)' y = 0,\ \forall\, j \in A(x^*). \tag{4.54}$$

Assume also that

$$\mu_j^* > 0, \qquad \forall\, j \in A(x^*). \tag{4.55}$$

Then x^* is a strict local minimum of f subject to

$$h(x) = 0 \qquad g(x) \leq 0.$$

Proof: (*Abbreviated*) It is straightforward to verify using our assumptions, that for the equivalent equality constrained problem (4.50), the second order sufficiency conditions are satisfied by $(x^*, z_1^*, \ldots, z_r^*)$, λ^*, μ^*, where

$$z_j^* = \left(-g_j(x^*)\right)^{1/2}.$$

Hence $(x^*, z_1^*, \ldots, z_r^*)$ is a strict local minimum for problem (4.50) and the result follows. **Q.E.D.**

The condition (4.55) is known as the *strict complementary slackness condition*. Exercise 4.3.7 provides a sharper version of the preceding sufficiency theorem, where the strict complementary slackness condition is somewhat weakened. However, this assumption cannot be entirely discarded as the following example shows.

Example 4.3.2

Consider the two-dimensional problem

$$\text{minimize } \tfrac{1}{2}(x_1^2 - x_2^2)$$
$$\text{subject to } x_2 \leq 0.$$

It can be verified that

$$x^* = (0,0), \qquad \mu^* = 0,$$

satisfy all the conditions of Prop. 4.3.2 except for the strict complementary slackness condition. It is seen that x^* is not a local minimum, because the cost can be decreased by moving from x^* along the feasible direction $(0, -1)$. This direction does not satisfy the condition (4.54). The difficulty here is that the constraint is active but *degenerate*, in the sense that its Lagrange multiplier is zero.

Using the transformation to an equality constrained problem, together with the corresponding sensitivity result of Prop. 4.2.2, we can also obtain the following proposition. The proof is left for the reader.

Proposition 4.3.3: (Sensitivity) Let x^* and λ^* be a local minimum and Lagrange multiplier, respectively, satisfying the second order sufficiency conditions of Prop. 4.3.2, and assume that x^* is a regular point. Consider the family of problems

$$\text{minimize } f(x)$$
$$\text{subject to } h(x) = u, \qquad (4.56)$$
$$g(x) \leq v,$$

parameterized by the vectors $u \in \Re^m$ and $v \in \Re^r$. Then there exists an open sphere S centered at $(u,v) = (0,0)$ such that for every $(u,v) \in S$ there is an $x(u,v) \in \Re^n$ and $\lambda(u,v) \in \Re^m$, $\mu(u,v) \in \Re^r$, which are a local minimum and associated Lagrange multiplier vectors of problem (4.56). Furthermore, $x(\cdot)$, $\lambda(\cdot)$, and $\mu(\cdot)$ are continuously differentiable in S and we have

$$x(0,0) = x^*, \qquad \lambda(0,0) = \lambda^*, \qquad \mu(0,0) = \mu^*.$$

In addition, for all $(u,v) \in S$, there holds

Sec. 4.3   Inequality Constraints

$$\nabla_u p(u,v) = -\lambda(u,v),$$
$$\nabla_v p(u,v) = -\mu(u,v),$$

where $p(u,v)$ is the optimal cost parameterized by (u,v),

$$p(u,v) = f\big(x(u,v)\big).$$

Sufficient Conditions and Lagrangian Minimization

The sufficient conditions that we have discussed so far involve second derivatives and Hessian positive definiteness assumptions. Our experience with unconstrained problems suggests that the first order Lagrange multiplier conditions together with convexity assumptions should also be sufficient for optimality. Indeed this is so, as we will demonstrate shortly. In fact we will not need to impose convexity or even differentiability assumptions explicitly. It turns out that a Lagrangian global minimization condition is sufficient. However, to verify this condition, one typically needs convexity assumptions in order to guarantee that candidate solutions x^* minimize globally the Lagrangian function.

Proposition 4.3.4: (General Sufficiency Condition) Consider the problem

$$\text{minimize} \quad f(x)$$
$$\text{subject to} \quad h_1(x) = 0, \ldots, h_m(x) = 0,$$
$$g_1(x) \leq 0, \ldots, g_r(x) \leq 0, \quad x \in X,$$

where f, h_i, and g_j are real-valued functions on $\Re^n$, and X is a given subset of $\Re^n$. Let x^* be a feasible vector which together with vectors $\lambda^* = (\lambda_1^*, \ldots, \lambda_m^*)$ and $\mu^* = (\mu_1^*, \ldots, \mu_r^*)$, satisfies

$$\mu_j^* \geq 0, \qquad j = 1, \ldots, r,$$
$$\mu_j^* = 0, \qquad \forall\, j \notin A(x^*),$$

and minimizes the Lagrangian function $L(x, \lambda^*, \mu^*)$ over $x \in X$:

$$x^* \in \arg\min_{x \in X} L(x, \lambda^*, \mu^*).$$

Then x^* is a global minimum of the problem.

Proof: The result is obtained from the relations:

$$f(x^*) = f(x^*) + \lambda^{*\prime} h(x^*) + \mu^{*\prime} g(x^*)$$
$$= \min_{x \in X} \{ f(x) + \lambda^{*\prime} h(x) + \mu^{*\prime} g(x) \}$$
$$\leq \inf_{x \in X, h(x)=0, g(x) \leq 0} \{ f(x) + \lambda^{*\prime} h(x) + \mu^{*\prime} g(x) \}$$
$$\leq \inf_{x \in X, h(x)=0, g(x) \leq 0} f(x),$$

where the first equality follows from the hypothesis, which implies that $h(x^*) = 0$ and $\mu^{*\prime} g(x^*) = 0$, and the last inequality follows from the nonnegativity of μ^*. **Q.E.D.**

In the case where f and g_j are differentiable convex functions, and h_i are linear, the Lagrangian function $L(x, \lambda^*, \mu^*)$ is convex as a function of x. It follows that for $X = \Re^n$, the Lagrangian minimization condition of the preceding proposition is equivalent to the first order condition

$$\nabla f(x^*) + \sum_{i=1}^{m} \lambda_i^* \nabla h_i(x^*) + \sum_{j=1}^{r} \mu_j^* \nabla g_j(x^*) = 0.$$

Thus, *in the presence of this convexity assumption, the first order optimality conditions are also sufficient.*

Note, however, that the proposition provides a more general sufficient condition because it does not require differentiability and convexity of f and g_j, and linearity of h_i, while X may be a strict subset of $\Re^n$.

4.3.3 Fritz John Optimality Conditions

We will now develop a generalization of the Karush-Kuhn-Tucker necessary conditions for problem (ICP), which does not require a regularity assumption. With the help of this generalization, we will derive all the necessary conditions obtained so far in this chapter, plus quite a few new ones.

To get a sense of the main idea, let us consider the case of the equality constrained problem

$$\text{minimize} \quad f(x)$$
$$\text{subject to} \quad h_1(x) = 0, \ldots, h_m(x) = 0.$$

We claim that an equivalent statement of the Lagrange multiplier theorem of Prop. 4.1.1 is that at a local minimum x^* (regular or not), there exist scalars $\mu_0^*, \lambda_1^*, \ldots, \lambda_m^*$, not all equal to 0, such that $\mu_0^* \geq 0$ and

$$\mu_0^* \nabla f(x^*) + \sum_{i=1}^{m} \lambda_i^* \nabla h_i(x^*) = 0. \tag{4.57}$$

Sec. 4.3   Inequality Constraints   387

To see the equivalence, note that there are two possibilities:

(a) x^* is regular, in which case Eq. (4.57) is satisfied with $\mu_0^* = 1$.

(b) x^* is not regular, in which case the gradients $\nabla h_i(x^*)$ are linearly dependent, and there exist $\lambda_1^*, \ldots, \lambda_m^*$, not all equal to 0, such that

$$\sum_{i=1}^{m} \lambda_i^* \nabla h_i(x^*) = 0,$$

so that Eq. (4.57) is satisfied with $\mu_0^* = 0$.

Necessary conditions that involve a nonnegative multiplier μ_0^* for the cost gradient, such as the one of Eq. (4.57), are known as *Fritz John necessary conditions*, and were first proposed in 1948 by John [Joh48]. We will first develop such conditions for the equality and inequality constrained problem (ICP). We will also consider some other related problems, in which case we will clearly indicate so. A simple adaptation of the penalty-based proof of the Lagrange multiplier theorem (cf. Section 4.1.1, and Props. 4.1.1, 4.3.1) yields the following proposition [a slightly stronger version is given in [BNO03] (Prop. 5.2.1)].

Proposition 4.3.5: (Fritz John Necessary Conditions) Let x^* be a local minimum of problem (ICP). Then there exists a scalar μ_0^* and multipliers $\lambda_1^*, \ldots, \lambda_m^*$ and $\mu_1^*, \ldots, \mu_r^*$, satisfying the following conditions:

(i) We have

$$\mu_0^* \nabla f(x^*) + \sum_{i=1}^{m} \lambda_i^* \nabla h_i(x^*) + \sum_{j=1}^{r} \mu_j^* \nabla g_j(x^*) = 0.$$

(ii) $\mu_j^* \geq 0$ for all $j = 0, 1, \ldots, r$.

(iii) $\mu_0^*, \lambda_1^*, \ldots, \lambda_m^*, \mu_1^*, \ldots, \mu_r^*$ are not all equal to 0.

(iv) In every neighborhood N of x^* there is an $x \in N$ such that $\lambda_i^* h_i(x) > 0$ for all i with $\lambda_i^* \neq 0$ and $\mu_j^* g_j(x) > 0$ for all j with $\mu_j^* \neq 0$. Moreover, if $(\lambda^*, \mu^*) \neq (0,0)$ this x can be chosen so that $f(x) < f(x^*)$.

Proof: We follow the line of argument of the penalty approach used in Section 4.1.1 and in Prop. 4.3.1. Consider the functions

$$g_j^+(x) = \max\{0, g_j(x)\}, \qquad j = 1, \ldots, r,$$

and for each $k = 1, 2, \ldots$, the "penalized" problem

$$\text{minimize } F^k(x) \equiv f(x) + \frac{k}{2}\sum_{i=1}^{m}\bigl(h_i(x)\bigr)^2 + \frac{k}{2}\sum_{j=1}^{r}\bigl(g_j^+(x)\bigr)^2 + \tfrac{1}{2}\|x - x^*\|^2$$

subject to $x \in S$,

where $S = \{x \mid \|x - x^*\| \leq \epsilon\}$, and $\epsilon > 0$ is such that $f(x^*) \leq f(x)$ for all feasible x with $x \in S$. The function $\bigl(g_j^+(x)\bigr)^2$ is continuously differentiable with gradient $2g_j^+(x)\nabla g_j(x)$, and if x^k minimizes $F^k(x)$ over S, an argument similar to the one used for the equality constraint case in Section 4.1.1 shows that $x^k \to x^*$ and that x^k is an interior point of S for all k greater than some $\bar{k}$. For such k, we have $\nabla F^k(x^k) = 0$, or

$$\nabla f(x^k) + \sum_{i=1}^{m}\xi_i^k \nabla h_i(x^k) + \sum_{j=1}^{r}\zeta_j^k \nabla g_j(x^k) + (x^k - x^*) = 0, \qquad (4.58)$$

where

$$\xi_i^k = k h_i(x^k), \qquad \zeta_j^k = k g_j^+(x^k). \qquad (4.59)$$

Denote,

$$\delta^k = \sqrt{1 + \sum_{i=1}^{m}(\xi_i^k)^2 + \sum_{j=1}^{r}(\zeta_j^k)^2}, \qquad (4.60)$$

$$\mu_0^k = \frac{1}{\delta^k}, \qquad \lambda_i^k = \frac{\xi_i^k}{\delta^k}, \quad i=1,\ldots,m, \qquad \mu_j^k = \frac{\zeta_j^k}{\delta^k}, \quad j=1,\ldots,r. \tag{4.61}$$

Then by dividing Eq. (4.58) with δ^k, we obtain

$$\mu_0^k \nabla f(x^k) + \sum_{i=1}^{m}\lambda_i^k \nabla h_i(x^k) + \sum_{j=1}^{r}\mu_j^k \nabla g_j(x^k) + \frac{1}{\delta^k}(x^k - x^*) = 0. \qquad (4.62)$$

Since by construction we have

$$(\mu_0^k)^2 + \sum_{i=1}^{m}(\lambda_i^k)^2 + \sum_{j=1}^{r}(\mu_j^k)^2 = 1, \qquad (4.63)$$

the sequence $\{\mu_0^k, \lambda_1^k, \ldots, \lambda_m^k, \mu_1^k, \ldots, \mu_r^k\}$ is bounded and must contain a subsequence that converges to some limit $\{\mu_0^*, \lambda_1^*, \ldots, \lambda_m^*, \mu_1^*, \ldots, \mu_r^*\}$. From Eq. (4.62), we see that this limit must satisfy condition (i), from Eqs. (4.59) and (4.61), it must satisfy condition (ii), and from Eq. (4.63), it must satisfy condition (iii). Finally, to show that condition (iv) is satisfied, let

$$I = \{i \mid \lambda_i^* \neq 0\}, \qquad J = \{j \mid \mu_j^* > 0\}.$$

Sec. 4.3   Inequality Constraints

Then, for all sufficiently large k within the index set of the convergent subsequence, we must have $\lambda_i^* \lambda_i^k > 0$ for all $i \in I$ and $\mu_j^* \mu_j^k > 0$ for all $j \in J$. Therefore, for these k, from Eqs. (4.59) and (4.61), we must have

$$\lambda_i^* h_i(x^k) > 0 \quad \forall\, i \in I, \qquad \mu_j^* g_j(x^k) > 0, \quad \forall\, j \in J.$$

Moreover, from the definition of x^k, we have $F^k(x^k) \leq F^k(x^*)$, or equivalently,

$$f(x^k) + \frac{k}{2}\sum_{i=1}^{m}(h_i(x^k))^2 + \frac{k}{2}\sum_{j=1}^{r}(g_j^+(x^k))^2 + \tfrac{1}{2}\|x^k - x^*\|^2 \leq f(x^*),$$

so if $I \cup J \neq \emptyset$, we have $f(x^k) < f(x^*)$. Since every neighborhood of x^* must contain some x^k from the subsequence, this proves condition (iv). **Q.E.D.**

Example 4.3.3

Consider the problem

$$\begin{aligned}
\text{minimize} \quad & f(x) = x_1 + x_2 \\
\text{subject to} \quad & g_1(x) = (x_1 - 1)^2 + x_2^2 - 1 \leq 0, \\
& g_2(x) = -(x_1 - 2)^2 - x_2^2 + 4 \leq 0.
\end{aligned}$$

This is an inequality constrained version of Example 4.1.1, whose geometry is illustrated in Fig. 4.1.2. It can be seen from this figure that the only feasible solution is $x^* = (0,0)$, which is thus the global minimum. The condition

$$\mu_0^* \nabla f(x^*) + \mu_1^* \nabla g_1(x^*) + \mu_2^* \nabla g_2(x^*) = 0$$

can be written as

$$\mu_0^* - 2\mu_1^* + 4\mu_2^* = 0, \qquad \mu_0^* = 0.$$

Thus, consistently with Example 4.1.1, where we found that no Lagrange multiplier exists, we have $\mu_0^* = 0$. It can be seen that any positive μ_1^* and μ_2^* with $\mu_1^* = 2\mu_2^*$ satisfy the Fritz John necessary conditions (i)-(iii), and from the geometry of the problem as shown in Fig. 4.1.2, it can be verified that condition (iv) is satisfied as well.

Statements of the Fritz John necessary conditions usually do not include condition (iv), which turns out to be important in the subsequent development. We call this condition the Fritz John *complementary violation* condition [CV for short, in analogy to the complementary slackness condition, $\mu_j^* g_j(x^*) = 0$ for all j, which is referred to as CS for short]. We

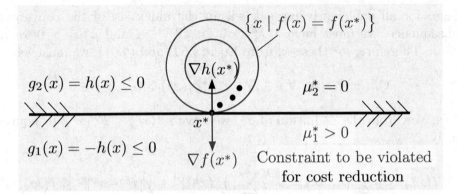

Figure 4.3.2. Illustration of the CV condition for a two-dimensional problem with a single linear constraint $h(x) = 0$, which is converted to the two inequality constraints $g_1(x) = -h(x) \leq 0$ and $g_2(x) = h(x) \leq 0$. While CS requires that $\mu_1^* \geq 0$ and $\mu_2^* \geq 0$, CV requires that $\mu_1^* > 0$ and $\mu_2^* = 0$ (we cannot violate simultaneously both constraints). Thus, while there are infinitely many multipliers satisfying CS, there is only one satisfying CV. This multiplier, through its sign pattern indicates that the constraint $-h(x) \leq 0$ should be violated for cost reduction. Moreover, its existence implies that we can take $\mu_0^* > 0$ in the Fritz John conditions and deduce the existence of a Lagrange multiplier: $\nabla f(x^*) + \lambda^* \nabla h(x^*) = 0$ if $\nabla h(x^*) \neq 0$ (i.e., x^* is regular).

note that the CV condition is stronger than (i.e., it implies) the CS condition. The reason is that if $\mu_j^* > 0$, by the CV condition we have $g_j(x) > 0$ for a vector x arbitrarily close to x^*, implying that $g_j(x^*) = 0$. In words, the CV condition asserts that *the constraints with nonzero multipliers can be simultaneously violated arbitrarily close to x^* in a way that the cost function is reduced while the constraint violations have the same signs as the corresponding multipliers*. This property is not shared by all the Lagrange multipliers corresponding to x^* (see Fig. 4.3.2). However, we will see later (see the subsequent Prop. 4.3.19) that if the set of Lagrange multipliers is nonempty, the Lagrange multiplier of minimum norm has this property.

The Importance of the CV Condition

To illustrate the significance of the CV condition in asserting the existence of Lagrange multipliers, suppose that we convert a problem with a single equality constraint, $\min_{h(x)=0} f(x)$, to the inequality constrained problem

$$\text{minimize } f(x)$$
$$\text{subject to } -h(x) \leq 0, \quad h(x) \leq 0.$$

We will argue that the CV condition (iv) is essential to obtain the Karush-Kuhn-Tucker conditions from the Fritz John conditions. Indeed the latter conditions assert the existence of nonnegative $\mu_0^*, \mu_1^*, \mu_2^*$, not all zero, such

Sec. 4.3 Inequality Constraints   391

that
$$\mu_0^* \nabla f(x^*) - \mu_1^* \nabla h(x^*) + \mu_2^* \nabla h(x^*) = 0. \tag{4.64}$$

The candidate multipliers that satisfy the above condition as well as the CS condition
$$-\mu_1^* h(x^*) = \mu_2^* h(x^*) = 0,$$

include those of the form
$$\mu_0^* = 0, \qquad \mu_1^* = \mu_2^* > 0, \tag{4.65}$$

so we cannot assert that $\mu_0^* > 0$. However, these multipliers fail the stronger CV condition (iv) of Prop. 4.3.5, which requires that if $\mu_0^* = 0$, we must have either $\mu_1^* \neq 0$ and $\mu_2^* = 0$ or $\mu_1^* = 0$ and $\mu_2^* \neq 0$ [since the constraints $-h(x) \leq 0$ and $h(x) \leq 0$ cannot be simultaneously violated; see Fig. 4.3.2]. Thus the candidate multipliers (4.65) are rejected by Prop. 4.3.5, and assuming that $\nabla h(x^*) \neq 0$ (i.e., x^* is regular), it follows from Eq. (4.64) that we cannot have $\mu_0^* = 0$. We can therefore divide Eq. (4.64) by μ_0^*, and obtain the familiar first order condition
$$\nabla f(x^*) + \lambda^* \nabla h(x^*) = 0$$

with
$$\lambda^* = \frac{\mu_2^* - \mu_1^*}{\mu_0^*}.$$

This deduction would not have been possible without the CV condition (iv), and explains why *with just the CS condition, the Fritz John conditions are analytically inadequate for the development of Lagrange multiplier theorems in some important settings* such as Prop. 4.3.1 (see also the following argument).

From the Fritz John to the Karush-Kuhn-Tucker Conditions

In our subsequent analysis we will consider a variety of assumptions and repeatedly apply the Fritz John conditions with a standard argument: in each case, we will verify that it is impossible to satisfy these conditions with $\mu_0^* = 0$, which in turn implies that $\mu_0^* > 0$. Then, by dividing all λ_i^* and μ_j^* with μ_0^*, we will recover the Karush-Kuhn-Tucker conditions with Lagrange multipliers equal to λ_i^*/μ_0^* and μ_j^*/μ_0^*. In fact, since the multipliers can be normalized by division with a positive scalar, it is sufficient to consider just two possibilities for μ_0^*: either $\mu_0^* = 0$ or $\mu_0^* = 1$.

As an example, suppose that x^* is regular as per the assumption of Prop. 4.3.1. This implies that we cannot have
$$\sum_{i=1}^m \lambda_i^* \nabla h_i(x^*) + \sum_{j=1}^r \mu_j^* \nabla g_j(x^*) = 0$$

for some λ_i^*, $i = 1, \ldots, m$, and μ_j^*, $j = 1, \ldots, r$, that are not all zero, so the Fritz John conditions cannot be satisfied with $\mu_0^* = 0$. Hence, we can take $\mu_0^* = 1$ and there exist λ_i^*, $i = 1, \ldots, m$, and μ_j^*, $j = 1, \ldots, r$, such that

$$\nabla f(x^*) + \sum_{i=1}^m \lambda_i^* \nabla h_i(x^*) + \sum_{j=1}^r \mu_j^* \nabla g_j(x^*) = 0, \qquad \mu_j^* \geq 0, \quad j = 1, \ldots, r.$$

Furthermore, in every neighborhood N of x^* there is an $x \in N$ such that $\lambda_i^* h_i(x) > 0$ for all i with $\lambda_i^* \neq 0$ and $\mu_j^* g_j(x) > 0$ for all j with $\mu_j^* \neq 0$. We have thus shown the following proposition, which represents a slightly more powerful form of the Karush-Kuhn-Tucker conditions of Prop. 4.3.1 [the CV condition (iv) of Prop. 4.3.5 has replaced the weaker CS condition, $\mu_j^* = 0$ if $j \notin A(x^*)$, of Prop. 4.3.1].

Proposition 4.3.6 (Strengthened Karush-Kuhn-Tucker Conditions) Let x^* be a local minimum of problem (ICP), and assume that x^* is regular. Then there exist Lagrange multipliers $\lambda_1^*, \ldots, \lambda_m^*$ and $\mu_1^*, \ldots, \mu_r^*$, satisfying the following conditions:

(i) We have

$$\nabla f(x^*) + \sum_{i=1}^m \lambda_i^* \nabla h_i(x^*) + \sum_{j=1}^r \mu_j^* \nabla g_j(x^*) = 0.$$

(ii) $\mu_j^* \geq 0$ for all $j = 1, \ldots, r$.

(iii) In every neighborhood N of x^* there is an $x \in N$ such that $\lambda_i^* h_i(x) > 0$ for all i with $\lambda_i^* \neq 0$ and $\mu_j^* g_j(x) > 0$ for all j with $\mu_j^* \neq 0$. Moreover, if $(\lambda^*, \mu^*) \neq (0,0)$ this x can be chosen so that $f(x) < f(x^*)$.

4.3.4 Constraint Qualifications and Pseudonormality

In many applications the regularity assumption of Prop. 4.3.6 is too strong. In this section we will use the Fritz John conditions of Prop. 4.3.5 to provide alternative assumptions that guarantee the existence of Lagrange multipliers. There are several assumptions of this type, and they are collectively known as *constraint qualifications*. The constraint qualifications developed in this section will all be shown to be special cases of a unifying constraint qualification, called *pseudonormality*.

Let us first consider in the following proposition the case where there are *linear* equality and *concave* (or, as a special case, linear) inequality constraints. The proof of the proposition relies strongly on the CV condition (iv) of Prop. 4.3.5, the key fact being that if at least one λ_i^* or μ_j^* is nonzero, there exists an x arbitrarily close to x^* such that

$$\sum_{i=1}^{m} \lambda_i^* h_i(x) + \sum_{j=1}^{r} \mu_j^* g_j(x) > 0.$$

We will see that this property forms the core of the notion of pseudonormality.

Proposition 4.3.7: (Linear/Concave Constraints) Let x^* be a local minimum of problem (ICP). Assume that the functions h_i are linear and the functions g_j are concave. Then x^* satisfies the necessary conditions of Prop. 4.3.6.

Proof: We consider a set of λ_i^*, $i = 1, \ldots, m$, and μ_j^*, $j = 0, 1, \ldots, r$, satisfying the Fritz John conditions, and we assume that $\mu_0^* = 0$. We will come to a contradiction, implying that we can take $\mu_0^* = 1$, thereby proving the proposition. Indeed, for any $x \in \Re^n$, we have, by the linearity of h_i and the concavity of g_j,

$$h_i(x) = h_i(x^*) + \nabla h_i(x^*)'(x - x^*), \qquad i = 1, \ldots, m,$$

$$g_j(x) \leq g_j(x^*) + \nabla g_j(x^*)'(x - x^*), \qquad j = 1, \ldots, r.$$

By multiplying these two relations with λ_i^* and μ_j^*, and by adding over i and j, respectively, we obtain

$$\sum_{i=1}^{m} \lambda_i^* h_i(x) + \sum_{j=1}^{r} \mu_j^* g_j(x) \leq \sum_{i=1}^{m} \lambda_i^* h_i(x^*) + \sum_{j=1}^{r} \mu_j^* g_j(x^*)$$

$$+ \left(\sum_{i=1}^{m} \lambda_i^* \nabla h_i(x^*) + \sum_{j=1}^{r} \mu_j^* \nabla g_j(x^*) \right)' (x - x^*)$$

$$= 0,$$

(4.66)

where the last equality holds because we have $\lambda_i^* h_i(x^*) = 0$ for all i and $\mu_j^* g_j(x^*) = 0$ for all j [by the Fritz John condition (iv)], and

$$\sum_{i=1}^{m} \lambda_i^* \nabla h_i(x^*) + \sum_{j=1}^{r} \mu_j^* \nabla g_j(x^*) = 0$$

[by the Fritz John condition (i)]. On the other hand, by the Fritz John condition (iii), we know that since $\mu_0^* = 0$, there exists some i for which

$\lambda_i^* \neq 0$ or some j for which $\mu_j^* > 0$. By the Fritz John condition (iv), there is an x satisfying $\lambda_i^* h_i(x) > 0$ for all i with $\lambda_i^* \neq 0$ and $\mu_j^* g_j(x) > 0$ for all j with $\mu_j^* > 0$. For this x, we have $\sum_{i=1}^m \lambda_i^* h_i(x) + \sum_{j=1}^r \mu_j^* g_j(x) > 0$, contradicting Eq. (4.66). **Q.E.D.**

The following proposition gives a classical result, which was shown in the paper [AHU61] for the case of inequality constraints only and was generalized to the case where there are equality constraints as well in [MaF67]. (An alternative proof is given in Exercise 4.3.5 for the case where there are no equality constraints; see also Exercise 4.3.18.)

Proposition 4.3.8: (Linear Independence/Interior Point Constraint Qualification) Let x^* be a local minimum of problem (ICP). Assume that the gradients $\nabla h_i(x^*)$, $i = 1, \ldots, m$, are linearly independent, and that there exists a vector d such that

$$\nabla h_i(x^*)'d = 0, \quad \forall\, i = 1, \ldots, m, \qquad \nabla g_j(x^*)'d < 0, \quad \forall\, j \in A(x^*).$$

Then x^* satisfies the necessary conditions of Prop. 4.3.6.

Proof: We consider a set of multipliers λ_i^*, $i = 1, \ldots, m$, μ_j^*, $j = 0, 1, \ldots, r$, satisfying the Fritz John conditions, and we assume that $\mu_0^* = 0$, so that

$$\sum_{i=1}^m \lambda_i^* \nabla h_i(x^*) + \sum_{j=1}^r \mu_j^* \nabla g_j(x^*) = 0. \qquad (4.67)$$

Since not all the λ_i^* and μ_j^* can be zero, we conclude that $\mu_j^* > 0$ for at least one $j \in A(x^*)$; otherwise we would have $\sum_{i=1}^m \lambda_i^* \nabla h_i(x^*) = 0$ with not all the λ_i^* equal to zero, and the linear independence of the gradients $\nabla h_i(x^*)$ would be violated. Since $\mu_j^* \geq 0$ for all j, with $\mu_j^* = 0$ for $j \notin A(x^*)$ and $\mu_j^* > 0$ for at least one j, we obtain

$$\sum_{i=1}^m \lambda_i^* \nabla h_i(x^*)'d + \sum_{j=1}^r \mu_j^* \nabla g_j(x^*)'d < 0,$$

where d is the vector of the hypothesis. This contradicts Eq. (4.67). Therefore, we must have $\mu_0^* > 0$, so that we can take $\mu_0^* = 1$. **Q.E.D.**

Note that the existence of a vector d satisfying the condition of Prop. 4.3.8 is equivalent to the sometimes more easily verifiable hypothesis that for each $j \in A(x^*)$, there exists a vector d^j such that

$$\nabla h_i(x^*)'d^j = 0, \quad \forall\, i = 1, \ldots, m,$$

Sec. 4.3 Inequality Constraints 395

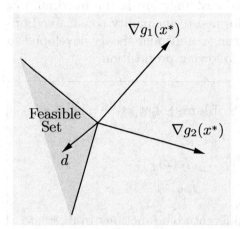

Figure 4.3.3. Illustration of the linear independence/interior point constraint qualification. There exists a vector d such that

$$\nabla g_j(x^*)'d < 0$$

for all $j \in A(x^*)$. As a result, if at least one of the μ_j^* is positive, we cannot have

$$\sum_{j=1}^{r} \mu_j^* \nabla g_j(x^*) = 0.$$

$$\nabla g_j(x^*)'d^j < 0, \qquad \nabla g_{\bar{j}}(x^*)'d^j \leq 0, \quad \forall\, \bar{j} \in A(x^*),\ \bar{j} \neq j.$$

Indeed, the vector $d = \sum_{j \in A(x^*)} d^j$ satisfies the condition of Prop. 4.3.8. Figure 4.3.3 illustrates the character of the linear independence/interior point constraint qualification.

The following proposition, due to Slater [Sla50], is really a special case of the preceding one, and was one of the first shown for inequality constraints.

Proposition 4.3.9: (Slater Constraint Qualification for Convex Inequalities) Let x^* be a local minimum of problem (ICP). Assume that the functions h_i are linear, the functions g_j are convex, and there exists a feasible vector $\bar{x}$ satisfying

$$g_j(\bar{x}) < 0, \qquad \forall\, j \in A(x^*).$$

Then x^* satisfies the necessary conditions of Prop. 4.3.6.

Proof: In view of the linearity of h_i, we can assume with no loss of generality that the gradients $\nabla h_i(x^*)$ are linearly independent; otherwise we can eliminate a sufficient number of equalities, while assigning to them zero Lagrange multipliers, to obtain a set of linearly independent equality constraints. Using the convexity of g_j,

$$0 > g_j(\bar{x}) \geq g_j(x^*) + \nabla g_j(x^*)'(\bar{x} - x^*) = \nabla g_j(x^*)'(\bar{x} - x^*), \qquad \forall\, j \in A(x^*),$$

and by the feasibility of $\bar{x}$ and the linearity of h_i,

$$0 = \nabla h_i(x^*)'(\bar{x} - x^*), \qquad i = 1, \ldots, m.$$

Now let $d = \bar{x} - x^*$ and apply the result of Prop. 4.3.8. **Q.E.D.**

Minimization problems where the cost function is the maximum of several functions can be readily transformed to inequality constrained optimization problems for which the Lagrange multiplier theory developed so far applies. In this way we obtain the following proposition.

Proposition 4.3.10: (Minimax Problems) Let x^* be a local minimum of the problem

$$\text{minimize } \max\{f_1(x), \ldots, f_p(x)\}$$
$$\text{subject to } h_1(x) = 0, \ldots, h_m(x) = 0,$$

where f_j and h_i are continuously differentiable functions from $\Re^n$ to $\Re$. Assume that the gradients $\nabla h_i(x^*)$, $i = 1, \ldots, m$, are linearly independent. Then there exists a vector $\lambda^* = (\lambda_1^*, \ldots, \lambda_m^*)$ and a vector $\mu^* = (\mu_1^*, \ldots, \mu_p^*)$ such that

(i)
$$\sum_{j=1}^{p} \mu_j^* \nabla f_j(x^*) + \sum_{i=1}^{m} \lambda_i^* \nabla h_i(x^*) = 0.$$

(ii)
$$\mu^* \geq 0, \qquad \sum_{j=1}^{p} \mu_j^* = 1.$$

(iii) For all $j = 1, \ldots, p$, if $\mu_j^* > 0$, then

$$f_j(x^*) = \max\{f_1(x^*), \ldots, f_p(x^*)\}.$$

(iv) In every neighborhood N of x^*, there is an $x \in N$ such that $\lambda_i^* h_i(x) > 0$ for all i with $\lambda_i^* \neq 0$.

Proof: We introduce an additional scalar variable z, and the problem

$$\begin{aligned}
&\text{minimize } z \\
&\text{subject to } h_1(x) = 0, \ldots, h_m(x) = 0, \qquad (4.68)\\
&\qquad\qquad f_1(x) \leq z, \ldots, f_p(x) \leq z.
\end{aligned}$$

Let

$$z^* = \max\{f_1(x^*), \ldots, f_p(x^*)\}.$$

It can be seen that (x^*, z^*) is a local minimum of the above problem, and satisfies the linear independence/interior point constraint qualification of

Prop. 4.3.8 with $d = (0, 1)$. The conclusion of that proposition, as applied to this problem, implies the desired assertions. **Q.E.D.**

The preceding proposition can be easily extended to the case where there are additional inequality constraints $g_j(x) \leq 0$, satisfying $\nabla g_j(x^*)'d < 0$ for some d (cf. Prop. 4.3.8). By adding these constraints to the ones of problem (4.68), and by applying Prop. 4.3.8, we obtain a corresponding Lagrange multiplier theorem.

Pseudonormality

We now introduce a characteristic property of the constraint set

$$\{x \mid h_1(x) = 0, \ldots, h_m(x) = 0, g_1(x) \leq 0, \ldots, g_r(x) \leq 0\},$$

which unifies the constraint qualifications discussed so far.

We say that a feasible vector x^* is *pseudonormal* if there is no set of scalars $\lambda_1, \ldots, \lambda_m, \mu_1, \ldots, \mu_r$, such that:

(i) $\sum_{i=1}^{m} \lambda_i \nabla h_i(x^*) + \sum_{j=1}^{r} \mu_j \nabla g_j(x^*) = 0$.

(ii) $\mu_j \geq 0$, $j = 1, \ldots, r$.

(iii) $\lambda_1, \ldots, \lambda_m, \mu_1, \ldots, \mu_r$ are not all equal to 0.

(iv) In every neighborhood N of x^* there is an $x \in N$ such that

$$\sum_{i=1}^{m} \lambda_i h_i(x) + \sum_{j=1}^{r} \mu_j g_j(x) > 0.$$

Thus if x^* is a pseudonormal local minimum, the Fritz John conditions of Prop. 4.3.5 cannot be satisfied with $\mu_0^* = 0$, so that there exist Lagrange multipliers satisfying the corresponding necessary conditions.

Pseudonormality admits an intuitive geometric interpretation. In particular, consider the case where there are no inequality constraints. Then it can be seen that x^* is pseudonormal if and only if one of the following two conditions holds:

(1) The gradients $\nabla h_i(x^*)$, $i = 1, \ldots, m$, are linearly independent.

(2) For every nonzero $\lambda = (\lambda_1, \ldots, \lambda_m)$ such that

$$\sum_{i=1}^{m} \lambda_i \nabla h_i(x^*) = 0,$$

the hyperplane through the origin with normal λ contains all vectors $h(x)$ for x within some sphere centered at x^* (see Fig. 4.3.4).

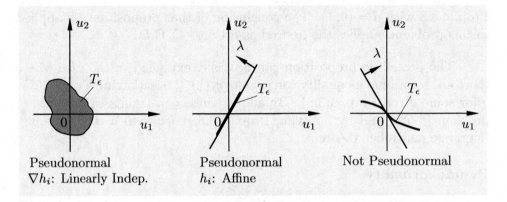

Figure 4.3.4. Geometrical interpretation of pseudonormality of x^* in the case where there are no inequality constraints. Consider the set

$$T_\epsilon = \{h(x) \mid \|x - x^*\| < \epsilon\}$$

for a small positive scalar ϵ. The vector x^* is pseudonormal if and only if either (1) the gradients $\nabla h_i(x^*)$, $i = 1, \ldots, m$, are linearly independent, or (2) for every nonzero λ satisfying

$$\sum_{i=1}^{m} \lambda_i \nabla h_i(x^*) = 0,$$

there is a small enough ϵ, such that the set T_ϵ lies in the hyperplane through the origin whose normal is λ.

Note that condition (2) above is satisfied in particular if the functions h_i are affine. To see this, note that if $h_i(x) = a_i'x - b_i$ and $\sum_{i=1}^{m} \lambda_i a_i = 0$, then we have $\lambda' h(x) = \sum_{i=1}^{m} \lambda_i a_i'(x - x^*) = 0$. It can also be argued that the only interesting special case where condition (2) is satisfied is when the functions h_i are affine (see Fig. 4.3.4). Thus, for the case of equality constraints only, pseudonormality is not much more general than the union of the two major constraint qualifications that guarantee that the constraint set admits Lagrange multipliers.

Figure 4.3.5 illustrates pseudonormality for the case where there are inequality constraints but no equality constraints. Clearly, from the definition of pseudonormality, we have that x^* is pseudonormal if the gradients $\nabla g_j(x^*)$, $j \in A(x^*)$, are linearly independent. It can also be seen from the figure that x^* is pseudonormal if the functions g_j, $j \in A(x^*)$, are concave.

We now note that pseudonormality is implicit in the proofs of Props. 4.3.6-4.3.9. Indeed by reviewing these proofs, it can be seen that x^* is pseudonormal under any one of the following conditions.

(a) x^* is a regular point.

Sec. 4.3 Inequality Constraints

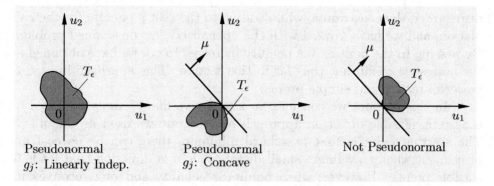

Pseudonormal
g_j: Linearly Indep.

Pseudonormal
g_j: Concave

Not Pseudonormal

Figure 4.3.5. Geometrical interpretation of pseudonormality of x^* in the case where all the constraints are inequalities that are active $[A(x^*) = \{1, \ldots, r\}]$. Consider the set

$$T_\epsilon = \big\{g(x) \mid \|x - x^*\| < \epsilon\big\}$$

for a small positive scalar ϵ. The vector x^* is pseudonormal if and only if either (1) the gradients $\nabla g_i(x^*)$, $j = 1, \ldots, r$, are linearly independent, or (2) for every nonzero $\mu \geq 0$ satisfying

$$\sum_{j=1}^{r} \mu_j \nabla g_j(x^*) = 0,$$

there is a small enough ϵ, such that the set T_ϵ does not cross into the positive open halfspace of the hyperplane through the origin whose normal is μ.

If the functions g_j are concave, the condition $\sum_{j=1}^{r} \mu_j \nabla g_j(x^*) = 0$ implies that x^* maximizes $\mu' g(x)$, so that

$$\mu' g(x) \leq \mu' g(x^*) = 0, \qquad \forall\, x \in \Re^n.$$

Therefore, as illustrated in the figure, the set T_ϵ does not cross into the positive open halfspace of the hyperplane through the origin whose normal is μ. This implies that x^* is pseudonormal.

(b) The constraint functions h_i are linear and the constraint functions g_j are concave.

(c) x^* satisfies the linear independence/interior point constraint qualification, as given in Prop. 4.3.8.

(d) x^* satisfies the Slater constraint qualification, as given in Prop. 4.3.9.

Thus pseudonormality unifies the constraint qualifications discussed so far, and embodies fundamental structure that when possessed by the constraint set, guarantees existence of Lagrange multipliers.

4.3.5 Abstract Set Constraints and the Tangent Cone

In our derivation of optimality conditions for inequality-constrained optimization problems, we have used a penalty function approach: we have disregarded the constraints, while adding to the cost a penalty for their violation, and we have worked with the "penalized" unconstrained problem. By passing to the limit as the penalty increases to ∞, we have obtained all the necessary conditions thus far in this section. This approach has led to powerful results and simple proofs.

In this section, we consider an alternative line of analysis, based on a descent/feasible direction approach like the one we used in Chapter 3. The starting point is that at a local minimum there can be no cost improvement when traveling a small distance along a direction that leads to feasible points. However, when nonlinear equality and/or nonconvex inequality constraints are involved, the notion of a feasible direction must be substantially modified before it can be used for analysis. This can be done through the notion of the *tangent cone* of the constraint at a feasible point x, a type of local approximation of the constraint set near x, which we will introduce shortly. The tangent cone can in turn be used to obtain optimality conditions as well as useful insights in many types of constrained problems, including some where the penalty approach does not work well.

Our aim in this section is to develop the tangent cone-based descent approach, and to use it to obtain some new results. These include necessary conditions for optimality in problems that involve an infinite number of inequality constraints.†

The Tangent Cone

Given a subset X of $\Re^n$ and a vector $x \in X$, a vector y is said to be a *tangent* of X at x if either $y = 0$ or there exists a sequence $\{x^k\} \subset X$ such that $x^k \neq x$ for all k and

$$x^k \to x, \qquad \frac{x^k - x}{\|x^k - x\|} \to \frac{y}{\|y\|}.$$

Thus a nonzero vector y is a tangent at x if it is possible to approach x with a sequence $\{x^k\} \subset X$ such that the normalized direction sequence

$$\frac{x^k - x}{\|x^k - x\|}$$

converges to $y/\|y\|$, the normalized direction of y.

The set of all tangents of X at x is easily seen to be a cone. It is called the *tangent cone* of X at x, and it is denoted by $T(x)$. Figure 4.3.6 illustrates the tangent cone with some examples. It can be shown that $T(x)$ is closed. In addition, if X is convex and does not consist of just the vector x, the cone $T(x)$ is equal to the closure of the set of feasible directions at x.

---

† The analysis in this section and the next is theoretically important, but will not be used further in the text. Thus it may be skipped at first reading.

Sec. 4.3 Inequality Constraints   401

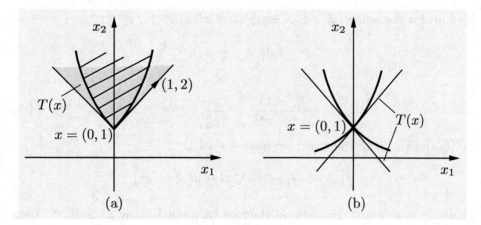

Figure 4.3.6. Examples of the tangent cone $T(x)$ of a set X at the vector $x = (0,1)$. In (a), we have

$$X = \left\{(x_1, x_2) \mid (x_1 + 1)^2 - x_2 \leq 0, \, (x_1 - 1)^2 - x_2 \leq 0\right\}.$$

Here X is convex and $T(x)$ is equal to the closure of the set of feasible directions at X. Note, however, that the vector $(1,2)$ belongs to $T(x)$ and also to the closure of the set of feasible directions, but is not a feasible direction. In (b), we have

$$X = \left\{(x_1, x_2) \mid \big((x_1 + 1)^2 - x_2\big)\big((x_1 - 1)^2 - x_2\big) = 0\right\}.$$

Here X is nonconvex, and $T(x)$ is closed but not convex.

A verification of these facts is given in Exercise 4.3.15; see also [BNO03], Prop. 4.6.2.

The importance of the tangent cone is due to the following proposition, which is a classical necessary optimality condition for the problem of minimizing a differentiable function f over a constraint set X.

Proposition 4.3.11: (Tangent Cone Condition for Optimality)
Let $f : \Re^n \mapsto \Re$ be a function and let X be a subset of $\Re^n$. Assume that f is continuously differentiable over an open set containing X, and let x^* be a local minimum of f over X. Then there is no descent direction within the tangent cone of X at x^*, i.e.,

$$\nabla f(x^*)'y \geq 0, \quad \forall \, y \in T(x^*). \tag{4.69}$$

Proof: Let y be a nonzero tangent of X at x^*. Then there exists a sequence

$\{\xi^k\}$ and a sequence $\{x^k\} \subset X$ such that $x^k \neq x^*$ for all k,

$$\xi^k \to 0, \qquad x^k \to x^*,$$

and

$$\frac{x^k - x^*}{\|x^k - x^*\|} = \frac{y}{\|y\|} + \xi^k. \tag{4.70}$$

By the mean value theorem, we have for all k

$$f(x^k) = f(x^*) + \nabla f(\tilde{x}^k)'(x^k - x^*),$$

where $\tilde{x}^k$ is a vector that lies on the line segment joining x^k and x^*. Using Eq. (4.70), the last relation can be written as

$$f(x^k) = f(x^*) + \frac{\|x^k - x^*\|}{\|y\|} \nabla f(\tilde{x}^k)' y^k, \tag{4.71}$$

where $y^k = y + \|y\|\xi^k$. If the tangent y is a descent direction of f at x^*, then $\nabla f(x^*)' y < 0$, and since $\tilde{x}^k \to x^*$ and $y^k \to y$, we obtain for all sufficiently large k, $\nabla f(\tilde{x}^k)' y^k < 0$ and [from Eq. (4.71)] $f(x^k) < f(x^*)$. This contradicts the local optimality of x^*. **Q.E.D.**

In many problems the condition (4.69) can be translated into more convenient necessary conditions. For example if X is a convex set, all the vectors of the form $x - x^*$, where $x \in X$, belong to $T(x^*)$. From Prop. 4.3.11 we thus obtain the familiar necessary condition

$$\nabla f(x^*)'(x - x^*) \geq 0, \qquad \forall \ x \in X,$$

which formed the basis for much of the analysis of Chapter 3.

Farkas' Lemma and Lagrange Multipliers

We will now use the tangent cone and the necessary condition of Prop. 4.3.11 as the basis for an alternative approach to Lagrange multipliers. We begin with Farkas' lemma, a basic fact relating to the geometry of polyhedral sets, which is useful in several constrained optimization contexts.

Farkas' lemma says that the cone C generated by r vectors $a_1, \ldots, a_r$,

$$C = \left\{ c \mid c = \sum_{j=1}^{r} \mu_j a_j, \ \mu_j \geq 0, \ j = 1, \ldots, r \right\},$$

consists of all vectors c satisfying

$$c'y \leq 0, \qquad \text{for all } y \text{ such that } a_j' y \leq 0, \ \forall \ j = 1, \ldots, r,$$

Sec. 4.3 Inequality Constraints

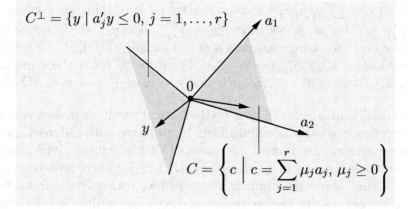

Figure 4.3.7. Geometric interpretation of Farkas' lemma. Consider the cone C generated by r vectors $a_1, \ldots, a_r$,

$$C = \left\{ c \;\middle|\; c = \sum_{j=1}^{r} \mu_j a_j,\; \mu_j \geq 0,\; j = 1, \ldots, r \right\},$$

and its polar cone $C^\perp = \{y \mid a_j' y \leq 0,\; j = 1, \ldots, r\}$. Farkas' lemma says that if c is such that

$$c'y \leq 0, \qquad \forall\, y \in C^\perp,$$

then c must belong to C (the reverse is also clearly true).

(see Fig. 4.3.7).

Farkas' lemma can be proved in several different ways. We give here a simple proof based on the Fritz John optimality conditions and the attendant Lagrange multiplier theorem for linear constraints (Prop. 4.3.7). In fact, this line of proof shows a slightly stronger version of Farkas' lemma, which includes a statement relating to the CV condition (this version is given in Exercise 4.3.22). For an alternative proof of Farkas' lemma, based on polyhedral convexity arguments, see Prop. 2.3.1 of [Ber09].

Proposition 4.3.12: (Farkas' Lemma) Let $a_1, \ldots, a_r$ be vectors in $\Re^n$. Then a vector c satisfies

$$c'y \leq 0, \qquad \text{for all } y \text{ such that } a_j' y \leq 0,\; \forall\, j = 1, \ldots, r,$$

if and only if there exist nonnegative scalars $\mu_1, \ldots, \mu_r$ such that

$$c = \mu_1 a_1 + \cdots + \mu_r a_r.$$

Proof: If $c = \sum_{j=1}^{r} \mu_j a_j$ for some scalars $\mu_j \geq 0$, then for every y satisfying $a_j' y \leq 0$ for all j, we have $c'y = \sum_{j=1}^{r} \mu_j a_j' y \leq 0$. Conversely, if c satisfies $c'y \leq 0$ for all y such that $a_j' y \leq 0$ for all j, then $y^* = 0$ minimizes $-c'y$ subject to $a_j' y \leq 0$, $j = 1, \ldots, r$. From Prop. 4.3.7, there must exist nonnegative scalars $\mu_1, \ldots, \mu_r$ such that $c = \mu_1 a_1 + \cdots + \mu_r a_r$. **Q.E.D.**

Farkas' lemma, viewed as a mathematical result, is in fact equivalent to the existence of Lagrange multipliers in problems with differentiable cost and linear constraints. Indeed, the proof of Farkas' lemma given above was based on the existence of Lagrange multipliers for such problems, and it turns out that a reverse argument is possible, i.e., starting from Farkas' lemma, we can assert the existence of Lagrange multipliers for linearly constrained problems. This is illustrated in Fig. 4.3.8 for the case where there are inequality constraints only (linear equality constraints can be handled by conversion into two linear inequality constraints). In fact, in several textbooks, the construction given in this figure is the basis for proofs of Lagrange multiplier theorems, starting from Farkas' lemma.

To formalize the use of Farkas' lemma just discussed, let us consider the equality and inequality constrained problem (ICP). We introduce the *cone of first order feasible variations* at x

$$V(x) = \{y \mid \nabla h_i(x)'y = 0, \ i = 1, \ldots, m, \ \nabla g_j(x)'y \leq 0, \ j \in A(x)\}.$$

We have the following proposition.

Proposition 4.3.13: (Necessary and Sufficient Condition for Existence of Lagrange Multipliers) Let x^* be a local minimum of problem (ICP). Then there exist Lagrange multipliers $\lambda_1^*, \ldots, \lambda_m^*$ and $\mu_1^*, \ldots, \mu_r^*$, satisfying

$$\nabla f(x^*) + \sum_{i=1}^{m} \lambda_i^* \nabla h_i(x^*) + \sum_{j=1}^{r} \mu_j^* \nabla g_j(x^*) = 0,$$

$$\mu_j^* \geq 0, \quad j = 0, 1, \ldots, r, \qquad \mu_j^* = 0, \quad j \notin A(x^*),$$

if and only if there is no descent direction within the cone of first order feasible variations $V(x^*)$, i.e.,

$$\nabla f(x^*)'y \geq 0, \qquad \forall\, y \in V(x^*). \tag{4.72}$$

Proof: Assume first that there are no equality constraints. The condition (4.72) is written as

$$\nabla f(x^*)'y \geq 0, \qquad \forall\, y \text{ with } \nabla g_j(x^*)'y \leq 0 \text{ and } j \in A(x^*),$$

Sec. 4.3 Inequality Constraints

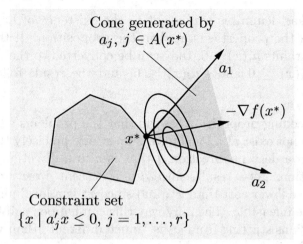

Figure 4.3.8. Proof of existence of Lagrange multipliers for linearly constrained problems, starting from Farkas' lemma. A local minimum x^* of f subject to $a_j'x \leq b_j$, $j = 1, \ldots, r$, is also a local minimum for the problem

$$\text{minimize } f(x)$$
$$\text{subject to } a_j'x \leq b_j, \quad j \in A(x^*),$$

where $A(x^*)$ is the set of active constraints at x^*. By Prop. 4.1.1,

$$\nabla f(x^*)'(x - x^*) \geq 0, \quad \forall\ x \text{ such that } a_j'x \leq b_j,\ j \in A(x^*).$$

Since a constraint $a_j'x \leq b_j$, $j \in A(x^*)$ can also be expressed as

$$a_j'(x - x^*) \leq 0,$$

the preceding condition is equivalent to

$$\nabla f(x^*)'y \geq 0, \quad \forall\ y \text{ such that } a_j'y \leq 0,\ j \in A(x^*).$$

From Farkas' lemma, it follows that there exist nonnegative scalars μ_j^*, $j \in A(x^*)$, such that

$$\nabla f(x^*) + \sum_{j \in A(x^*)} \mu_j^* a_j = 0.$$

By letting $\mu_j^* = 0$ for all $j \notin A(x^*)$, the desired Lagrange multiplier conditions

$$\nabla f(x^*) + \sum_{j=1}^{r} \mu_j^* a_j = 0,$$

$$\mu_j^* = 0, \quad \forall\ j \notin A(x^*),$$

follow.

which by Farkas' lemma is equivalent to the existence of Lagrange multipliers μ_j^* with the properties stated in the proposition. If there are some equality constraints $h_i(x) = 0$, they can be converted to the two inequality constraints $h_i(x) \leq 0$ and $-h_i(x) \leq 0$, and the result follows similarly. **Q.E.D.**

The preceding proposition asserts that the problems where Lagrange multipliers do not exist at a local minimum x^* are precisely those problems where there is a descent direction $\bar{d}$ that is also a direction of first order feasible variation. As a result, since $\bar{d}$ is a descent direction, points near x^* along $\bar{d}$ have lower cost than x^*, and since x^* is a local minimum, these points must be infeasible. The only way this can happen is if the true set of feasible variations starting from x^* is "much different" than what the cone of first order feasible variations $V(x^*)$ suggests (see Example 4.1.1 and Fig. 4.1.2 for one such situation). In order to fully understand this phenomenon, we introduce a concept that allows us to describe more precisely the local character of the feasible set around x^*.

Quasiregularity

Consider problem (ICP), i.e., the problem of minimizing $f(x)$ subject to $x \in C$, where
$$C = \{x \mid h_1(x) = 0, \ldots, h_m(x) = 0, g_1(x) \leq 0, \ldots, g_r(x) \leq 0\},$$
and f, h_i, and g_j are continuously differentiable functions. A vector $x \in C$ is said to be a *quasiregular* point if the tangent cone of C at x is equal to the cone of first order feasible variations at x:
$$T(x) = V(x).$$
Figure 4.3.6(b) provides an example where there is a point that is not quasiregular [at the point $x = (0,1)$, we have $\nabla h(x) = 0$, so $V(x)$ is equal to $\Re^2$ and strictly contains $T(x)$; every other vector in C is quasiregular]. See also Fig. 4.1.2 for another similar example.

By combining Props. 4.3.11 and 4.3.13, we obtain the following proposition.

Proposition 4.3.14: Let x^* be a local minimum of problem (ICP). If x^* is a quasiregular point, there exist Lagrange multipliers $\lambda_1^*, \ldots, \lambda_m^*$ and $\mu_1^*, \ldots, \mu_r^*$, satisfying

$$\nabla f(x^*) + \sum_{i=1}^{m} \lambda_i^* \nabla h_i(x^*) + \sum_{j=1}^{r} \mu_j^* \nabla g_j(x^*) = 0,$$

$$\mu_j^* \geq 0, \quad j = 1, \ldots, r, \qquad \mu_j^* = 0, \quad j \notin A(x^*).$$

Sec. 4.3   Inequality Constraints   407

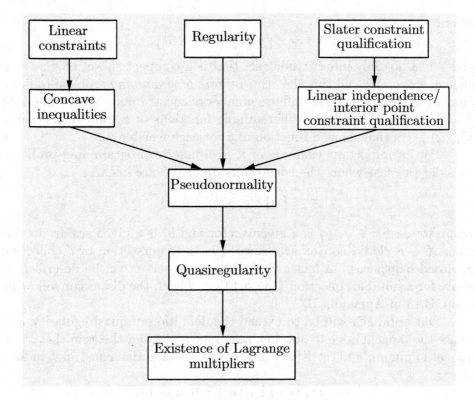

Figure 4.3.9. Relations between various conditions, which when satisfied at a local minimum x^*, guarantee existence of corresponding Lagrange multipliers.

It is possible to show that pseudonormality implies quasiregularity. This requires an elaborate proof, which will not be given; see [OzB04]. The reverse is not true; see Exercise 4.3.20 for a counterexample. Note that while quasiregularity is a more general constraint qualification, it is not verified as easily as pseudonormality in specific contexts of interest. Moreover, it breaks down in the more general case where there is an additional abstract constraint set as in the next section. Figure 4.3.9 summarizes the relations of various conditions that guarantee existence of Lagrange multipliers corresponding to a local minimum of f subject to a finite number of equality and inequality constraints. Each of the implications in this figure is one-directional.

Semi-Infinite Programming

We will now use the tangent cone-based analysis to derive optimality conditions for problems involving an infinite number of constraints. Consider a problem of the form

$$\begin{aligned} \text{minimize} \quad & f(x) \\ \text{subject to} \quad & x \in X, \end{aligned} \qquad (4.73)$$

where
$$X = \{x \mid g_j(x) \leq 0, j \in J\},$$
and J is a possibly infinite index set that is a compact subset of some Euclidean space. We call this the *semi-infinite programming* problem (a finite number of variables but an infinite number of constraints). We assume that f and g_j are continuously differentiable functions of x, and furthermore, $\nabla g_j(x)$ is continuous as a function of j for each feasible x.

For an important example of a semi-infinite programming problem, consider the case where the constraints $g_j(x) \leq 0$ are linear, i.e.,
$$X = \{x \mid a_j'x \leq b_j, j \in J\},$$
where for each $j \in J$, a_j is a given vector and b_j is a given scalar. In this case, X is a closed convex set, since it is the intersection of a collection of closed halfspaces. In fact any closed convex set X can be described as above for a suitable collection $\{(a_j, b_j) \mid j \in J\}$ (cf. the discussion following Prop. B.13 in Appendix B).

Our approach will be to extend the definition of quasiregularity, and to use the tangent cone to obtain a Lagrange multiplier theorem. Let x^* be a local minimum and let $A(x^*)$ denote the set of active constraint indices at x^*:
$$A(x^*) = \{j \mid g_j(x^*) = 0, j \in J\}.$$
The cone of first order feasible variations is
$$V(x^*) = \{y \mid \nabla g_j(x^*)'y \leq 0, \forall j \in A(x^*)\}.$$
Let $T(x^*)$ denote the tangent cone of X at x^*. We say that x^* is a *quasiregular* point if the following two conditions are satisfied:

(a) $T(x^*) = V(x^*)$.

(b) The cone $G(x^*)$ generated by the gradients $\nabla g_j(x^*)$, $j \in A(x^*)$ is closed, i.e., the set of all vectors of the form
$$\sum_{j \in \bar{A}} \mu_j \nabla g_j(x^*),$$

where $\bar{A}$ is a finite subset of $A(x^*)$ and $\mu_j \geq 0$ for all $j \in \bar{A}$, is closed.

Note that the cone $G(x^*)$ is necessarily closed if the index set $A(x^*)$ is finite; see Prop. B.15(b) in Appendix B. It can also be shown that $G(x^*)$ is closed if there exists a vector d such that
$$\nabla g_j(x^*)'d < 0, \qquad \forall j \in A(x^*),$$
(see the proof of the subsequent Prop. 4.3.16). We have the following proposition, whose proof makes use of the background material on cones in Section B.3 of Appendix B.

Sec. 4.3   Inequality Constraints

Proposition 4.3.15: (Lagrange Multiplier Theorem for Semi-Infinite Programming) Assume that x^* is a local minimum of f over X, and that x^* is quasiregular. Then either $\nabla f(x^*) = 0$ or else there exists an index set $J^* \subset A(x^*)$, and positive multipliers μ_j^*, $j \in J^*$, such that the gradients $\nabla g_j(x^*)$, $j \in J^*$, are linearly independent, and

$$\nabla f(x^*) + \sum_{j \in J^*} \mu_j^* \nabla g_j(x^*) = 0.$$

Proof: We first assert that we have

$$V(x^*)^\perp = G(x^*), \tag{4.74}$$

where $V(x^*)^\perp$ is the polar cone of $V(x^*)$ (see Section B.3 of Appendix B for the definition and various properties of polar cones). To see this, note that we have $V(x^*) = G(x^*)^\perp$, so that

$$V(x^*)^\perp = \bigl(G(x^*)^\perp\bigr)^\perp,$$

and since $G(x^*)$ is closed by the quasiregularity assumption, we have $G(x^*) = \bigl(G(x^*)^\perp\bigr)^\perp$ [Prop. B.15(a) in Appendix B].

Now, by Prop. 4.3.11, since x^* is a local minimum, we have $\nabla f(x^*)'y \geq 0$ for all $y \in T(x^*)$. By quasiregularity, we also have $T(x^*) = V(x^*)$, so that $\nabla f(x^*)'y \geq 0$ for all $y \in V(x^*)$. Hence, $-\nabla f(x^*)$ belongs to $V(x^*)^\perp$, and by Eq. (4.74), $-\nabla f(x^*)$ belongs to the cone $G(x^*)$. Thus, either $\nabla f(x^*) = 0$ or else, by Caratheodory's theorem for cones [Prop. B.6(a) in Appendix B], $-\nabla f(x^*)$ can be expressed as a positive combination of a collection of linearly independent vectors $\nabla g_j(x^*)$, $j \in J^*$. **Q.E.D.**

The following example illustrates the preceding proposition and highlights the significance of various properties of the index set J, including compactness.

Example 4.3.4

Consider the following problem in $\Re^2$:

$$\text{minimize} \quad -(x_1 + x_2)$$
$$\text{subject to} \quad x_1 \cos\phi + x_2 \sin\phi \leq 1, \quad \phi \in \Phi,$$

where Φ is a given compact subset of the set $[0, 2\pi]$. To visualize the constraint set, note that for each ϕ, the equation $x_1 \cos\phi + x_2 \sin\phi = 1$ specifies a line

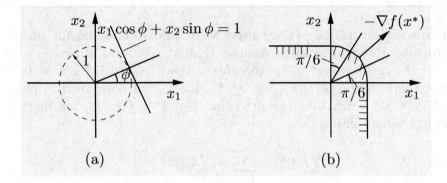

Figure 4.3.10. (a) The linear constraints in Example 4.3.4. (b) Constraint set and optimal solution for the case $\Phi = [0, \pi/6] \cup [\pi/3, \pi/2]$.

that is tangent to the unit circle, as shown in Fig. 4.3.10(a). It can be seen that since there can be at most one active constraint at any feasible vector, quasiregularity is satisfied everywhere.

Consider first the case where $\Phi = [0, 2\pi]$. Then the constraint set of the problem is the unit circle. In this case, the optimal solution is $x^* = (\sqrt{2}/2, \sqrt{2}/2)$, and it can be seen that the only constraint that is active corresponds to $\phi = \pi/4$. From the condition $\nabla f(x^*) + \sum_{j \in J^*} \mu_j^* \nabla g_j(x^*) = 0$, we see that the corresponding Lagrange multiplier is

$$\mu_\phi^* = \begin{cases} 0 & \text{if } \phi \neq \pi/4, \\ \sqrt{2} & \text{if } \phi = \pi/4. \end{cases}$$

Consider next the case where $\Phi = [0, \pi/6] \cup [\pi/3, \pi/2]$. Then the constraint set is as shown in Fig. 4.3.10(b). It can be seen from this figure that the optimal solution is $x^* = \big(1/(1+\sqrt{3}), 1/(1+\sqrt{3})\big)$. There are two constraints that are active, corresponding to $\phi = \pi/6$ and $\phi = \pi/3$. The corresponding Lagrange multiplier is

$$\mu_\phi^* = \begin{cases} 0 & \text{if } \phi \neq \pi/6, \pi/3, \\ \dfrac{2}{1+\sqrt{3}} & \text{if } \phi = \pi/6 \text{ or } \phi = \pi/3. \end{cases}$$

It is interesting to observe that if we exclude the points $\pi/6$ and $\pi/3$ from Φ, so that Φ is the *noncompact* set $[0, \pi/6] \cup (\pi/3, \pi/2]$, the optimal solution x^* will remain the same, but there will be no active constraints and corresponding Lagrange multipliers. This indicates the significance of the compactness assumption in the problem formulation.

Similarly, it can be seen that for any choice of the compact set Φ, there are at most two active constraints at the optimal solution, and the corresponding Lagrange multipliers are positive. We finally note that if we introduce some additional inequality constraints [such as for example the constraint $x_1 \geq x_2$ or the constraint $x_1 \leq 1/(1+\sqrt{3})$ in the case of Fig. 4.3.10(b)], it is still always possible to select a Lagrange multiplier vector, where at most two of the Lagrange multipliers are nonzero, even if the number of active constraints is larger than two. This is consistent with Prop. 4.3.15.

Sec. 4.3   Inequality Constraints   411

We now provide a condition that guarantees quasiregularity.

> **Proposition 4.3.16: (Constraint Qualification for Semi-Infinite Programming)** Let x^* be a local minimum of f over X, and assume that there exists a vector d such that
>
> $$\nabla g_j(x^*)'d < 0, \qquad \forall \, j \in A(x^*). \tag{4.75}$$
>
> Then x^* is quasiregular and the conclusions of Prop. 4.3.15 hold.

Proof: For any vector $y \in V(x^*)$, the vector $y + \alpha(d-y)$, where $0 < \alpha \le 1$, satisfies $\nabla g_j(x^*)'\big(y + \alpha(d-y)\big) < 0$ for all $j \in A(x^*)$, and based on this property, it can be shown that $y + \alpha(d - y)$ is a feasible direction at x^* so that it belongs to the tangent cone $T(x^*)$ (see Exercise 4.3.17). Hence $V(x^*)$ is a subset of the closure of $T(x^*)$, and since $T(x^*)$ is closed, we have $V(x^*) \subset T(x^*)$.

The reverse inclusion, $T(x^*) \subset V(x^*)$, is also true. To see this, let y be a nonzero tangent of X at x^*. Then there exists a sequence $\{\xi^k\}$ and a sequence $\{x^k\} \subset X$ such that $x^k \ne x^*$ for all k and

$$\xi^k \to 0, \qquad x^k \to x^*,$$

and

$$\frac{x^k - x^*}{\|x^k - x^*\|} = \frac{y}{\|y\|} + \xi^k.$$

By the mean value theorem, we have for all $j \in A(x^*)$ and k

$$0 \ge g_j(x^k) = g_j(x^*) + \nabla g_j(\tilde{x}^k)'(x^k - x^*) = \nabla g_j(\tilde{x}^k)'(x^k - x^*),$$

where $\tilde{x}^k$ is a vector that lies on the line segment joining x^k and x^*. This relation can be written as

$$0 \ge \frac{\|x^k - x^*\|}{\|y\|} \nabla g_j(\tilde{x}^k)'y^k,$$

or equivalently

$$\nabla g_j(\tilde{x}^k)'y^k \le 0,$$

where $y^k = y + \xi^k\|y\|$. Taking the limit as $k \to \infty$, we obtain

$$\nabla g_j(x^*)'y \le 0, \qquad \forall \, j \in A(x^*),$$

showing that $y \in V(x^*)$. Thus, we have $T(x^*) \subset V(x^*)$, completing the proof that $T(x^*) = V(x^*)$.

To show quasiregularity of x^*, there remains to show that the cone $G(x^*)$ generated by the set of gradients

$$R = \{\nabla g_j(x^*) \mid j \in A(x^*)\}$$

is closed. Let us first note that since $\nabla g_j(x^*)$ is continuous with respect to j, the set $A(x^*)$ is closed, and since J is compact, $A(x^*)$ is also compact. It follows that the set R, which is the image of the compact set $A(x^*)$ under a continuous mapping, is compact.

Consider any sequence $\{y^k\} \subset G(x^*)$ converging to some $y \neq 0$. We will show that $y \in G(x^*)$. Indeed, by Caratheodory's theorem (Prop. B.6 in Appendix B), for each k, there are indices $j_1^k, \ldots, j_{n+1}^k$ and nonnegative scalars $\beta_1^k, \ldots, \beta_{n+1}^k$ not all zero, such that

$$y^k = \beta_1^k \nabla g_{j_1^k}(x^*) + \cdots + \beta_{n+1}^k \nabla g_{j_{n+1}^k}(x^*).$$

Let

$$\beta^k = \beta_1^k + \cdots + \beta_{n+1}^k, \qquad \gamma_j^k = \frac{\beta_j^k}{\beta^k}, \quad j = j_1^k, \ldots, j_{n+1}^k.$$

Then we have

$$\frac{1}{\beta^k} y^k = \gamma_1^k \nabla g_{j_1^k}(x^*) + \cdots + \gamma_{n+1}^k \nabla g_{j_{n+1}^k}(x^*). \qquad (4.76)$$

The sequences $\nabla g_{j_1^k}(x^*), \ldots, \nabla g_{j_{n+1}^k}(x^*)$ lie in the compact set R, while the sequences $\gamma_1^k, \ldots, \gamma_{n+1}^k$ are nonnegative and satisfy $\sum_{m=1}^{n+1} \gamma_m^k = 1$. Therefore, from Eq. (4.76), we see that there is a subsequence $\{y^k/\beta^k\}_{k \in \mathcal{K}}$ that converges to a vector of the form

$$g = \gamma_1 \nabla g_{j_1}(x^*) + \cdots + \gamma_{n+1} \nabla g_{j_{n+1}}(x^*),$$

where the scalars $\gamma_1, \ldots, \gamma_{n+1}$ are nonnegative with $\sum_{m=1}^{n+1} \gamma_m = 1$, and the indices $j_1, \ldots, j_{n+1}$ belong to $A(x^*)$. In view of the assumption (4.75), we must have $g \neq 0$. This together with the fact $y^k \to y$ imply that the subsequence $\{\beta^k\}_{k \in \mathcal{K}}$ must converge to some $\beta > 0$, and that we must have $y = \beta g$. Since $g \in G(x^*)$ it follows that $y \in G(x^*)$. Therefore, $G(x^*)$ is closed. **Q.E.D.**

The assumption of the preceding proposition resembles the linear independence/interior point constraint qualification of Prop. 4.3.8, and in fact reduces to it for the case where there is only a finite number of inequality constraints and no equality constraints. A related assumption that resembles the Slater constraint qualification of Prop. 4.3.9 is that the functions g_j are convex and there exists a feasible vector $\bar{x}$ such that

$$g_j(\bar{x}) < 0, \qquad \forall \, j \in A(x^*).$$

It can then be seen that the vector $d = \bar{x} - x^*$ satisfies the assumptions of Prop. 4.3.15, so x^* is quasiregular under this assumption as well.

Finally, suppose that the function f and the functions g_j, $j \in J$, are all convex over $\Re^n$, and suppose that we have a feasible vector x^*, a nonempty and finite index set $J^* \subset A(x^*)$, and positive multipliers μ_j^*, $j \in J^*$, such that

$$\nabla f(x^*) + \sum_{j \in J^*} \mu_j^* \nabla g_j(x^*) = 0.$$

Then by the convexity of f and g_j, x^* minimizes $f(x) + \sum_{j \in J^*} \mu_j^* g_j(x)$ over $x \in \Re^n$, and the proof of Prop. 4.3.4 can be used to assert that x^* is a global minimum of f subject to $g_j(x) \leq 0$, $j \in J$. Thus, under convexity assumptions on f and g_j, the necessary condition of Prop. 4.3.15 is also sufficient for optimality.

Example 4.3.5

Given a compact set C in $\Re^n$, consider the problem of finding the smallest sphere that contains C. We formulate this as the minimax problem

$$\text{minimize} \ \max_{c \in C} \|x - c\|^2$$
$$\text{subject to} \ x \in \Re^n,$$

where x is the unknown center of the sphere. Since the cost function is strictly convex and coercive, we conclude that there exists a unique optimal solution x^*. By introducing an additional variable z, the problem is equivalently written as

$$\text{minimize} \ z$$
$$\text{subject to} \ x \in \Re^n, \quad \|x - c\|^2 \leq z, \quad \forall \ c \in C,$$

which is a semi-infinite programming problem with convex cost and constraints. It can be seen, that the assumption (4.75) of Prop. 4.3.15 is satisfied with $d = (0, 1)$, and taking also into account the preceding discussion, we conclude that x^* is the optimal solution if and only if the optimality conditions of Prop. 4.3.15 hold. After a little calculation, it then follows that there exist elements $c_1, \ldots, c_s$ of C, with $s \leq n + 1$, and corresponding positive multipliers $\mu_1^*, \ldots, \mu_s^*$, such that

$$\sum_{j=1}^{s} \mu_j^* = 1, \qquad x^* = \sum_{j=1}^{s} \mu_j^* c_j, \tag{4.77}$$

and the radius R of the optimal sphere satisfies

$$R = \|x^* - c_j\|, \qquad \forall \ j = 1, \ldots, s.$$

Let us now show that the radius satisfies the inequality

$$R \leq d\sqrt{\frac{s-1}{2s}}, \qquad (4.78)$$

due to [Joh48], where the scalar d is the diameter of the set $\{c_1, \ldots, c_s\}$:

$$d = \max_{i,j=1,\ldots,s} \|c_i - c_j\|.$$

Indeed, let

$$D_{ij} = \|c_i - c_j\|^2,$$

and note that

$$D_{ij} = \|c_i - c_j\|^2 = \|(c_i - x^*) - (c_j - x^*)\|^2 = 2R^2 - 2(c_i - x^*)'(c_j - x^*).$$

Multiplying this equation with $\mu_i^* \mu_j^*$, adding over i and j, and using the conditions (4.77), we have

$$\sum_{i,j} \mu_i^* \mu_j^* D_{ij} = 2R^2 \left(\sum_{j=1}^{s} \mu_j^*\right)^2 - \left(\sum_{i=1}^{s} \mu_i^*(c_i - x^*)\right)' \left(\sum_{j=1}^{s} \mu_j^*(c_j - x^*)\right)$$

$$= 2R^2.$$

Since $D_{ij} \leq d^2$ and $D_{ii} = 0$, we thus obtain

$$2R^2 \leq d^2 \sum_{i \neq j} \mu_i^* \mu_j^*.$$

Since

$$\sum_{j=1}^{s} \mu_j^* = 1,$$

it can be seen that

$$\sum_{j=1}^{s} (\mu_j^*)^2 \geq 1/s,$$

so that

$$\sum_{i \neq j} \mu_i^* \mu_j^* = 1 - \sum_{j=1}^{s} (\mu_j^*)^2 \leq 1 - \frac{1}{s} = \frac{s-1}{s}.$$

By combining the last two relations, we obtain the desired bound (4.78).

Sec. 4.3 Inequality Constraints

4.3.6 Abstract Set Constraints, Equality, and Inequality Constraints

Let us now consider a more general version of the equality/inequality constrained problem, where there is the additional constraint $x \in X$. In this section *we assume that X is closed and convex*. The analysis can be generalized to the case where X is nonconvex, albeit with considerable mathematical complication.† The Fritz John conditions of Prop. 4.3.5 are extended in the following proposition, where the gradient of the Lagrangian function, instead of being zero, satisfies the necessary condition for minimization over X (cf. Section 3.1).

Proposition 4.3.17: (Fritz John Conditions for an Additional Convex Set Constraint) Let x^* be a local minimum of the problem

$$\begin{aligned} \text{minimize} \quad & f(x) \\ \text{subject to} \quad & h_1(x) = 0, \ldots, h_m(x) = 0, \\ & g_1(x) \leq 0, \ldots, g_r(x) \leq 0, \quad x \in X, \end{aligned} \qquad (4.80)$$

where f, h_i, g_j are continuously differentiable functions from $\Re^n$ to $\Re$, and X is a closed convex set. Then there exist scalars $\lambda_1^*, \ldots, \lambda_m^*$ and $\mu_0^*, \mu_1^*, \ldots, \mu_r^*$, satisfying the following conditions:

---

† A more general version of the theory, is given in the papers [BeO02], [OzB04], and Chapter 5 of the book [BNO03], where the set X is assumed closed but possibly nonconvex. Then Prop. 4.3.17 holds with condition (i) replaced by

$$-\left(\mu_0^* \nabla f(x^*) + \sum_{i=1}^m \lambda_i^* \nabla h_i(x^*) + \sum_{j=1}^r \mu_j^* \nabla g_j(x^*)\right) \in N_X(x^*), \qquad (4.79)$$

where $N_X(x^*)$ is the so-called *normal cone* of X at x^*. This cone is prominent in the theory of nonsmooth analysis (see e.g., the book [RoW98]). It is defined as the set of all $z \in \Re^n$ such that there exist sequences $\{x^k\} \subset X$ and $\{z^k\} \subset \Re^n$ satisfying $x^k \to x^*$, $z^k \to z$, and $z_k' y \leq 0$, for all $y \in T(x^*)$, where $T(x^*)$ is the tangent cone of X at x^*.

When X is convex, the cone $N_X(x^*)$ can be shown to be equal to $T(x^*)^\perp$ [the polar cone of $T(x^*)$ as defined in Section B.3 of Appendix B], in which case the condition (4.79) is seen to be equivalent to condition (i) of Prop. 4.3.17. This, however, need not be true when X is nonconvex, in which case x^* need not be a stationary point of the Lagrangian function $L(\cdot, \lambda^*, \mu^*)$ [i.e., the Lagrangian may decrease starting from x^* along some direction $y \in T(x^*)$, cf. Prop. 4.3.11]. Then the significance of the multipliers λ^* and μ^* comes to question. Section 5.3.2, and Examples 5.3.1 and 5.3.2 of [BNO03] provide a discussion of this point.

(i) For all $x \in X$, we have

$$\left(\mu_0^* \nabla f(x^*) + \sum_{i=1}^m \lambda_i^* \nabla h_i(x^*) + \sum_{j=1}^r \mu_j^* \nabla g_j(x^*)\right)'(x - x^*) \geq 0.$$

(ii) $\mu_j^* \geq 0$ for all $j = 0, 1, \ldots, r$.

(iii) $\mu_0^*, \lambda_1^*, \ldots, \lambda_m^*, \mu_1^*, \ldots, \mu_r^*$ are not all equal to 0.

(iv) In every neighborhood N of x^* there is an $x \in N \cap X$ such that $\lambda_i^* h_i(x) > 0$ for all i with $\lambda_i^* \neq 0$ and $\mu_j^* g_j(x) > 0$ for all j with $\mu_j^* \neq 0$. Moreover, if $(\lambda^*, \mu^*) \neq (0, 0)$ this x can be chosen so that $f(x) < f(x^*)$.

Proof: The proof is similar to the one of Prop. 4.3.5. Following the same notation and penalty approach, the penalized cost $F^k(x)$ is minimized over $x \in X \cap S$, where $S = \{x \mid \|x - x^*\| \leq \epsilon\}$, and $\epsilon > 0$ is such that $f(x^*) \leq f(x)$ for all feasible x with $x \in S$, yielding a vector x^k. Similar to the proof of Prop. 4.3.5, we have that $x^k \to x^*$. For sufficiently large k, x^k is an interior point of S, so that

$$\nabla F^k(x^k)'(x - x^k) \geq 0, \qquad \forall\, x \in X,$$

or [cf. Eq. (4.58)]

$$\left(\nabla f(x^k) + \sum_{i=1}^m \xi_i^k \nabla h_i(x^k) + \sum_{j=1}^r \zeta_j^k \nabla g_j(x^k) + (x^k - x^*)\right)'(x - x^k) \geq 0,$$

for all $x \in X$, where [cf. Eq. (4.59)]

$$\xi_i^k = k h_i(x^k), \qquad \zeta_j^k = k g_j^+(x^k).$$

The proof from this point is essentially identical to the one of Prop. 4.3.5. **Q.E.D.**

A vector (λ^*, μ^*) that satisfies the Fritz John conditions (i)-(iii) of Prop. 4.3.17 with $\mu_0^* = 1$, as well as the CS condition $\mu_j^* g_j(x^*) = 0$ for all $j = 1, \ldots, r$, is called a *Lagrange multiplier* for problem (4.80). The following proposition gives an extension of the linear independence/interior point constraint qualification, which guarantees that we can take $\mu_0^* = 1$ in the preceding Fritz John conditions.

Sec. 4.3   Inequality Constraints

> **Proposition 4.3.18: (Constraint Qualification for an Additional Convex Set Constraint)** Let x^* be a local minimum of problem (4.80), and assume that the following two conditions hold:
>
> (1) There does not exist a nonzero vector $\lambda = (\lambda_1, \ldots, \lambda_m)$ such that
>
> $$\left(\sum_{i=1}^{m} \lambda_i \nabla h_i(x^*)\right)'(x - x^*) \geq 0, \qquad \forall\, x \in X.$$
>
> (2) There exists a feasible direction d of X at x^* such that
>
> $$\nabla h_i(x^*)'d = 0, \quad i = 1, \ldots, m, \qquad \nabla g_j(x^*)'d < 0, \quad \forall\, j \in A(x^*).$$
>
> Then the Fritz John conditions of Prop. 4.3.17 hold with $\mu_0^* = 1$.

Proof: Consider a set of multipliers λ_i^*, $i = 1, \ldots, m$, μ_j^*, $j = 0, 1, \ldots, r$, satisfying the Fritz John conditions of Prop. 4.3.17. Assume to arrive at a contradiction that $\mu_0^* = 0$, so that

$$\left(\sum_{i=1}^{m} \lambda_i^* \nabla h_i(x^*) + \sum_{j=1}^{r} \mu_j^* \nabla g_j(x^*)\right)'(x - x^*) \geq 0, \qquad \forall\, x \in X. \quad (4.81)$$

Since not all the λ_i^* and μ_j^* can be zero, we conclude that $\mu_j^* > 0$ for at least one $j \in A(x^*)$; otherwise condition (1) would be violated. Since $\mu_j^* \geq 0$ for all j, with $\mu_j^* = 0$ for $j \notin A(x^*)$ and $\mu_j^* > 0$ for at least one j, we obtain

$$\sum_{i=1}^{m} \lambda_i^* \nabla h_i(x^*)'d + \sum_{j=1}^{r} \mu_j^* \nabla g_j(x^*)'d < 0,$$

where d is the vector of condition (2). This contradicts Eq. (4.81). Therefore, we must have $\mu_0^* > 0$, so that we can take $\mu_0^* = 1$. **Q.E.D.**

In the case where the functions h_i are linear and the functions g_j are convex, it is possible to show that a Lagrange multiplier exists under assumptions that resemble the Slater constraint qualification of Prop. 4.3.9. Basically, condition (2) of Prop. 4.3.18 is replaced by the condition that there exists a feasible vector $\bar{x}$ satisfying $g_j(\bar{x}) < 0$ for all $j \in A(x^*)$. Here is an example where the conditions of Prop. 4.3.18 are violated and there exists no Lagrange multiplier.

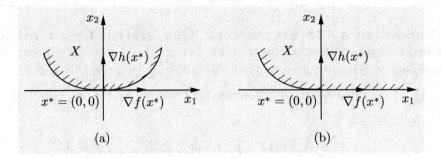

Figure 4.3.11. Constraint set X for the problems of Example 4.3.6.

Example 4.3.6

Consider the two-dimensional problem

$$\text{minimize } f(x) = x_1$$
$$\text{subject to } h(x) = x_2 = 0, \quad x \in X = \{(x_1, x_2) \mid x_1^2 \leq x_2\};$$

[see Fig. 4.3.11(a)]. The only feasible solution is $x^* = (0,0)$, but it can be graphically verified that there is no Lagrange multiplier λ^* such that

$$\big(\nabla f(x^*) + \lambda^* \nabla h(x^*)\big)'(x - x^*) = x_1 + \lambda^* x_2 \geq 0, \quad \forall\, x \in X.$$

Here, both conditions (1) and (2) of Prop. 4.3.18 are violated. By contrast, any scalars μ_0^*, λ^* with $\mu_0^* = 0$ and $\lambda^* > 0$ satisfy

$$\lambda^* \nabla h(x^*)'(x - x^*) = \lambda^* x_2 \geq 0, \quad \forall\, x \in X,$$

which is consistent with Prop. 4.3.17.

In a variation of this example, suppose instead that the set X is given by

$$X = \{(x_1, x_2) \mid x_1^2 \leq x_2,\, x_1 \leq 0\} \cup \{(x_1, x_2) \mid x_1 \geq 0,\, x_2 \geq 0\};$$

[see Fig. 4.3.11(b)]. The optimal solution is $x^* = (0,0)$, and again it can be seen that there is no Lagrange multiplier λ^* such that

$$\big(\nabla f(x^*) + \lambda^* \nabla h(x^*)\big)'(x - x^*) = x_1 + \lambda^* x_2 \geq 0, \quad \forall\, x \in X.$$

Here, condition (2) of Prop. 4.3.18 is satisfied, but condition (1), is violated.

Pseudonormality

We now indicate how to extend the definition of pseudonormality to the case of the constraint set

$$X \cap \{x \mid h_1(x) = 0, \ldots, h_m(x) = 0, g_1(x) \leq 0, \ldots, g_r(x) \leq 0\}.$$

We say that a feasible vector x^* is *pseudonormal* if there is no set of scalars $\lambda_1, \ldots, \lambda_m, \mu_1, \ldots, \mu_r$, such that:

(i) $\left(\sum_{i=1}^{m} \lambda_i \nabla h_i(x^*) + \sum_{j=1}^{r} \mu_j \nabla g_j(x^*)\right)'(x - x^*) \geq 0$ for all $x \in X$.

(ii) $\mu_j \geq 0$ for all $j = 1, \ldots, r$.

(iii) $\lambda_1, \ldots, \lambda_m, \mu_1, \ldots, \mu_r$ are not all equal to 0.

(iv) In every neighborhood N of x^* there is an $x \in N \cap X$ such that

$$\sum_{i=1}^{m} \lambda_i h_i(x) + \sum_{j=1}^{r} \mu_j g_j(x) > 0.$$

We note that if x^* is pseudonormal, there can be no scalars $\lambda_1, \ldots, \lambda_m$, $\mu_1, \ldots, \mu_r$, such that conditions (i)-(iii) above are satisfied, and in every neighborhood N of x^* there is an $x \in N \cap X$ such that $\lambda_i h_i(x) > 0$ for all i with $\lambda_i \neq 0$ and $\mu_j g_j(x) > 0$ for all j with $\mu_j \neq 0$. It follows that if x^* is a pseudonormal local minimum, the Fritz John conditions of Prop. 4.3.17 cannot be satisfied with $\mu_0^* = 0$, so that there exist Lagrange multipliers satisfying the corresponding necessary conditions. Moreover it can be verified that the constraint qualification of Prop. 4.3.18 implies pseudonormality of x^*. Thus pseudonormality unifies the constraint qualifications discussed so far, and embodies fundamental structure that when possessed by the constraint set, guarantees existence of Lagrange multipliers. For an extension of the notion of pseudonormality and corresponding Lagrange multiplier analysis, to the case where X is a closed but not necessarily convex set, we refer to [BeO02], [BNO03], and [OzB04].

Informative Lagrange Multipliers and Sensitivity

We noted that the Lagrange multipliers satisfying the CV condition provide a significant amount of sensitivity information by indicating through their sign pattern the direction in which to violate the constraints in order to effect a cost reduction. Such multipliers have been further analyzed in the paper [BeO02] and the book [BNO03], Section 5.3, and they have been called *informative*. Moreover, it was shown there that *if there exists at least one Lagrange multiplier, there exists one that is informative*. One of these informative multipliers is the Lagrange multiplier that has minimum norm.

This particular multiplier has an optimality property: it quantifies the optimal rate of cost reduction that can be achieved through violation of the constraints, and also provides the asymptotic direction of the corresponding constraint violation. This result, stated in the following proposition, was derived in the paper [Ber05c] to which we refer for the proof.

Proposition 4.3.19: (Sensitivity) Consider problem (4.80), and let x^* be a local minimum. Assume that the set of Lagrange multipliers is nonempty, and let (λ^*, μ^*) be the vector of minimum norm on this set. Then for every sequence $\{x^k\} \subset X$ of infeasible vectors such that $x^k \to x^*$, we have

$$f(x^*) - f(x^k) \le \|(\lambda^*, \mu^*)\| \|(h(x^k), g^+(x^k))\| + o(\|x^k - x^*\|), \quad (4.82)$$

where $g^+(x^k)$ is the vector with components

$$g_j^+(x^k) = \max\{0, g_j(x^k)\}, \qquad j = 1, \ldots, r.$$

Furthermore, if $(\lambda^*, \mu^*) \ne (0,0)$, the preceding inequality is sharp in the sense that there exists a sequence of infeasible vectors $\{x^k\} \subset X$ such that $x^k \to x^*$ and

$$\lim_{k \to \infty} \frac{f(x^*) - f(x^k)}{\|(h(x^k), g^+(x^k))\|} = \|(\lambda^*, \mu^*)\|. \qquad (4.83)$$

For this sequence, we have

$$\lim_{k \to \infty} \frac{h(x^k)}{\|(h(x^k), g^+(x^k))\|} = \frac{\lambda^*}{\|(\lambda^*, \mu^*)\|},$$

$$\lim_{k \to \infty} \frac{g^+(x^k)}{\|(h(x^k), g^+(x^k))\|} = \frac{\mu^*}{\|(\lambda^*, \mu^*)\|}.$$

The preceding proposition is consistent with the classical sensitivity interpretation of a Lagrange multiplier as the gradient of the primal function at 0 (cf. Prop. 4.3.3, which was obtained under second order sufficiency conditions). Here, however, we are not making enough assumptions for this stronger type of sensitivity interpretation to be valid, and the preceding proposition holds when there are multiple Lagrange multipliers, and also when the primal function has discontinuous directional derivatives (see [Ber05c] and [BOT06] for some examples).

EXERCISES

4.3.1

Solve the two-dimensional problem

$$\text{minimize } (x-a)^2 + (y-b)^2 + xy$$
$$\text{subject to } 0 \le x \le 1, \quad 0 \le y \le 1,$$

for all possible values of the scalars a and b.

4.3.2

Given a vector y, consider the problem

$$\text{maximize } y'x$$
$$\text{subject to } x'Qx \le 1,$$

where Q is a positive definite symmetric matrix. Show that the optimal value is $\sqrt{y'Q^{-1}y}$ and use this fact to establish the inequality

$$(x'y)^2 \le (x'Qx)(y'Q^{-1}y).$$

4.3.3

Solve Exercise 3.1.15 of Section 3.1 using Lagrange multipliers.

4.3.4

Solve Exercise 3.1.16 of Section 3.1 using Lagrange multipliers.

4.3.5 (Constraint Qualifications for Inequality Constraints) (www)

The purpose of this exercise is to explore a condition that implies all of the existence results for Lagrange multipliers that we have proved in this section for the case of inequality constraints only (unfortunately the approach does not generalize easily to nonlinear equality constraints). Consider the problem

$$\text{minimize } f(x)$$
$$\text{subject to } g_j(x) \le 0, \quad j = 1, \ldots, r.$$

For a feasible x, let $F(x)$ be the set of feasible directions at x,

$$F(x) = \{d \mid \text{for some } \bar{\alpha} > 0, \ g(x + \alpha d) \le 0 \text{ for all } \alpha \in [0, \bar{\alpha}]\}$$

and denote by $\overline{F}(x)$ the closure of $F(x)$. Let x^* be a local minimum. Show that:

(a)
$$\nabla f(x^*)'d \ge 0, \qquad \forall \ d \in \overline{F}(x^*).$$

(b) If we have $\overline{F}(x^*) = V(x^*)$, where

$$V(x^*) = \{d \mid \nabla g_j(x^*)'d \le 0, \ \forall \ j \in A(x^*)\},$$

then there exists a Lagrange multiplier vector $\mu^* \ge 0$ such that

$$\nabla f(x^*) + \sum_{j=1}^{r} \mu_j^* \nabla g_j(x^*) = 0, \qquad \mu_j^* = 0, \qquad \forall \ j \notin A(x^*).$$

Hint: Use part (a) and Farkas' Lemma (Prop. 4.3.12).

(c) The condition of part (b) holds if any one of the following conditions holds:

1. The functions g_j are linear.

2. ([AHU61]) There exists a vector d such that

$$\nabla g_j(x^*)'d < 0, \qquad \forall \ j \in A(x^*).$$

Hint: Let $\bar{d}$ satisfy $\nabla g_j(x^*)'\bar{d} \le 0$ for all $j \in A(x^*)$, and let $d_\gamma = \gamma d + (1 - \gamma)\bar{d}$. From the mean value theorem, we have for some $\epsilon \in [0, 1]$

$$g_j(x^* + \alpha d_\gamma) = g_j(x^*) + \alpha \nabla g_j(x^* + \epsilon \alpha d_\gamma)' d_\gamma$$
$$\le \alpha \big(\gamma \nabla g_j(x^* + \epsilon \alpha d_\gamma)'d + (1 - \gamma)\nabla g_j(x^* + \epsilon \alpha d_\gamma)'\bar{d}\big).$$

For a fixed γ, there exists an $\bar{\alpha} > 0$ such that the right-hand side above is nonpositive for $\alpha \in [0, \bar{\alpha}]$. Thus $d_\gamma \in F(x^*)$, for all γ and $\lim_{\gamma \to 0} d_\gamma = \bar{d}$. Hence $\bar{d} \in \overline{F}(x^*)$.

3. ([Sla50]) The functions g_j are convex and there exists a vector $\bar{x}$ satisfying

$$g_j(\bar{x}) < 0, \qquad \forall \ j \in A(x^*).$$

Hint: Let $d = \bar{x} - x^*$ and use condition 2.

4. The gradients $\nabla g_j(x^*)$, $j \in A(x^*)$, are linearly independent. *Hint*: Let d be such that $\nabla g_j(x^*)'d = -1$ for all $j \in A(x^*)$ and use condition 2.

(d) Consider the two-dimensional problem with two inequality constraints where

$$f(x_1, x_2) = x_1 + x_2,$$

Sec. 4.3 Inequality Constraints

$$g_1(x_1, x_2) = \begin{cases} -x_1^2 + x_2 & \text{if } x_1 \geq 0, \\ -x_1^4 + x_2 & \text{if } x_1 \leq 0, \end{cases}$$

$$g_2(x_1, x_2) = \begin{cases} x_1^4 - x_2 & \text{if } x_1 \geq 0, \\ x_1^2 - x_2 & \text{if } x_1 \leq 0. \end{cases}$$

Show that $x^* = (0,0)$ is a local minimum, but each of the conditions 1-4 of part (c) is violated and there exists no Lagrange multiplier vector.

(e) Consider the equality constrained problem $\min_{h(x)=0} f(x)$, and suppose that x^* is a local minimum such that $\nabla f(x^*) \neq 0$. Repeat part (d) for the case of the equivalent inequality constrained problem

$$\min_{\|h(x)\|^2 \leq 0} f(x).$$

4.3.6 (Gordan's Theorem of the Alternative) (www)

Let A be an $m \times n$ matrix. Then exactly one of the following two alternative conditions hold:

(i) There is an $x \in \Re^n$ such that $Ax < 0$ (i.e., all components of Ax are strictly negative).

(ii) There is a $\mu \in \Re^m$ such that $A'\mu = 0$, $\mu \neq 0$, and $\mu \geq 0$.

Hint: Assume that there exist x and μ satisfying simultaneously (i) and (ii), and reach a contradiction. Also, show that (i) and (ii) cannot fail to hold simultaneously by using Prop. 4.3.10 for a suitable minimax problem to prove that if (i) fails to hold, then (ii) must hold. *Note*: There other several important theorems (including Farkas' Lemma) that are of interest in optimization, and state that exactly one of two alternatives can hold in a given situation. For a systematic development of the Gordan, Farkas, and other theorems of the alternative, see [Ber09], Section 5.6.

4.3.7 (Sufficiency Conditions Without Strict Complementarity) (www)

Show that the sufficiency result of Prop. 4.3.2 holds if the strict complementary slackness condition (4.55) is removed, but the assumption on the Hessian of the Lagrangian is strengthened to read

$$y'\nabla_{xx}^2 L(x^*, \lambda^*, \mu^*) y > 0$$

for all $y \neq 0$ satisfying

$$\nabla h_i(x^*)'y = 0, \quad i = 1, \ldots, m,$$
$$\nabla g_j(x^*)'y = 0, \quad \forall\, j \in A(x^*) \text{ with } \mu_j^* > 0,$$
$$\nabla g_j(x^*)'y \leq 0, \quad \forall\, j \in A(x^*) \text{ with } \mu_j^* = 0.$$

Hint: Extend the proof of Prop. 4.2.1, based on the descent approach, given in Section 4.2.2. Show that the vector y in that proof satisfies the above conditions.

4.3.8

Find $x_1^*, x_2^*, \mu_0^*, \mu_1^*, \mu_2^*$ satisfying the Fritz John conditions for the following two-dimensional problems with inequality constraints:

(a) $f(x_1, x_2) = -x_1$, $\quad g_1(x_1, x_2) = x_2 - (1-x_1)^3$, $\quad g_2(x_1, x_2) = -x_2$.

(b) $f(x_1, x_2) = x_1 + x_2$, $\quad g_1(x_1, x_2) = x_1^2 + x_2^2 - 2$, $\quad g_2(x_1, x_2) = -x_1^2 - x_2^2 + 2$.

(c) $f(x_1, x_2) = x_1 + x_2$, $\quad g_1(x_1, x_2) = x_1^2 + x_2^2 - 2$, $\quad g_2(x_1, x_2) = x_1^2 + x_2^2 - 2$.

(d)
$$f(x_1, x_2) = x_1 + x_2,$$

$$g_1(x_1, x_2) = \begin{cases} -x_1^2 + x_2 & \text{if } x_1 \geq 0, \\ -x_1^4 + x_2 & \text{if } x_1 \leq 0, \end{cases}$$

$$g_2(x_1, x_2) = \begin{cases} x_1^4 - x_2 & \text{if } x_1 \geq 0, \\ x_1^2 - x_2 & \text{if } x_1 \leq 0. \end{cases}$$

4.3.9

Consider the equality constrained problem $\min_{h(x)=0} f(x)$, and suppose that x^* is a local minimum such that $h(x)$ is identically zero in a neighborhood of x^*. Use the Fritz John conditions to show that $\nabla f(x^*) = 0$.

4.3.10 (Relation of Fritz John and KKT multipliers) (www)

Consider the equality/inequality constrained problem

$$\begin{aligned} \text{minimize} \quad & f(x) \\ \text{subject to} \quad & h_1(x) = 0, \ldots, h_m(x) = 0, \\ & g_1(x) \leq 0, \ldots, g_r(x) \leq 0, \end{aligned}$$

let x^* be a (not necessarily regular) local minimum and assume that there exist multipliers $\lambda_1^*, \ldots, \lambda_m^*$ and $\mu_1^*, \ldots, \mu_r^*$ that satisfy the Karush-Kuhn-Tucker conditions of Prop. 4.3.1.

(a) Show by example that $\mu_0^* = 1$, together with $\lambda_1^*, \ldots, \lambda_m^*$ and $\mu_1^*, \ldots, \mu_r^*$, need not satisfy the Fritz John conditions of Prop. 4.3.5 (the CV condition may be violated).

(b) Show that there exist some $\lambda_1, \ldots, \lambda_m$ and $\mu_1, \ldots, \mu_r$ (possibly other than λ_i^* and μ_j^*), which together with $\mu_0 = 1$ satisfy all the Fritz John conditions of Prop. 4.3.5, and are such that the vectors $\nabla h_i(x^*)$ with $\lambda_i \neq 0$, and $\nabla g_j(x^*)$ with $\mu_j > 0$ are linearly independent. *Hint*: Express $\nabla f(x^*)$ as a linear combination of a minimal number of constraint gradients and show that the coefficients of the linear combination have the desired property.

4.3.11 (www)

Consider the problem of minimizing $f(x)$ subject to $g_j(x) \leq 0$, $j = 1, \ldots, r$, and assume that $g_1, \ldots, g_r$ are convex over $\Re^n$. Let x^* be a local minimum which together with multipliers $\mu_0^*, \mu_1^*, \ldots, \mu_r^*$ satisfies the Fritz John conditions of Prop. 4.3.5 with $\mu_0^* = 0$. Show that $g_j(x) = 0$ for all feasible vectors x and all j with $\mu_j^* > 0$. In particular, there exists an index j such that $g_j(x) = 0$ for all feasible vectors x. *Hint*: Write $\mu_j^* g_j(x) \geq \mu_j^* g_j(x^*) + \mu_j^* \nabla g_j(x^*)'(x - x^*)$ and add over j.

4.3.12 (An Alternative Fritz John Condition) (www)

Consider a version of Prop. 4.3.5, where the CV condition (iv) is replaced by the following: In every neighborhood N of x^* there is an $x \in N$ that simultaneously violates all the constraints corresponding to nonzero Lagrange multipliers. Moreover, if $(\lambda^*, \mu^*) \neq (0,0)$ this x can be chosen so that $f(x) < f(x^*)$. Show that this version of Prop. 4.3.5 is equivalent to the one given in the text, in the sense that each version of the proposition follows from the other. *Hint*: The condition above is clearly implied by the CV condition of Prop. 4.3.5 as given in the text. To show the reverse, replace each equality constraint $h_i(x) = 0$ with the two constraints $h_i(x) \leq 0$ and $-h_i(x) \leq 0$.

4.3.13 (Fritz John Conditions for Nondifferentiable Cost) (www)

Let x^* be a local minimum of the problem

$$\text{minimize } f(x)$$
$$\text{subject to } h_1(x) = 0, \ldots, h_m(x) = 0,$$
$$g_1(x) \leq 0, \ldots, g_r(x) \leq 0,$$

where f is a convex function, and h_i and g_j are continuously differentiable functions from $\Re^n$ to $\Re$.

(a) Adapt the proof of Prop. 4.3.5 to show that there exists a scalar μ_0^* and Lagrange multipliers $\lambda_1^*, \ldots, \lambda_m^*$ and $\mu_1^*, \ldots, \mu_r^*$, satisfying the following conditions:

   (i) The vector
   $$-\sum_{i=1}^m \lambda_i^* \nabla h_i(x^*) - \sum_{j=1}^r \mu_j^* \nabla g_j(x^*)$$
   is a subgradient of the function $\mu_0^* f(\cdot)$ at x^*.

   (ii) $\mu_j^* \geq 0$ for all $j = 0, 1, \ldots, r$.

   (iii) $\mu_0^*, \lambda_1^*, \ldots, \lambda_m^*, \mu_1^*, \ldots, \mu_r^*$ are not all equal to 0.

   (iv) In every neighborhood N of x^* there is an $x \in N$ such that $\lambda_i^* h_i(x) > 0$ for all i with $\lambda_i^* \neq 0$ and $\mu_j^* g_j(x) > 0$ for all j with $\mu_j^* \neq 0$. Moreover, if $(\lambda^*, \mu^*) \neq (0,0)$ this x can be chosen so that $f(x) < f(x^*)$.

Hint: Verify that if x^* is an unconstrained local minimum of $f_1(x) + f_2(x)$, where f_1 is convex and f_2 is continuously differentiable, then $-\nabla f_2(x^*)$ is a subgradient of f_1 at x^*. Use Prop. B.21 in Appendix B.

(b) Prove that one can take $\mu_0^* = 1$ if the constraint functions h_i and g_j are linear (cf. Prop. 4.3.7).

(c) Prove that one can take $\mu_0^* = 1$ under conditions that are analogous to the linear independence/interior point and Slater constraint qualifications (cf. Props. 4.3.8 and 4.3.9).

4.3.14 (www)

Consider the problem of finding a sphere of minimum radius that contains given vectors $y_1, \ldots, y_p$ in $\Re^n$. Formulate this problem as the minimax problem

$$\text{minimize} \quad \max\{\|x - y_1\|^2, \ldots, \|x - y_p\|^2\}$$
$$\text{subject to} \quad x \in \Re^n,$$

and write the corresponding optimality conditions of Prop. 4.3.10. Characterize the optimal solution and the Lagrange multipliers in the special case where $p = 3$.

4.3.15 (www)

Consider a subset X of $\Re^n$, and let x be a point in X. Show that $T(x)$, the tangent cone of X at x, has the following properties:

(a) $T(x)$ is a closed cone.

(b) If X is convex and contains at least one feasible direction at x, $T(x)$ is equal to the closure of the set of feasible directions at x.

4.3.16 (www)

Consider the subset of $\Re^n$ defined by

$$X = \{x \mid g_j(x) \leq 0, j = 1, \ldots, r\},$$

where $g_j : \Re^n \mapsto \Re$ are concave and differentiable functions over $\Re^n$. Show that every point in X is quasiregular.

4.3.17 (Feasible Directions for Infinitely Many Constraints) (www)

Consider the subset of $\Re^n$ defined by

$$X = \{x \mid g_j(x) \leq 0, j \in J\},$$

where J is an index set that is a compact subset of some Euclidean space. Assume that the g_j are real-valued continuously differentiable functions of x, and

Sec. 4.3   Inequality Constraints

furthermore, $\nabla g_j(x)$ is continuous as a function of j. Let x^* be a vector in X, and let $A(x^*) = \{j \mid g_j(x^*) = 0\}$. Let y be a vector such that $\nabla g_j(x^*)'y < 0$ for all $j \in A(x^*)$. Show that y is a feasible direction of X at x^* and belongs to the tangent cone $T(x^*)$. *Hint*: By the continuity of $\nabla g_j(x^*)$, there exists a neighborhood N of x^* and a neighborhood A of $A(x^*)$ (relative to J) such that $\nabla g_j(x)'y < 0$ for all $x \in N$ and $j \in A$. Furthermore, N can be chosen so that $g_j(x) < 0$ for all $x \in N$ and j in the compact set $J - A$. Choose $\delta > 0$ so that $x^* + \alpha y$ is in N when $0 \le \alpha \le \delta$. Show that $g_j(x^* + \alpha y) < 0$ for all α with $0 < \alpha \le \delta$.

4.3.18 (www)

Use the elimination approach to prove the existence of Lagrange multipliers under the linear independence/interior point and Slater constraint qualifications (cf. Props. 4.3.8 and 4.3.9) starting from the special case of this result where there are no equality constraints. *Hint*: As in Section 4.1.2, use the implicit function theorem to express m of the variables in terms of the remaining variables, while eliminating the equality constraints.

4.3.19 (Boundedness of the Set of Lagrange Multipliers [Gau77], [KlH98]) (www)

Let x^* be a local minimum of f subject to $h_i(x) = 0$, $i = 1,\ldots,m$, $g_j(x) \le 0$, $j = 1,\ldots,r$, where f, h_i, and g_j are continuously differentiable. Consider the set of pairs (λ^*,μ^*) that are Lagrange multipliers [i.e., satisfy $\mu^* \ge 0$, $\mu_j^* g_j(x^*) = 0$ for all j, and $\nabla L(x^*,\lambda^*,\mu^*) = 0$]. Show that this set is nonempty and bounded if and only if the linear independence/interior point constraint qualification (cf. Prop. 4.3.8) holds. *Hint*: Suppose that $\{\lambda^k,\mu^k\}$ is a sequence of Lagrange multipliers with $\|\lambda^k\| + \|\mu^k\| \to \infty$. Define

$$\beta^k = \frac{\lambda^k}{\|\lambda^k\|}, \qquad \gamma^k = \frac{\mu^k}{\|\mu^k\|},$$

and show that $\nabla h(x^*)\beta + \nabla g(x^*)\gamma = 0$ for all limit points (β,γ) of $\{\beta^k,\gamma^k\}$. Show that this is a contradiction of the linear independence/interior point constraint qualification. For the converse, use Gordan's theorem, given in Exercise 4.3.6. *Note*: Related results that apply to the case of an additional convex set constraint $x \in X$ (cf. Prop. 4.3.18) are given in Exercises 6.3.1 and 6.3.3.

4.3.20 (Pseudonormality, Quasinormality, and Quasiregularity) (www)

For the equality and inequality constraints of the problem (ICP) of Section 4.3, we say that a feasible vector x^* is *quasinormal* if there is no set of scalars $\lambda_1,\ldots,\lambda_m,\mu_1,\ldots,\mu_r$, such that:

(i) $\sum_{i=1}^{m}\lambda_i \nabla h_i(x^*) + \sum_{j=1}^{r}\mu_j \nabla g_j(x^*) = 0$.

(ii) $\mu_j \geq 0$, $\quad j = 1, \ldots, r$.

(iii) $\lambda_1, \ldots, \lambda_m, \mu_1, \ldots, \mu_r$ are not all equal to 0.

(vi) In every neighborhood N of x^* there is an $x \in N$ such that $\lambda_i h_i(x) > 0$ for all i with $\lambda_i \neq 0$ and $\mu_j g_j(x) > 0$ for all j with $\mu_j \neq 0$.

Thus if x^* is a quasinormal local minimum, the Fritz John conditions of Prop. 4.3.5 cannot be satisfied with $\mu_0^* = 0$, so that there exist corresponding Lagrange multipliers satisfying the corresponding necessary conditions. Moreover, it its clear that if x^* is pseudonormal it is also quasinormal (the reverse is not true).

Consider the case of two variables and two constraints, $h_1(x) = 0$ and $h_2(x) = 0$, given by

$$h_1(x) = x_2, \qquad h_2(x) = \begin{cases} x_1^4 \sin\left(\frac{1}{x_1}\right) - x_2 & \text{if } x_1 \neq 0, \\ 0 & \text{if } x_1 = 0. \end{cases}$$

Show that h_1 and h_2 are continuously differentiable, and that the vector $x^* = (0, 0)$ is quasiregular but not quasinormal or pseudonormal. *Note*: The paper [OzB04] explores the relations between pseudonormality, quasinormality, and quasiregularity in greater depth, including the case where there is an abstract set constraint $x \in X$.

4.3.21 (An Alternative Version of the Linear Independence/ Interior Point Constraint Qualification) (www)

Let x^* be a local minimum of the problem

$$\begin{aligned} \text{minimize} \quad & f(x) \\ \text{subject to} \quad & h_1(x) = 0, \ldots, h_m(x) = 0, \\ & g_1(x) \leq 0, \ldots, g_r(x) \leq 0, \end{aligned}$$

where h_i are linear functions, and f and g_j are continuously differentiable functions. Let J be the subset of indices $j \in A(x^*)$ such that g_j is a concave function. Assume that there exists a vector d such that

$$\nabla h_i(x^*)'d = 0, \quad i = 1, \ldots, m,$$

$$\nabla g_j(x^*)'d \leq 0, \quad \forall\, j \in J,$$

$$\nabla g_j(x^*)'d < 0, \quad \forall\, j \in A(x^*),\, j \notin J.$$

Then x^* satisfies the necessary conditions of Prop. 4.3.6. *Hint*: Combine the proofs of Props. 4.3.7 and 4.3.8.

4.3.22 (Enhanced Version of Farkas' Lemma [BNO03])

Let $a_1, \ldots, a_r$ and c be given vectors in $\Re^n$, and assume that $c \neq 0$. Show that we have

$$c'y \leq 0, \qquad \text{for all } y \text{ such that } a_j'y \leq 0, \ \forall \ j = 1, \ldots, r,$$

if and only if there exist nonnegative scalars $\mu_1, \ldots, \mu_r$ and a vector $\bar{y} \in \Re^n$ such that

$$c = \mu_1 a_1 + \cdots + \mu_r a_r, \qquad c'\bar{y} > 0,$$

$$a_j'\bar{y} > 0 \quad \text{for all } j \text{ with } \mu_j > 0,$$

$$a_j'\bar{y} \leq 0 \quad \text{for all } j \text{ with } \mu_j = 0.$$

Hint: Strengthen the proof of Prop. 4.3.12 to use the CV condition. *Note*: The classical version of Farkas' Lemma (Prop. 4.3.12) does not include the assertion on the existence of a $\bar{y}$ that satisfies $c'\bar{y} > 0$ while violating precisely those inequalities that correspond to positive multipliers. For a proof, see [BNO03], Prop. 5.4.2.

4.4 LINEAR CONSTRAINTS AND DUALITY

Problems with linear constraints have a remarkable property, which we derived in Prop. 4.3.7: they possess Lagrange multipliers always, even for local minima which are not regular. In this section we use this property to develop a duality theory for a problem with differentiable convex cost and linear constraints. In particular, we show that there is another problem, called *dual*, which has the same optimal value and has as optimal solutions the Lagrange multipliers of the original. This is an important relationship, which will be discussed extensively and generalized considerably in Chapter 6. Here, we develop some of the simpler aspects of duality theory, which follow easily from the developments so far in this chapter.

4.4.1 Convex Cost Function and Linear Constraints

As one expects based on the corresponding results of Section 4.1 and Prop. 4.3.4, if f is convex, the necessary optimality conditions for linear constraints can be proved to be sufficient. In fact a sharper necessary and sufficient condition can be derived, involving minimization of a Lagrangian function. This minimization may involve *any* subset of the inequality constraints, while the remaining constraints are taken into account using Lagrange multipliers, as shown in the following proposition. The flexibility of assigning Lagrange multipliers to only some of the constraints, while dealing with the other constraints explicitly is often very useful.

Proposition 4.4.1: (Optimality Conditions for Convex Cost and Linear Constraints) Consider the problem

$$\text{minimize } f(x)$$
$$\text{subject to } e_i'x = d_i, \quad i = 1, \ldots, m,$$
$$a_j'x \leq b_j, \quad j = 1, \ldots, r,$$

where e_i, a_j, and d_i, b_j are given vectors and scalars, respectively, and $f : \Re^n \mapsto \Re$ is convex and continuously differentiable. Let I be a subset of the index set $\{1, \ldots, m\}$, and J be a subset of the index set $\{1, \ldots, r\}$. Then x^* is a global minimum if and only if x^* is feasible and there exist scalars λ_i^*, $i \in I$, and μ_j^*, $j \in J$, such that

$$\mu_j^* \geq 0, \quad j \in J, \tag{4.84}$$

$$\mu_j^* = 0, \quad \forall\, j \in J \text{ with } j \notin A(x^*), \tag{4.85}$$

$$x^* \in \arg\min_{\substack{e_i'x = d_i,\ i \notin I \\ a_j'x \leq b_j,\ j \notin J}} \left\{ f(x) + \sum_{i \in I} \lambda_i^*(e_i'x - d_i) + \sum_{j \in J} \mu_j^*(a_j'x - b_j) \right\}. \tag{4.86}$$

Proof: Without loss of generality, we assume that all the constraints are inequalities (each linear equality constraint can be converted to two linear inequality constraints). Assume that x^* is a global minimum. Then by Prop. 4.3.7, there exist nonnegative scalars $\mu_1^*, \ldots, \mu_r^*$ such that

$$\mu_j^*(a_j'x^* - b_j) = 0, \quad j = 1, \ldots, r, \tag{4.87}$$

and

$$\nabla f(x^*) + \sum_{j=1}^{r} \mu_j^* a_j = 0.$$

Using the convexity of f, the last relation implies that

$$x^* \in \arg\min_{x \in \Re^n} \left\{ f(x) + \sum_{j=1}^{r} \mu_j^*(a_j'x - b_j) \right\}. \tag{4.88}$$

Combining Eqs. (4.87) and (4.88), we obtain

$$f(x^*) = \min_{x \in \Re^n} \left\{ f(x) + \sum_{j=1}^{r} \mu_j^*(a_j'x - b_j) \right\}. \tag{4.89}$$

Sec. 4.4 Linear Constraints and Duality

Since $\mu_j^* \geq 0$, it follows that

$$\mu_j^*(a_j'x - b_j) \leq 0 \quad \text{if} \quad a_j'x - b_j \leq 0,$$

so Eq. (4.89) implies that

$$f(x^*) \leq \min_{\substack{a_j'x \leq b_j \\ j \notin J}} \left\{ f(x) + \sum_{j=1}^r \mu_j^*(a_j'x - b_j) \right\}$$

$$= \inf_{\substack{a_j'x \leq b_j \\ j \notin J}} \left\{ f(x) + \sum_{j \in J} \mu_j^*(a_j'x - b_j) + \sum_{j \notin J} \mu_j^*(a_j'x - b_j) \right\} \quad (4.90)$$

$$\leq \inf_{\substack{a_j'x \leq b_j \\ j \notin J}} \left\{ f(x) + \sum_{j \in J} \mu_j^*(a_j'x - b_j) \right\}$$

$$\leq f(x^*) + \sum_{j \in J} \mu_j^*(a_j'x^* - b_j)$$

$$= f(x^*),$$

where the last equality holds by Eq. (4.87). Thus, x^* attains the minimum in the above relations, proving the desired Eq. (4.86).

Conversely, assume that x^* is feasible and there exist scalars μ_j^*, $j \in J$, satisfying conditions (4.84)-(4.86). Then, using Prop. 4.3.4, it follows that x^* is a global minimum. **Q.E.D.**

Note that given x^*, the vectors λ^* and μ^* satisfying conditions (4.84)-(4.86) in the preceding proof are independent of the index subsets I and J; they are the vectors λ^* and μ^* corresponding to $I = \{1, \ldots, m\}$ and $J = \{1, \ldots, r\}$, respectively, but they satisfy condition (4.86) for any subsets I and J. This is not surprising in view of the sensitivity interpretation of Lagrange multipliers.

Example 4.4.1 (Optimization Over a Simplex)

Consider the case where f is convex and the constraint set is the simplex

$$\left\{ x \; \middle| \; x \geq 0, \; \sum_{i=1}^n x_i = r \right\},$$

where $r > 0$ is a given scalar. We will use the optimality condition of Prop. 4.4.1 to recover the conditions obtained in Example 3.1.2 of Section 3.1.

By Prop. 4.4.1, if x^* is a global minimum, there exists a scalar λ^* such that

$$x^* \in \arg\min_{x \geq 0} \left\{ f(x) + \lambda^* \left(r - \sum_{i=1}^n x_i \right) \right\}. \quad (4.91)$$

By applying the optimality conditions for an orthant constraint (cf. Example 3.1.1 in Section 3.1), we obtain from Eq. (4.91)

$$\frac{\partial f(x^*)}{\partial x_i} \geq \lambda^*, \qquad i = 1, \ldots, n, \tag{4.92}$$

$$\frac{\partial f(x^*)}{\partial x_i} = \lambda^*, \qquad \text{if } x_i^* > 0. \tag{4.93}$$

We see therefore that all positive components x_i^* must have partial cost derivatives that are minimal and equal to the Lagrange multiplier λ^*. By Prop. 4.4.1, this condition is also sufficient for a feasible vector x^* to be optimal.

4.4.2 Duality Theory: A Simple Form for Linear Constraints

Consider the problem

$$\begin{aligned} \text{minimize} \quad & f(x) \\ \text{subject to} \quad & e_i' x = d_i, \quad i = 1, \ldots, m, \\ & a_j' x \leq b_j, \quad j = 1, \ldots, r, \quad x \in X, \end{aligned} \tag{P}$$

where e_i, a_j, and d_i, b_j are given vectors and scalars, respectively, $f : \Re^n \to \Re$ is a convex continuously differentiable function, and X is a polyhedral set, i.e., a set specified by a finite collection of linear equality and inequality constraints. We refer to problem (P) as the *primal problem*.

Define the Lagrangian function

$$L(x, \lambda, \mu) = f(x) + \sum_{i=1}^{m} \lambda_i (e_i' x - d_i) + \sum_{j=1}^{r} \mu_j (a_j' x - b_j).$$

Consider also the *dual function* defined by

$$q(\lambda, \mu) = \inf_{x \in X} L(x, \lambda, \mu). \tag{4.94}$$

The *dual problem* is

$$\begin{aligned} \text{maximize} \quad & q(\lambda, \mu) \\ \text{subject to} \quad & \lambda \in \Re^m, \quad \mu \geq 0. \end{aligned} \tag{D}$$

Note that if the polyhedral set X is bounded, then it is also compact (every polyhedral set is closed since it is the intersection of closed subspaces), and the infimum in Eq. (4.94) is attained for all (λ, μ) by Weierstrass' theorem (Prop. A.8 in Appendix A). Thus if X is bounded, the dual function takes real values. In general, however, $q(\lambda, \mu)$ can take the value $-\infty$. Thus in effect, the constraint set of the dual problem (D) is the set

$$Q = \{(\lambda, \mu) \mid \mu \geq 0, \; q(\lambda, \mu) > -\infty\}.$$

Sec. 4.4 Linear Constraints and Duality

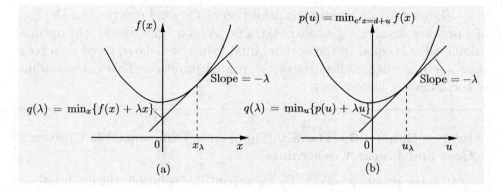

Figure 4.4.1. (a) Illustration of the dual function for the one-dimensional simple problem

$$\text{minimize } f(x)$$
$$\text{subject to } x = 0,$$

where x is a scalar. For a given λ, the dual value

$$q(\lambda) = \inf_x \{f(x) + \lambda x\}$$

is obtained by determining the point x_λ attaining the infimum and calculating $q(\lambda)$ as the crossing point of the vertical axis and the corresponding line of slope $-\lambda$ supporting the graph of f. It can be seen that the maximum value of $q(\lambda)$ is obtained when λ is equal to $-\nabla f(0)$, which is the Lagrange multiplier of the problem.

(b) Illustration of the dual function for the case of a single-constraint problem

$$\text{minimize } f(x)$$
$$\text{subject to } e'x = d_i,$$

where $x \in \Re^n$. This problem is equivalent to the one-dimensional problem

$$\text{minimize } p(u)$$
$$\text{subject to } u = 0,$$

where

$$p(u) = \inf_{\{x \in \Re^n | e'x = d_i + u\}} f(x).$$

(This is essentially the primal function; see the sensitivity theorem, Prop. 4.2.2.) The dual function can now be interpreted in terms of this problem as in (a). For the given λ, the dual value $q(\lambda)$ is obtained as the crossing point of the vertical axis and the corresponding line of slope $-\lambda$ supporting the graph of p. It can be seen that the maximum value of $q(\lambda)$ is obtained when λ is equal to $-\nabla p(0)$, which is the Lagrange multiplier of the problem (cf. the sensitivity theorem, Prop. 4.2.2).

Figure 4.4.1 illustrates the definition of the dual function for the case of a problem involving a single constraint. As the figure shows, the optimal primal value is equal to the optimal dual value, and the optimal dual solutions are Lagrange multipliers of the primal problem. This is the essence of the following basic result.

Proposition 4.4.2: (Duality Theorem - Differentiable Convex Cost and Linear Constraints)

(a) If the primal problem (P) has an optimal solution, the dual problem (D) also has an optimal solution and the corresponding optimal values are equal.

(b) In order for x^* to be an optimal primal solution and (λ^*, μ^*) to be an optimal dual solution, it is necessary and sufficient that x^* is primal feasible, $\mu^* \geq 0$, $\mu_j^* = 0$ for all $j \notin A(x^*)$, and

$$x^* \in \arg\min_{x \in X} L(x, \lambda^*, \mu^*). \tag{4.95}$$

Proof: (a) For all primal feasible x, and all $\lambda \in \Re^m$ and $\mu \geq 0$, we have $\lambda_i(e_i'x - d_i) = 0$ and $\mu_j'(a_j'x - b_j) \leq 0$, so that

$$q(\lambda, \mu) \leq f(x) + \sum_{i=1}^m \lambda_i(e_i'x - d_i) + \sum_{j=1}^r \mu_j(a_j'x - b_j) \leq f(x). \tag{4.96}$$

By taking the minimum of the right-hand side over all primal feasible x, we obtain

$$q(\lambda, \mu) \leq f(x^*), \quad \forall\, \lambda \in \Re^m,\ \mu \geq 0, \tag{4.97}$$

where x^* is a primal optimal solution. By Prop. 4.4.1, there exist $\lambda^* \in \Re^m$ and $\mu^* \geq 0$ such that $\mu_j^*(a_j'x^* - b_j) = 0$ for all j, and

$$x^* \in \arg\min_{x \in X} L(x, \lambda^*, \mu^*),$$

so by using also the definition (4.94) of q, we have

$$q(\lambda^*, \mu^*) = \inf_{x \in X} L(x, \lambda^*, \mu^*)$$
$$= L(x^*, \lambda^*, \mu^*)$$
$$= f(x^*) + \sum_{i=1}^m \lambda_i^*(e_i'x^* - d_i) + \sum_{j=1}^r \mu_j^*(a_j'x^* - b_j)$$
$$= f(x^*).$$

Sec. 4.4   Linear Constraints and Duality

By combining this equation with Eq. (4.97), we see that (λ^*, μ^*) is a dual optimal solution and that $q(\lambda^*, \mu^*) = f(x^*)$.

(b) If x^* is primal optimal and (λ^*, μ^*) is dual optimal, then $\mu^* \geq 0$ since (λ^*, μ^*) must be dual feasible. Furthermore, by part (a) we obtain

$$f(x^*) = q(\lambda^*, \mu^*),$$

which when combined with Eq. (4.96), yields

$$f(x^*) = L(x^*, \lambda^*, \mu^*) = q(\lambda^*, \mu^*) = \min_{x \in X} L(x, \lambda^*, \mu^*).$$

Conversely, the given conditions imply the above relation. Hence, since x^* is primal feasible and $\mu^* \geq 0$, Eq. (4.96) implies that x^* is primal optimal and (λ^*, μ^*) is dual optimal. **Q.E.D.**

The duality assertions of Prop. 4.4.2 were proved under hypotheses which are stronger than necessary. For example, the differentiability assumption on f is not needed – it turns out that convexity is sufficient. Furthermore, the existence of a primal optimal solution is not needed for the conclusion of Prop. 4.4.2(a) – it turns out that finiteness of the infimum of the primal cost is sufficient. However, to prove these and other extensions of the duality theorem, we need an approach that is not calculus-based like the one of this chapter, but is based instead on the geometry of convex sets; see Chapter 6.

Example 4.4.2 (The Dual of a Linear Program)

Consider the linear program

$$\begin{aligned}&\text{minimize} \quad c'x \\ &\text{subject to} \quad e_i' x = d_i, \quad i = 1, \ldots, m, \quad x \geq 0,\end{aligned} \quad \text{(LP)}$$

where c and e_i are given vectors in $\Re^n$, and d_i are given scalars. We consider the dual function

$$q(\lambda) = \inf_{x \geq 0} \left\{ \sum_{j=1}^{n} \left(c_j - \sum_{i=1}^{m} \lambda_i e_{ij} \right) x_j + \sum_{i=1}^{m} \lambda_i d_i \right\},$$

where e_{ij} is the jth component of the vector e_i. Clearly, if $c_j - \sum_{i=1}^{m} \lambda_i e_{ij} \geq 0$ for all j, the infimum above is attained for $x = 0$, and we have $q(\lambda) = \sum_{i=1}^{m} \lambda_i d_i$. On the other hand, if $c_j - \sum_{i=1}^{m} \lambda_i e_{ij} < 0$ for some j, we can make the expression in braces above arbitrarily small by taking x_j sufficiently large, so that $q(\lambda) = -\infty$ in this case. Thus, the dual problem is

$$\begin{aligned}&\text{maximize} \quad \sum_{i=1}^{m} \lambda_i d_i \\ &\text{subject to} \quad \sum_{i=1}^{m} \lambda_i e_{ij} \leq c_j, \quad j = 1, \ldots, n.\end{aligned} \quad \text{(DLP)}$$

By the optimality conditions of Prop. 4.4.2, (x^*, λ^*) is a primal and dual optimal solution pair if and only if x^* is primal feasible, and the Lagrangian optimality condition

$$x^* \in \arg\min_{x \geq 0} \left\{ \left(c - \sum_{i=1}^{m} \lambda_i^* e_i\right)' x + \sum_{i=1}^{m} \lambda_i^* d_i \right\}$$

holds [cf. Eq. (4.95)]. The above condition is equivalent to $c \geq \sum_{i=1}^{m} \lambda_i^* e_i$ (otherwise the minimum above would not be achieved at x^*), $x^* \geq 0$, and the following two relations:

$$x_j^* > 0 \quad \Rightarrow \quad \sum_{i=1}^{m} \lambda_i^* e_{ij} = c_j, \quad j = 1, \ldots, n, \quad (4.98)$$

$$\sum_{i=1}^{m} \lambda_i^* e_{ij} < c_j \quad \Rightarrow \quad x_j^* = 0, \quad j = 1, \ldots, n, \quad (4.99)$$

which are known as the *CS conditions for linear programming*.

Let us consider now the dual of the dual problem (DLP). We first convert this problem into the equivalent minimization problem

$$\text{minimize} \quad \sum_{i=1}^{m} (-d_i) \lambda_i$$

$$\text{subject to} \quad \sum_{i=1}^{m} \lambda_i e_{ij} \leq c_j, \quad j = 1, \ldots, n.$$

Assigning a Lagrange multiplier x_j to the jth inequality constraint, the dual function of this problem is given by

$$p(x) = \inf_{\lambda \in \Re^m} \left\{ \sum_{i=1}^{m} \left(\sum_{j=1}^{n} e_{ij} x_j - d_i \right) \lambda_i - \sum_{j=1}^{n} c_j x_j \right\}$$

$$= \begin{cases} -c'x & \text{if } e_i'x = d_i, \ i = 1, \ldots, m, \\ -\infty & \text{otherwise.} \end{cases}$$

The corresponding dual problem is

$$\text{maximize} \quad p(x)$$
$$\text{subject to} \quad x \geq 0,$$

or equivalently

$$\text{minimize} \quad c'x$$
$$\text{subject to} \quad e_i'x = d_i, \ i = 1, \ldots, m, \quad x \geq 0,$$

Sec. 4.4   Linear Constraints and Duality

which is identical to the primal problem (LP). We have thus shown that the duality is symmetric, i.e., *the dual of the dual problem (DLP) is the primal problem (LP)*.

The pair of primal and dual linear programs (LP) and (DLP) can also be written compactly in terms of the $m \times n$ matrix E having rows $e'_1, \ldots, e'_m$, and the vector d having components $d_1, \ldots, d_m$:

$$\min_{Ex=d,\ x\geq 0} c'x \quad \Longleftrightarrow \quad \max_{E'\lambda \leq c} d'\lambda.$$

Other linear programming duality relations that can be verified by the reader include the following:

$$\min_{A'x \geq b} c'x \quad \Longleftrightarrow \quad \max_{A\mu = c,\ \mu \geq 0} b'\mu,$$

$$\min_{A'x \geq b,\ x \geq 0} c'x \quad \Longleftrightarrow \quad \max_{A\mu \leq c,\ \mu \geq 0} b'\mu;$$

see Exercise 4.4.3.

Example 4.4.3 (The Dual of a Quadratic Program)

Consider the quadratic programming problem

$$\begin{aligned}\text{minimize} \quad & \tfrac{1}{2}x'Qx + c'x \\ \text{subject to} \quad & Ax \leq b,\end{aligned} \qquad \text{(QP)}$$

where Q is a given $n \times n$ positive definite symmetric matrix, A is a given $r \times n$ matrix, and $b \in \Re^r$ and $c \in \Re^n$ are given vectors. The dual function is

$$q(\mu) = \inf_{x \in \Re^n} \left\{ \tfrac{1}{2}x'Qx + c'x + \mu'(Ax - b) \right\}.$$

The infimum is attained for

$$x = -Q^{-1}(c + A'\mu),$$

and a straightforward calculation after substituting this expression in the preceding relation, yields

$$q(\mu) = -\tfrac{1}{2}\mu' AQ^{-1}A'\mu - \mu'(b + AQ^{-1}c) - \tfrac{1}{2}c'Q^{-1}c.$$

The dual problem, after dropping the constant $\tfrac{1}{2}c'Q^{-1}c$ and changing the minus sign to convert the maximization to a minimization, can be written as

$$\begin{aligned}\text{minimize} \quad & \tfrac{1}{2}\mu' P\mu + t'\mu \\ \text{subject to} \quad & \mu \geq 0,\end{aligned} \qquad \text{(DQP)}$$

where
$$P = AQ^{-1}A', \qquad t = b + AQ^{-1}c.$$

If μ^* is any dual optimal solution, then the optimal solution of the primal problem is
$$x^* = -Q^{-1}(c + A'\mu^*).$$

Note that the dual problem is also a quadratic program, but it has simpler constraints than the primal. Furthermore, if the row dimension r of A is smaller than its column dimension n, the dual problem is defined on a space of smaller dimension than the primal, and this can be algorithmically significant.

We finally note that when a convex quadratic program has a finite optimal value, it also has at least one primal optimal solution (cf. Prop. 3.1.3), so the existence assumption of Prop. 4.4.2(a) is satisfied. It follows that *for a convex quadratic program with a finite optimal value, both the primal and the dual programs have an optimal solution and their optimal values are equal.*

EXERCISES

4.4.1

A company has available for sale a quantity of Q units of a certain product that will be sold in n market outlets. For each $i = 1, \ldots, n$, the quantity d_i, which is demanded at outlet i, and the price of sale p_i are known. The company wishes to determine the quantities s_i^* with $0 \le s_i^* \le d_i$ to be sold at each outlet that maximize the revenue
$$\sum_{i=1}^{n} p_i s_i$$
from the sale. Assuming that
$$\sum_{i=1}^{n} d_i \ge Q, \qquad d_i > 0, \qquad p_i > 0, \qquad i = 1, \ldots, n,$$
show that there exists a cutoff price level y such that for each i, if $p_i > y$, then $s_i^* = d_i$, and if $p_i < y$, then $s_i^* = 0$. What happens if $p_i = y$? Describe a procedure for obtaining s_i^*.

4.4.2

Consider a problem of finding the best way to place bets totalling A dollars ($A > 0$) in a race involving n horses. Assume that we know the probability p_i, that the ith horse wins, and the amount $s_i > 0$ that the rest of the public is betting on the ith horse. The track keeps a proportion $1 - c$ of the total amount ($0 < 1 - c < 1$) and distributes the rest among the public in proportion to the amounts bet on the winning horse. Thus, if we bet x_i on the ith horse, we receive

$$c\left(A + \sum_{i=1}^{n} s_i\right) \frac{x_i}{s_i + x_i}$$

if the ith horse wins. The problem is to find x_i^*, $i = 1, \ldots, n$, which maximize the expected net return

$$c\left(A + \sum_{i=1}^{n} s_i\right) \left\{\sum_{i=1}^{n} \frac{p_i x_i}{s_i + x_i}\right\} - A,$$

or, equivalently,

$$\sum_{i=1}^{n} \frac{p_i x_i}{s_i + x_i}$$

subject to

$$\sum_{i=1}^{n} x_i = A, \qquad x_i \geq 0, \quad i = 1, \ldots, n.$$

Assume that

$$\frac{p_1}{s_1} > \frac{p_2}{s_2} > \cdots > \frac{p_n}{s_n},$$

and show that there exists a scalar λ^* such that the optimal solution is

$$x_i^* = \begin{cases} \sqrt{\frac{s_i p_i}{\lambda^*}} - s_i & \text{for } i = 1, \ldots, m^*, \\ 0 & \text{for } i = m^* + 1, \ldots, n, \end{cases}$$

where m^* is the largest index m for which

$$\frac{p_m}{s_m} \geq \lambda^*.$$

4.4.3 (Pairs of Dual Linear Programs) (www)

Verify the linear programming duality relations

$$\min_{A'x \geq b} c'x \quad \Longleftrightarrow \quad \max_{A\mu = c, \ \mu \geq 0} b'\mu,$$

$$\min_{A'x \geq b, \ x \geq 0} c'x \quad \Longleftrightarrow \quad \max_{A\mu \leq c, \ \mu \geq 0} b'\mu,$$

show that they are symmetric (i.e., the dual of the dual is the primal), and derive the corresponding CS conditions [cf. Eqs. (4.98) and (4.99)].

4.4.4 (Duality for Transportation Problems) (www)

Suppose that a quantity of a certain material must be shipped from m supply points to n demand points so as to minimize the total transportation cost. The supply at point i is denoted α_i and the demand at point j is denoted β_j. The unit transportation cost from i to j is a_{ij}. Letting x_{ij} denote the quantity shipped from supply point i to demand point j, the problem is

$$\text{minimize} \sum_{i,j} a_{ij} x_{ij}$$

$$\text{subject to}$$

$$\sum_{j=1}^{n} x_{ij} = \alpha_i, \quad i = 1, \ldots, m,$$

$$\sum_{i=1}^{m} x_{ij} = \beta_j, \quad j = 1, \ldots, n,$$

$$0 \leq x_{ij}, \quad \forall\ i,\ j,$$

where α_i and β_j are positive scalars, which for feasibility must satisfy

$$\sum_{i=1}^{m} \alpha_i = \sum_{j=1}^{n} \beta_j.$$

(a) Assign Lagrange multipliers to the equality constraints and derive the corresponding dual problem.

(b) Introduce a price p_j that demand point j will pay per unit delivered at j. Show that if x^* is an optimal solution of the transportation problem, there is a set of prices $\{p_j^* \mid j = 1, \ldots, n\}$ such that if $x_{ij}^* > 0$ then j offers maximum net profit for i, i.e.,

$$p_j^* - a_{ij} = \max_{k=1,\ldots,n} \{p_k^* - a_{ik}\}.$$

(c) Relate prices to dual feasible solutions, and show that every dual optimal solution has a property of the type described in (b) above.

4.4.5 (Duality and Zero Sum Games) (www)

Let A be an $n \times m$ matrix, and let X and Z be the unit simplices in $\Re^n$ and $\Re^m$, respectively:

$$X = \left\{ x \,\Big|\, \sum_{i=1}^{n} x_i = 1,\ x_i \geq 0,\ i = 1, \ldots, n \right\},$$

$$Z = \left\{ z \,\Big|\, \sum_{j=1}^{m} z_j = 1,\ z_j \geq 0,\ j = 1, \ldots, m \right\}.$$

Use linear programming duality to show that

$$\max_{z \in Z} \min_{x \in X} x' A z = \min_{x \in X} \max_{z \in Z} x' A z;$$

this is the classical *saddle point theorem* for zero sum games. *Hint*: For a fixed z,

$$\min_{x \in X} x' A z$$

is equal to the minimum component of the vector Az, so

$$\max_{z \in Z} \min_{x \in X} x' A z = \max_{z \in Z} \min\{(Az)_1, \ldots, (Az)_n\} = \max_{\xi e \leq Az, \, z \in Z} \xi, \quad (4.100)$$

where e is the unit vector in $\Re^n$ (all components are equal to 1). Similarly,

$$\min_{x \in X} \max_{z \in Z} x' A z = \min_{\zeta e \geq A' x, \, x \in X} \zeta. \quad (4.101)$$

Show that the linear programs in the right-hand sides of Eqs. (4.100) and (4.101) are dual to each other.

4.5 NOTES AND SOURCES

Section 4.1: The treatment of Lagrange multipliers using penalty functions was introduced in the nonlinear programming literature in the paper by McShane [McS73], and the book by Hestenes [Hes75]; see also Rockafellar [Roc93], who considered the case where there is an additional abstract set constraint (cf. Section 4.3.6). For an alternative proof of the Lagrange multiplier theorem, using an approach based on differentiable curves, see Hestenes [Hes75] and Luenberger [Lue84]. In this approach, starting from a local minimum we consider small variations along the (possibly nonlinear) constraint surface, and argue that the cost function cannot decrease asymptotically as the variations tend to zero. In particular, differentiable curves $x(t)$, parameterized by a scalar t, are constructed using the implicit function theorem so that they lie on the constraint surface and pass through x^*, i.e., $x^* = x(0)$. The unconstrained optimality condition

$$\left. \frac{df(x(t))}{dt} \right|_{t=0} = 0$$

is then applied to yield the first order Lagrange multiplier condition. This approach is insightful, but is considerably more complicated than the penalty approach we have followed, particularly for the case of inequality constraints.

Section 4.2: For textbook treatments of sensitivity, see Fiacco [Fia78], and Bonnans and Shapiro [BoS00]. For sensitivity analysis and related topics under weaker assumptions than the second order sufficiency conditions assumed here, see Dontchev and Jongen [DoJ86], Robinson [Rob87], Gauvin and Janin [GaJ88], Shapiro [Sha88], Auslender and Cominetti [AuC90], Bonnans [Bon92], King and Rockafellar [KiR92], Ioffe [Iof94], Bonnans, Cominetti, and Shapiro [BCS99], Bonnans and Shapiro [BoS00], and Dontchev and Rockafellar [DoR09]. The paper by Bertsekas [Ber05c] showed that when there are multiple Lagrange multipliers, the multiplier of minimum norm defines the optimal rate of improvement of the cost per unit constraint violation (see Prop. 4.3.19). A version of this result for the case of a convex programming problem will be given in Section 6.4.5.

Section 4.3: Lagrange multiplier theorems for inequality constraints became available considerably later than their equality constraint counterparts. Important early works are those of Karush [Kar39] (an unpublished MS thesis), John [Joh48], and Kuhn and Tucker [KuT51]. The survey by Kuhn [Kuh76] gives a historical account of the development of the subject.

There has been considerable effort to derive constraint qualifications guaranteeing the existence of Lagrange multipliers. Important examples are those of Arrow, Hurwicz, and Uzawa [AHU61], Dubovitskii and Milyutin [DuM65], Abadie [Aba67], Mangasarian and Fromovitz [MaF67], and Guignard [Gui69]. For textbook treatments see Mangasarian [Man69] and Hestenes [Hes75]. There has been much subsequent work on the subject, some of which addresses nondifferentiable problems, e.g., Gould and Tolle [GoT71], [GoT72], Bazaraa, Goode, and Shetty [BGS72], Ben-Tal and Zowe [BeZ82], Clarke [Cla83], Demjanov and Vasilév [DeV85], Mordukhovich [Mor88], and Rockafellar [Roc93]. For a more detailed treatment of our line of analysis in this book, including that case of nonconvex abstract set constraints, and the relation between the existence of Lagrange multipliers and the existence of exact nondifferentiable penalty functions, to be discussed in Section 5.3, see the book by Bertsekas, Nedić, and Ozdaglar [BNO03], Section 5.5.

Semi-infinite programming arises often in engineering applications, approximation theory, and optimal control. For further material on optimality conditions and computational methods, see the articles in the edited volume by Reemtsen and Ruckman [ReR98], and the book by Polak [Pol97].

Our treatment of the Fritz John theory, and its use in obtaining Lagrange multiplier theorems differs significantly from standard treatments. The starting point is the work of Hestenes [Hes75], who in turn relied on the penalty-based proof of the Karush-Kuhn-Tucker theorem given by McShane [McS73]. Hestenes introduced the notion of constraint quasinormality for the case where there is no abstract constraint set (see Exercise 4.3.20). The notion of pseudonormality and the extension of the Fritz John theory for the case where there is an abstract (possibly nonconvex) constraint set X were introduced by Bertsekas and Ozdaglar [BeO02], and are discussed in detail in the book [BNO03].

The Fritz John theory of this section becomes stronger in the case of a convex programming problem; see the paper by Bertsekas, Ozdaglar, and Tseng [BOT06]. For an illustrative result, consider the inequality constrained problem

$$\text{minimize } f(x)$$
$$\text{subject to } g_1(x) \leq 0, \ldots, g_r(x) \leq 0, \quad x \in X, \quad (4.102)$$

where $f, g_j : \Re^n \mapsto \Re$, $j = 1, \ldots, r$, are convex functions and X is a closed convex set. Assume further that the problem has an optimal solution, denoted x^*, and let $f^* = f(x^*)$ be the optimal value. We have the following analog of Prop. 4.3.5.

Proposition 4.5.1: (Fritz John Optimality Conditions for Convex Cost and Constraints) Under the preceding assumptions for problem (4.102), there exist a scalar μ_0^* and a vector $\mu^* = (\mu_1^*, \ldots, \mu_r^*)'$ satisfying the following conditions:

(i) $\mu_0^* f^* = \inf_{x \in X} \{\mu_0^* f(x) + \mu^{*\prime} g(x)\}$.

(ii) $\mu_j^* \geq 0$ for all $j = 0, 1, \ldots, r$.

(iii) $\mu_0^*, \mu_1^*, \ldots, \mu_r^*$ are not all equal to 0.

(iv) If $\mu^* \neq 0$, then there exists a sequence $\{x^k\} \subset X$ of infeasible points that converges to x^* and satisfies

$$f(x^k) \to f^*, \quad g^+(x^k) \to 0,$$

$$\frac{f^* - f(x^k)}{\|g^+(x^k)\|} \to \begin{cases} \|\mu^*\|/\mu_0^* & \text{if } \mu_0^* \neq 0, \\ \infty & \text{if } \mu_0^* = 0, \end{cases}$$

$$\frac{g^+(x^k)}{\|g^+(x^k)\|} \to \frac{\mu^*}{\|\mu^*\|},$$

where we denote by $g^+(x)$ the vector of constraint violations, i.e., the vector with components

$$g_j^+(x) = \max\{0, g_j(x)\}, \quad j = 1, \ldots, r, \ x \in X.$$

For the proof, see [BOT06], Prop. 2. Additional related results are given in [BOT06], including a sensitivity result that will be discussed in Section 6.4.5 (cf. Prop. 6.4.11).

Section 4.4: Duality theory has its origins in the work of von Neumann on zero sum games. This connection is illustrated in Exercise 4.4.5. The proof of linear programming duality was given by Gale, Kuhn, and Tucker [GKT51].

While the analysis of this section covers the important cases of linear programming and convex quadratic programming, it suffers from lack of generality: it requires differentiability of the cost function and linearity of the constraints. In Chapter 6, we will give a more general and powerful development of duality, and will provide a geometrical framework that allows insightful visualization of duality results. Moreover, in Chapter 7 we will use duality as a foundation for a variety of algorithms, which aim to exploit the special structure of large-scale problems.

5
Lagrange Multiplier Algorithms

Contents

5.1. Barrier and Interior Point Methods	p. 446
5.1.1. Path Following Methods for Linear Programming	p. 450
5.1.2. Primal-Dual Methods for Linear Programming	p. 458
5.2. Penalty and Augmented Lagrangian Methods	p. 469
5.2.1. The Quadratic Penalty Function Method	p. 471
5.2.2. Multiplier Methods – Main Ideas	p. 479
5.2.3. Convergence Analysis of Multiplier Methods	p. 488
5.2.4. Duality and Second Order Multiplier Methods	p. 492
5.2.5. Nonquadratic Augmented Lagrangians - The Exponential Method of Multipliers	p. 494
5.3. Exact Penalties – Sequential Quadratic Programming	p. 502
5.3.1. Nondifferentiable Exact Penalty Functions	p. 503
5.3.2. Sequential Quadratic Programming	p. 513
5.3.3. Differentiable Exact Penalty Functions	p. 520
5.4. Lagrangian Methods	p. 527
5.4.1. First-Order Lagrangian Methods	p. 528
5.4.2. Newton-Like Methods for Equality Constraints	p. 535
5.4.3. Global Convergence	p. 545
5.4.4. A Comparison of Various Methods	p. 548
5.5. Notes and Sources	p. 550

In this chapter, we consider several computational methods for problems with equality and inequality constraints. All of these methods use some form of Lagrange multiplier estimates and typically provide in the limit not just stationary points of the original problem but also associated Lagrange multipliers. Additional methods using Lagrange multipliers will also be discussed in the next two chapters after the development of duality theory.

The methods of this chapter are based on one of the following two ideas:

(a) *Using a penalty or a barrier function*. Here a constrained problem is converted into a sequence of unconstrained problems, which involve an added high cost either for infeasibility or for approaching the boundary of the feasible region via its interior. These methods are discussed in Sections 5.1-5.3, and include interior point linear programming methods based on the logarithmic barrier function, augmented Lagrangian methods, and sequential quadratic programming.

(b) *Solving the necessary optimality conditions*, viewing them as a system of equations and/or inequalities in the problem variables and the associated Lagrange multipliers. These methods are first discussed in Section 5.1.2 in a specialized linear programming context, and later in Section 5.4. For nonlinear programming problems, they guarantee only local convergence in their pure form; that is, they converge only when a good solution estimate is initially available. However, their convergence region can be enlarged by using various schemes that involve penalty and barrier functions.

The methods based on these two ideas turn out to be quite interconnected, and as an indication of this, we note the derivations of optimality conditions using penalty and augmented Lagrangian techniques in Chapter 4. Generally, the methods of this chapter are particularly well-suited for nonlinear constraints, because, contrary to the feasible direction methods of Chapter 3, they do not involve projections or direction finding subproblems, which tend to become more difficult when the constraints are nonlinear. Still, however, some of the methods of this chapter are very well suited for linear and quadratic programming problems, thus illustrating the power of blending descent, penalty/barrier, and Lagrange multiplier ideas within a common algorithmic framework.

5.1 BARRIER AND INTERIOR POINT METHODS

Barrier methods apply to inequality constrained problems of the form

$$\text{minimize } f(x) \qquad \qquad (5.1)$$
$$\text{subject to } x \in X, \qquad g_j(x) \leq 0, \quad j = 1, \ldots, r,$$

Sec. 5.1 Barrier and Interior Point Methods

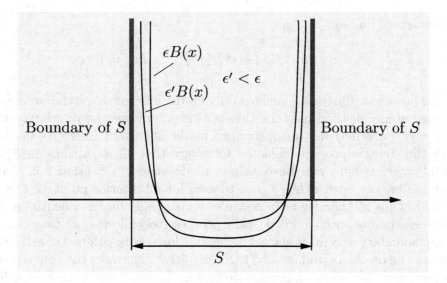

Figure 5.1.1. Form of a barrier function.

where X is a closed set, and $f : \Re^n \mapsto \Re$ and $g_j : \Re^n \mapsto \Re$ are given functions, with f continuous. The interior (relative to X) of the set defined by the inequality constraints is

$$S = \{x \in X \mid g_j(x) < 0, j = 1, \ldots, r\}.$$

We assume that S is nonempty and that any feasible point that is not in S can be approached arbitrarily closely by a vector from S; that is, given any feasible x and any $\delta > 0$, there exists $\tilde{x} \in S$ such that $\|\tilde{x} - x\| \leq \delta$. This property holds automatically if X and the constraint functions g_j are convex, as can be seen using the line segment principle [Prop. B.7(a) in Appendix B].

In barrier methods, we add to the cost a function $B(x)$ that is defined in the interior set S. This function, called the *barrier function*, is continuous and goes to ∞ as any one of the constraints $g_j(x)$ approaches 0 from negative values. The two most common examples of barrier functions are:

$$B(x) = -\sum_{j=1}^{r} \ln\{-g_j(x)\}, \qquad \text{logarithmic,}$$

$$B(x) = -\sum_{j=1}^{r} \frac{1}{g_j(x)}, \qquad \text{inverse.}$$

Note that both of these barrier functions are convex if all the constraint functions g_j are convex. Figure 5.1.1 illustrates the form of $B(x)$.

The barrier method is defined by introducing a parameter sequence $\{\epsilon^k\}$ with

$$0 < \epsilon^{k+1} < \epsilon^k, \quad k = 0, 1, \ldots, \qquad \epsilon^k \to 0.$$

It consists of finding

$$x^k \in \arg\min_{x \in S}\{f(x) + \epsilon^k B(x)\}, \qquad k = 0, 1, \ldots$$

Since the barrier function is defined only on the interior set S, the successive iterates of any method used for this minimization must be interior points. If $X = \Re^n$, one may use unconstrained methods such as Newton's method with the stepsize properly selected to ensure that all iterates lie in S; an initial interior point can be obtained as discussed in Section 3.2. Note that the barrier term $\epsilon^k B(x)$ goes to zero for all interior points $x \in S$ as $\epsilon^k \to 0$. Thus the barrier term becomes increasingly inconsequential as far as interior points are concerned, while progressively allowing x^k to get closer to the boundary of S (as it should if the solutions of the original constrained problem lie on the boundary of S). Figure 5.1.2 illustrates the convergence process, and the following proposition gives the main convergence result.

Proposition 5.1.1: Every limit point of a sequence $\{x^k\}$ generated by a barrier method is a global minimum of the original constrained problem (5.1).

Proof: Let $\{\bar{x}\}$ be the limit of a subsequence $\{x^k\}_{k \in K}$. If $\bar{x} \in S$, we have $\lim_{k \to \infty,\, k \in K} \epsilon^k B(x^k) = 0$, while if $\bar{x}$ lies on the boundary of S, we have by assumption $\lim_{k \to \infty,\, k \in K} B(x^k) = \infty$. In either case we obtain

$$\liminf_{k \to \infty} \epsilon^k B(x^k) \geq 0,$$

which implies that

$$\liminf_{k \to \infty,\, k \in K}\{f(x^k) + \epsilon^k B(x^k)\} = f(\bar{x}) + \liminf_{k \to \infty,\, k \in K}\{\epsilon^k B(x^k)\} \geq f(\bar{x}). \quad (5.2)$$

The vector $\bar{x}$ is a feasible point of the original problem (5.1), since $x^k \in S$ and X is a closed set. If $\bar{x}$ were not a global minimum, there would exist a feasible vector x^* such that $f(x^*) < f(\bar{x})$. Therefore, using the continuity of f and our assumption that x^* can be approached arbitrarily closely through the interior set S, there would also exist an interior point $\hat{x} \in S$ such that $f(\hat{x}) < f(\bar{x})$. We now have by the definition of x^k,

$$f(x^k) + \epsilon^k B(x^k) \leq f(\hat{x}) + \epsilon^k B(\hat{x}), \qquad k = 0, 1, \ldots,$$

which by taking the limit as $k \to \infty$ and $k \in K$, implies together with Eq. (5.2), that $f(\bar{x}) \leq f(\hat{x})$. This is a contradiction, thereby proving that $\bar{x}$ is a global minimum of the original problem. **Q.E.D.**

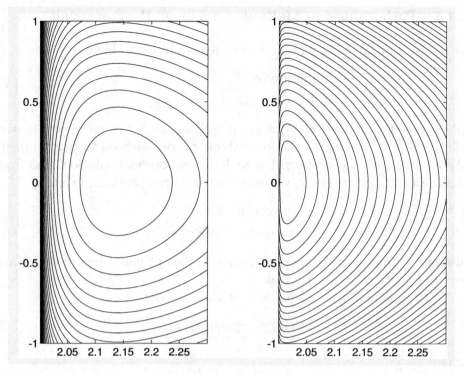

Figure 5.1.2. The convergence process of the barrier method for the problem

$$\text{minimize} \quad f(x) = \tfrac{1}{2}\left(x_1^2 + x_2^2\right)$$
$$\text{subject to} \quad 2 \leq x_1,$$

with optimal solution $x^* = (2, 0)$. For the case of the logarithmic barrier function $B(x) = -\ln(x_1 - 2)$, we have

$$x^k \in \arg\min_{x_1 > 2} \left\{ \tfrac{1}{2}\left(x_1^2 + x_2^2\right) - \epsilon^k \ln(x_1 - 2) \right\} = \left(1 + \sqrt{1 + \epsilon^k}, 0\right),$$

so as ϵ^k is decreased, the unconstrained minimum x^k approaches the constrained minimum $x^* = (2, 0)$. The figure shows the equal cost surfaces of

$$f(x) + \epsilon B(x)$$

for $\epsilon = 0.3$ (left side) and $\epsilon = 0.03$ (right side).

The logarithmic barrier function has been central to much research on methods that generate successive iterates lying in the interior set S. These methods are generically referred to as *interior point methods*, and have been extensively applied to linear and quadratic programming problems following the influential paper [Kar84]. We proceed to discuss the linear programming case in detail, using two different approaches based on the logarithmic barrier.

5.1.1 Path Following Methods for Linear Programming

In this section we consider the linear programming problem

$$\begin{aligned} \text{minimize} \quad & c'x \\ \text{subject to} \quad & Ax = b, \quad x \geq 0, \end{aligned} \quad \text{(LP)}$$

where $c \in \Re^n$ and $b \in \Re^m$ are given vectors, and A is an $m \times n$ matrix of rank m. We will adapt the logarithmic barrier method to this problem. We assume that the problem has at least one optimal solution, and from the theory of Section 4.4.2, we have that the dual problem, given by

$$\begin{aligned} \text{maximize} \quad & b'\lambda \\ \text{subject to} \quad & A'\lambda \leq c, \end{aligned} \quad \text{(DP)}$$

also has an optimal solution. Furthermore, the optimal values of the primal and the dual problem are equal.

The method involves finding for various $\epsilon > 0$,

$$x(\epsilon) \in \arg\min_{x \in S} F_\epsilon(x), \tag{5.3}$$

where

$$F_\epsilon(x) = c'x - \epsilon \sum_{i=1}^{n} \ln x_i,$$

and S is the interior set

$$S = \{x \mid Ax = b, \ x > 0\},$$

where $x > 0$ means that all the coordinates of x are strictly positive. We assume that S is nonempty and bounded. Since $-\ln x_i$ grows to ∞ as $x_i \to 0$, this assumption can be used together with Weierstrass' theorem (Prop. A.8 in Appendix A) to show that there exists at least one global minimum of $F_\epsilon(x)$ over S, which must be unique because the function F_ϵ can be seen to be strictly convex. Therefore, for each $\epsilon > 0$, $x(\epsilon)$ is uniquely defined by Eq. (5.3).

The Central Path

For given A, b, and c, as ϵ is reduced towards 0, $x(\epsilon)$ follows a trajectory that is known as the *central path*. Figure 5.1.3 illustrates the central path for various values of the cost vector c. Note the following:

(a) For fixed A and b, the central paths corresponding to different cost vectors c start at the same vector x_∞. This is the unique minimizing point over S of

$$-\sum_{i=1}^{n} \ln x_i,$$

Sec. 5.1 Barrier and Interior Point Methods   451

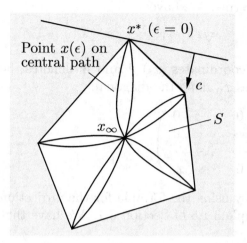

Figure 5.1.3. Central path trajectories $\{x(\epsilon) \mid 0 < \epsilon < \infty\}$ corresponding to ten different values of the cost vector c. All central paths start at the same vector, the *analytic center* x_∞, which corresponds to $\epsilon = \infty$,

$$x_\infty \in \arg\min_{x \in S} \left\{ -\sum_{i=1}^{n} \ln x_i \right\},$$

and end at optimal solutions of (LP).

corresponding to $\epsilon = \infty$, and is known as the *analytic center* of S.

(b) If c is such that (LP) has a unique optimal solution x^*, the central path ends at x^* [i.e., $\lim_{\epsilon \to 0} x(\epsilon) = x^*$]. This follows from Prop. 5.1.1, which implies that for every sequence $\{\epsilon^k\}$ with $\epsilon^k \to 0$, the corresponding sequence $\{x(\epsilon^k)\}$ converges to x^*.

(c) If c is such that (LP) has multiple optimal solutions, it can be shown that the central path ends at one of the optimal solutions [i.e., $\lim_{\epsilon \to 0} x(\epsilon)$ exists and is equal to some optimal solution of (LP)]. We will not prove this fact (see the end-of-chapter references).

Following Approximately the Central Path

The most straightforward way to implement the logarithmic barrier method is to use some iterative algorithm to minimize the function F_{ϵ^k} for each ϵ^k in a sequence $\{\epsilon^k\}$ with $\epsilon^k \downarrow 0$. This is equivalent to finding a sequence $\{x(\epsilon^k)\}$ of points on the central path. However, this approach is inefficient because it requires an infinite number of iterations to compute each point $x(\epsilon^k)$.

It turns out that a far more efficient approach is possible, whereby each minimization is done approximately through a few iterations (possibly only one) of the constrained version of Newton's method that was given in Section 3.3. For a fixed ϵ and a given $x \in S$, this method replaces x by

$$\tilde{x} = x + \alpha(\bar{x} - x),$$

where $\bar{x}$ is the pure Newton iterate defined as the optimal solution of the quadratic program in the vector z

$$\text{minimize } \nabla F_\epsilon(x)'(z - x) + \tfrac{1}{2}(z - x)'\nabla^2 F_\epsilon(x)(z - x)$$
$$\text{subject to } Az = b, \quad z \in \Re^n,$$

and α is a stepsize selected by some rule. We have

$$\nabla F_\epsilon(x) = c - \epsilon x^{-1}, \qquad \nabla^2 F_\epsilon(x) = \epsilon X^{-2},$$

where x^{-1} denotes the vector with coordinates $(x_i)^{-1}$ and X denotes the diagonal matrix with the coordinates x_i along the diagonal:

$$X = \begin{pmatrix} x_1 & 0 & \cdots & 0 \\ 0 & x_2 & \cdots & 0 \\ \cdots & \cdots & \cdots & \cdots \\ 0 & 0 & \cdots & x_n \end{pmatrix}.$$

We can obtain an expression for $\bar{x}$ by using the formula for the projection on a linear manifold given in Example 3.1.5 of Section 3.1. We have that the pure Newton iterate is

$$\bar{x} = x - \epsilon^{-1} X^2 \left(c - \epsilon x^{-1} - A'\lambda \right),$$

where

$$\lambda = (AX^2 A')^{-1} AX^2 \left(c - \epsilon x^{-1} \right).$$

These formulas can also be written as

$$\bar{x} = x - X q(x, \epsilon), \tag{5.4}$$

where

$$q(x, \epsilon) = \frac{Xz}{\epsilon} - e, \tag{5.5}$$

with e and z being the vectors

$$e = \begin{pmatrix} 1 \\ \vdots \\ 1 \end{pmatrix}, \qquad z = c - A'\lambda, \tag{5.6}$$

and

$$\lambda = (AX^2 A')^{-1} AX \left(Xc - \epsilon e \right). \tag{5.7}$$

Based on Eq. (5.4), we have $q(x, \epsilon) = X^{-1}(\bar{x} - x)$, so we may view the vector $q(x, \epsilon)$ as a transformed version of the Newton increment $(x - \bar{x})$ using the transformation matrix X^{-1} that maps the vector x into the vector e. Since $\bar{x}$ is the Newton step approximation to $x(\epsilon)$, we can consider $\|q(x, \epsilon)\|$ as a *measure of proximity* of the current point x to the point $x(\epsilon)$ on the central path. In particular, it can be seen that we have $q(x, \epsilon) = 0$ if and only if $x = x(\epsilon)$.

The key result to be shown shortly is that for convergence of the logarithmic barrier method, it is sufficient to stop the minimization of F_{ϵ^k} and decrease ϵ^k to ϵ^{k+1} once the current iterate x^k satisfies

$$\|q(x^k, \epsilon^k)\| < 1.$$

Sec. 5.1 Barrier and Interior Point Methods   453

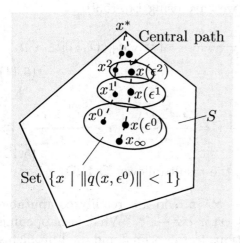

Figure 5.1.4. Following approximately the central path. For each ϵ^k, it is sufficient to carry out the minimization of F_{ϵ^k} up to where $\|q(x, \epsilon^k)\| < 1$.

Another way to phrase this result is that if a sequence of pairs $\{(x^k, \epsilon^k)\}$ satisfies

$$\|q(x^k, \epsilon^k)\| < 1, \qquad 0 < \epsilon^{k+1} < \epsilon^k, \quad k = 0, 1, \ldots, \qquad \epsilon^k \to 0,$$

then every limit point of $\{x^k\}$ is an optimal solution of (LP); see Fig. 5.1.4. The following proposition establishes this result.

Proposition 5.1.2: If $x > 0$, $Ax = b$, and $\|q(x, \epsilon)\| < 1$, then

$$c'x - f^* \leq c'x - b'\lambda \leq \epsilon(n + \|q(x, \epsilon)\|\sqrt{n}) \leq \epsilon(n + \sqrt{n}), \qquad (5.8)$$

where λ is given by Eq. (5.7), and f^* is the optimal value of (LP), i.e.,

$$f^* = \min_{Ay=b,\, y \geq 0} c'y.$$

Proof: Using the definition (5.5)-(5.7) of q, we can write the hypothesis $\|q(x, \epsilon)\| < 1$ as

$$\left\|\frac{X(c - A'\lambda)}{\epsilon} - e\right\| < 1. \qquad (5.9)$$

Thus the coordinates of $(X(c - A'\lambda)/\epsilon) - e$ must lie between -1 and 1, implying that the coordinates of $X(c - A'\lambda)$ are positive. Since the diagonal elements of X are positive, it follows that the coordinates of $c - A'\lambda$ are also positive. Hence $c \geq A'\lambda$, and for any optimal solution x^* of (LP), we obtain (using the fact $x^* \geq 0$)

$$f^* = c'x^* \geq \lambda' A x^* = \lambda' b. \qquad (5.10)$$

On the other hand, since $\|e\| = \sqrt{n}$, we have using Eq. (5.9),

$$e'\left(\frac{X(c - A'\lambda)}{\epsilon} - e\right) \leq \|e\| \left\|\frac{X(c - A'\lambda)}{\epsilon} - e\right\| = \sqrt{n}\,\|q(x,\epsilon)\| \leq \sqrt{n}, \tag{5.11}$$

and by using also Eq. (5.10),

$$e'\left(\frac{X(c - A'\lambda)}{\epsilon} - e\right) = \frac{x'(c - A'\lambda)}{\epsilon} - n = \frac{c'x - b'\lambda}{\epsilon} - n \geq \frac{c'x - f^*}{\epsilon} - n. \tag{5.12}$$

By combining Eqs. (5.11) and (5.12), the result follows. **Q.E.D.**

Note that from Eq. (5.8), $c'x - b'\lambda$ provides a readily computable upper bound to the (unknown) cost error $c'x - f^*$. What is happening here is that x and λ are feasible solutions to the primal and dual problems (LP) and (DP), respectively, and the common optimal value f^* lies between the corresponding primal and dual costs $c'x$ and $b'\lambda$.

Path-Following by Using Newton's Method

Since in order to implement the termination criterion $\|q(x,\epsilon)\| < 1$, we must calculate the pure Newton iterate $\bar{x} = x - Xq(x,\epsilon)$, it is natural to use a convergent version of Newton's method for approximate minimization of F_ϵ. This method replaces x by

$$\tilde{x} = x + \alpha(\bar{x} - x),$$

where α is a stepsize selected by the minimization rule or the Armijo rule (with unit initial stepsize) over the range of positive stepsizes such that $\tilde{x}$ is an interior point. We expect that for x sufficiently close to $x(\epsilon)$, the stepsize α can be taken equal to 1, so that the pure form of the method is used and a quadratic rate of convergence is obtained. The following proposition shows that the "termination set" $\{x \mid \|q(x,\epsilon)\| < 1\}$ is part of the region of quadratic convergence of the pure form of Newton's method.

> **Proposition 5.1.3:** If $x > 0$, $Ax = b$, and $\|q(x,\epsilon)\| < 1$, then the pure Newton iterate $\bar{x} = x - Xq(x,\epsilon)$ is an interior point, i.e., $\bar{x} \in S$. Furthermore, we have $\|q(\bar{x},\epsilon)\| < 1$ and in fact
>
> $$\|q(\bar{x},\epsilon)\| \leq \|q(x,\epsilon)\|^2. \tag{5.13}$$

Proof: Let us define

$$p = Xz/\epsilon = X(c - A'\lambda)/\epsilon,$$

so that $q(x,\epsilon) = p - e$ [cf. Eqs. (5.4) and (5.5)]. Since $\|p-e\| < 1$, we see that the coordinates of p satisfy $0 < p_i < 2$ for all i. We have $\bar{x} = x - X(p-e)$, so that

$$\bar{x}_i = (2 - p_i)x_i > 0$$

for all i, and since also $A\bar{x} = b$, it follows that $\bar{x}$ is an interior point.

It can be shown (Exercise 5.1.3) that the vector $\bar{\lambda}$ corresponding to $\bar{x}$ in the manner of Eq. (5.7) satisfies

$$\bar{\lambda} \in \arg\min_{\xi \in \Re^m} \left\| \frac{\bar{X}(c - A'\xi)}{\epsilon} - e \right\|,$$

where $\bar{X}$ is the diagonal matrix with $\bar{x}_i$ along the diagonal. Hence,

$$\|q(\bar{x},\epsilon)\| = \left\| \frac{\bar{X}(c - A'\bar{\lambda})}{\epsilon} - e \right\| \leq \left\| \frac{\bar{X}(c - A'\lambda)}{\epsilon} - e \right\| = \|\bar{X}X^{-1}p - e\|.$$

Since $\bar{x} = 2x - Xp$, we have

$$\bar{X}X^{-1}p = (2X - XP)X^{-1}p = 2p - Pp,$$

where P is the diagonal matrix with p_i along the diagonal. The last two relations yield

$$\|q(\bar{x},\epsilon)\|^2 \leq \|2p - Pp - e\|^2 \leq \sum_{i=1}^n \left(2p_i - p_i^2 - 1\right)^2 = \sum_{i=1}^n (p_i - 1)^4$$

$$\leq \left(\sum_{i=1}^n (p_i - 1)^2\right)^2 = \|p - e\|^4 = \|q(x,\epsilon)\|^4.$$

This proves the result. **Q.E.D.**

The preceding proposition shows that if $\|q(x,\epsilon)\|$ is substantially less than 1, then a single pure Newton step, changing x to $\bar{x}$, reduces $\|q(x,\epsilon)\|$ by a substantial factor [cf. Eq. (5.13)]. Thus, we expect that if $\bar{\epsilon}$ is not much smaller than ϵ and $\|q(x,\epsilon)\|$ is substantially less than 1, then $\|q(\bar{x},\bar{\epsilon})\|$ will also be substantially less than 1. This means that by carefully selecting the ϵ-reduction factor $\epsilon^{k+1}/\epsilon^k$ in combination with an appropriately small termination tolerance for the first minimization of F_ϵ ($k = 0$), we can execute all the subsequent approximate minimizations of F_{ϵ^k} ($k \geq 1$) in a single pure Newton step; see Fig. 5.1.5. One possibility is, given ϵ^k and x^k such that

$$\|q(x^k, \epsilon^k)\| \leq 1,$$

to obtain x^{k+1} by a single Newton step and then to select ϵ^{k+1} so that $\|q(x^{k+1}, \epsilon^{k+1})\|$ is minimized. This minimization can be done in closed

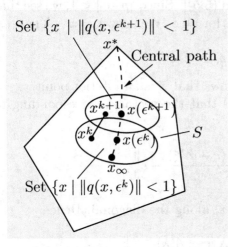

Figure 5.1.5. Following approximately the central path by using a single Newton step for each ϵ^k. If ϵ^k is close to ϵ^{k+1} and x^k is close to the central path, one expects that x^{k+1} obtained from x^k by a single pure Newton step will also be close to the central path.

form because $\|q(x,\epsilon)\|$ is quadratic in $1/\epsilon$ [cf. Eq. (5.5)]. Another possibility is shown in the next proposition.

Proposition 5.1.4: Suppose that $x > 0$, $Ax = b$, and that $\|q(x,\epsilon)\| \leq \gamma$ for some $\gamma < 1$. For any $\delta \in (0, n^{1/2})$, let $\bar{\epsilon} = (1 - \delta n^{-1/2})\epsilon$. Then

$$\|q(\bar{x}, \bar{\epsilon})\| \leq \frac{\gamma^2 + \delta}{1 - \delta n^{-1/2}}.$$

In particular, if

$$\delta \leq \frac{\gamma(1-\gamma)}{1+\gamma}, \tag{5.14}$$

we have $\|q(\bar{x}, \bar{\epsilon})\| \leq \gamma$.

Proof: Let $\theta = \delta n^{-1/2}$. We have using Eq. (5.5)

$$q(\bar{x}, \bar{\epsilon}) = \frac{\bar{X}z}{\bar{\epsilon}} - e = \frac{\bar{X}z}{(1-\theta)\epsilon} - e = \frac{q(\bar{x}, \epsilon) + e}{1 - \theta} - e = \frac{1}{1-\theta}\left(q(\bar{x}, \epsilon) + \theta e\right).$$

Thus, using also Eq. (5.13),

$$\|q(\bar{x}, \bar{\epsilon})\| \leq \frac{1}{1-\theta}\left(\|q(\bar{x}, \epsilon)\| + \theta\|e\|\right)$$

$$= \frac{1}{1-\theta}\left(\|q(\bar{x}, \epsilon)\| + \theta n^{1/2}\right)$$

$$\leq \frac{1}{1-\theta}\left(\|q(x, \epsilon)\|^2 + \delta\right)$$

$$\leq \frac{\gamma^2 + \delta}{1 - \theta}.$$

Sec. 5.1 Barrier and Interior Point Methods 457

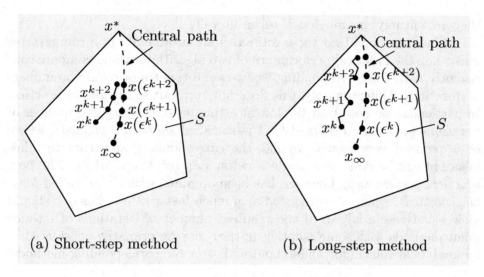

(a) Short-step method   (b) Long-step method

Figure 5.1.6. Following approximately the central path by decreasing ϵ^k slowly as in (a) or quickly as in (b). In (a) a single Newton step is required in each approximate minimization at the expense of a large number of approximate minimizations.

Finally, Eq. (5.14) can be written as $(\gamma^2 + \delta)/(1 - \delta) \leq \gamma$, which, in combination with the relation just proved, implies that $\|q(\bar{x}, \bar{\epsilon})\| \leq \gamma$. **Q.E.D.**

Note that in the preceding proposition one can maintain x very close to the central path ($\gamma << 1$) provided one takes δ to be very small [cf. Eq. (5.14)], or equivalently, one uses an ϵ-reduction factor $1 - \delta n^{-1/2}$ that is very close to 1. Unfortunately, even when γ is close to 1, in order to guarantee the single-step attainment of the tolerance $\|q(x, \epsilon)\| < \gamma$, it is still necessary to decrease ϵ very slowly. In particular, since we must take $\delta < 1$ in order for $\|q(\bar{x}, \bar{\epsilon})\| < \gamma$ [cf. Eq. (5.14)], the reduction factor $\bar{\epsilon}/\epsilon$ must exceed $1 - n^{-1/2}$, which is very close to 1. This means that, even though each approximate minimization after the first will require a single Newton step, a very large number of approximate minimizations will be needed to attain an acceptable accuracy. Thus, it may be more efficient in practice to decrease ϵ^k at a faster rate, while accepting the possibility of multiple Newton steps before switching from ϵ^k to ϵ^{k+1}, as illustrated in Fig. 5.1.6.

The preceding results form the basis for worst-case estimates of the number of Newton iterations required to reduce the error $c'x^k - f^*$ below some tolerance, where x^k is obtained by approximate minimization of F_{ϵ^k} using different termination criteria and reduction factors $\epsilon^{k+1}/\epsilon^k$. Exercise 5.1.5 provides a sample of this type of analysis. Many researchers consider a low estimate of number of iterations a good indicator of algorithmic performance. We note, however, that the worst-case estimates that have been obtained for interior point methods are so unrealistically high that they are entirely meaningless if taken literally.

One may hope that these estimates are meaningful in a comparative sense, i.e., the practical performance of two algorithms would compare consistently with the corresponding worst-case estimates of required numbers of iterations. Unfortunately, this does not turn out to be true in practice. In particular, we note that the lowest estimates of the required number of iterations have been obtained for the so-called *short-step methods*, where ϵ^k is reduced very slowly so that the corresponding approximate minimization can be done in a single Newton step (cf. Prop. 5.1.4). The best practical performance, however, has been obtained with the so-called *long-step methods*, where ϵ^k is reduced at a much faster rate. Thus one should view worst-case analyses of the required number of iterations of interior point methods with some skepticism; they may be primarily considered as an analytical vehicle for understanding better the corresponding methods.

Quadratic and Convex Programming

The logarithmic barrier method in conjunction with Newton's method can also be fruitfully applied to the convex programming problem

$$\text{minimize } f(x)$$
$$\text{subject to } Ax = b, \quad x \geq 0,$$

where $f : \Re^n \mapsto \Re$ is a convex function. The implementation of the method benefits from the extensive experience that has been accumulated from the linear programming case. For the special case of the quadratic programming problem

$$\text{minimize } c'x + \tfrac{1}{2}x'Qx$$
$$\text{subject to } Ax = b, \quad x \geq 0,$$

with Q positive semidefinite, the performance and the analysis of the method are similar to that for linear programs. We refer to the end-of-chapter references for a detailed treatment.

5.1.2 Primal-Dual Methods for Linear Programming

We will now discuss an alternative interior point method for solving the linear program

$$\text{minimize } c'x$$
$$\text{subject to } Ax = b, \quad x \geq 0, \tag{LP}$$

and its dual problem,

$$\text{maximize } b'\lambda$$
$$\text{subject to } A'\lambda \leq c. \tag{DP}$$

Here as in the preceding section, $c \in \Re^n$ and $b \in \Re^m$ are given vectors, and A is an $m \times n$ matrix of rank m.

Sec. 5.1   Barrier and Interior Point Methods

The logarithmic barrier method of the preceding section involves the sequential minimization of

$$F_{\epsilon^k}(x) = c'x - \epsilon^k \sum_{i=1}^{n} \ln x_i,$$

where S is the interior set

$$S = \{x \mid Ax = b, \, x > 0\},$$

and $\{\epsilon^k\}$ is a sequence that decreases to 0. This minimization is done approximately, using one or more Newton iterations.

We will now consider a related approach: applying Newton's method for solving the system of optimality conditions of the problem of minimizing $F_{\epsilon^k}(\cdot)$ over S. The salient features of this approach are:

(a) Only one Newton iteration is carried out for each value of ϵ^k.

(b) The Newton iterations generate a sequence of primal and dual solution pairs (x^k, λ^k), corresponding to a sequence of barrier parameters ϵ^k that converge to 0.

(c) For every k, the pair (x^k, λ^k) is such that x^k is an interior point of the positive orthant, i.e., $x^k > 0$, while λ^k is an interior point of the dual feasible region, i.e.,

$$c - A'\lambda^k > 0.$$

(However, x^k need not be primal-feasible, i.e., it need not satisfy the equation $Ax = b$ as it does in the path-following approach of Section 5.1.1.)

(d) Global convergence is enforced by using as merit function the expression

$$P^k = x^{k'}z^k + \|Ax^k - b\|, \qquad (5.15)$$

where z^k is the vector

$$z^k = c - A'\lambda^k.$$

The expression (5.15) consists of two nonnegative terms: the first term is $x^{k'}z^k$, which is positive (since $x^k > 0$ and $z^k > 0$) and can be written as

$$x^{k'}z^k = x^{k'}(c - A'\lambda^k) = c'x^k - b'\lambda^k + (b - Ax^k)'\lambda^k.$$

Thus when x^k is primal-feasible ($Ax^k = b$), $x^{k'}z^k$ is equal to the duality gap, that is, the difference between the primal and the dual costs, $c'x^k - b'\lambda^k$. The second term is the norm of the primal constraint violation $\|Ax^k - b\|$. In the method to be described, neither of the terms $x^{k'}z^k$ and $\|Ax^k - b\|$ may increase at each iteration, so that $P^{k+1} \leq P^k$ (and typically $P^{k+1} < P^k$) for all k. If we can show that $P^k \to 0$, then asymptotically both the duality gap and the primal constraint violation will be driven to zero. Thus every limit point of $\{(x^k, \lambda^k)\}$ will be a pair of primal and dual optimal solutions, in view of the duality relation

$$\min_{Ax=b,\, x \geq 0} c'x = \max_{A'\lambda \leq c} b'\lambda,$$

shown in Section 4.4.2.

Let us write the necessary and sufficient conditions for (x, λ) to be a (global) minimum-Lagrange multiplier pair for the problem of minimizing the barrier function $F_\epsilon(x)$ subject to $Ax = b$. They are

$$c - \epsilon x^{-1} - A'\lambda = 0, \qquad Ax = b, \tag{5.16}$$

where x^{-1} denotes the vector with coordinates $(x_i)^{-1}$. Let z be the vector

$$z = c - A'\lambda,$$

and note that λ is dual feasible if and only if $z \geq 0$.

Using the vector z, we can write the first condition of Eq. (5.16) as $z - \epsilon x^{-1} = 0$ or, equivalently, $XZe = \epsilon e$, where X and Z are the diagonal matrices with the coordinates of x and z, respectively, along the diagonal, and e is the vector with unit coordinates:

$$X = \begin{pmatrix} x_1 & 0 & \cdots & 0 \\ 0 & x_2 & \cdots & 0 \\ \cdots & \cdots & \cdots & \cdots \\ 0 & 0 & \cdots & x_n \end{pmatrix}, \quad Z = \begin{pmatrix} z_1 & 0 & \cdots & 0 \\ 0 & z_2 & \cdots & 0 \\ \cdots & \cdots & \cdots & \cdots \\ 0 & 0 & \cdots & z_n \end{pmatrix}, \quad e = \begin{pmatrix} 1 \\ 1 \\ \vdots \\ 1 \end{pmatrix}.$$

Thus the optimality conditions (5.16) can be written in the equivalent form

$$XZe = \epsilon e, \tag{5.17}$$

$$Ax = b, \tag{5.18}$$

$$z + A'\lambda = c. \tag{5.19}$$

Given (x, λ, z) satisfying $z + A'\lambda = c$, and such that $x > 0$ and $z > 0$, a Newton iteration for solving this system is

$$x(\alpha, \epsilon) = x + \alpha \Delta x, \tag{5.20}$$

$$\lambda(\alpha, \epsilon) = \lambda + \alpha \Delta \lambda,$$

$$z(\alpha, \epsilon) = z + \alpha \Delta z,$$

Sec. 5.1 Barrier and Interior Point Methods

where α is a stepsize such that $0 < \alpha \leq 1$ and

$$x(\alpha, \epsilon) > 0, \qquad z(\alpha, \epsilon) > 0,$$

and the pure Newton step $(\Delta x, \Delta \lambda, \Delta z)$ solves the linearized version of the system (5.17)-(5.19)

$$X\Delta z + Z\Delta x = -v, \tag{5.21}$$

$$A\Delta x = b - Ax, \tag{5.22}$$

$$\Delta z + A'\Delta \lambda = 0, \tag{5.23}$$

with v defined by

$$v = XZe - \epsilon e. \tag{5.24}$$

After a straightforward calculation, it can be verified that the solution of the linearized system (5.21)-(5.23) can be written as

$$\Delta \lambda = \left(AZ^{-1}XA'\right)^{-1}\left(AZ^{-1}v + b - Ax\right), \tag{5.25}$$

$$\Delta z = -A'\Delta \lambda, \tag{5.26}$$

$$\Delta x = -Z^{-1}v - Z^{-1}X\Delta z.$$

Note that $\lambda(\alpha, \epsilon)$ is dual feasible, since from Eq. (5.23) and the condition $z + A'\lambda = c$, we see that $z(\alpha, \epsilon) + A'\lambda(\alpha, \epsilon) = c$. Note also that if $\alpha = 1$, i.e., a pure Newton step is used, $x(\alpha, \epsilon)$ is primal feasible, since from Eq. (5.22) we have $A(x + \Delta x) = b$.

Merit Function Improvement

We will now evaluate the changes in the constraint violation and the merit function induced by the Newton iteration. By using Eqs. (5.20) and (5.22), the new constraint violation is given by

$$Ax(\alpha, \epsilon) - b = Ax + \alpha A\Delta x - b = Ax + \alpha(b - Ax) - b = (1-\alpha)(Ax - b). \tag{5.27}$$

Thus, since $0 < \alpha \leq 1$, the new norm of constraint violation $\|Ax(\alpha, \epsilon) - b\|$ is always no larger than the old one. Furthermore, if x is primal-feasible ($Ax = b$), the new iterate $x(\alpha, \epsilon)$ is also primal-feasible.

The inner product

$$g = x'z \tag{5.28}$$

after the iteration becomes

$$\begin{aligned} g(\alpha, \epsilon) &= x(\alpha, \epsilon)'z(\alpha, \epsilon) \\ &= (x + \alpha\Delta x)'(z + \alpha\Delta z) \\ &= x'z + \alpha(x'\Delta z + z'\Delta x) + \alpha^2 \Delta x'\Delta z. \end{aligned} \tag{5.29}$$

From Eqs. (5.22) and (5.26) we have

$$\Delta x' \Delta z = (Ax - b)' \Delta \lambda,$$

while by premultiplying Eq. (5.21) with e' and using the definition (5.24) for v, we obtain

$$x' \Delta z + z' \Delta x = -e'v = n\epsilon - x'z.$$

By substituting the last two relations in Eq. (5.29) and by using also the expression (5.28) for g, we see that

$$g(\alpha, \epsilon) = g - \alpha(g - n\epsilon) + \alpha^2 (Ax - b)' \Delta \lambda. \qquad (5.30)$$

Let us now denote by P and $P(\alpha, \epsilon)$ the value of the merit function (5.15) before and after the iteration, respectively. We have by using the expressions (5.27) and (5.30),

$$\begin{aligned} P(\alpha, \epsilon) &= g(\alpha, \epsilon) + \|Ax(\alpha, \epsilon) - b\| \\ &= g - \alpha(g - n\epsilon) + \alpha^2 (Ax - b)' \Delta \lambda + (1 - \alpha)\|Ax - b\|, \end{aligned}$$

or

$$P(\alpha, \epsilon) = P - \alpha \big(g - n\epsilon + \|Ax - b\|\big) + \alpha^2 (Ax - b)' \Delta \lambda.$$

Thus if ϵ is chosen to satisfy

$$\epsilon < \frac{g}{n}$$

and α is chosen to be small enough so that the second order term $\alpha^2(Ax - b)'\Delta\lambda$ is dominated by the first order term $\alpha(g - n\epsilon)$, the merit function will be improved as a result of the iteration.

A General Class of Primal-Dual Algorithms

Let us consider now the general class of algorithms of the form

$$x^{k+1} = x(\alpha^k, \epsilon^k), \qquad \lambda^{k+1} = \lambda(\alpha^k, \epsilon^k), \qquad z^{k+1} = z(\alpha^k, \epsilon^k),$$

where α^k and ϵ^k are positive scalars such that

$$x^{k+1} > 0, \qquad z^{k+1} > 0, \qquad \epsilon^k < \frac{g^k}{n},$$

where g^k is the inner product

$$g^k = x^{k\prime} z^k + (Ax^k - b)' \lambda^k,$$

and α^k is such that the merit function P^k is reduced. Initially we must have $x^0 > 0$, and $z^0 = c - A'\lambda^0 > 0$ (such a point can often be easily found; otherwise an appropriate reformulation of the problem is necessary for which we refer to the specialized literature). These methods have been called *primal-dual*. As we have seen, they are really Newton-like methods for approximate solution of the system of optimality conditions (5.16), supplemented by a stepsize procedure that guarantees that the merit function P^k is improved at each iteration.

It can be shown that it is possible to choose α^k and ϵ^k so that the merit function is not only reduced at each iteration, but also converges to zero. Furthermore, with suitable choices of α^k and ϵ^k, algorithms with good theoretical properties, such as polynomial complexity and superlinear convergence, can be derived. The main convergence analysis ideas rely on a primal-dual version of the central path discussed in Section 5.1.1, and some of the associated path following concepts. We refer to the research monograph [Wri97a] for a detailed discussion.

With properly chosen sequences α^k and ϵ^k, and appropriate implementation, the practical performance of the primal-dual methods has been shown to be excellent. The choice

$$\epsilon^k = \frac{g^k}{n^2},$$

leading to the relation

$$g^{k+1} = (1 - \alpha^k + \alpha^k/n)g^k$$

for feasible x^k, has been suggested as a good practical rule. Usually, when x^k has already become feasible, α^k is chosen as $\theta\tilde{\alpha}^k$, where θ is a factor very close to 1 (say 0.999), and $\tilde{\alpha}^k$ is the maximum stepsize α that guarantees that $x(\alpha, \epsilon^k) \geq 0$ and $z(\alpha, \epsilon^k) \geq 0$

$$\tilde{\alpha}^k = \min\left\{\min_{i=1,\ldots,n}\left\{\frac{x_i^k}{-\Delta x_i} \mid \Delta x_i < 0\right\}, \min_{i=1,\ldots,n}\left\{\frac{z_i^k}{-\Delta z_i} \mid \Delta z_i < 0\right\}\right\}.$$

When x^k is not feasible, the choice of α^k must also be such that the merit function is improved. In some works, a different stepsize for the x update than for the (λ, z) update has been suggested. The stepsize for the x update is near the maximum stepsize α that guarantees $x(\alpha, \epsilon^k) \geq 0$, and the stepsize for the (λ, z) update is near the maximum stepsize α that guarantees $z(\alpha, \epsilon^k) \geq 0$. There are a number of additional practical issues related to implementation, for which we refer to the specialized literature.

Predictor-Corrector Variants

We mentioned briefly in Section 1.4.3 the variation of Newton's method where the Hessian is evaluated periodically every $p > 1$ iterations in order to economize in iteration overhead. When $p = 2$ and the problem is to solve the system $g(x) = 0$, where $g : \Re^n \mapsto \Re^n$, this variation of Newton's method takes the form

$$\hat{x}^k = x^k - \left(\nabla g(x^k)'\right)^{-1} g(x^k), \tag{5.31}$$

$$x^{k+1} = \hat{x}^k - \left(\nabla g(x^k)'\right)^{-1} g(\hat{x}^k). \tag{5.32}$$

Thus, given x^k, this iteration performs a regular Newton step to obtain $\hat{x}^k$, and then an approximate Newton step from $\hat{x}^k$, using, however, the already available Jacobian inverse $\left(\nabla g(x^k)'\right)^{-1}$. It can be shown that if $x^k \to x^*$, the order of convergence of the error $\|x^k - x^*\|$ is cubic, i.e.,

$$\limsup_{k \to \infty} \frac{\|x^{k+1} - x^*\|}{\|x^k - x^*\|^3} < \infty,$$

under the same assumptions that the ordinary Newton's method ($p = 1$) attains a quadratic order of convergence; see [OrR70], p. 315. Thus, the price for the 50% saving in Jacobian evaluations and inversions is a small degradation of the convergence rate over the ordinary Newton's method (which attains a quartic order of convergence when two successive ordinary Newton steps are counted as one).

Two-step Newton methods such as the iteration (5.31), (5.32), when applied to the system of optimality conditions (5.17)-(5.19) for the linear program (LP) are known as *predictor-corrector* methods (the name comes from their similarity with predictor-corrector methods for solving differential equations). They operate as follows:

Given (x, z, λ) with

$$x > 0, \qquad z = c - A'\lambda > 0,$$

the *predictor iteration* [cf. Eq. (5.31)], solves for $\left(\Delta \hat{x}, \Delta \hat{z}, \Delta \hat{\lambda}\right)$ the system

$$X \Delta \hat{z} + Z \Delta \hat{x} = -\hat{v}, \tag{5.33}$$

$$A \Delta \hat{x} = b - Ax, \tag{5.34}$$

$$\Delta \hat{z} + A' \Delta \hat{\lambda} = 0, \tag{5.35}$$

with $\hat{v}$ defined by

$$\hat{v} = XZe - \hat{\epsilon}e, \tag{5.36}$$

[cf. Eqs. (5.21)-(5.24)].

The *corrector iteration* [cf. Eq. (5.32)], solves for $\left(\Delta \bar{x}, \Delta \bar{z}, \Delta \bar{\lambda}\right)$ the system

$$X \Delta \bar{z} + Z \Delta \bar{x} = -\bar{v}, \tag{5.37}$$

$$A \Delta \bar{x} = b - A(x + \Delta \hat{x}), \tag{5.38}$$

Sec. 5.1   Barrier and Interior Point Methods

$$\Delta \bar{z} + A' \Delta \bar{\lambda} = 0, \tag{5.39}$$

with $\bar{v}$ defined by

$$\bar{v} = (X + \Delta \hat{X})(Z + \Delta \hat{Z})e - \bar{\epsilon}e, \tag{5.40}$$

where $\Delta \hat{X}$ and $\Delta \hat{Z}$ are the diagonal matrices with the components of $\Delta \hat{x}$ and $\Delta \hat{z}$ along the diagonal, respectively. Here $\hat{\epsilon}$ and $\bar{\epsilon}$ are the barrier parameters corresponding to the two iterations.

The *composite Newton direction* is

$$\Delta x = \Delta \hat{x} + \Delta \bar{x},$$

$$\Delta z = \Delta \hat{z} + \Delta \bar{z},$$

$$\Delta \lambda = \Delta \hat{\lambda} + \Delta \bar{\lambda},$$

and the corresponding iteration is

$$x(\alpha, \epsilon) = x + \alpha \Delta x,$$

$$\lambda(\alpha, \epsilon) = \lambda + \alpha \Delta \lambda,$$

$$z(\alpha, \epsilon) = z + \alpha \Delta z,$$

where α is a stepsize such that $0 < \alpha \leq 1$ and

$$x(\alpha, \epsilon) > 0, \qquad z(\alpha, \epsilon) > 0.$$

Adding Eqs. (5.33)-(5.35) and Eqs. (5.37)-(5.39), we obtain

$$X(\Delta \hat{z} + \Delta \bar{z})z + Z(\Delta \hat{x} + \Delta \bar{x}) = -\hat{v} - \bar{v}, \tag{5.41}$$

$$A(\Delta \hat{x} + \Delta \bar{x})x = b - Ax + b - A(x + \Delta \hat{x}), \tag{5.42}$$

$$\Delta \hat{z} + \Delta \bar{z} + A'(\Delta \hat{\lambda} + \Delta \bar{\lambda}) = 0, \tag{5.43}$$

We now use the fact

$$b - A(x + \Delta \hat{x}) = 0$$

[cf. Eq. (5.34)], and we also use Eqs. (5.40) and (5.33) to write

$$\bar{v} = (X + \Delta \hat{X})(Z + \Delta \hat{Z})e - \bar{\epsilon}e$$

$$= XZe + \Delta \hat{X} Z e + X \Delta \hat{Z} e + \Delta \hat{X} \Delta \hat{Z} e - \bar{\epsilon}e$$

$$= XZe + Z \Delta \hat{x} + X \Delta \hat{z} + \Delta \hat{X} \Delta \hat{Z} e - \bar{\epsilon}e$$

$$= XZe - \hat{v} + \Delta \hat{X} \Delta \hat{Z} e - \bar{\epsilon}e.$$

Substituting in Eqs. (5.41)-(5.43), we see that the composite Newton direction

$$(\Delta x, \Delta z, \Delta \lambda) = (\Delta \hat{x} + \Delta \bar{x}, \Delta \hat{z} + \Delta \bar{z}, \Delta \hat{\lambda} + \Delta \bar{\lambda})$$

is obtained by solving the following system of equations:

$$X\Delta z + Z\Delta x = -XZe - \Delta \hat{X}\Delta \hat{Z}e + \bar{\epsilon}e, \qquad (5.44)$$

$$A\Delta x = b - Ax, \qquad (5.45)$$

$$\Delta z + A'\Delta \lambda = 0. \qquad (5.46)$$

To implement the predictor-corrector method, we need to solve the system (5.33)-(5.36) for some value of $\hat{\epsilon}$ to obtain $(\Delta \hat{X}, \Delta \hat{Z})$, and then to solve the system (5.44)-(5.46) for some value of $\bar{\epsilon}$ to obtain $(\Delta x, \Delta z, \Delta \lambda)$. It is important to note here that most of the work needed for the first system, namely the factorization of the matrix

$$AZ^{-1}XA'$$

in Eq. (5.25), need not be repeated when solving the second system, so that solving both systems requires relatively little extra work over solving the first one.

In an implementation that has proved successful in practice, $\hat{\epsilon}$ is taken to be zero. Furthermore, $\bar{\epsilon}$ is chosen on the basis of the solution of the first system according to the formula

$$\bar{\epsilon} = \left(\frac{\hat{g}}{x'z}\right)^2 \frac{\hat{g}}{n},$$

where $\hat{g}$ is the duality gap that would result from a feasibility-restricted primal-dual step given by

$$\hat{g} = (x + \alpha_P \Delta \hat{x})'(z + \alpha_D \Delta \hat{z}),$$

where

$$\alpha_P = \theta \min_{i=1,\ldots,n} \left\{ \frac{x_i^k}{-\Delta x_i} \;\middle|\; \Delta x_i < 0 \right\},$$

$$\alpha_D = \theta \min_{i=1,\ldots,n} \left\{ \frac{z_i^k}{-\Delta z_i} \;\middle|\; \Delta z_i < 0 \right\}.$$

and θ is a factor very close to 1 (say 0.999). We refer to the specialized literature for further details [Meh92], [LMS92], [Wri97a], [Ye97].

EXERCISES

5.1.1

Consider the linear program

$$\text{minimize} \quad x_1 + 2x_2 + 3x_3$$
$$\text{subject to} \quad x_1 + x_2 + x_3 = 1, \quad x \geq 0.$$

(a) Sketch on paper the central path. Write a computer program to implement a short-step and a long-step path-following method based on Newton's method for this problem. Compare the number of Newton steps for a given solution accuracy for the starting points $x^0 = (.8, .15, .05)$ and $x^0 = (.1, .2, .7)$.

(b) Write a computer program to implement a primal-dual interior point method and its predictor-corrector variant, and solve the problem for $\lambda^0 = 0$ and x^0 as in part (a).

5.1.2

Given x, show how to find an $\epsilon > 0$ that minimizes $\|q(x, \epsilon)\|$ [cf. Eq. (5.5)]. How would you use this idea to accelerate convergence in a short-step path-following method?

5.1.3

Show that the vector λ of Eq. (5.7) satisfies

$$\lambda \in \arg\min_{\xi \in \Re^m} \left\| \frac{X(c - A'\xi)}{\epsilon} - e \right\|.$$

5.1.4

Let $\delta = \|q(x, \epsilon)\|$, where $q(x, \epsilon)$ is the scaled Newton step defined by Eq. (5.5), and assume that $\delta < 1$. Show that

$$\|X^{-1}(x - x(\epsilon))\| \leq \frac{\delta}{1 - \delta},$$

$$F_\epsilon(x) - F_\epsilon\big(x(\epsilon)\big) \leq \frac{\delta^2}{1 - \delta^2},$$

$$|c'x - c'x(\epsilon)| \leq \frac{\delta(1 + \delta)\epsilon}{1 - \delta} \sqrt{n}.$$

5.1.5 (Complexity of the Short-Step Method [Tse89])

The purpose of this exercise is to show that the number of iterations required by the short-step logarithmic barrier method to achieve a given accuracy is proportional to $\sqrt{n}$. Consider the linear programming problem (LP), a vector $x^0 \in S$, and a sequence $\{\epsilon^k\}$ such that $\|q(x^0,\epsilon^0)\| \leq 1/2$ and $\epsilon^{k+1} = (1-\theta)\epsilon^k$, where $\theta = 1/(6n^{1/2})$ (cf. Prop. 5.1.4). Let x^{k+1} be generated from x^k by a pure Newton step aimed at minimizing F_{ϵ^k}. For a given integer r, let $\bar{k}$ be the smallest integer k such that $-\ln(n\epsilon^k) \geq r$ and let $r^0 = -\ln(n\epsilon^0)$. Show that

$$\bar{k} \leq 6(r - r^0)\sqrt{n}$$

and

$$c'x^{\bar{k}} - f^* \leq \frac{3}{2}e^{-r}.$$

Note: We have assumed here that a vector x^0 with $\|q(x^0,\epsilon^0)\| \leq 1/2$ is available. It is possible to show that such a point can be found in a number of Newton steps that is proportional to $\sqrt{n}$.

5.1.6 (The Dual Problem as an Equality Constrained Problem)

Consider the dual problem

$$\begin{aligned}\text{maximize} \quad & b'\lambda \\ \text{subject to} \quad & A'\lambda \leq c,\end{aligned} \qquad \text{(DP)}$$

and its equivalent version

$$\begin{aligned}\text{maximize} \quad & b'\lambda \\ \text{subject to} \quad & A'\lambda + z = c, \quad z \geq 0,\end{aligned}$$

that involves the vector of additional variables z. Let P_A be the matrix that projects a vector x onto the nullspace of the matrix A, and note that using the analysis of Example 3.1.5 in Section 3.1, we have

$$P_A = I - A'(AA')^{-1}A.$$

Show that the dual linear program (DP) is equivalent to the linear program

$$\begin{aligned}\text{minimize} \quad & \bar{x}'z \\ \text{subject to} \quad & P_A z = P_A c, \quad z \geq 0,\end{aligned} \qquad (5.47)$$

where $\bar{x}$ is any primal feasible vector, i.e., $A\bar{x} = b$, $\bar{x} \geq 0$.

5.1.7 (Dual Central Path)

Consider the dual problem (DP). Using its equivalent reformulation (5.47) of Exercise 5.1.6, it is seen that the appropriate definition of the central path of the dual problem is

$$z(\epsilon) \in \arg \min_{P_A z = P_A c,\, z > 0} \left\{ \bar{x}'z - \sum_{i=1}^{n} \ln z_i \right\},$$

where $\bar{x}$ is any primal feasible vector. Show that the primal and dual central paths are related by

$$z(\epsilon) = \epsilon x(\epsilon)^{-1},$$

and that the corresponding duality gap satisfies

$$c'x(\epsilon) - b'\lambda(\epsilon) = n\epsilon,$$

where $\lambda(\epsilon)$ is any vector such that $A'\lambda(\epsilon) = c - z(\epsilon)$.

5.2 PENALTY AND AUGMENTED LAGRANGIAN METHODS

The basic idea in penalty methods is to eliminate some or all of the constraints and add to the cost function a penalty term that prescribes a high cost to infeasible points. Associated with these methods is a penalty parameter c that determines the severity of the penalty and as a consequence, the extent to which the resulting unconstrained problem approximates the original constrained problem. As c takes higher values, the approximation becomes increasingly accurate. We focus attention primarily on the popular quadratic penalty function. Some other penalty functions, including the exponential, are discussed in Section 5.2.5.

Consider first the equality constrained problem

$$\begin{aligned} \text{minimize} \quad & f(x) \\ \text{subject to} \quad & h(x) = 0, \quad x \in X, \end{aligned} \quad (5.48)$$

where $f : \Re^n \mapsto \Re$, $h : \Re^n \mapsto \Re^m$ are given functions, and X is a given subset of $\Re^n$. Much of our analysis in this section will focus on the case where $X = \Re^n$, and x^* together with a Lagrange multiplier vector λ^* satisfies the sufficient optimality conditions of Prop. 4.2.1. At the center of our development is the *augmented Lagrangian function* $L_c : \Re^n \times \Re^m \mapsto \Re$, introduced in Section 4.2 and given by

$$L_c(x, \lambda) = f(x) + \lambda' h(x) + \frac{c}{2} \|h(x)\|^2,$$

where c is a positive penalty parameter.

There are two mechanisms by which unconstrained minimization of $L_c(\cdot, \lambda)$ can yield points close to x^*:

(a) *By taking λ close to λ^**. Indeed, as shown in Section 4.2.1, if c is higher than a certain threshold, then for some $\gamma > 0$ and $\epsilon > 0$ we have

$$L_c(x, \lambda^*) \geq L_c(x^*, \lambda^*) + \frac{\gamma}{2}\|x - x^*\|^2, \qquad \forall\, x \text{ with } \|x - x^*\| < \epsilon,$$

so that x^* is a strict unconstrained local minimum of the augmented Lagrangian $L_c(\cdot, \lambda^*)$ corresponding to λ^*. This suggests that if λ is close to λ^*, a good approximation to x^* can be found by unconstrained minimization of $L_c(\cdot, \lambda)$.

(b) *By taking c large*. Indeed for high c, there is high cost for infeasibility, so the unconstrained minima of $L_c(\cdot, \lambda)$ will be nearly feasible. Since $L_c(x, \lambda) = f(x)$ for feasible x, we expect that $L_c(x, \lambda) \approx f(x)$ for nearly feasible x. Therefore, we can also expect to obtain a good approximation to x^* by unconstrained minimization of $L_c(\cdot, \lambda)$ when c is large.

Example 5.2.1

Consider the two-dimensional problem

$$\text{minimize } f(x) = \tfrac{1}{2}(x_1^2 + x_2^2)$$
$$\text{subject to } x_1 = 1,$$

with optimal solution $x^* = (1, 0)$ and corresponding Lagrange multiplier $\lambda^* = -1$. The augmented Lagrangian is

$$L_c(x, \lambda) = \tfrac{1}{2}(x_1^2 + x_2^2) + \lambda(x_1 - 1) + \tfrac{c}{2}(x_1 - 1)^2,$$

and by setting its gradient to zero we can verify that its unique unconstrained minimum $x(\lambda, c)$ has coordinates given by

$$x_1(\lambda, c) = \frac{c - \lambda}{c + 1}, \qquad x_2(\lambda, c) = 0. \tag{5.49}$$

Thus, we have for all $c > 0$,

$$\lim_{\lambda \to \lambda^*} x_1(\lambda, c) = x_1(-1, c) = 1 = x_1^*, \qquad \lim_{\lambda \to \lambda^*} x_2(\lambda^*, c) = 0 = x_2^*,$$

showing that as λ is chosen close to λ^*, the unconstrained minimum of $L_c(\cdot, \lambda)$ approaches the constrained minimum (see Fig. 5.2.1).

Using Eq. (5.49), we also have for all λ,

$$\lim_{c \to \infty} x_1(\lambda, c) = 1 = x_1^*, \qquad \lim_{c \to \infty} x_2(\lambda, c) = 0 = x_2^*,$$

showing that as c increases, the unconstrained minimum of $L_c(\cdot, \lambda)$ approaches the constrained minimum (see Fig. 5.2.2).

Sec. 5.2 Penalty and Augmented Lagrangian Methods

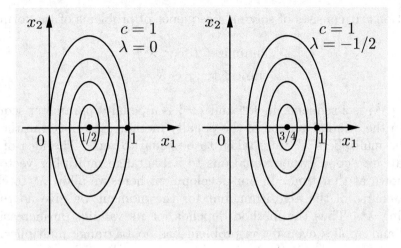

Figure 5.2.1. Equal cost surfaces of the augmented Lagrangian

$$L_c(x, \lambda) = \tfrac{1}{2}(x_1^2 + x_2^2) + \lambda(x_1 - 1) + \tfrac{c}{2}(x_1 - 1)^2,$$

of Example 5.2.1, for $c = 1$ and two different values of λ. The unconstrained minimum of $L_c(\cdot, \lambda)$ approaches the constrained minimum $x^* = (1, 0)$ as $\lambda \to \lambda^* = -1$.

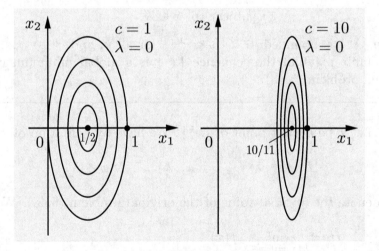

Figure 5.2.2. Equal cost surfaces of the augmented Lagrangian

$$L_c(x, \lambda) = \tfrac{1}{2}(x_1^2 + x_2^2) + \lambda(x_1 - 1) + \tfrac{c}{2}(x_1 - 1)^2,$$

of Example 5.2.1, for $\lambda = 0$ and two different values of c. The unconstrained minimum of $L_c(\cdot, \lambda)$ approaches the constrained minimum $x^* = (1, 0)$ as $c \to \infty$.

5.2.1 The Quadratic Penalty Function Method

The quadratic penalty function method is motivated by the preceding considerations. It consists of solving a sequence of problems of the form

$$\text{minimize } L_{c^k}(x, \lambda^k)$$
$$\text{subject to } x \in X,$$

where $\{\lambda^k\}$ is a sequence in $\Re^m$ and $\{c^k\}$ is a penalty parameter sequence.

In the original version of the penalty method introduced in the early 60s, the multipliers λ^k were taken to be equal to zero. The idea of using λ^k that are "good" approximations to a Lagrange multiplier vector was not known at that time. In our development here, we allow λ^k to change in the course of the algorithm, but for the moment we give no rule for updating λ^k. Thus the method depends for its validity on increasing c^k to ∞, and applies even if the problem has no Lagrange multiplier. The following proposition is the basic convergence result.

Proposition 5.2.1: Assume that f and h are continuous functions, that X is a closed set, and that the constraint set $\{x \in X \mid h(x) = 0\}$ is nonempty. For $k = 0, 1, \ldots$, let x^k be a global minimum of the problem

$$\text{minimize } L_{c^k}(x, \lambda^k)$$
$$\text{subject to } x \in X,$$

where $\{\lambda^k\}$ is bounded, $0 < c^k < c^{k+1}$ for all k, and $c^k \to \infty$. Then every limit point of the sequence $\{x^k\}$ is a global minimum of the original problem (5.48).

Proof: Let $\bar{x}$ be a limit point of $\{x^k\}$. We have by definition of x^k

$$L_{c^k}(x^k, \lambda^k) \leq L_{c^k}(x, \lambda^k), \qquad \forall \, x \in X. \tag{5.50}$$

Let f^* denote the optimal value of the original problem (5.48). We have

$$f^* = \inf_{h(x)=0,\, x \in X} f(x)$$
$$= \inf_{h(x)=0,\, x \in X} \left\{ f(x) + \lambda^{k\prime} h(x) + \frac{c^k}{2} \|h(x)\|^2 \right\}$$
$$= \inf_{h(x)=0,\, x \in X} L_{c^k}(x, \lambda^k).$$

Hence, by taking the infimum of the right-hand side of Eq. (5.50) over $x \in X$, $h(x) = 0$, we obtain

$$L_{c^k}(x^k, \lambda^k) = f(x^k) + \lambda^{k\prime} h(x^k) + \frac{c^k}{2} \|h(x^k)\|^2 \leq f^*.$$

Sec. 5.2   Penalty and Augmented Lagrangian Methods   **473**

The sequence $\{\lambda^k\}$ is bounded and hence it has a limit point $\bar{\lambda}$. Without loss of generality, we may assume that $\lambda^k \to \bar{\lambda}$. By taking upper limit in the relation above and by using the continuity of f and h, we obtain

$$f(\bar{x}) + \bar{\lambda}'h(\bar{x}) + \limsup_{k\to\infty} \frac{c^k}{2} \|h(x^k)\|^2 \leq f^*. \tag{5.51}$$

Since $\|h(x^k)\|^2 \geq 0$ and $c^k \to \infty$, it follows that $h(x^k) \to 0$ and

$$h(\bar{x}) = 0, \tag{5.52}$$

for otherwise the left-hand side of Eq. (5.51) would equal ∞, while $f^* < \infty$ (since the constraint set is assumed nonempty). Since X is a closed set, we also obtain that $\bar{x} \in X$. Hence, $\bar{x}$ is feasible, and since from Eqs. (5.51) and (5.52) we have $f(\bar{x}) \leq f^*$, it follows that $\bar{x}$ is optimal. **Q.E.D.**

Lagrange Multiplier Estimates – Inexact Minimization

The preceding convergence result assumes implicitly that the minimum of the augmented Lagrangian is found exactly. On the other hand, unconstrained minimization methods are usually terminated when the cost gradient is sufficiently small, but not necessarily zero. In particular, when $X = \Re^n$, and f and h are differentiable, the algorithm for solving the unconstrained problem

$$\text{minimize } L_{c^k}(x, \lambda^k)$$
$$\text{subject to } x \in \Re^n,$$

will typically be terminated at a point x^k satisfying

$$\|\nabla_x L_{c^k}(x^k, \lambda^k)\| \leq \epsilon^k,$$

where ϵ^k is some small scalar. We address this situation in the next proposition, where we show in addition that we can usually obtain a Lagrange multiplier vector as a by-product of the computation.

Proposition 5.2.2: Assume that $X = \Re^n$, and f and h are continuously differentiable. For $k = 0, 1, \ldots$, let x^k satisfy

$$\|\nabla_x L_{c^k}(x^k, \lambda^k)\| \leq \epsilon^k,$$

where $\{\lambda^k\}$ is bounded, and $\{\epsilon^k\}$ and $\{c^k\}$ satisfy

$$0 < c^k < c^{k+1}, \quad \forall\, k, \quad c^k \to \infty,$$

$$0 \leq \epsilon^k, \quad \forall\, k, \quad \epsilon^k \to 0.$$

> Assume that a subsequence $\{x^k\}_K$ converges to a vector x^* such that $\nabla h(x^*)$ has rank m. Then
>
> $$\{\lambda^k + c^k h(x^k)\}_K \to \lambda^*,$$
>
> where λ^* is a vector satisfying, together with x^*, the first order necessary conditions
>
> $$\nabla f(x^*) + \nabla h(x^*)\lambda^* = 0, \qquad h(x^*) = 0.$$

Proof: Without loss of generality we assume that the entire sequence $\{x^k\}$ converges to x^*. Define for all k

$$\tilde{\lambda}^k = \lambda^k + c^k h(x^k).$$

We have

$$\nabla_x L_{c^k}(x^k, \lambda^k) = \nabla f(x^k) + \nabla h(x^k)\bigl(\lambda^k + c^k h(x^k)\bigr) = \nabla f(x^k) + \nabla h(x^k)\tilde{\lambda}^k. \tag{5.53}$$

Since $\nabla h(x^*)$ has rank m, $\nabla h(x^k)$ has rank m for all k that are sufficiently large. Without loss of generality, we assume that $\nabla h(x^k)$ has rank m for all k. Then, by multiplying Eq. (5.53) with

$$\bigl(\nabla h(x^k)'\nabla h(x^k)\bigr)^{-1}\nabla h(x^k)',$$

we obtain

$$\tilde{\lambda}^k = \bigl(\nabla h(x^k)'\nabla h(x^k)\bigr)^{-1}\nabla h(x^k)'\bigl(\nabla_x L_{c^k}(x^k, \lambda^k) - \nabla f(x^k)\bigr). \tag{5.54}$$

The hypothesis implies that $\nabla_x L_{c^k}(x^k, \lambda^k) \to 0$, so Eq. (5.54) yields

$$\tilde{\lambda}^k \to \lambda^*,$$

where

$$\lambda^* = -\bigl(\nabla h(x^*)'\nabla h(x^*)\bigr)^{-1}\nabla h(x^*)'\nabla f(x^*).$$

Using again the fact $\nabla_x L_{c^k}(x^k, \lambda^k) \to 0$ and Eq. (5.53), we see that

$$\nabla f(x^*) + \nabla h(x^*)\lambda^* = 0.$$

Since $\{\lambda^k\}$ is bounded and $\lambda^k + c^k h(x^k) \to \lambda^*$, it follows that $\{c^k h(x^k)\}$ is bounded. Since $c^k \to \infty$, we must have $h(x^k) \to 0$ and we conclude that $h(x^*) = 0$. **Q.E.D.**

Practical Behavior – Ill-Conditioning

Let us now consider the practical behavior of the quadratic penalty method, assuming that $X = \Re^n$, and f and h are continuously differentiable. Suppose that the kth unconstrained minimization of $L_{c^k}(x, \lambda^k)$ is terminated when
$$\left\|\nabla_x L_{c^k}(x^k, \lambda^k)\right\| \leq \epsilon^k, \tag{5.55}$$
where $\epsilon^k \to 0$. There are three possibilities:

(a) The method breaks down because an x^k satisfying the condition (5.55) cannot be found.

(b) A sequence $\{x^k\}$ satisfying the condition (5.55) for all k is obtained, but it either has no limit points, or for each of its limit points x^* the matrix $\nabla h(x^*)$ has linearly dependent columns.

(c) A sequence $\{x^k\}$ satisfying the condition (5.55) for all k is found and it has a limit point x^* such that $\nabla h(x^*)$ has rank m. Then, by Prop. 5.2.2, x^* together with λ^* [the corresponding limit point of $\{\lambda^k + c^k h(x^k)\}$] satisfies the first order necessary conditions for optimality.

Possibility (a) usually occurs when $L_{c^k}(\cdot, \lambda^k)$ is unbounded below as discussed following Prop. 5.2.1.

Possibility (b) usually occurs when $L_{c^k}(\cdot, \lambda^k)$ is bounded below, but the original problem has no feasible solution. Typically then the penalty term dominates as $k \to \infty$, and the method usually converges to an infeasible vector x^*, which is a stationary point of the function $\|h(x)\|^2$. This means that
$$\nabla h(x^*) h(x^*) = \tfrac{1}{2} \nabla \{\|h(x^*)\|^2\} = 0,$$
implying that $\nabla h(x^*)$ has linearly dependent columns.

Possibility (c) is the normal case, where the unconstrained minimization algorithm terminates successfully for each k and $\{x^k\}$ converges to a feasible vector, which is also regular. It is of course possible (although unusual in practice) that $\{x^k\}$ converges to a local minimum x^*, which is not regular. Then, if there is no Lagrange multiplier vector corresponding to x^*, the sequence $\{\lambda^k + c^k h(x^k)\}$ diverges and has no limit point.

Extensive practical experience shows that the penalty function method is on the whole quite reliable and usually converges to at least a local minimum of the original problem. Whenever it fails, this is usually due to the increasing difficulty of the minimization

$$\text{minimize } L_{c^k}(x, \lambda^k)$$
$$\text{subject to } x \in X,$$

as $c^k \to \infty$. In particular, let us assume that $X = \Re^n$, and f and h are twice differentiable. Then, according to the convergence rate analysis of Section

1.3, the degree of difficulty depends on the ratio of largest to smallest eigenvalue (the condition number) of the Hessian matrix $\nabla^2_{xx} L_{c^k}(x^k, \lambda^k)$, and this ratio tends to increase with c^k. We illustrate this by means of an example. A proof is outlined in Exercise 5.2.8; see also [Ber82a], p. 102.

Example 5.2.2

Consider the problem of Example 5.2.1:

$$\text{minimize } f(x) = \tfrac{1}{2}(x_1^2 + x_2^2)$$
$$\text{subject to } x_1 = 1.$$

The augmented Lagrangian is

$$L_c(x, \lambda) = \tfrac{1}{2}(x_1^2 + x_2^2) + \lambda(x_1 - 1) + \tfrac{c}{2}(x_1 - 1)^2,$$

and its Hessian is

$$\nabla^2_{xx} L_c(x, \lambda) = I + c \begin{pmatrix} 1 \\ 0 \end{pmatrix} \begin{pmatrix} 1 & 0 \end{pmatrix} = \begin{pmatrix} 1+c & 0 \\ 0 & 1 \end{pmatrix}.$$

The ratio of largest to smallest eigenvalue of the Hessian is $1+c$ and tends to ∞ as $c \to \infty$. The associated ill-conditioning can also be observed from the narrow level sets of the augmented Lagrangian for large c in Fig. 5.2.2.

To overcome ill-conditioning, it is recommended to use a Newton-like method for minimizing $L_{c^k}(\cdot, \lambda^k)$, as well as double precision arithmetic to deal with roundoff errors. It is common to adopt as a starting point for minimizing $L_{c^k}(\cdot, \lambda^k)$ the last point x^{k-1} of the previous minimization. In order, however, for x^{k-1} to be near a minimizing point of $L_{c^k}(\cdot, \lambda^k)$, it is necessary that c^k is close to c^{k-1}. This in turn requires that the rate of increase of the penalty parameter c^k should be relatively small. There is a basic tradeoff here. If c^k is increased at a fast rate, then $\{x^k\}$ converges faster, but the likelihood of ill-conditioning is greater. Usually, a sequence $\{c^k\}$ generated by $c^{k+1} = \beta c^k$ with β in the range $[4, 10]$ works well if a Newton-like method is used for minimizing $L_{c^k}(\cdot, \lambda^k)$; otherwise, a smaller value of β may be needed. Some trial and error may be needed to choose the initial penalty parameter c^0, since there is no safe guideline on how to determine this value. For an indication of this, note that if the problem functions f and h are scaled by multiplication with a scalar $s > 0$, then c^0 should be divided by s to maintain the same condition number for the Hessian of the augmented Lagrangian.

Inequality Constraints

The simplest way to treat inequality constraints in the context of the quadratic penalty method, is to convert them to equality constraints by

Sec. 5.2   Penalty and Augmented Lagrangian Methods

using squared additional variables. We have already used this device in our discussion of optimality conditions for inequality constraints in Section 4.3.2.

Consider the problem

$$\begin{aligned}\text{minimize} \quad & f(x) \\ \text{subject to} \quad & h_1(x) = 0, \ldots, h_m(x) = 0, \\ & g_1(x) \leq 0, \ldots, g_r(x) \leq 0.\end{aligned} \quad (5.56)$$

As discussed in Section 4.3.2, we can convert this problem to the equality constrained problem

$$\begin{aligned}\text{minimize} \quad & f(x) \\ \text{subject to} \quad & h_1(x) = 0, \ldots, h_m(x) = 0, \\ & g_1(x) + z_1^2 = 0, \ldots, g_r(x) + z_r^2 = 0,\end{aligned} \quad (5.57)$$

where $z_1, \ldots, z_r$ are additional variables. The quadratic penalty method for this problem involves unconstrained minimizations of the form

$$\min_{x,z} \bar{L}_c(x, z, \lambda, \mu) = f(x) + \lambda' h(x) + \frac{c}{2}\|h(x)\|^2$$
$$+ \sum_{j=1}^r \left\{ \mu_j \left(g_j(x) + z_j^2\right) + \frac{c}{2} \left|g_j(x) + z_j^2\right|^2 \right\},$$

for various values of λ, μ, and c. This type of minimization can be done by first minimizing $\bar{L}_c(x, z, \lambda, \mu)$ with respect to z, obtaining

$$L_c(x, \lambda, \mu) = \min_z \bar{L}_c(x, z, \lambda, \mu),$$

and then by minimizing $L_c(x, \lambda, \mu)$ with respect to x. A key observation is that *the first minimization with respect to z can be carried out in closed form for each fixed x*, thereby yielding a closed form expression for $L_c(x, \lambda, \mu)$.

Indeed, we have

$$\min_z \bar{L}_c(x, z, \lambda, \mu) = f(x) + \lambda' h(x) + \frac{c}{2}\|h(x)\|^2$$
$$+ \sum_{j=1}^r \min_{z_j} \left\{ \mu_j \left(g_j(x) + z_j^2\right) + \frac{c}{2} \left|g_j(x) + z_j^2\right|^2 \right\},$$
$$(5.58)$$

and the minimization with respect to z_j in the last term is equivalent to

$$\min_{u_j \geq 0} \left\{ \mu_j \left(g_j(x) + u_j\right) + \frac{c}{2} \left|g_j(x) + u_j\right|^2 \right\}.$$

The function in braces above is quadratic in u_j. Its constrained minimum is $u_j^* = \max\{0, \hat{u}_j\}$, where $\hat{u}_j$ is the unconstrained minimum at which the derivative, $\mu_j + c(g_j(x) + \hat{u}_j)$, is zero. Thus,

$$u_j^* = \max\left\{0, -\left(\frac{\mu_j}{c} + g_j(x)\right)\right\}.$$

Denoting

$$g_j^+(x, \mu, c) = \max\left\{g_j(x), -\frac{\mu_j}{c}\right\}, \qquad (5.59)$$

we have $g_j(x) + u_j^* = g_j^+(x, \mu, c)$. Substituting this expression in Eq. (5.58), we obtain a closed form expression for $L_c(x, \lambda, \mu) = \min_z \bar{L}_c(x, z, \lambda, \mu)$ given by

$$\begin{aligned}
L_c(x, \lambda, \mu) &= f(x) + \lambda' h(x) + \frac{c}{2}\|h(x)\|^2 \\
&\quad + \sum_{j=1}^{r}\left\{\mu_j g_j^+(x, \mu, c) + \frac{c}{2}\left(g_j^+(x, \mu, c)\right)^2\right\}.
\end{aligned} \qquad (5.60)$$

After some calculation, left for the reader, we can also write this expression as

$$\begin{aligned}
L_c(x, \lambda, \mu) &= f(x) + \lambda' h(x) + \frac{c}{2}\|h(x)\|^2 \\
&\quad + \frac{1}{2c}\sum_{j=1}^{r}\left\{\left(\max\{0, \mu_j + cg_j(x)\}\right)^2 - \mu_j^2\right\},
\end{aligned} \qquad (5.61)$$

and we can view it as the augmented Lagrangian function for the inequality constrained problem (5.56).

Note that the penalty term

$$\frac{1}{2c}\left(\max\{0, \mu_j + cg_j(x)\}\right)^2 - \mu_j^2$$

corresponding to the jth inequality constraint in Eq. (5.61) is continuously differentiable in x if g_j is continuously differentiable (see Fig. 5.2.3). However, its Hessian matrix is discontinuous for all x such that $g_j(x) = -\mu_j/c$; this may cause some difficulties in the minimization of $L_c(x, \lambda, \mu)$ and motivates alternative augmented Lagrangian methods for inequality constraints (see Section 5.2.5).

To summarize, the quadratic penalty method for the inequality constrained problem (5.56) consists of a sequence of minimizations of the form

$$\begin{aligned}
&\text{minimize} \quad L_{c^k}(x, \lambda^k, \mu^k) \\
&\text{subject to} \quad x \in X,
\end{aligned}$$

Sec. 5.2   Penalty and Augmented Lagrangian Methods   479

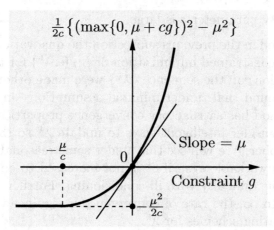

Figure 5.2.3. Form of the quadratic penalty function for inequality constraints.

where $L_c(x, \lambda^k, \mu^k)$ is given by Eq. (5.60) or Eq. (5.61), $\{\lambda^k\}$ and $\{\mu^k\}$ are sequences in $\Re^m$ and $\Re^r$, respectively, and $\{c^k\}$ is a positive penalty parameter sequence. Since this method is equivalent to the equality-constrained method applied to the corresponding equality-constrained problem (5.57), our convergence result of Prop. 5.2.1 applies with the obvious modifications.

Furthermore, if $X = \Re^n$, f, h, and g are continuously differentiable, and the generated sequence $\{x^k\}$ converges to a local minimum x^* which is also regular, application of Prop. 5.2.2 to the equivalent equality constrained problem (5.57) shows that the sequences

$$\{\lambda_i^k + c^k h_i(x^k)\}, \qquad \max\{0, \mu_j^k + c^k g_j(x^k)\} \qquad (5.62)$$

converge to the corresponding Lagrange multipliers λ_i^* and μ_j^* [for the jth inequality constraint, the Lagrange multiplier estimate is

$$\mu_j^k + c^k g_j^+(x^k, \mu^k, c^k),$$

which is equal to $\max\{0, \mu_j^k + c^k g_j(x^k)\}$ in view of the form (5.59) for g_j^+].

5.2.2   Multiplier Methods – Main Ideas

Let us return to the case where $X = \Re^n$ and the problem has only equality constraints,

$$\text{minimize } f(x)$$
$$\text{subject to } h(x) = 0.$$

We mentioned earlier that optimal solutions of this problem can be well approximated by unconstrained minima of the augmented Lagrangian $L_c(\cdot, \lambda)$ under two types of circumstances:

(a) The vector λ is close to a Lagrange multiplier.

(b) The penalty parameter c is large.

We analyzed in the previous subsection the quadratic penalty method consisting of unconstrained minimization of $L_{c^k}(\cdot, \lambda^k)$ for a sequence $c^k \to \infty$. No assumptions on the sequence $\{\lambda^k\}$ were made other than boundedness. Still, we found that, under minimal assumptions on f and h (continuity), the method has satisfactory convergence properties (Prop. 5.2.1).

We now consider intelligent ways to update λ^k so that it tends to a Lagrange multiplier. We will see that under some reasonable assumptions, this approach is workable even if c^k is not increased to ∞, thereby alleviating much of the difficulty with ill-conditioning. Furthermore, even when c^k is increased to ∞, the rate of convergence is significantly enhanced by using good updating schemes for λ^k.

The Method of Multipliers

A first update formula for λ^k in the quadratic penalty method is

$$\lambda^{k+1} = \lambda^k + c^k h(x^k). \tag{5.63}$$

The rationale is provided by Prop. 5.2.2, which shows that, if the generated sequence $\{x^k\}$ converges to a local minimum x^* that is regular, then $\{\lambda^k + c^k h(x^k)\}$ converges to the corresponding Lagrange multiplier λ^*.

The quadratic penalty method with the preceding update formula for λ^k is known as the *method of multipliers* (also called *augmented Lagrangian method*). There are a number of interesting convergence and rate of convergence results regarding this method, which will be given shortly. We first illustrate the method with some examples.

Example 5.2.3 (A Convex Problem)

Consider again the problem of Examples 5.2.1 and 5.2.2:

$$\text{minimize } f(x) = \tfrac{1}{2}(x_1^2 + x_2^2)$$
$$\text{subject to } x_1 = 1,$$

with optimal solution $x^* = (1, 0)$ and Lagrange multiplier $\lambda^* = -1$. The augmented Lagrangian is

$$L_c(x, \lambda) = \tfrac{1}{2}(x_1^2 + x_2^2) + \lambda(x_1 - 1) + \tfrac{c}{2}(x_1 - 1)^2.$$

The vectors x^k generated by the method of multipliers minimize $L_{c^k}(\cdot, \lambda^k)$ and are given by

$$x^k = \left(\frac{c^k - \lambda^k}{c^k + 1}, 0\right).$$

Sec. 5.2 Penalty and Augmented Lagrangian Methods 481

Using this expression, the multiplier updating formula (5.63) can be written as

$$\lambda^{k+1} = \lambda^k + c^k \left(\frac{c^k - \lambda^k}{c^k + 1} - 1 \right) = \frac{\lambda^k}{c^k + 1} - \frac{c^k}{c^k + 1},$$

or by introducing the Lagrange multiplier $\lambda^* = -1$,

$$\lambda^{k+1} - \lambda^* = \frac{\lambda^k - \lambda^*}{c^k + 1}.$$

From this formula, it can be seen that

(a) $\lambda^k \to \lambda^* = -1$ and $x^k \to x^* = (1, 0)$ for every nondecreasing sequence $\{c^k\}$ [since the scalar $1/(c^k+1)$ multiplying $\lambda^k - \lambda^*$ in the above formula is always less than one].

(b) The convergence rate becomes faster as c^k becomes larger; in fact the error sequence $\{|\lambda^k - \lambda^*|\}$ converges superlinearly if $c^k \to \infty$.

Note that it is not necessary to increase c^k to ∞, although doing so results in a better convergence rate.

Example 5.2.4 (A Nonconvex Problem)

Consider the problem

$$\text{minimize } \tfrac{1}{2}(-x_1^2 + x_2^2)$$
$$\text{subject to } x_1 = 1$$

with optimal solution $x^* = (1, 0)$ and Lagrange multiplier $\lambda^* = 1$. The augmented Lagrangian is given by

$$L_c(x, \lambda) = \tfrac{1}{2}(-x_1^2 + x_2^2) + \lambda(x_1 - 1) + \tfrac{c}{2}(x_1 - 1)^2.$$

The vector x^k minimizing $L_{c^k}(x, \lambda^k)$ is given by

$$x^k = \left(\frac{c^k - \lambda^k}{c^k - 1}, 0 \right). \tag{5.64}$$

For this formula to be correct, however, it is necessary that $c^k > 1$; for $c^k < 1$ the augmented Lagrangian has no minimum, and the same is true for $c^k = 1$ unless $\lambda^k = 1$. The multiplier updating formula (5.63) can be written using Eq. (5.64) as

$$\lambda^{k+1} = \lambda^k + c^k \left(\frac{c^k - \lambda^k}{c^k - 1} - 1 \right) = -\frac{\lambda^k}{c^k - 1} + \frac{c^k}{c^k - 1},$$

or by introducing the Lagrange multiplier $\lambda^* = 1$,

$$\lambda^{k+1} - \lambda^* = -\frac{\lambda^k - \lambda^*}{c^k - 1}. \tag{5.65}$$

From this iteration, it can be seen that similar conclusions to those of the preceding example can be drawn. In particular, it is not necessary to increase c^k to ∞ to obtain convergence, although doing so results in a better convergence rate. However, there is a difference: whereas in the preceding example, convergence was guaranteed for all positive sequences $\{c^k\}$, in the present example, the minimizing points exist only if $c^k > 1$. Here, c^k plays a convexification role: once it exceeds the threshold value of 1 the penalty term convexifies the augmented Lagrangian, thus compensating for the nonconvexity of the cost function. Moreover, it is seen from Eq. (5.65) that to obtain convergence, the penalty parameter c^k must eventually exceed 2, so that the scalar

$$\frac{-1}{c^k - 1}$$

multiplying λ^k has absolute value less than one. The need for c^k to exceed twice the value of the convexification threshold is a fundamental characteristic of multiplier methods when applied to nonconvex problems, as we will see shortly.

Geometric Interpretation of the Method of Multipliers

The conclusions from the preceding two examples hold in considerable generality. We first provide a geometric interpretation. Assume that f and h are twice differentiable and let x^* be a local minimum of f over $h(x) = 0$. Assume also that x^* is regular and together with its associated Lagrange multiplier vector λ^* satisfies the second order sufficiency conditions of Prop. 4.2.1. Then the assumptions of the sensitivity theorem (Prop. 4.2.2) are satisfied and we can consider the *primal function*

$$p(u) = \min_{h(x)=u} f(x),$$

defined for u in an open sphere centered at $u = 0$. [The minimization above is understood to be local in an open sphere within which x^* is the unique local minimum of f over $h(x) = 0$ (cf. Prop. 4.2.2).] Note that we have

$$p(0) = f(x^*), \qquad \nabla p(0) = -\lambda^*,$$

(cf. Prop. 4.2.2). The primal function is illustrated in Fig. 5.2.4.

We can break down the minimization of $L_c(\cdot, \lambda)$ into two stages, first minimizing over all x such that $h(x) = u$ with u fixed, and then minimizing over all u. Thus,

$$\min_x L_c(x, \lambda) = \min_u \min_{h(x)=u} \left\{ f(x) + \lambda' h(x) + \frac{c}{2} \|h(x)\|^2 \right\}$$
$$= \min_u \left\{ p(u) + \lambda' u + \frac{c}{2} \|u\|^2 \right\},$$

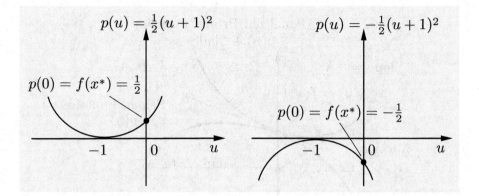

Figure 5.2.4. Illustration of the primal function. In (a) we show the primal function

$$p(u) = \min_{x_1 - 1 = u} \tfrac{1}{2}(x_1^2 + x_2^2)$$

for the problem of Example 5.2.3. In (b) we show the primal function

$$p(u) = \min_{x_1 - 1 = u} \tfrac{1}{2}(-x_1^2 + x_2^2)$$

for the problem of Example 5.2.4. The latter primal function is not convex because the cost function is not convex on the subspace that is orthogonal to the constraint set (this observation can be generalized).

where the minimization above is understood to be local in a neighborhood of $u = 0$. This minimization can be interpreted as shown in Fig. 5.2.5. The minimum is attained at the point $u(\lambda, c)$ for which the gradient of $p(u) + \lambda'u + \tfrac{c}{2}\|u\|^2$ is zero, or, equivalently,

$$\nabla\left\{p(u) + \frac{c}{2}\|u\|^2\right\}\bigg|_{u=u(\lambda,c)} = -\lambda.$$

Thus, the minimizing point $u(\lambda, c)$ is obtained as shown in Fig. 5.2.5. We also have

$$\min_x L_c(x, \lambda) - \lambda'u(\lambda, c) = p(u(\lambda, c)) + \frac{c}{2}\|u(\lambda, c)\|^2,$$

so the tangent hyperplane to the graph of $p(u) + \tfrac{c}{2}\|u\|^2$ at $u(\lambda, c)$ (which has "slope" $-\lambda$) intersects the vertical axis at the value $\min_x L_c(x, \lambda)$ as shown in Fig. 5.2.5. It can be seen that if c is sufficiently large, then the function

$$p(u) + \lambda'u + \frac{c}{2}\|u\|^2$$

is convex in a neighborhood of the origin. Furthermore, for λ close to λ^* and large c, the value $\min_x L_c(x, \lambda)$ is close to $p(0) = f(x^*)$.

Figure 5.2.6 provides a geometric interpretation of the multiplier iteration

$$\lambda^{k+1} = \lambda^k + c^k h(x^k).$$

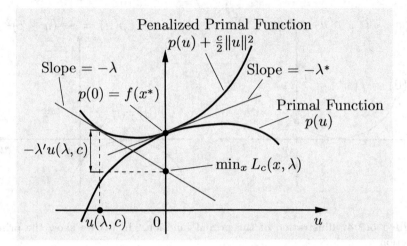

Figure 5.2.5. Geometric interpretation of minimization of the augmented Lagrangian. The value $\min_x L_c(x, \lambda)$ is the point where the tangent hyperplane to the graph of $p(u) + \frac{c}{2}\|u\|^2$ at $u(\lambda, c)$ (which has "slope" $-\lambda$) intersects the vertical axis. This point is close to $p(0) = f(x^*)$ if either λ is close to λ^* or c is large (or both).

To understand this figure, note that if x^k minimizes $L_{c^k}(\cdot, \lambda^k)$, then by the preceding analysis the vector u^k given by $u^k = h(x^k)$ minimizes $p(u) + \lambda^{k'}u + \frac{c^k}{2}\|u\|^2$. Hence,

$$\nabla\left\{p(u) + \frac{c^k}{2}\|u\|^2\right\}\bigg|_{u=u^k} = -\lambda^k,$$

and

$$\nabla p(u^k) = -(\lambda^k + c^k u^k) = -\left(\lambda^k + c^k h(x^k)\right).$$

It follows that the next multiplier λ^{k+1} is

$$\lambda^{k+1} = \lambda^k + c^k h(x^k) = -\nabla p(u^k),$$

as shown in Fig. 5.2.6. The figure shows that if λ^k is sufficiently close to λ^* and/or c^k is sufficiently large, the next multiplier λ^{k+1} will be closer to λ^* than λ^k is. Furthermore, c^k need not be increased to ∞ in order to obtain convergence; it is sufficient that c^k eventually exceeds some threshold level. The figure also shows that if $p(u)$ is linear, convergence to λ^* will be achieved in one iteration.

In summary, the geometric interpretation of the method of multipliers just presented suggests the following:

(a) If c is large enough so that $cI + \nabla^2 p(0)$ is positive definite, then the "penalized primal function"

$$p(u) + \frac{c}{2}\|u\|^2$$

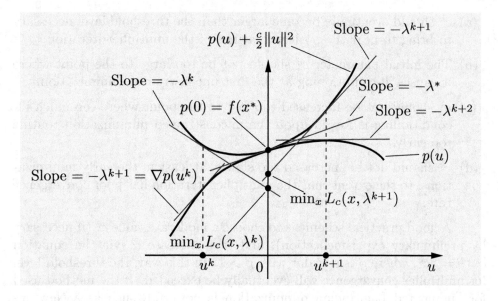

Figure 5.2.6. Geometric interpretation of the first order multiplier iteration. The figure shows the process of obtaining λ^{k+1} from λ^k, assuming a constant penalty parameter $c^k = c^{k+1} = c$.

is convex within a sphere centered at $u = 0$. Furthermore, a local minimum of the augmented Lagrangian $L_c(x, \lambda)$ that is near x^* exists if λ is close enough to λ^*. The reason is that $\nabla^2_{xx} L_c(x^*, \lambda^*)$ is positive definite if and only if $cI + \nabla^2 p(0)$ is positive definite, a fact the reader may wish to verify as an exercise.

(b) If c^k is sufficiently large [the threshold can be shown to be twice the value of c needed to make $cI + \nabla^2 p(0)$ positive definite; see Exercise 5.2.4], then

$$\|\lambda^{k+1} - \lambda^*\| \le \|\lambda^k - \lambda^*\|$$

and $\lambda^k \to \lambda^*$.

(c) Convergence can be obtained even if c^k is not increased to ∞.

(d) As c^k is increased, the rate of convergence of λ^k improves.

(e) If $\nabla^2 p(0) = 0$, the convergence is very fast.

These conclusions will be formalized in Section 5.2.3.

Computational Aspects – Choice of Parameters

In addition to addressing the problem of ill-conditioning, an important practical question in the method of multipliers is how to select the initial multiplier λ^0 and the penalty parameter sequence. Clearly, in view of the interpretations given earlier, any prior knowledge should be exploited to select λ^0 as close as possible to λ^*. The main considerations to be kept in mind for selecting the penalty parameter sequence are the following:

(a) c^k should eventually become larger than the threshold level necessary to bring to bear the positive features of the multiplier iteration.

(b) The initial parameter c^0 should not be too large to the point where it causes ill-conditioning at the first unconstrained minimization.

(c) c^k should not be increased too fast to the point where too much ill-conditioning is forced upon the unconstrained minimization routine too early.

(d) c^k should not be increased too slowly, at least in the early minimizations, to the extent that the multiplier iteration has poor convergence rate.

A good practical scheme is to choose a moderate value c^0 (if necessary by preliminary experimentation), and then increase c^k via the equation $c^{k+1} = \beta c^k$, where β is a scalar with $\beta > 1$. In this way, the threshold level for multiplier convergence will eventually be exceeded. If the method used for augmented Lagrangian minimization is powerful, such as a Newton-like method, fairly large values of β (say $\beta \in [5, 10]$) are recommended; otherwise, smaller values of β may be necessary, depending on the method's ability to deal with ill-conditioning.

Another reasonable parameter adjustment scheme is to increase c^k by multiplication with a factor $\beta > 1$ only if the constraint violation as measured by $\|h(x(\lambda^k, c^k))\|$ is not decreased by a factor $\gamma < 1$ over the previous minimization; i.e.,

$$c^{k+1} = \begin{cases} \beta c^k & \text{if } \|h(x^k)\| > \gamma \|h(x^{k-1})\|, \\ c^k & \text{if } \|h(x^k)\| \leq \gamma \|h(x^{k-1})\|. \end{cases}$$

A choice such as $\gamma = 0.25$ is typically recommended.

Still another possibility is an adaptive scheme that uses a different penalty parameter c_i^k for each constraint $h_i(x) = 0$, and increases by a certain factor the penalty parameters of the constraints that are violated most. For example, increase c_i^k if the constraint violation as measured by $|h_i(x^k)|$ is not decreased by a certain factor over $|h_i(x^{k-1})|$.

Inexact Minimization of the Augmented Lagrangian

In practice the minimization of $L_{c^k}(x, \lambda^k)$ is typically terminated early. For example, it may be terminated at a point x^k satisfying

$$\|\nabla_x L_{c^k}(x^k, \lambda^k)\| \leq \epsilon^k,$$

where $\{\epsilon^k\}$ is a positive sequence converging to zero. Then it is still appropriate to use the multiplier update

$$\lambda^{k+1} = \lambda^k + c^k h(x^k),$$

although in theory, some of the linear convergence rate results to be given shortly will not hold any more. This deficiency does not seem to be important in practice, but can also be corrected by using the alternative termination criterion

$$\|\nabla_x L_{c^k}(x^k, \lambda^k)\| \leq \min\{\epsilon^k, \gamma^k \|h(x^k)\|\},$$

where $\{\epsilon^k\}$ and $\{\gamma^k\}$ are positive sequences converging to zero; for an analysis see the author's works [Ber75b], [Ber76a], and [Ber82a], Section 2.5.

In Section 5.4, we will see that with certain safeguards, it is possible to terminate the minimization of the augmented Lagrangian after a few Newton steps (possibly only one), and follow it by a second order multiplier update of the type that will be discussed later in this section. Such algorithmic strategies give rise to some of the most effective methods using Lagrange multipliers.

Inequality Constraints

To treat inequality constraints $g_j(x) \leq 0$ in the context of the method of multipliers, we convert them into equality constraints $g_j(x) + z_j^2 = 0$, using the additional variables z_j [cf. problems (5.56) and (5.57)]. In particular, the multiplier update formulas are

$$\lambda^{k+1} = \lambda^k + c^k h(x^k),$$

$$\mu_j^{k+1} = \max\left\{0, \mu_j^k + c^k g_j(x^k)\right\}, \qquad j = 1, \ldots, r,$$

[cf. Eq. (5.62)], where x^k minimizes the augmented Lagrangian

$$L_{c^k}(x, \lambda^k, \mu^k) = f(x) + \lambda^{k'} h(x) + \frac{c}{2} \|h(x)\|^2$$

$$+ \frac{1}{2c^k} \sum_{j=1}^{r} \left\{ \left(\max\{0, \mu_j^k + c^k g_j(x)\}\right)^2 - (\mu_j^k)^2 \right\}.$$

Many problems encountered in practice involve two-sided constraints of the form

$$\alpha_j \leq g_j(x) \leq \beta_j,$$

where α_j and β_j are given scalars. Each two-sided constraint could of course be separated into two one-sided constraints. This would require, however, the assignment of two multipliers per two-sided constraint, and is somewhat wasteful, since we know that at a solution at least one of the two multipliers will be zero. It turns out that there is an alternative approach that requires only *one multiplier per two-sided constraint* (see Exercise 5.2.7).

Partial Elimination of Constraints

In the preceding multiplier algorithms, all the equality and inequality constraints are eliminated by means of a penalty. In some cases it is convenient to eliminate only part of the constraints, while retaining the remaining constraints explicitly. A typical example is a problem of the form

$$\text{minimize } f(x)$$
$$\text{subject to } h(x) = 0, \quad x \geq 0.$$

While, in addition to $h(x) = 0$, it is possible to eliminate by means of a penalty the bound constraints $x \geq 0$, it is often desirable to handle these constraints explicitly by a gradient projection or two-metric projection method of the type discussed in Sections 3.3 and 3.4, respectively.

More generally, a method of multipliers with partial elimination of constraints for the problem

$$\text{minimize } f(x)$$
$$\text{subject to } h(x) = 0, \quad g(x) \leq 0,$$

consists of finding x^k that solves the problem

$$\text{minimize } f(x) + \lambda^{k'} h(x) + \frac{c}{2}\|h(x)\|^2$$
$$\text{subject to } g(x) \leq 0,$$

followed by the multiplier iteration

$$\lambda^{k+1} = \lambda^k + c^k h(x^k).$$

In fact it is not essential that just the equality constraints are eliminated by means of a penalty above. Any mixture of equality and inequality constraints can be eliminated by means of a penalty and a multiplier, while the remaining constraints can be explicitly retained. For a detailed treatment of partial elimination of constraints, we refer to [Ber77], [Ber82a], Section 2.4, [Dun91a], and [Dun93b].

5.2.3 Convergence Analysis of Multiplier Methods

We now discuss the convergence properties of multiplier methods and substantiate the conclusions derived informally earlier. We focus attention throughout on the equality constrained problem

$$\text{minimize } f(x)$$
$$\text{subject to } h(x) = 0,$$

and on a particular local minimum x^*. We assume that x^* *is regular and together with a Lagrange multiplier vector λ^* satisfies the second order sufficiency conditions of Prop. 4.2.1.* In view of our earlier treatment of inequality constraints by conversion to equalities, our analysis readily carries over to the case of mixed equality and inequality constraints, under the second order sufficiency conditions of Prop. 4.2.1.

The convergence results described in this section can be strengthened considerably under additional convexity assumptions on the problem. This is discussed further in Section 7.3; see also [Ber82a], Chapter 5.

Error Bounds for Local Minima of the Augmented Lagrangian

A first basic issue is whether local minima x^k of the augmented Lagrangian $L_{c^k}(\cdot, \lambda^k)$ exist, so that the method itself is well-defined. We have shown that for the local minimum-Lagrange multiplier pair (x^*, λ^*) there exist scalars $\bar c > 0$, $\gamma > 0$, and $\epsilon > 0$, such that

$$L_c(x, \lambda^*) \geq L_c(x^*, \lambda^*) + \frac{\gamma}{2}\|x - x^*\|^2, \qquad \forall\, x \text{ with } \|x - x^*\| < \epsilon, \text{ and } c \geq \bar c,$$

(cf. the discussion following Lemma 3.2.1). It is thus reasonable to infer that if λ is close to λ^*, there should exist a local minimum of $L_c(\cdot, \lambda)$ close to x^* for every $c \geq \bar c$. More precisely, for a fixed $c \geq \bar c$, we can show this by considering the system of equations

$$\nabla_x L_c(x, \lambda) = \nabla f(x) + \nabla h(x)\big(\lambda + ch(x)\big) = 0,$$

and by using the implicit function theorem in a neighborhood of (x^*, λ^*). [This can be done because the Jacobian of the system with respect to x is $\nabla_x^2 L_c(x^*, \lambda^*)$, and is positive definite since $c \geq \bar c$.] Thus, for λ sufficiently close to λ^*, there is an unconstrained local minimum $x(\lambda, c)$ of $L_c(\cdot, \lambda)$, which is defined via the equation

$$\nabla f\big(x(\lambda, c)\big) + \nabla h\big(x(\lambda, c)\big)\big(\lambda + ch(x(\lambda, c))\big) = 0.$$

A closer examination of the preceding argument shows that for application of the implicit function theorem it is not essential that λ be close to λ^* but rather that the vector $\lambda + ch\big(x(\lambda, c)\big)$ be close to λ^*. Proposition 5.2.2 indicates that for any λ, if c is sufficiently large and $x(\lambda, c)$ minimizes $L_c(x, \lambda)$, the vector $\lambda + ch\big(x(\lambda, c)\big)$ is close to λ^*. This suggests that *there should exist a local minimum of $L_c(\cdot, \lambda)$ close to x^* even for λ that are far from λ^*, provided c is sufficiently large.* This can indeed be shown. In fact it turns out that for existence of the local minimum $x(\lambda, c)$, what is really important is that the ratio $\|\lambda - \lambda^*\|/c$ be sufficiently small. However, proving this simultaneously for the entire range of values $c \in [\bar c, \infty)$ is not easy. The following proposition, due to [Ber82a], provides a precise mathematical statement of the existence result, together with some error estimates that quantify the rate of convergence. The proof requires the introduction of the variables $\tilde\lambda = \lambda + ch(x)$, and $t = (\lambda - \lambda^*)/c$, together with the system of equations $\nabla_x L_0(x, \tilde\lambda) = 0$ and an analysis based on a more refined form of the implicit function theorem than the one given in Appendix A.

Proposition 5.2.3: Let $\bar c$ be a positive scalar such that

$$\nabla^2_{xx} L_{\bar c}(x^*, \lambda^*) > 0.$$

There exist positive scalars δ, ϵ, and M such that:

(a) For all (λ, c) in the set $D \subset \Re^{m+1}$ defined by

$$D = \{(\lambda, c) \mid \|\lambda - \lambda^*\| < \delta c,\ \bar c \leq c\}, \qquad (5.66)$$

the problem
$$\text{minimize}\quad L_c(x, \lambda)$$
$$\text{subject to}\quad \|x - x^*\| < \epsilon$$

has a unique solution denoted $x(\lambda, c)$. The function $x(\cdot, \cdot)$ is continuously differentiable in the interior of D, and for all $(\lambda, c) \in D$, we have

$$\|x(\lambda, c) - x^*\| \leq M \frac{\|\lambda - \lambda^*\|}{c}.$$

(b) For all $(\lambda, c) \in D$, we have

$$\|\tilde\lambda(\lambda, c) - \lambda^*\| \leq M \frac{\|\lambda - \lambda^*\|}{c},$$

where
$$\tilde\lambda(\lambda, c) = \lambda + ch\bigl(x(\lambda, c)\bigr).$$

(c) For all $(\lambda, c) \in D$, the matrix $\nabla^2_{xx} L_c\bigl(x(\lambda, c), \lambda\bigr)$ is positive definite and the matrix $\nabla h\bigl(x(\lambda, c)\bigr)$ has rank m.

Proof: See [Ber82a], p. 108.

Figure 5.2.7 shows the set D of pairs (λ, c) within which the conclusions of Prop. 5.2.3 are valid [cf. Eq. (5.66)]. It can be seen that, for any λ, there exists a c_λ such that (λ, c) belongs to D for every $c \geq c_\lambda$. The estimate δc on the allowable distance of λ from λ^* grows linearly with c

[compare with Eq. (5.66)]. In particular problems, the actual allowable distance may grow at a higher than linear rate, and in fact it is possible that for every λ and $c > 0$ there exists a unique global minimum of $L_c(\cdot, \lambda)$. (Take for instance the scalar problem $\min\{x^2 \mid x = 0\}$.) However, it is shown by example in [Ber82a], p. 111, that the estimate of a linear order of growth cannot be improved.

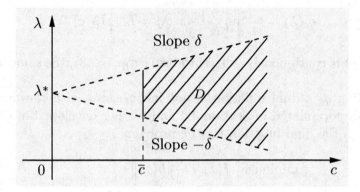

Figure 5.2.7. Illustration of the set

$$D = \left\{(\lambda, c) \mid \|\lambda - \lambda^*\| < \delta c, \ \bar{c} \leq c\right\}$$

within which the conclusions of Prop. 5.2.3 are valid.

Convergence and Rate of Convergence

Proposition 5.2.3 yields both a convergence and a convergence rate result for the multiplier iteration

$$\lambda^{k+1} = \lambda^k + c^k h(x^k).$$

It shows that if the generated sequence $\{\lambda^k\}$ is bounded [this can be enforced if necessary by leaving λ^k unchanged if $\lambda^k + c^k h(x^k)$ does not belong to a prespecified bounded open set known to contain λ^*], the penalty parameter c^k is sufficiently large after a certain index [so that $(\lambda^k, c^k) \in D$], and after that index, minimization of $L_{c^k}(\cdot, \lambda^k)$ yields the local minimum $x^k = x(\lambda^k, c^k)$ closest to x^*, then we obtain $x^k \to x^*$, $\lambda^k \to \lambda^*$. Furthermore, the rate of convergence of the error sequences $\{\|x^k - x^*\|\}$ and $\{\|\lambda^k - \lambda^*\|\}$ is linear, and it is superlinear if $c^k \to \infty$.

It is possible to conduct a more refined convergence and rate of convergence analysis that supplements Prop. 5.2.3. This analysis quantifies the threshold level of the penalty parameter for convergence to occur and gives a precise estimate of the linear convergence rate. We refer to the book [Ber82a] for an extensive discussion; see also Exercise 5.2.4.

5.2.4 Duality and Second Order Multiplier Methods

Let $\bar{c}$, δ, and ϵ be as in Prop. 5.2.3, and define for (λ, c) in the set

$$D = \{(\lambda, c) \mid \|\lambda - \lambda^*\| < \delta c, \; \bar{c} \leq c\}$$

the *dual function* q_c by

$$q_c(\lambda) = \min_{\|x - x^*\| < \epsilon} L_c(x, \lambda) = L_c\big(x(\lambda, c), \lambda\big). \tag{5.67}$$

Since $x(\cdot, c)$ is continuously differentiable (Prop. 5.2.3), the same is true for q_c.

Calling q_c a dual function is not inconsistent with the duality theory already formulated in Section 4.4 and further developed in Chapter 6. Indeed q_c is the dual function for the problem

$$\text{minimize} \quad f(x) + \frac{c}{2}\|h(x)\|^2$$

$$\text{subject to} \quad \|x - x^*\| < \epsilon, \quad h(x) = 0,$$

which for $c \geq \bar{c}$, has x^* as its unique optimal solution and λ^* as the corresponding Lagrange multiplier.

We compute the gradient of q_c with respect to λ. From Eq. (5.67), we have

$$\nabla q_c(\lambda) = \nabla_\lambda x(\lambda, c) \nabla_x L_c\big(x(\lambda, c), \lambda\big) + h\big(x(\lambda, c)\big).$$

Since $\nabla_x L_c\big(x(\lambda, c), \lambda\big) = 0$, we obtain

$$\nabla q_c(\lambda) = h\big(x(\lambda, c)\big), \tag{5.68}$$

and since $x(\cdot, c)$ is continuously differentiable, the same is true for ∇q_c.

Next we compute the Hessian $\nabla^2 q_c$. Differentiating ∇q_c, as given by Eq. (5.68), with respect to λ, we obtain

$$\nabla^2 q_c(\lambda) = \nabla_\lambda x(\lambda, c) \nabla h\big(x(\lambda, c)\big). \tag{5.69}$$

We also have, for all (λ, c) in the set D,

$$\nabla_x L_c\big(x(\lambda, c), \lambda\big) = 0.$$

Differentiating with respect to λ, we obtain

$$\nabla_\lambda x(\lambda, c) \nabla^2_{xx} L_c\big(x(\lambda, c), \lambda\big) + \nabla^2_{\lambda x} L_c\big(x(\lambda, c), \lambda\big) = 0,$$

and since

$$\nabla^2_{\lambda x} L_c\big(x(\lambda, c), \lambda\big) = \nabla h\big(x(\lambda, c)\big)',$$

it follows that
$$\nabla_\lambda x(\lambda, c) = -\nabla h\big(x(\lambda, c)\big)' \big(\nabla^2_{xx} L_c(x(\lambda, c), \lambda)\big)^{-1}.$$

Substitution in Eq. (5.69) yields the formula

$$\nabla^2 q_c(\lambda) = -\nabla h\big(x(\lambda, c)\big)' \big(\nabla^2_{xx} L_c(x(\lambda, c), \lambda)\big)^{-1} \nabla h\big(x(\lambda, c)\big). \qquad (5.70)$$

Since $\nabla^2_{xx} L_c(x(\lambda, c), \lambda)$ is positive definite and $\nabla h(x(\lambda, c))$ has rank m for $(\lambda, c) \in D$ (cf. Prop. 5.2.3), it follows from Eq. (5.70) that $\nabla^2 q_c(\lambda)$ is negative definite for all $(\lambda, c) \in D$, so that q_c is concave within the set $\{\lambda \mid \|\lambda - \lambda^*\| \leq \delta c\}$. Furthermore, using Eq. (5.68), we have, for all $c \geq \bar{c}$,

$$\nabla q_c(\lambda^*) = h\big(x(\lambda^*, c)\big) = h(x^*) = 0.$$

Thus, *for every* $c \geq \bar{c}$, λ^* *maximizes* $q_c(\lambda)$ *over the set* $\{\lambda \mid \|\lambda - \lambda^*\| < \delta c\}$. Also in view of Eq. (5.68), the multiplier update formula can be written as

$$\lambda^{k+1} = \lambda^k + c^k \nabla q_{c^k}(\lambda^k), \qquad (5.71)$$

so *it is a steepest ascent iteration for maximizing* q_{c^k}. When $c^k = c$ for all k, then Eq. (5.71) is the constant stepsize steepest ascent method

$$\lambda^{k+1} = \lambda^k + c \nabla q_c(\lambda^k)$$

for maximizing q_c.

The Second Order Method of Multipliers

In view of the interpretation of the multiplier iteration as a steepest ascent method, it is natural to consider the Newton-like iteration

$$\lambda^{k+1} = \lambda^k - \big(\nabla^2 q_{c^k}(\lambda^k)\big)^{-1} \nabla q_{c^k}(\lambda^k),$$

for maximizing the dual function. In view of the gradient and Hessian formulas (5.68) and (5.70), this iteration can be written as

$$\lambda^{k+1} = \lambda^k + (B^k)^{-1} h(x^k),$$

where

$$B^k = \nabla h(x^k)' \big(\nabla^2_{xx} L_{c^k}(x^k, \lambda^k)\big)^{-1} \nabla h(x^k)$$

and x^k minimizes $L_{c^k}(\cdot, \lambda^k)$.

An alternative form, which turns out to be more appropriate when the minimization of the augmented Lagrangian is inexact is given by

$$\lambda^{k+1} = \lambda^k + (B^k)^{-1} \big(h(x^k) - \nabla h(x^k)' \big(\nabla^2_{xx} L_{c^k}(x^k, \lambda^k)\big)^{-1} \nabla_x L_{c^k}(x^k, \lambda^k)\big). \qquad (5.72)$$

When the augmented Lagrangian is minimized exactly, $\nabla_x L_{c^k}(x^k, \lambda^k) = 0$, and the two forms are equivalent.

To provide motivation for iteration (5.72), let us consider Newton's method for solving the system of necessary conditions

$$\nabla_x L_c(x, \lambda) = \nabla f(x) + \nabla h(x)(\lambda + ch(x)) = 0, \qquad h(x) = 0.$$

In this method, we linearize the above system around the current iterate (x^k, λ^k), and we obtain the next iterate (x^{k+1}, λ^{k+1}) from the solution of the linearized system

$$\begin{pmatrix} \nabla_{xx}^2 L_{c^k}(x^k, \lambda^k) & \nabla h(x^k) \\ \nabla h(x^k)' & 0 \end{pmatrix} \begin{pmatrix} x^{k+1} - x^k \\ \lambda^{k+1} - \lambda^k \end{pmatrix} = - \begin{pmatrix} \nabla_x L_{c^k}(x^k, \lambda^k) \\ h(x^k) \end{pmatrix}.$$

It is straightforward to verify (a derivation will be given in Section 5.4.2) that λ^{k+1} is given by Eq. (5.72), while

$$x^{k+1} = x^k - \left(\nabla_{xx}^2 L_{c^k}(x^k, \lambda^k)\right)^{-1} \nabla_x L_{c^k}(x^k, \lambda^{k+1}).$$

This justifies the use of the extra term

$$\nabla h(x^k)' \left(\nabla_{xx}^2 L_{c^k}(x^k, \lambda^k)\right)^{-1} \nabla_x L_{c^k}(x^k, \lambda^k)$$

in Eq. (5.72) when the minimization of the augmented Lagrangian is inexact.

5.2.5 Nonquadratic Augmented Lagrangians - The Exponential Method of Multipliers

One of the drawbacks of the method of multipliers when applied to inequality constrained problems is that the corresponding augmented Lagrangian function is not twice differentiable even if the cost and constraint functions are. As a result, serious difficulties can arise when Newton-like methods are used to minimize the augmented Lagrangian, particularly for polyhedral-type problems. This motivates alternative twice differentiable augmented Lagrangians to handle inequality constraints, which we now describe.

Consider the problem

$$\begin{aligned} \text{minimize} \quad & f(x) \\ \text{subject to} \quad & g_1(x) \leq 0, \ldots, g_r(x) \leq 0. \end{aligned}$$

We introduce a method of multipliers characterized by a twice differentiable penalty function $\psi : \Re \mapsto \Re$ with the following properties:

(i) $\nabla^2 \psi(t) > 0$ for all $t \in \Re$,

(ii) $\psi(0) = 0$, $\nabla \psi(0) = 1$,

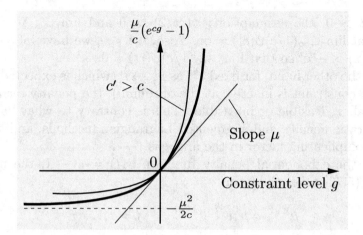

Figure 5.2.8. The penalty term of the exponential method of multipliers. The slope at 0 is μ, regardless of the value of x.

(iii) $\lim_{t \to -\infty} \psi(t) > -\infty$,

(iv) $\lim_{t \to -\infty} \nabla \psi(t) = 0$ and $\lim_{t \to \infty} \nabla \psi(t) = \infty$.

A simple and interesting special case is the *exponential penalty function*

$$\psi(t) = e^t - 1,$$

(see Fig. 5.2.8).

The method consists of the sequence of unconstrained minimizations

$$x^k \in \arg\min_{x \in \Re^n} \left\{ f(x) + \sum_{j=1}^{m} \frac{\mu_j^k}{c_j^k} \psi\big(c_j^k g_j(x)\big) \right\},$$

followed by the multiplier iterations

$$\mu_j^{k+1} = \mu_j^k \nabla \psi\big(c_j^k g_j(x^k)\big), \qquad j = 1, \ldots, r. \tag{5.73}$$

Here $\{c_j^k\}$ is a positive penalty parameter sequence for each j, and the initial multipliers μ_j^0 are arbitrary positive numbers.

Note that for fixed $\mu_j^k > 0$, the "penalty" term

$$\frac{\mu_j^k}{c_j^k} \psi\big(c_j^k g_j(x)\big)$$

tends (as $c_j^k \to \infty$) to ∞ for all infeasible x $[g_j(x) > 0]$ and to zero for all feasible x $[g_j(x) \leq 0]$. To see this, note that by convexity of ψ, we have $\psi(ct) \geq \psi(ct/2) + (ct/2)\nabla\psi(ct/2)$, from which we obtain

$$\frac{1}{c}\psi(ct) \geq \frac{1}{c}\psi(ct/2) + \frac{t}{2}\nabla\psi(ct/2).$$

Thus if $t > 0$, the assumptions $\psi(ct/2) > 0$ and $\lim_{\tau \to \infty} \nabla \psi(\tau) = \infty$ imply that $\lim_{c \to \infty}(1/c)\psi(ct) = \infty$. Also, if $t < 0$, we have $\inf_{c > 0} \psi(ct) = \inf_{\tau < 0} \psi(\tau) > -\infty$, so that $\lim_{c \to \infty}(1/c)\psi(ct) = 0$.

On the other hand, for fixed c_j^k, as $\mu_j^k \to 0$ (which is expected to occur if the jth constraint is inactive at the optimum), the penalty term goes to zero for all x, feasible or infeasible. This is contrary to what happens in the quadratic penalty and augmented Lagrangian methods, and turns out to be a complicating factor in the analysis.

For the exponential penalty function $\psi(t) = e^t - 1$, the multiplier iteration (5.73) takes the form

$$\mu_j^{k+1} = \mu_j^k e^{c_j^k g_j(x^k)}, \qquad j = 1, \ldots, r.$$

Another interesting method, known as the *modified barrier method*, is based on the following version of the logarithmic barrier function

$$\psi(t) = -\ln(1 - t),$$

for which the multiplier iteration (5.73) takes the form

$$\mu_j^{k+1} = \frac{\mu_j^k}{1 - c_j^k g_j(x^k)}, \qquad j = 1, \ldots, r.$$

This method is not really a special case of the generic method (5.73) because the penalty function ψ is defined only on the set $(-\infty, 1)$, but it shares the same qualitative characteristics as the generic method.

Practical Implementation

Two practical points regarding the exponential method are worth mentioning. The first is that the exponential terms in the augmented Lagrangian function can easily become very large with an attendant computer overflow. [The modified barrier method has a similar and even more serious difficulty: it tends to ∞ as $c_j^k g_j(x^k) \to 1$.] One way to deal with this disadvantage is to define $\psi(t)$ as the exponential $e^t - 1$ only for t in an interval $(-\infty, A]$, where A is such that e^A is within the floating point range of the computer; outside the interval $(-\infty, A]$, $\psi(t)$ can be defined as any function such that the properties required of ψ, including twice differentiability, are maintained over the entire real line. For example ψ can be a quadratic function with parameters chosen so that $\nabla^2 \psi$ is continuous at the splice point A.

The second point is that it makes sense to introduce a different penalty parameter c_j^k for the jth constraint and to let c_j^k depend on the current values of the corresponding multiplier μ_j^k via

$$c_j^k = \frac{w^k}{\mu_j^k}, \qquad j = 1, \ldots, r, \qquad (5.74)$$

where $\{w^k\}$ is a positive scalar sequence with $w^k \leq w^{k+1}$ for all k. The reason can be seen by using the series expansion of the exponential term to write

$$\frac{\mu_j}{c}\left(e^{cg_j(x)} - 1\right) = \frac{\mu_j}{c}\left(cg_j(x) + \frac{c^2}{2}(g_j(x))^2 + \frac{c^3}{3!}(g_j(x))^3 + \cdots\right).$$

If the jth constraint is active at the eventual limit, the terms of order higher than quadratic can be neglected and we can write approximately

$$\frac{\mu_j}{c}\left(e^{cg_j(x)} - 1\right) \approx \mu_j g_j(x) + \frac{c\mu_j}{2}g_j(x)^2.$$

Thus the exponential augmented Lagrangian term becomes similar to the quadratic term, except that the role of the penalty parameter is played by the product $c\mu_j$. This motivates the use of selection rules such as Eq. (5.74). A similar penalty selection rationale applies also to other penalty functions in the class.

Generally the convergence analysis of the exponential method of multipliers and other methods in the class, under second order sufficiency conditions, turns out to be not much more difficult than for the quadratic method (see the references). The convergence results available are as powerful as for the quadratic method. The practical performances of the exponential and the quadratic method of multipliers are roughly comparable for problems where second order differentiability of the augmented Lagrangian function turns out to be of no concern. The exponential method has an edge for problems where the lack of second order differentiability in the quadratic method causes difficulties.

EXERCISES

5.2.1

Consider the problem

$$\text{minimize } f(x) = \tfrac{1}{2}\left(x_1^2 - x_2^2\right) - 3x_2$$
$$\text{subject to } x_2 = 0.$$

(a) Calculate the optimal solution and the Lagrange multiplier.

(b) For $k = 0, 1, 2$ and $c^k = 10^{k+1}$ calculate and compare the iterates of the quadratic penalty method with $\lambda^k = 0$ for all k and the method of multipliers with $\lambda^0 = 0$.

(c) Draw the figure that interprets geometrically the two methods (cf. Figs. 5.2.5 and 5.2.6) for this problem, and plot the iterates of the two methods on this figure for $k = 0, 1, 2$.

(d) Suppose that c is taken to be constant in the method of multipliers. For what values of c would the augmented Lagrangian have a minimum and for what values of c would the method converge?

5.2.2

Consider the problem

$$\text{minimize} \quad f(x) = \tfrac{1}{2}\left(x_1^2 + |x_2|^\rho\right) + 2x_2$$
$$\text{subject to} \quad x_2 = 0,$$

where $\rho > 1$.

(a) Calculate the optimal solution and the Lagrange multiplier.

(b) Write a computer program to calculate the iterates of the multiplier method with $\lambda^0 = 0$, and $c^k = 1$ for all k. Confirm computationally that the rate of convergence is sublinear if $\rho = 1.5$, linear if $\rho = 2$, and superlinear if $\rho = 3$.

(c) Give a heuristic argument why the rate of convergence is sublinear if $\rho < 2$, linear if $\rho = 2$, and superlinear if $\rho > 2$. What happens in the limit where $\rho = 1$?

5.2.3

Consider the problem of Exercise 5.2.1. Verify that the second order method of multipliers converges in one iteration provided c is sufficiently large, and estimate the threshold value for c.

5.2.4 (Convergence Threshold and Convergence Rate of the Method of Multipliers) (www)

Consider the quadratic problem

$$\text{minimize} \quad \tfrac{1}{2}x'Qx$$
$$\text{subject to} \quad Ax = b,$$

where Q is symmetric and A is an $m \times n$ matrix of rank m. Let f^* be the optimal value of the problem and assume that the problem has a unique minimum x^* with associated Lagrange multiplier λ^*. Verify that for sufficiently large c, the penalized dual function is

$$q_c(\lambda) = -\tfrac{1}{2}(\lambda - \lambda^*)'A(Q + cA'A)^{-1}A'(\lambda - \lambda^*) + f^*.$$

(Use the quadratic programming duality theory of Section 4.4.2 to show that q_c is a quadratic function and to derive the Hessian matrix of q_c. Then use the fact that q_c is maximized at λ^* and that its maximum value is f^*.) Consider the first order method of multipliers.

(a) Use the theory of Section 1.3 to show that for all k

$$\|\lambda^{k+1} - \lambda^*\| \leq r^k \|\lambda^k - \lambda^*\|,$$

where

$$r^k = \max\{|1 - c^k E_{c^k}|, |1 - c^k e_{c^k}|\}$$

and E_c and e_c denote the maximum and the minimum eigenvalues of the matrix $A(Q + cA'A)^{-1}A'$.

(b) Assume that Q is invertible. Using the matrix identity

$$\left(I + c^k AQ^{-1}A'\right)^{-1} = I - c^k A(Q + c^k A'A)^{-1}A'$$

(cf. Section A.3 in Appendix A), relate the eigenvalues of the matrix $A(Q + c^k A'A)^{-1}A'$ with those of the matrix $AQ^{-1}A'$. Show that if $\gamma_1, \ldots, \gamma_m$ are the eigenvalues of $(AQ^{-1}A')^{-1}$, we have

$$r^k = \max_{i=1,\ldots,m} \left|\frac{\gamma_i}{\gamma_i + c^k}\right|.$$

(c) Show that the method converges to λ^* if $c > \bar{c}$, where $\bar{c} = 0$ if $\gamma_i \geq 0$ for all i, and $\bar{c} = -2\min\{\gamma_1, \ldots, \gamma_m\}$ otherwise.

5.2.5 (Stepsize Analysis of the Method of Multipliers) (www)

Consider the problem of Exercise 5.2.4. Use the results of that exercise to analyze the convergence and rate of convergence of the generalized method of multipliers

$$\lambda^{k+1} = \lambda^k + \alpha^k (Ax^k - b),$$

where α^k is a positive stepsize. Show in particular that if Q is positive definite and $c^k = c$ for all k, convergence is guaranteed if $\delta \leq \alpha^k \leq 2c$ for all k, where δ is some positive scalar. (For a solution and related analysis, see [Ber75d] and [Ber82a], p. 126.)

5.2.6

A weakness of the quadratic penalty method is that the augmented Lagrangian may not have a global minimum. As an example, show that the scalar problem

$$\text{minimize} \quad -x^4$$
$$\text{subject to} \quad x = 0$$

has the unique global minimum $x^* = 0$ but its augmented Lagrangian

$$L_{c^k}(x, \lambda^k) = -x^4 + \lambda^k x + \frac{c^k}{2} x^2$$

has no global minimum for every c^k and λ^k. To overcome this difficulty, consider a penalty function of the form

$$\frac{c}{2} \|h(x)\|^2 + \|h(x)\|^\rho,$$

where $\rho > 4$, instead of $(c/2)\|h(x)\|^2$. Show that $L_{c^k}(x, \lambda^k)$ has a global minimum for every λ^k and $c^k > 0$.

5.2.7 (Two-Sided Inequality Constraints [Ber77], [Ber82a])

The purpose of this exercise is to show how to treat two-sided inequality constraints by using a *single* multiplier per constraint. Consider the problem

$$\text{minimize } f(x)$$
$$\text{subject to } \alpha_j \leq g_j(x) \leq \beta_j, \qquad j = 1, \ldots, r,$$

where $f : \Re^n \mapsto \Re$ and $g_j : \Re^n \mapsto \Re$ are given functions, and α_j and β_j, $j = 1, \ldots, r$, are given scalars with $\alpha_j < \beta_j$. The method consists of sequential minimizations of the form

$$\text{minimize } f(x) + \sum_{j=1}^{r} P_j\big(g_j(x), \mu_j^k, c^k\big)$$
$$\text{subject to } x \in \Re^n,$$

where

$$P_j\big(g_j(x), \mu_j^k, c^k\big) = \min_{u_j \in [g_j(x) - \beta_j, \, g_j(x) - \alpha_j]} \left\{ \mu_j^k u_j + \frac{c^k}{2} |u_j|^2 \right\}.$$

Each of these minimizations is followed by the multiplier iteration

$$\mu_j^{k+1} = \begin{cases} \mu_j^k + c^k\big(g_j(x^k) - \beta_j\big) & \text{if } \mu_j^k + c^k\big(g_j(x^k) - \beta_j\big) > 0, \\ \mu_j^k + c^k\big(g_j(x^k) - \alpha_j\big) & \text{if } \mu_j^k + c^k\big(g_j(x^k) - \alpha_j\big) < 0, \\ 0 & \text{otherwise,} \end{cases}$$

where x^k is a minimizing vector. Justify the method by introducing artificial variables u_j, by converting the problem to the equivalent form

$$\text{minimize } f(x)$$
$$\text{subject to } \alpha_j \leq g_j(x) - u_j \leq \beta_j, \qquad u_j = 0, \quad j = 1, \ldots, r,$$

and by applying a multiplier method for this problem, where only the constraints $u_j = 0$ are eliminated by means of a quadratic penalty function (partial elimination of constraints).

5.2.8 (Proof of Ill-Conditioning as $c^k \to \infty$)

Consider the quadratic penalty method ($c^k \to \infty$) for the equality constrained problem of minimizing $f(x)$ subject to $h(x) = 0$, and assume that the generated sequence converges to a local minimum x^* that is also a regular point. Show that the condition number of the Hessian $\nabla^2_{xx} L_{c^k}(x^k, \lambda^k)$ tends to ∞. *Hint*: We have

$$\nabla^2_{xx} L_{c^k}(x^k, \lambda^k) = \nabla^2_{xx} L_0(x^k, \tilde{\lambda}^k) + c^k \nabla h(x^k) \nabla h(x^k)',$$

where $\tilde{\lambda}^k = \lambda^k + c^k h(x^k)$. The minimum eigenvalue $m(x^k, \lambda^k, c^k)$ of this matrix satisfies

$$\begin{aligned} m(x^k, \lambda^k, c^k) &= \min_{\|z\|=1} z' \nabla^2_{xx} L_{c^k}(x^k, \lambda^k) z \\ &\leq \min_{\|z\|=1, \nabla h(x^k)'z=0} z' \nabla^2_{xx} L_{c^k}(x^k, \lambda^k) z \\ &= \min_{\|z\|=1, \nabla h(x^k)'z=0} z' \nabla^2_{xx} L_0(x, \tilde{\lambda}^k) z. \end{aligned}$$

The maximum eigenvalue $M(x^k, \lambda^k, c^k)$ satisfies

$$\begin{aligned} M(x^k, \lambda^k, c^k) &= \max_{\|z\|=1} z' \nabla^2_{xx} L_{c^k}(x^k, \lambda^k) z \\ &\geq \min_{\|z\|=1} z' \nabla^2_{xx} L_0(x^k, \tilde{\lambda}^k) z + c^k \max_{\|z\|=1} z' \nabla h(x^k) \nabla h(x^k)' z. \end{aligned}$$

Use Prop. 5.2.2 to argue that $\nabla h(x^k)$ has rank m for sufficiently large k, and hence

$$\lim_{c^k \to \infty} \frac{M(x^k, \lambda^k, c^k)}{m(x^k, \lambda^k, c^k)} = \infty.$$

5.2.9 (www)

Let $\{x^k\}$ be a sequence generated by the logarithmic barrier method. Formulate conditions under which the sequences $\{-\epsilon^k/g_j(x^k)\}$ converge to corresponding Lagrange multipliers. *Hint*: Compare with the corresponding result of Prop. 5.2.2 for the quadratic penalty function.

5.2.10

State and prove analogs of Props. 5.2.1 and 5.2.2 for the case where the penalty function

$$\frac{c}{2}\|h(x)\|^2 + \|h(x)\|^\rho$$

with $\rho > 1$ is used in place of the quadratic $\frac{c}{2}\|h(x)\|^2$.

5.2.11 (Primal-Dual Methods not Using a Penalty) (www)

This exercise shows that an important dual ascent method, to be discussed in Section 7.2 (see also the end of Section 5.4.1), turns out to be equivalent to the first order method of multipliers for an artificial problem. Consider minimizing $f(x)$ subject to $h(x) = 0$, where f and h are twice continuously differentiable, and let x^* be a local minimum that is a regular point. Let λ^* be the associated Lagrange multiplier vector, and assume that the Hessian $\nabla^2_{xx} L(x^*, \lambda^*)$ of the (ordinary) Lagrangian

$$L(x, \lambda) = f(x) + \lambda' h(x)$$

is positive definite. (Note that this is stronger than what is required by the second order sufficiency conditions.) Consider the iteration

$$\lambda^{k+1} = \lambda^k + \alpha h(x^k),$$

where α is a positive scalar stepsize and x^k minimizes $L(x, \lambda^k)$ (the minimization is local in a suitable neighborhood of x^*). Show that there exists a threshold $\bar{\alpha} > 0$ and a sphere centered at λ^* such that if λ^0 belongs to this sphere and $\alpha < \bar{\alpha}$, then λ^k converges to λ^*. Consider first the case where f is quadratic and h is linear, and sketch an analysis for the more general case. *Hint*: Even though there is no penalty parameter here, the method can be viewed as a method of multipliers for the artificial problem

$$\text{minimize } f(x) - \frac{\alpha}{2}\|h(x)\|^2$$
$$\text{subject to } h(x) = 0.$$

Use the threshold of Exercise 5.2.4(c) to verify that $\bar{\alpha}$ can be taken to be equal to twice the minimum eigenvalue of $\nabla h(x^*)' \big(\nabla^2_{xx} L(x^*, \lambda^*)\big)^{-1} \nabla h(x^*)$. For analysis along this line, see [Ber82a], Section 2.6.

5.3 EXACT PENALTIES – SEQUENTIAL QUADRATIC PROGRAMMING

In this section we consider penalty methods that are *exact* in the sense that they require only one unconstrained minimization to obtain an optimal solution of the original constrained problem. We will use exact penalties as the basis for a broad class of algorithms, called *sequential quadratic programming*. These algorithms may also be viewed within the context of the Lagrangian methods of Section 5.4, and will be reencountered there.

To get a sense of how this is possible, consider the equality constrained problem

$$\text{minimize } f(x)$$
$$\text{subject to } h_i(x) = 0, \quad i = 1, \ldots, m, \tag{5.75}$$

where f and h_i are continuously differentiable, and let

$$L(x, \lambda) = f(x) + \lambda' h(x)$$

Sec. 5.3   Exact Penalties – Sequential Quadratic Programming   503

be the corresponding Lagrangian function. Then by minimizing the function

$$P(x,\lambda) = \|\nabla_x L(x,\lambda)\|^2 + \|h(x)\|^2 \qquad (5.76)$$

over $(x,\lambda) \in \Re^{n+m}$ we can obtain local minima-Lagrange multiplier pairs (x^*, λ^*) satisfying the first order necessary conditions

$$\nabla_x L(x^*, \lambda^*) = 0, \qquad h(x^*) = 0.$$

We may view $P(x, \lambda)$ as an *exact penalty function*, i.e., a function whose unconstrained minima are (or strongly relate to) optimal solutions and/or Lagrange multipliers of a constrained problem.

The exact penalty function $P(x, \lambda)$ of Eq. (5.76) has (in effect) been used extensively in the special case where $m = n$ and the problem is to solve the system of constraint equations $h(x) = 0$ (in this case, any cost function f may be used). However, in the case where $m < n$, $P(x, \lambda)$ has significant drawbacks because it does not discriminate between local minima and local maxima, and it may also have local minima $(\bar{x}, \bar{\lambda})$ that are not global and do not satisfy the necessary optimality conditions, i.e., $P(\bar{x}, \bar{\lambda}) > 0$. There are, however, more sophisticated exact penalty functions that do not have these drawbacks, as we will see later.

We may distinguish between *differentiable* and *nondifferentiable* exact penalty functions. The former have the advantage that they can be minimized by the unconstrained methods we have already studied in Chapters 1 and 2. The latter involve nondifferentiabilities, so the methods of Chapters 1 and 2 are not directly applicable. We will develop special algorithms, called *sequential quadratic programming methods*, for their minimization.

Nondifferentiable exact penalty methods have been more popular in practice than their differentiable counterparts, and they will receive most of our attention. On the other hand, differentiable exact penalty methods have some interesting advantages; see the monograph [Ber82a] (Section 4.3), which treats extensively both types of methods.

5.3.1 Nondifferentiable Exact Penalty Functions

Our first objective in this section is to show that solutions of the equality constrained problem (5.75) are related to solutions of the (nondifferentiable) unconstrained problem

$$\text{minimize} \ \ f(x) + cP(x)$$
$$\text{subject to} \ \ x \in \Re^n,$$

where $c > 0$ and P is the nondifferentiable penalty function defined by

$$P(x) = \max_{i=1,\ldots,m} |h_i(x)|.$$

We develop the main argument for equality constraints, then generalize to include inequality constraints, and finally state the result in Prop. 5.3.1.

Indeed let x^* be a local minimum, which is a regular point and satisfies together with a corresponding Lagrange multiplier vector λ^*, the second order sufficiency conditions of Prop. 4.2.1. Consider also the *primal function* p defined in a neighborhood of the origin by

$$p(u) = \min\{f(x) \mid h(x) = u, \|x - x^*\| < \epsilon\},$$

where $\epsilon > 0$ is some scalar; see the sensitivity theorem (Prop. 4.2.2). Then if we locally minimize $f + cP$ around x^*, we can split the minimization in two: first minimize over all x satisfying $h(x) = u$ and then minimize over all possible u. We have

$$\inf_{\|x-x^*\|<\epsilon} \left\{ f(x) + c \max_{i=1,\ldots,m} |h_i(x)| \right\}$$

$$= \inf_{u \in U_\epsilon} \inf_{\{x \mid h(x)=u, \|x-x^*\|<\epsilon\}} \left\{ f(x) + c \max_{i=1,\ldots,m} |h_i(x)| \right\}$$

$$= \inf_{u \in U_\epsilon} p_c(u),$$

where

$$p_c(u) = p(u) + c \max_{i=1,\ldots,m} |u_i|,$$

$$U_\epsilon = \{u \mid h(x) = u \text{ for some } x \text{ with } \|x - x^*\| < \epsilon\}.$$

We may view p_c as a *penalized primal function*. We will now show that for large enough c, p_c has a local minimum at $u = 0$; cf. Fig. 5.3.1.

Since, according to the sensitivity theorem, we have $\nabla p(0) = -\lambda^*$, we can use the mean value theorem to write for each u in a neighborhood of the origin

$$p(u) = p(0) - u'\lambda^* + \tfrac{1}{2} u' \nabla^2 p(\bar{\alpha} u) u,$$

where $\bar{\alpha}$ is some scalar in $[0, 1]$. Thus

$$p_c(u) = p(0) - \sum_{i=1}^{m} u_i \lambda_i^* + c \max_{i=1,\ldots,m} |u_i| + \tfrac{1}{2} u' \nabla^2 p(\bar{\alpha} u) u. \tag{5.77}$$

Assume that c is sufficiently large so that for some $\gamma > 0$,

$$c \geq \sum_{i=1}^{m} |\lambda_i^*| + \gamma.$$

Then it follows that

$$c \max_{i=1,\ldots,m} |u_i| \geq \left(\sum_{i=1}^{m} |\lambda_i^*| + \gamma \right) \max_{i=1,\ldots,m} |u_i|$$

$$\geq \sum_{i=1}^{m} u_i \lambda_i^* + \gamma \max_{i=1,\ldots,m} |u_i|.$$

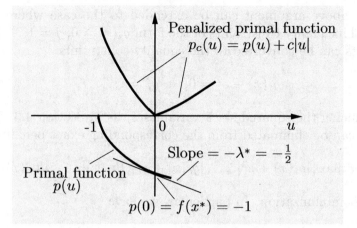

Figure 5.3.2. Illustration of how for c large enough, $u = 0$ is a strict local minimum of $p_c(u) = p(u) + c \max_{i=1,\ldots,m} |u_i|$. The figure corresponds to the two-dimensional problem where $f(x) = x_1$ and $h(x) = x_1^2 + x_2^2 - 1$. The optimal solution and Lagrange multiplier are $x^* = (-1, 0)$ and $\lambda^* = 1/2$, respectively. The primal function is defined for $u \geq -1$ and is given by

$$p(u) = \min_{x_1^2 + x_2^2 - 1 = u} x_1 = -\sqrt{1+u}.$$

Note that $\nabla p(0) = \lambda^*$ and that we must have $c > \lambda^*$ in order for the nondifferentiable penalty function to be exact.

Using this relation in Eq. (5.77), we obtain

$$p_c(u) \geq p(0) + \gamma \max_{i=1,\ldots,m} |u_i| + \tfrac{1}{2} u' \nabla^2 p(\bar{\alpha} u) u.$$

For u sufficiently close to zero, the last term is dominated by the next to last term, so

$$p_c(u) > p(0) = p_c(0)$$

for all $u \neq 0$ in a neighborhood N of the origin. Hence $u = 0$ is a strict local minimum of p_c as shown in Fig. 5.3.2. Since we have

$$f(x) + cP(x) \geq p(u) + c \max_{i=1,\ldots,m} |u_i| = p_c(u)$$

for all $u \neq 0$ in the neighborhood N and x such that $h(x) = u$ with $\|x - x^*\| < \epsilon$, and we also have $p_c(0) = f(x^*)$, we obtain that for all x in a neighborhood of x^* with $P(x) > 0$,

$$f(x) + cP(x) > f(x^*).$$

Since x^* is a strict local minimum of f over all x with $P(x) = 0$, it follows that x^* is a strict local minimum of $f + cP$, provided that $c > \sum_{i=1}^{m} |\lambda_i^*|$.

The above argument can be extended to the case where there are additional inequality constraints of the form $g_j(x) \leq 0$, $j = 1, \ldots, r$. These constraints can be converted to the equality constraints

$$g_j(x) + z_j^2 = 0, \qquad j = 1, \ldots, r,$$

by introducing the squared slack variables z_j as in Section 4.3. The slack variables can be eliminated from the corresponding exact penalty function

$$f(x) + c \max\{|g_1(x) + z_1^2|, \ldots, |g_r(x) + z_r^2|, |h_1(x)|, \ldots, |h_m(x)|\} \quad (5.78)$$

by explicit minimization. In particular, we have

$$\min_{z_j} |g_j(x) + z_j^2| = \max\{0, g_j(x)\},$$

so minimization of the exact penalty function (5.78) is equivalent to minimization of the function

$$f(x) + c \max\{0, g_1(x), \ldots, g_r(x), |h_1(x)|, \ldots, |h_m(x)|\}.$$

Thus, by repeating the earlier argument given for equality constraints, we have the following proposition.

Proposition 5.3.1: Let x^* be a local minimum of the problem

minimize $f(x)$
subject to $h_i(x) = 0$, $i = 1, \ldots, m$, $\qquad g_j(x) \leq 0$, $j = 1, \ldots, r$,

which is regular and satisfies together with corresponding Lagrange multiplier vectors λ^* and μ^*, the second order sufficiency conditions of Prop. 4.3.2. Then, if

$$c > \sum_{i=1}^{m} |\lambda_i^*| + \sum_{j=1}^{r} \mu_j^*,$$

the vector x^* is a strict unconstrained local minimum of $f + cP$, where

$$P(x) = \max\{0, g_1(x), \ldots, g_r(x), |h_1(x)|, \ldots, |h_m(x)|\}.$$

An example illustrating the above proposition is given in Fig. 5.3.3. The proof of the proposition was relatively simple but made assumptions that are stronger than necessary. There are related results that do not

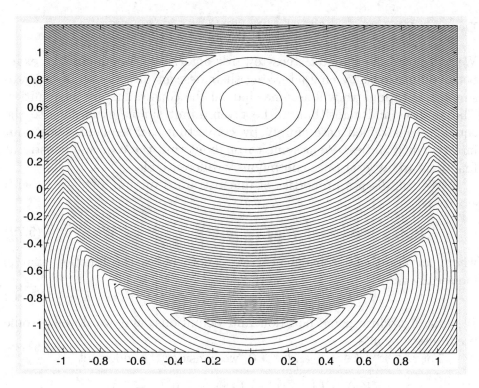

Figure 5.3.3. Equal cost surfaces of the function $f + cP$ for the two-dimensional problem where
$$f(x) = x_1, \qquad h(x) = x_1^2 + x_2^2 - 1$$
(cf. Fig. 5.3.2). For c greater than the Lagrange multiplier $\lambda^* = 1/2$, the optimal solution $x^* = (-1, 0)$ is a local minimum of $f + cP$. This is not so for $c < \lambda^*$. The figure corresponds to $c = 0.8$.

require second order differentiability assumptions. In particular, it is shown in Exercise 5.3.4 that if c is sufficiently large, a regular local minimum x^* is a "stationary" point of $f + cP$ (in a sense to be made precise shortly). The reverse is not necessarily true. In particular, there may exist local minima of $f + cP$ that do not correspond to constrained local minima of f for any c; see Exercise 5.3.1. There is also a more refined analysis that requires just first order differentiability; see [BNO03], Section 5.5.

Finally, let us note that under convexity assumptions, where f is convex, h_i are linear, g_j are convex, and there is an additional convex abstract set constraint, the results that connect global minima of the original problem and global minima of corresponding nondifferentiable exact penalty functions are more powerful. In particular, the analysis is not tied to a specific local minimum x^*, and there no need for the type of differentiability, sufficiency, and regularity assumptions that we are using in Prop. 5.3.1; see [Ber75a], and the textbook accounts [BNO03], Section 7.3, and [Ber15a], Section 1.5.

Descent Directions of Exact Penalties

We will now take the first steps towards algorithms for minimizing exact penalty functions, by characterizing their descent directions. In order to simplify notation, we will assume that all the constraints are inequalities. The analysis and algorithms to be given admit simple extensions to the equality constrained case simply by converting each equality constraint into two inequalities. We assume throughout that the cost and constraint functions are at least once continuously differentiable.

We will discuss properties of unconstrained minima of $f + cP$, with $c > 0$ and

$$P(x) = \max\{g_0(x), g_1(x), \ldots, g_r(x)\}, \qquad \forall\, x \in \Re^n, \tag{5.79}$$

where for notational convenience, we denote by g_0 the function that is identically zero:

$$g_0(x) \equiv 0, \qquad x \in \Re^n. \tag{5.80}$$

We first introduce some notation and definitions, and we develop some preliminary results.

For $x \in \Re^n$, $d \in \Re^n$, and $c > 0$, we consider the index set

$$J(x) = \{j \mid g_j(x) = P(x),\; j = 0, 1, \ldots, r\},$$

and we denote

$$\theta_c(x; d) = \max\{\nabla f(x)'d + c\nabla g_j(x)'d \mid j \in J(x)\}.$$

The function θ_c plays the role that the gradient would play if $f + cP$ were differentiable. In particular, the function

$$f(x) + cP(x) + \theta_c(x; d)$$

may be viewed as a linear approximation of $f + cP$ for variations d around x; see Fig. 5.3.4.

Since at an unconstrained local minimum x^*, $f + cP$ cannot decrease along any direction, the preceding interpretation of θ_c motivates us to call a vector x^* a *stationary point of* $f + cP$ if for all $d \in \Re^n$ there holds

$$\theta_c(x^*; d) \geq 0.$$

The following proposition shows that local minima of $f + cP$ must be stationary points of $f + cP$. Furthermore, *descent directions of* $f + cP$ *at a nonstationary point* x *can be obtained from the following convex quadratic program, in* $(d, \xi) \in \Re^{n+1}$:

$$\begin{aligned}
\text{minimize} \quad & \nabla f(x)'d + \tfrac{1}{2}d'Hd + c\xi \\
\text{subject to} \quad & (d, \xi) \in \Re^{n+1}, \qquad g_j(x) + \nabla g_j(x)'d \leq \xi, \quad j = 0, 1, \ldots, r,
\end{aligned} \tag{5.81}$$

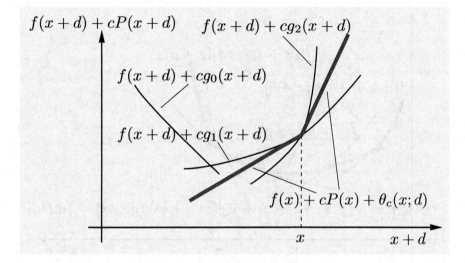

Figure 5.3.4. Illustration of $\theta_c(x;d)$ at a point x. It is the first order estimate of the variation

$$f(x+d) + cP(x+d) - f(x) - cP(x)$$

of $f + cP$ around x. Here the index set $J(x)$ is $\{1, 2\}$.

where $c > 0$ and H is a positive definite symmetric matrix.

To understand the role of this quadratic program, note that for a fixed d, the minimum with respect to ξ is attained at

$$\xi = \max_{j=0,1,\ldots,r} \{g_j(x) + \nabla g_j(x)'d\}.$$

Thus, by eliminating the variable ξ, and by adding to the cost the constant term $f(x)$, we can write the quadratic program (5.81) in the alternative form

$$\text{minimize} \max_{j=0,1,\ldots,r} \{f(x) + cg_j(x) + \nabla f(x)'d + c\nabla g_j(x)'d\} + \tfrac{1}{2} d'Hd$$

subject to $d \in \Re^n$.

(5.82)

For small $\|d\|$, the maximum over j is attained for $j \in J(x)$, so we can substitute $P(x)$ in place of $g_j(x)$. The cost function then takes the form

$$f(x) + cP(x) + \theta_c(x;d) + \tfrac{1}{2} d'Hd,$$

so locally, for d near zero, the problem (5.82) can be viewed as minimization of a quadratic approximation of $f + cP$ around x; see Fig. 5.3.5. Also since the cost function of problem (5.82) is strictly convex in d, its optimal solution is unique, implying also that the quadratic program (5.81) has a unique optimal solution (d, ξ).

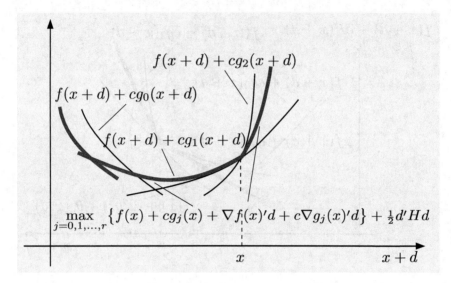

Figure 5.3.5. Illustration of the cost function

$$\max_{j=0,1,\ldots,r} \left\{ f(x) + cg_j(x) + \nabla f(x)'d + c\nabla g_j(x)'d \right\} + \tfrac{1}{2}d'Hd$$

of the quadratic program (5.82). For small $\|d\|$ this function takes the form

$$f(x) + cP(x) + \theta_c(x;d) + \tfrac{1}{2}d'Hd,$$

and is a quadratic approximation of $f + cP$ around x. It can be seen that by minimizing this function over d we obtain a direction of descent of $f + cP$ at x.

Proposition 5.3.2: (Descent Directions of $f + cP$)

(a) For all $x \in \Re^n$, $d \in \Re^n$, and $\alpha > 0$, we have

$$f(x+\alpha d) + cP(x+\alpha d) - f(x) - cP(x) = \alpha \theta_c(x;d) + o(\alpha), \quad (5.83)$$

where $\lim_{\alpha \to 0^+} o(\alpha)/\alpha = 0$. As a result, if $\theta_c(x;d) < 0$, then d is a descent direction, i.e., there exists $\bar{\alpha} > 0$ such that

$$f(x+\alpha d) + cP(x+\alpha d) < f(x) + cP(x), \quad \forall \, \alpha \in (0, \bar{\alpha}].$$

Moreover, a local minimum of $f + cP$ is a stationary point.

(b) If f and g_j are convex functions, then a stationary point of $f + cP$ is also a global minimum of $f + cP$.

(c) For any $x \in \Re^n$ and positive definite symmetric H, if (d, ξ) is the optimal solution of the quadratic program (5.81), then

Sec. 5.3 Exact Penalties – Sequential Quadratic Programming 511

$$\theta_c(x;d) \leq -d'Hd. \qquad (5.84)$$

(d) A vector x is a stationary point of $f + cP$ if and only if the quadratic program (5.81) has $\{d = 0, \xi = P(x)\}$ as its optimal solution.

Proof: (a) We have for all $\alpha > 0$ and $j \in J(x)$,

$$f(x+\alpha d) + cg_j(x+\alpha d) = f(x) + \alpha \nabla f(x)'d + c\big(g_j(x) + \alpha \nabla g_i(x)'d\big) \\ + o_j(\alpha),$$

where $\lim_{\alpha \to 0+} o_j(\alpha)/\alpha = 0$. Hence, by using the fact $g_j(x) = P(x)$ for all $j \in J(x)$,

$$f(x+\alpha d) + c\max\{g_j(x+\alpha d) \mid j \in J(x)\} \\ = f(x) + \alpha \nabla f(x)'d + c\max\{g_j(x) + \alpha \nabla g_j(x)'d \mid j \in J(x)\} + o(\alpha) \\ = f(x) + cP(x) + \alpha \theta_c(x;d) + o(\alpha),$$

where $\lim_{\alpha \to 0+} o(\alpha)/\alpha = 0$. We have, for all α that are sufficiently small,

$$\max\{g_j(x+\alpha d) \mid j \in J(x)\} = \max\{g_j(x+\alpha d) \mid j = 0, 1, \ldots, r\} \\ = P(x+\alpha d).$$

Combining the two above relations, we obtain

$$f(x+\alpha d) + cP(x+\alpha d) = f(x) + cP(x) + \alpha \theta_c(x;d) + o(\alpha),$$

which is Eq. (5.83).

If x^* is a local minimum of $f + cP$, then by Eq. (5.83), we have, for all d and $\alpha > 0$ such that $\|d\|$ and α are sufficiently small,

$$\alpha \theta_c(x^*;d) + o(\alpha) \geq 0.$$

Dividing by α and taking the limit as $\alpha \to 0$, we obtain $\theta_c(x^*;d) \geq 0$, so x^* is a stationary point of $f + cP$.

(b) By convexity, we have [cf. Prop. B.3(a) in Appendix B] for all j and $x \in \Re^n$,

$$f(x) + cg_j(x) \geq f(x^*) + cg_j(x^*) + \big(\nabla f(x^*) + c\nabla g_j(x^*)\big)'(x - x^*).$$

Taking the maximum over j, we obtain

$$f(x) + cP(x) \geq \max_{j=0,1,\ldots,r} \big\{f(x^*) + cg_j(x^*) + \big(\nabla f(x^*) + c\nabla g_j(x^*)\big)'(x - x^*)\big\}.$$

For a sufficiently small scalar ϵ and for all x with $\|x-x^*\| < \epsilon$, the maximum above is attained for some $j \in J(x^*)$. Since $g_j(x^*) = P(x^*)$ for all $j \in J(x^*)$, we obtain for all x with $\|x - x^*\| < \epsilon$,

$$f(x) + cP(x) \geq f(x^*) + cP(x^*) + \theta_c(x^*; x - x^*) \geq f(x^*) + cP(x^*),$$

where the last inequality holds because x^* is a stationary point of $f + cP$. Hence x^* is a local minimum of $f + cP$, and in view of the convexity of $f + cP$, x^* is a global minimum.

(c) We have $g_j(x) + \nabla g_j(x)'d \leq \xi$ for all j. Since $g_j(x) = P(x)$ for all $j \in J(x)$, it follows that $\nabla g_j(x)'d \leq \xi - P(x)$ for all $j \in J(x)$ and therefore using the definition of θ_c we have

$$\theta_c(x; d) \leq \nabla f(x)'d + c(\xi - P(x)). \tag{5.85}$$

Let $\{\mu_j\}$ be a set of Lagrange multipliers for the quadratic program (5.81). The optimality conditions yield

$$\nabla f(x) + Hd + \sum_{j=0}^{r} \mu_j \nabla g_j(x) = 0, \tag{5.86}$$

$$c - \sum_{j=0}^{r} \mu_j = 0, \tag{5.87}$$

$$g_j(x) + \nabla g_j(x)'d \leq \xi, \quad \mu_j \geq 0, \quad j = 0, 1, \ldots, r,$$
$$\mu_j\big(g_j(x) + \nabla g_j(x)'d - \xi\big) = 0, \quad j = 0, 1, \ldots, r.$$

By adding the last equation over all j and using Eq. (5.87), we have

$$\sum_{j=0}^{r} \mu_j \nabla g_j(x)'d = \sum_{j=0}^{r} \mu_j \xi - \sum_{j=0}^{r} \mu_j g_j(x)$$

$$\geq \sum_{j=0}^{r} \mu_j \big(\xi - \max_{m=0,1,\ldots,r} g_m(x)\big)$$

$$= \sum_{j=0}^{r} \mu_j \big(\xi - P(x)\big)$$

$$= c(\xi - P(x)).$$

Combining this equation with Eq. (5.86) we obtain

$$\nabla f(x)'d + d'Hd + c(\xi - P(x)) \leq 0, \tag{5.88}$$

which in conjunction with Eq. (5.85) yields

$$\theta_c(x; d) + d'Hd \leq 0,$$

thus proving Eq. (5.84).

(d) We have that x is a stationary point of $f + cP$ if and only if $\theta_c(x; d) \geq 0$ for all d, which by Eq. (5.84) is true if and only if $\{d = 0,\ \xi = P(x)\}$ is the optimal solution of the quadratic program (5.81). **Q.E.D.**

5.3.2 Sequential Quadratic Programming

We now introduce an iterative descent algorithm for minimizing the exact penalty function $f+cP$. It is called the *linearization algorithm* or *sequential quadratic programming*. Like the gradient projection method, it calculates the descent direction by solving a quadratic programming subproblem of the form (5.81). The algorithm is given by

$$x^{k+1} = x^k + \alpha^k d^k,$$

where α^k is a nonnegative scalar stepsize, and d^k is a direction obtained by solving the quadratic program in (d, ξ)

$$\begin{aligned}\text{minimize} \quad & \nabla f(x^k)'d + \tfrac{1}{2} d' H^k d + c\xi \\ \text{subject to} \quad & g_j(x^k) + \nabla g_j(x^k)'d \leq \xi, \qquad j = 0, 1, \ldots, r,\end{aligned} \qquad (5.89)$$

where H^k is a positive definite symmetric matrix. Proposition 5.3.2(c) implies that the solution d is a descent direction of $f + cP$ at x^k.

The initial vector x^0 is arbitrary and the stepsize α^k is chosen by any one of the stepsize rules listed below:

(a) *Minimization rule*: Here α^k is chosen so that

$$f(x^k + \alpha^k d^k) + cP(x^k + \alpha^k d^k) = \min_{\alpha \geq 0}\{f(x^k + \alpha d^k) + cP(x^k + \alpha d^k)\}.$$

(b) *Limited minimization rule*: Here a fixed scalar $s > 0$ is selected and α^k is chosen so that

$$f(x^k + \alpha^k d^k) + cP(x^k + \alpha^k d^k) = \min_{\alpha \in [0,s]}\{f(x^k + \alpha d^k) + cP(x^k + \alpha d^k)\}.$$

(c) *Armijo rule*: Here fixed scalars s, β, and σ with $s > 0$, $\beta \in (0, 1)$, and $\sigma \in (0, \tfrac{1}{2})$, are selected, and we set $\alpha^k = \beta^{m_k} s$, where m_k is the first nonnegative integer m for which

$$f(x^k) + cP(x^k) - f(x^k + \beta^m s d^k) - cP(x^k + \beta^m s d^k) \geq \sigma \beta^m s d^{k\prime} H^k d^k. \qquad (5.90)$$

It can be shown that if $d^k \neq 0$, the Armijo rule will yield a stepsize after a finite number of arithmetic operations. To see this, note that by Prop. 5.3.2(a) and Eq. (5.84), we have for all $\alpha > 0$,

$$f(x^k) + cP(x^k) - f(x^k + \alpha d^k) - cP(x^k + \alpha d^k) = -\alpha \theta_c(x^k; d^k) + o(\alpha)$$
$$\geq \alpha d^{k\prime} H^k d^k + o(\alpha).$$

We then obtain

$$f(x^k) + cP(x^k) - f(x^k + \alpha d^k) - cP(x^k + \alpha d^k) \geq \sigma \alpha d^{k\prime} H^k d^k, \qquad \forall \, \alpha \in (0, \bar{\alpha}],$$

where $\bar{\alpha} > 0$ is such that we have $(1 - \sigma)\alpha d^{k'}H^k d^k + o(\alpha) \geq 0$ for all $\alpha \in (0, \bar{\alpha}]$. Therefore, if $d^k \neq 0$, there is an integer m such that the Armijo test (5.90) is passed, while if $d^k = 0$, by Prop. 5.3.2(d), x^k is a stationary point of $f + cP$.

We have the following convergence result. Its proof is patterned after the corresponding proof for gradient methods for unconstrained minimization (cf. Prop. 1.2.1 in Section 1.2), but is considerably more complicated due to the constraints.

Proposition 5.3.3: Let $\{x^k\}$ be a sequence generated by the linearization algorithm, where the stepsize α^k is chosen by the minimization rule, the limited minimization rule, or the Armijo rule. Assume that there exist positive scalars γ and Γ such that

$$\gamma \|z\|^2 \leq z' H^k z \leq \Gamma \|z\|^2, \qquad \forall\, z \in \Re^n, \qquad k = 0, 1, \ldots,$$

(this condition corresponds to the assumption of a gradient-related direction sequence in unconstrained optimization). Then every limit point of $\{x^k\}$ is a stationary point of $f + cP$.

Proof: We argue by contradiction. Assume that a subsequence $\{x^k\}_K$ generated by the algorithm using the Armijo rule converges to a vector $\bar{x}$ that is not a stationary point of $f + cP$. Since $f(x^k) + cP(x^k)$ is monotonically decreasing, we have

$$f(x^k) + cP(x^k) \to f(\bar{x}) + cP(\bar{x})$$

and hence also

$$f(x^k) + cP(x^k) - f(x^{k+1}) - cP(x^{k+1}) \to 0.$$

By the definition of the Armijo rule, we have

$$f(x^k) + cP(x^k) - f(x^{k+1}) - cP(x^{k+1}) \geq \sigma \alpha^k d^{k'} H^k d^k.$$

Hence

$$\alpha^k d^{k'} H^k d^k \to 0. \tag{5.91}$$

Since for $k \in K$, d^k is the optimal solution of the quadratic program (5.89), we must have for some set of Lagrange multipliers $\{\mu_j^k\}$ and all $k \in K$,

$$\nabla f(x^k) + \sum_{j=0}^{r} \mu_j^k \nabla g_j(x^k) + H^k d^k = 0, \qquad c = \sum_{j=0}^{r} \mu_j^k, \tag{5.92}$$

Sec. 5.3   Exact Penalties – Sequential Quadratic Programming

$$\mu_j^k \geq 0, \qquad \mu_j^k\big(g_j(x^k) + \nabla g_j(x^k)'d^k - \xi^k\big) = 0, \qquad j = 0, 1, \ldots, r, \quad (5.93)$$

where

$$\xi^k = \max_{j=0,1,\ldots,r} \big\{g_j(x^k) + \nabla g_j(x^k)'d^k\big\}.$$

The relations $c = \sum_{j=0}^{r} \mu_j^k$ and $\mu_j^k \geq 0$ imply that the subsequences $\{\mu_j^k\}$ are bounded. Hence, without loss of generality, we may assume that for some μ_j, $j = 0, 1, \ldots, r$, we have

$$\{\mu_j^k\}_K \to \bar{\mu}_j, \qquad j = 0, 1, \ldots, r. \qquad (5.94)$$

Using the assumption $\gamma\|z\|^2 \leq z'H^k z \leq \Gamma\|z\|^2$, we may also assume without loss of generality that

$$\{H^k\}_K \to \bar{H} \qquad (5.95)$$

for some positive definite matrix $\bar{H}$.

Now from the fact $\alpha^k d^{k'} H^k d^k \to 0$ [cf. Eq. (5.91)], we see that there are two possibilities. Either

$$\liminf_{k\to\infty,\ k\in K} \|d^k\| = 0, \qquad (5.96)$$

or else

$$\liminf_{k\to\infty,\ k\in K} \alpha^k = 0, \qquad \liminf_{k\to\infty,\ k\in K} \|d^k\| > 0. \qquad (5.97)$$

If Eq. (5.96) holds, then we may assume without loss of generality that $\{d^k\}_K \to 0$, and by taking the limit in Eqs. (5.92) and (5.93), and using Eq. (5.94), we have

$$\nabla f(\bar{x}) + \sum_{j=0}^{r} \bar{\mu}_j \nabla g_j(\bar{x}) = 0, \qquad c = \sum_{j=0}^{r} \bar{\mu}_j,$$

$$\bar{\mu}_j \geq 0, \qquad \bar{\mu}_j\big(g_j(\bar{x}) - \xi\big) = 0, \qquad j = 0, 1, \ldots, r,$$

where $\xi = \max_{j=0,1,\ldots,r} g_j(\bar{x})$. Hence the quadratic program (5.81) corresponding to $\bar{x}$ has $\{d = 0,\ \xi = P(\bar{x})\}$ as its optimal solution. From Prop. 5.3.2(d), it follows that $\bar{x}$ is a stationary point of $f + cP$, thus contradicting the hypothesis made earlier.

It will thus suffice to assume that Eq. (5.97) holds and to arrive at a contradiction. We may assume without loss of generality that

$$\{\alpha^k\}_K \to 0.$$

Since Eqs. (5.92), (5.94), and (5.95) show that $\{d^k\}_K$ is a bounded sequence, we may also assume without loss of generality that

$$\{d^k\}_K \to \bar{d},$$

where $\bar{d}$ is some vector that cannot be zero in view of Eq. (5.97).

Since $\{\alpha^k\}_K \to 0$, it follows, in view of the definition of the Armijo rule, that the initial stepsize s will be reduced at least once for all $k \in K$ after some index $\bar{k}$. This means that for all $k \in K$, $k \geq \bar{k}$,

$$f(x^k) + cP(x^k) - f(x^k + \bar{\alpha}^k d^k) - cP(x^k + \bar{\alpha}^k d^k) < \sigma \bar{\alpha}^k d^{k\prime} H^k d^k, \quad (5.98)$$

where $\bar{\alpha}^k = \alpha^k/\beta$.

Define for all k and d,

$$\zeta^k(d) = \nabla f(x^k)'d + c \max_{j \in J(x^k)} \{g_j(x^k) + \nabla g_j(x^k)'d\} - cP(x^k),$$

and restrict attention to $k \in K$, $k \geq \bar{k}$, that are sufficiently large so that $\bar{\alpha}^k \leq 1$, $J(x^k) \subset J(\bar{x})$, and $J(x^k + \bar{\alpha}^k d^k) \subset J(\bar{x})$. We will show that

$$f(x^k) + cP(x^k) - f(x^k + \bar{\alpha}^k d^k) - cP(x^k + \bar{\alpha}^k d^k) = -\zeta^k(\bar{\alpha}^k d^k) + o(\bar{\alpha}^k), \quad (5.99)$$

where

$$\lim_{k \to \infty, k \in K} \frac{o(\bar{\alpha}^k)}{\bar{\alpha}^k} = 0. \quad (5.100)$$

Indeed, we have

$$f(x^k + \bar{\alpha}^k d^k) = f(x^k) + \bar{\alpha}^k \nabla f(x^k)'d^k + o_0(\bar{\alpha}^k \|d^k\|)$$

$$g_j(x^k + \bar{\alpha}^k d^k) = g_j(x^k) + \bar{\alpha}^k \nabla g_j(x^k)'d^k + o_j(\bar{\alpha}^k \|d^k\|), \quad j \in J(x^k),$$

where $o_j(\cdot)$ are functions satisfying $\lim_{k \to \infty} o_j(\bar{\alpha}^k \|d^k\|)/\bar{\alpha}^k = 0$. Adding and taking the maximum over $j \in J(x)$, and using the fact $J(x^k + \bar{\alpha}^k d^k) \subset J(\bar{x})$ [implying that $P(x^k + \bar{\alpha}^k d^k) = \max_j g_j(x^k + \bar{\alpha}^k d^k)$], we obtain for sufficiently large k,

$$f(x^k + \bar{\alpha}^k d^k) + cP(x^k + \bar{\alpha}^k d^k) = f(x^k) + \bar{\alpha}^k \nabla f(x^k)'d^k$$
$$+ c \max_{j \in J(x^k)} \{g_j(x^k) + \bar{\alpha}^k \nabla g_j(x^k)'d^k\} + o(\bar{\alpha}^k \|d^k\|)$$
$$= f(x^k) + cP(x^k) + \zeta^k(\bar{\alpha}^k d^k) + o(\bar{\alpha}^k),$$

thus proving Eq. (5.99).

We also claim that

$$-\frac{\zeta^k(\bar{\alpha}^k d^k)}{\bar{\alpha}^k} \geq -\zeta^k(d^k) \geq d^{k\prime} H^k d^k. \quad (5.101)$$

Indeed, let (d^k, ξ^k) be the optimal solution of the quadratic program

$$\text{minimize } \nabla f(x^k)'d + \tfrac{1}{2} d' H^k d + c\xi$$
$$\text{subject to } g_j(x^k) + \nabla g_j(x^k)'d \leq \xi, \quad j = 0, 1, \ldots, r.$$

Sec. 5.3   Exact Penalties – Sequential Quadratic Programming   517

We have
$$\xi^k = \max_{j=0,1,\ldots,r} \{g_j(x^k) + \nabla g_j(x^k)'d^k\}$$
$$\geq \max_{j \in J(x)} \{g_j(x^k) + \nabla g_j(x^k)'d^k\}$$
$$= \frac{\zeta^k(d^k) - \nabla f(x^k)'d^k}{c} + P(x^k).$$

On the other hand, in the proof of Prop. 5.3.2(c) we showed [cf. Eq. (5.88)] that
$$c(\xi^k - P(x^k)) - \nabla f(x^k)'d^k \geq d^{k'} H^k d^k.$$

The last two equations, together with the relation $\zeta^k(\bar{\alpha}^k d) \leq \bar{\alpha}^k \zeta^k(d)$, which follows from the convexity of $\zeta^k(\cdot)$, prove Eq. (5.101).

By dividing Eq. (5.98) with $\bar{\alpha}^k$ and by combining it with Eq. (5.99), we obtain
$$\sigma d^{k'} H^k d^k > -\frac{\zeta^k(\bar{\alpha}^k d)}{\bar{\alpha}^k} + \frac{o(\bar{\alpha}^k)}{\bar{\alpha}^k},$$
which in view of Eq. (5.101), yields
$$(1-\sigma)d^{k'} H^k d^k + \frac{o(\bar{\alpha}^k)}{\bar{\alpha}^k} < 0.$$

Since $\{H^k\}_K \to \bar{H}$, $\{d^k\}_K \to \bar{d}$, $\bar{H}$ is positive definite, $\bar{d} \neq 0$, and $o(\bar{\alpha}^k)/\bar{\alpha}^k \to 0$ [cf. Eq. (5.100)], we obtain a contradiction. This completes the proof of the proposition for the case of the Armijo rule.

Consider now the minimization rule and let $\{x^k\}_K$ converge to a vector $\bar{x}$, which is not a stationary point of $f + cP$. Let $\tilde{x}^{k+1}$ be the point that would be generated from x^k via the Armijo rule and let $\tilde{\alpha}^k$ be the corresponding stepsize. We have
$$f(x^k) - f(x^{k+1}) \geq f(x^k) - f(\tilde{x}^{k+1}) \geq \sigma \tilde{\alpha}^k d^{k'} H^k d^k.$$

By replacing α^k by $\tilde{\alpha}^k$ in the arguments of the earlier proof, we obtain a contradiction. This line of argument establishes that any stepsize rule that gives a larger reduction in the value of $f + cP$ at each iteration than the Armijo rule inherits its convergence properties, so it also proves the proposition for the limited minimization rule. **Q.E.D.**

Application to Constrained Optimization Problems

Given the inequality constrained problem

$$\begin{aligned}&\text{minimize } f(x)\\&\text{subject to } g_j(x) \leq 0, \quad j=1,\ldots,r,\end{aligned} \quad (5.102)$$

we can attempt its solution by using the linearization algorithm to minimize the corresponding exact penalty function $f+cP$ for a value of c that exceeds the threshold $\sum_{j=1}^{r} \mu_j^*$ (cf. Prop. 5.3.1).

There are a number of complex implementation issues here. One difficulty is that we may not know a suitable threshold value for c. Under these circumstances, a possible approach is to choose an initial value c^0 for c and increase it as necessary at each iteration k if the algorithm indicates that the current value c^k is inadequate. An important question is to decide on the conditions that would prompt an increase of c^k. The most common approach is based on trying to solve the quadratic program

$$\begin{aligned} \text{minimize} \quad & \nabla f(x^k)'d + \tfrac{1}{2}d'H^k d \\ \text{subject to} \quad & g_j(x^k) + \nabla g_j(x^k)'d \leq 0, \qquad j = 1,\ldots,r, \end{aligned} \tag{5.103}$$

which differs from the direction finding quadratic program

$$\begin{aligned} \text{minimize} \quad & \nabla f(x^k)'d + \tfrac{1}{2}d'H^k d + c\xi \\ \text{subject to} \quad & g_j(x^k) + \nabla g_j(x^k)'d \leq \xi, \qquad j = 0,1,\ldots,r, \end{aligned} \tag{5.104}$$

of the linearization method in that ξ has been set to zero. If program (5.103) has a feasible solution, then it must have a unique optimal solution d^k and at least one set of Lagrange multipliers $\mu_1^k,\ldots,\mu_r^k$ (since its cost function is a strictly convex quadratic and its constraints are linear). It can then be verified by checking the corresponding optimality conditions (Exercise 5.3.3) that for all $c > \sum_{j=1}^{r} \mu_j^k$, the pair $\{d^k, \xi = 0\}$ is the optimal solution of the quadratic program (5.104). Thus the direction d^k can be used as a direction of descent for minimizing $f + c^k P$, where $\sum_{j=1}^{r} \mu_j^k$ provides an underestimate for the appropriate value for c^k. The penalty parameter may be updated by

$$c^k = \max\left\{ c^{k-1}, \sum_{j=1}^{r} \mu_j^k + \gamma \right\},$$

where γ is some positive scalar. Note that by the optimality conditions of the quadratic program (5.103) we have approximately, for small $\|d^k\|$,

$$\nabla f(x^k) + \sum_{j=1}^{r} \mu_j^k \nabla g_j(x^k) \approx 0, \qquad \mu_j^k g_j(x^k) \approx 0, \qquad j = 1,\ldots,r,$$

so near convergence, the scalars μ_j^k are approximately equal to Lagrange multipliers of the constrained problem (5.102). This is consistent with the strategy of setting c^k at a somewhat higher value than $\sum_{j=1}^{r} \mu_j^k$. If on the other hand, the quadratic program (5.103) has no feasible solution, one can set $c^k = c^{k-1}$ and obtain a direction of descent d^k for the quadratic program (5.104).†

One of the drawbacks of this approach is that the value of the penalty parameter c^k may increase rapidly during the early stages of the algorithm, while during the final stage of the algorithm a much smaller value of c^k may be adequate. A large value of c^k results in very sharp corners of the surfaces of equal cost of the penalized cost $f + c^k P$ along the boundary of the constraint set, and can have a substantial adverse effect on the effectiveness of the stepsize procedure and thus on algorithmic progress (see Fig. 5.3.3). In this connection, it is interesting to note that if the system

$$g_j(x^k) + \nabla g_j(x^k)'d \leq 0, \qquad j = 1, \ldots, r,$$

is feasible, then the direction d^k obtained from the quadratic program (5.103) is independent of c^k, while the stepsize α^k depends strongly on c^k. For this reason, it may be important to provide schemes that allow for the reduction of c^k if circumstances appear to be favorable. The details of this can become quite complicated and we refer to the book [Ber82a] for a discussion of some possibilities.

An important question relates to the choice of the matrices H^k. In unconstrained minimization, one tries to employ a stepsize $\alpha^k = 1$ together with matrices H^k that approximate the Hessian of the cost function at a solution. A natural analog for the constrained case would be to choose H^k close to the Hessian of the Lagrangian function

$$L(x, \mu) = f(x) + \mu'g(x),$$

evaluated at (x^*, μ^*); a justification for this is provided in the next section, where it is shown that the direction d^k calculated by the linearization algorithm, with this choice of H^k, can be viewed as a Newton step.

There are two difficulties relating to such an approach. The first is that $\nabla^2_{xx} L(x^*, \mu^*)$ may not be positive definite. Actually this is not as serious as might appear. As we discuss more fully in the next section, what is important is that H^k approximate closely $\nabla^2_{xx} L(x^*, \mu^*)$ only on the

---

† It is possible that because of the constraint nonlinearities the quadratic program (5.103) has no feasible solution. This will not happen if the constraint functions g_j are convex and the original inequality constrained problem (5.102) has at least one feasible solution, say $\bar{x}$; it can be seen that the vector $\bar{d}^k = \bar{x} - x^k$ is a feasible solution of the quadratic program (5.103) since

$$g_j(x^k) + \nabla g_j(x^k)'(\bar{x} - x^k) \leq g_j(\bar{x}) \leq 0$$

by Prop. B.3(a) of Appendix B. Usually, even if the constraints are nonconvex, the quadratic problem (5.103) is feasible, provided the constrained problem (5.102) is feasible.

subspace tangent to the active constraints. Under second order sufficiency assumptions on (x^*, μ^*), this can be done with positive definite H^k, since then $\nabla^2_{xx} L(x^*, \mu^*)$ is positive definite on this subspace.

The second difficulty relates to the fact that even if we were to choose H^k equal to the (generally unknown) matrix $\nabla^2_{xx} L(x^*, \mu^*)$ and even if this matrix is positive definite, it may happen that arbitrarily close to x^* a stepsize $\alpha^k = 1$ is not acceptable by the algorithm because it does not decrease the value of $f + cP$; this can happen even for very simple problems (see Exercise 5.3.9). The book [Ber82a] (p. 290) discusses this point in detail, and introduces modifications to the basic linearization algorithm that allow a superlinear convergence rate.

Extension to Equality Constraints

The development given earlier for inequality constraints can be extended to the case of additional equality constraints simply by converting each equality constraint $h_i(x) = 0$ to the two inequalities

$$h_i(x) \le 0, \quad -h_i(x) \le 0.$$

For example, the direction finding quadratic program of the linearization method is

$$\begin{aligned} \text{minimize} \quad & \nabla f(x^k)' d + \tfrac{1}{2} d' H^k d + c\xi \\ \text{subject to} \quad & g_j(x^k) + \nabla g_j(x^k)' d \le \xi, \quad j = 0, 1, \ldots, r, \\ & \left| h_i(x^k) + \nabla h_i(x^k)' d \right| \le \xi, \quad i = 1, \ldots, m. \end{aligned}$$

This program yields a descent direction for the exact penalty function

$$f(x) + c \max\{0, g_1(x), \ldots, g_r(x), |h_1(x)|, \ldots, |h_m(x)|\},$$

and can be used as a basis for an algorithm similar to the one developed for inequality constraints.

5.3.3 Differentiable Exact Penalty Functions

We now discuss briefly differentiable exact penalty functions for the equality constrained problem

$$\begin{aligned} \text{minimize} \quad & f(x) \\ \text{subject to} \quad & h_i(x) = 0, \quad i = 1, \ldots, m. \end{aligned} \tag{5.105}$$

We assume that f and h_i are twice continuously differentiable. Furthermore, we assume that the matrix $\nabla h(x)$ has rank m for all x, although much of the following analysis can be conducted assuming $\nabla h(x)$ has rank m in a suitable open subset of $\Re^n$. Motivated by the exact penalty function

$$\|\nabla_x L(x,\lambda)\|^2 + \|h(x)\|^2$$

discussed earlier, we consider the function

$$P_c(x,\lambda) = L(x,\lambda) + \tfrac{1}{2}\|W(x)\nabla_x L(x,\lambda)\|^2 + \tfrac{c}{2}\|h(x)\|^2, \qquad (5.106)$$

where

$$L(x,\lambda) = f(x) + \lambda' h(x)$$

is the Lagrangian function, c is a positive parameter, and $W(x)$ is any continuously differentiable $m \times n$ matrix such that the $m \times m$ matrix $W(x)\nabla h(x)$ is nonsingular for all x.

The idea here is that by introducing the Lagrangian $L(x,\lambda)$ in the penalized cost P_c, we build a preference towards local minima rather than local maxima. The use of the matrix function $W(x)$ cannot be motivated easily, but will be justified by subsequent developments. Two examples of choices of $W(x)$ that turn out to be useful are

$$W(x) = \nabla h(x)', \qquad (5.107)$$

$$W(x) = \bigl(\nabla h(x)'\nabla h(x)\bigr)^{-1}\nabla h(x)'. \qquad (5.108)$$

The monograph [Ber82a] discusses in greater detail the role of the matrix $W(x)$ and also considers a different type of method whereby $W(x)$ is taken equal to the identity matrix.

Let us write $W(x)$ in the form

$$W(x) = \begin{pmatrix} w_1(x)' \\ \vdots \\ w_m(x)' \end{pmatrix},$$

where $w_i : \Re^n \mapsto \Re^n$ are some functions, and let $e_1, \ldots, e_m$ be the columns of the $m \times m$ identity matrix. It is then straightforward to verify that

$$\nabla_x P_c = \nabla_x L + \left(\nabla^2_{xx} L W' + \sum_{i=1}^m \nabla w_i \nabla_x L e_i'\right) W \nabla_x L + c \nabla h h, \qquad (5.109)$$

$$\nabla_\lambda P_c = h + \nabla h' W' W \nabla_x L, \qquad (5.110)$$

where all functions and gradients in the above expressions are evaluated at the typical pair (x,λ).

It can be seen that if (x^*, λ^*) is a local minimum-Lagrange multiplier pair of the original problem (5.105), then (x^*, λ^*) is also a stationary point of $P_c(x, \lambda)$, i.e.,

$$\nabla_x P_c(x^*, \lambda^*) = 0, \qquad \nabla_\lambda P_c(x^*, \lambda^*) = 0.$$

Under appropriate conditions, the reverse assertions are possible, namely that stationary points (x^*, λ^*) of $P_c(x, \lambda)$ satisfy the first order necessary conditions for the original constrained optimization problem. There are several results of this type, of which the following is typical. We outline the proof in Exercise 5.3.7 and we also refer to [Ber82a] for an extensive analysis.

Proposition 5.3.4: For every compact subset $X \times \Lambda$ of $\Re^{n+m}$ there exists a $\bar{c} > 0$ such that for all $c \geq \bar{c}$, every stationary point (x^*, λ^*) of P_c that belongs to $X \times \Lambda$ satisfies the first order necessary conditions

$$\nabla_x L(x^*, \lambda^*) = 0, \qquad \nabla_\lambda L(x^*, \lambda^*) = 0.$$

Differentiable Exact Penalty Functions Depending Only on x

One approach to minimizing $P_c(x, \lambda)$ is to first minimize it with respect to λ and then minimize it with respect to x. To simplify the subsequent formulas, let us focus on the function

$$W(x) = \left(\nabla h(x)' \nabla h(x)\right)^{-1} \nabla h(x)'$$

of Eq. (5.108). For this function, $W(x)\nabla h(x)$ is equal to the identity matrix and from Eq. (5.106) we have

$$P_c(x, \lambda) = f(x) + \lambda' h(x) + \tfrac{1}{2}\|W(x)\nabla f(x) + \lambda\|^2 + \tfrac{c}{2}\|h(x)\|^2. \quad (5.111)$$

We can minimize explicitly this function with respect to λ by setting

$$\nabla_\lambda P_c(x, \lambda) = h(x) + W(x)\nabla f(x) + \lambda = 0.$$

Substituting λ from this equation into Eq. (5.111), we obtain

$$\min_\lambda P_c(x, \lambda) = f(x) + \hat{\lambda}(x)' h(x) + \frac{c-1}{2}\|h(x)\|^2,$$

where

$$\hat{\lambda}(x) = -W(x)\nabla f(x).$$

Sec. 5.3 Exact Penalties – Sequential Quadratic Programming 523

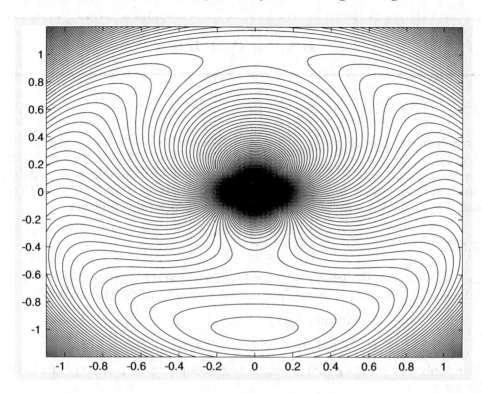

Figure 5.3.6. Equal cost surfaces of the differentiable exact penalty function $\hat{P}_c(x)$ for the two-dimensional problem where

$$f(x) = x_1, \qquad h(x) = x_1^2 + x_2^2 - 1$$

(cf. Figs. 5.3.2 and 5.3.3). The figure corresponds to $c = 2$. Note that there is a singularity at $(0,0)$, which is a nonregular point at which $\hat{\lambda}(x)$ is undefined. The function $\hat{P}_c(x)$ takes arbitrarily large and arbitrarily small values sufficiently close to $(0,0)$. This type of singularity can be avoided by using a modification of the exact penalty function (see Exercise 5.3.8).

Replacing $c - 1$ by c, it is seen that the function

$$\hat{P}_c(x) = f(x) + \hat{\lambda}(x)'h(x) + \frac{c}{2}\|h(x)\|^2 \qquad (5.112)$$

is an exact penalty function, inheriting its properties from the exact penalty function $P_c(x, \lambda)$. Figure 5.3.6 illustrates the function $\hat{P}_c(x)$ for the same example problem that we used to illustrate the nondifferentiable exact penalty function $f + cP$ in Fig. 5.3.3. It can be seen that the two exact penalty functions $\hat{P}_c$ and $f + cP$ have quite different structures, including that $\hat{P}_c$ is not defined at nonregular points. For a detailed analysis of algorithms for minimization of $\hat{P}_c$ (including superlinearly converging Newton-like methods), the associated convergence and implementation issues, and extensions to inequality constraints we refer to the monograph [Ber82a] (Section 4.3.3).

EXERCISES

5.3.1

Consider a one-dimensional problem with two inequality constraints where $f(x) = 0$, $g_1(x) = -x$, $g_2(x) = 1 - x^2$. Show that for all c, $x = (1/2)(1 - \sqrt{5})$ and $x = 0$ are stationary points of $f + cP$, where P is the nondifferentiable exact penalty function (5.79)-(5.80), but are infeasible for the constrained problem. Plot $P(x)$ and discuss the behavior of the linearization method for this problem.

5.3.2

Let H be a positive definite symmetric matrix. Show that the pair (x^*, μ^*) satisfies the first order necessary conditions of Prop. 4.3.1 for the problem

$$\text{minimize } f(x)$$
$$\text{subject to } g_j(x) \leq 0, \quad j = 1, \ldots, r,$$

if and only if $(0, \mu^*)$ is a global minimum-Lagrange multiplier pair of the quadratic program

$$\text{minimize } \nabla f(x^*)'d + \tfrac{1}{2}d'Hd$$
$$\text{subject to } g_j(x^*) + \nabla g_j(x^*)'d \leq 0, \quad j = 1, \ldots, r.$$

(See [Ber82a], Section 4.1 for a solution.)

5.3.3

Show that if (d, μ) is a global minimum-Lagrange multiplier pair of the quadratic program

$$\text{minimize } \nabla f(x)'d + \tfrac{1}{2}d'Hd$$
$$\text{subject to } g_j(x) + \nabla g_j(x)'d \leq 0, \quad j = 1, \ldots, r,$$

where H is positive definite symmetric, and

$$c \geq \sum_{j=1}^{r} \mu_j,$$

then $(d, \xi = 0, \bar{\mu})$ is a global minimum-Lagrange multiplier pair of the quadratic program

$$\text{minimize } \nabla f(x)'d + \tfrac{1}{2}d'Hd + c\xi$$
$$\text{subject to } (x, \xi) \in \Re^{n+1}, \quad g_j(x) + \nabla g_j(x)'d \leq \xi, \quad j = 0, 1, \ldots, r,$$

where $\bar{\mu}_j = \mu_j$ for $j = 1, \ldots, r$, $\bar{\mu}_0 = c - \sum_{j=1}^{r} \mu_j$, and $g_0(x) \equiv 0$. (See [Ber82a], Section 4.1 for a solution.)

5.3.4

Show that if the pair (x^*, μ^*) satisfies the first order necessary conditions of Prop. 4.3.1 for the problem

$$\text{minimize } f(x)$$
$$\text{subject to } g_j(x) \leq 0, \quad j = 1, \ldots, r,$$

then x^* is a stationary point of $f + cP$ for all $c \geq \sum_{j=1}^{r} \mu_j^*$. *Hint*: Combine the results of Exercises 5.3.2 and 5.3.3.

5.3.5

Show that when the constraints are linear, the linearization method based on the quadratic program (5.103) is equivalent to one of the gradient projection methods of Section 3.3.

5.3.6

For the one-dimensional problem of minimizing $(1/6)x^3$ subject to $x = 0$, consider the differentiable exact penalty function $P_c(x, \lambda)$ of Eq. (5.106) with $W(x)$ given by Eq. (5.107) or Eq. (5.108). Show that it has two stationary points: the pairs $(0, 0)$ and $\big(c-1, (1-c^2)/2\big)$. Are both of these local minima of $P_c(x, \lambda)$? Discuss how your analysis is consistent with Prop. 5.3.4.

5.3.7

Prove Prop. 5.3.4. *Hint*: By Eq. (5.110), the condition $\nabla_\lambda P_c = 0$ at some point of $X \times \Lambda$ implies $W \nabla_x L = -(\nabla h' W')^{-1} h$. If at this point $\nabla_x P_c = 0$ also holds, we obtain after some calculation

$$0 = W \nabla_x P_c$$
$$= \left\{ cW \nabla h - \left(I + W \left(\nabla_{xx}^2 L W' + \sum_{i=1}^{m} \nabla w_i \nabla_x L e_i' \right) \right) (\nabla h' W')^{-1} \right\} h.$$

Show that there exists $\bar{c} > 0$ such that for all $c \geq \bar{c}$ and stationary points within $X \times \Lambda$, the matrix within braces is nonsingular, implying that at such points $h = 0$. Conclude that we also have $W \nabla_x L = 0$ so that from Eq. (5.109), $\nabla_x L = 0$.

5.3.8 (Dealing with Singularities [Ber82a], p. 215)

A difficulty with the penalty function $P_c(x, \lambda)$ of Eq. (5.106) is the assumption that the matrix $\nabla h(x)$ has rank m for all x. When this assumption is violated, the λ-dependent terms of P_c may be unbounded below. Furthermore, the function $\hat{P}_c$ of Eq. (5.112) is undefined at some points and singularities of the type shown in Fig. 5.3.3 at $x = 0$ may arise. To deal with this difficulty, introduce the following modified version of P_c:

$$P_{c,\tau}(x, \lambda) = L(x, \lambda) + \tfrac{1}{2}\left\|\nabla h(x)\nabla_x L(x, \lambda)\right\|^2 + \tfrac{c+\tau\|\lambda\|^2}{2}\left\|h(x)\right\|^2,$$

where τ is an additional positive parameter.

(a) Show that $P_{c,\tau}(x, \lambda)$ is bounded from below if the function $f(x)+(c/2)\left\|h(x)\right\|^2$ is bounded from below.

(b) Obtain a corresponding differentiable penalty function depending only on x, by minimizing $P_{c,\tau}(x, \lambda)$ with respect to λ.

(c) Plot the contours of this function for the problem of Fig. 5.3.6 and verify that the singularity exhibited in that figure does not occur.

5.3.9 (Maratos' Effect [Mar78])

This example illustrates a fundamental difficulty in attaining superlinear convergence when using the nondifferentiable exact penalty function for monitoring descent. (This difficulty does not arise for differentiable exact penalty functions; see [Ber82a], pp. 271-277.) Consider the problem

$$\text{minimize } f(x) = x_1$$
$$\text{subject to } h(x) = x_1^2 + x_2^2 - 1 = 0,$$

with optimal solution $x^* = (-1, 0)$ and Lagrange multiplier $\lambda^* = 1/2$ (see Figs. 5.3.2, 5.3.3, and 5.3.6). For any x, let (d, λ) be an optimal solution-Lagrange multiplier pair of the problem

$$\text{minimize } \nabla f(x) + \tfrac{1}{2}d'\nabla^2_{xx}L(x^*, \lambda^*)d$$
$$\text{subject to } h(x) + \nabla h(x)'d = 0.$$

(Note that d is the Newton direction; see also the next section.) Show that for all c,

$$f(x+d) + c\left|h(x+d)\right| - f(x) - c\left|h(x)\right| = \lambda h(x) - c\left|h(x)\right| + (c - \lambda^*)\|d\|^2.$$

Conclude that for $c > 2\lambda^*$, there are points x arbitrarily close to x^* for which the exact penalty function $f(x) + c\left|h(x)\right|$ is not reduced by a pure Newton step. (For a solution of the exercise and for a broader discussion of this phenomenon, see [Ber82a], p. 290.)

5.4 LAGRANGIAN METHODS

In this section we consider the direct solution of the system of necessary optimality conditions of the equality constrained problem

$$\text{minimize } f(x)$$
$$\text{subject to } h(x) = 0.$$

Thus we view the optimality conditions

$$\nabla f(x) + \nabla h(x)\lambda = 0, \qquad h(x) = 0, \tag{5.113}$$

as a system of $(n + m)$ nonlinear equations with $(n + m)$ unknowns, the vectors x and λ. We refer to this as the *Lagrangian system*, and we will aim to solve it with a variety of first and second order methods, called *Lagrangian methods*. We will also discuss supplementary schemes, based on merit function descent, which offer improved convergence guarantees. Moreover, we will provide alternative or modified forms of the optimality conditions to accommodate inequality constraints and approximations.

An important example of the Lagrangian approach is the primal-dual methodology for linear programming that we have discussed in Section 5.1.2. In that case, we considered solution of the system of primal and dual optimality conditions (5.16), Newton-like solution methods, and the use of the merit function (5.15) to enforce global convergence. Here our approach is more closely related to traditional lines of analysis of algorithms for solution of differentiable systems of nonlinear equations.

We will first consider Lagrangian algorithms for the system (5.113) that have the generic form

$$x^{k+1} = G(x^k, \lambda^k), \qquad \lambda^{k+1} = H(x^k, \lambda^k), \tag{5.114}$$

where $G : \Re^{n+m} \mapsto \Re^n$ and $H : \Re^{n+m} \mapsto \Re^n$ are continuously differentiable functions. Since this iteration can only converge to a pair (x^*, λ^*) such that

$$x^* = G(x^*, \lambda^*), \qquad \lambda^* = H(x^*, \lambda^*),$$

the functions G and H must be chosen so that local minima-Lagrange multiplier pairs satisfy the above equations.

We start with a first order method, which does not require second derivatives. We then consider Newton-like methods, and various ways to implement them. The difficulty with all these methods, as well as with most other methods for solving nonlinear systems of equations, is that they guarantee only local convergence, i.e., convergence from a starting point that is sufficiently close to a solution. To enlarge the region of convergence, it is necessary to use some type of line search based on the improvement of some merit function. While the existence of such a function for the Lagrangian system (5.113) is not obvious, we will see that a number of functions, such as the augmented Lagrangian function, the exact penalty functions of Section 5.3, and other functions can serve as the basis for globally convergent versions.

5.4.1 First Order Lagrangian Methods

The simplest Lagrangian method for solving the system of optimality conditions (5.113) is given by

$$x^{k+1} = x^k - \alpha \nabla_x L(x^k, \lambda^k), \tag{5.115}$$

$$\lambda^{k+1} = \lambda^k + \alpha h(x^k), \tag{5.116}$$

where L is the Lagrangian function

$$L(x, \lambda) = f(x) + \lambda' h(x)$$

and $\alpha > 0$ is a scalar stepsize. To motivate this method, consider the function

$$P(x, \lambda) = \tfrac{1}{2} \left\| \nabla_x L(x, \lambda) \right\|^2 + \tfrac{1}{2} \left\| h(x) \right\|^2.$$

This function is minimized at a local minimum-Lagrange multiplier pair, so it can be viewed as an exact penalty function (cf. the discussion of Section 5.3).

Let us consider the direction

$$d(x^k, \lambda^k) = \left(-\nabla_x L(x^k, \lambda^k), h(x^k) \right)$$

used in the first order iteration (5.115)-(5.116) and derive conditions under which it is a descent direction of the exact penalty function $P(x, \lambda)$. We have

$$\nabla P(x, \lambda) = \begin{pmatrix} \nabla^2_{xx} L(x, \lambda) \nabla_x L(x, \lambda) + \nabla h(x) h(x) \\ \nabla h(x)' \nabla_x L(x, \lambda) \end{pmatrix},$$

so that

$$\begin{aligned} d(x^k, \lambda^k)' \nabla P(x^k, \lambda^k) &= -\nabla_x L(x^k, \lambda^k)' \big(\nabla^2_{xx} L(x^k, \lambda^k) \nabla_x L(x^k, \lambda^k) \\ &\quad + \nabla h(x^k) h(x^k) \big) + h(x^k)' \nabla h(x^k)' \nabla_x L(x^k, \lambda^k) \\ &= -\nabla_x L(x^k, \lambda^k)' \nabla^2_{xx} L(x^k, \lambda^k) \nabla_x L(x^k, \lambda^k). \end{aligned}$$

If the Hessian of the Lagrangian $\nabla^2_{xx} L(x^k, \lambda^k)$ is positive definite, we see that $d(x^k, \lambda^k)$ is a descent direction of the exact penalty function P [assuming that $\nabla_x L(x^k, \lambda^k) \neq 0$]. Note, however, that *positive definiteness of $\nabla^2_{xx} L(x^k, \lambda^k)$ is essential and is a stronger requirement than the second order sufficiency conditions of Section 4.2.*

To analyze the convergence of the first order iteration (5.115)-(5.116), we cannot quite use the global convergence methodology for descent methods of Chapters 1 and 2 because $d(x^k, \lambda^k)$ need not be a descent direction of the exact penalty function P when far from (x^*, λ^*). Thus we need some new tools, and to this end, we develop a general result on the local convergence of methods for solving systems of nonlinear equations. A pair (x^*, λ^*) is said to be a *point of attraction* of the iteration (5.114) if there exists an open set $S \subset \Re^{n+m}$ such that if $(x^0, \lambda^0) \in S$, then the sequence $\{(x^k, \lambda^k)\}$ generated by the iteration belongs to S and converges to (x^*, λ^*). The following proposition is very useful for our purposes.

Proposition 5.4.1: Let $G : \Re^{n+m} \mapsto \Re^n$ and $H : \Re^{n+m} \mapsto \Re^n$ be continuously differentiable functions. Assume that (x^*, λ^*) satisfies

$$x^* = G(x^*, \lambda^*), \qquad \lambda^* = H(x^*, \lambda^*),$$

and that all eigenvalues of the $(n+m) \times (n+m)$ matrix

$$R^* = \begin{pmatrix} \nabla_x G(x^*, \lambda^*) & \nabla_x H(x^*, \lambda^*) \\ \nabla_\lambda G(x^*, \lambda^*) & \nabla_\lambda H(x^*, \lambda^*) \end{pmatrix} \qquad (5.117)$$

lie strictly within the unit circle of the complex plane. Then (x^*, λ^*) is a point of attraction of the iteration

$$x^{k+1} = G(x^k, \lambda^k), \qquad \lambda^{k+1} = H(x^k, \lambda^k), \qquad (5.118)$$

and when the generated sequence $\{(x^k, \lambda^k)\}$ converges to (x^*, λ^*), the rate of convergence of $\|x^k - x^*\|$ and $\|\lambda^k - \lambda^*\|$ is linear.

Proof: Denote $y = (x, \lambda)$, $y^k = (x^k, \lambda^k)$, $y^* = (x^*, \lambda^*)$, and consider the function $M : \Re^{n+m} \mapsto \Re^{n+m}$ given by $M(y) = \big(G(x, \lambda), H(x, \lambda)\big)$. By the mean value theorem, we have for any two vectors y and $\tilde{y}$,

$$M(\tilde{y}) - M(y) = R'(\tilde{y} - y),$$

where R is the matrix having as ith column the gradient $\nabla M_i(\hat{y}^i)$ of the ith component of M evaluated at some vector $\hat{y}^i$ on the line segment connecting y and $\tilde{y}$. By taking $\tilde{y}$ and y sufficiently close to y^*, we can make R as close to the matrix R^* of Eq. (5.117) as desired, and therefore we can make the eigenvalues of the transpose R' lie within the unit circle [the eigenvalues of R and R' coincide by Prop. A.13(f) of Appendix A]. It follows from Prop. A.15 of Appendix A that there exists a norm $\|\cdot\|$ and an open sphere S with respect to that norm centered at (x^*, λ^*) such that, within S, the induced matrix norm of R' is less than $1 - \epsilon$ where ϵ is some positive scalar. Since

$$\|M(\tilde{y}) - M(y)\| \leq \|R'\| \, \|\tilde{y} - y\|,$$

it follows that within the sphere S, the mapping M is a contraction as defined in Appendix A. The result then follows from the contraction mapping theorem (Prop. A.26 in Appendix A). **Q.E.D.**

We now prove the local convergence of the first order Lagrangian iteration (5.115)-(5.116).

Proposition 5.4.2: Assume that f and h are twice continuously differentiable, and let (x^*, λ^*) be a local minimum-Lagrange multiplier pair. Assume also that x^* is regular and that the matrix $\nabla^2_{xx} L(x^*, \lambda^*)$ is positive definite. Then there exists $\bar{\alpha} > 0$, such that for all $\alpha \in (0, \bar{\alpha}]$, (x^*, λ^*) is a point of attraction of iteration (5.115)-(5.116), and if the generated sequence $\{(x^k, \lambda^k)\}$ converges to (x^*, λ^*), then the rate of convergence of $\|x^k - x^*\|$ and $\|\lambda^k - \lambda^*\|$ is linear.

Proof: The proof consists of showing that, for α sufficiently small, the hypothesis of Prop. 5.4.1 is satisfied. Indeed for $\alpha > 0$, consider the mapping $M_\alpha : \Re^{n+m} \mapsto \Re^{n+m}$ defined by

$$M_\alpha(x, \lambda) = \begin{pmatrix} x - \alpha \nabla_x L(x, \lambda) \\ \lambda + \alpha \nabla_\lambda L(x, \lambda) \end{pmatrix}.$$

Clearly $(x^*, \lambda^*) = M_\alpha(x^*, \lambda^*)$, and we have

$$\nabla M_\alpha(x^*, \lambda^*)' = I - \alpha B, \tag{5.119}$$

where

$$B = \begin{pmatrix} \nabla^2_{xx} L(x^*, \lambda^*) & \nabla h(x^*) \\ -\nabla h(x^*)' & 0 \end{pmatrix}. \tag{5.120}$$

We will show that the real part of each eigenvalue of B is strictly positive, and then the result will follow from Eq. (5.119) by using Prop. 5.4.1. For any complex vector y, denote by $\hat{y}$ its complex conjugate, and for any complex number γ, denote by $Re(\gamma)$ its real part. Let β be an eigenvalue of B, and let $(z, w) \neq 0$ be a corresponding eigenvector where z and w are complex vectors of dimension n and m, respectively. We have

$$Re\left\{(\hat{z}' \quad \hat{w}') B \begin{pmatrix} z \\ w \end{pmatrix}\right\} = Re\left\{\beta(\hat{z}' \quad \hat{w}') \begin{pmatrix} z \\ w \end{pmatrix}\right\} = Re(\beta)(\|z\|^2 + \|w\|^2), \tag{5.121}$$

while at the same time, by using Eq. (5.120),

$$Re\left\{(\hat{z}' \quad \hat{w}') B \begin{pmatrix} z \\ w \end{pmatrix}\right\} = Re\{\hat{z}' \nabla^2_{xx} L(x^*, \lambda^*) z + \hat{z}' \nabla h(x^*) w - \hat{w}' \nabla h(x^*)' z\}. \tag{5.122}$$

Since for any real $n \times m$ matrix Q, we have

$$Re\{\hat{z}' Q' w\} = Re\{\hat{w}' Q z\},$$

it follows from Eqs. (5.121) and (5.122) that

$$Re\{\hat{z}'\nabla^2_{xx}L(x^*,\lambda^*)z\} = Re\left\{(\hat{z}' \quad \hat{w}')B\begin{pmatrix}z\\w\end{pmatrix}\right\} = Re(\beta)(\|z\|^2 + \|w\|^2). \tag{5.123}$$

Since for any positive definite matrix A, we have

$$Re\{\hat{z}'Az\} > 0, \quad \forall\, z \neq 0,$$

it follows from Eq. (5.123) and the positive definiteness assumption on $\nabla^2_{xx}L(x^*,\lambda^*)$ that either $Re(\beta) > 0$ or else $z = 0$. But if $z = 0$, the equation

$$B\begin{pmatrix}z\\w\end{pmatrix} = \beta\begin{pmatrix}z\\w\end{pmatrix}$$

yields

$$\nabla h(x^*)w = 0.$$

Since $\nabla h(x^*)$ has rank m, it follows that $w = 0$. This contradicts our earlier assumption that $(z,w) \neq 0$. Consequently, we must have $Re(\beta) > 0$. **Q.E.D.**

We note that by appropriately scaling the vectors x and λ, we can show that the result of Prop. 5.4.2 holds also for the more general iteration

$$x^{k+1} = x^k - \alpha D\nabla_x L(x^k,\lambda^k), \tag{5.124}$$

$$\lambda^{k+1} = \lambda^k + \alpha E h(x^k), \tag{5.125}$$

where D and E are any positive definite symmetric matrices of appropriate dimension [reduce the preceding iteration to an iteration of the form (5.115)-(5.116) using a change of variables $x = D^{1/2}y$ and $\lambda = E^{1/2}\mu$; cf. the discussion of Section 1.3.2].

Augmented Lagrangian Convexification

We noted that positive definiteness of $\nabla^2_{xx}L(x^*,\lambda^*)$ is necessary for the validity of the Lagrangian method (5.115)-(5.116) and its scaled version (5.124)-(5.125). This essentially requires that the problem has a "locally convex" structure in the neighborhood of x^*. On the other hand when this structure is not present, we can remedy the situation by convexification, through the use of an augmented Lagrangian of the form

$$L_c(x,\lambda) = f(x) + \lambda'h(x) + \frac{c}{2}\|h(x)\|^2. \tag{5.126}$$

In particular, we may apply the method to the equivalent problem

$$\text{minimize } f(x) + \frac{c}{2}\|h(x)\|^2$$

$$\text{subject to } h(x) = 0,$$

where c is chosen sufficiently large to ensure that the corresponding matrix

$$\nabla^2_{xx} L_c(x^*, \lambda^*) = \nabla^2_{xx} L(x^*, \lambda^*) + c \nabla h(x^*) \nabla h(x^*)'$$

is positive definite [assuming of course that (x^*, λ^*) satisfy the second order sufficiency conditions so that the augmented Lagrangian theory applies; cf. Lemma 4.2.1 and the subsequent discussion].

The method (5.124)-(5.125) applied to the preceding equivalent problem takes the form

$$x^{k+1} = x^k - \alpha D \nabla_x L_c(x^k, \lambda^k), \tag{5.127}$$

$$\lambda^{k+1} = \lambda^k + \alpha E h(x^k), \tag{5.128}$$

where D and E are positive definite symmetric matrices of appropriate dimension. By using Prop. 5.4.2, we see that this method converges linearly to (x^*, λ^*), assuming that the second order sufficiency conditions are satisfied, c is sufficiently large to ensure that $\nabla^2_{xx} L_c(x^*, \lambda^*)$ is positive definite, the starting pair (x^0, λ^0) is sufficiently close to (x^*, λ^*), and α is sufficiently small.

Lagrangian Method in the Space of Primal Variables

An important observation for our purposes is that given a good approximation of a local minimum x^* that is a regular point [$\nabla h(x^*)$ has rank m], we can obtain analytically a good approximation of the associated Lagrange multiplier λ^*. One way to do this is to use the function $\hat{\lambda}$ defined by

$$\hat{\lambda}(x) = \big(\nabla h(x)' \nabla h(x)\big)^{-1} \big(h(x) - \nabla h(x)' \nabla f(x)\big), \tag{5.129}$$

for all x such that $\nabla h(x)$ has rank m. Indeed, by multiplying the necessary condition $\nabla f(x^*) + \nabla h(x^*) \lambda^* = 0$ with $\nabla h(x^*)'$, we obtain $\nabla h(x^*)' \nabla f(x^*) + \nabla h(x^*)' \nabla h(x^*) \lambda^* = 0$, so that

$$\lambda^* = -\big(\nabla h(x^*)' \nabla h(x^*)\big)^{-1} \nabla h(x^*)' \nabla f(x^*).$$

Thus, using the fact $h(x^*) = 0$ in Eq. (5.129), we obtain $\hat{\lambda}(x^*) = \lambda^*$. Since $\hat{\lambda}(\cdot)$ is a continuous function, it follows that $\hat{\lambda}(x)$ is near λ^* if x is near x^*.

One benefit of this observation is to alleviate the requirement for a good initial choice of (x^0, λ^0) in the preceding Lagrangian methods (cf. Prop. 5.4.2). If a good initial choice x^0 is available, we can obtain a good initial choice λ^0 from $\lambda^0 = \hat{\lambda}(x^0)$.

Carrying this idea further, we may consider a Lagrangian method where λ^k is taken to be equal to $\hat{\lambda}(x^k)$ rather than updated according to Eq. (5.116) or (5.125), leading to the algorithm

$$x^{k+1} = x^k - \alpha \nabla_x L\big(x^k, \hat{\lambda}(x^k)\big), \tag{5.130}$$

Sec. 5.4   Lagrangian Methods   533

which iterates exclusively within the space of the vector x. To show local convergence to x^* of this iteration for sufficiently small stepsize α, it is necessary that the matrix
$$I - \alpha G(x^*)$$
where
$$G(x) = \nabla\bigl(\nabla_x L(x, \hat{\lambda}(x))\bigr),$$
has eigenvalues strictly within the unit circle (cf. Prop. 5.4.1).

Indeed, remarkably, it can be shown that if (x^*, λ^*) satisfy the second order sufficiency conditions of Prop. 4.2.1 (in addition to x^* being a regular point), the eigenvalues of $G(x^*)$ are all real-valued and positive, so the eigenvalues of $I - \alpha G(x^*)$ lie within the unit circle for sufficiently small α. The proof is fairly complicated, and is given in Prop. 4.26 of [Ber82a]. A superlinearly converging Newton-like Lagrangian method that operates exclusively within the space of x is also described in [Ber82a] (Prop. 4.27).

An interesting observation is that [assuming that $\nabla h(x^k)$ has rank m] we can obtain both $\hat{\lambda}(x^k)$ and $\nabla_x L(x^k, \hat{\lambda}(x^k))$ by solving the quadratic program
$$\begin{aligned}\text{minimize} \quad & \nabla f(x^k)'d + \tfrac{1}{2}\|d\|^2 \\ \text{subject to} \quad & h(x^k) + \nabla h(x^k)'d = 0.\end{aligned} \qquad (5.131)$$
Indeed the optimality conditions for this program are
$$\nabla f(x^k) + \nabla h(x^k)\lambda + d = 0, \qquad h(x^k) + \nabla h(x^k)'d = 0,$$
and it can be seen that the unique Lagrange multiplier vector is $\hat{\lambda}(x^k)$, while the unique optimal solution is
$$d(x^k) = -\nabla_x L\bigl(x^k, \hat{\lambda}(x^k)\bigr).$$
As a special case we note that if h is linear and x^k is a feasible point, $-\nabla_x L(x^k, \hat{\lambda}(x^k))$ is equal to $-\nabla f(x^k)$ projected onto the feasible set, so the iteration reduces to a gradient projection iteration. In the more general case, where h is linear but x^k is infeasible, $-\nabla_x L(x^k, \hat{\lambda}(x^k))$ has two components; one component is $-\nabla f(x^k)$ projected onto the feasible set and aims at cost function reduction, while the other component is orthogonal to the feasible set, and aims at infeasibility reduction (for sufficiently small α). Note also that based on the quadratic programming implementation (5.131), the method is related to the linearization method (5.103) of Section 5.3.2.

A noteworthy fact here is that *positive definiteness of the matrix* $\nabla^2_{xx}L(x^*, \lambda^*)$ *is not required*. Convergence is guaranteed assuming just the second order sufficiency conditions of Prop. 4.2.1 and regularity of x^*. To get a sense of this, we note that *the iteration (5.130) when applied to the equivalent problem*
$$\begin{aligned}\text{minimize} \quad & f_c(x) \stackrel{\text{def}}{=} f(x) + \frac{c}{2}\|h(x)\|^2 \\ \text{subject to} \quad & h(x) = 0,\end{aligned}$$

is independent of the value of c, and embodies any desired amount of augmented Lagrangian convexification. Indeed it can be seen that the solution of the quadratic program (5.131) is not affected if ∇f is replaces by ∇f_c, since subject to the constraint $h(x^k) + \nabla h(x^k)'d = 0$, the inner products $\nabla f(x^k)'d$ and $\nabla f_c(x^k)'d$ differ by the constant $c\|h(x^k)\|^2$.

Let us finally note that the iteration (5.130) involves the calculation of $\hat{\lambda}(x^k)$, which may be significant extra overhead, particularly if h is nonlinear and/or the number of constraints m is large. Under favorable conditions, however, the method is viable and applies to problems that do not have the "locally convex" structure [positive definiteness of $\nabla^2_{xx} L(x^*, \lambda^*)$], which is required for the Lagrangian iteration (5.115)-(5.116).

Decomposition and Parallelization in Separable Problems

The Lagrangian methods and their variations in this section are well-suited for separable problems of the form

$$\text{minimize} \quad \sum_{i=1}^{n} f_i(x_i)$$
$$\text{subject to} \quad \sum_{i=1}^{n} h_{ij}(x_i) = 0, \quad j = 1, \ldots, m, \tag{5.132}$$

where $f_i : \Re \mapsto \Re$ and $h_{ij} : \Re \mapsto \Re$ are twice continuously differentiable scalar functions. Problems of this type arise naturally in many contexts and they will be discussed in greater detail in the context of convex programming in Section 6.1.5, and in a variety of algorithmic contexts in Chapter 7.

Here we wish to point out the mechanism by which Lagrangian methods can exploit the structure of separable problems. Indeed the Lagrangian function of the problem takes the form

$$L(x, \lambda) = \sum_{i=1}^{n} \left\{ f_i(x_i) + \sum_{j=1}^{m} \lambda_j h_{ij}(x_i) \right\},$$

and is separable with respect to x_i. As a result, the Lagrangian method (5.115)-(5.116) takes the form

$$x_i^{k+1} = x_i^k - \alpha \left(\frac{\partial f_i(x_i^k)}{\partial x_i} + \sum_{j=1}^{m} \lambda_j^k \frac{\partial h_{ij}(x_i^k)}{\partial x_i} \right), \quad i = 1, \ldots, n, \tag{5.133}$$

$$\lambda_j^{k+1} = \lambda_j^k + \alpha \sum_{i=1}^{n} h_{ij}(x_i^k), \quad j = 1, \ldots, m. \tag{5.134}$$

Note that the iteration (5.133) decomposes with respect to the coordinates x_i, and is well-suited for parallel computation. For example, one may consider a parallel computing system with n processors, each updating a single scalar coordinate x_i according to iteration (5.133), and another (central) processor updating the multiplier vector λ according to Eq. (5.134). The ith processor communicates with the central processor, sending its current values x_i^k or $h_{ij}(x_i^k)$, $j = 1, \ldots, m$, while receiving from the central processor the current value of λ^k.

For a small enough stepsize α, this parallel algorithmic process is convergent under the conditions of Prop. 5.4.2 [including the requirement that the Hessian of the Lagrangian $\nabla_{xx}^2 L(x^*, \lambda^*)$ is positive definite]. Also, a certain amount of asynchronism can be allowed into the algorithm, based on the totally and partially asynchronous guidelines of Section 2.5. Moreover, the iteration (5.130) admits similar parallelization.

Let us finally note an alternative method for separable problems, which uses a Lagrangian minimization in place of the gradient iteration (5.133). The method also requires that the Hessian of the Lagrangian $\nabla_{xx}^2 L(x^*, \lambda^*)$ is positive definite, and has the form

$$x_i^{k+1} = \arg\min_{x_i} \left\{ f_i(x_i) + \sum_{j=1}^{m} \lambda_j^k h_{ij}(x_i) \right\}, \qquad i = 1, \ldots, n,$$

$$\lambda_j^{k+1} = \lambda_j^k + \alpha \sum_{i=1}^{n} h_{ij}(x_i^k), \qquad j = 1, \ldots, m,$$

where the minimization is assumed to be local in a neighborhood of x^*. This method can be understood by viewing it as a method of multipliers for the problem

$$\text{minimize} \quad f(x) - \frac{\alpha}{2} \|h(x)\|^2$$

$$\text{subject to} \quad h(x) = 0.$$

In particular, for convergence, α must not exceed twice the minimum eigenvalue of $\nabla h(x^*)' \left(\nabla_{xx}^2 L(x^*, \lambda^*) \right)^{-1} \nabla h(x^*)$; see Exercise 5.2.11.

5.4.2 Newton-Like Methods for Equality Constraints

We will now turn to second order methods for solving the Lagrangian system

$$\nabla f(x) + \nabla h(x)\lambda = 0, \qquad h(x) = 0,$$

by viewing it as the vector equation

$$\nabla L(x, \lambda) = 0.$$

Newton's method for this equation is

$$x^{k+1} = x^k + \Delta x^k, \qquad \lambda^{k+1} = \lambda^k + \Delta \lambda^k, \tag{5.135}$$

where $(\Delta x^k, \Delta \lambda^k) \in \Re^{n+m}$ is obtained by solving the system

$$\nabla^2 L(x^k, \lambda^k) \begin{pmatrix} \Delta x^k \\ \Delta \lambda^k \end{pmatrix} = -\nabla L(x^k, \lambda^k), \qquad (5.136)$$

the linearized version of the optimality condition $\nabla L(x, \lambda) = 0$.

We say that (x^{k+1}, λ^{k+1}) is *well-defined* by the Newton iteration (5.135)-(5.136) if the matrix $\nabla^2 L(x^k, \lambda^k)$ is invertible. Note that if x^* is a local minimum that is regular and together with a Lagrange multiplier λ^* satisfies the second order sufficiency condition of Prop. 4.2.1, then $\nabla^2 L(x^*, \lambda^*)$ is invertible; this was shown as part of the proof of the sensitivity theorem (Prop. 4.2.2). As a result, $\nabla^2 L(x, \lambda)$ is invertible in a neighborhood of (x^*, λ^*), and within this neighborhood, points generated by the Newton iteration are well-defined. In the subsequent discussion, when stating various local convergence properties of the Newton iteration in connection with such a pair, we implicitly restrict the iteration within a neighborhood where it is well-defined.

The local convergence properties of the method can be inferred from the results of Section 1.4. For purposes of convenient reference, we provide the corresponding result in the following proposition.

Proposition 5.4.3: Let x^* be a strict local minimum that is regular and satisfies together with a corresponding Lagrange multiplier vector λ^* the second order sufficiency conditions of Prop. 4.2.1. Then (x^*, λ^*) is a point of attraction of the Newton iteration (5.135)-(5.136). Furthermore, if the generated sequence converges to (x^*, λ^*), the rate of convergence of $\{\|(x^k, \lambda^k) - (x^*, \lambda^*)\|\}$ is superlinear (at least order two if $\nabla^2 f$ and $\nabla^2 h_i$, $i = 1, \ldots, m$, are Lipschitz continuous in a neighborhood of x^*).

Proof: Use Prop. 1.4.1 of Section 1.4. **Q.E.D.**

The Newton iteration (5.135)-(5.136) has a rich structure that can be used to provide interesting implementations, three of which we discuss below.

A First Implementation of Newton's Method

Let us write the gradient and Hessian of the Lagrangian function as

$$\nabla L(x^k, \lambda^k) = \begin{pmatrix} \nabla_x L(x^k, \lambda^k) \\ h(x^k) \end{pmatrix}, \qquad \nabla^2 L(x^k, \lambda^k) = \begin{pmatrix} H^k & N^k \\ N^{k\prime} & 0 \end{pmatrix},$$

where

$$H^k = \nabla^2_{xx} L(x^k, \lambda^k), \qquad N^k = \nabla h(x^k).$$

Sec. 5.4   Lagrangian Methods   537

Thus, the Newton system (5.136) takes the form

$$\begin{pmatrix} H^k & N^k \\ N^{k'} & 0 \end{pmatrix} \begin{pmatrix} \Delta x^k \\ \Delta \lambda^k \end{pmatrix} = - \begin{pmatrix} \nabla_x L(x^k, \lambda^k) \\ h(x^k) \end{pmatrix}. \qquad (5.137)$$

Let us assume that H^k is *invertible and* N^k *has rank* m. Then we can provide a more explicit expression for the Newton iteration. Indeed the Newton system (5.137) can be written as

$$H^k \Delta x^k + N^k \Delta \lambda^k = -\nabla_x L(x^k, \lambda^k), \qquad (5.138)$$

$$N^{k'} \Delta x^k = -h(x^k). \qquad (5.139)$$

By multiplying the first equation with $N^{k'}(H^k)^{-1}$ and by using the second equation, it follows that

$$-h(x^k) + N^{k'}(H^k)^{-1} N^k \Delta \lambda^k = -N^{k'}(H^k)^{-1} \nabla_x L(x^k, \lambda^k).$$

Since N^k has rank m, the matrix $N^{k'}(H^k)^{-1} N^k$ is nonsingular, and we obtain

$$\lambda^{k+1} - \lambda^k = \Delta \lambda^k = \left(N^{k'}(H^k)^{-1} N^k \right)^{-1} \left(h(x^k) - N^{k'}(H^k)^{-1} \nabla_x L(x^k, \lambda^k) \right). \qquad (5.140)$$

We have

$$\begin{aligned} \nabla_x L(x^k, \lambda^k) &= \nabla f(x^k) + N^k \lambda^k \\ &= \nabla f(x^k) + N^k \lambda^{k+1} - N^k \Delta \lambda^k \\ &= \nabla_x L(x^k, \lambda^{k+1}) - N^k \Delta \lambda^k, \end{aligned}$$

and by using this equation to substitute $\nabla_x L(x^k, \lambda^k)$ in Eqs. (5.138) and (5.140), we finally obtain the two-step iteration

$$\lambda^{k+1} = \left(N^{k'}(H^k)^{-1} N^k \right)^{-1} \left(h(x^k) - N^{k'}(H^k)^{-1} \nabla f(x^k) \right), \qquad (5.141)$$

$$x^{k+1} = x^k - (H^k)^{-1} \nabla_x L(x^k, \lambda^{k+1}). \qquad (5.142)$$

This is a first implementation of the Newton iteration (under the assumption that H^k is invertible and N^k has rank m). It has the advantage that it requires the solution of systems of dimension at most n [as opposed to $n + m$, which is the dimension of $\nabla^2 L(x^k, \lambda^k)$].

An Implementation of Newton's Method Based on Augmented Lagrangian Functions

We will now derive another way to write the system of equations (5.138)-(5.139). It is based on the observation that, for every scalar c, we have from Eq. (5.139),

$$c N^k N^{k'} \Delta x = -c N^k h(x^k),$$

which added to Eq. (5.138), yields

$$(H^k + cN^kN^{k'})\Delta x^k + N^k(\Delta\lambda^k + ch(x^k)) = -\nabla_x L(x^k, \lambda^k). \quad (5.143)$$

Equivalently, by writing $\Delta\lambda^k = \lambda^{k+1} - \lambda^k$,

$$(H^k + cN^kN^{k'})\Delta x^k = -\nabla_x L(x^k, \lambda^{k+1} + ch(x^k)) = -\nabla_x L_c(x^k, \lambda^{k+1}), \quad (5.144)$$

where L_c is the augmented Lagrangian function

$$L_c(x, \lambda) = L(x, \lambda) + \frac{c}{2}\|h(x)\|^2.$$

Also if $(H^k + cN^kN^{k'})^{-1}$ exists, by multiplying Eq. (5.143) with $N^{k'}(H^k + cN^kN^{k'})^{-1}$, we obtain

$$N^{k'}(H^k + cN^kN^{k'})^{-1}N^k(\Delta\lambda^k + ch(x^k))$$
$$= -N^{k'}\Delta x^k - N^{k'}(H^k + cN^kN^{k'})^{-1}\nabla_x L(x^k, \lambda^k),$$

which by writing $\Delta\lambda^k = \lambda^{k+1} - \lambda^k$ and $N^{k'}\Delta x^k = -h(x^k)$ [cf. Eq. (5.139)], yields

$$N^{k'}(H^k + cN^kN^{k'})^{-1}N^k(\lambda^{k+1} - \lambda^k + ch(x^k))$$
$$= h(x^k) - N^{k'}(H^k + cN^kN^{k'})^{-1}\nabla_x L(x^k, \lambda^k),$$

or by using the fact $\nabla_x L(x^k, \lambda^k) = \nabla f(x^k) + N^k\lambda^k$,

$$N^{k'}(H^k + cN^kN^{k'})^{-1}N^k(\lambda^{k+1} + ch(x^k))$$
$$= h(x^k) - N^{k'}(H^k + cN^kN^{k'})^{-1}\nabla f(x^k). \quad (5.145)$$

Thus, from Eqs. (5.144) and (5.145), we obtain the following equivalent form of Newton's method

$$\lambda^{k+1} = -ch(x^k) + \left(N^{k'}(H^k + cN^kN^{k'})^{-1}N^k\right)^{-1} \\ \left(h(x^k) - N^{k'}(H^k + cN^kN^{k'})^{-1}\nabla f(x^k)\right), \quad (5.146)$$

$$x^{k+1} = x^k - (H^k + cN^kN^{k'})^{-1}\nabla_x L_c(x^k, \lambda^{k+1}). \quad (5.147)$$

An advantage that this implementation may offer over the one of Eqs. (5.141)-(5.142) (which corresponds to $c = 0$) is that the matrix H^k may not be invertible while $H^k + cN^kN^{k'}$ may be invertible for some values of c. For example, for c sufficiently large, we have that $H^k + cN^kN^{k'}$ is not only invertible but also positive definite near (x^*, λ^*) (cf. Lemma 4.2.1 in Section 4.2). An additional benefit of this property is that it allows us to differentiate between local minima and local maxima (near a local maximum-Lagrange multiplier pair, $H^k + cN^kN^{k'}$ is not likely to be

Sec. 5.4   Lagrangian Methods

positive definite for any positive value of c). Note that positive definiteness of $H^k + cN^k N^{k'}$ can be easily detected if the Cholesky factorization method is used for solving the various linear systems of equations in Eqs. (5.146) and (5.147) (cf. the discussion of Section 1.4).

Another property, which is particularly useful for enlarging the region of convergence of Newton's method (see Section 5.4.3), can be inferred from Eq. (5.147): if c is large enough so that $H^k + cN^k N^{k'}$ is positive definite, $x^{k+1} - x^k$ is a descent direction of the augmented Lagrangian function $L_c(\cdot, \lambda^{k+1})$ at x^k.

An Implementation of Newton's Method Based on Quadratic Programming

We will now derive a third implementation of Newton's method. It is based on the observation that the Newton system (5.138)-(5.139) is written as

$$\nabla f(x^k) + H^k \Delta x^k + N^k \lambda^{k+1} = 0, \qquad h(x^k) + N^{k'} \Delta x^k = 0,$$

which are the necessary optimality conditions for $(\Delta x^k, \lambda^{k+1})$ to be a global minimum-Lagrange multiplier pair of the quadratic program

$$\begin{aligned}\text{minimize} \quad & \nabla f(x^k)' \Delta x + \tfrac{1}{2} \Delta x' H^k \Delta x \\ \text{subject to} \quad & h(x^k) + N^{k'} \Delta x = 0.\end{aligned} \tag{5.148}$$

Thus we can obtain $(\Delta x^k, \lambda^{k+1})$ by solving this problem.

This implementation is not particularly useful for practical purposes but provides an interesting connection with the linearization method of Section 5.3. This connection can be made more explicit by noting that the solution Δx^k of the quadratic program (5.148) is unaffected if H^k is replaced by any matrix of the form $H^k + cN^k N^{k'}$, where c is a scalar, thereby obtaining the program

$$\begin{aligned}\text{minimize} \quad & \nabla f(x^k)' \Delta x + \tfrac{1}{2} \Delta x' (H^k + cN^k N^{k'}) \Delta x \\ \text{subject to} \quad & h(x^k) + N^{k'} \Delta x = 0.\end{aligned} \tag{5.149}$$

To see that problems (5.148) and (5.149) have the same solution Δx^k, simply note that they have the same constraints while their cost functions differ by the term $c\Delta x' N^k N^{k'} \Delta x$, which is equal to $c\|h(x^k)\|^2$ and is therefore constant. Near a local minimum-Lagrange multiplier pair (x^*, λ^*) satisfying the sufficiency conditions, we have that $H^k + cN^k N^{k'}$ is positive definite if c is sufficiently large (Lemma 4.2.1 in Section 4.2), and the quadratic program (5.149) is positive definite.

We see therefore that, under these circumstances, the Newton iteration can be viewed in effect as a special case of the linearization method of Section 5.3 with a constant unity stepsize, and scaling matrix

$$\bar{H}^k = H^k + cN^k N^{k'},$$

where c is any scalar for which $\bar{H}^k$ is positive definite.

Merit Functions and Descent Properties of Newton's Method

Since we would like to improve the global convergence properties of Newton's method, it is interesting to look for appropriate merit functions, i.e., functions for which $(x^{k+1} - x^k)$ is a descent direction at x^k or $(x^{k+1} - x^k, \lambda^{k+1} - \lambda^k)$ is a descent direction at (x^k, λ^k). By this, we mean functions F such that for a sufficiently small positive scalar $\bar{\alpha}$, we have

$$F\big(x^k + \alpha(x^{k+1} - x^k)\big) < F(x^k), \qquad \forall \, \alpha \in (0, \bar{\alpha}],$$

or

$$F\big(x^k + \alpha(x^{k+1} - x^k), \lambda^k + \alpha(\lambda^{k+1} - \lambda^k)\big) < F(x^k, \lambda^k), \qquad \forall \, \alpha \in (0, \bar{\alpha}].$$

The following proposition shows that there are several possible merit functions.

Proposition 5.4.4: (Merit Functions for Lagrangian Methods)
Let x^* be a local minimum that is a regular point and satisfies together with a corresponding Lagrange multiplier vector λ^* the second order sufficiency conditions of Prop. 4.2.1. There exists a neighborhood S of (x^*, λ^*) such that if $(x^k, \lambda^k) \in S$ and $x^k \neq x^*$, then (x^{k+1}, λ^{k+1}) is well-defined by the Newton iteration and the following hold:

(a) There exists a scalar $\bar{c}$ such that for all $c \geq \bar{c}$, the vector $(x^{k+1} - x^k)$ is a descent direction at x^k for the exact penalty function

$$f(x) + c \max_{i=1,\ldots,m} |h_i(x)|. \qquad (5.150)$$

(b) The vector $(x^{k+1} - x^k, \lambda^{k+1} - \lambda^k)$ is a descent direction at (x^k, λ^k) for the exact penalty function

$$P(x, \lambda) = \tfrac{1}{2} \|\nabla L(x, \lambda)\|^2.$$

Furthermore, given any scalar $r > 0$, there exists a $\delta > 0$ such that if

$$\|(x^k - x^*, \lambda^k - \lambda^*)\| < \delta,$$

we have

$$P(x^{k+1}, \lambda^{k+1}) \leq r P(x^k, \lambda^k). \qquad (5.151)$$

(c) For every scalar c such that $H^k + cN^k N^{k\prime}$ is positive definite, the vector $(x^{k+1} - x^k)$ is a descent direction at x^k of the augmented Lagrangian function $L_c(\cdot, \lambda^{k+1})$.

Sec. 5.4   Lagrangian Methods   541

Proof: (a) Take $\bar{c} > 0$ sufficiently large and a neighborhood S of (x^*, λ^*), which is sufficiently small, so that for $(x^k, \lambda^k) \in S$, the matrix $H^k + \bar{c} N^k N^{k'}$ is positive definite. Since Δx^k is the solution of the quadratic program (5.149), it follows from Prop. 5.3.2 that if $x^k \neq x^*$, then Δx^k is a descent direction of the exact penalty function (5.150) for all $c \geq \bar{c}$.

(b) We have

$$\begin{pmatrix} x^{k+1} - x^k \\ \lambda^{k+1} - \lambda^k \end{pmatrix} = -\nabla^2 L(x^k, \lambda^k)^{-1} \nabla L(x^k, \lambda^k)$$

and

$$\nabla P(x^k, \lambda^k) = \nabla^2 L(x^k, \lambda^k) \nabla L(x^k, \lambda^k),$$

so

$$\big((x^{k+1} - x^k)', (\lambda^{k+1} - \lambda^k)'\big) \nabla P(x^k, \lambda^k) = -\|\nabla L(x^k, \lambda^k)\|^2 < 0,$$

and the descent property follows.

From Prop. 1.4.1, we have that, given any $\bar{r} > 0$, there exists a $\bar{\delta} > 0$ such that for $\|(x^k - x^*, \lambda^k - \lambda^*)\| < \bar{\delta}$, we have

$$\|(x^{k+1} - x^*, \lambda^{k+1} - \lambda^*)\| \leq \bar{r} \|(x^k - x^*, \lambda^k - \lambda^*)\|. \quad (5.152)$$

For every (x, λ), we have, by the mean value theorem,

$$\nabla L(x, \lambda) = B \begin{pmatrix} x - x^* \\ \lambda - \lambda^* \end{pmatrix},$$

where each row of B is the corresponding row of $\nabla^2 L$ evaluated at a point between (x, λ) and (x^*, λ^*). Since $\nabla^2 L(x^*, \lambda^*)$ is invertible, it follows that there is an $\epsilon > 0$ and scalars $\mu > 0$ and $M > 0$ such that for $\|(x - x^*, \lambda - \lambda^*)\| < \epsilon$, we have

$$\mu \|(x - x^*, \lambda - \lambda^*)\| \leq \|\nabla L(x, \lambda)\| \leq M \|(x - x^*, \lambda - \lambda^*)\|. \quad (5.153)$$

From Eqs. (5.152) and (5.153), it follows that for each $\bar{r} > 0$ there exists $\delta > 0$ such that, for $\|(x^k - x^*, \lambda^k - \lambda^*)\| < \delta$,

$$\|\nabla L(x^{k+1}, \lambda^{k+1})\| \leq (M\bar{r}/\mu) \|\nabla L(x^k, \lambda^k)\|,$$

or, equivalently,

$$P(x^{k+1}, \lambda^{k+1}) \leq (M^2 \bar{r}^2 / \mu^2) P(x^k, \lambda^k).$$

Given $r > 0$, we take $\bar{r} = (\mu/M)\sqrt{r}$ in the preceding relation, and Eq. (5.151) follows.

(c) We have shown that $x^{k+1} - x^k = -(H^k + cN^k N^{k'})^{-1} \nabla_x L_c(x^k, \lambda^{k+1})$ [cf. Eq. (5.147)], which implies the conclusion. **Q.E.D.**

It is also possible to use the differentiable exact penalty functions of Section 5.3.3 as merit functions for Newton's method. The verification of this is somewhat tedious, so we refer to [Ber82a], p. 219, and [Ber82c] for an analysis. Moreover, it can be shown that differentiable exact penalty functions, while more complicated, have an interesting advantage over the nondifferentiable penalty function (5.150): they are not susceptible to the Maratos' effect discussed in Exercise 5.3.9 of Section 5.3, and they allow superlinear convergence of Newton-like methods without any complex modifications. For this analysis, we refer to the paper [Ber80b] (see also [Ber82a], pp. 271-279).

Variations of Newton's Method

There are a number of variations of Newton's method, which aim at some beneficial effect, and are obtained by introducing some extra terms in the left-hand side of the Newton system. These variations have the general form

$$x^{k+1} = x^k + \Delta x^k, \qquad \lambda^{k+1} = \lambda^k + \Delta \lambda^k,$$

where

$$\left(\nabla^2 L(x^k, \lambda^k) + V^k(x^k, \lambda^k)\right) \begin{pmatrix} \Delta x^k \\ \Delta \lambda^k \end{pmatrix} = -\nabla L(x^k, \lambda^k),$$

with the extra term $V^k(x^k, \lambda^k)$ being "small" enough relative to $\nabla^2 L(x^k, \lambda^k)$, so that the eigenvalues of the matrix

$$I - \left(\nabla^2 L(x^k, \lambda^k) + V^k(x^k, \lambda^k)\right)^{-1} \nabla^2 L(x^k, \lambda^k)$$

are within the unit circle and the convergence result of Prop. 5.4.1 applies. In the case where the extra term $V^k(x^k, \lambda^k)$ converges to zero, superlinear convergence is attained; otherwise, the rate of convergence is linear (cf. Prop. 5.4.1).

An interesting approximation of Newton's method is obtained by adding a term $-(1/c^k)\Delta\lambda^k$ in the left-hand side of the equation $N^{k'}\Delta x^k = -h(x^k)$, where c^k is a positive parameter, so that Δx^k and $\Delta\lambda^k$ are obtained by solving the system

$$H^k \Delta x^k + N^k \Delta \lambda^k = -\nabla_x L(x^k, \lambda^k), \tag{5.154}$$

$$N^{k'} \Delta x^k - (1/c^k)\Delta \lambda^k = -h(x^k). \tag{5.155}$$

As $c^k \to \infty$, the system asymptotically becomes identical to the one corresponding to Newton's method.

Sec. 5.4 Lagrangian Methods

We can show that the system (5.154)-(5.155) has a unique solution if either $(H^k)^{-1}$ or $(H^k + c^k N^k N^{k'})^{-1}$ exists. Indeed when $(H^k)^{-1}$ exists, we can write explicitly the solution. By multiplying Eq. (5.154) by $N^{k'}(H^k)^{-1}$ and by using Eq. (5.155), we obtain

$$(1/c^k)\Delta\lambda^k - h(x^k) + N^{k'}(H^k)^{-1}N^k\Delta\lambda^k = -N^{k'}(H^k)^{-1}\nabla_x L(x^k, \lambda^k),$$

from which

$$\Delta\lambda^k = \left((1/c^k)I + N^{k'}(H^k)^{-1}N^k\right)^{-1}\left(h(x^k) - N^{k'}(H^k)^{-1}\nabla_x L(x^k, \lambda^k)\right)$$

and

$$\lambda^{k+1} = \lambda^k + \left((1/c^k)I + N^{k'}(H^k)^{-1}N^k\right)^{-1}\left(h(x^k) - N^{k'}(H^k)^{-1}\nabla_x L(x^k, \lambda^k)\right).$$

From Eq. (5.154), we then obtain

$$x^{k+1} = x^k - (H^k)^{-1}\nabla_x L(x^k, \lambda^{k+1}).$$

Also if $(H^k + c^k N^k N^{k'})^{-1}$ exists, by multiplying Eq. (5.155) with $c^k N^k$ and adding the resulting equation to Eq. (5.154), we obtain

$$(H^k + c^k N^k N^{k'})\Delta x^k = -\nabla_x L(x^k, \lambda^k) - c^k N^k h(x^k),$$

and finally,

$$x^{k+1} = x^k - (H^k + c^k N^k N^{k'})^{-1}\nabla L_{c^k}(x^k, \lambda^k), \tag{5.156}$$

where L_{c^k} is the augmented Lagrangian function. Furthermore, from Eq. (5.155), we obtain

$$\lambda^{k+1} = \lambda^k + c^k\left(h(x^k) + N^{k'}(x^{k+1} - x^k)\right). \tag{5.157}$$

Note that the preceding development shows that N^k *need not have rank m in order for the system (5.154)-(5.155) to have a unique solution, while this is not true for the Newton iteration*. Thus by introducing the term $(1/c^k)\Delta\lambda^k$ in the second equation, we avoid potential difficulties due to linear dependence of the constraint gradients.

The preceding analysis suggests that if c^k is taken sufficiently large, then the approximate Newton iteration (5.156)-(5.157) should converge locally to a local minimum-Lagrange multiplier pair (x^*, λ^*) under the same conditions as the exact Newton iteration (cf. Prop. 5.4.3). Furthermore, the rate of convergence should be superlinear if $c^k \to \infty$. The proof of this, using Prop. 5.4.1, is straightforward, but is tedious and will not be given; see [Ber82a], pp. 240-243, where some variations of the method of Eqs. (5.156)-(5.157) are also discussed.

Another type of approximate Newton's method is obtained by introducing extra terms in the right-hand side (rather than the left-hand side) of the Newton system. For local convergence to (x^*, λ^*), it is essential that the extra terms tend to zero. The primal-dual methods for linear programming of Section 5.1.2 are of this type.

Connection with the First Order Method of Multipliers

From Eq. (5.156) it is seen that if $H^k + c^k N^k N^{k'}$ is positive definite, then $(x^{k+1} - x^k)$ is a descent direction for the augmented Lagrangian function $L_{c^k}(\cdot, \lambda^k)$. Furthermore, if the constraint functions h_i are linear, then Eq. (5.157) can be written as

$$\lambda^{k+1} = \lambda^k + c^k h(x^{k+1}), \qquad (5.158)$$

while if in addition f is quadratic and $H^k + c^k N^k N^{k'}$ is positive definite, then from Eq. (5.156), x^{k+1} is the unique minimizing point of the augmented Lagrangian $L_{c^k}(\cdot, \lambda^k)$. Hence, it follows that *if the constraints are linear and the cost function is quadratic, then the iteration (5.156)-(5.157) is equivalent to the first order method of multipliers of Section 5.2.*

In the more general case where the constraints are nonlinear, it is natural to consider the iteration

$$x^{k+1} = x^k - \nabla^2_{xx} L_{c^k}(x^k, \lambda^k)^{-1} \nabla_x L_{c^k}(x^k, \lambda^k), \qquad (5.159)$$

followed by the first order multiplier iteration

$$\lambda^{k+1} = \lambda^k + c^k h(x^{k+1}). \qquad (5.160)$$

This is simply the first order multiplier iteration where the minimization of the augmented Lagrangian is replaced by a *single* pure Newton step, a method known as the *diagonalized method of multipliers*.

Note that for c^k large and x^k close to x^*, the Hessian $\nabla^2_{xx} L_{c^k}(x^k, \lambda^k)$ is nearly equal to $H^k + c^k N^k N^{k'}$, and $h(x^{k+1})$ is nearly equal to $h(x^k) + N^{k'}(x^{k+1} - x^k)$. Thus the diagonalized method of multipliers (5.159)-(5.160) can be viewed as an approximation to the variation of Newton's method (5.156)-(5.157) discussed earlier. This suggests that if c^k is taken larger than some threshold for all k, then the method should converge locally to a local minimum-Lagrange multiplier pair (x^*, λ^*) under the conditions of Prop. 5.4.3. This is indeed true, and the proof may be found in [Tap77] and [Ber82a], pp. 241-243, where it is also shown that the rate of convergence is superlinear if $c^k \to \infty$.

Extension to Inequality Constraints

Let us consider the inequality constrained problem

$$\text{minimize } f(x)$$
$$\text{subject to } g_j(x) \leq 0, \qquad j = 1, \ldots, r,$$

and focus on a local minimum x^* that is regular and together with a Lagrange multiplier μ^*, satisfies the second order sufficiency conditions of Prop. 4.3.2.

We can develop a Newton method for this problem, which is an extension of the quadratic programming implementation given earlier for equality constraints [cf. Eq. (5.148)]. This method is also similar to the constrained version of Newton's method for convex constraint sets given in Section 3.3 (in fact the two methods coincide when all the constraints are linear). In particular, given (x^k, μ^k), we obtain (x^{k+1}, μ^{k+1}) as an optimal solution-Lagrange multiplier pair of the quadratic program

$$\text{minimize} \quad \nabla f(x^k)'(x - x^k) + \tfrac{1}{2}(x - x^k)'\nabla^2_{xx}L(x^k, \mu^k)(x - x^k)$$
$$\text{subject to} \quad g_j(x^k) + \nabla g_j(x^k)'(x - x^k) \leq 0, \qquad j = 1, \ldots, r.$$

It is possible to show that there exists a neighborhood S of (x^*, μ^*) such that if (x^k, μ^k) is within S, then (x^{k+1}, μ^{k+1}) is uniquely defined as an optimal solution-Lagrange multiplier pair within S (an application of the implicit function theorem is needed to formalize this statement). Furthermore, (x^k, μ^k) converges to (x^*, μ^*) superlinearly. The details of this development are quite complex, and we refer to the book [Ber82a], Section 4.4.3, and the literature cited at the end of the chapter for further material.

5.4.3 Global Convergence

In order to enlarge the region of convergence of Lagrangian methods, it is necessary to combine them with some other method that has satisfactory global convergence properties. We refer to such a method as a *global method*. The main idea here is to construct a method that when sufficiently close to a local minimum switches automatically to a fast Lagrangian method, while when far away from such a point switches automatically to the global method, which is designed to make steady progress towards approaching the set of optimal solutions. The Lagrangian method can be any method that updates both vectors x and λ, based on the guidelines of this section. Prime candidates for use as global methods are multiplier methods and exact penalty methods.

There are many possibilities for combining global and Lagrangian methods, and the suitability of any one of these depends strongly on the problem at hand. For this reason, our main purpose in this section is not to develop and recommend specific algorithms, but rather to focus on the main guidelines for harmoniously interfacing global and Lagrangian methods while retaining the advantages of both. We note that combinations of global and Lagrangian methods, which involve augmented Lagrangian functions for convexification purposes, underlie several practical algorithms, including some popular nonlinear programming codes [CGT92], [MuS87].

Once a global and a Lagrangian method have been selected, the main issue to be settled is the choice of what we shall call the *switching rule* and the *acceptance rule*. The switching rule determines at each iteration, on the basis of certain tests, whether a switch should be made to the Lagrangian method. The tests depend on the information currently available, and their purpose is to decide whether an iteration of the Lagrangian method has a reasonable chance of success. As an example, such tests might include verification that ∇h has rank m and that $\nabla^2_{xx}L$ is positive definite on the subspace $\{y \mid \nabla h'y = 0\}$. We hasten to add here that these tests should not require excessive computational overhead. In some cases a switch might be made without any test at all, subject only to the condition that the Lagrangian iteration is well-defined.

The acceptance rule determines whether the results of the Lagrangian iteration will be accepted as they are, whether they will be modified, or whether they will be rejected completely and a switch will be made back to the global method. Typically, acceptance of the results of the Lagrangian iteration is based on reducing the value of a suitable merit function.

Combinations with Multiplier Methods

One possibility for enlarging the region of convergence of Lagrangian methods is to combine them with the first or second order methods of multipliers discussed in Section 5.2. The resulting methods tend to be very reliable since they inherit the robustness of the method of multipliers. At the same time they typically require fewer iterations to converge within the same accuracy than pure methods of multipliers.

To convey the general idea, let us discuss a few of the many possibilities. The simplest one is to switch to a Lagrangian method at the end of each (perhaps approximate) unconstrained minimization of a method of multipliers and continue using the Lagrangian method as long as the value of the exact penalty function $\|\nabla L\|^2$ is being decreased by a certain factor at each iteration. If satisfactory progress in decreasing $\|\nabla L\|^2$ is not observed, a switch back to the method of multipliers is made. Another possibility is to attempt a switch to a Lagrangian method at each iteration. As an example, consider a method for the equality constrained problem, which combines Newton's method for unconstrained minimization of the augmented Lagrangian together with the approximate Newton/Lagrangian iterations (5.156)-(5.160), which correspond to the method of multipliers.

At iteration k, we have x^k, λ^k, and a penalty parameter c^k. We also have a positive scalar w^k, which represents a target value of the exact penalty function $\|\nabla L\|^2$ that must be attained in order to accept the Lagrangian iteration, and a positive scalar ϵ^k that controls the accuracy of the unconstrained minimization of the method of multipliers. At the kth iteration, we determine x^{k+1}, λ^{k+1}, w^{k+1}, and ϵ^{k+1} as follows:

We form the Cholesky factorization $L^k L^{k'}$ of $\nabla^2_{xx}L_{c^k}(x^k, \lambda^k)$ as in Section 1.4. In the process, we modify $\nabla^2_{xx}L_{c^k}(x^k, \lambda^k)$ if it is not "sufficiently positive definite" (compare with Section 1.4). We then find the Newton direction

$$d^k = -(L^k L^{k'})^{-1} \nabla_x L_{c^k}(x^k, \lambda^k), \qquad (5.161)$$

and if $\nabla^2_{xx} L_{c^k}(x^k, \lambda^k)$ was found "sufficiently positive definite" during the factorization process, we also carry out the Lagrangian iteration [compare with Eqs. (5.159) and (5.160)]:

$$\bar{x}^k = x^k + d^k, \tag{5.162}$$

$$\bar{\lambda}^k = \lambda^k + c^k h(\bar{x}^k). \tag{5.163}$$

[The analog of Eq. (5.157) could also be used in place of Eq. (5.163).]
    If

$$\left\| \nabla L(\bar{x}^k, \bar{\lambda}^k) \right\|^2 \leq w^k,$$

then we accept the Lagrangian iteration and we set

$$x^{k+1} = \bar{x}^k, \qquad \lambda^{k+1} = \bar{\lambda}^k, \qquad c^{k+1} = c^k, \qquad \epsilon^{k+1} = \epsilon^k,$$

$$w^{k+1} = \gamma \left\| \nabla L(\bar{x}^k, \bar{\lambda}^k) \right\|^2,$$

where γ is a fixed scalar with $0 < \gamma < 1$.

Otherwise, we do not accept the results of the Lagrangian iteration, that is we do not update λ^k. Instead we revert to minimization of the augmented Lagrangian $L_{c^k}(\cdot, \lambda^k)$ by performing an Armijo-type line search. In particular, we set

$$x^{k+1} = x^k + \alpha^k d^k,$$

where the stepsize is obtained as

$$\alpha^k = \beta^{m_k},$$

where m_k is the first nonnegative integer m such that

$$L_{c^k}(x^k, \lambda^k) - L_{c^k}(x^k + \beta^m d^k, \lambda^k) \geq -\sigma \beta^m d^{k'} \nabla_x L_{c^k}(x^k, \lambda^k),$$

and β and σ are fixed scalars with $\beta \in (0, 1)$ and $\sigma \in (0, \frac{1}{2})$. If

$$\left\| \nabla_x L_{c^k}(x^{k+1}, \lambda^k) \right\| \leq \epsilon^k,$$

implying termination of the current unconstrained minimization, we do the ordinary first order multiplier iteration, setting

$$\lambda^{k+1} = \lambda^k + c^k h(x^k), \tag{5.164}$$

$$\epsilon^{k+1} = \gamma \epsilon^k, \qquad c^{k+1} = r c^k, \qquad w^{k+1} = \gamma \left\| \nabla L(x^{k+1}, \lambda^{k+1}) \right\|^2,$$

where r is a fixed scalar with $r > 1$. If

$$\left\| \nabla_x L_{c^k}(x^{k+1}, \lambda^k) \right\| > \epsilon^k,$$

we set

$$\lambda^{k+1} = \lambda^k, \qquad \epsilon^{k+1} = \epsilon^k, \qquad c^{k+1} = c^k, \qquad w_{k=1} = w^k,$$

and proceed with the next iteration.

An alternative combined algorithm is obtained by using, in place of the first order iteration (5.163), the second order iteration

$$\bar{\lambda}^k = \left(N^{k'}(H^k + c^k N^k N^{k'})^{-1} N^k\right)^{-1}$$
$$\left(h(\tilde{x}^k) - N^{k'}(H^k + c^k N^k N^{k'})^{-1} \nabla f(\tilde{x}^k)\right) - c^k h(\tilde{x}^k),$$

where $\tilde{x}^k$ is obtained by a pure Newton step

$$\tilde{x}^k = x^k + d^k = x^k - (L^k L^{k'})^{-1} \nabla_x L(x^k, \lambda^k),$$

[cf. Eq. (5.161)]. This corresponds to the second implementation of Newton's method of Eqs. (5.146)-(5.147)]. One could then obtain the vector $\bar{x}^k$ by a line search on the augmented Lagrangian $L_{c^k}(\cdot, \bar{\lambda}^k)$ along the direction d^k. The first order multiplier update (5.164) could also be replaced by a second order update. This combination of Newton's method and the second order multiplier method has outstanding rate of convergence properties, particularly if relatively good starting points are known. The combination given earlier based on the first order multiplier updates (5.163)-(5.164) is simpler, particularly if second derivatives are hard to compute and/or a quasi-Newton approximation is used in Eq. (5.161) in place of the inverse Hessian of the augmented Lagrangian $(L^k L^{k'})^{-1}$.

5.4.4 A Comparison of Various Methods

Quite a few barrier, penalty, and Lagrange multiplier methods were given in this chapter, so it is worth reflecting on their suitability for different types of problems. Even though it is hard to provide reliable guidelines, one may at least delineate the relative strengths and weaknesses of the various methods in specific practical contexts.

The barrier methods of Section 5.1, generally must solve a sequence of minimization problems that are increasingly ill-conditioned. This is a disadvantage relative to the multiplier methods of Section 5.2, whose sequence of minimization problems need not be ill-conditioned, and also relative to the exact penalty methods of Section 5.3, which require solution of only one minimization problem. However, for linear and for quadratic programs there is special structure that makes the logarithmic barrier and also the primal-dual interior point methods of Section 5.1.2 preferable to multiplier and exact penalty methods, from the theoretical and apparently the practical point of view. Whether, there are other important classes of problems for which this is also true, is an open question.

Multiplier methods are excellent general purpose constrained optimization methods. Their main advantages are simplicity and robustness. They rely on well-developed unconstrained optimization technology, and they require fewer assumptions for their validity relative to their competitors. In particular, they can deal with nonlinear equality constraints, and they do not require the existence of second derivatives and the regularity of the generated iterates (although they can be made more efficient when second derivatives can be used and when the iterates are regular). For these reasons some of the most popular software packages for solving nonlinear programming problems are based on multiplier methods.

The main disadvantage of multiplier methods relative to exact penalty methods is that they require a sequence of unconstrained minimizations as opposed to a single minimization. This disadvantage can be ameliorated by making the minimizations inexact or by combining the multiplier method with a Lagrangian method as described in Section 5.4.3. Still, practice has shown that minimization of an exact penalty function by a Newton-like method can require substantially fewer iterations relative to a multiplier method. Note, however, that each of these iterations may require a potentially costly subproblem (as in the linearization method) or may require complex calculations (as in differentiable exact penalty methods).

Generally, both multiplier methods and exact penalty methods require some trial and error to obtain appropriate values of the penalty parameter and also to ensure that there are no difficulties with ill-conditioning. However, multiplier methods typically are easier to "tune" than exact penalty methods, and deal more comfortably with the absence of a good starting point. Thus, if only a limited number of optimization runs are required in a given practical problem after development of the optimization code, one is typically better off with a method of multipliers than with an exact penalty method. If on the other hand, repetitive solution of the same problem with minor variations is envisioned, solution time is an issue, and the associated overhead per iteration is reasonable, one may prefer to use an exact penalty method.

EXERCISES

5.4.1

Consider the problem

$$\text{minimize} \quad -x_1 x_2$$
$$\text{subject to} \quad x_1 + x_2 = 2.$$

Write the implementation (5.141)-(5.142) of Newton's method, and show that it finds the optimal solution in a single iteration, regardless of the starting point. Write also the approximate implementation (5.156)-(5.157) for the starting point $x^0 = (0,0)$, $\lambda^0 = 0$, and for the two values $c = 10^{-2}$ and $c = 10^2$. How does the error after a single iteration depend on c?

5.4.2

Use Prop. 5.4.1 to derive a local convergence result for the approximate implementation (5.156)-(5.157) of Newton's method. Do the same for iteration (5.159)-(5.160).

5.5 NOTES AND SOURCES

Section 5.1: The logarithmic barrier method dates to the work of Frisch in the middle 50s [Fri56]. Other barrier methods have been proposed by Carroll [Car61]. An important early reference on penalty and barrier methods is the book by Fiacco and McCormick [FiM68]. Properties of the central path are investigated by McLinden [McL80], Sonnevend [Son86], Bayer and Lagarias [BaL89], and Guler [Gul94]. For surveys of interior point methods for linear programming, which give many additional references, see Gonzaga [Gon92], den Hertog [Her94], and Forsgren, Gill, and Wright [FGW02]. The line of analysis that we use is due to Tseng [Tse89], which also gives a computational complexity result along the lines of Exercise 5.1.5 (see also Gonzaga [Gon91]).

There has been a lot of effort to apply interior point methods to nonlinear problems, such as quadratic programming (Anstreicher, den Hertog, Roos, and Terlaky [AHR93], Wright [Wri96]), linear complementarity problems (Kojima, Meggido, Noma, and Yoshise [KMN91], Tseng [Tse92], Wright [Wri93c]), matrix inequalities (Alizadeh [Ali92], [Ali95], Nesterov and Nemirovskii [NeN94], Vandenberghe and Boyd [VaB95]), general convex programming problems (Wright [Wri92], Kortanek and Zhu [KoZ93], Anstreicher and Vial [AnV94], Jarre and Saunders [JaS95], Kortanek and Zhu [KoZ95]), and optimal control (Wright [Wri93b]). The research monographs by Nesterov and Nemirovskii [NeN94], Wright [Wri97a], Ye [Ye97], and Renegar [Ren01] are devoted to interior point methods for linear, quadratic, and convex programming.

Among interior point algorithms for linear programming, primal-dual methods are widely considered as generally the most effective; see e.g., McShane, Monma, and Shanno [MMS91]. They were introduced for linear programming through the study of the primal-dual central path by Megiddo [Meg88]. They were turned into path-following algorithms in the subsequent papers by Kojima, Mizuno, and Yoshise [KMY89], and Monteiro and Adler [MoA89a]. Their convergence, rate of convergence, and computational complexity are discussed in Zhang and Tapia [ZhT92], [ZTP93], [ZhT93], Potra [Pot94], and Tapia, Zhang, and Ye [TZY95]. The predictor-corrector variant was proposed by Mehrotra [Meh92]. There have been several extensions to broader classes of problems, including quadratic programming (Monteiro and Adler [MoA89b]), and linear complementarity (Wright [Wri94] and Tseng [Tse95b]). The research monograph by Wright [Wri97a] provides an extensive development of primal-dual interior point methods.

Section 5.2: The method of multipliers for equality constraints was independently proposed by Hestenes [Hes69], Powell [Pow69], and Haarhoff and Buys [HaB70]. These references contained very little analysis and empirical evidence, but much subsequent work established the convergence properties of the method and proposed various extensions. Surveys by Bertsekas [Ber76b] and Rockafellar [Roc76c] summarize the work up to 1976, and the author's research monograph [Ber82a] provides a detailed analysis and many references. The global convergence of the method is discussed by Poljak and Tretjakov [PoT73b], Bertsekas [Ber76a], Polak and Tits [PoT80a], Bertsekas [Ber82a], and Conn, Gould, and Toint [CGT91]. Global and superlinear convergence results for second order methods of multipliers, which are analogous to Prop. 5.2.3, are given in [Ber82a], Section 2.3.2.

The class of nonquadratic penalty methods for inequality constraints, given in Section 5.2.5, was introduced by Kort and Bertsekas [KoB72], with a special focus on the exponential method of multipliers. The convergence properties of the sequence $\{x^k\}$ generated by this method are quite intricate and are discussed by Bertsekas [Ber82a], Tseng and Bertsekas [TsB93], and Iusem [Ius99] for convex problems. For nonconvex problems under second order sufficiency conditions, the convergence analysis follows the pattern of the corresponding analysis for the quadratic method of multipliers (see Nguyen and Strodiot [NgS79]). The exponential method was applied to the solution of systems of nonlinear inequalities by Bertsekas ([Ber82a], Section 5.1.3), and Schnabel [Sch82]. The modified barrier method was proposed and developed by Polyak; see [Pol92] and the references given therein.

For subsequent research on the exponential penalty, the modified barrier, and other related methods that use nonquadratic penalty functions; see Freund [Fre91], Guler [Gul92], Teboulle [Teb92], Chen and Teboulle [ChT93], Tseng and Bertsekas [TsB93], Eckstein [Eck94a], Iusem, Svaiter, and Teboulle [IST94], Iusem and Teboulle [IuT95], Ben-Tal and Zibulevsky [BeZ97], Polyak and Teboulle [PoT97], Wei, Qi, and Birge [WQB98], and Iusem [Ius99].

Section 5.3: Nondifferentiable exact penalty methods were first proposed by Zangwill [Zan67b]; see also Han and Mangasarian [HaM79], who survey the subject and give many references. The linearization method is due to Pschenichny [Psc70] (see also Pschenichny and Danilin [PsD76]). It was independently derived later by Han [Han77]. Convergence rate issues and modifications to improve the convergence rate of sequential quadratic programming algorithms and to avoid the Maratos' effect (Exercise 5.3.9) are discussed in Boggs, Tolle, and Wang [BTW82], Coleman and Conn [CoC82a], [CoC82b], Gabay [Gab82], Mayne and Polak [MaP82], Panier and Tits [PaT91], Bonnans, Panier, Tits, and Zhou [BPT92], and Bonnans [Bon89b], [Bon94]. Combinations of the linearization method and the two-metric projection method of Section 3.4 have been proposed by Heinkenschloss [Hei96]. Note that since the linearization method can minimize the nondifferentiable exact penalty function $f(x) + cP(x)$, it can also be used to minimize

$$P(x) = \max\{g_0(x), g_1(x), \ldots, g_r(x)\},$$

which is a typical case of a minimax problem.

Exact differentiable penalty methods involving only x were introduced by Fletcher [Fle70b]. Exact differentiable penalty methods involving both x and λ were introduced by DiPillo and Grippo [DiG79]. The relation between these two types of methods, their utility for sequential quadratic programming, and a number of variations were derived by Bertsekas [Ber82c] (see also [Ber82a], Section 4.3, which contains a detailed convergence analysis). Extensions to inequality constraints are given by Glad and Polak [GlP79], and Bertsekas [Ber82a]; see also Mukai and Polak [MuP75], Boggs and Tolle [BoT80], and DiPillo and Grippo [DiG89]. Differentiable exact penalty functions are used by Nazareth [Naz96], and Nazareth and Qi [NaQ96] to extend the region of convergence of Newton-like methods for solving systems of nonlinear equations.

Section 5.4: First order Lagrangian methods were introduced by Arrow, Hurwicz, and Uzawa [AHU58]. They were also analyzed by Poljak [Pol70], and Psenichnyi and Danilin [PsD75], whom we follow in our presentation. Combinations of Lagrangian methods with the proximal algorithm were proposed in more general form for convex programming problems by Chen and Teboulle [ChT94], together with applications in decomposition of separable problems.

A Newton-like Lagrangian method for inequality constraints was proposed by Wilson [Wil63]. For this method, a superlinear convergence rate was established by Robinson [Rob74] under second order sufficiency conditions, including strict complementarity. Superlinear convergence results for a variant of the method were shown under weaker conditions by Wright [Wri98] and Hager [Hag99]. Combinations of Lagrangian methods with first order methods of multipliers were given by Glad [Gla79]. An extensive discussion of Newton-like Lagrangian methods is given in the author's research monograph [Ber82a].

6

Duality and Convex Programming

Contents

6.1. Duality and Dual Problems . . . . p. 554
   6.1.1. Geometric Multipliers . . . . p. 556
   6.1.2. The Weak Duality Theorem . . . . p. 561
   6.1.3. Primal and Dual Optimal Solutions . . . . p. 566
   6.1.4. Treatment of Equality Constraints . . . . p. 568
   6.1.5. Separable Problems and their Geometry . . . . . p. 570
   6.1.6. Additional Issues About Duality . . . . p. 575
6.2. Convex Cost – Linear Constraints . . . . p. 582
6.3. Convex Cost – Convex Constraints . . . . p. 589
6.4. Conjugate Functions and Fenchel Duality . . . . . . p. 598
   6.4.1. Conic Programming . . . . p. 604
   6.4.2. Monotropic Programming . . . . p. 612
   6.4.3. Network Optimization . . . . p. 617
   6.4.4. Games and the Minimax Theorem . . . . p. 620
   6.4.5. The Primal Function and Sensitivity Analysis . . . p. 623
6.5. Discrete Optimization and Duality . . . . p. 630
   6.5.1. Examples of Discrete Optimization Problems . . . p. 631
   6.5.2. Branch-and-Bound . . . . p. 639
   6.5.3. Lagrangian Relaxation . . . . p. 648
6.6. Notes and Sources . . . . p. 660

αγεωμετρητος μηδεις εισιτω

(No one ignorant of geometry shall enter)

Door inscription at Plato's Academy

Our analysis of earlier chapters was primarily calculus-based, comparing local minima to their close neighbors by using first and second derivatives of the cost and the constraints. In this chapter we use a fundamentally different line of analysis. In particular, we will be concerned exclusively with global minima, as opposed to local minima, and at least in the beginning, we will make no differentiability or even continuity assumptions on the cost and constraint functions.

An important consequence of the generality of our framework is that much of the theory that we will develop is very broadly applicable. For example, it is useful for discrete optimization problems, such as integer programming, as we will show in Section 6.5. On the other hand, the power of the theory is enhanced as the problem becomes more structured. In particular, the strongest results of this chapter apply only when the cost and constraints are convex. In the next section we develop a general and intuitive geometrical framework for duality, which illustrates the basic concepts, the results that can be expected under convexity conditions, the associated theoretical pitfalls, and the conditions needed to obtain these results.

6.1 DUALITY AND DUAL PROBLEMS

The central notion of this chapter is *duality*. We have already encountered a simple version of duality theory in Section 4.4, which was developed under differentiability assumptions on the cost function and linearity assumptions on the constraints. This theory covered, among others, linear and convex quadratic programming. Our analysis in this chapter will be deeper and more powerful, making full use of intuitive geometrical notions such as hyperplanes, and their convex set support and separation properties. Because our development is geometry-based, the duality results that we develop and their proofs can be easily visualized. As an example, we formulate two abstract problems, dual to each other, which will prove particularly relevant in the context of the subsequent analysis. Let S be a subset of $\Re^n$ and consider the following problems.

(a) *Min common point problem*: Here, among all points that are common to both S and the nth axis, we want to find the one whose nth coordinate is minimum.

Sec. 6.1   Duality and Dual Problems   555

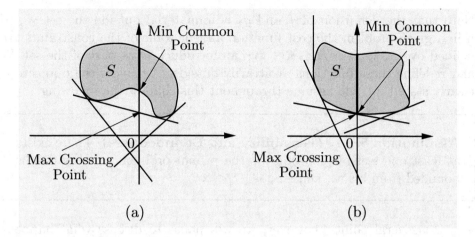

Figure 6.1.1. Illustration of the optimal values of the min common point and max crossing point problems. In (a), the two optimal values are not equal. In (b), the set S, when "extended upwards" along the nth axis, yields the set

$$\bar{S} = \{\bar{x} \mid \text{for some } x \in S, \; \bar{x}_n \geq x_n, \; \bar{x}_i = x_i, \forall \, i = 1, \ldots, n-1\},$$

which is convex. As a result, the two optimal values are equal. This fact, when suitably formalized, will form the basis for a number of duality results in Sections 6.2 and 6.3.

(b) *Max crossing point problem*: Here we consider all hyperplanes that intersect the nth axis and support the set S from "below" (see Fig. 6.1.1). We want to find the maximum point of interception of the nth axis with such a hyperplane.

Figure 6.1.1 shows that the optimal value of the max crossing point problem is no larger than the optimal value of the min common point problem, and that under favorable circumstances the two optimal values are equal.

We first focus on problems that involve inequality constraints. The extension to equality-constrained problems is simple, and will be discussed in Section 6.1.5. In particular, we consider the problem

$$\text{minimize } f(x)$$
$$\text{subject to } x \in X, \quad g_j(x) \leq 0, \quad j = 1, \ldots, r,$$

where $f : \Re^n \mapsto \Re$, $g_j : \Re^n \mapsto \Re$ are given functions, and X is a subset of $\Re^n$. We also write the constraints $g_j(x) \leq 0$ compactly as $g(x) \leq 0$, where

$$g(x) = \big(g_1(x), \ldots, g_r(x)\big).$$

We refer to this problem as the *primal problem* and we denote by f^* its optimal value:

$$f^* = \inf_{\substack{x \in X \\ g_j(x) \leq 0, \, j=1,\ldots,r}} f(x).$$

Note that the definition of f and g_j is immaterial outside the set X, so if in a given problem the cost function and/or some of the constraints are defined over a domain $D \subset \Re^n$, we can introduce D as part of the set X, and redefine these functions arbitrarily outside D. Unless the opposite is clearly stated, we will assume throughout this chapter the following:

Assumption 5.1.1: (Feasibility and Boundedness) There exists at least one feasible solution for the primal problem and the cost is bounded from below, i.e., $-\infty < f^* < \infty$.

At several points, however, we will pause to discuss what happens when this assumption is violated.

6.1.1 Geometric Multipliers

We want to define a notion of multiplier that is not tied to a specific local or global minimum, and does not assume differentiability of the cost and constraint functions. To this end, we draw motivation from the case where $X = \Re^n$, and f and g_j are convex and differentiable. Then, if x^* is a global minimum and a regular point, the associated Lagrange multiplier vector $\mu^* = (\mu_1^*, \ldots, \mu_r^*) \geq 0$ satisfies $\nabla_x L(x^*, \mu^*) = 0$ and $\mu_j^* g_j(x^*) = 0$ for all j, so that

$$f^* = f(x^*) = \min_{x \in \Re^n} L(x, \mu^*), \tag{6.1}$$

where $L : \Re^{n+r} \mapsto \Re$ is the Lagrangian function

$$L(x, \mu) = f(x) + \sum_{j=1}^{r} \mu_j g_j(x) = f(x) + \mu' g(x).$$

Taking the Lagrangian condition (6.1) as a point of departure, we are led to the following definition.

Definition 5.1.1: A vector $\mu^* = (\mu_1^*, \ldots, \mu_r^*)$ is said to be a *geometric multiplier vector* (or simply *geometric multiplier*) for the primal problem if

$$\mu_j^* \geq 0, \qquad j = 1, \ldots, r,$$

and

$$f^* = \inf_{x \in X} L(x, \mu^*).$$

As indicated by the preceding discussion, there is a connection between geometric and Lagrange multipliers in the case where X, f, and

Sec. 6.1   Duality and Dual Problems   557

g_j are convex, and f and g_j are continuously differentiable. Then, it can be shown that given an optimal solution x^*, the set of Lagrange multipliers corresponding to x^* (as defined in Section 4.3) is equal to the set of geometric multipliers. However, even under convexity assumptions, it is possible that the problem has no optimal solution and hence no Lagrange multipliers, while the set of geometric multipliers may be nonempty. As an example, consider the one-dimensional convex problem of minimizing e^x subject to the single inequality constraint $x \leq 0$, with optimal value $f^* = 0$; it has the geometric multiplier $\mu^* = 0$ (according to Definition 5.1.1), but it has no optimal solution.

To visualize the definition of a geometric multiplier, as well as other concepts related to duality, it is useful to consider hyperplanes in the space of constraint-cost pairs $\big(g(x), f(x)\big)$ (viewed as vectors in $\Re^{r+1}$). We recall from Section B.2 in Appendix B that a hyperplane H in $\Re^{r+1}$ is specified by a linear equation involving a nonzero vector (μ, μ_0) (called the *normal vector of H*), where $\mu \in \Re^r$ and $\mu_0 \in \Re$, and by a constant c as follows:

$$H = \big\{(z, w) \mid z \in \Re^r, w \in \Re, \mu_0 w + \mu' z = c\big\}.$$

Any vector $(\bar{z}, \bar{w})$ that belongs to the hyperplane H specifies the constant c as

$$c = \mu_0 \bar{w} + \mu' \bar{z}.$$

Thus the hyperplane with given normal (μ, μ_0) that passes through a given vector $(\bar{z}, \bar{w})$ is the set of (z, w) that satisfy the equation

$$\mu_0 w + \mu' z = \mu_0 \bar{w} + \mu' \bar{z}.$$

This hyperplane defines two halfspaces: the *positive halfspace*

$$H^+ = \big\{(z, w) \mid \mu_0 w + \mu' z \geq \mu_0 \bar{w} + \mu' \bar{z}\big\}$$

and the *negative halfspace*

$$H^- = \big\{(z, w) \mid \mu_0 w + \mu' z \leq \mu_0 \bar{w} + \mu' \bar{z}\big\}.$$

Hyperplanes with normals (μ, μ_0) where $\mu_0 \neq 0$ are referred to as *nonvertical* (their normal has a nonzero last component). A nonvertical hyperplane can be *normalized* by dividing its normal vector by μ_0, and assuming this is done, we have $\mu_0 = 1$.

The above definitions will now be used to interpret geometric multipliers as normalized nonvertical hyperplanes with a certain orientation to the set of all constraint-cost pairs as x ranges over X, i.e., the subset of $\Re^{r+1}$

$$S = \big\{(g(x), f(x)) \mid x \in X\big\}. \qquad (6.2)$$

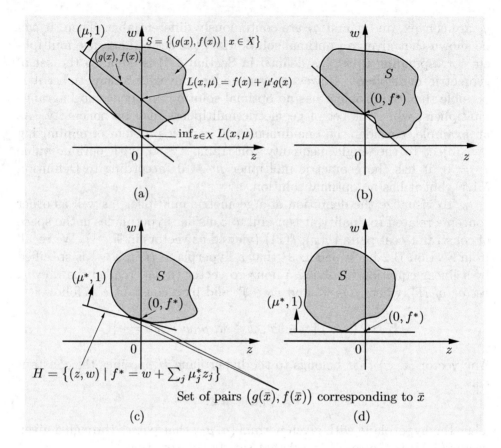

Figure 6.1.2. (a) Geometrical interpretation of the Lagrangian value $L(x,\mu)$ as the level of interception of the vertical axis with the hyperplane with normal $(\mu, 1)$ that passes through $(g(x), f(x))$. (b) A case where there is no geometric multiplier. Here, there is no hyperplane that passes through the point $(0, f^*)$ and contains S in its positive halfspace. (c) and (d) Illustration of a geometric multiplier vector μ^* in the cases $\mu^* \neq 0$ and $\mu^* = 0$, respectively. It defines a hyperplane H that has normal $(\mu^*, 1)$, passes through $(0, f^*)$, and contains S in its positive halfspace. Note that the common points of H and S (if any) are the pairs $(g(\bar{x}), f(\bar{x}))$ corresponding to points $\bar{x}$ that minimize the Lagrangian $L(x, \mu^*)$ over $x \in X$. In (c), the point $(0, f^*)$ belongs to S and corresponds to optimal primal solutions for which the constraint is active (as well as to some infeasible primal solutions). In (d), the point $(0, f^*)$ does not belong to S and corresponds to an optimal primal solution for which the constraint is inactive.

We have the following lemma, which is graphically illustrated in Fig. 6.1.2.

Visualization Lemma:

(a) The hyperplane with normal $(\mu, 1)$ that passes through a vector $(g(x), f(x))$ intercepts the vertical axis $\{(0, w) \mid w \in \Re\}$ at the level $L(x, \mu)$.

> (b) Among all hyperplanes with normal $(\mu, 1)$ that contain in their positive halfspace the set S of Eq. (6.2), the highest attained level of interception of the vertical axis is $\inf_{x \in X} L(x, \mu)$.
>
> (c) μ^* is a geometric multiplier if and only if $\mu^* \geq 0$ and among all hyperplanes with normal $(\mu^*, 1)$ that contain in their positive halfspace the set S, the highest attained level of interception of the vertical axis is f^*.

Proof: (a) By the preceding discussion, the hyperplane with normal $(\mu, 1)$ that passes through $(g(x), f(x))$ is the set of (z, w) satisfying

$$w + \mu'z = f(x) + \mu'g(x) = L(x, \mu).$$

The only vector in the vertical axis satisfying this equation is $(0, L(x, \mu))$.

(b) The hyperplane with normal $(\mu, 1)$ that intercepts the vertical axis at level c is the set of vectors (z, w) that satisfy the equation

$$w + \mu'z = c,$$

and this hyperplane contains S in its positive halfspace if and only if

$$L(x, \mu) = f(x) + \mu'g(x) \geq c, \qquad \forall\, x \in X.$$

Therefore the maximum point of interception is $c^* = \inf_{x \in X} L(x, \mu)$.

(c) Follows from the definition of a geometric multiplier and part (b). **Q.E.D.**

The hyperplane with normal $(\mu, 1)$ that attains the highest level of interception of the vertical axis, as in part (b) of the visualization lemma, is seen to "support" the set S (the notion of a supporting hyperplane is defined more precisely in Section B.2 of Appendix B). Figure 6.1.3 gives some examples where there exist one or more geometric multipliers. Figure 6.1.4 shows cases where there is no geometric multiplier.

If a geometric multiplier μ^* is known, then all optimal solutions x^* can be obtained by minimizing the Lagrangian $L(x, \mu^*)$ over $x \in X$, as indicated in Fig. 6.1.2(c) and shown in the following proposition. However, there may be vectors that minimize $L(x, \mu^*)$ over $x \in X$ but do not satisfy the inequality constraints $g(x) \leq 0$ [cf. Figs. 6.1.2(c) and 6.1.3(a)].

> **Proposition 6.1.1:** Let μ^* be a geometric multiplier. Then x^* is a global minimum of the primal problem if and only if x^* is feasible and
>
> $$x^* \in \arg\min_{x \in X} L(x, \mu^*), \qquad \mu_j^* g_j(x^*) = 0, \qquad j = 1, \ldots, r. \qquad (6.3)$$

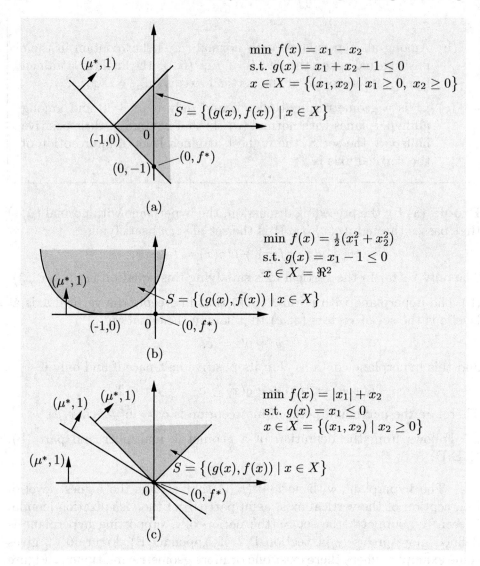

Figure 6.1.3. Examples where there exists at least one geometric multiplier. In (a), there is a unique geometric multiplier, $\mu^* = 1$. In (b), there is a unique geometric multiplier, $\mu^* = 0$. In (c), the set of geometric multipliers is the interval $[0, 1]$.

Proof: If x^* is a global minimum, then x^* is feasible and furthermore,

$$f^* = f(x^*) \geq f(x^*) + \sum_{j=1}^{r} \mu_j^* g_j(x^*) = L(x^*, \mu^*) \geq \inf_{x \in X} L(x, \mu^*), \quad (6.4)$$

where the first inequality follows from the definition of a geometric multiplier ($\mu^* \geq 0$) and the feasibility of x^* [$g(x^*) \leq 0$]. Using again the definition of geometric multiplier, we have $f^* = \inf_{x \in X} L(x, \mu^*)$, so that equality holds throughout in Eq. (6.4), implying Eq. (6.3).

Sec. 6.1 Duality and Dual Problems   561

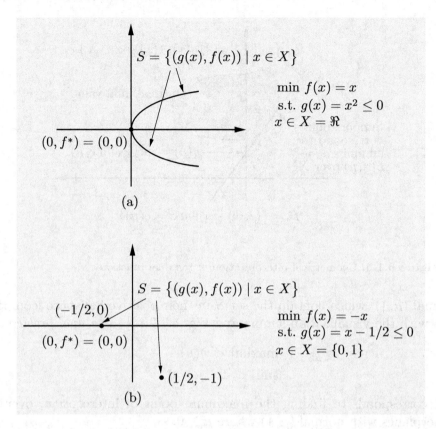

Figure 6.1.4. Examples where there exists no geometric multiplier. In (a), a geometric multiplier does not exist because the only hyperplane that supports S at $(0, f^*)$ is vertical. In (b), we have an integer programming problem, where the set X is discrete. Here there is no geometric multiplier because there is no hyperplane that supports S and passes through $(0, f^*)$.

Conversely, if x^* is feasible and Eq. (6.3) holds, we have, using also the definition of a geometric multiplier,

$$f(x^*) = f(x^*) + \sum_{j=1}^{r} \mu_j^* g_j(x^*) = L(x^*, \mu^*) = \min_{x \in X} L(x, \mu^*) = f^*,$$

so x^* is a global minimum. **Q.E.D.**

6.1.2 The Weak Duality Theorem

We introduce the *dual function* q, which is defined for $\mu \in \Re^r$ by

$$q(\mu) = \inf_{x \in X} L(x, \mu).$$

This definition is illustrated in Fig. 6.1.5, where $q(\mu)$ is interpreted as the highest point of interception with the vertical axis over all hyperplanes with

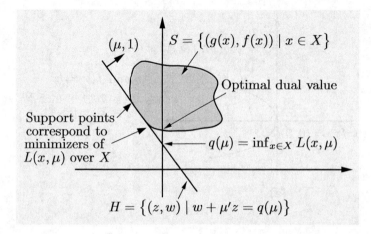

Figure 6.1.5. Geometrical interpretation of the dual function.

normal $(\mu, 1)$, which contain the set S in their positive halfspace [compare also with the visualization lemma and Fig. 6.1.2(a)]. The *dual problem* is

$$\text{maximize } q(\mu)$$
$$\text{subject to } \mu \geq 0,$$

and corresponds to finding the maximum point of interception, over all hyperplanes with normal $(\mu, 1)$ where $\mu \geq 0$.

Note that $q(\mu)$ may be equal to $-\infty$ for some μ. In this case, we effectively have the additional constraint $\mu \in D$ in the dual problem, where D, called the *domain* of q, is the set of μ for which $q(\mu)$ is finite:

$$D = \{\mu \mid q(\mu) > -\infty\}.$$

In fact, we may have $q(\mu) = -\infty$ for all $\mu \geq 0$, in which case the dual optimal value

$$q^* = \sup_{\mu \geq 0} q(\mu),$$

is equal to $-\infty$ (situations where this can happen will be discussed later).

Regardless of the structure of the cost and constraints of the primal problem, the dual problem has nice convexity properties, as shown by the following proposition.

Proposition 6.1.2: The domain D of the dual function q is convex and q is concave over D.

Proof: For any x, μ, $\bar{\mu}$, and $\alpha \in [0, 1]$, we have

$$L(x, \alpha\mu + (1-\alpha)\bar{\mu}) = \alpha L(x, \mu) + (1-\alpha)L(x, \bar{\mu}).$$

Sec. 6.1   Duality and Dual Problems

Taking the infimum over all $x \in X$, we obtain

$$\inf_{x \in X} L(x, \alpha\mu + (1-\alpha)\bar{\mu}) \geq \alpha \inf_{x \in X} L(x, \mu) + (1-\alpha) \inf_{x \in X} L(x, \bar{\mu})$$

or

$$q(\alpha\mu + (1-\alpha)\bar{\mu}) \geq \alpha q(\mu) + (1-\alpha)q(\bar{\mu}).$$

Therefore if μ and $\bar{\mu}$ belong to D, the same is true for $\alpha\mu + (1-\alpha)\bar{\mu}$, so D is convex. Furthermore, q is concave over D. **Q.E.D.**

The concavity of q can also be verified by observing that q is defined as the infimum over $x \in X$ of the collection of the concave (in fact linear) functions $L(x, \cdot)$; such a function is concave by Prop. B.2(d) in Appendix B.

Another important property is that the optimal dual value is always an underestimate of the optimal primal value. This is evident from the geometric interpretation of Fig. 6.1.5 and is formally shown in the next proposition.

Proposition 6.1.3: (Weak Duality Theorem) We have

$$q^* \leq f^*.$$

Proof: For all $\mu \geq 0$, and $x \in X$ with $g(x) \leq 0$, we have

$$q(\mu) = \inf_{z \in X} L(z, \mu) \leq f(x) + \sum_{j=1}^{r} \mu_j g_j(x) \leq f(x),$$

so

$$q^* = \sup_{\mu \geq 0} q(\mu) \leq \inf_{x \in X,\, g(x) \leq 0} f(x) = f^*.$$

Q.E.D.

If $q^* = f^*$ we say that *there is no duality gap* and if $q^* < f^*$ we say that *there is a duality gap*. Note that if there exists a geometric multiplier μ^*, the weak duality theorem ($q^* \leq f^*$) and the definition of a geometric multiplier $[f^* = q(\mu^*) \leq q^*]$ imply that there is no duality gap. However, the converse is not true. In particular, it is possible that no geometric multiplier exists even though there is no duality gap [cf. Fig. 6.1.4(a)]; in this case the dual problem does not have an optimal solution, as implied by the following proposition.

Proposition 6.1.4:

(a) If there is no duality gap, the set of geometric multipliers is equal to the set of optimal dual solutions.

(b) If there is a duality gap, the set of geometric multipliers is empty.

Proof: By definition, a vector $\mu^* \geq 0$ is a geometric multiplier if and only if $f^* = q(\mu^*) \leq q^*$, which by the weak duality theorem, holds if and only if there is no duality gap and μ^* is a dual optimal solution. **Q.E.D.**

The preceding results are illustrated in Figs. 6.1.6 and 6.1.7 for the problems of Figs. 6.1.3 and 6.1.4, respectively.

Duality theory is most useful when there is no duality gap. To guarantee that there is no duality gap and that a geometric multiplier exists, it is typically necessary to impose various types of convexity conditions on the cost and the constraints of the primal problem; we will develop some of these conditions in Sections 6.2-6.4. However, even with a duality gap, the dual problem can be useful, as indicated by the following example.

Example 6.1.1 (Integer Programming, and Branch-and-Bound)

Many important practical optimization problems of the form

$$\text{minimize } f(x)$$
$$\text{subject to } x \in X, \quad g_j(x) \leq 0, \quad j = 1, \ldots, r,$$

have a finite constraint set X. An example is *integer programming*, where the coordinates of x must be integers from a bounded range (usually 0 or 1). An important special case is the linear 0-1 integer programming problem

$$\text{minimize } c'x$$
$$\text{subject to } Ax \leq b, \quad x_i = 0 \text{ or } 1, \quad i = 1, \ldots, n.$$

A principal approach for solving such problems is the *branch-and-bound method*, which will be described in Section 6.5. This method relies on obtaining lower bounds to the optimal cost of restricted problems of the form

$$\text{minimize } f(x)$$
$$\text{subject to } x \in \tilde{X}, \quad g_j(x) \leq 0, \quad j = 1, \ldots, r,$$

where $\tilde{X}$ is a subset of X; for example in the 0-1 integer case where X specifies that all x_i should be 0 or 1, $\tilde{X}$ may be the set of all 0-1 vectors x such that one or more coordinates x_i are restricted to satisfy $x_i = 0$ for all $x \in \tilde{X}$ or

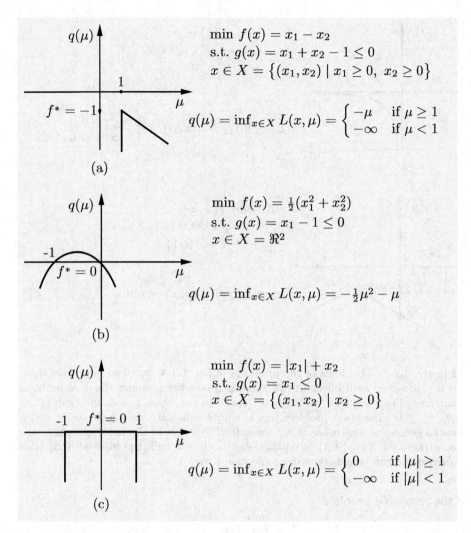

Figure 6.1.6. The dual functions of the problems of Fig. 6.1.3. In all these problems, there is no duality gap and the set of geometric multipliers is equal to the set of dual optimal solutions [cf. Prop. 6.1.4(a)]. In (a), there is a unique dual optimal solution $\mu^* = 1$. In (b), there is a unique dual optimal solution $\mu^* = 0$; at the corresponding primal optimal solution ($x^* = 0$) the constraint is inactive. In (c), the set of dual optimal solutions is $\{\mu^* \mid 0 \leq \mu^* \leq 1\}$.

$x_i = 1$ for all $x \in \tilde{X}$. These lower bounds can often be obtained by finding a dual-feasible (possibly dual-optimal) solution μ of this problem and the corresponding dual value

$$q(\mu) = \inf_{x \in \tilde{X}} \left\{ f(x) + \sum_{j=1}^{r} \mu_j g_j(x) \right\},$$

which by the weak duality theorem, is a lower bound to the optimal value of

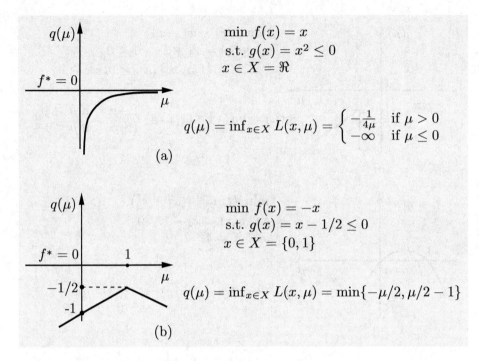

Figure 6.1.7. The duals of the problems of Fig. 6.1.4. In these problems, there is no geometric multiplier. In (a), there is no duality gap and the dual problem has no optimal solution [cf. Prop. 6.1.4(a)]. In (b), there is a duality gap ($f^* - q^* = 1/2$). The dual problem has a unique optimal solution $\mu^* = 1$, which is the geometric multiplier of a "convexified" version of the problem, where the constraint set $X = \{0, 1\}$ is replaced by its convex hull, the interval $[0, 1]$ (this observation can be generalized; see Section 6.5).

the restricted problem

$$\min_{x \in \tilde{X},\, g(x) \leq 0} f(x).$$

One is interested in finding lower bounds that are as tight as possible, so the usual approach is to start with some dual feasible solution and iteratively improve it by using some algorithm. A major difficulty here is that the dual function $q(\mu)$ is typically nondifferentiable, so the methods developed so far cannot be used. We will develop special methods for optimization of nondifferentiable cost functions in the next chapter. Note that in many problems of interest that have favorable structure, the value $q(\mu)$ can be calculated easily, and we will see in the next chapter that other quantities, such as subgradients, which are needed for application of nondifferentiable optimization methods, are also easily obtained together with $q(\mu)$.

6.1.3 Primal and Dual Optimal Solutions

There are powerful characterizations of primal and dual optimal solution pairs, given in the following two propositions. Note, however, that these characterizations are useful only if there is no duality gap, since otherwise there is no geometric multiplier [cf. Prop. 6.1.4(b)], even if the dual problem has an optimal solution.

Proposition 6.1.5: (Optimality Conditions) A pair (x^*, μ^*) is an optimal solution-geometric multiplier pair if and only if

$$x^* \in X, \quad g(x^*) \leq 0, \qquad \text{(Primal Feasibility)}, \qquad (6.5)$$

$$\mu^* \geq 0, \qquad \text{(Dual Feasibility)}, \qquad (6.6)$$

$$x^* \in \arg\min_{x \in X} L(x, \mu^*), \qquad \text{(Lagrangian Optimality)}, \qquad (6.7)$$

$$\mu_j^* g_j(x^*) = 0, \quad j = 1, \ldots, r, \qquad \text{(Complementary Slackness)}. \qquad (6.8)$$

Proof: If (x^*, μ^*) is an optimal solution-geometric multiplier pair, then x^* is primal feasible and μ^* is dual feasible. Equations (6.7) and (6.8) follow from Prop. 6.1.1.

Conversely, using Eqs. (6.5)-(6.8), we obtain

$$f^* \leq f(x^*) = L(x^*, \mu^*) = \min_{x \in X} L(x, \mu^*) = q(\mu^*) \leq q^*.$$

Using the weak duality theorem (Prop. 6.1.3), we see that equality holds throughout in the preceding relation. It follows that x^* is primal optimal and μ^* is dual optimal, while there is no duality gap. **Q.E.D.**

Proposition 6.1.6: (Saddle Point Theorem) A pair (x^*, μ^*) is an optimal solution-geometric multiplier pair if and only if $x^* \in X$, $\mu^* \geq 0$, and (x^*, μ^*) is a saddle point of the Lagrangian, in the sense that

$$L(x^*, \mu) \leq L(x^*, \mu^*) \leq L(x, \mu^*), \qquad \forall\, x \in X, \; \mu \geq 0. \qquad (6.9)$$

Proof: If (x^*, μ^*) is an optimal solution-geometric multiplier pair, then x^* is primal optimal and μ^* is dual optimal, so that $x^* \in X$, $\mu^* \geq 0$, and Eq. (6.7) holds, thereby proving the right-hand side of Eq. (6.9). Furthermore, for all $\mu \geq 0$, using the fact $g(x^*) \leq 0$, we have $L(x^*, \mu) \leq f(x^*)$ and in view of Eqs. (6.7) and (6.8), we obtain

$$L(x^*, \mu) \leq f(x^*) = L(x^*, \mu^*),$$

proving the left-hand side of Eq. (6.9).

Conversely, assume that $x^* \in X$, $\mu^* \geq 0$, and Eq. (6.9) holds. We will show that the four conditions (6.5)-(6.8) of Prop. 6.1.5 hold. We have

$$\sup_{\mu \geq 0} L(x^*, \mu) = \sup_{\mu \geq 0} \left\{ f(x^*) + \sum_{j=1}^{r} \mu_j g_j(x^*) \right\} = \begin{cases} f(x^*) & \text{if } g(x^*) \leq 0, \\ \infty & \text{otherwise.} \end{cases}$$

Therefore, from the left-hand side of Eq. (6.9), we obtain that $g(x^*) \leq 0$, and $f(x^*) = L(x^*, \mu^*)$. Thus, Eqs. (6.5) and (6.6) hold, and we have $\sum_{j=1}^{r} \mu_j^* g_j(x^*) = 0$. Since $\mu^* \geq 0$ and $g(x^*) \leq 0$, we obtain $\mu_j^* g_j(x^*) = 0$ for all j, and taking into account the right-hand side of Eq. (6.9), we see that Eq. (6.7) holds. From Prop. 6.1.5 it follows that (x^*, μ^*) is an optimal solution-geometric multiplier pair. **Q.E.D.**

Infeasible or Unbounded Primal Problem

We now consider what happens when our standing Assumption 5.1.1 (feasibility and boundedness) does not hold.

Suppose that the primal problem is unbounded, i.e., $f^* = -\infty$. Then, it is seen that the proof of the weak duality theorem still applies and that $q(\mu) = -\infty$ for all $\mu \geq 0$. As a result the dual problem is infeasible.

Suppose now that X is nonempty but the primal problem is infeasible, i.e., the set $\{x \in X \mid g(x) \leq 0\}$ is empty. Then, by convention, we write

$$f^* = \infty.$$

The dual function $q(\mu) = \inf_{x \in X} L(x, \mu)$ satisfies $q(\mu) < \infty$ for all μ, but it is possible that $q^* = \infty$, in which case the dual problem is unbounded. It is also possible, however, that $q^* < \infty$ or even that $q^* = -\infty$, as the examples of Fig. 6.1.8 show. For linear and quadratic programs, it will be shown in Section 6.2 that $-\infty < f^* < \infty$ implies $f^* = q^*$. However, even for linear programs, it is possible that both the primal and the dual are infeasible, i.e., $f^* = \infty$ and $q^* = -\infty$ (see Exercise 6.1.5).

6.1.4 Treatment of Equality Constraints

The theory developed so far in this section can be extended to handle additional equality constraints of the form $h_i(x) = 0$, $i = 1, \ldots, m$. A constraint of this type can be converted into the two inequality constraints

$$h_i(x) \leq 0, \qquad -h_i(x) \leq 0.$$

In particular, consider the problem

minimize $f(x)$

subject to $x \in X$, $\quad g_j(x) \leq 0$, $j = 1, \ldots, r$, $\quad h_i(x) = 0$, $i = 1, \ldots m$.

Sec. 6.1 Duality and Dual Problems 569

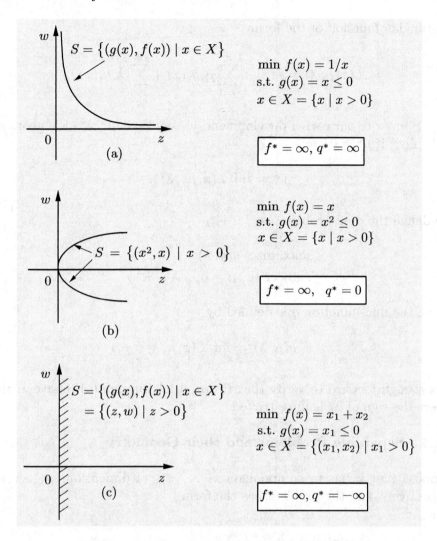

Figure 6.1.8. Examples where X is nonempty but the primal problem is infeasible. In (a), we have $f^* = q^* = \infty$. In (b), we have $f^* = \infty$ and $-\infty < q^* < \infty$. In (c), we have $f^* = \infty$ and $q^* = -\infty$.

If we replace each equality constraint by two inequalities as above, we see that a vector $(\mu_1^*, \ldots, \mu_r^*, \lambda_1^+, \lambda_1^-, \ldots, \lambda_m^+, \lambda_m^-)$ is a geometric multiplier if $\mu_j^* \geq 0$ for all j, $\lambda_i^+ \geq 0$ and $\lambda_i^- \geq 0$, for all i, and

$$f^* = \inf_{x \in X} \left\{ f(x) + \sum_{j=1}^{r} \mu_j^* g_j(x) + \sum_{i=1}^{m} (\lambda_i^+ - \lambda_i^-) h_i(x) \right\},$$

where λ_i^+ corresponds to the constraint $h_i(x) \leq 0$ and λ_i^- corresponds to the constraint $-h_i(x) \leq 0$. By replacing $\lambda_i^+ - \lambda_i^-$ with a single scalar λ_i^* with unrestricted sign in the above expression, we are led to introduce a

Lagrangian function of the form

$$L(x,\mu,\lambda) = f(x) + \sum_{j=1}^{r} \mu_j g_j(x) + \sum_{i=1}^{m} \lambda_i h_i(x).$$

Similar to our earlier development, we say that (μ^*, λ^*) is a geometric multiplier, if $\mu^* \geq 0$ and

$$f^* = \inf_{x \in X} L(x, \mu^*, \lambda^*).$$

We define the dual problem as

$$\text{maximize } q(\mu, \lambda)$$
$$\text{subject to } \mu \geq 0, \quad \lambda \in \Re^m,$$

where the dual function q is defined by

$$q(\mu, \lambda) = \inf_{x \in X} L(x, \mu, \lambda).$$

It is straightforward to verify that Props. 6.1.1 through 6.1.6 have analogs where the sign of λ_i is unrestricted.

6.1.5 Separable Problems and their Geometry

Suppose that x has m components $x_1, \ldots, x_m$ of dimensions $n_1, \ldots, n_m$, respectively, and the problem has the form

$$\text{minimize } \sum_{i=1}^{m} f_i(x_i)$$
$$\text{subject to } \sum_{i=1}^{m} g_{ij}(x_i) \leq 0, \quad j = 1, \ldots, r,$$
$$x_i \in X_i, \quad i = 1, \ldots, m,$$

where $f_i : \Re^{n_i} \mapsto \Re$ and $g_{ij} : \Re^{n_i} \mapsto \Re$ are given functions, and X_i are given subsets of $\Re^{n_i}$. We call such a problem *separable*, and we note that if the constraints $\sum_{i=1}^{m} g_{ij}(x_i) \leq 0$ were not present, it would be possible to decompose this problem into m independent subproblems. This motivates us to consider the following dual problem that involves geometric multipliers for these constraints:

$$\text{maximize } q(\mu)$$
$$\text{subject to } \mu \geq 0,$$

Sec. 6.1   Duality and Dual Problems

where the dual function is given by

$$q(\mu) = \inf_{\substack{x_i \in X_i \\ i=1,\ldots,m}} \left\{ \sum_{i=1}^{m} \left(f_i(x_i) + \sum_{j=1}^{r} \mu_j g_{ij}(x_i) \right) \right\} = \sum_{i=1}^{m} q_i(\mu),$$

and

$$q_i(\mu) = \inf_{x_i \in X_i} \left\{ f_i(x_i) + \sum_{j=1}^{r} \mu_j g_{ij}(x_i) \right\}, \qquad i = 1,\ldots,m.$$

Note that the minimization involved in the calculation of the dual function has been decomposed into m simpler minimizations. These minimizations are often conveniently done either analytically or computationally, in which case the dual function can be easily evaluated. This is the key advantageous structure of separable problems.

The separable structure is additionally helpful when the cost and/or the constraints are not convex. In particular, in this case *the duality gap turns out to be relatively small and can often be shown to diminish to zero relative to the optimal primal value as the number m of separable terms increases*. As a result, one can often obtain a near-optimal primal solution, starting from a dual-optimal solution, without resorting to costly branch-and-bound procedures (cf. Example 6.1.1).

The small duality gap size is a consequence of the structure of the set of constraint-cost pairs

$$S = \{(g(x), f(x)) \mid x \in X\}.$$

In the case of a separable problem, this set can be written as a vector sum of m sets in $\Re^{r+1}$, one for each separable term, i.e.,

$$S = S_1 + \cdots + S_m,$$

where

$$S_i = \{(g_i(x_i), f_i(x_i)) \mid x_i \in X_i\},$$

and $g_i : \Re^{n_i} \mapsto \Re^r$ is the function $g_i(x_i) = (g_{i1}(x_i), \ldots, g_{im}(x_i))$. Generally, a set S that is the vector sum of a large number of possibly nonconvex but roughly similar sets "tends to be convex" in the sense that any vector in the convex hull of S can be closely approximated by a vector in S. As a result, the duality gap tends to be relatively small, as illustrated in Fig. 6.1.9. An example is given in Exercise 6.1.7. The analytical substantiation is based on a theorem by Shapley and Folkman (see Exercise 6.1.13). In particular, it is shown in [Ber82a], Section 5.6.1, under some reasonable assumptions, that the duality gap satisfies

$$f^* - q^* \leq (r+1) \max_{i=1,\ldots,m} \rho_i,$$

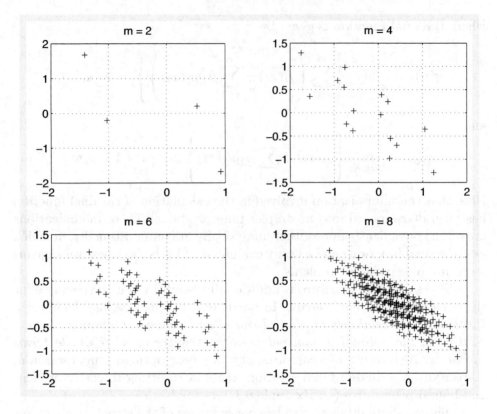

Figure 6.1.9. Illustration of the set of constraint-cost pairs

$$S = \{(g(x), f(x)) \mid x \in X\}$$

for the case of a separable problem. It can be expressed as the vector sum

$$S = S_1 + \cdots + S_m,$$

where

$$S_i = \{(g_i(x_i), f_i(x_i)) \mid x_i \in X_i\}.$$

The figure shows S for the integer programming problem

$$\text{minimize} \quad \frac{1}{m} \sum_{i=1}^{m} r_i x_i$$

$$\text{subject to} \quad \frac{1}{m} \sum_{i=1}^{m} w_i x_i \leq 0.25, \quad x_i = -1 \text{ or } 1,$$

for various values of m, and for r_i and w_i chosen randomly from the range [-2,0] and [0,2], respectively, according to a uniform distribution. It can be seen that the duality gap diminishes relative to the optimal primal value. For an analytical substantiation, based on the Shapley-Folkman theorem, see [Ber82a], Section 5.6.1, or [Ber09], Section 5.7 for a more abstract version.

where for each i, ρ_i is a nonnegative scalar that depends on the structure of the functions f_i, g_{ij}, $j = 1, \ldots, r$, and the set X_i. This estimate suggests that as $m \to \infty$, the duality gap is bounded. Thus, if $|f^*| \to \infty$ as $m \to \infty$, the "relative" duality gap $(f^* - q^*)/|f^*|$ diminishes to 0 as $m \to \infty$. See also [Ber09], Section 5.7, for a more abstract version of this analysis, which includes an estimate of the duality gap in zero-sum games.

Let us close this section with a simple but important example of a separable problem.

Example 6.1.2 (Separable Problem with a Single Constraint)

Suppose that at least A units of a certain commodity (for example, electric power) must be produced by n independent production units at minimum cost. Let x_i denote the amount produced by the ith unit and let $f_i(x_i)$ be the corresponding production cost. The problem is

$$\text{minimize} \quad \sum_{i=1}^{n} f_i(x_i)$$

$$\text{subject to} \quad \sum_{i=1}^{n} x_i \geq A, \qquad \alpha_i \leq x_i \leq \beta_i, \qquad i = 1, \ldots, n,$$

where α_i and β_i are known lower and upper bounds on the production capacity of the ith unit.

We assume that each f_i is continuously differentiable and convex, and that $\sum_{i=1}^{n} \beta_i \geq A$, so that the problem has at least one feasible solution. By Prop. 6.1.5, (x^*, μ^*) is an optimal solution-geometric multiplier pair if and only if x^* is primal feasible, $\mu^* \geq 0$, $\mu^* \left(\sum_{i=1}^{n} x_i^* - A \right) = 0$, and the Lagrangian optimality condition

$$x^* \in \arg \min_{\substack{\alpha_i \leq x_i \leq \beta_i \\ i=1,\ldots,n}} \left\{ \sum_{i=1}^{n} \bigl(f_i(x_i) - \mu^* x_i\bigr) + \mu^* A \right\}$$

holds. Thanks to the separable structure of the problem, the last condition can be decomposed into the n conditions

$$x_i^* \in \arg \min_{\alpha_i \leq x_i \leq \beta_i} \bigl\{ f_i(x_i) - \mu^* x_i \bigr\}, \qquad i = 1, \ldots, n.$$

Using the differentiability and convexity of f_i, it is seen that the preceding condition is equivalent to the following three conditions

$$\nabla f_i(x_i^*) = \mu^*, \qquad \text{if } \alpha_i < x_i^* < \beta_i, \tag{6.10}$$

$$\nabla f_i(x_i^*) \geq \mu^*, \qquad \text{if } \alpha_i = x_i^*, \tag{6.11}$$

$$\nabla f_i(x_i^*) \leq \mu^*, \qquad \text{if } x_i^* = \beta_i. \tag{6.12}$$

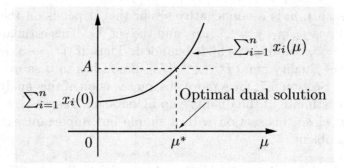

Figure 6.1.10. Graphical approach for solving the production allocation problem. We plot $\sum_{i=1}^{n} x_i(\mu)$ as an increasing function of $\mu \geq 0$ based on the solution of Eqs. (6.13)-(6.15), and we find μ^* for which

$$\sum_{i=1}^{n} x_i(\mu^*) = A.$$

The figure shows the case where the vector $x^*(0)$ corresponding to $\mu^* = 0$ satisfies $\sum_{i=1}^{n} x_i(0) < A$. If instead $\sum_{i=1}^{n} x_i(0) \geq A$, then $x(0)$ is optimal.

The gradient $\nabla f_i(x_i)$ is also called the *marginal production cost* of the ith unit. Thus, at an optimum, all units not operating at the upper or lower bound must operate at points of equal marginal production cost.

The computational solution of the problem typically involves duality. For example, in many situations the dual problem can be written in a convenient closed form and it can be solved by a one-dimensional search (cf. Appendix C). In an alternative but related approach, one may try to find a solution (x^*, μ^*) of the Lagrangian optimality conditions of Eqs. (6.10)-(6.12) and the primal feasibility condition

$$\sum_{i=1}^{n} x_i^* \geq A, \qquad \alpha_i \leq x_i^* \leq \beta_i, \quad i = 1, \ldots, n.$$

Let us assume for convenience that each f_i is strictly convex on $[\alpha_i, \beta_i]$, so that for every μ, the equation

$$\nabla f_i(x_i) = \mu$$

has a unique solution for x_i. Then the set of relations

$$\nabla f_i(x_i) = \mu, \qquad \text{if } \alpha_i < x_i < \beta_i, \tag{6.13}$$

$$\nabla f_i(x_i) \geq \mu, \qquad \text{if } \alpha_i = x_i, \tag{6.14}$$

$$\nabla f_i(x_i) \leq \mu, \qquad \text{if } x_i = \beta_i, \tag{6.15}$$

[cf. Eqs. (6.10)-(6.12)] has a unique solution, denoted $x_i(\mu)$. In view of the convexity of f_i, it is seen that $\sum_{i=1}^{n} x_i(\mu)$ increases as μ increases. Thus, we may plot $\sum_{i=1}^{n} x_i(\mu)$ versus μ. The point of interception of the resulting graph with the level A yields the optimal dual solution μ^* as shown in Fig. 6.1.10.

Sec. 6.1   Duality and Dual Problems   575

In the preceding example, we may view the dual solution method illustrated in Fig. 6.1.10 as a *coordination mechanism*. A coordinator sets the price of the commodity to be produced at the optimal dual value μ^*. The production units then respond by acting in their own self-interest, setting their production levels to the values

$$x_i^* \in \arg\min_{\alpha_i \leq x_i \leq \beta_i} \{f_i(x_i) - \mu^* x_i\}, \qquad i = 1, \ldots, n,$$

that minimize their net production cost (cost of production minus revenue from sale of the product). This interpretation of separable problem duality can be extended to the case where multiple commodities are being produced, there is a demand inequality constraint for each commodity, and a coordinator sets a separate price for each commodity.

6.1.6 Additional Issues About Duality

Mathematical programming duality is a broadly applicable subject with a rich theory, and most of the remainder of the book revolves around its analytical and algorithmic aspects. Here is a preview and brief summary of some of the topics that we will address in the sequel:

(a) *Conditions under which there is no duality gap and there exists a geometric multiplier.* Convexity of the cost function and the constraints is essentially a prerequisite for this, but additional technical conditions, similar to the constraint qualifications encountered in Chapter 4, are needed. We focus on two cases, which parallel some of the corresponding conditions discussed in Section 4.3.5.

   (1) The constraints $g_j(x) \leq 0$ are linear and the constraint set X is polyhedral. This case is discussed in Section 6.2.

   (2) The constraints $g_j(x) \leq 0$ are convex and possess a common interior point $\bar{x} \in X$, satisfying $g_j(\bar{x}) < 0$ for all j (cf. the Slater constraint qualification discussed in Section 4.3.5). This case is discussed in Section 6.3.

(b) *Specially structured dual problems.* Here, specific assumptions are made on the form of the cost functions and the constraints, and strengthened versions of duality theorems and/or other specialized results are accordingly obtained. A principal example is Fenchel duality, developed in Section 6.4, which relies on the theory of conjugate functions to derive an alternative framework for obtaining duality results in specially structured problems. Examples of such problems include:

   (1) *Conic programming.* This class of problems, discussed in Section 6.4.1, involves a constraint set of the form $X \cap C$, where X

is convex and C is a cone. Important special cases are *second order cone programming* and *semidefinite programming*, which correspond to a linear cost function, some linear constraints, and some special types of cones.

(2) *Monotropic programming.* This class of problems, discussed in Section 6.4.2, involve a separable cost and a subspace constraint, and include linear and quadratic programming problems. The duality results here are symmetric and very strong. In fact, monotropic programming problems form the largest class of nonlinear programming problems with a duality theory that is as sharp as the one for linear and quadratic programs.

(3) *Network optimization problems* with a single commodity and no side constraints. This is a practically important class, including problems such as shortest path, assignment, max-flow, etc. The duality theory here is a special case of monotropic programming duality, and is discussed in Section 6.4.3.

(c) *Exact penalty function methodology.* There is a strong and insightful relation between the nondifferentiable exact penalty functions discussed in Section 5.3 and Fenchel duality theory, to be discussed in Section 6.4. This relation provides the connecting link between a seemingly disparate set of topics relating to penalty and augmented Lagrangian methods, sensitivity, and smoothing approaches for nondifferentiable optimization. For a discussion of some of these connections, see [BNO03], Section 7.3, and [Ber15a], Section 1.5.

(d) *Discrete optimization methodology.* Discrete optimization problems are often approached via duality, as indicated earlier in Example 6.1.1. In particular, the solution of dual problems provides lower bounds to the corresponding optimal primal values, which can be used in the context of the branch-and-bound technique to be discussed in Section 6.5.

(e) *Dual algorithmic methodology.* There are several important methods for large scale optimization that rely on duality. In effect these methods try to solve the dual problem directly, while exploiting the available special structure. Several dual algorithmic approaches will be discussed in Chapter 7.

(f) *Infinite dimensional problems.* Since duality theory is based on geometrical constructions, it is not surprising that it has far-reaching application to problems with optimization variables taking values in infinite dimensional linear spaces, where convexity notions can be readily defined. Extensions of this type can be challenging and complicated, but the advanced reader can verify that the entire framework, analysis, and results presented so far in this section can be used

Sec. 6.1   Duality and Dual Problems   577

in an infinite dimensional context with essentially no modification, as long as the number of inequality constraints is finite. In particular, the definitions and derivations given in this section for the problem

$$\text{minimize } f(x)$$
$$\text{subject to } x \in X, \quad g_j(x) \leq 0, \quad j = 1, \ldots, r,$$

apply nearly verbatim to the case where x belongs to an arbitrary linear vector space rather than the Euclidean space $\Re^n$. Furthermore, while we will not go into the development of the corresponding theory, the advanced reader can verify that the proof of the geometric multiplier theorem of Prop. 6.3.1 under the Slater constraint qualification (to be given later in Section 6.3) also carries through.

EXERCISES

6.1.1

Find the sets of all optimal solutions and all geometric multipliers, and sketch the dual function for the following two-dimensional convex programming problems:

$$\text{minimize } x_1$$
$$\text{subject to } |x_1| + |x_2| \leq 1, \quad x \in X = \Re^2,$$

and

$$\text{minimize } x_1$$
$$\text{subject to } |x_1| + |x_2| \leq 1, \quad x \in X = \{x \mid |x_1| \leq 1, |x_2| \leq 1\}.$$

6.1.2

Consider the problem

$$\text{minimize } f(x) = 10x_1 + 3x_2$$
$$\text{subject to } 5x_1 + x_2 \geq 4, \quad x_1, x_2 = 0 \text{ or } 1.$$

(a) Sketch the set of constraint-cost pairs

$$\{(4 - 5x_1 - x_2, 10x_1 + 3x_2) \mid x_1, x_2 = 0 \text{ or } 1\}.$$

(b) Sketch the dual function.

(c) Solve the problem and its dual, and relate the solutions to your sketch in part (a).

6.1.3

Derive the dual of the projection problem

$$\text{minimize } \|z - x\|^2$$
$$\text{subject to } Ax = 0,$$

where the $m \times n$ matrix A and the vector $z \in \Re^n$ are given. Show that the dual problem is also a problem of projection on a subspace.

6.1.4 (www)

Use duality to show that in three-dimensional space, the (minimum) distance from the origin to a line is equal to the maximum over all (minimum) distances of the origin from planes that contain the line.

6.1.5

Show that the dual of the (infeasible) linear program

$$\text{minimize } x_1 - x_2$$
$$\text{subject to } x \in X = \{x \mid x_1 \geq 0, x_2 \geq 0\}, \quad x_1 + 1 \leq 0, \ 1 - x_1 - x_2 \leq 0,$$

is the (infeasible) linear program

$$\text{maximize } \mu_1 + \mu_2$$
$$\text{subject to } \mu_1 \geq 0, \ \mu_2 \geq 0, \ -\mu_1 + \mu_2 - 1 \leq 0, \ \mu_2 + 1 \leq 0.$$

6.1.6

Consider the optimal control problem of minimizing $\sum_{i=1}^{N-1} |u_i|^2$ subject to

$$x_{i+1} = A_i x_i + B_i u_i, \quad i = 0, \ldots, N-1,$$
$$x_0 : \text{given},$$
$$x_N \geq c,$$

where c is a given vector. Show that a dual problem is of the form

$$\text{minimize } \mu' Q \mu + \mu' d$$
$$\text{subject to } \mu \geq 0,$$

where Q is an appropriate $n \times n$ matrix (n is the dimension of x_N) and $d \in \Re^n$ is an appropriate vector.

Sec. 6.1   Duality and Dual Problems

6.1.7 (Duality Gap of the Knapsack Problem) (www)

Given objects $i = 1, \ldots, n$, with positive weights w_i and values v_i, we want to assemble a subset of the objects such that the sum of the weights of the subset does not exceed a given $A > 0$, and the sum of the values of the subset is maximized.

(a) Show that this problem can be written as

$$\text{maximize} \quad \sum_{i=1}^{n} v_i x_i$$
$$\text{subject to} \quad \sum_{i=1}^{n} w_i x_i \leq A, \qquad x_i \in \{0, 1\}, \quad i = 1, \ldots, n.$$

(b) Use a graphical procedure to solve the dual problem where a multiplier is assigned to the constraint $\sum_{i=1}^{n} w_i x_i \leq A$.

(c) Let f^* and q^* be the optimal values of the primal and dual problems, respectively. Show that

$$0 \leq q^* - f^* \leq \max_{i=1,\ldots,n} v_i.$$

(d) Consider the problem where A is multiplied by a positive integer k and each object is replaced by k replicas of itself, while the object weights and values stay the same. Let $f^*(k)$ and $q^*(k)$ be the corresponding optimal primal and dual values. Show that

$$\frac{q^*(k) - f^*(k)}{f^*(k)} \leq \frac{1}{k} \frac{\max_{i=1,\ldots,n} v_i}{f^*},$$

so that the relative value of the duality gap tends to 0 as $k \to \infty$.

6.1.8 (Sensitivity) (www)

Consider the family of problems

$$\text{minimize} \quad f(x)$$
$$\text{subject to} \quad x \in X, \qquad g_j(x) \leq u_j, \quad j = 1, \ldots, r,$$

where $u = (u_1, \ldots, u_r)$ is a vector parameterizing the right-hand side of the constraints. Given two distinct values $\bar{u}$ and $\tilde{u}$ of u, let $\bar{f}$ and $\tilde{f}$ be the corresponding optimal values, and let $\bar{\mu}$ and $\tilde{\mu}$ be corresponding geometric multipliers. Assume that $-\infty < \bar{f} < \infty$ and $-\infty < \tilde{f} < \infty$. Show that

$$(\tilde{u} - \bar{u})' \tilde{\mu} \leq \bar{f} - \tilde{f} \leq (\tilde{u} - \bar{u})' \bar{\mu}.$$

6.1.9 (Hoffman's Bound) (www)

Consider the parametric program

$$\text{minimize } \|z - x\|$$
$$\text{subject to } Ax \leq b + y, \quad x \in \Re^n,$$

where y is a parameter vector taking values in a subset Y of $\Re^m$, z is a parameter vector taking values in $\Re^n$, A is a given $m \times n$ matrix, and b is a given vector in $\Re^m$. Let $f^*(y, z)$ be the optimal value of this program and assume that we have $f^*(y, z) < \infty$ for all $y \in Y$ and $z \in \Re^n$.

(a) Show that it is possible to select for each $y \in Y$ and $z \in \Re^n$ a geometric multiplier $\mu^*(y, z)$ such that the set $\{\mu^*(y, z) \mid y \in Y, z \in \Re^n\}$ is bounded. Furthermore, we have

$$f^*(y, z) \leq \mu^*(y, z)'(Az - b - y)^+, \quad \forall \, y \in Y, \, z \in \Re^n,$$

where $(v)^+$ denotes the vector with components $\max\{0, v_i\}$ if v_i are the components of the vector v.

(b) Use part (a) to show that there exists a constant c such that

$$f^*(y, z) \leq c \|(Az - b - y)^+\|, \quad \forall \, y \in Y, \, z \in \Re^n.$$

(For a generalization involving nonlinear inequality constraints as well as a set constraint $x \in X$, see [Ber99].)

6.1.10 (Upper Bounds to the Optimal Dual Value [Ber99]) (www)

Consider the problem

$$\text{minimize } f(x)$$
$$\text{subject to } x \in X, \quad g_j(x) \leq 0, \quad j = 1, \ldots, r,$$

where X is a nonempty subset of $\Re^n$, and $f : \Re^n \mapsto \Re$, $g_j : \Re^n \mapsto \Re$ are given functions. Let f^* and q^* be the optimal primal and dual values, respectively:

$$f^* = \inf_{\substack{x \in X \\ g_j(x) \leq 0, \, j=1,\ldots,r}} f(x), \qquad q^* = \sup_{\mu \geq 0} q(\mu),$$

where $q : \Re^r \mapsto [-\infty, \infty)$ is the dual function $q(\mu) = \inf_{x \in X} L(x, \mu)$. Assume that we have two vectors x_F and x_I from X such that:

(1) x_F is feasible, i.e., $g(x_F) \leq 0$.

(2) x_I is infeasible, i.e., $g_j(x_I) > 0$ for at least one j.

(3) $f(x_I) < f(x_F)$.

Let

$$\Gamma = \inf\{\gamma \geq 0 \mid g_j(x_I) \leq -\gamma g_j(x_F), \, j = 1, \ldots, r\}.$$

Show that

$$q^* \leq \frac{\Gamma}{\Gamma + 1} f(x_F) + \frac{1}{\Gamma + 1} f(x_I),$$

and that this inequality provides a tighter upper bound to q^* than $f(x_F)$.

6.1.11 (www)

The purpose of this exercise is to illustrate how to prove that there is no duality gap for a given problem by establishing that there is no duality gap for a potentially simpler problem. Consider the problem

$$\text{minimize } f(x)$$
$$\text{subject to } x \in X, \quad g_j(x) \leq 0, \quad j = 1, \ldots, r, \qquad (6.16)$$

where X is a nonempty subset of $\Re^n$, and $f : \Re^n \mapsto \Re$, $g_j : \Re^n \mapsto \Re$ are some functions. Consider a partition of the index set $\{1, \ldots, r\}$ into two complementary subsets J and $\bar{J}$. Assume that there is a geometric multiplier $\{\mu_j^* \mid j \in J\}$ for the problem

$$\text{minimize } f(x)$$
$$\text{subject to } x \in \bar{X}, \quad g_j(x) \leq 0, \quad j \in J,$$

where

$$\bar{X} = \{x \in X \mid g_j(x) \leq 0, j \in \bar{J}\}.$$

Show that if the problem

$$\text{minimize } \left\{ f(x) + \sum_{j \in J} \mu_j^* g_j(x) \right\}$$
$$\text{subject to } x \in X, \quad g_j(x) \leq 0, \quad j \in \bar{J}, \qquad (6.17)$$

has no duality gap, the same is true for problem (6.16). Furthermore, if $\{\mu_j^* \mid j \in \bar{J}\}$ is a geometric multiplier for problem (6.17), then $\{\mu_j^* \mid j = 1, \ldots, r\}$ is a geometric multiplier for problem (6.16).

6.1.12 (Extended Representation) (www)

Consider the problem

$$\text{minimize } f(x)$$
$$\text{subject to } x \in X, \quad g_j(x) \leq 0, \quad j = 1, \ldots, r, \qquad (6.18)$$

where X is a nonempty subset of $\Re^n$, and $f : \Re^n \mapsto \Re$, $g_j : \Re^n \mapsto \Re$ are some functions. Assume that X has the form

$$X = \tilde{X} \cap \{x \mid \tilde{g}_j(x) \leq 0, j = 1, \ldots, \tilde{r}\},$$

where $\tilde{X}$ is a nonempty subset of $\Re^n$ and $\tilde{g}_j : \Re^n \mapsto \Re$ are some functions. Consider the problem where the constraints $\tilde{g}_j(x) \leq 0$ are lumped together with the constraints $g_j(x) \leq 0$, and $\tilde{X}$ is the abstract constraint set:

$$\text{minimize } f(x)$$
$$\text{subject to } x \in \tilde{X}, \quad g_j(x) \leq 0, \quad j = 1, \ldots, r, \quad \tilde{g}_j(x) \leq 0, \quad j = 1, \ldots, \tilde{r},$$
$$(6.19)$$

(a) Show that if problem (6.19) has no duality gap, the same is true for problem (6.18).

(b) Show that if μ_j^*, $j = 1, \ldots, r$, and $\tilde{\mu}_j^*$, $j = 1, \ldots, \tilde{r}$, are geometric multipliers for problem (6.19), then μ_j^*, $j = 1, \ldots, r$, are geometric multipliers for problem (6.18).

6.1.13 (Shapley-Folkman Theorem)

Let X_i, $i = 1, \ldots, m$, be nonempty subsets of $\Re^n$ and let $X = X_1 + \cdots + X_m$. Assume that $m > n$. Then every vector $x \in \text{conv}(X)$ can be represented as $x = x_1 + \cdots + x_m$, where $x_i \in \text{conv}(X_i)$ for all $i = 1, \ldots, m$, and $x_i \in X_i$ for at least $m - n$ indices i. Abbreviated proof ([Zho93]): Let $x = \sum_{i=1}^{m} y_i$ with $y_i \in \text{conv}(X_i)$ and $x_i \in X_i$, and let

$$y_i = \sum_{j=1}^{t_i} a_{ij} y_{ij},$$

where $a_{ij} > 0$, $\sum_{j=1}^{t_i} a_{ij} = 1$, and $y_{ij} \in X_i$ for all pairs (i, j). Consider the following vectors of $\Re^{n+m}$:

$$z' = (x', 1, 1, \ldots, 1),$$

$$z'_{1j} = (y'_{1j}, 1, 0, \ldots, 0),$$

$$\cdots$$

$$z'_{mj} = (y'_{mj}, 0, 0, \ldots, 1),$$

so that $z = \sum_{i=1}^{m} \sum_{j=1}^{t_i} a_{ij} z_{ij}$. Use Caratheodory's theorem for cones [Prop. B.6(a), in Appendix B] to write $z = \sum_{i=1}^{m} \sum_{j=1}^{t_i} b_{ij} z_{ij}$, where all the coefficients b_{ij} are nonnegative and at most $n+m$ of them are strictly positive. In particular, we have

$$x = \sum_{i=1}^{m} \sum_{j=1}^{t_i} b_{ij} x_{ij}, \qquad \sum_{j=1}^{t_i} b_{ij} = 1, \qquad \forall\ i = 1, \ldots, m.$$

Let $x_i = \sum_{j=1}^{t_i} b_{ij} y_{ij}$, so that $x = x_1 + \cdots + x_m$. Since for each i, at least one of $b_{i1}, \ldots, b_{it_i}$ must be positive and at most $n + m$ of the b_{ij} are positive, it follows that for at least $m - n$ indices i, we have $b_{ik} = 1$ for some k and $b_{ij} = 0$ for all $j \neq k$.

6.2 CONVEX COST – LINEAR CONSTRAINTS

We saw in Sections 4.3.5 and 4.4 that problems with differentiable cost and linear constraints are special because they possess Lagrange multipliers, even for local minima that are not regular points. We traced this fact

Sec. 6.2   Convex Cost – Linear Constraints   583

to properties of polyhedral convexity and we discussed a connection with Farkas' lemma. It turns out that these same properties are responsible for powerful duality results for the linearly constrained problem

$$\text{minimize } f(x)$$
$$\text{subject to } x \in X, \quad e_i'x - d_i = 0, \quad i = 1, \ldots, m, \qquad (6.20)$$
$$a_j'x - b_j \leq 0, \quad j = 1, \ldots, r,$$

where $f : \Re^n \mapsto \Re$ is a convex (not necessarily differentiable) function and X is a polyhedral subset of $\Re^n$. Here is the main result of this section.

Proposition 6.2.1: (Strong Duality Theorem - Linear Constraints) Let the cost function f of problem (6.20) be convex over $\Re^n$ and let the set X be polyhedral. Assume that the optimal value f^* is finite. Then there is no duality gap and there exists at least one geometric multiplier.

Before we prove the proposition, let us note that we have already given a proof for the special case where f is not only convex but also continuously differentiable over $\Re^n$, and there exists a primal optimal solution; see Prop. 4.4.2 in Section 4.4.2. On the other hand, the proof of Prop. 6.2.1 in the general case, to be given shortly, is considerably more challenging.

Let us also note that convexity of f over X is not enough for Prop. 6.2.1 to hold; it is essential that f be convex over the entire space $\Re^n$, as the following example shows.

Example 6.2.1 (A Convex Problem with a Duality Gap)

Consider the two-dimensional problem

$$\text{minimize } f(x)$$
$$\text{subject to } x_1 = 0, \quad x \in X = \{x \mid x \geq 0\},$$

where

$$f(x) = e^{-\sqrt{x_1 x_2}}, \quad \forall \, x \in X,$$

and $f(x)$ is arbitrarily defined for $x \notin X$. Here it can be verified that f is convex over X (its Hessian is positive definite in the interior of X). Since for feasibility, we must have $x_1 = 0$, we see that $f^* = 1$. On the other hand, for all $\mu \geq 0$ we have

$$q(\mu) = \inf_{x \geq 0} \left\{ e^{-\sqrt{x_1 x_2}} + \mu x_1 \right\} = 0,$$

since the expression in braces is nonnegative for $x \geq 0$ and can approach zero by taking $x_1 \to 0$ and $x_1 x_2 \to \infty$. It follows that $q^* = 0$. Thus, there is a duality gap, $f^* - q^* = 1$. The difficulty here is that $f(x)$ is not defined as a convex function over $\Re^2$.

Proposition 6.2.1 asserts that the dual problem has an optimal solution when the optimal value is finite and the constraint set is nonempty and polyhedral. If this constraint set is also bounded, then the primal problem has an optimal solution by Weierstrass' theorem (Prop. A.8 in Appendix A), since f is a convex function and therefore is also continuous (Prop. B.9 in Appendix B). In the important special case of a linear or a convex quadratic program, however, this boundedness assumption is not needed to guarantee that the primal problem has an optimal solution, as has been shown in Prop. 3.1.5. We pointed this out at the end of Section 4.4, and we state it here as a proposition for completeness.

Proposition 6.2.2: (Linear and Quadratic Programming Duality) Let the cost function f of problem (6.20) be convex quadratic and let the set X be polyhedral. Assume that the optimal value f^* is finite. Then, the primal and dual problems have optimal solutions, and there is no duality gap.

Proof: Combine Props. 6.2.1 and 3.1.3. **Q.E.D.**

Proof of Duality Theorem for Linear Constraints

The proof of Prop. 6.2.1 is based on a version of Farkas' lemma, where some of the linearity assumptions are relaxed. The lemma uses the notion of *relative interior* of a convex set, discussed in Section B.1 of Appendix B. In particular, if C is a convex subset of $\Re^n$, and aff(C) is the affine hull of C (the intersection of all linear manifolds containing C), the relative interior of C, denoted ri(C), is the set of all $x \in C$ for which there exists an $\epsilon > 0$ such that all $z \in$ aff(C) with $\|z - x\| < \epsilon$ are contained in C, i.e., ri(C) is the interior of C relative to aff(C). Every nonempty convex set has a nonempty relative interior [Prop. B.7(b) in Appendix B].

Proposition 6.2.3: (Nonlinear Farkas' Lemma) Let C be a convex subset of $\Re^n$ and let $F : C \mapsto \Re$ be a function that is convex over C. Let also $a_j \in \Re^n$ and $b_j \in \Re$, $j = 1, \ldots, r$, be given vectors and scalars, respectively. Assume that the set

$$S = \{x \mid a_j'x \leq b_j, \ j = 1, \ldots, r\} \qquad (6.21)$$

Sec. 6.2   Convex Cost – Linear Constraints   585

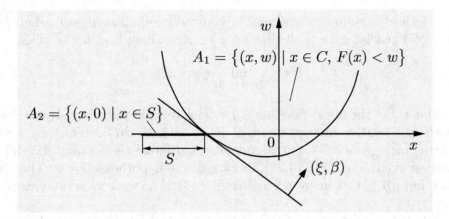

Figure 6.2.1. Separating hyperplane argument used for the proof of the nonlinear Farkas' lemma (Prop. 6.2.3).

contains a vector in the relative interior of C, and that

$$F(x) \geq 0, \quad \forall\, x \in S \cap C.$$

Then, there exist scalars $\mu_j \geq 0$, $j = 1, \ldots, r$, such that

$$F(x) + \sum_{j=1}^{r} \mu_j(a_j' x - b_j) \geq 0, \quad \forall\, x \in C. \qquad (6.22)$$

Proof: Consider the convex subsets A_1 and A_2 of $\Re^{n+1}$ defined by

$$A_1 = \{(x,w) \mid x \in C,\ F(x) < w\}, \qquad A_2 = \{(x,0) \mid x \in S\},$$

(cf. Fig. 6.2.1). In view of the hypothesis $F(x) \geq 0$ for all $x \in S \cap C$, we see that A_1 and A_2 are disjoint, while A_2 is polyhedral. Hence, by the proper separation theorem [Prop. B.14(b) of Appendix B], there exists a hyperplane that separates A_1 and A_2, and does not contain A_1, i.e., a vector $(\xi, \beta) \neq (0, 0)$ such that

$$\xi' z \leq \xi' y + \beta w, \quad \forall\, (y, w) \in A_1,\ z \in S, \qquad (6.23)$$

$$\inf_{(y,w) \in A_1} \{\xi' y + \beta w\} < \sup_{(y,w) \in A_1} \{\xi' y + \beta w\}. \qquad (6.24)$$

In view of the definition of A_1, it is seen that $\beta \geq 0$, since otherwise it would be possible to decrease the right-hand side in Eq. (6.23) without bound. We will now show that $\beta > 0$.

Indeed, assume to arrive at a contradiction, that $\beta = 0$. Let $\bar{x} \in S \cap \text{ri}(C)$ and let $\bar{w}$ be such that $(\bar{x}, \bar{w}) \in A_1$. Then, Eq. (6.23) yields

$$\xi'\bar{x} \le \inf_{(y,w)\in A_1} \xi'y \le \xi'\bar{x}.$$

It follows that the linear function $\xi'y$ attains its minimum over A_1 at $(\bar{x}, \bar{w})$, which is a relative interior point of A_1 since $\bar{x} \in \text{ri}(C)$. Thus, by Prop. B.18(a) of Appendix B, $\xi'y$ is constant and equal to $\xi'\bar{x}$ over A_1. This, however, contradicts Eq. (6.24), in view of the hypothesis $\beta = 0$. Thus, we must have $\beta > 0$, and by normalizing (ξ, β) if necessary, we may assume that $\beta = 1$.

Using the definition (6.21) of the set S and Eq. (6.23), we obtain

$$\sup_{a'_j x - b_j \le 0,\ j=1,\ldots,r} \xi'x \le \inf_{x \in C} \big\{ F(x) + \xi'x \big\}. \tag{6.25}$$

The linear program in the left-hand side above has finite optimal value, since S is nonempty, so by Prop. 3.1.5, it has an optimal solution x^*. By the Lagrange multiplier theorem for linear constraints (Prop. 4.4.1), it follows that there exist nonnegative scalars $\mu_1, \ldots, \mu_r$ such that

$$\xi = \sum_{j=1}^{r} \mu_j a_j, \tag{6.26}$$

$$\mu_j = 0, \qquad \forall\ j \text{ with } a'_j x^* < b_j. \tag{6.27}$$

Taking the inner product of both sides of Eq. (6.26) with x^* and using Eq. (6.27), we obtain

$$\xi'x^* = \sum_{j=1}^{r} \mu_j a'_j x^* = \sum_{j=1}^{r} \mu_j b_j. \tag{6.28}$$

Using Eqs. (6.26) and (6.28) in Eq. (6.25), we obtain

$$\sum_{j=1}^{r} \mu_j b_j \le \inf_{x \in C} \left\{ F(x) + \left(\sum_{j=1}^{r} \mu_j a_j \right)' x \right\}$$

or equivalently,

$$0 \le \inf_{x \in C} \left\{ F(x) + \sum_{j=1}^{r} \mu_j (a'_j x - b_j) \right\}.$$

This proves the desired relation (6.22). **Q.E.D.**

We are now ready to prove Prop. 6.2.1.

Sec. 6.2   Convex Cost – Linear Constraints

Proof of Prop. 6.2.1: Without loss of generality, we assume that there are no equality constraints, so we are dealing with the problem

$$\text{minimize } f(x)$$
$$\text{subject to } x \in X, \quad a'_j x - b_j \leq 0, \quad j = 1, \ldots, r,$$

(each equality constraint can be converted into two inequality constraints, as discussed in Section 6.1.4). Let X be expressed in terms of linear inequalities as

$$X = \{x \mid a'_j x - b_j \leq 0, \ j = r+1, \ldots, p\},$$

where p is an integer with $p > r$. By applying the nonlinear Farkas' lemma (Prop. 6.2.3) with

$$C = \Re^n, \quad S = \{x \mid a'_j x - b_j \leq 0, \ j = 1, \ldots, p\}, \quad F(x) = f(x) - f^*,$$

we see that there exist $\mu_1, \ldots, \mu_p$ with $\mu_j \geq 0$ for all j, such that

$$f^* \leq f(x) + \sum_{j=1}^{p} \mu_j (a'_j x - b_j), \quad \forall \, x \in \Re^n.$$

Moreover for $x \in X$, we have

$$\mu_j(a'_j x - b_j) \leq 0, \quad \forall \, j = r+1, \ldots, p,$$

so the preceding two relations yield

$$f^* \leq f(x) + \sum_{j=1}^{r} \mu_j (a'_j x - b_j), \quad \forall \, x \in X,$$

from which

$$f^* \leq \inf_{x \in X} L(x, \mu) = q(\mu) \leq q^*.$$

Using the weak duality theorem (Prop. 6.1.3), it follows that μ is a geometric multiplier and that there is no duality gap. **Q.E.D.**

We finally note a version of the strong duality theorem of Prop. 6.2.1, which relaxes the requirement that X be a polyhedral set (see Exercise 6.2.2). Instead, X is assumed to be the intersection of a polyhedral set and a general convex set C. However, there is an additional assumption that there exists a feasible solution, which is a relative interior point of C (this is needed in order to fulfill the relative interior point condition of the nonlinear Farkas' lemma).

EXERCISES

6.2.1

Consider the problem

$$\text{minimize} \sum_{i=0}^{m} f_i(x)$$
$$\text{subject to } x \in X_i, \quad i = 0, 1, \ldots, m,$$

where $f_i : \Re^n \mapsto \Re$ are convex functions and X_i are bounded polyhedral subsets of $\Re^n$ with nonempty intersection. Show that a dual problem is given by

$$\text{maximize} \quad q_0(\lambda_1 + \cdots + \lambda_m) + \sum_{i=1}^{m} q_i(\lambda_i)$$
$$\text{subject to } \lambda_i \in \Re^n, \quad i = 1, \ldots, m,$$

where the functions $q_i : \Re^n \mapsto \Re$ are given by

$$q_0(\lambda) = \min_{x \in X_0} \{f_0(x) - \lambda' x\},$$

$$q_i(\lambda) = \min_{x \in X_i} \{f_i(x) + \lambda' x\}, \quad i = 1, \ldots, m.$$

Show also that the primal and dual problems have optimal solutions, and that there is no duality gap.

6.2.2 (A Stronger Version of the Duality Theorem) (www)

Show that the conclusions of Prop. 6.2.1 hold if in problem (6.20), the assumption that X is a polyhedral set and $f : \Re^n \mapsto \Re$ is convex over $\Re^n$ is replaced by the following assumptions:

(1) X is the intersection of a polyhedral set P and a convex set C.

(2) $f : \Re^n \mapsto \Re$ is convex over C.

(3) There exists a feasible solution of the problem that belongs to the relative interior of C.

Use Example 6.2.1 to demonstrate the need for assumption (3). *Hint*: Express P as

$$P = \{x \mid a_j' x - b_j \leq 0, \ j = r+1, \ldots, p\},$$

where p is an integer with $p > r$. Apply the nonlinear Farkas' lemma (Prop. 6.2.3) and proceed as in the proof of Prop. 6.2.1.

6.3 CONVEX COST – CONVEX CONSTRAINTS

We now consider the nonlinearly constrained problem

$$\text{minimize } f(x)$$
$$\text{subject to } x \in X, \quad g_j(x) \leq 0, \quad j = 1, \ldots, r, \tag{6.29}$$

under convexity assumptions. In particular, we will assume the following.

Assumption 6.3.1: (Convexity and Interior Point) Problem (6.29) is feasible and its optimal value f^* is finite. Furthermore, the set X is a convex subset of $\Re^n$ and the functions $f : \Re^n \mapsto \Re$, $g_j : \Re^n \mapsto \Re$ are convex over X. In addition, there exists a vector $\bar{x} \in X$ such that

$$g_j(\bar{x}) < 0, \quad \forall\, j = 1, \ldots, r. \tag{6.30}$$

The interior point condition (6.30) is known as the *Slater constraint qualification* [Sla50] (compare also with the analysis of Section 4.3.5). The following is the main result of this section.

Proposition 6.3.1: (Strong Duality Theorem - Inequality Constraints) Let Assumption 6.3.1 hold for problem (6.29). Then there is no duality gap and there exists at least one geometric multiplier.

Proof: Consider the subset of $\Re^{r+1}$ given by

$$A = \big\{(z_1, \ldots, z_r, w) \mid \text{there exists } x \in X \text{ such that}$$
$$g_j(x) \leq z_j,\ j = 1, \ldots, r,\ f(x) \leq w\big\},$$

(cf. Fig. 6.3.1). We first show that A is convex. To this end, we consider vectors $(z, w) \in A$ and $(\tilde{z}, \tilde{w}) \in A$, and we show that their convex combinations lie in A. The definition of A implies that for some $x \in X$ and $\tilde{x} \in X$, we have

$$f(x) \leq w, \quad g_j(x) \leq z_j, \quad j = 1, \ldots, r,$$
$$f(\tilde{x}) \leq \tilde{w}, \quad g_j(\tilde{x}) \leq \tilde{z}_j, \quad j = 1, \ldots, r.$$

For any $\alpha \in [0, 1]$, we multiply these relations with α and $1-\alpha$, respectively, and add. By using the convexity of f and g_j, we obtain

$$f\big(\alpha x + (1-\alpha)\tilde{x}\big) \leq \alpha f(x) + (1-\alpha)f(\tilde{x}) \leq \alpha w + (1-\alpha)\tilde{w},$$

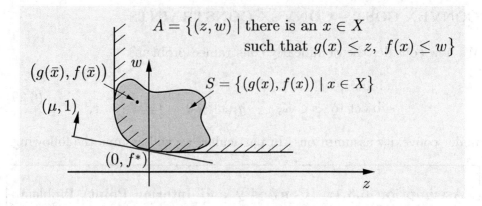

Figure 6.3.1 Illustration of the set

$$S = \{(g(x), f(x)) \mid x \in X\}$$

and the set

$$A = \{(z_1, \ldots, z_r, w) \mid \text{there exists } x \in X \text{ such that}$$
$$g_j(x) \leq z_j, \ j = 1, \ldots, r, \ f(x) \leq w\},$$

used in the proof of Prop. 6.3.1. The idea of the proof is to show that A is convex and that $(0, f^*)$ is not an interior point of A. It follows that there exists a hyperplane passing through $(0, f^*)$ and supporting A. This hyperplane is used to construct a geometric multiplier.

$$g_j\big(\alpha x + (1-\alpha)\tilde{x}\big) \leq \alpha g_j(x) + (1-\alpha)g_j(\tilde{x}) \leq \alpha z_j + (1-\alpha)\tilde{z}_j, \quad j = 1, \ldots, r.$$

In view of the convexity of X, we have $\alpha x + (1-\alpha)\tilde{x} \in X$, so these equations imply that the convex combination of (z, w) and $(\tilde{z}, \tilde{w})$, i.e., $\big(\alpha z + (1-\alpha)\tilde{z}, \alpha w + (1-\alpha)\tilde{w}\big)$, belongs to A. This proves the convexity of A.

We next observe that $(0, f^*)$ is not an interior point of A; otherwise, for some $\epsilon > 0$, the point $(0, f^* - \epsilon)$ would belong to A, contradicting the definition of f^* as the primal optimal value. Therefore, by Prop. B.11 of Appendix B, there exists a hyperplane passing through $(0, f^*)$ and containing A in one of the two corresponding halfspaces. In particular, there exists a vector $(\mu, \beta) \neq (0, 0)$ such that

$$\beta f^* \leq \beta w + \mu' z, \quad \forall \ (z, w) \in A. \tag{6.31}$$

This equation implies that

$$\beta \geq 0, \quad \mu_j \geq 0, \quad \forall \ j = 1, \ldots, r, \tag{6.32}$$

since for each $(z, w) \in A$, we have that $(z, w + \gamma) \in A$ and $(z_1, \ldots, z_j + \gamma, \ldots, z_r, w) \in A$ for all $\gamma > 0$ and j.

Sec. 6.3   Convex Cost – Convex Constraints   591

We now claim that $\beta > 0$. Indeed, if this were not so, we would have $\beta = 0$ and Eq. (6.31) would imply that $0 \leq \mu'z$ for all $(z, w) \in A$. Since, $\big(g_1(\bar{x}), \ldots, g_r(\bar{x}), f(\bar{x})\big) \in A$, we would obtain

$$0 \leq \sum_{j=1}^{r} \mu_j g_j(\bar{x}),$$

which in view of $\mu \geq 0$ [cf. Eq. (6.32)] and the assumption $g_j(\bar{x}) < 0$ for all j, implies that $\mu = 0$. This means that $(\mu, \beta) = (0, 0)$, arriving at a contradiction. Thus, we must have $\beta > 0$ and by dividing if necessary the vector (μ, β) by β, we may assume that $\beta = 1$. Thus, since $\big(g(x), f(x)\big) \in A$ for all $x \in X$, Eq. (6.31) yields

$$f^* \leq f(x) + \mu'g(x), \qquad \forall\, x \in X. \tag{6.33}$$

Taking the infimum over $x \in X$ and using the fact $\mu \geq 0$, we obtain

$$f^* \leq \inf_{x \in X} \big\{f(x) + \mu'g(x)\big\} = q(\mu) \leq q^*,$$

where q^* is the optimal dual value. From the weak duality theorem (Prop. 6.1.3), it follows that μ is a geometric multiplier for problem (6.29) and that there is no duality gap. **Q.E.D.**

The one-dimensional problem

$$\text{minimize } f(x) = x$$
$$\text{subject to } g(x) = x^2 \leq 0, \qquad x \in X = \Re,$$

[cf. Fig. 6.1.4(a)], for which there is no geometric multiplier, shows that the interior point condition of Assumption 6.3.1 is necessary in Prop. 6.3.1. In this problem there is no duality gap. Exercise 6.3.9 provides a two-dimensional example where the interior point condition is violated and there is a duality gap.

Mixed Convex and Linear Constraints

Consider now the case where in addition to the convex inequality constraints in problem (6.29), there are linear constraints

$$\text{minimize } f(x)$$
$$\text{subject to } x \in X, \ e_i'x - d_i = 0, \ i = 1, \ldots, m,$$
$$g_j(x) \leq 0, \ j = 1, \ldots, \bar{r}, \quad a_j'x - b_j \leq 0, \ j = \bar{r}+1, \ldots, r. \tag{6.34}$$

Even after each equality constraint $e_i'x - d_i = 0$ is converted to the two inequality constraints $e_i'x - d_i \leq 0$ and $-e_i'x + d_i \leq 0$, Prop. 6.3.1 cannot be applied to this problem, because the interior point condition of Assumption 6.3.1 is violated. To cover this case, we develop a related result, which involves the following assumption.

Assumption 6.3.2: (For the Mixed Constraints Problem (6.34))
The optimal value f^* of problem (6.34) is finite, and the following hold:

(1) The set X is the intersection of a polyhedral set and a convex set C.

(2) The functions $f : \Re^n \mapsto \Re$ and $g_j : \Re^n \mapsto \Re$ are convex over C.

(3) There exists a feasible vector $\bar{x}$ such that $g_j(\bar{x}) < 0$ for all $j = 1, \ldots, \bar{r}$.

(4) There exists a vector that satisfies the linear constraints [but not necessarily the constraints $g_j(x) \leq 0$, $j = 1, \ldots, \bar{r}$], and belongs to X and to the relative interior of C.

Proposition 6.3.2: (Strong Duality Theorem - Mixed Convex and Linear Constraints) Let Assumption 6.3.2 hold for problem (6.34). Then there is no duality gap and there exists at least one geometric multiplier.

Proof: Using Prop. 6.3.1, we argue that there exist $\mu_j^* \geq 0$, $j = 1, \ldots, \bar{r}$, such that

$$f^* = \inf_{\substack{x \in X,\, a_j'x - b_j \leq 0,\, j=\bar{r}+1,\ldots,r \\ e_i'x - d_i = 0,\, i=1,\ldots,m}} \left\{ f(x) + \sum_{j=1}^{\bar{r}} \mu_j^* g_j(x) \right\}.$$

Then we apply the result of Exercise 6.2.2 to the minimization problem in the right-hand side of the above equation to show that there exist λ_i^*, $i = 1, \ldots, m$, and $\mu_j^* \geq 0$, $j = \bar{r} + 1, \ldots, r$, such that

$$f^* = \inf_{x \in X} \left\{ f(x) + \sum_{i=1}^{m} \lambda_i^*(e_i'x - d_i) + \sum_{j=\bar{r}+1}^{r} \mu_j^*(a_j'x - b_j) + \sum_{j=1}^{\bar{r}} \mu_j^* g_j(x) \right\}.$$

Thus, the vector $(\lambda_1^*, \ldots, \lambda_m^*, \mu_1^*, \ldots, \mu_r^*)$ is a geometric multiplier. **Q.E.D.**

To see the need for the relative interior condition in Assumption 6.3.2, consider the problem of Example 6.2.1, where there is a duality gap. This problem satisfies all the conditions of Assumption 6.3.2 with $C = X$, except that the feasible set and the relative interior of X have no common point.

EXERCISES

6.3.1 (Boundedness of the Set of Geometric Multipliers) (www)

Consider problem (6.29) assuming that X is convex, f and g_j are convex over X, and f^* is finite. Show that the set of geometric multipliers is nonempty and bounded if and only if there exists an $\bar{x} \in X$ such that $g_j(\bar{x}) < 0$ for all j. *Hint*: Show that if $\bar{x} \in X$ is such that $g_j(\bar{x}) < 0$ for all j, then for every geometric multiplier μ, we have

$$\sum_{j=1}^{r} \mu_j \leq \frac{f(\bar{x}) - f^*}{\min_{j=1,\ldots,r}\{-g_j(\bar{x})\}}.$$

Conversely, assume that there is no vector $\bar{x} \in X$ such that $g_j(\bar{x}) < 0$ for all j. Use the line of proof of Prop. 6.3.1 to show that there exists a nonzero vector $\gamma \geq 0$ such that $\gamma'g(x) \geq 0$ for all $x \in X$, and conclude that the set of geometric multipliers must be unbounded.

6.3.2 (Optimality Conditions for Nonconvex Cost) (www)

Consider the problem

$$\text{minimize } f(x)$$
$$\text{subject to } x \in X, \quad g_j(x) \leq 0, \quad j = 1, \ldots, r,$$

where f is continuously differentiable, and $X \subset \Re^n$ is a closed convex set. Let x^* be a local minimum.

(a) Assume that the functions g_j are convex over X, and that there exists a vector $\bar{x} \in X$ such that $g_j(\bar{x}) < 0$ for all $j = 1, \ldots, r$. Show that there exists a vector $\mu^* \geq 0$ such that

$$x^* \in \arg\min_{x \in X} \left\{ \nabla f(x^*)'x + \sum_{j=1}^{r} \mu_j^* g_j(x) \right\}, \quad \mu_j^* g_j(x^*) = 0, \quad j = 1, \ldots, r.$$

(b) Assume that the functions g_j are continuously differentiable, and that there exists a feasible direction d of X at x^* such that

$$\nabla g_j(x^*)'d < 0, \quad \forall\, j \in A(x^*).$$

Show that there exists a vector $\mu^* \geq 0$ such that

$$x^* \in \arg\min_{x \in X} \nabla_x L(x^*, \mu^*)'x, \quad \mu_j^* g_j(x^*) = 0, \quad j = 1, \ldots, r.$$

6.3.3 (Boundedness of the Set of Lagrange Multipliers for Nonconvex Constraints) (www)

Consider the problem

$$\text{minimize } f(x)$$
$$\text{subject to } x \in X, \quad g_j(x) \leq 0, \quad j = 1, \ldots, r,$$

where f and the g_j are continuously differentiable, and $X \subset \Re^n$ is a closed convex set. Let x^* be a local minimum, and let M^* be the set of vectors $\mu \geq 0$ such that

$$x^* \in \arg\min_{x \in X} \nabla_x L(x^*, \mu)' x, \quad \mu' g(x^*) = 0.$$

Show that M^* is nonempty and bounded if and only if there exists a feasible direction d of X at x^* such that

$$\nabla g_j(x^*)' d < 0, \quad \forall\, j \in A(x^*).$$

Hint: Combine Prop. 4.3.12 and Exercise 6.3.1.

6.3.4 (Characterization of Pareto Optimality) (www)

A decisionmaker wishes to choose a vector $x \in X$, which keeps the values of *two* cost functions $f_1 : \Re^n \mapsto \Re$ and $f_2 : \Re^n \mapsto \Re$ reasonably small. Since a vector x^* minimizing simultaneously both f_1 and f_2 over X need not exist, he/she decides to settle for a *Pareto optimal solution*, i.e., a vector $x^* \in X$ with the property that there does not exist any vector $\bar{x} \in X$ that is strictly better than x^*, in the sense that either

$$f_1(\bar{x}) \leq f_1(x^*), \quad f_2(\bar{x}) < f_2(x^*)$$

or

$$f_1(\bar{x}) < f_1(x^*), \quad f_2(\bar{x}) \leq f_2(x^*).$$

(a) Show that if x^* is a vector in X, and λ_1^* and λ_2^* are two positive scalars such that

$$\lambda_1^* f_1(x^*) + \lambda_2^* f_2(x^*) = \min_{x \in X} \{\lambda_1^* f_1(x) + \lambda_2^* f_2(x)\},$$

then x^* is a Pareto optimal solution.

(b) Assume that X is convex and f_1, f_2 are convex over X. Show that if x^* is a Pareto optimal solution, then there exist non-negative scalars λ_1^*, λ_2^*, not both zero, such that

$$\lambda_1^* f_1(x^*) + \lambda_2^* f_2(x^*) = \min_{x \in X} \{\lambda_1^* f_1(x) + \lambda_2^* f_2(x)\}.$$

Hint: Consider the set

$$A = \{(z_1, z_2) \mid \text{there exists } x \in X \text{ such that } f_1(x) \leq z_1,\ f_2(x) \leq z_2\}$$

and show that it is a convex set. Use hyperplane separation arguments.

(c) Generalize the results of (a) and (b) to the case where there are m cost functions rather than two.

6.3.5 (Directional Convexity) (www)

Consider the following problem, which involves a pair of optimization variables (x, u):

$$\text{minimize } f(x, u)$$
$$\text{subject to } h(x, u) = 0, \qquad u \in U,$$

where $f : \Re^{n+s} \mapsto \Re$ and $h : \Re^{n+s} \mapsto \Re^m$ are given functions, and U is a subset of $\Re^s$. Assume that:

(1) The following subset of $\Re^{m+1}$

$$\big\{(z, w) \mid \text{there exists } (x, u) \in \Re^{n+s} \text{ such that } h(x, u) = z,\ f(x, u) \leq w,\ u \in U\big\}$$

is convex. (This property is known as *directional convexity*.)

(2) There exists an $\epsilon > 0$ such that for all $z \in \Re^m$ with $\|z\| < \epsilon$, we have $h(x, u) = z$ for some $u \in U$ and $x \in \Re^n$.

Show that there exists $\lambda \in \Re^m$ such that

$$\inf_{x \in \Re^n,\ u \in U} \big\{f(x, u) + \lambda' h(x, u)\big\} = f^*,$$

where f^* is the optimal value of the problem. Furthermore, if (x^*, u^*) is an optimal solution, we have

$$u^* \in \arg\min_{u \in U} \big\{f(x^*, u) + \lambda' h(x^*, u)\big\}.$$

If in addition, f and h are continuously differentiable with respect to x for each $u \in U$, we have

$$\nabla_x f(x^*, u^*) + \nabla_x h(x^*, u^*)\lambda = 0.$$

Hint: Follow the proof of Prop. 6.3.1.

6.3.6 (Directional Convexity and Optimal Control) (www)

Consider finding sequences $u = (u_0, u_1, \ldots, u_{N-1})$ and $x = (x_1, x_2, \ldots, x_N)$, which minimize

$$g_N(x_N) + \sum_{i=0}^{N-1} g_i(x_i, u_i),$$

subject to the constraints

$$x_{i+1} = f_i(x_i, u_i), \quad i = 0, \ldots, N-1, \qquad x_0 : \text{ given},$$

$$u_i \in U_i \subset \Re^m, \quad i = 0, \ldots, N-1.$$

We assume that the functions g_i and f_i are continuously differentiable with respect to x_i for each $u_i \in U_i$, and we also assume that the set

$$\left\{(z, w) \;\middle|\; x_{i+1} = f_i(x_i, u_i) + z_i,\ g_N(x_N) + \sum_{i=0}^{N-1} g_i(x_i, u_i) \leq w,\ u_i \in U_i \right\}$$

is convex; this is a directional convexity assumption (refer to the preceding exercise). Furthermore, there exists $\epsilon > 0$ such that for all $z = (z_0, \ldots, z_{N-1})$ with $\|z\| < \epsilon$, we have $x_{i+1} = f_i(x_i, u_i) + z_i$ for some x_{i+1}, $u_i \in U_i$, and all i. Show that if u^* and x^* are optimal, then the minimum principle

$$u_i^* \in \arg\min_{u_i \in U_i} H_i(x_i^*, u_i, p_{i+1}^*), \qquad \forall\, i = 0, \ldots, N-1,$$

holds, where

$$H_i(x_i, u_i, p_{i+1}) = g_i(x_i, u_i) + p_{i+1}' f_i(x_i, u_i)$$

is the Hamiltonian function, and the vectors $p_1^*, \ldots, p_N^*$ are obtained from the adjoint equation

$$p_i^* = \nabla_{x_i} H_i(x_i^*, u_i^*, p_{i+1}^*), \qquad i = 1, \ldots, N-1,$$

with the terminal condition

$$p_N^* = \nabla g_N(x_N^*).$$

6.3.7

Consider the problem of finding a circle of minimum radius that contains r points $y_1, \ldots, y_r$ in the plane, i.e., find x and R that minimize R subject to $\|x - y_j\| \leq R$ for all $j = 1, \ldots, r$, where x is the center of the circle under optimization.

(a) Introduce multipliers μ_j, $j = 1, \ldots, r$, for the constraints, and show that the dual problem has an optimal solution and there is no duality gap.

(b) Show that calculating the dual function at some $\mu \geq 0$ involves the calculation of a Weber point of $y_1, \ldots, y_r$ with weights $\mu_1, \ldots, \mu_r$, i.e., the solution of the problem $\min_{x \in \Re^2} \sum_{j=1}^{r} \mu_j \|x - y_j\|$ (see Exercise 1.1.6).

(c) Show that the dual problem is equivalent to a geometric problem, involving the Varignon frame with *equal* string lengths, and shown in Fig. 6.3.2 (compare with Fig. 1.1.9). This problem is to distribute one unit of weight to the r strings, so that the potential energy of the system at equilibrium is maximized.

(d) Show by example that there may exist an optimal solution μ^* of the geometric problem of part (c) and an optimal solution x^* of the Weber problem corresponding to μ^*, such that x^* is not the center of a minimum radius circle containing the points $y_1, \ldots, y_r$. Explain how this can happen.

6.3.8 (Inconsistent Systems of Convex Inequalities) (www)

Let $g_j : \Re^n \mapsto \Re$, $j = 1, \ldots, r$, be functions that are convex over the nonempty convex set $X \subset \Re^n$. Show that the system

$$g_j(x) < 0, \qquad j = 1, \ldots, r,$$

Sec. 1.1   Convex Cost - Convex Constraints   597

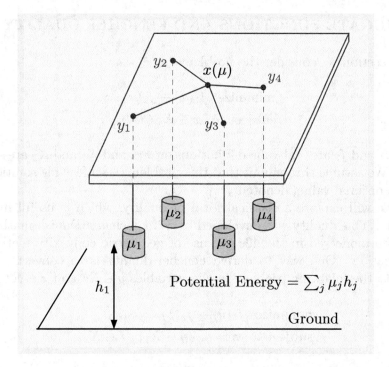

Figure 6.3.2. Varignon frame with equal string lengths (cf. Exercise 6.3.7).

has no solution within X if and only if there exists a vector $\mu \in \Re^r$ such that

$$\sum_{j=1}^{r} \mu_j = 1, \qquad \mu \geq 0,$$

$$\mu' g(x) \geq 0, \qquad \forall \, x \in X.$$

(Note that this result generalizes Gordan's theorem of the alternative; see Exercise 4.3.6.) *Hint*: Consider the problem

$$\text{minimize } y$$
$$\text{subject to } x \in X, \quad y \in \Re, \quad g_j(x) \leq y, \quad j = 1, \ldots, r.$$

6.3.9 (Duality Gap Example) www

Consider the two-dimensional problem

$$\text{minimize } e^{x_2}$$
$$\text{subject to } \|x\| - x_1 \leq 0, \qquad x \in X = \Re^2.$$

Calculate f^* and q^*, and show that there is a duality gap.

6.4 CONJUGATE FUNCTIONS AND FENCHEL DUALITY

In this section we consider the problem

$$\begin{aligned}\text{minimize } & f_1(x) - f_2(x) \\ \text{subject to } & x \in X_1 \cap X_2,\end{aligned} \quad (6.35)$$

where f_1 and f_2 are real-valued functions on $\Re^n$, and X_1 and X_2 are subsets of $\Re^n$. We assume throughout that this problem has a feasible solution and a finite optimal value, denoted f^*.†

We will explore a classical form of duality, which is useful in many contexts. This duality was developed by W. Fenchel, a Danish mathematician who pioneered in the 40s the use of geometric convexity methods in optimization. One way to derive Fenchel duality is to convert problem (6.35) to the following problem in the variables $y \in \Re^n$ and $z \in \Re^n$

$$\begin{aligned}\text{minimize } & f_1(y) - f_2(z) \\ \text{subject to } & z = y, \ y \in X_1, \ z \in X_2,\end{aligned} \quad (6.36)$$

and to dualize the constraint $z = y$. The dual function is

$$\begin{aligned}q(\lambda) &= \inf_{y \in X_1, z \in X_2} \{f_1(y) - f_2(z) + (z-y)'\lambda\} \\ &= \inf_{z \in X_2} \{z'\lambda - f_2(z)\} + \inf_{y \in X_1} \{f_1(y) - y'\lambda\}\end{aligned}$$

or, equivalently,

$$q(\lambda) = g_2(\lambda) - g_1(\lambda), \quad (6.37)$$

where the functions $g_1 : \Re^n \mapsto (-\infty, \infty]$ and $g_2 : \Re^n \mapsto [-\infty, \infty)$ are defined by

$$g_1(\lambda) = \sup_{x \in X_1} \{x'\lambda - f_1(x)\}, \quad (6.38)$$

$$g_2(\lambda) = \inf_{x \in X_2} \{x'\lambda - f_2(x)\}. \quad (6.39)$$

The function g_1 is known as the *conjugate convex function* corresponding to the pair (f_1, X_1), while the function g_2 is known as the *conjugate concave function* corresponding to (f_2, X_2) (see Fig. 6.4.1). It is

---

† Much of the literature dealing with the material of the present section treats the functions f_1 and f_2 as extended real-valued functions, which are defined over $\Re^n$ but take the value ∞ outside the sets X_1 and X_2, respectively. There are notational advantages to this format, but it is simpler for our purposes to maintain the framework of real-valued functions, while keeping explicit the (effective) domains of f_1, f_2, and other related functions.

Sec. 6.4 Conjugate Functions and Fenchel Duality

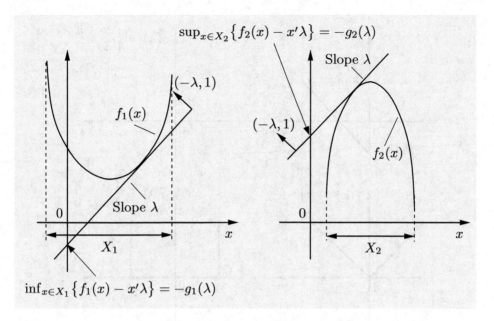

Figure 6.4.1. Visualization of the conjugate convex function

$$g_1(\lambda) = \sup_{x \in X_1} \{x'\lambda - f_1(x)\}$$

and the conjugate concave function

$$g_2(\lambda) = \inf_{x \in X_2} \{x'\lambda - f_2(x)\}.$$

The point of interception of the vertical axis with the hyperplane that has normal $(-\lambda, 1)$ and supports the graph of (f_1, X_1) is $\inf_{x \in X_1}\{f_1(x) - x'\lambda\}$, which by definition is equal to $-g_1(\lambda)$. This is illustrated in the figure on the left. The figure on the right gives a similar interpretation of $g_2(\lambda)$.

straightforward to show the convexity of the sets Λ_1 and Λ_2 over which g_1 and g_2 are finite,

$$\Lambda_1 = \{\lambda \mid g_1(\lambda) < \infty\}, \qquad \Lambda_2 = \{\lambda \mid g_2(\lambda) > -\infty\}, \tag{6.40}$$

(compare with the proof of convexity of the domain of the dual function in Prop. 6.1.2). Furthermore, g_1 is convex over Λ_1 and g_2 is concave over Λ_2. Figure 6.4.2 shows some examples of conjugate convex functions.

The dual problem is given by [cf. Eqs. (6.37)-(6.40)]

$$\begin{aligned} \text{maximize} \quad & g_2(\lambda) - g_1(\lambda) \\ \text{subject to} \quad & \lambda \in \Lambda_1 \cap \Lambda_2. \end{aligned} \tag{6.41}$$

By applying Prop. 6.1.5 to the constrained optimization problem (6.36), we obtain the following necessary and sufficient conditions for optimality.

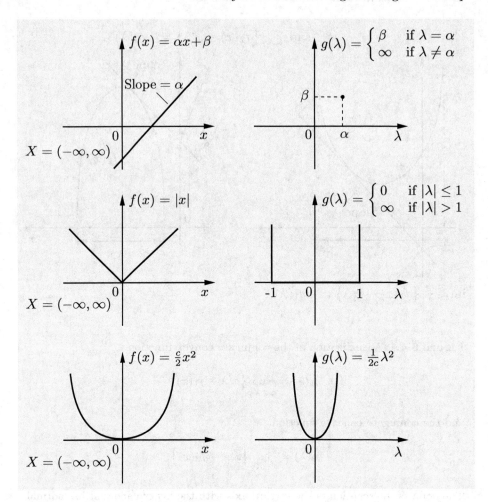

Figure 6.4.2. Some examples of conjugate convex functions.

Proposition 6.4.1: (Fenchel Optimality Conditions) A pair (x^*, λ^*) is an optimal solution pair for problems (6.35) and (6.41) if and only if

$$x^* \in X_1 \cap X_2, \quad \text{(primal feasibility)}, \tag{6.42}$$

$$\lambda^* \in \Lambda_1 \cap \Lambda_2, \quad \text{(dual feasibility)}, \tag{6.43}$$

$$x^* \in \arg\max_{x \in X_1} \{x'\lambda^* - f_1(x)\}, \quad x^* \in \arg\min_{x \in X_2} \{x'\lambda^* - f_2(x)\},$$

(Lagrangian optimality). (6.44)

The duality between the problems (6.35) and (6.41), and the Lagrangian optimality condition (6.44) are illustrated in Fig. 6.4.3. Note that

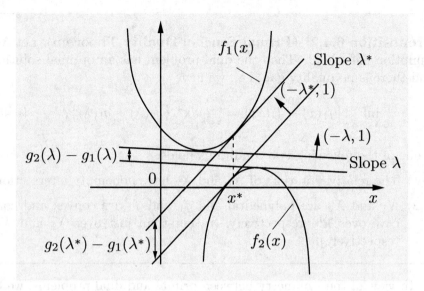

Figure 6.4.3. Illustration of the Fenchel duality theory. The dual value

$$g_2(\lambda) - g_1(\lambda)$$

becomes maximal when hyperplanes that are normal to $(-\lambda, 1)$ support the graphs of f_1 and f_2 at a common point x^*, which solves the primal problem [cf. the Lagrangian optimality condition (6.44)].

if $X_1 = \Re^n$ and f_1 is differentiable, the Lagrangian optimality condition (6.44) shows that

$$\lambda^* = \nabla f_1(x^*),$$

while if $X_2 = \Re^n$ and f_2 is differentiable, it shows that

$$\lambda^* = \nabla f_2(x^*).$$

As in Sections 6.2 and 6.3, to assert that there is no duality gap, we need some convexity assumptions.

Assumption 6.4.1: The sets X_1, X_2 are convex. Furthermore, the function f_1 is convex over X_1, and the function f_2 is concave over X_2.

Under Assumption 6.4.1, by applying Props. 6.2.1 and 6.3.2, together with the fact

$$\text{ri}(X_1 \times X_2) = \text{ri}(X_1) \times \text{ri}(X_2),$$

(cf. Prop. B.8 of Appendix B), to problem (6.36), we obtain the following:

> **Proposition 6.4.2: (Primal Fenchel Duality Theorem)** Let Assumption 6.4.1 hold. Then the dual problem has an optimal solution and there is no duality gap, i.e., we have
>
> $$\inf_{x \in X_1 \cap X_2} \{f_1(x) - f_2(x)\} = \max_{\lambda \in \Lambda_1 \cap \Lambda_2} \{g_2(\lambda) - g_1(\lambda)\}, \qquad (6.45)$$
>
> if one of the following two conditions holds:
>
> (1) The relative interiors of X_1 and X_2 have nonempty intersection.
>
> (2) X_1 and X_2 are polyhedral and f_1 and f_2 are convex and concave over $\Re^n$, respectively, (rather than just over X_1 and X_2, respectively).

In view of the symmetry between primal and dual problems, we may apply the preceding theorem to the dual problem. We will state this theorem but not fully prove it, as we will not need it in the sequel. If the epigraphs $\{(x,w) \mid f_1(x) \leq w\}$ and $\{(x,w) \mid f_2(x) \geq w\}$ are closed and convex subsets of $\Re^{n+1}$, it is possible to show that the conjugate convex function of (g_1, Λ_1) is

$$\sup_{\lambda \in \Lambda_1} \{\lambda'x - g_1(\lambda)\} = \begin{cases} f_1(x) & \text{if } x \in X_1, \\ \infty & \text{if } x \notin X_1, \end{cases}$$

and the conjugate concave function of (g_2, X_2) is

$$\inf_{\lambda \in \Lambda_2} \{\lambda'x - g_2(\lambda)\} = \begin{cases} f_2(x) & \text{if } x \in X_2, \\ -\infty & \text{if } x \notin X_2, \end{cases}$$

where Λ_1 and Λ_2 are the sets over which g_1 and g_2 take finite values; this is a major theorem of convex analysis, which states that the conjugate of the conjugate of a closed proper convex function is the original (see e.g., [Ber09], Prop. 1.6.1; we will not prove the theorem here). Thus, by applying Prop. 6.4.2, we obtain the following:

> **Proposition 6.4.3: (Dual Fenchel Duality Theorem)** Assume that the epigraphs $\{(x,w) \mid x \in X_1, f_1(x) \leq w\}$ and $\{(x,w) \mid x \in X_2, f_2(x) \geq w\}$ are closed convex sets. Then there exists an optimal primal solution and there is no duality gap [i.e., we have Eq. (6.45)] if one of the following two conditions holds:
>
> (1) The relative interiors of Λ_1 and Λ_2 have nonempty intersection.
>
> (2) Λ_1 and Λ_2 are polyhedral, and g_1 and g_2 can be extended to a real-valued convex and concave function, respectively, which are defined over the entire space $\Re^n$.

Note, that similar to the case of Eqs. (6.42)-(6.44), we have that (x^*, λ^*) are a primal and dual optimal solution pair if and only if the primal and dual feasibility conditions (6.42) and (6.43) hold, together with the alternative Lagrangian optimality condition

$$\lambda^* \in \arg\max_{\lambda \in \Lambda_1}\{\lambda' x^* - g_1(\lambda)\}, \qquad \lambda^* \in \arg\min_{\lambda \in \Lambda_2}\{\lambda' x^* - g_2(\lambda)\}. \qquad (6.46)$$

Indeed, the Lagrangian optimality conditions (6.44) and (6.46) are equivalent, since they are both equivalent to the condition

$$x^{*\prime}\lambda^* = f_1(x^*) + g_1(\lambda^*) = f_2(x^*) + g_2(\lambda^*).$$

Moreover, if $\Lambda_1 = \Re^n$ and g_1 is differentiable, the Lagrangian optimality condition (6.46) shows that

$$x^* = \nabla g_1(\lambda^*),$$

while if $\Lambda_2 = \Re^n$ and g_2 is differentiable, it shows that

$$x^* = \nabla g_2(\lambda^*).$$

An Extended Form of the Fenchel Duality Theorem

Let us consider a slightly more general form of the Fenchel duality framework. It involves the problem

$$\text{minimize} \quad f_1(x) + f_2(Ax)$$
$$\text{subject to} \quad x \in X, \quad Ax \in Z,$$

where A is an $m \times n$ matrix, $f_1 : \Re^n \mapsto \Re$ and $f_2 : \Re^m \mapsto \Re$ are convex functions, and X and Z are convex subsets of $\Re^n$ and $\Re^m$, respectively. This may be viewed as a special case of the earlier form (6.35), with $f_2(x)$ replaced by the function $-f_2(Ax)$, X_1 replaced by X, and X_2 replaced by $\{x \mid Ax \in Z\}$.

The most straightforward way to obtain the corresponding Fenchel duality theorems is to essentially repeat the derivation given earlier for the case where A is the identity. In particular, we consider the following equivalent constrained optimization problem in the variables $x \in \Re^n$ and $z \in \Re^m$:

$$\text{minimize} \quad f_1(x) + f_2(z)$$
$$\text{subject to} \quad x \in X, \quad z \in Z, \quad z = Ax.$$

Viewing this as a convex programming problem with the linear equality constraint $z = Ax$, the dual function is

$$q(\lambda) = \inf_{x \in X,\ z \in Z}\{f_1(x) + f_2(z) + \lambda'(z - Ax)\}$$
$$= \inf_{x \in X}\{f_1(x) - \lambda' Ax\} + \inf_{z \in Z}\{f_2(z) + \lambda' z\}.$$

The dual problem is to maximize $q(\lambda)$ over $\lambda \in \Re^m$. After a sign change to convert it to a minimization problem, the dual problem takes the form

$$\text{minimize} \quad g_1(A'\lambda) + g_2(-\lambda)$$
$$\text{subject to} \quad \lambda \in \Re^m,$$

where

$$g_1(A'\lambda) = \sup_{x \in X} \{\lambda' Ax - f_1(x)\}, \qquad g_2(-\lambda) = \sup_{z \in Z} \{-\lambda' z - f_2(z)\}.$$

The functions g_1 and g_2 are the convex conjugate functions corresponding to the pairs (f_1, X) and (f_2, Z), respectively. Equivalently, we can write the problem as

$$\begin{aligned}\text{minimize} \quad & g_1(A'\lambda) + g_2(-\lambda) \\ \text{subject to} \quad & A'\lambda \in \Lambda_1, \ -\lambda \in \Lambda_2,\end{aligned} \qquad (6.47)$$

where Λ_1 and Λ_2 are the sets over which g_1 and g_2 take finite values.

The Fenchel duality relation, which holds under the corresponding relative interior assumptions, states that there is no duality gap. Moreover (x^*, λ^*) is a primal and dual optimal solution pair if and only if the following Lagrangian optimality conditions hold:

$$x^* \in \arg\min_{x \in X} \{f_1(x) - x'A'\lambda^*\}, \qquad (6.48)$$

and

$$Ax^* \in \arg\min_{z \in Z} \{f_2(z) + z'\lambda^*\}. \qquad (6.49)$$

In the next few subsections, we will discuss some applications of Fenchel duality.

6.4.1 Conic Programming

An important problem structure, which can be analyzed as a special case of the Fenchel duality framework is *conic programming*. This is the problem

$$\begin{aligned}\text{minimize} \quad & f(x) \\ \text{subject to} \quad & x \in X \cap C,\end{aligned} \qquad (6.50)$$

where C is a closed convex cone in $\Re^n$, X is a convex set, and $f : \Re^n \mapsto \Re$ is a convex function over X, with epigraph $\{(x, w) \mid x \in X, f(x) \leq w\}$ that is a closed set.

Indeed, let us apply the Fenchel duality framework (6.35) with the definitions

$$f_1(x) = f(x), \quad X_1 = X, \quad f_2(x) \equiv 0, \quad X_2 = C.$$

Sec. 6.4 Conjugate Functions and Fenchel Duality

The corresponding conjugates are

$$g_1(\lambda) = \sup_{x \in X} \{\lambda'x - f(x)\}, \qquad g_2(\lambda) = \sup_{x \in C} \lambda'x = \begin{cases} 0 & \text{if } \lambda \in C^*, \\ \infty & \text{if } \lambda \notin C^*, \end{cases}$$

where

$$C^* = \{\lambda \mid \lambda'x \leq 0, \, \forall \, x \in C\}$$

is the *polar cone* of C (see Section B.3 of Appendix B). The dual problem (6.47) is

$$\begin{aligned} \text{minimize} \quad & g(\lambda) \\ \text{subject to} \quad & \lambda \in \Lambda \cap \hat{C}, \end{aligned} \tag{6.51}$$

where g is the conjugate of (f, X), Λ is the set where g takes finite values, $\Lambda = \{\lambda \mid g(\lambda) < \infty\}$, and $\hat{C}$ is the negative polar cone (also called the *dual cone* of C):

$$\hat{C} = -C^* = \{\lambda \mid \lambda'x \geq 0, \, \forall \, x \in C\}.$$

The strong duality relation $f^* = q^*$ can be written as

$$\inf_{x \in X \cap C} f(x) = - \inf_{\lambda \in \Lambda \cap \hat{C}} g(\lambda).$$

Note the symmetry between primal and dual problems.

The following proposition translates the conditions of the primal Fenchel duality theorem (Prop. 6.4.2) to guarantee that there is no duality gap and that the dual problem has an optimal solution.

Proposition 6.4.4: (Conic Duality Theorem) Assume that the optimal value of the primal conic problem (6.50) is finite, and that $\text{ri}(X) \cap \text{ri}(C) \neq \emptyset$. Then, there is no duality gap and the dual problem (6.51) has an optimal solution.

Using the symmetry of the primal and dual problems, we also obtain that there is no duality gap and the primal problem (6.50) has an optimal solution if the optimal value of the dual conic problem (6.51) is finite, if C and the epigraph $\{(x, w) \mid x \in X, f(x) \leq w\}$ are closed, and $\text{ri}(\Lambda) \cap \text{ri}(\hat{C}) \neq \emptyset$ (cf. Prop. 6.4.3). It is also possible to derive primal and dual optimality conditions by translating the optimality conditions of the Fenchel duality framework (cf. Prop. 6.4.1).

Linear-Conic Problems

An important special case of conic programming, called *linear-conic problem*, arises when X is an affine set, $X = b + S$, where b is a given vector and S is a subspace, and f is linear over X, i.e.,

$$f(x) = c'x, \qquad \forall \, x \in b + S,$$

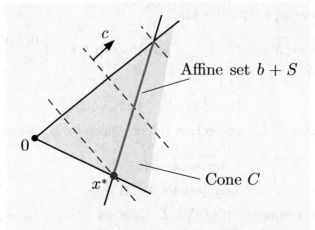

Figure 6.4.4. Illustration of a linear-conic problem: minimizing a linear function $c'x$ over the intersection of an affine set $b + S$ and a convex cone C.

where c is a given vector. Then the primal problem can be written as

$$\text{minimize} \quad c'x$$
$$\text{subject to} \quad x - b \in S, \quad x \in C; \tag{6.52}$$

see Fig. 6.4.4.

To derive the dual problem, we note that

$$g(\lambda) = \sup_{x - b \in S} (\lambda - c)'x$$
$$= \sup_{y \in S} (\lambda - c)'(y + b)$$
$$= \begin{cases} (\lambda - c)'b & \text{if } \lambda - c \in S^\perp, \\ \infty & \text{if } \lambda - c \notin S. \end{cases}$$

It can be seen that the dual problem $\min_{\lambda \in \Lambda \cap \hat{C}} g(\lambda)$ [cf. Eq. (6.51)], after discarding the superfluous term $c'b$ from the cost, can be written as

$$\text{minimize} \quad b'\lambda$$
$$\text{subject to} \quad \lambda - c \in S^\perp, \quad \lambda \in \hat{C}, \tag{6.53}$$

where $\hat{C}$ is the dual cone:

$$\hat{C} = \{\lambda \mid x'\lambda \geq 0, \, \forall \, x \in C\}.$$

The following proposition translates the conditions of Prop. 6.4.4 to the linear-conic duality context.

Sec. 6.4   Conjugate Functions and Fenchel Duality   607

> **Proposition 6.4.5: (Linear-Conic Duality Theorem)** Assume that the primal problem (6.52) has finite optimal value. Assume further that $(b + S) \cap \text{ri}(C) \neq \emptyset$. Then, there is no duality gap and the dual problem has an optimal solution.

Special Forms of Linear-Conic Problems

The primal and dual linear-conic problems have been placed in the elegant symmetric form (6.52) and (6.53). There are also other useful formats that parallel and generalize similar formats in linear programming. For example, we have the following dual problem pairs:

$$\min_{Ax=b,\ x\in C} c'x \quad \Longleftrightarrow \quad \max_{c-A'\lambda\in \hat{C}} b'\lambda, \qquad (6.54)$$

$$\min_{Ax-b\in C} c'x \quad \Longleftrightarrow \quad \max_{A'\lambda=c,\ \lambda\in \hat{C}} b'\lambda, \qquad (6.55)$$

where A is an $m \times n$ matrix, and $x \in \Re^n$, $\lambda \in \Re^m$, $c \in \Re^n$, $b \in \Re^m$.

To verify the duality relation (6.54), let $\bar{x}$ be any vector such that $A\bar{x} = b$, and let us write the primal problem on the left in the primal conic form (6.52) as

minimize $c'x$

subject to $x - \bar{x} \in \text{N}(A)$, $x \in C$,

where $\text{N}(A)$ is the nullspace of A. The corresponding dual conic problem (6.53) is to solve for μ the problem

$$\begin{aligned}&\text{minimize}\quad \bar{x}'\mu \\ &\text{subject to}\quad \mu - c \in \text{N}(A)^\perp,\quad \mu \in \hat{C}.\end{aligned} \qquad (6.56)$$

Since $\text{N}(A)^\perp$ is equal to $\text{Ra}(A')$, the range of A', the constraints of problem (6.56) can be equivalently written as $c - \mu \in -\text{Ra}(A') = \text{Ra}(A')$, $\mu \in \hat{C}$, or

$$c - \mu = A'\lambda, \quad \mu \in \hat{C},$$

for some $\lambda \in \Re^m$. Making the change of variables $\mu = c - A'\lambda$, the dual problem (6.56) can be written as

minimize $\bar{x}'(c - A'\lambda)$

subject to $c - A'\lambda \in \hat{C}$.

By discarding the constant $\bar{x}'c$ from the cost function, using the fact $A\bar{x} = b$, and changing from minimization to maximization, we see that this dual problem is equivalent to the one in the right-hand side of the duality pair (6.54). The duality relation (6.55) is proved similarly.

We next discuss two important special cases of conic programming: *second order cone programming* and *semidefinite programming*. These problems involve two different special cones, and an explicit definition of the affine set constraint. They arise in a variety of applications, and their computational difficulty in practice tends to lie between that of linear and quadratic programming on one hand, and general convex programming on the other hand.

Second Order Cone Programming

Let us consider the linear-conic problem (6.55), with the cone

$$C = \left\{ (x_1, \ldots, x_n) \;\middle|\; x_n \geq \sqrt{x_1^2 + \cdots + x_{n-1}^2} \right\},$$

which is known as the *second order cone* (see Fig. 6.4.5). The dual cone is

$$\hat{C} = \{ y \mid 0 \leq y'x, \; \forall \; x \in C \} = \left\{ y \;\middle|\; 0 \leq \inf_{\|(x_1,\ldots,x_{n-1})\| \leq x_n} y'x \right\},$$

and it can be shown that $\hat{C} = C$. This property is referred to as *self-duality* of the second order cone, and is fairly evident from Fig. 6.4.5. For a proof, we write

$$\inf_{\|(x_1,\ldots,x_{n-1})\| \leq x_n} y'x = \inf_{x_n \geq 0} \left\{ y_n x_n + \inf_{\|(x_1,\ldots,x_{n-1})\| \leq x_n} \sum_{i=1}^{n-1} y_i x_i \right\}$$

$$= \inf_{x_n \geq 0} \left\{ y_n x_n - \|(y_1, \ldots, y_{n-1})\| \, x_n \right\}$$

$$= \begin{cases} 0 & \text{if } \|(y_1, \ldots, y_{n-1})\| \leq y_n, \\ -\infty & \text{otherwise}, \end{cases}$$

where the second equality follows because the minimum of the inner product of a vector $z \in \Re^{n-1}$ with vectors in the unit ball of $\Re^{n-1}$ is $-\|z\|$. Combining the preceding two relations, we have

$$y \in \hat{C} \quad \text{if and only if} \quad 0 \leq y_n - \|(y_1, \ldots, y_{n-1})\|,$$

so $\hat{C} = C$.

The second order cone programming problem (SOCP for short) is

$$\begin{aligned} & \text{minimize} && c'x \\ & \text{subject to} && A_i x - b_i \in C_i, \; i = 1, \ldots, m, \end{aligned} \quad (6.57)$$

Sec. 6.4   Conjugate Functions and Fenchel Duality   609

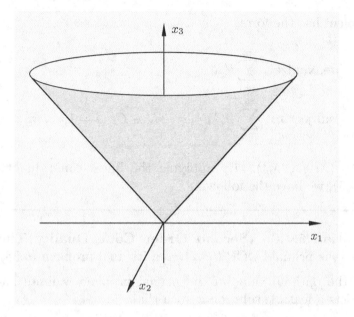

Figure 6.4.5. The second order cone

$$C = \left\{ (x_1, \ldots, x_n) \mid x_n \geq \sqrt{x_1^2 + \cdots + x_{n-1}^2} \right\},$$

in $\Re^3$.

where $x \in \Re^n$, c is a vector in $\Re^n$, and for $i = 1, \ldots, m$, A_i is an $n_i \times n$ matrix, b_i is a vector in $\Re^{n_i}$, and C_i is the second order cone of $\Re^{n_i}$. It is seen to be a special case of the primal problem in the left-hand side of the duality relation (6.55), where

$$A = \begin{pmatrix} A_1 \\ \vdots \\ A_m \end{pmatrix}, \quad b = \begin{pmatrix} b_1 \\ \vdots \\ b_m \end{pmatrix}, \quad C = C_1 \times \cdots \times C_m.$$

Note that linear inequality constraints of the form $a_i'x - b_i \geq 0$ can be written as

$$\begin{pmatrix} 0 \\ a_i' \end{pmatrix} x - \begin{pmatrix} 0 \\ b_i \end{pmatrix} \in C_i,$$

where C_i is the second order cone of $\Re^2$. As a result, linear-conic problems involving second order cones contain as special cases linear programming problems.

We now observe that from the right-hand side of the duality relation (6.55), and the self-duality relation $C = \hat{C}$, the corresponding dual linear-conic problem has the form

$$\text{maximize} \quad \sum_{i=1}^{m} b'_i \lambda_i$$
$$\text{subject to} \quad \sum_{i=1}^{m} A'_i \lambda_i = c, \quad \lambda_i \in C_i, \ i = 1, \ldots, m, \quad (6.58)$$

where $\lambda = (\lambda_1, \ldots, \lambda_m)$. By applying the linear-conic duality theorem (Prop. 6.4.5), we have the following.

Proposition 6.4.6: (Second Order Cone Duality Theorem) Consider the primal SOCP (6.57), and its dual problem (6.58).

(a) If the optimal value of the primal problem is finite and there exists a feasible solution $\bar{x}$ such that

$$A_i \bar{x} - b_i \in \text{int}(C_i), \quad i = 1, \ldots, m,$$

then there is no duality gap, and the dual problem has an optimal solution.

(b) If the optimal value of the dual problem is finite and there exists a feasible solution $\bar{\lambda} = (\bar{\lambda}_1, \ldots, \bar{\lambda}_m)$ such that

$$\bar{\lambda}_i \in \text{int}(C_i), \quad i = 1, \ldots, m,$$

then there is no duality gap, and the primal problem has an optimal solution.

Note that while the linear-conic duality theorem requires a relative interior point condition, the preceding proposition requires an interior point condition. The reason is that the second order cone has nonempty interior, so its relative interior coincides with its interior.

Semidefinite Programming

We now consider the linear-conic problem (6.54) with C being the cone of positive semidefinite and symmetric matrices. This is called the *positive semidefinite cone*. To define the problem, we view the space of symmetric $n \times n$ matrices as the space $\Re^{n^2}$ with the inner product

$$<X, Y> = \text{trace}(XY) = \sum_{i=1}^{n} \sum_{j=1}^{n} x_{ij} y_{ij}.$$

The interior of C can be seen to be the set of positive definite matrices.

The dual cone is

$$\hat{C} = \{Y \mid \text{trace}(XY) \geq 0, \forall\ X \in C\},$$

and it can be shown that $\hat{C} = C$, i.e., C is self-dual. Indeed, if $Y \notin C$, there exists a vector $v \in \Re^n$ such that

$$0 > v'Yv = \text{trace}(vv'Y).$$

Hence the positive semidefinite matrix $X = vv'$ satisfies $0 > \text{trace}(XY)$, so $Y \notin \hat{C}$ and it follows that $C \supset \hat{C}$. Conversely, let $Y \in C$, and let X be any positive semidefinite matrix. We can express X as

$$X = \sum_{i=1}^{n} \lambda_i e_i e_i',$$

where λ_i are the nonnegative eigenvalues of X, and e_i are corresponding orthonormal eigenvectors. Then,

$$\text{trace}(XY) = \text{trace}\left(Y \sum_{i=1}^{n} \lambda_i e_i e_i'\right) = \sum_{i=1}^{n} \lambda_i e_i' Y e_i \geq 0.$$

It follows that $Y \in \hat{C}$, and $C \subset \hat{C}$.

The semidefinite programming problem (SDP for short) is to minimize a linear function of a symmetric matrix over the intersection of an affine set with the positive semidefinite cone. It has the form

$$\begin{aligned}
\text{minimize} \quad & <D, X> \\
\text{subject to} \quad & <A_i, X> = b_i, \quad i = 1, \ldots, m, \quad X \in C,
\end{aligned} \tag{6.59}$$

where $D, A_1, \ldots, A_m$, are given $n \times n$ symmetric matrices, and $b_1, \ldots, b_m$, are given scalars. It is seen to be a special case of the primal problem in the left-hand side of the duality relation (6.54).

We can view SDP as a problem with linear cost, linear constraints, and a convex set constraint. Then, similar to the case of SOCP, it can be verified that the dual problem (6.53), as given by the right-hand side of the duality relation (6.54), takes the form

$$\begin{aligned}
\text{maximize} \quad & b'\lambda \\
\text{subject to} \quad & D - (\lambda_1 A_1 + \cdots + \lambda_m A_m) \in C,
\end{aligned} \tag{6.60}$$

where $b = (b_1, \ldots, b_m)$ and the maximization is over the vector $\lambda = (\lambda_1, \ldots, \lambda_m)$. By applying the linear-conic duality theorem (Prop. 6.4.5), we have the following proposition.

Proposition 6.4.7: (Semidefinite Duality Theorem) Consider the primal SDP (6.59), and its dual problem (6.60).

(a) If the optimal value of the primal problem is finite and there exists a primal-feasible solution, which is positive definite, then there is no duality gap, and the dual problem has an optimal solution.

(b) If the optimal value of the dual problem is finite and there exist scalars $\bar\lambda_1, \ldots, \bar\lambda_m$ such that $D - (\bar\lambda_1 A_1 + \cdots + \bar\lambda_m A_m)$ is positive definite, then there is no duality gap, and the primal problem has an optimal solution.

It can be shown that the SOCP can be cast as an SDP problem. Thus SDP involves a more general structure than SOCP. This is consistent with the practical observation that the latter problem is generally more amenable to computational solution.

6.4.2 Monotropic Programming

Monotropic programming problems have the generic form

$$\text{minimize} \quad \sum_{i=1}^{n} f_i(x_i) \qquad \text{(MP)}$$
$$\text{subject to} \quad x \in S, \quad x_i \in X_i, \quad i = 1, \ldots, n,$$

where

(a) X_i are convex intervals of real numbers.

(b) S is a subspace of $\Re^n$.

The following analysis extends easily to the case where S is a linear manifold, i.e. the corresponding constraint is $Ax = b$ for some $m \times n$ matrix A and some vector $b \in \Re^n$ (Exercises 6.4.1 and 6.4.2, or [Ber09], Section 5.3.5). We assume throughout that this problem has at least one feasible solution and a finite optimal value, denoted f^*.

Let us apply Fenchel duality to problem (MP) with the following identifications:

(a) The functions f_1 and f_2 in the Fenchel duality framework are

$$\sum_{i=1}^{n} f_i(x_i)$$

and the identically zero function, respectively.

(b) The sets X_1 and X_2 in the Fenchel duality framework are the Cartesian product
$$X_1 \times \cdots \times X_n$$
and the subspace S, respectively.

The corresponding conjugate concave and convex functions g_2 and g_1 in the Fenchel duality framework are

$$\inf_{x \in S} \lambda' x = \begin{cases} 0 & \text{if } \lambda \in S^\perp, \\ -\infty & \text{if } \lambda \notin S^\perp, \end{cases}$$

where $S^\perp$ is the orthogonal subspace of S, and

$$\sup_{x_i \in X_i} \left\{ \sum_{i=1}^n (x_i \lambda_i - f_i(x_i)) \right\} = \sum_{i=1}^n g_i(\lambda_i),$$

where, for each i,
$$g_i(\lambda_i) = \sup_{x_i \in X_i} \{ x_i \lambda_i - f_i(x_i) \}$$

is the conjugate convex function of (f_i, X_i). After a sign change which converts it to a minimization problem, the Fenchel dual problem (6.41) becomes

$$\text{minimize} \quad \sum_{i=1}^n g_i(\lambda_i) \qquad \text{(DMP)}$$
$$\text{subject to} \quad \lambda \in S^\perp, \quad \lambda_i \in \Lambda_i, \quad i = 1, \ldots, n,$$

where Λ_i is the domain of g_i,

$$\Lambda_i = \{ \lambda_i \mid g_i(\lambda_i) < \infty \}.$$

Here, the dual problem has an optimal solution and there is no duality gap if the functions f_i are convex and one of the following two conditions holds:

(1) The relative interiors of S and $X_1 \times \cdots \times X_n$ intersect.

(2) The intervals X_i are closed and the functions f_i are convex over the entire real line.

These conditions correspond to the two conditions for no duality gap in Prop. 6.4.2. At least one of these conditions is satisfied by the great majority of practical monotropic programs.

Finally, let us consider the Lagrangian optimality condition, which together with primal and dual feasibility, guarantees that x^* is primal optimal, λ^* is dual optimal, and there is no duality gap. This condition takes the form [cf. Eq. (6.44)]

$$x_i^* \in \arg\max_{x_i \in X_i} \{ x_i \lambda_i^* - f_i(x_i) \}, \qquad \forall \, i = 1, \ldots, n. \tag{6.61}$$

We can express this condition in terms of the right and left derivatives of the function f_i over the set X_i. Denote

$$f_i^+(x_i) = \begin{cases} \infty & \text{if } x_i \text{ is a right endpoint of } X_i, \\ \lim_{\gamma \downarrow 0} \dfrac{f_i(x_i + \gamma) - f_i(x_i)}{\gamma} & \text{otherwise,} \end{cases}$$

$$f_i^-(x_i) = \begin{cases} -\infty & \text{if } x_i \text{ is a left endpoint of } X_i, \\ \lim_{\gamma \uparrow 0} \dfrac{f_i(x_i + \gamma) - f_i(x_i)}{\gamma} & \text{otherwise.} \end{cases}$$

Then the Lagrangian optimality condition (6.61) can be written as

$$f_i^-(x_i^*) \leq \lambda_i^* \leq f_i^+(x_i^*), \qquad \forall \; i = 1, \ldots, n,$$

as illustrated in Fig. 6.4.6.

The theory of monotropic programming can be developed under more refined assumptions than the ones of the Fenchel duality theorems, and consequently sharper duality results can be obtained. In fact, in monotropic programming equality of the primal and dual optimal values can be shown under conditions as weak as for linear programs. We state the corresponding result below, and for the proof we refer to the books [Roc84] and [Ber98], which develop monotropic programming in detail.

Proposition 6.4.8: Assume that for all $i = 1, \ldots, n$, the epigraphs $\{(x_i, w) \mid x \in X_i, \, f_i(x_i) \leq w\}$ are closed and convex. Then, if there exists at least one feasible solution to the primal problem (MP), or at least one feasible solution to the dual problem (DMP), the optimal primal and dual costs are equal.

Note that part of the assertion of the above proposition is that if the primal problem is feasible but unbounded, then the dual problem is infeasible (the optimal costs of both problems are equal to $-\infty$), and that if the dual problem is feasible but unbounded, the primal problem is infeasible (the optimal costs of both problems are equal to ∞).

Extended Monotropic Programming

We will now discuss an extension of the monotropic programming problem, which allows multidimensional components x_i. In particular, we consider the problem

$$\text{minimize} \quad \sum_{i=1}^{m} f_i(x_i) \tag{6.62}$$

$$\text{subject to} \quad x \in S, \quad x_i \in X_i, \; i = 1, \ldots, m,$$

Sec. 6.4 Conjugate Functions and Fenchel Duality

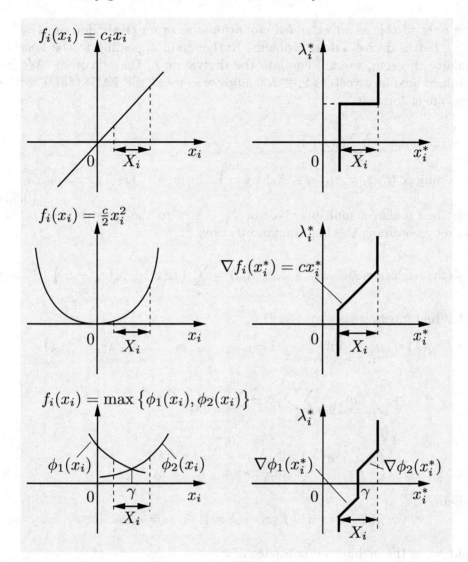

Figure 6.4.6. Illustration of the Lagrangian optimality condition

$$f_i^-(x_i^*) \leq \lambda_i^* \leq f_i^+(x_i^*), \qquad \forall\ i = 1, \ldots, n,$$

for several special cases of the function f_i. To satisfy this condition, the pair (x_i^*, λ_i^*) should lie on the thick-lined graph.

where $x \stackrel{\text{def}}{=} (x_1, \ldots, x_m) \in \Re^{n_1 + \cdots + n_m}$ is the optimization vector, with components $x_i \in \Re^{n_i}$, $i = 1, \ldots, m$, and

X_i is a convex subset of $\Re^{n_i}$.

$f_i : \Re^{n_i} \mapsto \Re$ is convex over X_i.

S is a subspace of $\Re^{n_1 + \cdots + n_m}$.

We refer to this as an *extended monotropic program* (EMP for short).

Let us derive a dual problem. Rather than appealing to the Fenchel duality theorem, we will emulate the derivation of that theorem. We introduce auxiliary vectors $z_i \in \Re^{n_i}$ and we convert the EMP (6.62) to the equivalent form

$$\text{minimize} \quad \sum_{i=1}^{m} f_i(z_i)$$
$$\text{subject to} \quad z_i = x_i, \; z_i \in X_i, \; i = 1, \ldots, m, \quad (x_1, \ldots, x_m) \in S. \tag{6.63}$$

We then assign a multiplier vector $\lambda_i \in \Re^{n_i}$ to the constraint $z_i = x_i$, thereby obtaining the Lagrangian function

$$L(x_1, \ldots, x_m, z_1, \ldots, z_m, \lambda_1, \ldots, \lambda_m) = \sum_{i=1}^{m} \big(f_i(z_i) + \lambda_i'(x_i - z_i)\big). \tag{6.64}$$

The dual function is

$$q(\lambda) = \inf_{(x_1,\ldots,x_m) \in S, \; z_i \in X_i} L(x_1, \ldots, x_m, z_1, \ldots, z_m, \lambda_1, \ldots, \lambda_m)$$

$$= \inf_{(x_1,\ldots,x_m) \in S} \sum_{i=1}^{m} \lambda_i' x_i + \sum_{i=1}^{m} \inf_{z_i \in X_i} \{f_i(z_i) - \lambda_i' z_i\}$$

$$= \begin{cases} \sum_{i=1}^{m} q_i(\lambda_i) & \text{if } (\lambda_1, \ldots, \lambda_m) \in S^\perp, \\ -\infty & \text{otherwise,} \end{cases}$$

where

$$q_i(\lambda_i) = \inf_{z_i \in X_i} \{f_i(z_i) - \lambda_i' z_i\}, \quad i = 1, \ldots, m,$$

and $S^\perp$ is the orthogonal subspace of S.

Note that since q_i can be written as

$$q_i(\lambda_i) = -\sup_{z_i \in X_i} \{\lambda_i' z_i - f_i(z_i)\},$$

it follows that $-q_i$ is the conjugate convex function of (f_i, X_i). The dual problem is

$$\text{maximize} \quad \sum_{i=1}^{m} q_i(\lambda_i) \tag{6.65}$$
$$\text{subject to} \quad (\lambda_1, \ldots, \lambda_m) \in S^\perp, \; \lambda_i \in \Lambda_i, \; i = 1, \ldots, m,$$

where $\Lambda_i = \{\lambda_i \mid q_i(\lambda_i) > -\infty\}$. Thus, with a change of sign to convert maximization to minimization, the dual problem has the same form as the primal.

Sec. 6.4   *Conjugate Functions and Fenchel Duality*   617

Since EMP in the form (6.63) can be viewed as a special case of the convex programming problem with equality constraints, it is possible to obtain optimality conditions as special cases of the corresponding results derived in Section 6.1. In particular, it can be seen that a pair (x, λ) satisfies the Lagrangian optimality condition of Prop. 6.1.5, applied to the Lagrangian (6.64), if and only if x_i attains the infimum in the equation

$$q_i(\lambda_i) = \inf_{z_i \in X_i} \{f_i(z_i) - \lambda_i' z_i\}, \qquad i = 1, \ldots, m.$$

Thus, by working out the details, we obtain the following.

Proposition 6.4.9: (EMP Optimality Conditions) There holds $-\infty < q^* = f^* < \infty$, and $(x_1^*, \ldots, x_m^*, \lambda_1^*, \ldots, \lambda_m^*)$ are an optimal primal and dual solution pair of the EMP problem if and only if

$$(x_1^* \ldots, x_m^*) \in S, \qquad (\lambda_1^* \ldots, \lambda_m^*) \in S^\perp,$$

and

$$x_i^* \in \arg\min_{x_i \in X_i} \{f_i(x_i) - x_i' \lambda_i^*\}, \quad i = 1, \ldots, m. \qquad (6.66)$$

There is also an interesting duality theory relating to EMP, which is similar to the one for the monotropic programming problem. We refer to [Ber10a] or [Ber15a], Chapter 6, for this theory, and to [Ber15a], Chapter 4, for algorithmic applications.

6.4.3 Network Optimization

We now consider an important special case of a monotropic programming problem. Consider a directed graph consisting of a set $\mathcal{N}$ of nodes and a set $\mathcal{A}$ of directed arcs. Letting x_{ij} be the flow of the arc (i, j), the problem is to minimize

$$\sum_{(i,j) \in \mathcal{A}} f_{ij}(x_{ij})$$

subject to the constraints

$$\sum_{\{j | (i,j) \in \mathcal{A}\}} x_{ij} - \sum_{\{j | (j,i) \in \mathcal{A}\}} x_{ji} = 0, \qquad \forall\, i \in \mathcal{N}, \qquad (6.67)$$

and

$$b_{ij} \leq x_{ij} \leq c_{ij}, \qquad \forall\, (i,j) \in \mathcal{A}, \qquad (6.68)$$

where b_{ij}, c_{ij} are given scalars, and $f_{ij} : \Re \mapsto \Re$ are convex functions.

The constraints (6.67), known as the *conservation of flow constraints*, specify that the total inflow into each node must be equal to the total outflow from that node. This formulation of the network optimization problem, where the right-hand side in the constraint (6.67) is zero for all nodes i, is called the *circulation* format. In a different formulation, the right-hand side of the constraint (6.67) may involve nonzero "supplies" or "demands" for some nodes i, but it is possible to convert the problem to the circulation format by introducing an artificial node to which all supplies and demands accumulate (see e.g., [Ber98]). In some formulations, the arc flow bounds (6.68) may not be present. The following development can be easily modified in this case, provided the directional derivatives of each arc cost function f_{ij} tend to $-\infty$ and ∞ as x_{ij} tends to $-\infty$ and ∞, respectively.

The problem is in the form of the monotropic program (MP), where the subspace S corresponds to the conservation of flow constraints (6.67). The orthogonal subspace $S^\perp$ is given by

$$S^\perp = \{\lambda \mid \lambda_{ij} = p_i - p_j \text{ for all } (i,j) \in \mathcal{A} \text{ and some scalars } p_m, m \in \mathcal{N}\},$$

so the dual problem is

$$\text{minimize} \sum_{(i,j) \in \mathcal{A}} g_{ij}(\lambda_{ij})$$

subject to $\lambda_{ij} = p_i - p_j$ for all $(i,j) \in \mathcal{A}$ and some scalars p_i, $i \in \mathcal{N}$,

where

$$g_{ij}(\lambda_{ij}) = \max_{b_{ij} \leq x_{ij} \leq c_{ij}} \{\lambda_{ij} x_{ij} - f_{ij}(x_{ij})\}.$$

Based on the results for monotropic programming that we outlined in Section 6.4.2, it can be seen that if the network flow problem is feasible, there exists an optimal dual solution and there is no duality gap.

For network problems, it is customary to write the dual problem in terms of the variables p_i (also known as the node *prices* or *potentials*) in the following equivalent form:

$$\text{maximize} \sum_{(i,j) \in \mathcal{A}} q_{ij}(p_i - p_j)$$

subject to no constraint on p_i, $i \in \mathcal{N}$,

where

$$q_{ij}(p_i - p_j) = \min_{b_{ij} \leq x_{ij} \leq c_{ij}} \{f_{ij}(x_{ij}) - (p_i - p_j) x_{ij}\}.$$

The form of the dual arc functions q_{ij} is illustrated in Fig. 6.4.7 for the case of a linear primal cost. Figure 6.4.8 illustrates the Lagrangian optimality condition

$$f_{ij}^-(x_{ij}^*) \leq p_i^* - p_j^* \leq f_{ij}^+(x_{ij}^*), \qquad \forall\ (i,j), \tag{6.69}$$

Sec. 6.4 Conjugate Functions and Fenchel Duality

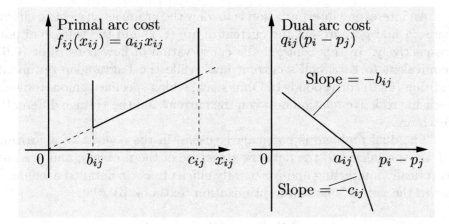

Figure 6.4.7. Illustration of primal and dual arc cost functions of the network optimization problem when the primal cost is linear.

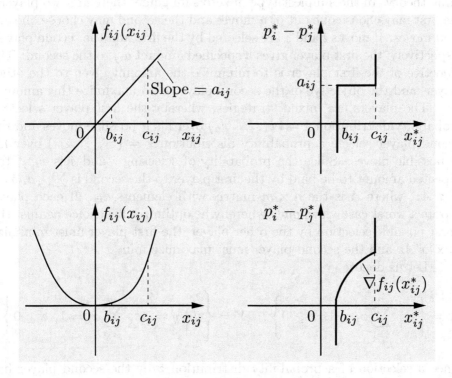

Figure 6.4.8. Illustration of the Lagrangian optimality condition

$$f_{ij}^-(x_{ij}^*) \leq p_i^* - p_j^* \leq f_{ij}^+(x_{ij}^*),$$

for some special cases of the network optimization problem. To satisfy this condition, the pair $(x_{ij}^*, p_i^* - p_j^*)$ should lie on the thick-lined graph.

in the network context.

An interesting interpretation is to view the graph as an electric circuit, where x_{ij} and p_i represent the current of arc (i,j) and the voltage of node i, respectively. In this context, the conservation of flow constraint (6.67) is equivalent to Kirchhoff's current law, while the Lagrangian optimality condition (6.69) corresponds to Ohm's law, which specifies a monotonically increasing resistive relation between the current and the voltage differential of each arc.

The dual problem is particularly useful in the context of algorithms that are specialized to the network structure of the problem, and are both theoretically interesting and practically efficient. For a detailed account we refer to the author's network optimization textbook [Ber98].

6.4.4 Games and the Minimax Theorem

As another application of Fenchel duality, we develop a classical result of game theory. In the simplest type of zero sum game, there are two players: the first may choose one out of n moves and the second may choose one out of m moves. If moves i and j are selected by the first and the second player, respectively, the first player gives a specified amount a_{ij} to the second. The objective of the first player is to minimize the amount given to the other player, and the objective of the second player is to maximize this amount.

The players use mixed strategies, whereby the first player selects a probability distribution $x = (x_1, \ldots, x_n)$ over his n possible moves and the second player selects a probability distribution $z = (z_1, \ldots, z_m)$ over his m possible moves. Since the probability of selecting i and j is $x_i z_j$, the expected amount to be paid by the first player to the second is $\sum_{i,j} a_{ij} x_i z_j$ or $x'Az$, where A is the $n \times m$ matrix with elements a_{ij}. If each player adopts a worst case viewpoint, whereby he optimizes his choice against the worst possible selection by the other player, the first player must minimize $\max_z x'Az$ and the second player must maximize $\min_x x'Az$.

Let us denote

$$X = \left\{ x \ \Big| \ \sum_{i=1}^{n} x_i = 1, \ x \geq 0 \right\}, \quad Y = \left\{ y \ \Big| \ y = Az, \ \sum_{j=1}^{m} z_j = 1, \ z \geq 0 \right\}.$$

Since a selection of a probability distribution z by the second player has the same effect on the payoff as a selection of the vector $y = Az$ from the set Y, the first player's optimal value (or guaranteed upper bound on the expected amount he must pay) is

$$\min_{x \in X} \max_{y \in Y} x'y.$$

This is also the minimum amount that the first player would pay if the second player chooses his move optimally with knowledge of the first player's

Sec. 6.4 Conjugate Functions and Fenchel Duality

probability distribution x. The second player's optimal value (or guaranteed lower bound on the expected amount he will receive) is

$$\max_{y \in Y} \min_{x \in X} x'y.$$

This is also the maximum amount that the second player would obtain if the first player chooses his move optimally with knowledge of the second player's probability distribution z.

The classical result (due to von Neumann [Neu28]) is that these two optimal values are equal, implying that there is an amount that can be meaningfully viewed as the value of the game for its participants. The following proposition uses Fenchel duality to prove this result, and to extend it to the case where X and Y are arbitrary convex and compact sets. More refined theorems may be found in [Roc70], in [BNO03], Section 2.6, and in [Ber09], Section 5.5. Exercise 4.4.5 deals with a simple but fundamental case where X and Y are polyhedral sets.

Proposition 6.4.10: (Minimax Theorem) Let X and Y be nonempty, convex, and compact subsets of $\Re^n$. Then

$$\min_{x \in X} \max_{y \in Y} x'y = \max_{y \in Y} \min_{x \in X} x'y.$$

Proof: We apply the Fenchel duality theory, with the identifications

$$f_1(x) = \max_{y \in Y} x'y, \qquad f_2(x) \equiv 0, \qquad X_1 = \Re^n, \qquad X_2 = X. \qquad (6.70)$$

For each x, the maximum above is attained by Weierstrass' theorem (Prop. A.8 in Appendix A), since Y is compact. Hence $f_1(x)$ is finite for all x and convex over $\Re^n$ [Prop. B.2(d) in Appendix B], so that Assumption 6.4.1 is satisfied. Furthermore, the relative interiors of X_1 and X_2 intersect, because $\mathrm{ri}(X_1) = \Re^n$ and $\mathrm{ri}(X_2)$ is nonempty [the relative interior of every nonempty convex set is nonempty; see Prop. B.7(b) in Appendix B]. Therefore, there is no duality gap and we have [cf. Eq. (6.45)]

$$\min_{x \in X} \max_{y \in Y} x'y = \max_{\lambda \in \Lambda_1 \cap \Lambda_2} \{g_2(\lambda) - g_1(\lambda)\}, \qquad (6.71)$$

where g_1, g_2 and Λ_1, Λ_2 are given by Eqs. (6.38)-(6.40). Using the identifications (6.70), and the definitions of the conjugates [cf. Eqs. (6.38) and (6.39)], we have

$$g_2(\lambda) = \inf_{x \in X} \lambda'x, \qquad (6.72)$$

$$g_1(\lambda) = \sup_{x \in \Re^n} \{\lambda'x - f_1(x)\}$$

$$= \sup_{x \in \Re^n} \left\{\lambda'x - \max_{y \in Y} x'y\right\}$$

$$= -\inf_{x \in \Re^n} \left\{\max_{y \in Y} x'y - x'\lambda\right\} \quad (6.73)$$

$$= -\inf_{x \in \Re^n} \max_{y \in Y} x'(y - \lambda).$$

If $\lambda \in Y$, then for every $x \in \Re^n$, we have

$$\max_{y \in Y} x'(y - \lambda) \geq 0, \qquad x \in \Re^n;$$

i.e. $\max_{y \in Y} x'(y - \lambda)$ is at least as large as zero, which is the value of $x'(y - \lambda)$ obtained for $y = \lambda$. Thus, the minimum of $\max_{y \in Y} x'(y - \lambda)$ is attained for $x = 0$ and we have

$$\inf_{x \in \Re^n} \max_{y \in Y} x'(y - \lambda) = 0, \quad \text{if } \lambda \in Y. \quad (6.74)$$

If $\lambda \notin Y$, then because Y is convex and compact, there exists a hyperplane passing through λ, not intersecting Y, and such that Y is contained in one of the corresponding halfspaces; that is, there exists $\bar{x} \in \Re^n$ such that

$$\bar{x}'y < \bar{x}'\lambda, \qquad \forall\, y \in Y.$$

Therefore, since Y is compact,

$$\max_{y \in Y} \bar{x}'(y - \lambda) < 0,$$

so by taking $x = \alpha \bar{x}$ for α arbitrarily large, we see that

$$\inf_{x \in \Re^n} \max_{y \in Y} x'(y - \lambda) = -\infty, \quad \text{if } \lambda \notin Y. \quad (6.75)$$

Combining Eqs. (6.73)-(6.75), we see that

$$g_1(\lambda) = \begin{cases} 0 & \text{if } \lambda \in Y, \\ \infty & \text{if } \lambda \notin Y. \end{cases}$$

Using this equation together with Eqs. (6.71) and (6.72), we obtain

$$\min_{x \in X} \max_{y \in Y} x'y = \max_{\lambda \in Y} \min_{x \in X} x'\lambda,$$

which by replacing λ with y, gives the desired result. **Q.E.D.**

6.4.5 The Primal Function and Sensitivity Analysis

Consider the inequality constrained problem

$$\begin{aligned}\text{minimize } & f(x) \\ \text{subject to } & x \in X, \quad g_j(x) \leq 0, \quad j = 1, \ldots, r,\end{aligned} \qquad (6.76)$$

where $f : \Re^n \mapsto \Re$ and $g_j : \Re^n \mapsto \Re$, $j = 1, \ldots, r$, are given functions, and X is a given subset of $\Re^n$. We introduce the family of problems where the right-hand side of the inequality constraints is perturbed by a vector $u = (u_1, \ldots, u_r)$, and we let $p(u)$ denote the corresponding optimal value, i.e.,

$$p(u) = \inf_{\substack{x \in X, g_j(x) \leq u_j, \\ j=1,\ldots,r}} f(x).$$

The function $p : \Re^r \mapsto [-\infty, \infty]$ is called the *primal function* of the problem. We encountered versions of this function in Section 4.2.3, in the context of sensitivity, and in Section 5.2, in the context of methods of multipliers.

Let P be the *domain* of p, i.e., the set of all u for which the constraint set $\{x \in X \mid g_j(x) \leq u_j, j = 1, \ldots, r\}$ is nonempty and, consequently, $p(u) < \infty$,

$$P = \{u \mid p(u) < \infty\}.$$

There are some interesting connections between the primal and dual functions. In particular, let us write for every $\mu \geq 0$,

$$q(\mu) = \inf_{x \in X} \left\{ f(x) + \sum_{j=1}^r \mu_j g_j(x) \right\}$$

$$= \inf_{\{(u,x)\mid u \in P, x \in X, g_j(x) \leq u_j, j=1,\ldots,r\}} \left\{ f(x) + \sum_{j=1}^r \mu_j g_j(x) \right\}$$

$$= \inf_{\{(u,x)\mid u \in P, x \in X, g_j(x) \leq u_j, j=1,\ldots,r\}} \left\{ f(x) + \sum_{j=1}^r \mu_j u_j \right\}$$

$$= \inf_{u \in P} \left\{ \inf_{\substack{x \in X, g_j(x) \leq u_j, \\ j=1,\ldots,r}} \left\{ f(x) + \sum_{j=1}^r \mu_j u_j \right\} \right\},$$

and finally

$$q(\mu) = \inf_{u \in P} \{p(u) + \mu' u\}, \qquad \forall \, \mu \geq 0. \qquad (6.77)$$

Thus, we have $q(\mu) = -\sup_{u \in P}\{(-\mu)'u - p(u)\}$, implying that

$$q(\mu) = -h(-\mu), \qquad \forall \, \mu \geq 0,$$

where h is the conjugate convex function of (p, P).

Generally, neither the primal function nor its domain have any convexity properties. However, if X is convex, f and g_j are convex over X, and $p(u) > -\infty$ for all $u \in P$, it can be shown that P is a convex set and p is convex over P. To verify this, take any $u_1, u_2 \in P$, $\alpha \in [0, 1]$, and $\epsilon > 0$, and choose $x_1, x_2 \in X$ such that $g(x_i) \leq u_i$ and $f(x_i) \leq p(u_i) + \epsilon$, $i = 1, 2$. Then, by using the convexity of X, f, and g_j, it is seen that

$$p(\alpha u_1 + (1 - \alpha)u_2) \leq \alpha p(u_1) + (1 - \alpha)p(u_2) + \epsilon.$$

This implies that $\alpha u_1 + (1 - \alpha)u_2 \in P$, so that P is convex, and also, by taking the limit as $\epsilon \to 0$, that p is convex over P.

Under the preceding convexity assumptions, we can show that the structure of the primal function around $u = 0$ determines the existence of geometric multipliers. In particular, if μ^* is a geometric multiplier for problem (6.76), then we have $q(\mu^*) = f^* = p(0)$, and from Eq. (6.77) we see that

$$p(0) \leq p(u) + u'\mu^*, \qquad \forall\, u \in \Re^r. \qquad (6.78)$$

This implies that $-\mu^*$ is a subgradient of $p(u)$ at 0 (see Appendix B for the definition and basic properties of subgradients). Conversely, if the above equation holds for some μ^*, then since $p(u)$ is monotonically nonincreasing with respect to the coordinates of u, we have $\mu^* \geq 0$ [otherwise the right side of Eq. (6.78) would be unbounded below]. Furthermore, from Eqs. (6.78) and (6.77), it follows that

$$f^* = p(0) \leq \inf_{u \in P}\{p(u) + u'\mu^*\} = q(\mu^*),$$

which, in view of the weak duality theorem (Prop. 6.1.3), implies that equality holds throughout in the preceding relation, and that μ^* is a geometric multiplier. In conclusion, if the problem is feasible, and p is convex and real-valued over P, we have

$-\mu^*$ is a subgradient of p at $u = 0$
$\iff \mu^*$ is a geometric multiplier of the problem $\min_{x \in X,\, g(x) \leq 0} f(x)$.

This result is consistent with the sensitivity analyses of Sections 4.2 and 4.3, and characterizes the subdifferential of p at $u = 0$. More generally, by replacing the constraint $g(x) \leq 0$ with a constraint $g(x) \leq u$, where $u \in P$, we see that if p is convex and real-valued over P, we have

$-\mu$ is a subgradient of p at $u \in P$
$\iff \mu$ is a geometric multiplier of the problem $\min_{x \in X,\, g(x) \leq u} f(x)$.

For further discussion of the relation between the primal and dual functions, and their subgradients, we refer to [BNO03], Sections 6.5.3 and

6.5.4, and [Ber09], Example 5.4.2. In particular, it can be shown that if p is convex and 0 is in the relative interior of P, then the geometric multiplier of minimum norm is the direction along which the directional derivative of p is minimized, so it provides the direction along which the right-hand side of the constraints should be perturbed to effect the maximum rate of improvement in cost function value.

We will present a more precise as well as more general version of the sensitivity result just mentioned. Consider problem (6.76) and the dual problem

$$\text{maximize } q(\mu) = \inf_{x \in X} \{f(x) + \mu'g(x)\} \qquad (6.79)$$
$$\text{subject to } \mu \geq 0,$$

without any assumptions of f, g_j, and X. Then, denoting $g^+(x)$ the vector of constraint violations, i.e., the vector with components

$$g_j^+(x) = \max\{0, g_j(x)\}, \qquad j = 1, \ldots, r, \ x \in X,$$

we have for all $x \in X$, and $\mu \geq 0$,

$$q(\mu) - f(x) \leq f(x) + \mu'g(x) - f(x) = \mu'g(x) \leq \mu'g^+(x) \leq \|\mu\|\,\|g^+(x)\|.$$

In particular, if μ^* is a dual optimal solution of minimum norm, by setting $\mu = \mu^*$ in the preceding relation, we have

$$f^* - f(x) \leq (f^* - q^*) + \|\mu^*\|\,\|g^+(x)\|, \qquad \forall\ x \in X, \qquad (6.80)$$

where f^* and q^* are the optimal primal and dual values. This means that if we change x within X to violate the constraints by an amount $\|g^+(x)\|$, the cost function cannot improve by an amount that exceeds the duality gap $f^* - q^*$ plus an amount that grows at a rate $\|\mu^*\|$ (per unit constraint violation). The following proposition from [BOT06] states that in fact this upper bound in cost function improvement is sharp under convexity assumptions.

Proposition 6.4.11: (Sensitivity) Let f, g_j, $j = 1, \ldots, r$, be convex, let X be closed and convex, and let f^* and q^* be finite. Assume further that the dual problem has at least one optimal solution, and let μ^* be the optimal solution of minimum norm. Then if $\mu^* \neq 0$, there exists a sequence $\{x^k\} \subset X$ of infeasible points such that

$$f(x^k) \to q^*, \qquad g^+(x^k) \to 0,$$

$$\frac{q^* - f(x^k)}{\|g^+(x^k)\|} \to \|\mu^*\|, \qquad \frac{g^+(x^k)}{\|g^+(x^k)\|} \to \frac{\mu^*}{\|\mu^*\|}.$$

We refer to [BOT06] (Prop. 5) for the proof of the proposition and related analysis. Note that in addition to showing that the inequality (6.80) is asymptotically sharp, the proposition shows that along the sequence $\{x^k\}$, the inequality constraints are violated in proportion to the size of the corresponding components of μ^*. If there is no duality gap ($f^* = q^*$), μ^* is the optimal rate of cost improvement (per unit constraint violation). If there is duality gap ($f^* > q^*$), the cost improvement is at the same rate but also includes the additive term $f^* - q^*$.

EXERCISES

6.4.1

Consider the problem

$$\text{minimize} \quad \sum_{i=1}^{n} f_i(x_i)$$
$$\text{subject to} \quad Ax = b, \quad x_i \in X_i, \quad i = 1, \ldots, n.$$

Convert this problem into a monotropic programming problem by introducing an additional vector z that is constrained by the equation $z = b$. Show that a dual problem is given by

$$\text{minimize} \quad \sum_{i=1}^{n} g_i(\lambda_i) - b'\xi$$
$$\text{subject to} \quad \lambda = A'\xi, \quad \lambda_i \in \Lambda_i, \quad i = 1, \ldots, n,$$

where $\Lambda_i = \{\lambda_i \mid g_i(\lambda_i) < \infty\}$.

6.4.2

Consider the variation of the monotropic programming problem

$$\text{minimize} \quad \sum_{i=1}^{n} f_i(x_i)$$
$$\text{subject to} \quad x \in M, \quad x_i \in X_i, \; i = 1, \ldots, n,$$

where M is a linear manifold given by $M = \bar{x} + S$, where $\bar{x}$ is a given vector and S is a subspace. Derive a dual problem by translating the origin to the vector $\bar{x}$.

6.4.3 (Smoothing of Nondifferentiabilities [Ber75c], [Ber77], [Ber82a], [Pap81], [Pol79]) (www)

A simple and effective technique to handle nondifferentiabilities in the cost or the constraints of optimization problems is to replace them by smooth approximations and then use gradient-based algorithms. This exercise develops a general technique for deriving such approximations. Let $\gamma : \Re^n \mapsto \Re$ be a convex real-valued function with conjugate convex function given by

$$g(\lambda) = \sup_{u \in \Re^n} \{u'\lambda - \gamma(u)\}.$$

For each $t \in \Re^n$, define

$$P_c(t) = \inf_{u \in \Re^n} \{\gamma(t-u) + \tfrac{c}{2}\|u\|^2\},$$

where c is a positive constant.

(a) Use the Fenchel duality theorem to show that

$$P_c(t) = \sup_{\lambda \in \Re^n} \{t'\lambda - g(\lambda) - \tfrac{1}{2c}\|\lambda\|^2\}.$$

Show also that P_c is convex and differentiable, and that the gradient $\nabla P_c(t)$ is the unique vector λ attaining the supremum in the above equation. *Hint:* Interpret $P_c(t)$ as the primal function of a suitable problem.

(b) Derive $g(\lambda)$ and $P_c(t)$ for the following functions, the first three of which are defined on $\Re$,

$$\gamma(t) = \max\{0, t\},$$
$$\gamma(t) = |t|,$$
$$\gamma(t) = \frac{s}{2}t^2, \quad (s \text{ is a positive scalar}),$$
$$\gamma(t) = \max\{t_1, \ldots, t_n\}.$$

(c) Show that $P_c(t)$ is real-valued and that for all t, $\lim_{c \to \infty} P_c(t) = \gamma(t)$, so the accuracy of the approximation can be controlled using c. Verify graphically this property for the first three functions in (b).

(d) Repeat (a) and prove results analogous to (b) and (c), when P_c is instead defined by

$$P_c(t) = \inf_{u \in \Re^n} \{\gamma(t-u) + y'u + \tfrac{c}{2}\|u\|^2\},$$

where y is some fixed vector in $\Re^n$.

6.4.4

Consider a closed convex set X in $\Re^n$ and a vector $z \notin X$. Use Fenchel duality to show that the minimum distance from z to X is equal to the maximum distance from z to a hyperplane that separates z from X. Generalize this result by replacing z with a closed convex set Z.

6.4.5 (www)

Consider the problem

$$\text{minimize } f(x)$$
$$\text{subject to } \|x\| \leq 1,$$

where the function $f : \Re^n \mapsto \Re$ is given by

$$f(x) = \max_{y \in Y} x'y,$$

and Y is a convex and compact set. Show that this problem can be solved as a problem of projection on the set Y.

6.4.6 (Quadratically Constrained Quadratic Problems [LVB98]) (www)

Consider the quadratically constrained quadratic problem

$$\text{minimize } x'P_0 x + 2q_0' x + r_0$$
$$\text{subject to } x'P_i x + 2q_i' x + r_i \leq 0, \quad i = 1, \ldots, p,$$

where $P_0, P_1, \ldots, P_p$ are symmetric positive definite matrices. Show that the problem can be converted to a second order cone programming problem and derive the corresponding dual problem. *Hint*: Consider the equivalent problem

$$\text{minimize } \left\| P_0^{1/2} x + P_0^{-1/2} q_0 \right\|$$
$$\text{subject to } \left\| P_i^{1/2} x + P_i^{-1/2} q_i \right\| \leq (r_i - q_i' P_i^{-1} q_i)^{1/2}, \quad i = 1, \ldots, p.$$

6.4.7 (Minimizing the Sum or the Maximum of Norms [LVB98]) (www)

Consider the problems

$$\text{minimize } \sum_{i=1}^{p} \|F_i x + g_i\| \tag{6.81}$$
$$\text{subject to } x \in \Re^n,$$

and

$$\text{minimize } \max_{i=1,\ldots,p} \|F_i x + g_i\|$$
$$\text{subject to } x \in \Re^n,$$

where F_i and g_i are given matrices and vectors, respectively. Convert these problems to second order cone programming problems and derive the corresponding dual problems.

Sec. 6.4   Conjugate Functions and Fenchel Duality   629

6.4.8 (Complex l_1 and l_∞ Approximation, and Second Order Cone Programming [LVB98]) (www)

Consider the complex l_1 approximation problem

$$\text{minimize} \quad \|Ax - b\|_1$$
$$\text{subject to} \quad x \in \mathcal{C}^n,$$

where $\mathcal{C}^n$ is the set of n-dimensional vectors with complex components. Show that it is a special case of problem (6.81) and derive the corresponding dual problem. Repeat for the complex l_∞ approximation problem

$$\text{minimize} \quad \|Ax - b\|_\infty$$
$$\text{subject to} \quad x \in \mathcal{C}^n.$$

6.4.9 (An Extension of the Minimax Theorem)

Show that

$$\min_{x \in X} \max_{y \in Y} \{x'y + F(x) - G(y)\} = \max_{y \in Y} \min_{x \in X} \{x'y + F(x) - G(y)\},$$

where X and Y are convex and compact subsets of $\Re^n$, F is a continuous convex function over X, and G is a continuous convex function over Y.

6.4.10 (Convexification of Nonconvex Problems [Ber79a])

Consider the problem

$$\text{minimize} \quad f(x)$$
$$\text{subject to} \quad Ax = b$$

where $f(x)$ is a quadratic but not necessarily positive semidefinite function, and assume that it has a unique optimal solution x^*. Consider the iteration

$$x^{k+1} \in \arg\min_{Ax=b} \left\{ f(x) + \frac{1}{2c}\|x - x^k\|^2 \right\}, \qquad (6.82)$$

where $c > 0$, and show that it converges to x^*. What is the connection with the proximal algorithm? *Note:* The idea of this exercise extends to problems where f is nonquadratic and the constraint $Ax = b$ is replaced by a nonlinear constraint $h(x) = 0$. The algorithm (6.82) is useful if there is a particularly advantageous method for finding the minimum in Eq. (6.82), which, however, requires convexity of the cost function $f(x) + (1/2c)\|x - x^k\|^2$. This requirement can be met by choosing c sufficiently small. Prominent examples are methods that rely on separability of the cost and the constraints (see Chapter 7). For an analysis in the case of nonquadratic cost and a nonlinear equality constraint, assuming x^* satisfies second order sufficiency conditions, see [Ber79a].

6.4.11 (Strong Duality for One-Dimensional Problems) (www)

Consider a one-dimensional problem (x is a scalar) of the form

$$\text{minimize } f(x)$$
$$\text{subject to } x \in X, \quad g_j(x) \leq 0, \quad j = 1, \ldots, r,$$

where X is convex, and f and g_j are convex and lower semicontinuous when their domain is restricted to X. Assuming that $-\infty < f^* < \infty$, show that there is no duality gap. (Based on unpublished joint work with P. Tseng.)

6.5 DISCRETE OPTIMIZATION AND DUALITY

Many optimization problems, in addition to the usual equality and inequality constraints, involve integer constraints. For example, the variables x_i may be constrained to be 0 or 1. Problems where some or all of the optimization variables take values in a finite set are generally referred to as *discrete optimization problems*. There is a large variety of practical problems of this type, arising for example in scheduling, resource allocation, and engineering design. The methodology for their solution is quite diverse, but an important subset of this methodology relies on the solution of continuous optimization subproblems, as well as on duality. In this section, we explore this interface between discrete and continuous optimization, and we illustrate it through analysis, algorithms, and paradigms.

We first discuss, in Section 6.5.1, some examples of discrete optimization problems that are suitable for the use of duality, and provide vehicles for illustrating the methodology to be discussed later. Then, in Section 6.5.2, we describe the branch-and-bound method, which is in principle capable of producing an optimal solution to an integer-constrained problem. This method relies on upper and lower bound estimates of the optimal cost of various problems that are derived from the given problem. Usually, the upper bounds are obtained with various heuristics, while the lower bounds are obtained through *integer constraint relaxation* (neglecting the integer constraints) or through *Lagrangian relaxation* (using the weak duality theorem).

In Section 6.5.3, we elaborate on the methods for obtaining lower bounds and we discuss in detail the Lagrangian relaxation method. This method requires the optimization of nondifferentiable functions, and some of the major relevant algorithms, subgradient and cutting plane methods, will be discussed in Chapter 7. The effectiveness of duality-based methods for discrete optimization is greatly enhanced when there is a small duality gap. As discussed in Section 6.1.5, separability, the structural property that most enhances duality, also tends to diminish the size of the duality gap in relative terms, particularly when the number of variables is much larger than the number of constraints that couple the variables.

The material in this section is fairly elementary. It is based on the main duality ideas of Section 6.1, but does not rely on the material of the subsequent Sections 6.2-6.4, so it can be read directly after reading Section 6.1. However, it does require some background on directed graphs and associated optimization concepts, since many of the interesting discrete optimization problems involve a graph structure. The author's text on the subject [Ber98] provides an introductory treatment, as do several other sources given at the end of the chapter.

6.5.1 Examples of Discrete Optimization Problems

There is a very large variety of integer-constrained optimization problems. Furthermore, small changes in the problem formulation can often make a significant difference in the character of the solution. As a result, it is not easy to provide a taxonomy of the major problems of interest. It is helpful, however, to study in some detail a few representative examples that can serve as paradigms when dealing with other problems that have similar structure.

Example 6.5.1 (Integer-Constrained Network Optimization)

Let us consider network optimization problems, various forms of which we have already discussed several times so far. In particular, in Section 3.1, we discussed convex multicommodity flow problems arising in communication and transportation networks, and in Section 6.4.3 we considered a convex network problem that is a special case of a monotropic programming problem. Here we will focus on the combinatorial aspects of linear cost network flow problems where integer constraints are involved.

Consider a directed graph consisting of a set $\mathcal{N}$ of nodes and a set $\mathcal{A}$ of directed arcs. The flow x_{ij} of each arc (i,j) *is constrained to be an integer*. The problem is to minimize

$$\sum_{(i,j)\in\mathcal{A}} a_{ij} x_{ij} \tag{6.83}$$

subject to the constraints

$$\sum_{\{j|(i,j)\in\mathcal{A}\}} x_{ij} - \sum_{\{j|(j,i)\in\mathcal{A}\}} x_{ji} = s_i, \quad \forall\, i \in \mathcal{N}, \tag{6.84}$$

and

$$b_{ij} \le x_{ij} \le c_{ij}, \quad \forall\, (i,j) \in \mathcal{A}, \tag{6.85}$$

where a_{ij} are given scalars, and s_i, b_{ij}, and c_{ij} are given integers.

We refer to the constraints (6.84) and (6.85) as the *conservation of flow constraints*, and the *flow bound constraints*, respectively. The conservation of flow constraint at a node i expresses the requirement that the difference between the total flows coming into and out of i must be equal to the given amount s_i, which may be viewed as a *supply* provided by node i to the outside world.

For a typical application of the minimum cost flow problem, think of the nodes as locations (cities, warehouses, or factories) where a certain product is produced or consumed. Think of the arcs as transportation links between the locations, each with transportation cost a_{ij} per unit transported. The problem then is to move the product from the production points to the consumption points at minimum cost while observing the capacity constraints of the transportation links. However, the minimum cost flow problem has many applications that are well beyond the transportation context just described, as will be seen from the following examples. These examples illustrate how some important discrete/combinatorial problems can be modeled as minimum cost flow problems, and highlight the important connection between continuous and discrete network optimization.

Let us consider the classical *shortest path problem*. Here, to each arc (i, j) of a graph, we assign a scalar cost a_{ij} and we define the cost of a path consisting of a sequence of arcs $(i_1, i_2), (i_2, i_3), \ldots, (i_{n-1}, i_n)$ to be the sum of the costs of the arcs

$$\sum_{k=1}^{n-1} a_{i_k i_{k+1}}.$$

Given an origin node s and a destination node t, the shortest path problem is to find a path that starts at s, ends at t, and has minimum cost.

It is possible to cast this problem as the following network optimization problem with integer constraints:

$$\text{minimize} \quad \sum_{(i,j) \in \mathcal{A}} a_{ij} x_{ij}$$

$$\text{subject to} \quad \sum_{\{j | (i,j) \in \mathcal{A}\}} x_{ij} - \sum_{\{j | (j,i) \in \mathcal{A}\}} x_{ji} = \begin{cases} 1 & \text{if } i = s, \\ -1 & \text{if } i = t, \\ 0 & \text{otherwise,} \end{cases} \quad (6.86)$$

$$x_{ij} = 0 \text{ or } 1, \quad \forall \, (i,j) \in \mathcal{A}.$$

Indeed, because of the form of the conservation of flow constraint, it can be seen that the feasible solutions x are associated with the paths P from s to t via the one-to-one correspondence

$$x_{ij} = \begin{cases} 1 & \text{if } (i,j) \text{ belongs to } P, \\ 0 & \text{otherwise.} \end{cases}$$

Furthermore, the cost of x is the length of the corresponding path P. Thus, the path corresponding to an optimal solution of the network optimization problem (6.86) is shortest.

Another interesting combinatorial optimization problem is the *assignment problem*. Here we have n persons and n objects, which we have to match on a one-to-one basis. There is a benefit or value a_{ij} for matching person i

with object j, and we want to assign persons to objects so as to maximize the total benefit. There is also a restriction that person i can be assigned to object j only if (i,j) belongs to a given set of pairs $\mathcal{A}$. Mathematically, we want to find a set of person-object pairs $(1, j_1), \ldots, (n, j_n)$ from $\mathcal{A}$ such that the objects $j_1, \ldots, j_n$ are all distinct, and the total benefit

$$\sum_{i=1}^{n} a_{ij_i}$$

is maximized.

To formulate the assignment problem as a network optimization problem with integer constraints, we introduce the graph shown in Fig. 6.5.1. Here, there are $2n$ nodes divided into two groups: n corresponding to persons and n corresponding to objects. Also, for every possible pair $(i,j) \in \mathcal{A}$, there is an arc connecting person i with object j. The variable x_{ij} is the flow of arc (i,j), and is constrained to be either 1 or 0, indicating that person i is or is not assigned to person j, respectively. The constraint that each person/node i must be assigned to some object can be expressed as

$$\sum_{\{j|(i,j)\in\mathcal{A}\}} x_{ij} = 1,$$

while the constraint that each object/node j must be assigned to some person can be expressed as

$$\sum_{\{i|(i,j)\in\mathcal{A}\}} x_{ij} = 1.$$

Finally, we may view $(-a_{ij})$ as the cost coefficient of the arc (i,j) (by reversing the sign of a_{ij}, we convert the problem from a maximization to a minimization problem). Thus the problem is

$$\begin{aligned}
\text{minimize} \quad & \sum_{(i,j)\in\mathcal{A}} (-a_{ij})x_{ij} \\
\text{subject to} \quad & \sum_{\{j|(i,j)\in\mathcal{A}\}} x_{ij} = 1, \quad \forall\, i=1,\ldots,n, \\
& \sum_{\{i|(i,j)\in\mathcal{A}\}} x_{ij} = 1, \quad \forall\, j=1,\ldots,n, \\
& x_{ij} = 0 \text{ or } 1, \quad \forall\, (i,j) \in \mathcal{A},
\end{aligned} \qquad (6.87)$$

which is a special case of the network optimization problem (6.83)-(6.85).

The most important property of the network optimization problem (6.83)-(6.85) is that *the integer constraints can be neglected*. By this we mean that the *relaxed* problem, i.e., the linear program of minimizing the cost $\sum_{(i,j)} a_{ij}x_{ij}$ subject to the conservation of flow and bound constraints (6.84) and (6.85), but *without* the integer constraints, has the same optimal value as the integer-constrained original. This remarkable fact will be shown as a consequence of a structural property of the network problem, called *unimodularity*, which is discussed in the following example.

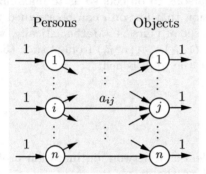

Figure 6.5.1. The graph representation of an assignment problem.

Example 6.5.2 (Unimodular Problems)

An important fact about linear programs of the form

$$\text{minimize } a'x$$
$$\text{subject to } Ex = d, \quad b \leq x \leq c,$$

is that if they attain an optimal solution, they attain one among the set of extreme points of the constraint polyhedron $P = \{x \mid Ex = d, b \leq x \leq c\}$. This is the fundamental theorem of linear programming, given as Prop. B.20(c) in Appendix B. Thus if we knew that all the extreme points of P are integer, we would be assured that the above linear program and its integer-constrained version, where the components of x are further constrained to be integer, share some optimal solutions. Furthermore, given the integer-constrained version of the problem, we could neglect the integer constraints, solve the resulting linear program with a method that finds an optimal extreme point (such as the simplex method), and thus solve the original integer-constrained version.

It turns out that the extreme points of the polyhedron $P = \{x \mid Ex = d, b \leq x \leq c\}$ are integer provided that the components of E, b, c, and d are integer, and the matrix E has a property called *total unimodularity*. To introduce this property, let us say that a square matrix A with integer components is *unimodular* if its determinant is 0, 1, or -1. Unimodularity can be used to assert the integrality of solutions of linear systems of equations. To see this, note that if A is invertible and unimodular, the components of the inverse matrix A^{-1} are integer, because by Kramer's rule, they are equal to a polynomial in the components of A divided by the determinant of A, which is 1 or -1 by the unimodularity and invertibility of A. Therefore, the unique solution x of the system $Ax = b$ is integer for every integer vector b.

A rectangular matrix with integer components is called *totally unimodular* if each of its square submatrices is unimodular. Using the property of unimodular matrices just described, we can show the integrality of all the extreme points (vertices) of a polyhedral set of the form $\{x \mid Ex = d, b \leq x \leq c\}$, where E is totally unimodular, and b, c, and d are vectors with integer components. This follows from standard linear programming theory, which asserts that every extreme point $\hat{x}$ of this polyhedral set is the solution of an equation of the form $\hat{E}\hat{x} = \hat{d}$, where $\hat{E}$ is a square invertible submatrix of E and $\hat{d}$ is a subvector of d (see Prop. B.20 of Appendix B).

It turns out that the conservation of flow constraints (6.84) of the network optimization problem (6.83)-(6.85) involve a matrix that is totally unimodular, and as a result, the extreme points of the associated constraint polyhedron are integer. This matrix, denoted E is the, so-called, *arc incidence matrix* of the underlying graph. It has a row for each node and a column for each arc. The component corresponding to the ith row and a given arc is a 1 if the arc is outgoing from i, is a -1 if the arc is incoming to i, and is a 0 otherwise. Indeed, we can show that the determinant of each square submatrix of E is 0, 1, or -1 by induction on the dimension of the submatrix. In particular, the submatrices of dimension 1 of E are the scalar components of E, which are 0, 1, or -1. Suppose that the determinant of each square submatrix of dimension $n \geq 1$ is 0, 1, or -1. Consider a square submatrix of dimension $n+1$. If this matrix has a column with all components 0, the matrix is singular, and its determinant is 0. If the matrix has a column with a single nonzero component (a 1 or a -1), by expanding its determinant along that component and using the induction hypothesis, we see that the determinant is 0, 1, or -1. Finally, if each column of the matrix has two components (a 1 and a -1), the sum of its rows is 0, so the matrix is singular, and its determinant is 0.

The network optimization problem (6.83)-(6.85) is the most important example of an integer programming problem with polyhedral constraints that has the unimodular property. However, there are other interesting examples for which we refer to the exercises and specialized sources, such as Schrijver [Sch86], and Nemhauser and Wolsey [NeW88].

Example 6.5.3 (Generalized Assignment and Facility Location Problems)

Consider a problem of assigning m jobs to n machines. If job i is performed at machine j, it costs a_{ij} and requires t_{ij} time units. We want to find a minimum cost assignment of the jobs to the machines, given the total available time T_j at machine j. We assume that each job must be performed in its entirety at a single machine, and this introduces a combinatorial character to the problem.

Let us introduce for each job i and machine j, a variable x_{ij} which takes the value 1 or 0 depending on whether the job is or is not assigned to machine j, respectively. Then the problem takes the form

$$\text{minimize} \quad \sum_{i=1}^{m} \sum_{j=1}^{n} a_{ij} x_{ij}$$

$$\text{subject to} \quad \sum_{j=1}^{n} x_{ij} = 1, \quad i = 1, \ldots, m,$$

$$\sum_{i=1}^{m} t_{ij} x_{ij} \leq T_j, \quad j = 1, \ldots, n,$$

$$x_{ij} = 0 \text{ or } 1, \quad i = 1, \ldots, m, \ j = 1, \ldots, n.$$

The constraint $\sum_{j=1}^{n} x_{ij} = 1$ specifies that each job i must be assigned to some machine, and the constraint $\sum_{i=1}^{m} t_{ij} x_{ij} \leq T_j$ specifies that the total working time of machine j must not exceed the available T_j.

When there is only one machine, the index j is superfluous, and the problem becomes equivalent to the classical *knapsack problem* (cf. Exercise 6.1.7). Here we want to place in a knapsack the most valuable subcollection out of a given collection of objects, subject to a total weight constraint

$$\sum_{i=1}^{m} w_i x_i \leq T,$$

where T is a given total weight threshold, w_i is the weight of object i, and x_i is a variable which is 1 or 0 depending on whether the ith object is placed in the knapsack or not. The value to be maximized is $\sum_{i=1}^{m} v_i x_i$, where v_i is the value of the ith object.

For another variant of the generalized assignment problem, let us introduce a setup cost b_j for using machine j, i.e., for having $\sum_{i=1}^{m} x_{ij} > 0$. We then obtain an integer-constrained problem of the form

$$\text{minimize} \quad \sum_{i=1}^{m} \sum_{j=1}^{n} a_{ij} x_{ij} + \sum_{j=1}^{n} b_j y_j$$

$$\text{subject to} \quad \sum_{j=1}^{n} x_{ij} = 1, \quad i = 1, \ldots, m,$$

$$\sum_{i=1}^{m} t_{ij} x_{ij} \leq T_j y_j, \quad j = 1, \ldots, n,$$

$$x_{ij} = 0 \text{ or } 1, \quad i = 1, \ldots, m, \; j = 1, \ldots, n,$$

$$y_j = 0 \text{ or } 1, \quad j = 1, \ldots, n,$$

where $y_j = 1$ indicates that the jth machine is used. When $t_{ij} = 1$ for all pairs (i, j), this problem is also known as the *facility location problem*. Within this context, we must select a subset of locations from a given candidate set, and place in each of these locations a "facility" that will serve the needs of certain "clients" up to a given capacity bound. The 0-1 decision variable y_j corresponds to selecting location j for facility placement. To make the connection with the generalized assignment problem, associate clients with jobs, and locations with machines.

Example 6.5.4 (Traveling Salesman Problem)

An important model for scheduling a sequence of operations is the classical traveling salesman problem. This is perhaps the most studied of all combinatorial optimization problems. In addition to its use as a practical model, it has served as a testbed for a large variety of formal and heuristic approaches in discrete optimization.

Sec. 6.5   Discrete Optimization and Duality   637

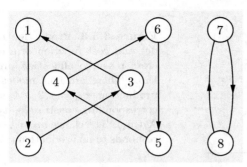

Figure 6.5.2. Example of an infeasible solution of a traveling salesman problem where all the constraints are satisfied except for the connectivity constraint (6.88). This solution may have been obtained by solving an $N \times N$ assignment problem and consists of multiple cycles [(1,2,3), (4,5,6), and (7,8) in the figure]. The arcs of the cycles correspond to the assigned pairs (i,j) in the assignment problem.

In a colloquial description of the problem, a salesman wants to find a minimum cost tour that visits each of N given cities exactly once and returns to the starting city. Let a_{ij} be the cost of going from city i to city j, and let x_{ij} be a variable that takes the value 1 if the salesman visits city j immediately following city i, and the value 0 otherwise. Then the problem is

$$\text{minimize} \quad \sum_{i=1}^{N} \sum_{\substack{j=1,\ldots,N \\ j \neq i}} a_{ij} x_{ij}$$

$$\text{subject to} \quad \sum_{\substack{j=1,\ldots,N \\ j \neq i}} x_{ij} = 1, \quad i = 1,\ldots,N,$$

$$\sum_{\substack{i=1,\ldots,N \\ i \neq j}} x_{ij} = 1, \quad j = 1,\ldots,N,$$

$$x_{ij} = 0 \text{ or } 1, \quad \forall\, (i,j) \in \mathcal{A},$$

plus the additional constraint that the set of arcs $\{(i,j) \mid x_{ij} = 1\}$ forms a connected tour. The last constraint can be expressed as

$$\sum_{i \in S,\, j \notin S} (x_{ij} + x_{ji}) \geq 2, \quad \forall \text{ nonempty proper subsets } S \text{ of cities}. \quad (6.88)$$

It turns out that if this constraint were not present, the problem would be much easier. In particular, it would be an assignment problem (assign the N cities to the N successor cities in a salesman tour), which is a relatively easy problem as discussed in Example 6.5.1. Unfortunately, however, these constraints are essential, since without them, there would be feasible solutions involving multiple disconnected cycles, as illustrated in Fig. 6.5.2. Nonetheless, with the aid of duality, the corresponding assignment problem can form the basis for solution of the traveling salesman problem.

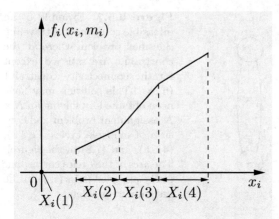

Figure 6.5.3. Production modes and cost function of a discrete resource allocation problem. There are four modes of production, $m_i = 1, 2, 3, 4$, and corresponding constraints, $x_i \in X_i(m_i)$. The choice $m_i = 1$ corresponds to no production ($x_i = 0$).

Example 6.5.5 (Separable Resource Allocation Problems)

Consider a problem of optimally producing a given amount of product using n production units. We assume that the production and cost characteristics of the units may change abruptly as the production level changes; for example there may be a setup or startup cost that is incurred at an arbitrarily small (but positive) level of production. This situation leads to a "mixed" discrete/continuous problem formulation.

Let x_i be the amount produced by the ith unit, where $i = 1, \ldots, n$. Let there be a finite set M_i of production modes associated with the ith production unit. The set of possible productions levels in production mode $m_i \in M_i$ of the ith unit is denoted $X_i(m_i)$, so that the amount produced by the ith unit must satisfy $x_i \in X_i(m_i)$ when the production mode n_i is selected (see Fig. 6.5.3).

Let $f_i(x_i, m_i)$ be the cost of producing amount x_i via the production mode m_i. The problem then is to

$$\text{minimize} \quad \sum_{i=1}^{n} f_i(x_i, m_i)$$

$$\text{subject to} \quad m_i \in M_i, \qquad x_i \in X(m_i).$$

There are also some additional constraints, which we generically represent in the form

$$\sum_{i=1}^{n} g_{ij}(x_i, m_i) \leq 0, \qquad j = 1, \ldots, r,$$

where the function g_{ij} represents the contribution of the ith unit in the jth constraint. These constraints specify that the units must collectively meet certain requirements (such as satisfying a given demand d, i.e., $\sum_{i=1}^{m} x_i \geq d$), security constraints (such as a lower bound on the number of units involved in production), etc.

An important characteristic of this problem is that it has a separable structure. As discussed in Section 6.1.5, this structure is particularly well-suited for the application of duality for two reasons: first, the computation of the dual function values is facilitated through decomposition, and second, the duality gap tends to be small in relative terms, particularly as the dimension n increases. In what follows, we will illustrate how these properties enhance the application of various solution algorithms.

6.5.2 Branch-and-Bound

The branch-and-bound method implicitly enumerates all the feasible solutions, using optimization problems that involve no integer constraints. The method can be very time-consuming, but is in principle capable of yielding an exactly optimal solution. The idea of the method is to partition the feasible set into smaller subsets, and then calculate certain bounds on the attainable cost within some of the subsets to eliminate from further consideration other subsets. This idea is encapsulated in the following simple observation.

Bounding Principle

Given the problem of minimizing $f(x)$ over $x \in X$, and two subsets $Y_1 \subset X$ and $Y_2 \subset X$, suppose that we have bounds

$$\underline{f}_1 \leq \min_{x \in Y_1} f(x), \qquad \bar{f}_2 \geq \min_{x \in Y_2} f(x).$$

Then, if $\bar{f}_2 \leq \underline{f}_1$, the solutions in Y_1 may be disregarded since their cost cannot be smaller than the cost of the best solution in Y_2.

The branch-and-bound method uses suitable upper and lower bounds, and the bounding principle to eliminate from consideration substantial portions of the feasible set. To describe the method, consider a general discrete optimization problem

$$\text{minimize} \quad f(x)$$
$$\text{subject to} \quad x \in X,$$

where the feasible set X is a *finite* set. The branch-and-bound algorithm uses an acyclic graph known as the *branch-and-bound tree*, which corresponds to a progressively finer partition of X. In particular, the nodes of this graph correspond to a collection $\mathcal{X}$ of subsets of X, which is such that:

1. $X \in \mathcal{X}$ (i.e., the set of all solutions is a node).
2. If x is a feasible solution, then $\{x\} \in \mathcal{X}$ (i.e., each solution viewed as a singleton set is a node).

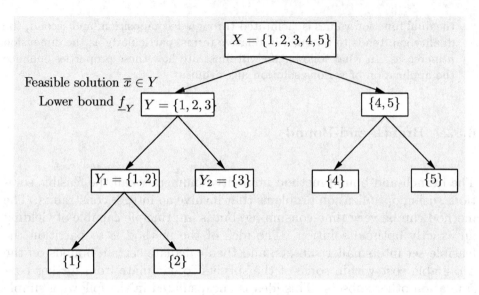

Figure 6.5.4. Illustration of a branch-and-bound tree. Each node Y (a subset of the feasible set X), except those consisting of a single solution, is partitioned into several other nodes (subsets) $Y_1, \ldots, Y_n$. The original feasible set is divided repeatedly into subsets until no more division is possible. For each node/subset Y of the tree, one may compute a lower bound $\underline{f}_Y$ to the optimal cost of the corresponding restricted subproblem $\min_{x \in Y} f(x)$, and a feasible solution $\bar{x} \in Y$, whose cost can serve as an upper bound to the optimal cost $\min_{x \in X} f(x)$ of the original problem. The idea is to use these bounds to economize computation by eliminating nodes of the tree that cannot contain an optimal solution.

3. If a set $Y \in \mathcal{X}$ contains more than one solution $x \in X$, then there exist disjoint sets $Y_1, \ldots, Y_n \in \mathcal{X}$ such that

$$\bigcup_{i=1}^n Y_i = Y.$$

The set Y is called the *parent* of $Y_1, \ldots, Y_n$, and the sets $Y_1, \ldots, Y_n$ are called the *children* or *descendants* of Y.

4. Each set in $\mathcal{X}$ other than X has a parent.

The collection of sets $\mathcal{X}$ defines the branch-and-bound tree as in Fig. 6.5.4. In particular, this tree has the set of all feasible solutions X as its root node and the singleton solutions $\{x\}$, $x \in X$, as terminal nodes. The arcs of the graph are those that connect parents Y and their children Y_i.

The key assumption in the branch-and-bound method is that for every nonterminal node Y, there is an algorithm that calculates:

(a) A lower bound $\underline{f}_Y$ to the minimum cost over Y

$$\underline{f}_Y \leq \min_{x \in Y} f(x).$$

(b) A feasible solution $\bar{x} \in Y$, whose cost $f(\bar{x})$ can serve as an upper bound to the minimum cost over Y (as well as over X).

These bounds are used to save computation by discarding the nodes/subsets of the tree that have no chance of containing a solution that is better than the best currently available. In particular, the algorithm selects nodes Y from the branch-and-bound tree, and checks whether the lower bound $\underline{f}_Y$ exceeds the best available upper bound [the minimal cost $f(\bar{x})$ over all feasible solutions $\bar{x}$ found so far]. If this is so, we know that Y cannot contain an optimal solution, so all its descendant nodes in the tree need not be considered further.

To organize the search through the tree, the algorithm maintains a node list called OPEN, and also maintains a scalar called UPPER, which is equal to the minimal cost over feasible solutions found so far. Initially, OPEN contains just X, and UPPER is equal to ∞ or to the cost $f(\bar{x})$ of some feasible solution $\bar{x} \in X$.

Branch-and-Bound Algorithm

Step 1: Remove a node Y from OPEN. For each child Y_j of Y, do the following: Find the lower bound $\underline{f}_{Y_j}$ and a feasible solution $\bar{x} \in Y_j$. If

$$\underline{f}_{Y_j} < \text{UPPER},$$

place Y_j in OPEN. If in addition

$$f(\bar{x}) < \text{UPPER},$$

set

$$\text{UPPER} = f(\bar{x})$$

and mark $\bar{x}$ as the best solution found so far.

Step 2: (Termination Test) If OPEN is nonempty, go to step 1. Otherwise, terminate; the best solution found so far is optimal.

A node Y_j that is not placed in OPEN in Step 1 is said to be *fathomed*. Such a node cannot contain a better solution than the best solution found so far, since the corresponding lower bound $\underline{f}_{Y_j}$ is not smaller than UPPER. Therefore nothing is lost when we drop this node from further consideration and forego the examination of its descendants. Regardless of how many nodes are fathomed, the branch-and-bound algorithm is guaranteed to examine either explicitly or implicitly (through fathoming) all the terminal nodes, which are the singleton solutions. As a result, it will terminate with an optimal solution. Note that a small (near-optimal) value of UPPER and tight lower bounds $\underline{f}_{Y_j}$ contribute to the quick fathoming of large portions of the branch-and-bound tree, so it is important to obtain good bounds as early as possible.

In a popular variant of the algorithm, termination is accelerated at the expense of obtaining a solution that is suboptimal within some tolerance $\epsilon > 0$. In particular, if we replace the test

$$\underline{f}_{Yj} < \text{UPPER}$$

with

$$\underline{f}_{Yj} < \text{UPPER} - \epsilon$$

in Step 1, the best solution obtained upon termination is guaranteed to be within ϵ of optimality.

Other variations of branch-and-bound relate to the method for selecting a node from OPEN in Step 1. For example, a possible strategy is to choose the node with minimal lower bound; alternatively, one may choose the node containing the best solution found so far. In fact *it is neither practical nor necessary to generate a priori the branch-and-bound tree*. Instead, one may adaptively decide on the order and the manner in which the nodes are partitioned into descendants based on the progress of the algorithm.

Branch-and-bound typically uses continuous optimization problems (without integer constraints) to obtain lower bounds to the optimal costs of the restricted problems $\min_{x \in Y} f(x)$ and to construct corresponding feasible solutions. For example, suppose that our original problem has a convex cost function, and a feasible set X that consists of convex inequality constraints, *plus the additional constraint that all optimization variables must be 0 or 1*. Then a restricted subset Y may specify that the values of some given subset of variables are fixed at 0 or at 1, while the remaining variables may take either the value 0 or the value 1. A lower bound to the restricted optimal cost $\min_{x \in Y} f(x)$ is then obtained by relaxing the 0-1 constraint on the latter variables, thereby allowing them to take any value in the interval $[0, 1]$ and resulting in a convex problem with inequality constraints. Thus the solution by branch-and-bound of a problem with convex cost and inequality constraints *plus* additional integer constraints requires the solution of many convex problems with inequality constraints but *without* integer constraints.

Example 6.5.6 (Facility Location Problems)

Let us consider the facility location problem introduced in Example 6.5.3, which involves m clients and n locations. By $x_{ij} = 1$ (or $x_{ij} = 0$) we indicate that client i is assigned to location j at a cost a_{ij} (or is not assigned, respectively). We also introduce a 0-1 integer variable y_j to indicate (with $y_j = 1$)

Sec. 6.5   Discrete Optimization and Duality   643

that a facility is placed at location j at a cost b_j. The problem is

$$\text{minimize} \quad \sum_{i=1}^{m}\sum_{j=1}^{n} a_{ij}x_{ij} + \sum_{j=1}^{n} b_j y_j$$

$$\text{subject to} \quad \sum_{j=1}^{n} x_{ij} = 1, \quad i = 1, \ldots, m,$$

$$\sum_{i=1}^{m} x_{ij} \leq T_j y_j, \quad j = 1, \ldots, n,$$

$$x_{ij} = 0 \text{ or } 1, \quad \forall\, (i,j),$$

$$y_j = 0 \text{ or } 1, \quad j = 1, \ldots, n,$$

where T_j is the maximum number of customers that can be served by a facility at location j.

The solution of the problem by branch-and-bound involves the partition of the feasible set X into subsets. The choice of subsets is somewhat arbitrary, but it is convenient to select subsets of the form

$$X(J_0, J_1) = \big\{(x,y) : \text{feasible} \mid y_j = 0,\ \forall\, j \in J_0,\ y_j = 1,\ \forall\, j \in J_1\big\},$$

where J_0 and J_1 are disjoint subsets of the index set $\{1,\ldots,n\}$ of facility locations. Thus, $X(J_0, J_1)$ is the subset of feasible solutions such that:

a facility is placed at the locations in J_1,

no facility is placed at the locations in J_0,

a facility may or may not be placed at the remaining locations.

For each node/subset $X(J_0, J_1)$, we may obtain a lower bound and a feasible solution by solving the linear program where all integer constraints are relaxed except that the variables $y_j,\ j \in J_0 \cup J_1$ are fixed at either 0 or 1.

As an illustration, let us work out the example shown in Fig. 6.5.5, which involves 3 clients and 2 locations. The facility capacities at the two locations are $T_1 = T_2 = 3$. The cost coefficients a_{ij} and b_j are shown next to the corresponding arcs. The optimal solution corresponds to $y_1 = 0$ and $y_2 = 1$, i.e., placing a facility only in location 2 and serving all the clients at that facility. The corresponding optimal cost is

$$f^* = 5.$$

Let us apply the branch-and-bound algorithm using the tree shown in Fig. 6.5.5. We first consider the top node $(J_0 = \emptyset, J_1 = \emptyset)$, where neither y_1 nor y_2 is fixed at 0 or 1. The lower bound $\underline{f}_Y$ is obtained by solving the (relaxed) linear program

$$\text{minimize} \quad (2x_{11} + x_{12}) + (2x_{21} + x_{22}) + (x_{31} + 2x_{32}) + 3y_1 + y_2$$

$$\text{subject to} \quad x_{11} + x_{12} = 1, \quad x_{21} + x_{22} = 1, \quad x_{31} + x_{32} = 1,$$

$$x_{11} + x_{21} + x_{31} \leq 3y_1, \quad x_{12} + x_{22} + x_{32} \leq 3y_2,$$

$$0 \leq x_{ij} \leq 1, \quad \forall\, (i,j),$$

$$0 \leq y_1 \leq 1, \quad 0 \leq y_2 \leq 1.$$

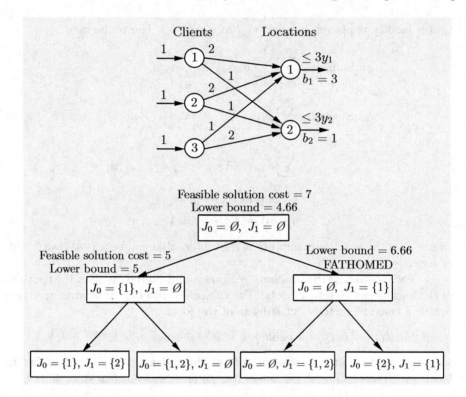

Figure 6.5.5. Branch-and-bound solution of a facility location problem with 3 clients and 2 locations. The facility capacities at the two locations are $T_1 = T_2 = 3$. The cost coefficients a_{ij} and b_j are shown next to the corresponding arcs. The relaxed problem for the top node ($J_0 = \emptyset, J_1 = \emptyset$), corresponding to relaxing all the integer constraints, is solved first, obtaining the lower and upper bounds shown. Then the relaxed problem corresponding to the left node ($J_0 = \{1\}, J_1 = \emptyset$) is solved, obtaining the lower and upper bounds shown. Finally, the relaxed problem corresponding to the right node ($J_0 = \emptyset, J_1 = \{1\}$) is solved, obtaining a lower bound that is higher than the current value of UPPER. As a result this node can be fathomed, and its descendants need not be considered further.

The optimal solution of this program can be obtained by a linear programming algorithm and can be verified to be

$$x_{ij} = \begin{cases} 1 & \text{if } (i,j) = (1,2), (2,2), (3,1), \\ 0 & \text{otherwise,} \end{cases}$$

$$y_1 = 1/3, \qquad y_2 = 2/3.$$

The corresponding optimal cost (lower bound) is

$$\underline{f}_Y = 4.66.$$

A feasible solution of the original problem is obtained by rounding the fractional values of y_1 and y_2 to

$$\bar{y}_1 = 1, \qquad \bar{y}_2 = 1,$$

Sec. 6.5 Discrete Optimization and Duality

and the associated cost is 7. Thus, we set

$$\text{UPPER} = 7,$$

and we place in OPEN the two descendants $\left(J_0 = \{1\}, J_1 = \emptyset\right)$ and $\left(J_0 = \emptyset, J_1 = \{1\}\right)$, corresponding to fixing y_1 at 0 and at 1, respectively.

We proceed with the left branch of the branch-and-bound tree, and consider the node $\left(J_0 = \{1\}, J_1 = \emptyset\right)$, corresponding to fixing y_1 as well as the corresponding variables x_{11}, x_{21}, and x_{31} to 0. The associated (relaxed) linear program is

$$\begin{aligned}
\text{minimize} \quad & x_{12} + x_{22} + 2x_{32} + y_2 \\
\text{subject to} \quad & x_{12} = 1, \quad x_{22} = 1, \quad x_{32} = 1, \\
& x_{12} + x_{22} + x_{32} \leq 3y_2, \\
& 0 \leq x_{12} \leq 1, \quad 0 \leq x_{22} \leq 1, \quad 0 \leq x_{32} \leq 1, \\
& 0 \leq y_2 \leq 1.
\end{aligned}$$

The optimal solution (in fact the only feasible solution) of this program is

$$x_{ij} = \begin{cases} 1 & \text{if } (i,j) = (1,2), (2,2), (3,2), \\ 0 & \text{otherwise}, \end{cases}$$

$$y_2 = 1,$$

and the corresponding optimal cost (lower bound) is

$$\underline{f}_Y = 5.$$

The optimal solution of the relaxed problem is integer, and its cost, 5, is lower than the current value of UPPER, so we set

$$\text{UPPER} = 5.$$

The two descendants, $\left(J_0 = \{1\}, J_1 = \{2\}\right)$ and $\left(J_0 = \{1,2\}, J_1 = \emptyset\right)$, corresponding to fixing y_2 at 1 and at 0, respectively, are placed in OPEN.

We proceed with the right branch of the branch-and-bound tree, and consider the node $\left(J_0 = \emptyset, J_1 = \{1\}\right)$, corresponding to fixing y_1 to 1. The associated (relaxed) linear program is

$$\begin{aligned}
\text{minimize} \quad & (2x_{11} + x_{12}) + (2x_{21} + x_{22}) + (x_{31} + 2x_{32}) + 3 + y_2 \\
\text{subject to} \quad & x_{11} + x_{12} = 1, \quad x_{21} + x_{22} = 1, \quad x_{31} + x_{32} = 1, \\
& x_{11} + x_{21} + x_{31} \leq 3, \quad x_{12} + x_{22} + x_{32} \leq 3y_2, \\
& 0 \leq x_{ij} \leq 1, \quad \forall\, (i,j), \\
& 0 \leq y_2 \leq 1.
\end{aligned}$$

The optimal solution of this program can be verified to be

$$x_{ij} = \begin{cases} 1 & \text{if } (i,j) = (1,2), (2,2), (3,1), \\ 0 & \text{otherwise}, \end{cases}$$

$$y_2 = 2/3,$$

and the corresponding optimal cost (lower bound) is

$$\underline{f}_Y = 6.66.$$

This is larger than the current value of UPPER, so the node can be fathomed, and its two descendants are not placed in OPEN.

We conclude that one of the two descendants of the left node, $(J_0 = \{1\}, J_1 = \{2\})$ and $(J_0 = \{1,2\}, J_1 = \emptyset)$ (the only nodes in OPEN), contains the optimal solution. We can proceed to solve the relaxed linear programs corresponding to these two nodes, and obtain the optimal solution. However, there is also a shortcut here: since these are the only two remaining nodes and the upper bound corresponding to these nodes coincides with the lower bound, we can conclude that the lower bound is equal to the optimal cost and the corresponding integer solution ($y_1 = 0, y_2 = 1$) is optimal.

Generally, for the success of the branch-and-bound approach it is important that the lower bounds are as tight as possible, because this facilitates the fathoming of nodes, and leads to fewer restricted problem solutions. On the other hand, the tightness of the bounds strongly depends on how the problem is formulated as an integer programming problem. There may be several possible formulations, some of which are "stronger" than others in the sense that they provide better bounds within the branch-and-bound context. The following example provides an illustration.

Example 6.5.7 (Facility Location – Alternative Formulation)

Consider the following alternative formulation of the preceding facility location problem

$$\text{minimize} \quad \sum_{i=1}^{m}\sum_{j=1}^{n} a_{ij} x_{ij} + \sum_{j=1}^{n} b_j y_j$$

$$\text{subject to} \quad \sum_{j=1}^{n} x_{ij} = 1, \quad i = 1, \ldots, m,$$

$$\sum_{i=1}^{m} x_{ij} \leq T_j y_j, \quad j = 1, \ldots, n,$$

$$x_{ij} \leq y_j, \quad \forall\, (i,j),$$

$$x_{ij} = 0 \text{ or } 1, \quad \forall\, (i,j),$$

$$y_j = 0 \text{ or } 1, \quad j = 1, \ldots, n.$$

This formulation involves a lot more constraints, but is in fact superior to the one given earlier (cf. Example 6.5.6). The reason is that, once we relax the 0-1 constraints on x_{ij} and y_j, the constraints

$$\sum_{i=1}^{m} x_{ij} \leq T_j y_j$$

Sec. 6.5 Discrete Optimization and Duality   647

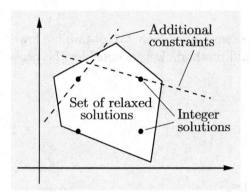

Figure 6.5.6. Illustration of the effect of additional constraints. They do not affect the set of feasible integer solutions, but they reduce the set of "relaxed solutions," i.e., those x that satisfy all the constraints except for the integer constraints. This results in improved lower bounds and a faster branch-and-bound solution.

of Example 6.5.6 define a larger region than the constraints $\sum_{i=1}^{m} x_{ij} \leq T_j y_j$ and $x_{ij} \leq y_j$ of the present example. As a result, the lower bounds obtained by relaxing some of the 0-1 constraints are tighter in the alternative formulation just given, thereby enhancing the effectiveness of the branch-and-bound method. In fact, it can be verified that for the example of Fig. 6.5.5, by relaxing the 0-1 constraints in the stronger formulation of the present example, we obtain the correct optimal integer solution at the very first node of the branch-and-bound tree.

The preceding example illustrates an important fact: *it is possible to accelerate the branch-and-bound solution of a problem by adding constraints that do not affect the set of feasible integer solutions*. Such constraints, referred to as *valid inequalities*, may improve the lower bounds obtained by relaxing the 0-1 constraints. Basically, when the integer constraints are relaxed, one obtains a superset of the feasible set of integer solutions, so with more constraints, the corresponding superset becomes smaller and approximates better the true feasible set (see Fig. 6.5.6). Thus one should strive to select a problem formulation and/or try to generate additional valid inequalities so that when the integer constraints are relaxed, the feasible set is as small as possible. In particular, this leads to a variant of branch-and-bound, called *branch-and-cut*, which generates suitable valid inequalities at several nodes of the branch-and-bound tree, together with the associated bounds. We refer to the surveys [JRT95], [LuB96], and [CaF97] for accounts on this subject.

We note that the subject of characterizing the feasible set of an integer programming problem, and approximating it tightly with a polyhedral set has received extensive attention. In particular, there is a lot of theory and accumulated practical knowledge on characterizing the feasible set in specific problem contexts; see the references cited at the end of the chapter. A further discussion of branch-and-bound is beyond our scope. We refer to sources on linear and combinatorial optimization, which also describe many applications; see the end-of-chapter references.

6.5.3 Lagrangian Relaxation

In this section, we consider an important approach for obtaining lower bounds to use in the branch-and-bound method. Let us consider the problem

$$\begin{aligned}\text{minimize} \quad & f(x) \\ \text{subject to} \quad & g_j(x) \leq 0, \quad j = 1, \ldots, r, \\ & x \in X,\end{aligned}$$

where X is a *finite* set of vectors in $\Re^n$.

In the Lagrangian relaxation approach, we eliminate the constraints $g_j(x) \leq 0$ by using nonnegative multipliers μ_j, and by forming the Lagrangian function

$$L(x, \mu) = f(x) + \sum_{j=1}^{r} \mu_j g_j(x).$$

As in Section 6.1, the dual function is

$$q(\mu) = \min_{x \in X} L(x, \mu)$$

and the dual problem is

$$\begin{aligned}\text{maximize} \quad & q(\mu) \\ \text{subject to} \quad & \mu \geq 0.\end{aligned}$$

A key fact here is that for any $\mu \geq 0$, *the dual value $q(\mu)$ provides a lower bound to the optimal primal value f^**, and so does the optimal dual value q^*. This is the weak duality theorem (Prop. 6.1.3). Thus solving the dual problem (even approximately) provides a lower bound that can be used in the context of the branch-and-bound procedure.

The minimization of $L(x, \mu)$ over $x \in X$ is facilitated when the cost function f, the constraint functions g_j, and the set X have a special structure, such as a separable structure, since then one can take advantage of the associated decomposition, as discussed in Section 6.1.5. More generally, the minimization of $L(x, \mu)$ is facilitated when by eliminating the constraints $g_j(x) \leq 0$ using multipliers, one obtains an easily solvable problem. A major case of this type is when the cost and constraint functions f and g_j are linear, and the finite set X is the constraint set of an integer-constrained network optimization problem discussed in Example 6.5.1. Then the minimization of $L(x, \mu)$ over X can be done with fast specialized network flow algorithms such as shortest path, max-flow, assignment, transportation, and minimum cost flow algorithms (see network optimization books, such as the author's [Ber98], for an extensive discussion).

Example 6.5.8 (Facility Location − Lagrangian Relaxation)

Consider the facility location problem as formulated in Example 6.5.7:

$$\text{minimize} \quad \sum_{i=1}^{m}\sum_{j=1}^{n} a_{ij}x_{ij} + \sum_{j=1}^{n} b_j y_j$$

$$\text{subject to} \quad \sum_{j=1}^{n} x_{ij} = 1, \quad i = 1, \ldots, m,$$

$$\sum_{i=1}^{m} x_{ij} \leq T_j y_j, \quad j = 1, \ldots, n,$$

$$x_{ij} \leq y_j, \quad \forall\, (i,j),$$

$$x_{ij} = 0 \text{ or } 1, \quad \forall\, (i,j),$$

$$y_j = 0 \text{ or } 1, \quad j = 1, \ldots, n.$$

The solution of the problem by branch-and-bound involves the partition of the feasible set into subsets whereby some of the integer variables are fixed at the 0 or 1 value, while the 0-1 constraints on the remaining variables are replaced by interval constraints. In particular, at the typical subproblem/node of the branch-and-bound tree we select two disjoint subsets J_0 and J_1 of the index set $\{1, \ldots, n\}$. These subsets correspond to:

placing a facility at the locations in J_1,

placing no facility at the locations in J_0,

leaving open the placement of a facility at the remaining locations.

We must then obtain a lower bound and a feasible solution of the restricted integer program where the variables y_j, $j \in J_0 \cup J_1$ have been fixed at either 0 or 1:

$$\text{minimize} \quad \sum_{i=1}^{m}\sum_{j=1}^{n} a_{ij}x_{ij} + \sum_{j=1}^{n} b_j y_j \tag{6.89}$$

$$\text{subject to} \quad \sum_{j=1}^{n} x_{ij} = 1, \quad i = 1, \ldots, m,$$

$$\sum_{i=1}^{m} x_{ij} \leq T_j y_j, \quad j \notin J_0,$$

$$x_{ij} \leq y_j, \quad j \notin J_0,$$

$$x_{ij} = 0 \text{ or } 1, \quad \forall\, (i,j) \text{ with } j \notin J_0,$$

$$x_{ij} = 0, \quad \forall\, (i,j) \text{ with } j \in J_0,$$

$$y_j = 0 \text{ or } 1, \quad \forall\, j \notin J_0 \cup J_1,$$

$$y_j = 0, \quad \forall\, j \in J_0, \qquad y_j = 1, \quad \forall\, j \in J_1.$$

Here are three ways to obtain a lower bound. The first is based on constraint relaxation, as in Examples 6.5.6 and 6.5.7, and the last two are based on Lagrangian relaxation:

(a) We replace the 0-1 integer constraints on the variables x_{ij} and y_j by $[0,1]$ interval constraints. The resulting linear problem is then solved by standard linear programming methods.

(b) We eliminate the constraints $\sum_{i=1}^{m} x_{ij} \leq T_j y_j$, $j \notin J_0$, using multipliers $\mu_j \geq 0$, and the constraints $x_{ij} \leq y_j$, $j \notin J_0$, using multipliers $\xi_{ij} \geq 0$. The corresponding Lagrangian function,

$$L(x,y,\mu,\xi) = \sum_{i=1}^{m}\sum_{j \notin J_0}(a_{ij}+\mu_j+\xi_{ij})x_{ij} + \sum_{j \notin J_0}\left(b_j - \mu_j T_j - \sum_{i=1}^{m}\xi_{ij}\right)y_j,$$

is then minimized over the remaining constraints. This minimization is decoupled with respect to x and y. Minimization over y_j yields optimal values

$$\hat{y}_j = \begin{cases} 0 & \text{if } j \in J_0, \text{ or } b_j - \mu_j T_j - \sum_{i=1}^{m}\xi_{ij} \geq 0 \text{ and } j \notin J_1, \\ 1 & \text{otherwise.} \end{cases}$$

Minimization over x_{ij} can be done by finding for each i, an index $j_i \notin J_0$ such that

$$a_{ij_i} + \mu_{j_i} + \xi_{ij_i} \in \arg\min_{j \notin J_0}\{a_{ij}+\mu_j+\xi_{ij}\},$$

and then obtaining the optimal values

$$\hat{x}_{ij} = \begin{cases} 0 & \text{if } j \neq j_i, \\ 1 & \text{if } j = j_i. \end{cases}$$

The value $L(\hat{x},\hat{y},\mu,\xi)$, which is the dual value corresponding to μ and ξ, is a lower bound to the optimal cost of the restricted integer program (6.89). This lower bound can be strengthened by maximizing $L(\hat{x},\hat{y},\mu,\xi)$ over $\mu \geq 0$ and $\xi \geq 0$, thereby obtaining the optimal dual value/lower bound (we discuss later the issues associated with this maximization).

(c) We eliminate only the constraints, $x_{ij} \leq y_j$, $j \notin J_0$, using multipliers $\xi_{ij} \geq 0$. The corresponding Lagrangian function,

$$L(x,y,\xi) = \sum_{i=1}^{m}\sum_{j \notin J_0}(a_{ij}+\xi_{ij})x_{ij} + \sum_{j \notin J_0}\left(b_j - \sum_{i=1}^{m}\xi_{ij}\right)y_j,$$

is then minimized over the remaining constraints for x. This minimization is decoupled with respect to x and y. Minimization over y_j yields optimal values

$$\hat{y}_j = \begin{cases} 0 & \text{if } j \in J_0, \text{ or } b_j - \sum_{i=1}^{m}\xi_{ij} \geq 0 \text{ and } j \notin J_1, \\ 1 & \text{otherwise.} \end{cases}$$

The minimization over x_{ij} does not decompose over the clients i as in case (b) above. Instead it has the form

$$\text{minimize} \quad \sum_{i=1}^{m} \sum_{j \notin J_0} (a_{ij} + \xi_{ij}) x_{ij}$$

$$\text{subject to} \quad \sum_{j=1}^{n} x_{ij} = 1, \quad i = 1, \ldots, m,$$

$$\sum_{i=1}^{m} x_{ij} \leq T_j, \quad \forall\, j \text{ with } \hat{y}_j = 1,$$

$$x_{ij} = 0 \text{ or } 1, \quad \forall\, (i,j) \text{ with } \hat{y}_j = 1,$$

$$x_{ij} = 0, \quad \forall\, (i,j) \text{ with } \hat{y}_j = 0.$$

This problem is an integer-constrained network optimization problem (cf. Example 6.5.1), known as a *transportation problem*, which can be solved efficiently with specialized algorithms. The value $L(\hat{x}, \hat{y}, \xi)$, is the dual value corresponding to ξ, and is a lower bound to the optimal cost of the restricted integer program (6.89). The optimal lower bound is obtained by maximizing $L(\hat{x}, \hat{y}, \xi)$ over $\xi \geq 0$.

Note that the three lower bounds involve qualitatively different computations. In particular, in cases (b) and (c), which involve Lagrangian relaxation, an easy optimization problem is solved many times in the context of the algorithm that maximizes the dual function. By contrast, in constraint relaxation [case (a)], a single optimization problem is solved to obtain the lower bound, but the solution of this problem is complicated by the presence of extra constraints.

Regarding the quality of the three bounds, it turns out that the lower bound obtained by constraint relaxation, and the two lower bounds obtained by Lagrangian relaxation, are all equal thanks to the linearity of the cost function, as we will demonstrate shortly in some generality.

Example 6.5.10 (Traveling Salesman Problem – Lagrangian Relaxation)

Consider the traveling salesman problem of Example 6.5.4. Here, we want to find a minimum cost tour in a complete graph where the cost of arc (i,j) is denoted a_{ij}. We formulate this as the discrete optimization problem

$$\text{minimize} \quad \sum_{i=1}^{N} \sum_{\substack{j=1,\ldots,N \\ j \neq i}} a_{ij} x_{ij}$$

$$\text{subject to} \quad \sum_{\substack{j=1,\ldots,N \\ j \neq i}} x_{ij} = 1, \quad i = 1, \ldots, N, \quad (6.90)$$

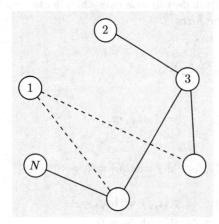

Figure 6.5.7. Illustration of a 1-tree. It consists of a tree that spans nodes $2, \ldots, N$, *plus* two arcs that are incident to node 1.

$$\sum_{\substack{i=1,\ldots,N \\ i \neq j}} x_{ij} = 1, \qquad j = 1, \ldots, N, \tag{6.91}$$

$$x_{ij} = 0 \text{ or } 1, \qquad \forall \ (i,j) \in \mathcal{A}, \tag{6.92}$$

the subgraph with node-arc set $\bigl(\mathcal{N}, \{(i,j) \mid x_{ij} = 1\}\bigr)$ is connected. (6.93)

We may express the connectivity constraint (6.93) in several different ways, leading to different Lagrangian relaxation and branch-and-bound algorithms. One of the most successful formulations is based on the notion of a *1-tree*, which consists of a tree that spans nodes $2, \ldots, N$, *plus* two arcs that are incident to node 1. Equivalently, a 1-tree is a connected subgraph that contains a single cycle passing through node 1 (see Fig. 6.5.7). Note that if the conservation of flow constraints (6.90) and (6.91), and the integer constraints (6.92) are satisfied, then the connectivity constraint (6.93) is equivalent to the constraint that the subgraph $\bigl(\mathcal{N}, \{(i,j) \mid x_{ij} = 1\}\bigr)$ is a 1-tree.

Let X_1 be the set of all x with $0-1$ components, and such that the subgraph $\bigl(\mathcal{N}, \{(i,j) \mid x_{ij} = 1\}\bigr)$ is a 1-tree. Let us consider a Lagrangian relaxation approach based on elimination of the conservation of flow equations. Assigning multipliers u_i and v_j to the constraints (6.90) and (6.91), respectively, the Lagrangian function is

$$L(x, u, v) = \sum_{i,j, i \neq j} (a_{ij} + u_i + v_j) x_{ij} - \sum_{i=1}^{N} u_i - \sum_{j=1}^{N} v_j.$$

The minimization of the Lagrangian is over all 1-trees, leading to the problem

$$\min_{x \in X_1} \left\{ \sum_{i,j, i \neq j} (a_{ij} + u_i + v_j) x_{ij} \right\}.$$

If we view $a_{ij} + u_i + v_j$ as a *modified cost* of arc (i,j), this minimization is quite easy. It is equivalent to obtaining a tree of minimum modified cost that spans the nodes $2, \ldots, N$, and then adding two arcs that are incident to node 1 and have minimum modified cost. The minimum cost spanning tree problem can be easily solved using a variety of efficient algorithms such as the Prim-Dijkstra algorithm (see e.g., [BeT97], [Ber98], [Mur92], [PaS82]).

The preceding examples illustrate an important characteristic of Lagrangian relaxation: *as we eliminate the troublesome constraints $g_j(x) \leq 0$, the problem is often simplified enough so that we also essentially eliminate the integer constraints.* This should be contrasted with the integer constraint relaxation approach, where we eliminate just the integer constraints, while leaving the other inequality constraints unaffected.

Comparing Lagrangian and Constraint Relaxation

Let us now compare the lower bounds obtained by relaxing the integer constraints and by dualizing the inequality constraints. We focus on the special case of a convex cost function $f(x)$ and the constraints

$$a'_j x \leq b_j, \qquad j = 1, \ldots, r,$$

$$x_i \in X_i, \qquad i = 1, \ldots, n,$$

where X_i are given finite subsets of the real line, and a_j and b_j are given vectors and scalars, respectively. Let f^* denote the optimal primal cost,

$$f^* = \inf_{\substack{a'_j x \leq b_j,\, j=1,\ldots,r \\ x_i \in X_i,\, i=1,\ldots,n}} f(x),$$

and let q^* denote the optimal dual cost,

$$q^* = \sup_{\mu \geq 0} q(\mu) = \sup_{\mu \geq 0} \inf_{x_i \in X_i,\, i=1,\ldots,n} L(x, \mu), \qquad (6.94)$$

where

$$L(x, \mu) = f(x) + \sum_{j=1}^{r} \mu_j (a'_j x - b_j)$$

is the Lagrangian function. Let $\hat{X}_i$ denote the interval which is the convex hull of the set X_i, and denote by $\hat{f}$ the optimal cost of the problem, where each set X_i is replaced by $\hat{X}_i$,

$$\hat{f} = \inf_{\substack{a'_j x \leq b_j,\, j=1,\ldots,r \\ x_i \in \hat{X}_i,\, i=1,\ldots,n}} f(x). \qquad (6.95)$$

Note that $\hat{f}$ is the lower bound obtained by constraint relaxation.

We will show that

$$\hat{f} \leq q^*,$$

so that *for a convex cost and linear inequality constraints, the lower bound q^* obtained by Lagrangian relaxation is no worse that the lower bound $\hat{f}$ obtained by constraint relaxation.* Indeed by using Prop. 6.2.1, we see that problem (6.95) has no duality gap, so $\hat{f}$ is equal to the corresponding dual cost, which is

$$\hat{q} = \sup_{\mu \geq 0} \inf_{x_i \in \hat{X}_i,\, i=1,\ldots,n} L(x, \mu). \tag{6.96}$$

By comparing Eqs. (6.94) and (6.96), we see that $\hat{q} \leq q^*$, so we conclude that

$$\hat{f} = \hat{q} \leq q^*.$$

Let us now focus on the special case where f is linear. Then, the Lagrangian function $L(\cdot, \mu)$ is also linear and the minimization in the two dual problems (6.94) and (6.96) yields identical results, so that $\hat{f} = \hat{q} = q^*$. Thus, *when the cost and the inequality constraints are linear, the lower bounds obtained by Lagrangian and constraint relaxation are equal.*

Weaknesses of Lagrangian Relaxation

Finally, let us point out that the Lagrangian relaxation method has several weaknesses. First, the minimization of $L(x, \mu)$ over the set X may yield an x that violates some of the constraints $g_j(x) \leq 0$, so it may be necessary to adjust this x for feasibility using heuristics. There is not much one can say in generality about such heuristics because they are typically problem-dependent.

Second, the maximization of the dual function $q(\mu)$ over $\mu \geq 0$ may be quite nontrivial for a number of reasons, including the fact that q is typically nondifferentiable. In Chapter 7, we will discuss the algorithmic methodology for solving the dual problem, including the subgradient and cutting plane methods, which have enjoyed a great deal of popularity for nondifferentiable optimization.

EXERCISES

6.5.1

Apply branch-and-bound to solve the problem

$$\text{minimize } 5x_1 + 2x_2 + x_3$$

$$\text{subject to } x_1 + 2x_2 + 2x_3 \geq 3, \quad x_1, x_2, x_3 \in \{0, 1\}.$$

6.5.2

Apply branch-and-bound to solve the problem of Exercise 6.1.2. Obtain lower bounds using Lagrangian relaxation and constraint relaxation, and verify that they are identical.

6.5.3 (Separable Problems with Integer/Simplex Constraints) (www)

Consider the problem

$$\text{minimize} \quad \sum_{j=1}^{n} f_j(x_j)$$
$$\text{subject to} \quad \sum_{j=1}^{n} x_j \leq A,$$
$$x_j \in \{0, 1, \ldots, m_j\}, \qquad j = 1, \ldots, n,$$

where A and $m_1, \ldots, m_n$ are given positive integers, and each function f_j is convex over the interval $[0, m_j]$. Consider an iterative algorithm (due to [IbK88]) that starts at $(0, \ldots, 0)$ and maintains a feasible vector $(x_1, \ldots, x_n)$. At the typical iteration, we consider the set of indices $J = \{j \mid x_j < m_j\}$. If J is empty or $\sum_{j=1}^{n} x_j = A$, the algorithm terminates. Otherwise, we find an index $\bar{j} \in J$ that maximizes $f_j(x_j) - f_j(x_j + 1)$. If $f_{\bar{j}}(x_{\bar{j}}) - f_{\bar{j}}(x_{\bar{j}} + 1) \leq 0$, the algorithm terminates. Otherwise, we increase $x_{\bar{j}}$ by one unit, and go to the next iteration. Show that upon termination, the algorithm yields an optimal solution. *Note:* The book [IbK88] contains a lot of material on this problem, and addresses the issues of efficient implementation.

6.5.4 (Monotone Discrete Problems [WuB01]) (www)

Consider the problem

$$\text{minimize} \quad f(x)$$
$$\text{subject to} \quad g_i(x) \leq 0, \qquad i = 1, \ldots, n,$$
$$x_i \in X_i, \qquad i = 1, \ldots, n,$$

where the functions $f : \Re^n \mapsto \Re$ and $g_i : \Re^n \mapsto \Re$ are given, and the sets X_i are *finite* subsets of real numbers. Let X denote the Cartesian product $X_1 \times \cdots \times X_n$ and let $e_i = (0, \ldots, 0, 1, 0, \ldots, 0)'$ denote the ith unit vector. Assume that for all $x = (x_1, \ldots, x_n) \in X$, all $i = 1, \ldots, n$, and all $\xi > x_i$, we have

$$f\big(x + (\xi - x_i)e_i\big) \geq f(x),$$
$$g_i\big(x + (\xi - x_i)e_i\big) < g_i(x),$$
$$g_j\big(x + (\xi - x_i)e_i\big) \geq g_j(x), \qquad \forall\, j \neq i.$$

[In words, this assumption says that f is monotone nondecreasing in all coordinate directions, and g_i is monotone decreasing in the ith direction and monotone nondecreasing in all other directions. It is satisfied for example if f and g_i are linear of the form

$$f(x) = \sum_{i=1}^{n} c_i x_i, \qquad g_i(x) = \sum_{j=1}^{n} a_{ij} x_j - b_i, \quad i = 1, \ldots, n,$$

where for all i, $c_i \geq 0$, $a_{ii} < 0$, and $a_{ij} \geq 0$, for $j \neq i$.]

(a) Assume that x^* is an optimal solution of the problem. Consider the following algorithm: Start with an $x^0 \in X$ satisfying $x^0 \leq x^*$ (for example, set $x_i^0 = \min X_i$). Given x^k, stop if $g_i(x^k) \leq 0$ for all $i = 1, \ldots, n$; otherwise, select any i such that $g_i(x^k) > 0$ and set

$$x^{k+1} = x^k + (\xi - x_i^k) e_i,$$

where ξ is the smallest element of X_i such that $g_i\bigl(x^k + (\xi - x_i^k)e_i\bigr) \leq 0$ (see Fig. 6.5.8). Show that the algorithm is well-defined and terminates after a finite number of iterations with an optimal solution. Under what conditions will this optimal solution be equal to x^*? *Hint*: Verify that $x^k \leq x^*$ for all k.

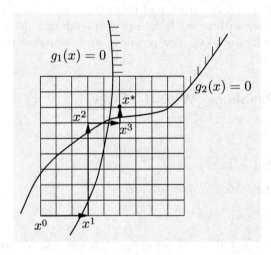

Figure 6.5.8. Illustration of the algorithm of Exercise 6.5.4 in two dimensions. The grid points correspond to the elements of $X_1 \times X_2$.

(b) Show that regardless of the starting point, the algorithm will terminate in a finite number of iterations even if the problem is infeasible. How should the starting point be chosen so that the algorithm can detect infeasibility?

(c) Consider the more general problem where there are inequality constraints $g_i(x) \leq 0$, $i \in I$, for only a subset I of the set of indices $\{1, \ldots, n\}$. Describe how this case can be reduced to the case where $I = \{1, \ldots, n\}$.

6.5.5 (www)

Show that an $m \times n$ matrix E is totally unimodular if and only if every subset J of $\{1, \ldots, n\}$ can be partitioned into two subsets J_1 and J_2 such that

$$\left| \sum_{j \in J_1} e_{ij} - \sum_{j \in J_2} e_{ij} \right| \leq 1, \qquad \forall \ i = 1, \ldots, m.$$

6.5.6 (www)

Let E be a matrix with entries -1, 0, or 1, and at most two nonzero entries in each of its columns. Show that E is totally unimodular if and only if the rows of E can be divided into two subsets such that for each column with two nonzero entries, the following hold: if the two nonzero entries in the column have the same sign, their rows are in different subsets, and if they have the opposite sign, their rows are in the same subset.

6.5.7 (www)

Let E be a matrix with entries that are either 0 or 1. Suppose that in each column, all the entries that are equal to 1 appear consecutively. Show that E is totally unimodular.

6.5.8 (Birkhoff's Theorem for Doubly Stochastic Matrices)

A *doubly stochastic* $n \times n$ matrix $X = \{x_{ij}\}$ is a matrix such that the elements of each of its rows and columns are nonnegative, and add to one. Thus the set of doubly stochastic matrices can be identified with the polyhedral set S of $\Re^{n^2}$ specified by

$$\sum_{j=1}^{n} x_{ij} = 1, \quad \forall \ i, \qquad \sum_{i=1}^{n} x_{ij} = 1, \quad \forall \ j,$$

$$x_{ij} \geq 0, \qquad \forall \ i, j.$$

A *permutation matrix* is a doubly stochastic matrix whose elements are either one or zero.

(a) Show that the extreme points of S are the permutation matrices.

(b) Verify that the matrix specifying S is totally unimodular.

(c) Show that every doubly stochastic matrix X can be written as $\sum_{i=1}^{k} \gamma_i X_i^*$, where X_i^* are permutation matrices and $\gamma_i \geq 0$, $\sum_{i=1}^{k} \gamma_i = 1$.

6.5.9 (Constructing Valid Inequalities [Gom58])

Consider the polyhedral set

$$P = \left\{ x \geq 0 \;\middle|\; \sum_{i=1}^{n} a_i x_i \leq b \right\},$$

where $a_1, \ldots, a_n$ and b are given vectors in $\Re^n$, and $x_1, \ldots, x_n$ are the coordinates of x. Let u be any vector with nonnegative coordinates. Show that the following steps produce progressively tighter inequalities, which are satisfied by all $x \in P$ that are integer.

(1) $\sum_{j=1}^{n} u' a_j x_j \leq u' b$.

(2) $\sum_{j=1}^{n} \lfloor u' a_j \rfloor x_j \leq u' b$.

(3) $\sum_{j=1}^{n} \lfloor u' a_j \rfloor x_j \leq \lfloor u' b \rfloor$.

Note: This procedure is important both for theoretical representations of valid inequalities and for use in algorithms (see e.g., [Wol98], p. 120).

6.5.10 (www)

The purpose of this problem is to show that potentially better bounds can be obtained from Lagrangian relaxation by dualizing as few constraints as possible. Consider the problem

$$\begin{aligned} & \text{minimize} \quad f(x) \\ & \text{subject to} \quad x \in X, \quad g_j(x) \leq 0, \quad j = 1, \ldots, r, \end{aligned} \quad (6.97)$$

where X is a subset of $\Re^n$, and $f : \Re^n \mapsto \Re$, $g_j : \Re^n \mapsto \Re$ are given functions. We assume that the problem is feasible and that its optimal value f^* satisfies $f^* > -\infty$.

Let $\bar{r}$ be an integer with $1 \leq \bar{r} < r$, consider the set

$$\bar{X} = \{x \in X \mid g_{\bar{r}+1}(x) \leq 0, \ldots, g_r(x) \leq 0\},$$

and the problem

$$\begin{aligned} & \text{minimize} \quad f(x) \\ & \text{subject to} \quad x \in \bar{X}, \quad g_j(x) \leq 0, \quad j = 1, \ldots, \bar{r}. \end{aligned} \quad (6.98)$$

Let q^* and $\bar{q}^*$ be the optimal dual values of problems (6.97) and (6.98), respectively.

(a) Show that $q^* \leq \bar{q}^* \leq f^*$.

(b) Show that if problem (6.97) has no duality gap, the same is true for problem (6.98). Furthermore, if $(\mu_1^*, \ldots, \mu_r^*)$ is a geometric multiplier for problem (6.97), then $(\mu_1^*, \ldots, \mu_{\bar{r}}^*)$ is a geometric multiplier for problem (6.98).

(c) Construct an example to show that it is possible that $q^* < \bar{q}^* = f^*$.

6.5.11 (Integer Approximations of Feasible Solutions)

(a) Given a feasible solution x of the network optimization problem of Example 6.5.1, show that there exists an integer feasible solution $\bar{x}$ satisfying

$$|x_{ij} - \bar{x}_{ij}| < 1, \qquad \forall \, (i,j) \in \mathcal{A}.$$

Hint: For each arc (i,j), define the integer flow bounds

$$\bar{b}_{ij} = \lfloor x_{ij} \rfloor, \qquad \bar{c}_{ij} = \lceil x_{ij} \rceil.$$

(b) Generalize the result of part (a) for the case of the unimodular problem of Example 6.5.2.

6.5.12 (Bilinear Cost Function)

Consider a cost function $f : \Re^n \mapsto \Re$, which is linear in each variable when all other variables are held fixed, i.e., for each i, and $\bar{x}_1, \ldots, \bar{x}_{i-1}, \bar{x}_{i+1}, \ldots, \bar{x}_n$, the function $f(\bar{x}_1, \ldots, \bar{x}_{i-1}, x_i, \bar{x}_{i+1}, \ldots, \bar{x}_n)$ is linear as a function of x_i. Show that the minimum of f over a set of the form

$$X = \{x \mid \alpha_i \le x_i \le \beta_i, \, i = 1, \ldots, n\}$$

is attained at an extreme point of X. What is the significance of this when the scalars α_i and β_i are integer?

6.5.13 (Multicommodity Flow Problems)

Contrary to the case of single commodity network flow problems (Example 6.5.1), multicommodity flow problems with integer data need not have integer optimal solutions. To see this, one may use an example involving six nodes, $1, \ldots, 6$, arranged in a ring. Each arc of the ring is bidirectional, and its flow is constrained to lie between 0 and 1 in each direction. Let node 1 send 1 unit of flow to node 4, node 3 send 1 unit of flow to node 6, and node 5 send 1 unit of flow to node 2. Show that there is a feasible solution, but there is no integer feasible solution.

6.5.14 (Integer and Concave Quadratic Programming)

The purpose of this exercise is to show that a 0-1 linear-integer programming problem can be converted to a noninteger minimization problem with concave quadratic cost. Consider the problem

$$\text{minimize} \quad a'x$$
$$\text{subject to} \quad Ax \le b, \qquad x_i \in \{0, 1\}, \quad i = 1, \ldots, n,$$

and the problem

$$\text{minimize} \quad a'x + c\sum_{i=1}^{n} x_i(1-x_i)$$
$$\text{subject to} \quad Ax \leq b, \quad 0 \leq x_i \leq 1, \quad i=1,\ldots,n,$$

where $a \in \Re^n$, $b \in \Re^m$, A is an $m \times n$ matrix, and c is a positive scalar parameter. Show that if c is sufficiently large, the two problems have the same optimal solutions. *Hint*: The minimum of the second problem is attained at one of the extreme points of the polyhedral set

$$\{x \mid Ax \leq b, 0 \leq x_i \leq 1, i=1,\ldots,n\},$$

in view of the concavity of the cost function (cf. Prop. B.19).

6.5.15

Consider the problem

$$\text{minimize} \quad x'Qx + a'x$$
$$\text{subject to} \quad x_i \in \{0,1\}, \quad i=1,\ldots,n,$$

and the problem

$$\text{minimize} \quad x'Qx + a'x + c\sum_{i=1}^{n} x_i(1-x_i)$$
$$\text{subject to} \quad 0 \leq x_i \leq 1, \quad i=1,\ldots,n,$$

where $a \in \Re^n$, Q is a symmetric $n \times n$ matrix, and c is a scalar parameter. Show that if c is sufficiently large so that $Q - cI$ is nonpositive definite, the second problem has an optimal solution, which is also an optimal solution of the first.

6.6 NOTES AND SOURCES

Section 6.1: An extensive account of duality theory for convex programming is given in the book by Rockafellar [Roc70]. The line of analysis of this chapter descends from this book, although it is more visually oriented.

The min-common point and max-crossing point problems were used as intuitive illustrations of duality in earlier editions of this book. They were developed into a systematic analytical framework for duality analysis in joint research with A. Nedić and A. Ozdaglar, which is described in the book [BNO03], and also in the author's book [Ber09]. The approach used there is to obtain a handful of broadly applicable theorems within the min-common/max-crossing framework, and then specialize them to particular problem settings (constrained optimization, Fenchel duality, minimax problems, theorems of the alternative, etc).

Estimates of the duality gap for separable problems are given by Aubin and Ekeland [AuE76], and under a different set of assumptions, by Bertsekas [Ber82a], Section 5.6.1. Related estimates of the duality gap for more general problems, including zero-sum games, are given in [Ber09], Section 5.7.

Section 6.2: The line of analysis for convex programming problems with polyhedral constraint structure descends from Rockafellar [Roc70]. This analysis is generalized to the broader min-common/max-crossing duality framework in the books [BNO03], Section 3.5.2, and [Ber09], Section 4.5. It is also specialized to zero-sum games in [BNO03], Section 3.5.3.

Section 6.3: This section descends from Slater's classical paper [Sla50]. The analysis is extended to the min-common/max-crossing duality framework in the books [BNO03], Section 2.6.2, and [Ber09], Section 4.4, where it is also applied to zero-sum games. For elaboration on Assumption 6.3.1, see Jeyakumar and Wolkowicz [JeW92]. The use of directional convexity in duality theory and optimality conditions for discrete-time optimal control (Exercise 6.3.6) is discussed by Cannon, Cullum, and Polak [CCP70].

Section 6.4: The notion of a conjugate convex function was developed by Fenchel [Fen49], [Fen51], where the relation with the classical Legendre transformation was noted. The Fenchel duality theorems were also derived in these references. A detailed development, including extensions, was given by Rockafellar [Roc70]; see also the books by Luenberger [Lue69], Zalinescu [Zal02], Auslender and Teboulle [AuT03], Bertsekas, Nedić, and Ozdaglar [BNO03], and Bertsekas [Ber09].

Conic programming duality has been known since the early days of convex programming (see e.g., Rockafellar [Roc70], Th. 31.4). The book by Ben-Tal and Nemirovski [BeN01] focuses on conic programming; see also the 2005 class notes by Nemirovski (on line), and the book by Boyd and Vanderberghe [BoV04], which describes many applications.

Monotropic programming has been developed by Rockafellar [Roc67], [Roc81], following the work of Minty [Min60], who treated the special case where the constraint subspace is specified by conservation of flow constraints in a network (cf. Section 6.4.3). An extensive account of monotropic programming theory, associated algorithms, and applications to network optimization are given in the book by Rockafellar [Roc84]. A more condensed version of this theory, including a proof of the main duality result (Prop. 6.4.8), is given in the author's network optimization book [Ber98], Section 9.7. Primal-dual algorithms for monotropic programming are given by Rockafellar [Roc84], and Tseng and Bertsekas [TsB90].

Extended monotropic programming was developed by the author in

[Ber10a] (see also [Ber15a], Chapter 6). A strong duality theorem, based on a constraint qualification involving ϵ-subdifferentials, may be found in these references, along with applications to several types of specially structured convex programming problems.

The sensitivity result of Prop. 6.4.11 is due to Bertsekas, Ozdaglar, and Tseng [BOT06]. This reference also discusses an enhanced form of the Fritz John conditions of Section 4.3.3, adapted to convex programming.

Section 6.5: There is a great variety of integer constrained optimization problems, and the associated methodological and applications literature is vast. For textbook treatments, some of which are restricted to network optimization, see Lawler [Law76], Zoutendijk [Zou76], Papadimitriou and Steiglitz [PaS82], Minoux [Min86], Schrijver [Sch86], [Sch93], Nemhauser and Wolsey [NeW88], Bazaraa, Jarvis, and Sherali [BJS90], Bogart [Bog90], Murty [Mur92], Pulleyblank, Cook, Cunningham, and Schrijver [PCC93], Cameron [Cam94], Floudas [Flo95], Bertsimas and Tsitsiklis [BeT97], Bertsekas [Ber98], Cook, Cunningham, Pulleyblank, and Schrijver [CCP98], Wolsey [Wol98], Sherali and Adams [ShA99], and Horst, Pardalos, and Thoai [HPT00].

Volumes 7 and 8 of the Handbooks in OR/MS, edited by Ball, Magnanti, Monma, and Nemhauser [BMM95a], [BMM95b], are devoted to network theory and applications, and include several excellent survey papers with large bibliographies; see also the survey by Ahuja, Magnanti, and Orlin [AMO91]. O'hEigeartaigh, Lenstra, and Rinnoy Kan [OLR85] provide an extensive bibliography on combinatorial optimization.

The paper by Dantzig, Fulkerson, and Johnson [DFJ54] introduced branch-and-bound method in the context of the traveling salesman problem. This paper was followed by Croes [Cro58], Eastman [Eas58], and Land and Doig [LaD60], who considered versions of the branch-and-bound method in the context of various integer programming problems. Balas and Toth [BaT85], Nemhauser and Wolsey [NeW88], and Wolsey [Wol98] provide extensive surveys of branch-and-bound and its variations. Lagrangian relaxation for discrete optimization was suggested by Held and Karp [HeK70], [HeK71], and was applied to traveling salesman problems.

In many practical discrete optimization problems, the branch-and-bound method is too time-consuming for exact optimal solution, so it can only be used as an approximation scheme. There are other approaches, which do not offer the theoretical guarantees of branch-and-bound, but are much faster in practice. Local search methods, such as genetic algorithms, tabu search, simulated annealing, and rollout algorithms are among the most popular of such alternative possibilities. For some textbook references that also include bibliographies, see Aarts and Lenstra [AaL97], Bertsekas [Ber98], [Ber05a], [Ber12], Bertsimas and Tsitsiklis [BeT97], Glover and Laguna [GlL97], Goldberg [Gol89], and Korst, Aarts, and Korst [KAK89].

7

Dual Methods

Contents

7.1. Dual Derivatives and Subgradients	p. 666
7.2. Dual Ascent Methods for Differentiable Dual Problems	p. 673
7.2.1. Coordinate Ascent for Quadratic Programming	p. 673
7.2.2. Separable Problems and Primal Strict Convexity	p. 675
7.2.3. Partitioning and Dual Strict Concavity	p. 677
7.3. Proximal and Augmented Lagrangian Methods	p. 682
7.3.1. The Method of Multipliers as a Dual Proximal Algorithm	p. 682
7.3.2. Entropy Minimization and Exponential Method of Multipliers	p. 686
7.3.3. Incremental Augmented Lagrangian Methods	p. 687
7.4. Alternating Direction Methods of Multipliers	p. 691
7.4.1. ADMM Applied to Separable Problems	p. 699
7.4.2. Connections Between Augmented Lagrangian-Related Methods	p. 703
7.5. Subgradient-Based Optimization Methods	p. 709
7.5.1. Subgradient Methods	p. 709
7.5.2. Approximate and Incremental Subgradient Methods	p. 714
7.5.3. Cutting Plane Methods	p. 717
7.5.4. Ascent and Approximate Ascent Methods	p. 724
7.6. Decomposition Methods	p. 735
7.6.1. Lagrangian Relaxation of the Coupling Constraints	p. 736
7.6.2. Decomposition by Right-Hand Side Allocation	p. 739
7.7. Notes and Sources	p. 742

In this chapter we consider dual methods, i.e., algorithms for solving dual problems. As a starting point, let us focus on the primal problem

$$\text{minimize } f(x)$$
$$\text{subject to } x \in X, \quad g_j(x) \leq 0, \quad j = 1, \ldots, r, \qquad \text{(P)}$$

and its dual

$$\text{maximize } q(\mu)$$
$$\text{subject to } \mu \geq 0, \qquad \text{(D)}$$

where $f : \Re^n \mapsto \Re$, $g_j : \Re^n \mapsto \Re$ are given functions, X is a subset of $\Re^n$, and

$$q(\mu) = \inf_{x \in X} L(x, \mu) = \inf_{x \in X} \{f(x) + \mu' g(x)\}$$

is the dual function (cf. Section 6.1). We may also consider equality constrained versions of the primal problem, with or without inequality constraints.

As we embark on the study of dual methods, it is worth reflecting on the potential incentives for solving the dual problem in place of the primal. These are:

(a) The dual is a concave problem (concave cost, convex constraint set). By contrast, the primal need not be convex.

(b) The dual may have smaller dimension and/or simpler constraints than the primal.

(c) If there is no duality gap and the dual is solved exactly to yield a geometric multiplier μ^*, all optimal primal solutions can be obtained by minimizing the Lagrangian $L(x, \mu^*)$ over $x \in X$ [however, there may be additional minimizers of $L(x, \mu^*)$ that are primal-infeasible]. Furthermore, if the dual is solved approximately to yield an approximate geometric multiplier μ, and x_μ minimizes $L(x, \mu)$ over $x \in X$, then by applying Prop. 6.1.5, it can be seen that x_μ also solves the problem

$$\text{minimize } f(x)$$
$$\text{subject to } x \in X, \quad g_j(x) \leq g_j(x_\mu), \quad j = 1, \ldots, r.$$

Thus if the constraint violations $g_j(x_\mu)$ are not much larger than zero, x_μ may be an acceptable practical solution.

(d) Even if there is a duality gap, for every $\mu \geq 0$, the dual value $q(\mu)$ is a lower bound to the optimal primal value (the weak duality theorem; Prop. 6.1.3). This lower bound may be useful in the context of discrete optimization and branch-and-bound procedures (cf. Section 6.5.2).

We should also consider some of the difficulties in solving the dual problem. The most important ones are the following:

(a) Evaluating the dual function at any μ requires minimization of the Lagrangian $L(x, \mu)$ over $x \in X$. In effect, this restricts the utility of dual methods to problems where this minimization can either be done in closed form or else is relatively simple; for example, when there is special structure that allows decomposition, as in the separable problems of Section 6.1.5 and the monotropic programming problems of Section 6.4.2.

(b) In many types of problems, the dual function is nondifferentiable, in which case the algorithms of Chapters 1, 2, 3, and 5 do not apply.

(c) Even if we find an optimal dual solution μ^*, it may be difficult to obtain a primal feasible vector x from the minimization of $L(x, \mu^*)$ over $x \in X$ as required by the primal-dual optimality conditions of Prop. 6.1.5, since this minimization can also yield primal-infeasible vectors.

Another important point regarding large-scale optimization problems is that there are several different ways to introduce duality in their solution. For example an alternative strategy to take advantage of separability, often called *partitioning*, is to divide the variables in two subsets, and minimize first with respect to one subset while taking advantage of whatever simplification may arise by fixing the variables in the other subset. In particular, consider the problem

$$\text{minimize} \quad F(x) + G(z) \qquad (7.1)$$
$$\text{subject to} \quad x \in X, \ z \in Z, \ Ax + Bz = d,$$

where $F : \Re^n \mapsto \Re$ and $G : \Re^m \mapsto \Re$ are convex functions, X and Z are convex subsets of $\Re^n$ and $\Re^m$, respectively, and A is an $r \times n$ matrix, B is an $r \times m$ matrix, and $d \in \Re^r$ is a given vector. The problem can be written as

$$\text{minimize} \quad F(x) + \inf_{Bz=d-Ax,\ z \in Z} G(z)$$
$$\text{subject to} \quad x \in X,$$

or

$$\text{minimize} \quad F(x) + H(d - Ax)$$
$$\text{subject to} \quad x \in X,$$

where $H(\cdot)$ is the primal function of the minimization problem involving z above:

$$H(u) = \inf_{Bz=u,\ z \in Z} G(z).$$

Assuming no duality gap, this primal function and its subgradients can be calculated using the corresponding dual function and associated geometric multipliers, as discussed in Section 6.4.5. We will discuss this approach in Section 7.2.

Still another way to use duality in algorithms is based on the use of augmented Lagrangians, in the spirit of the method of multipliers of Section 5.2. In Section 7.3, we will use Fenchel duality to show that under convexity assumptions, the method of multipliers can be viewed as a dual version of the proximal algorithm of Section 3.6. In Section 7.4, we will focus on the problem (7.1) and we will discuss the *alternating direction method of multipliers* (ADMM for short), which algorithmically decouples the variables x and z. It is based on minimization of the augmented Lagrangian

$$L_c(x, z, \lambda) = F(x) + G(z) + \lambda'(Ax + Bz - d) + \frac{c}{2}\|Ax + Bz - d\|^2,$$

where $c > 0$, and takes the form

$$x^{k+1} \in \arg\min_{x \in X} L_c(x, z^k, \lambda^k),$$

$$z^{k+1} \in \arg\min_{z \in Z} L_c(x^{k+1}, z, \lambda^k),$$

$$\lambda^{k+1} = \lambda^k + c(Ax^{k+1} + Bz^{k+1} - d).$$

We will see in Section 7.4 that the ADMM has strong convergence properties and many applications. Note that both the method of multipliers and ADMM operate simultaneously in the spaces of primal and dual variables, and this is reminiscent of the Lagrangian methods of Section 5.4. However, while the Lagrangian methods require differentiability of the cost and constraints, and guarantee local (possibly superlinear) convergence, the methods of the present chapter require convexity, converge globally, but guarantee at most linear convergence (at least the versions discussed here).

Naturally, the differentiability properties of the dual function is a very important determinant of the type of dual method that is appropriate for a given problem. We consequently develop first these properties in Section 7.1. In Section 7.2, we consider methods for the case where q is differentiable, and discuss the applicability of various gradient-based methods from Chapters 1-3. In Sections 7.3 and 7.4 we discuss methods based on augmented Lagrangian minimizations, and we discuss their connections to proximal algorithms. In Sections 7.5 and 7.6, we consider dealing with a nondifferentiable dual cost by using subgradients in place of gradients, and we discuss specific methods for large problems with special structure. The author's book on convex optimization algorithms [Ber15a] provides a more extensive account of the material of this chapter, including convergence analysis that is briefly mentioned or covered selectively in exercises here.

7.1 DUAL DERIVATIVES AND SUBGRADIENTS

In this section, we will use some properties of subgradients of a convex or a concave function, developed in Appendix B, to characterize various differentiability properties that are of interest in the context of duality. We have already characterized in Section 6.4.5 the subgradients of the primal function in terms of optimal solutions of the dual problem, assuming no duality gap. A consequence of this is the essential equivalence of differentiability of the primal function at the origin, and the uniqueness of solution of the dual problem. We now focus on differentiability properties of the dual function (without making any convexity assumptions on the primal problem).

For a given $\mu \in \Re^r$, suppose that x_μ minimizes the Lagrangian $L(x, \mu)$ over $x \in X$,

$$x_\mu \in \arg\min_{x \in X} L(x, \mu) = \arg\min_{x \in X} \{f(x) + \mu' g(x)\}.$$

An important fact for our purposes is that $g(x_\mu)$ *is a subgradient of the dual function q at μ*, i.e.,

$$q(\bar{\mu}) \leq q(\mu) + (\bar{\mu} - \mu)' g(x_\mu), \qquad \forall\, \bar{\mu} \in \Re^r. \tag{7.2}$$

To see this, we use the definition of q and x_μ to write for all $\bar{\mu} \in \Re^r$,

$$\begin{aligned} q(\bar{\mu}) &= \inf_{x \in X} \{f(x) + \bar{\mu}' g(x)\} \\ &\leq f(x_\mu) + \bar{\mu}' g(x_\mu) \\ &= f(x_\mu) + \mu' g(x_\mu) + (\bar{\mu} - \mu)' g(x_\mu) \\ &= q(\mu) + (\bar{\mu} - \mu)' g(x_\mu). \end{aligned}$$

Note that this calculation is valid for all $\mu \in \Re^r$ for which there is a minimizing vector x_μ, regardless of whether $\mu \geq 0$.

What is particularly important here is that we need to compute x_μ anyway in order to evaluate the dual function at μ, so typically *a subgradient $g(x_\mu)$ is obtained at little additional cost*. Moreover, many of the dual methods to be discussed solve the dual problem by computing the dual function value and a subgradient at a sequence of vectors $\{\mu^k\}$. It is not necessary to compute the set of *all* subgradients at μ^k in these methods; a single subgradient is sufficient.

Differentiable Dual Function

Even though the full set of subgradients at a point is not needed for the application of the methods of this chapter, it is still useful to characterize this set. For example it is important to derive conditions under which q is differentiable. We know from our preceding discussion that if q is differentiable at μ, there can be at most one value of $g(x_\mu)$ corresponding to vectors $x_\mu \in X$ minimizing $L(x, \mu)$. This suggests that q is everywhere differentiable (as well as real-valued and concave) if for all μ, $L(x, \mu)$ is minimized at a unique $x_\mu \in X$. Indeed this can be inferred from the convexity results developed in Appendix B, under some assumptions. In particular, we have the following proposition.

Proposition 7.1.1: Let X be a compact set, and let f and g be continuous over X. Assume also that for every $\mu \in \Re^r$, $L(x,\mu)$ is minimized over $x \in X$ at a unique point x_μ. Then, q is everywhere continuously differentiable and

$$\nabla q(\mu) = g(x_\mu), \qquad \forall\, \mu \in \Re^r.$$

Proof: To assert the uniqueness of the subgradient of q at μ, apply Danskin's theorem (Prop. B.22 in Appendix B) with the identifications $x \sim z$, $X \sim Z$, $\mu \sim x$, and $-L(x,\mu) \sim \phi(x,z)$. The assumptions of this theorem are satisfied because X is compact, while $L(x,\mu)$ is continuous as a function of x and concave (in fact linear) as a function of μ. The continuity of the dual gradient ∇q follows from Prop. B.21(c) in Appendix B. **Q.E.D.**

Note that if the constraint functions g_j are linear, X is convex and compact, and f is strictly convex, then the assumptions of Prop. 7.1.1 are satisfied and the dual function q is differentiable. We will focus on this and other related cases in the next section, where we discuss methods for differentiable dual functions.

Twice Differentiable Dual Function

Under particularly favorable circumstances, the dual function may be twice differentiable, which opens up the possibility of using Newton's method to solve the dual problem.

To illustrate this, let us assume the following:

(a) $X = \Re^n$.

(b) f and g_j are twice continuously differentiable convex functions.

(c) There exists a primal and dual optimal solution pair (x^*, μ^*) such that

$$\nabla^2_{xx} L(x^*, \mu^*) : \text{ positive definite.} \qquad (7.3)$$

Consider the system of n equations

$$\nabla_x L(x, \mu) = 0, \qquad (7.4)$$

with the $(n+r)$ unknowns (x, μ). The pair (x^*, μ^*) is a solution of this system, and the Jacobian with respect to x at this solution is $\nabla^2_{xx} L(x^*, \mu^*)$,

Sec. 7.1 Dual Derivatives and Subgradients

which is positive definite by assumption (c). Therefore, the implicit function theorem (Prop. A.25 in Appendix A) applies and asserts that there exist neighborhoods S_{x^*} and S_{μ^*} of x^* and μ^*, respectively, such that Eq. (7.4) has a unique solution x_μ within S_{x^*} for every $\mu \in S_{\mu^*}$, i.e.,

$$\nabla_x L(x_\mu, \mu) = 0, \qquad \forall\, \mu \in S_{\mu^*}. \tag{7.5}$$

For $\mu \geq 0$, the Lagrangian $L(x, \mu)$ is convex as a function of x, so it follows that x_μ minimizes $L(x, \mu)$ over $x \in \Re^n$ for all fixed $\mu \in S_{\mu^*}$ with $\mu \geq 0$, and the dual function satisfies

$$q(\mu) = \min_{x \in \Re^n} L(x, \mu) = L(x_\mu, \mu), \qquad \forall\, \mu \in S_{\mu^*} \cap \{\mu \mid \mu \geq 0\}. \tag{7.6}$$

We can derive the Hessian matrix of q by differentiating twice the above equation. Here, we will use the implicit function theorem to infer that x_μ is continuously differentiable as a function of μ with an $n \times r$ gradient matrix denoted ∇x_μ. By differentiating Eq. (7.6) and by using the fact $\nabla_x L(x_\mu, \mu) = 0$ [cf. Eq. (7.5)], we obtain

$$\nabla q(\mu) = \nabla x_\mu \nabla_x L(x_\mu, \mu) + g(x_\mu) = g(x_\mu),$$

which, incidentally, is consistent with the dual function differentiability result of Prop. 7.1.1 [cf. Eq. (7.5)]. By differentiating again this relation, we have

$$\nabla^2 q(\mu) = \nabla x_\mu \nabla g(x_\mu), \tag{7.7}$$

while by differentiating the relation $\nabla_x L(x_\mu, \mu) = 0$, we obtain

$$\nabla x_\mu \nabla^2_{xx} L(x_\mu, \mu) + \nabla g(x_\mu)' = 0. \tag{7.8}$$

By taking the neighborhood S_{μ^*} sufficiently small if needed, we can assert that

$$\nabla^2_{xx} L(x_\mu, \mu) : \text{ positive definite}, \qquad \forall\, \mu \in S_{\mu^*},$$

so that $\left(\nabla^2_{xx} L(x_\mu, \mu)\right)^{-1}$ exists. From Eq. (7.8) we see that

$$\nabla x_\mu = -\nabla g(x_\mu)' \left(\nabla^2_{xx} L(x_\mu, \mu)\right)^{-1},$$

which by substitution in Eq. (7.7), yields the desired Hessian of the dual function:

$$\nabla^2 q(\mu) = -\nabla g(x_\mu)' \left(\nabla^2_{xx} L(x_\mu, \mu)\right)^{-1} \nabla g(x_\mu). \tag{7.9}$$

Nondifferentiable Dual Function

In many important cases the dual function is nondifferentiable. In particular, this is typically true when X is a discrete set, as for example in integer programming. Then the continuity and compactness assumptions of Prop. 7.1.1 are satisfied, but there typically exist some μ for which $L(x, \mu)$ has multiple minima, leading to nondifferentiabilities. In fact, it can be shown that *if there exists a duality gap, the dual function is nondifferentiable at every dual optimal solution*; see Exercise 7.1.1. Thus, nondifferentiabilities tend to arise at the most interesting points and cannot be ignored in dual methods.

An important special case of a nondifferentiable dual function is when q is polyhedral, i.e., it has the form

$$q(\mu) = \min_{i \in I} \{a_i' \mu + b_i\}, \tag{7.10}$$

where I is a finite index set, and $a_i \in \Re^r$ and b_i are given vectors and scalars, respectively. This case arises, for example, when dealing with a discrete problem where X is a finite set [the elements of the set I in Eq. (7.10) correspond to the elements of X]. The set of all subgradients of q is then the convex hull of the vectors a_i for which the minimum is attained in Eq. (7.10), as stated in the following proposition.

Proposition 7.1.2: Let the dual function q be polyhedral of the form (7.10), and for every $\mu \in \Re^r$, let I_μ be the set of indices attaining the minimum in Eq. (7.10),

$$I_\mu = \{i \in I \mid a_i' \mu + b_i = q(\mu)\}.$$

The set of all subgradients of q at μ is given by

$$\partial q(\mu) = \left\{ g \;\middle|\; g = \sum_{i \in I_\mu} \xi_i a_i,\; \xi_i \geq 0,\; \sum_{i \in I_\mu} \xi_i = 1 \right\}.$$

Proof: Apply Danskin's theorem [Prop. B.22(b) in Appendix B]. **Q.E.D.**

Even though a subgradient may not be a direction of ascent at points μ where $q(\mu)$ is nondifferentiable, it still maintains an important property of the gradient: *it makes an angle less than 90 degrees with all ascent directions at μ*, i.e., all the vectors $\alpha(\bar{\mu} - \mu)$ such that $\alpha > 0$ and $q(\bar{\mu}) > q(\mu)$. In particular, *a small move from μ along any subgradient at μ decreases the* distance to any maximizer μ^* of q. This property follows from Eq. (7.2) and is illustrated in Fig. 7.1.1. It will form the basis for a number of dual methods that use subgradients (see Section 7.5).

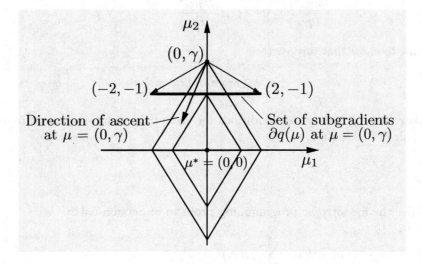

Figure 7.1.1. Illustration of the set of all subgradients $\partial q(\mu)$ at $\mu = (0, 1)$ of the function
$$q(\mu) = -2|\mu_1| - |\mu_2|.$$
By expressing this function as
$$q(\mu) = \min\{-2\mu_1 - \mu_2, 2\mu_1 - \mu_2, -2\mu_1 + \mu_2, 2\mu_1 + \mu_2\},$$
we see that $\partial q(\mu)$ is the convex hull of the gradients of the functions attaining the above minimum (cf. Prop. 7.1.2). Thus, for $\mu = (0, \gamma)$, $\gamma > 0$, the minimum is attained by the first two functions having gradients $(-2, -1)$ and $(2, -1)$. As a result, $\partial q(\mu)$ is as shown in the figure. Note the characteristic property of a subgradient at μ; it makes an angle less than 90 degrees with every ascent direction at μ. As a result, a small move along a subgradient decreases the distance from the maximizing point $\mu^* = (0, 0)$.

EXERCISES

7.1.1

This exercise shows that nondifferentiabilities of the dual function often tend to arise at the most interesting points and thus cannot be ignored. Consider problem (P) and assume that for all $\mu \geq 0$, the infimum of the Lagrangian $L(x, \mu)$ over X

is attained by at least one $x_\mu \in X$. Show that if there is a duality gap, then the dual function $q(\mu) = \inf_{x \in X} L(x, \mu)$ is nondifferentiable at every dual optimal solution. *Hint*: If q is differentiable at a dual optimal solution μ^*, we must have

$$\frac{\partial q(\mu^*)}{\partial \mu_j} \leq 0, \qquad \mu_j^* \frac{\partial q(\mu^*)}{\partial \mu_j} = 0, \qquad \forall\ j.$$

Use this to show that any vector

$$x_{\mu^*} \in \arg\min_{x \in X} L(x, \mu^*)$$

together with μ^* satisfy the conditions for an optimal solution-geometric multiplier pair of Section 6.1.

7.1.2

Consider the monotropic programming problem of Section 6.4.2:

$$\text{minimize} \quad \sum_{i=1}^{n} f_i(x_i)$$

$$\text{subject to } x \in S, \qquad x_i \in X_i, \quad i = 1, \ldots, n.$$

Assume that the sets X_i are closed intervals, and that the functions $f_i : \Re \mapsto \Re$ are strictly convex and satisfy

$$\lim_{|x_i| \to \infty} f_i(x_i) = \infty.$$

Use Danskin's theorem (Prop. B.22 in Appendix B) to show that the dual function has the form

$$\sum_{i=1}^{n} g_i(\lambda_i)$$

and is real-valued and differentiable, where g_i is the conjugate of (f_i, X_i),

$$g_i(\lambda_i) = \sup_{x_i \in X_i} \{x_i \lambda_i - f_i(x_i)\}.$$

Moreover, $\partial g_i(\lambda_i)/\partial \lambda_i$ is the unique scalar attaining the supremum in the preceding equation.

7.1.3

Let $g(\lambda)$ be the convex conjugate function of a pair (f, X) (cf. Section 6.4):

$$g(\lambda) = \sup_{x \in X} \{\lambda' x - f(x)\}, \qquad \lambda \in \Re^n.$$

Show that if x_λ attains the supremum above, then x_λ is a subgradient of g at λ.

7.2 DUAL ASCENT METHODS FOR DIFFERENTIABLE DUAL PROBLEMS

If the dual problem is differentiable, we can apply the methodology for unconstrained and simply constrained problems of Chapters 1-3. For example, the use of scaled steepest ascent, gradient projection, and Newton-like methods is straightforward using the formulas for the first and the second derivatives developed in Section 7.1. In particular, if the Hessian matrix $\nabla^2 q$ is positive definite in a neighborhood of a dual optimal solution [cf. Eq. (7.9)], the problem may be converted to an equivalent problem, for which the first and second order methods of multipliers are equivalent to gradient and Newton methods, thereby bringing to bear the analysis of Section 5.2 (see Exercise 5.2.11). In some cases, even methods requiring line search, such as the conjugate gradient method, become possible thanks to the special structure of the dual function (see Exercise 7.2.1).

Coordinate ascent methods are often particularly interesting in a dual context, because of the convenient structure of the constraint $\mu \geq 0$ of the dual problem. We first discuss such methods for the case of a strictly convex quadratic programming problem. We then consider various special structures and we delineate the type of assumptions that are needed in order to use methods that rely on differentiability.

7.2.1 Coordinate Ascent for Quadratic Programming

Consider the quadratic programming problem

$$\begin{aligned} \text{minimize} \quad & \tfrac{1}{2}x'Qx + c'x \\ \text{subject to} \quad & Ax \leq b, \end{aligned} \qquad (7.11)$$

where Q is a given $n \times n$ positive definite symmetric matrix, A is a given $m \times n$ matrix, and $b \in \Re^m$ and $c \in \Re^n$ are given vectors. Based on the duality theory developed in Section 4.4, the dual of the quadratic programming problem (7.11) is given by

$$\begin{aligned} \text{minimize} \quad & \tfrac{1}{2}\mu'P\mu + t'\mu \\ \text{subject to} \quad & \mu \geq 0, \end{aligned}$$

where

$$P = AQ^{-1}A', \qquad t = b + AQ^{-1}c.$$

It is shown in Section 4.4 that if μ^* solves the dual problem, then $x^* = -Q^{-1}(c + A'\mu^*)$ solves the primal problem (7.11). The dual problem has a simple constraint set, so it may be easier to solve than the primal. Furthermore, it turns out to be particularly well suited to the use of parallel algorithms; see e.g. the books [BeT89] and [CeZ97].

Let a_j denote the jth column of A'. We assume that a_j is nonzero for all j (if $a_j = 0$, then the corresponding constraint $a_j'x \le b_j$ is meaningless and can be eliminated). Since Q is symmetric and positive definite, the jth diagonal element of P, given by $p_{jj} = a_j'Q^{-1}a_j$, is positive. This means that for every j, the dual cost function is strictly convex along the jth coordinate. Therefore, the unique attainment assumption of Prop. 3.7.1 is satisfied and it is possible to use the coordinate ascent method. Because the dual cost is quadratic, the minimization with respect to μ along each coordinate can be done analytically, and the iteration can be written explicitly.

The first partial derivative of the dual cost function with respect to μ_j is given by
$$t_j + \sum_{k=1}^{m} p_{jk}\mu^k,$$
where p_{jk} and t_j are the corresponding elements of the matrix P and the vector t, respectively. Setting the derivative to zero, we see that the unconstrained minimum of the dual cost along the jth coordinate starting from μ is attained at $\tilde{\mu}_j$ given by
$$\tilde{\mu}_j = -\frac{1}{p_{jj}}\left(t_j + \sum_{k \ne j} p_{jk}\mu^k\right) = \mu_j - \frac{1}{p_{jj}}\left(t_j + \sum_{k=1}^{m} p_{jk}\mu^k\right).$$

Taking into account the nonnegativity constraint $\mu_j \ge 0$, we see that the jth coordinate update has the form
$$\mu_j := \max\{0, \tilde{\mu}_j\} = \max\left\{0, \mu_j - \frac{1}{p_{jj}}\left(t_j + \sum_{k=1}^{m} p_{jk}\mu^k\right)\right\}. \tag{7.12}$$

The matrix A often has a sparse structure in practice, and one would like to take advantage of this structure. Unfortunately, the matrix $P = AQ^{-1}A'$ typically has a less advantageous sparsity structure than A. Furthermore, it may be undesirable to calculate and store the elements of P, particularly when m is large. It turns out that the coordinate ascent iteration (7.12) can be performed without explicit knowledge of the elements p_{jk} of the matrix P; only the elements of the matrix AQ^{-1} are needed instead. To see how this can be done, consider the vector
$$\nu = -A'\mu. \tag{7.13}$$
We have
$$P\mu = AQ^{-1}A'\mu = -AQ^{-1}\nu,$$
and the jth component of this vector equation yields
$$\sum_{k=1}^{m} p_{jk}\mu^k = -w_j'\nu, \tag{7.14}$$

Sec. 7.2   *Dual Ascent Methods for Differentiable Dual Problems*   **675**

where w'_j is the jth row of AQ^{-1}. We also have

$$p_{jj} = w'_j a_j, \qquad (7.15)$$

where a_j is the jth column of A'. The coordinate ascent iteration (7.12) can now be written, using Eqs. (7.14) and (7.15), as

$$\mu_i := \begin{cases} \max\left\{0, \mu_j - \dfrac{1}{w'_j a_j}(t_j - w'_j \nu)\right\} & \text{if } i = j, \\ \mu_i & \text{if } i \neq j, \end{cases}$$

or, equivalently,

$$\mu := \mu - \min\left\{\mu_j, \frac{1}{w'_j a_j}(t_j - w'_j \nu)\right\} e_j, \qquad (7.16)$$

where e_j is the jth unit vector (all its elements are 0 except for the jth, which is 1). The corresponding iteration for the vector ν of Eq. (7.13) is obtained by multiplication of Eq. (7.16) with $-A'$ yielding

$$\nu := \nu + \min\left\{\mu_j, \frac{1}{w'_j a_j}(t_j - w'_j \nu)\right\} a_j. \qquad (7.17)$$

The coordinate ascent method can now be summarized as follows. Initially, μ is any vector in the nonnegative orthant and $\nu = -A'\mu$. At each iteration, a coordinate index j is chosen, and μ and ν are iterated using Eqs. (7.16) and (7.17).

7.2.2 Separable Problems and Primal Strict Convexity

We will now consider a strictly convex version of the separable problem discussed in Section 6.1.5:

$$\begin{aligned}
\text{minimize} \quad & \sum_{i=1}^{m} f_i(x_i) \\
\text{subject to} \quad & \sum_{i=1}^{m} g_{ij}(x_i) \leq 0, \qquad j = 1, \ldots, r, \\
& x_i \in X_i, \qquad i = 1, \ldots, m,
\end{aligned} \qquad (7.18)$$

where $f_i : \Re^{n_i} \mapsto \Re$ are strictly convex functions, x_i are the components of x, g_{ij} are given convex functions on $\Re^{n_i}$, and X_i are given convex and compact subsets of $\Re^{n_i}$. We derived the dual problem

$$\begin{aligned}
\text{maximize} \quad & q(\mu) \\
\text{subject to} \quad & \mu \geq 0,
\end{aligned}$$

where the dual function is given by

$$q(\mu) = \inf_{\substack{x_i \in X_i \\ i=1,\ldots,m}} \left\{ \sum_{i=1}^{m} \left(f_i(x_i) + \sum_{j=1}^{r} \mu_j g_{ij}(x_i) \right) \right\} = \sum_{i=1}^{m} q_i(\mu),$$

and

$$q_i(\mu) = \inf_{x_i \in X_i} \left\{ f_i(x_i) + \sum_{j=1}^{r} \mu_j g_{ij}(x_i) \right\}, \qquad i = 1,\ldots,m. \qquad (7.19)$$

Note that since the dual function q is the sum of n functions, $q = \sum_{i=1}^{n} q_i$, the dual problem is suitable for the incremental gradient and Newton methods of Section 2.4, if the first and second order derivatives of q_i, respectively, exist.

By applying Prop. 7.1.1, we see that strict convexity of f_i typically implies that the dual function is continuously differentiable, and that if the minimum in Eq. (7.19) is attained at the point $x_i(\mu)$, the partial derivative of q with respect to μ_j is given by

$$\frac{\partial q(\mu)}{\partial \mu_j} = \sum_{i=1}^{m} g_{ij}\big(x_i(\mu)\big),$$

where $x(\mu) = \{x_1(\mu),\ldots,x_m(\mu)\}$. Since the dual function is differentiable, we can apply methods considered in Chapters 1-3, such as coordinate ascent and gradient projection.

The application of these methods is enhanced by the separable structure of the problem, which in turn greatly facilitates the evaluation of the dual function, and allows the use of distributed, asynchronous, and incremental gradient-like methods. In this chapter, we will not focus much on the possibilities for the use of a parallel computing system, but we refer instead to the books [BeT89], [CeZ97], which describe a number of related algorithms.

Example 7.2.1 (Price Coordination and Hierarchical Decomposition)

Ascent methods for solving the dual of the separable problem (7.18) admit an interesting interpretation in the context of optimization of a system consisting of m subsystems, where the cost function f_i is associated with the operations of the ith subsystem, and there are r resources to be shared by the subsystems. Here x_i is viewed as a local decision vector that influences the cost of the ith subsystem only, and each of the constraints

$$\sum_{i=1}^{m} g_{ij}(x_i) \leq 0$$

is viewed as a restriction of the availability of the jth resource. In particular, $g_{ij}(x_i)$ accounts for the portion of the jth resource consumed by the ith subsystem when its decision vector is x_i.

We may interpret the dual variable μ_j as a price of the jth resource, which is set by a coordinator. Given a price vector μ, the subsystems respond by setting their local decision vectors to values $x_i(\mu)$ that minimize their total cost

$$f_i(x_i) + \sum_{j=1}^{r} \mu_j g_{ij}(x_i),$$

including the cost of the resources at the given price vector μ. A dual ascent algorithm can be viewed as an iterative adjustment by the coordinator, which aims to set the prices at levels such that the subsystems, acting in their own best interest, minimize the overall system cost subject to the given resource constraints.

Another important consequence of separability is that the Hessian of the Lagrangian is a diagonal matrix. This facilitates the use of Newton's method in the case where q is twice differentiable, and $X_i = \Re^{n_i}$. In particular, assuming that (x^*, μ^*) is a primal and dual optimal solution pair, that the components x_i are scalar, and that the positive definiteness condition (7.3) is satisfied, we have from Eq. (7.9)

$$\nabla^2 q(\mu) = -\sum_{i=1}^{n} \frac{G_i' G_i}{\nabla^2_{x_i x_i} L} = -\sum_{i=1}^{n} \frac{G_i' G_i}{\nabla^2 f_i + \sum_{j=1}^{r} \mu_j \nabla^2 g_{ij}},$$

where G_i is the row vector $(\partial g_{i1}/\partial x_i, \ldots, \partial g_{ir}/\partial x_i)$, and all derivatives are evaluated at x_μ. Thus, since $\nabla^2_{x_i x_i} L$ is a scalar, the calculation of $\nabla^2 q$ does not involve a matrix inverse, thereby greatly simplifying the use of Newton's method for solving the dual problem.

Unfortunately, in many convex separable problems the functions f_i are not strictly convex, so even if there is no duality gap, the dual function may not be differentiable. In such cases it is possible to approach the problem via the proximal ideas of Section 3.6, and introduce strict convexity by adding to the primal cost function $\sum_{i=1}^{m} f_i(x_i)$ the corresponding strictly convex proximal term. Thus the problem is solved by solving a sequence of strictly convex separable problems, each of which has a differentiable dual problem.

7.2.3 Partitioning and Dual Strict Concavity

We now consider a different way to take advantage of separability, which is based on the partitioning approach outlined in the introduction to this chapter. Our starting point is the problem

$$\begin{aligned} \text{minimize} \quad & F(x) + G(z) \\ \text{subject to} \quad & x \in X, \ z \in Z, \ Ax + Bz = d, \end{aligned} \quad (7.20)$$

where $F : \Re^n \mapsto \Re$ and $G : \Re^m \mapsto \Re$ are convex functions, X and Z are convex subsets of $\Re^n$ and $\Re^m$, respectively, A is an $r \times n$ matrix, B is an $r \times m$ matrix, and $d \in \Re^r$ is a given vector [cf. problem (7.1)]. The optimization variables are x and z, and they are linked through the constraint $Ax + Bz = d$.

The idea here is to eliminate z by expressing its optimal value as a function of x. In particular, we first consider optimization with respect to z for a fixed value of x, i.e.,

$$\text{minimize } G(z)$$
$$\text{subject to } Bz = d - Ax, \quad z \in Z, \quad (7.21)$$

and then minimize with respect to x. This leads to the problem

$$\text{minimize } F(x) + H(d - Ax)$$
$$\text{subject to } x \in X, \quad (7.22)$$

where

$$H(u) = \inf_{z \in Z,\, Bz = u} G(z). \quad (7.23)$$

If x^* is an optimal solution of this problem and z^* is an optimal solution of problem (7.21) when $x = x^*$, then (x^*, z^*) is an optimal solution of the original problem (7.20). We call problem (7.22)-(7.23) the *master problem*. Note that the function H is related to the primal function of problem (7.21). The following proposition summarizes its differentiability properties.

Proposition 7.2.1: Assume that problem (7.21) has an optimal solution and at least one geometric multiplier for each $x \in X$.

(a) The set of subgradients of the function H at $d - Ax$ is the set of all $-\lambda$, where λ is a geometric multiplier of the problem (7.21) corresponding to the constraint $Bz = d - Ax$.

(b) We have
$$H(u) = \sup_{\lambda \in \Re^r} \{\tilde{q}(\lambda) - \lambda' u\}, \quad (7.24)$$
where
$$\tilde{q}(\lambda) = \inf_{z \in Z} \{G(z) + \lambda' Bz\}.$$

Furthermore, H is differentiable if $\tilde{q}$ is strictly concave over the set $\{\lambda \mid \tilde{q}(\lambda) > -\infty\}$.

Proof: Part (a) follows from the analysis of Section 6.4.5. For part (b), we use the fact that $\tilde{q}(\lambda) - \lambda' u$ is the dual function for the minimization defining H, so H is differentiable at u if there is a unique λ attaining the maximum in Eq. (7.24). **Q.E.D.**

The master problem (7.22)-(7.23) can be solved using an iterative method that updates x based on gradients or subgradients of H, obtained by solving problems of the form (7.21), as shown by Prop. 7.2.1. In some special cases, H can be explicitly calculated; see Exercise 7.2.3.

Let us now focus on the case where $z = (z_1, \ldots, z_m)$, G and Z have the separable forms

$$G(z) = \sum_{j=1}^{m} G_j(z_j), \qquad Z = \{z \mid z_j \in Z_j, j = 1, \ldots, m\},$$

and the constraint $Ax + Bz = d$ also has the separable form

$$A_j x + B_j z_j = d_j, \qquad j = 1, \ldots, m.$$

Then the master problem becomes

$$\text{minimize} \quad F(x) + \sum_{j=1}^{m} H_j(d_j - A_j x)$$

$$\text{subject to} \quad x \in X,$$

where

$$H_j(u_j) = \inf_{z_j \in Z_j, \, B_j z_j = u_j} G_j(z_j), \qquad j = 1, \ldots, m.$$

Note that this problem has an additive cost function, and it can be solved using the incremental methods discussed in Section 2.4, provided H_j are differentiable functions and the problem is unconstrained.†

Problems of the preceding form arise often in practice. A representative example is *stochastic programming*, an optimization model with many applications (see e.g. the books [KaW94], [Pre95], [BiL97], [Fra12], and [ShD14], and the computational surveys [Rus97] and [Sah04]). This model can also be viewed as a stochastic optimal control problem and addressed by dynamic programming methods (see e.g., [VaW89], [Ber05a]), but here we will adopt a nonlinear programming approach, as described in the following example.

---

† If the functions H_j are nondifferentiable and/or some of the constraint sets X and Z_j are strict subsets of the corresponding Euclidean spaces, the incremental subgradient methods to be discussed in Section 7.5.2 apply.

Example 7.2.2 (Stochastic Programming and Incremental Solution Methods)

In a two-stage version of the stochastic programming problem, a vector $x \in X$ is selected, a random event occurs that has m possible outcomes $w_1, \ldots, w_m$, and then another vector $z \in Z$ is selected with knowledge of the outcome that occurred. Thus for optimization purposes, we need to specify a different vector $z_j \in Z$ for each outcome w_j. The problem is to minimize the expected cost

$$F(x) + \sum_{j=1}^{m} \pi_j G_j(z_j),$$

where $G_j(z_j)$ is the cost associated with the occurrence of w_j and π_j is the corresponding probability. There are also constraints $A_j x + B_j z_j = d_j$, $j = 1, \ldots, m$, which couple x and z_j, so the problem becomes a special case of the problem formulated above.

The master problem is to minimize over $x \in X$,

$$F(x) + \sum_{j=1}^{m} H_j(d_j - A_j x),$$

where H_j is given by

$$H_j(u_j) = \inf_{z_j \in Z_j,\, B_j z_j = u_j} \pi_j G_j(z_j), \qquad j = 1, \ldots, m.$$

By Prop. 7.2.1(b), H_j differentiable if the function

$$\tilde{q}_j(\lambda_j) = \inf_{z_j \in Z_j} \left\{ \pi_j G_j(z_j) + \lambda_j' B_j z_j \right\}$$

is strictly concave. In this case the incremental gradient methods of Section 2.4 and their constrained extensions apply.

EXERCISES

7.2.1 (Line Search in Dual Ascent Methods [Pan84])

Consider the convex programming problem $\min_{Ax=b,\, x \in X} f(x)$ and its dual function $q(\lambda)$. Given a vector λ and a direction d, suppose that we want to find a stepsize α^* by the line search

$$\alpha^* \in \arg\max_{\alpha \geq 0} q(\lambda + \alpha d).$$

Show that α^* is a geometric multiplier for the single-constraint problem

$$\text{minimize} \quad f(x) + \lambda' A x$$
$$\text{subject to} \quad d'(Ax - b) \leq 0, \quad x \in X.$$

7.2.2

Consider the separable problem of Section 7.2.2 for the case where each x_i is not a scalar coordinate but rather is an m-dimensional vector. Write a form of $\nabla^2 q$ that involves inversion of order m.

7.2.3 (Partitioning and Thevenin Decomposition [Ber96a])

Consider the problem of minimizing $F(x) + G(z)$ subject to $x \in X$, $z \in Z$, and $Ax + Bz = d$, where $F : \Re^n \mapsto \Re$ and $G : \Re^m \mapsto \Re$ are convex functions, X and Z are convex subsets of $\Re^n$ and $\Re^m$, respectively, A is an $r \times n$ matrix, B is an $r \times m$ matrix, and $d \in \Re^r$ is a given vector [cf. problem (7.20)]. Assume that the matrix B has rank r, and that

$$G(z) = \tfrac{1}{2} z'Rz + w'z, \qquad Z = \{z \mid Dz = d\},$$

where R is a positive definite symmetric $m \times m$ matrix, D is a given matrix, and d, w are given vectors. Assume further that the constraint set $\{z \mid Bz = d - Ax, Dz = d\}$ is nonempty for all x.

(a) Show that the function H appearing in the master problem (7.22) is given by

$$H(d - Ax) = \min_{Bz = d-Ax,\, z \in Z} G(z) = \tfrac{1}{2}(Ax - b)'M(Ax - b) + \gamma,$$

where M is an $r \times r$ positive definite symmetric matrix, γ is a constant, and

$$b = d - B\bar{z},$$

with

$$\bar{z} \in \arg\min_{z \in Z} G(z).$$

Furthermore, the vector

$$M(Ax - b)$$

is the unique geometric multiplier of the problem

$$\text{minimize } G(z)$$
$$\text{subject to } Bz = d - Ax, \qquad z \in Z,$$

[cf. Eq. (7.21)] associated with the constraint $Bz = d - Ax$.

(b) Suppose that the matrix A has rank r. Show how to calculate the columns of M by solving r problems of the form (7.21) with x equal to each of r vectors such that the corresponding vectors $Ax - b$ are linearly independent.

Note: Thevenin's theorem is a celebrated result from electric circuit theory. For a discussion of its connection with optimization and the results of this exercise, see the author's paper [Ber96a] and the textbook [Ber98] (Example 9.3).

7.3 PROXIMAL AND AUGMENTED LAGRANGIAN METHODS

In this section we will explore an important equivalence relation, within a convex programming context, between the proximal algorithm of Section 3.6 and the method of multipliers of Section 5.2.2. Glimpses of this connection may be obtained from the duality analysis of Section 5.2.4. In the presence of convexity, we will show that this connection is stronger, and can be based on the Fenchel duality theorem, and the conjugacy relation between the primal and the dual functions discussed in Section 6.4.5.

7.3.1 The Method of Multipliers as a Dual Proximal Algorithm

Consider the constrained minimization problem

$$\begin{aligned} \text{minimize} \quad & f(x) \\ \text{subject to} \quad & x \in X, \quad Ax = b, \end{aligned} \tag{7.25}$$

where $f : \Re^n \mapsto \Re$ is a convex function, X is a closed convex set, A is an $m \times n$ matrix, and $b \in \Re^m$.† Consider also the corresponding primal and dual functions

$$p(u) = \inf_{x \in X,\, Ax - b = u} f(x), \qquad q(\lambda) = \inf_{x \in X} \{f(x) + \lambda'(Ax - b)\}.$$

Let us apply the proximal algorithm of Section 3.6 to the dual problem of maximizing q (reversing signs and replacing minimization by maximization as necessary). It has the form

$$\lambda^{k+1} \in \arg\max_{\lambda \in \Re^m} \left\{ q(\lambda) - \frac{1}{2c^k} \|\lambda - \lambda^k\|^2 \right\}. \tag{7.26}$$

In what follows, we assume that $\{c^k\}$ is bounded below by a positive number, so that by Prop. 3.6.3, the sequence $\{q(\lambda^k)\}$ converges to the optimal dual value, and $\{\lambda^k\}$ converges to an optimal dual solution, provided such a solution exists. We will use the Fenchel duality theorem to write the above maximization in an equivalent form involving the conjugates of the two functions involved.

Indeed, the convex conjugate of $\frac{1}{2c^k}\|\lambda - \lambda^k\|^2$ is the function $\lambda^{k'}u + \frac{c^k}{2}\|u\|^2$, as can be verified with a straightforward calculation. Moreover, according to the discussion of Section 6.4.5, q and p are conjugates of each

---

† We focus on linear equality constraints for convenience, but the analysis can be extended to convex inequality constraints as well. In particular, as in Section 5.2.2, we may argue by converting a linear inequality constraint of the form $a_j'x \le b_j$ to a linear equality constraint $a_j'x + z_j = b_j$, and a slack variable constraint $z_j \ge 0$ that can be absorbed into the set X.

other, in the sense that $-q(-\lambda)$ is the conjugate convex function of $p(u)$. Thus by the Fenchel duality theorem (with the necessary sign adjustments, and assuming the conditions of the theorem are fulfilled), the maximization in Eq. (7.26) can be equivalently done by performing the minimization

$$u^{k+1} \in \arg\min_{u \in \Re^m} \left\{ p(u) + \lambda^{k'} u + \frac{c^k}{2} \|u\|^2 \right\}, \tag{7.27}$$

(cf. Prop. 6.4.2), and by following it with the Lagrangian optimality equation

$$\lambda^{k+1} = \lambda^k + c^k u^{k+1}; \tag{7.28}$$

(cf. the third condition of Prop. 6.4.1).

While the minimization in Eq. (7.27) involves the unknown primal function p, it can be conveniently written in terms of the augmented Lagrangian function

$$L_c(x, \lambda) = f(x) + \lambda'(Ax - b) + \frac{c}{2} \|Ax - b\|^2, \qquad x \in \Re^n, \ \lambda \in \Re^m.$$

In particular, we can use the definition of p to write the minimization (7.27) as

$$\inf_{u \in \Re^m} \left\{ \inf_{x \in X, \ Ax - b = u} \{f(x)\} + \lambda^{k'} u + \frac{c^k}{2} \|u\|^2 \right\}$$

$$= \inf_{u \in \Re^m} \inf_{x \in X, \ Ax - b = u} \left\{ f(x) + \lambda^{k'}(Ax - b) + \frac{c^k}{2} \|Ax - b\|^2 \right\}$$

$$= \inf_{x \in X} \left\{ f(x) + \lambda^{k'}(Ax - b) + \frac{c^k}{2} \|Ax - b\|^2 \right\}$$

$$= \inf_{x \in X} L_{c^k}(x, \lambda^k).$$

The minimizing u and x in this calculation are related by the equation

$$u^{k+1} = Ax^{k+1} - b,$$

where x^{k+1} is any vector that minimizes $L_{c^k}(\cdot, \lambda^k)$ over X. We assume that such vectors exist. The existence of the minimizing u^{k+1} is guaranteed if the epigraph of p, $\{(u, w) \mid u \in \Re^r, w \in \Re, p(u) \leq w\}$, is closed, since then the minimization (7.27) has a unique solution [cf. Props. 1.1.2 and B.10(b)]. The existence of the minimizing x^{k+1} is not guaranteed, and must be either assumed or verified independently; for example by showing that either f is coercive or that $A'A$ is invertible, in which case $L_c(\cdot, \lambda)$ has a nonempty and compact set of minima [cf. Prop. B.10(b)].

Thus the iteration (7.27)-(7.28) [which is equivalent to the proximal algorithm (7.26), applied to maximization of the dual function q] can be written as

$$\lambda^{k+1} = \lambda^k + c^k(Ax^{k+1} - b), \tag{7.29}$$

where
$$x^{k+1} \in \arg\min_{x \in X} L_{c^k}(x, \lambda^k). \tag{7.30}$$

This is precisely the first order method of multipliers of Section 5.2.2.

One consequence of this connection is that the convergence properties of the multiplier method for convex programming problems can be derived from the corresponding properties of the proximal algorithm (cf. Section 3.6). Assuming that there exists a dual optimal solution, so that $\{\lambda^k\}$ converges to such a solution, we also claim that every limit point of the generated sequence $\{x^k\}$ is an optimal solution of the primal problem (7.25). To see this, note that using Eq. (7.29), we obtain

$$c^k(Ax^{k+1} - b) \to 0, \qquad Ax^{k+1} - b \to 0.$$

Furthermore, we have

$$L_{c^k}(x^{k+1}, \lambda^k) = \min_{x \in X}\left\{f(x) + \lambda^{k'}(Ax - b) + \frac{c^k}{2}\|Ax - b\|^2\right\}.$$

The preceding relations yield

$$\limsup_{k \to \infty} f(x^{k+1}) = \limsup_{k \to \infty} L_{c^k}(x^{k+1}, \lambda^k) \le f(x), \quad \forall\, x \in X \text{ with } Ax = b,$$

so if $x^* \in X$ is a limit point of $\{x^k\}$, we obtain

$$f(x^*) \le f(x), \qquad \forall\, x \in X \text{ with } Ax = b,$$

as well as $Ax^* = b$ (in view of $Ax^{k+1} - b \to 0$). Therefore any limit point x^* of the generated sequence $\{x^k\}$ is an optimal solution of the primal problem (7.25).

Note that there is no guarantee that $\{x^k\}$ has a limit point, and indeed the dual sequence $\{\lambda^k\}$ will converge to a dual optimal solution if one exists, even if the primal problem (7.25) does not have an optimal solution. As an example, the reader may verify that for the two-dimensional/single constraint problem

$$\text{minimize} \quad f(x) = e^{x_1}$$
$$\text{subject to} \quad x_1 + x_2 = 0, \qquad x_1 \in \Re, \quad x_2 \ge 0,$$

the dual optimal solution is $\lambda^* = 0$, but there is no primal optimal solution. For this problem, the augmented Lagrangian algorithm will generate sequences $\{\lambda^k\}$ and $\{x^k\}$ such that $\lambda^k \to 0$ and $x_1^k \to -\infty$.

Another property of the multiplier method, which can be derived from the convergence properties of the proximal algorithm, is that it converges to an optimal dual solution in a finite number of iterations for problems

Sec. 7.3   Proximal and Augmented Lagrangian Methods

with polyhedral cost functions and linear constraints (including linear programs), assuming at least one dual solution exists. Moreover it converges in a single iteration, if the initial penalty parameter c^0 is sufficiently large. This property, first shown in [Ber75a], is a direct consequence of Prop. 3.6.5.

One difficulty with the multiplier method is that even if the cost function is separable, the augmented Lagrangian $L_c(\cdot, \lambda)$ is typically nonseparable because it involves the quadratic term $\|Ax - b\|^2$. With some reformulation, however, it is possible to preserve a good deal of the separable structure through the use of coordinate descent ideas, as described in the following example.

Example 7.3.1: (Additive Cost Problems)

Consider the problem

$$\text{minimize} \quad \sum_{i=1}^{m} f_i(x)$$
$$\text{subject to} \quad x \in \cap_{i=1}^{m} X_i,$$

where $f_i : \Re^n \mapsto \Re$ are convex functions and X_i are closed, convex sets with nonempty intersection. There are several important special cases of this problem, such as regularized regression, classification, and maximum likelihood estimation (see e.g., [Ber15a], Section 1.3).

We introduce additional artificial variables z_i, $i = 1, \ldots, m$, we consider the equivalent problem

$$\text{minimize} \quad \sum_{i=1}^{m} f_i(z_i)$$
$$\text{subject to} \quad x = z_i, \quad z_i \in X_i, \quad i = 1, \ldots, m,$$

and we apply the multiplier method to eliminate the constraints

$$z_i = x,$$

using corresponding multiplier vectors λ_i. The method takes the form

$$\lambda_i^{k+1} = \lambda_i^k + c^k(x^{k+1} - z_i^{k+1}), \quad i = 1, \ldots, m, \qquad (7.31)$$

where x^{k+1} and z_i^{k+1}, $i = 1, \ldots, m$, solve the problem

$$\text{minimize} \quad \sum_{i=1}^{m} \left(f_i(z_i) + \lambda_i^{k'}(x - z_i) + \frac{c^k}{2} \|x - z_i\|^2 \right)$$
$$\text{subject to} \quad x \in \Re^n, \quad z_i \in X_i, \quad i = 1, \ldots, m.$$

Note that there is coupling between x and all the vectors z_i, so this problem cannot be decomposed into separate minimizations with respect to some of the variables. On the other hand, the problem has a Cartesian product constraint set, and a structure that is suitable for the application of block coordinate descent methods that cyclically minimize the cost function, one component at a time. In particular, we can consider a method that minimizes the augmented Lagrangian with respect to $x \in \Re^n$ with the iteration

$$x := \frac{\sum_{i=1}^{m} z_i}{m} - \frac{\sum_{i=1}^{m} \lambda_i^k}{mc^k}, \qquad (7.32)$$

then minimizes the augmented Lagrangian with respect to $z_i \in X_i$, with the iteration

$$z_i \in \arg\min_{z_i \in X_i} \left\{ f_i(z_i) - \lambda_i^{k'} z_i + \frac{c^k}{2} \|x - z_i\|^2 \right\}, \qquad i = 1, \ldots, m, \qquad (7.33)$$

and repeats until convergence to a minimum of the augmented Lagrangian, which is then followed by a multiplier update of the form (7.31). Of course, the coordinate descent method can be very slow, because of ill-conditioning, particularly if a large value of penalty parameter c^k is used (which enhances in turn the convergence rate of the method of multipliers, as discussed in Section 5.2, and more generally the proximal algorithm as discussed in Section 3.6).

In the preceding example the minimization of the augmented Lagrangian exploits the problem structure, yet requires an infinite number of cyclic minimization iterations of the form (7.32)-(7.33), before the multiplier update (7.31) can be performed. Actually, exact convergence to a minimum of the augmented Lagrangian is not necessary, only a limited number of minimization cycles in x and z_i may be performed prior to a multiplier update. In particular, there are versions of methods of multipliers, with sound convergence properties, which allow for inexact minimization of the augmented Lagrangian, subject to certain termination criteria, as noted in Section 5.2.

In Section 7.3.3 we will consider additional methods to take advantage of the separable structure in methods that use augmented Lagrangians. Moreover, in Section 7.4, we will discuss the alternating direction method of multipliers (ADMM for short), a somewhat different type of method, which is based on augmented Lagrangian ideas and performs only one minimization cycle in x and z_i before updating the multiplier vectors.

7.3.2 Entropy Minimization and Exponential Method of Multipliers

The duality correspondence between the proximal and augmented Lagrangian algorithms can be expanded considerably by introducing more general penalty functions. As an illustration, let us consider the problem

minimize $f(x)$

subject to $x \in X, \qquad g_j(x) \leq 0, \quad j = 1, \ldots, r,$

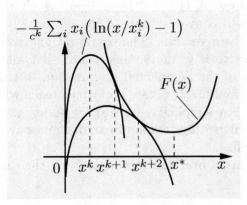

Figure 7.3.1. Illustration of the entropy minimization algorithm, applied to the generic problem of minimizing a convex function $F : \Re^n \mapsto \Re$ over a convex set X, which is a nonempty closed convex subset of the nonnegative orthant.

where $f : \Re^n \mapsto \Re$, $g_j : \Re^n \mapsto \Re$ are convex functions, and X is a convex subset of $\Re^n$.

Suppose that we use an exponential penalty function in place of the quadratic. Then the earlier Fenchel duality-based line of analysis can be used to show that the dual counterpart of the exponential method of multipliers of Section 5.2.5 is a variant of the proximal algorithm of Section 3.6, called *entropy minimization algorithm*. This algorithm, when applied to the generic problem of minimizing a convex function $F : \Re^n \mapsto \Re$ over a convex set X, which is a convex strict subset of the nonnegative orthant,

$$\text{minimize} \quad F(x)$$
$$\text{subject to} \quad x \in X,$$

takes the form

$$x^{k+1} \in \arg\min_{x \in X} \left\{ F(x) + \frac{1}{c^k} \sum_{i=1}^{n} x_i \left(\ln\left(\frac{x_i}{x_i^k}\right) - 1 \right) \right\};$$

see Fig. 7.3.1.

The convergence properties of the entropy minimization algorithm are quite similar to those of the proximal algorithm of Section 3.6. However, the analysis is more complicated, because when one of the coordinates x_i^k tends to zero, the fraction of the corresponding entropy term tends to infinity. We refer to the literature cited at the end of the chapter.

7.3.3 Incremental Augmented Lagrangian Methods

Let us now consider the separable constrained optimization problem

$$\text{minimize} \quad \sum_{i=1}^{m} f_i(x_i)$$

$$\text{subject to} \quad x_i \in X_i, \ i = 1, \ldots, m, \quad \sum_{i=1}^{m} (A_i x_i - b_i) = 0.$$

Here $f_i : \Re^{n_i} \mapsto \Re$ are convex functions (n_i is a positive integer, which may depend on i), X_i are nonempty closed convex subsets of $\Re^{n_i}$, A_i are given $r \times n_i$ matrices, and $b_i \in \Re^r$ are given vectors. The optimization vector is $x = (x_1, \ldots, x_m)$, and our objective is to derive algorithms that allow decomposition in the minimization of the augmented Lagrangian, so that m separate augmented Lagrangian minimizations are performed, each with respect to a single component x_i. For simplicity, we restrict ourselves to linear equality constraints, but the algorithms extend in an obvious way to convex inequality constraints as well.

By decomposing the minimization over the components x_i, the dual function q can be expressed as

$$q(\lambda) = \sum_{i=1}^{m} q_i(\lambda),$$

where q_i is the concave function

$$q_i(\lambda) = \inf_{x_i \in X_i} \{f_i(x_i) + \lambda'(A_i x_i - b_i)\}, \qquad i = 1, \ldots, m, \qquad (7.34)$$

similar to Section 7.2.2. The additive form of the dual function q makes it suitable for application of incremental methods. In the case where the components q_i are differentiable [which is true if the infimum in the definition (7.34) is attained uniquely for all λ], we may use the incremental proximal methods discussed in Section 3.6.2. The component gradients $\nabla q_i(\lambda)$ can be determined as $A_i x_i - b_i$, where x_i minimizes in Eq. (7.34) (cf. the discussion of Section 7.1).

The incremental proximal algorithm discussed in Section 3.6.2 takes the form

$$\lambda^{k+1} = \arg\max_{\lambda \in \Re^r} \left\{ q_{i_k}(\lambda) - \frac{1}{2c^k} \|\lambda - \lambda^k\|^2 \right\},$$

where i_k is the index of the dual function component selected for iteration at time k. Similar to the analysis that connects proximal and multiplier iterations (cf. Section 7.3.1), it can be given as a two-step process, whereby the single component minimization is performed

$$x_{i_k}^{k+1} \in \arg\min_{x_{i_k} \in X_{i_k}} \left\{ f_{i_k}(x_{i_k}) + \lambda^{k'}(A_{i_k} x_{i_k} - b_{i_k}) + \frac{c^k}{2} \|A_{i_k} x_{i_k} - b_{i_k}\|^2 \right\},$$

$$(7.35)$$

while the others are kept unchanged, $x_i^{k+1} = x_i^k$ for all $i \neq i_k$, and λ is updated according to

$$\lambda^{k+1} = \lambda^k + c^k(A_{i_k} x_{i_k}^{k+1} - b_{i_k}). \qquad (7.36)$$

Note that the augmented Lagrangian minimization (7.35) is decoupled with respect to the components x_i. This overcomes the major limitation of the augmented Lagrangian approach for separable problems, at the

Sec. 7.3 Proximal and Augmented Lagrangian Methods

expense of more restrictive operational requirements: for convergence, the parameter sequence $\{c^k\}$ should be chosen to be diminishing, as discussed in Section 3.6.2. Unfortunately, this runs contrary to our earlier suggestion for (nonincremental) methods of multipliers as applied to nonconvex problems (cf. Section 5.2.2). We recommended there an increasing or constant sequence $\{c^k\}$, both to ensure sufficient convexification and to accelerate convergence. The next algorithm aims to guarantee convergence with a constant (but sufficiently small) c^k under some conditions.

Incremental Aggregated Version

Let us consider an aggregated version of the algorithm (7.35)-(7.36), which is patterned after the corresponding incremental aggregated proximal algorithm discussed in Section 3.6.2. Similarly, this algorithm can be implemented in terms of augmented Lagrangian minimizations: at iteration k, it selects a component index i_k, and updates the single component x_{i_k} according to

$$x_{i_k}^{k+1} \in \arg\min_{x_{i_k} \in X_{i_k}} \left\{ f_{i_k}(x_{i_k}) + \lambda^{k'} A_{i_k} x_{i_k} + \frac{c^k}{2} \left\| A_{i_k} x_{i_k} + \sum_{i \neq i_k} A_i x_i^k - b \right\|^2 \right\} \tag{7.37}$$

while keeping the others unchanged, $x_i^{k+1} = x_i^k$ for all $i \neq i_k$. Then it updates λ according to

$$\lambda^{k+1} = \lambda^k + c^k \left(\sum_{i=1}^m A_i x_i^{k+1} - b \right). \tag{7.38}$$

The difference from the nonaggregated algorithm (7.35)-(7.36) is that the penalty term in the minimization in Eq. (7.37) involves the components x_i^k, with $i \neq i_k$, which are fixed at their current values. Using this penalty term in place of the one in the minimization Eq. (7.35) may be more effective. Indeed, by using the results for incremental proximal methods, which were discussed in Section 3.6.2, one may prove a linear convergence result for the algorithm (7.37)-(7.38) (under appropriate assumptions, which include that c^k should be sufficiently small).

Note that as in other incremental methods, all component indexes in both aggregated and nonaggregated methods, should be selected for minimization in Eq. (7.35) or Eq. (7.37) with equal long-term frequency. The author's paper [Ber15b] contains further discussion and analysis, and surveys augmented Lagrangian-like methods for separable problems.

Incremental Aggregated Augmented Lagrangian Methods for Inequality Constraints

The preceding development applies to problem (7.25) that has equality constraints. Let us now consider incremental aggregated augmented Lagrangian methods for the inequality-constrained separable constrained optimization problem

$$\text{minimize} \quad \sum_{i=1}^{m} f_i(x_i)$$

$$\text{subject to} \quad x_i \in X_i, \ i=1,\ldots,m, \quad \sum_{i=1}^{m} g_{ij}(x_i) \leq 0, \ j=1,\ldots,r,$$

where f_i and g_{ij} are convex real-valued functions, and X_i are convex sets. In analogy with Sections 5.2.5 and 7.3.2, we may use a nonquadratic penalty ψ for the inequality constraints, such as the exponential. This method maintains a vector $\mu^k > 0$ and operates as follows.

Select a component index i_k, and update the single component x_{i_k} according to

$$x_{i_k}^{k+1} \in \arg\min_{x_{i_k} \in X_{i_k}} \left\{ f_{i_k}(x_{i_k}) + \sum_{j=1}^{r} \frac{\mu_j^k}{c_j^k} \psi \left(c_j^k \left(g_{i_k j}(x_{i_k}) + \sum_{i \neq i_k} g_{ij}(x_i^k) \right) \right) \right\},$$

while keeping the others unchanged, $x_i^{k+1} = x_i^k$ for all $i \neq i_k$. Then update μ according to

$$\mu_j^{k+1} = \mu_j^k \nabla \psi \left(c_j^k \sum_{i=1}^{m} g_{ij}(x_i^{k+1}) \right), \quad j=1,\ldots,r.$$

The exponential method corresponds to

$$\psi(t) = \exp(t) - 1,$$

cf. Section 5.2.5. Again we refer to [Ber15b] for discussion and analysis.

EXERCISES

7.3.1 (Partial Proximal Algorithm [BeT94], [IbF96]) (www)

The purpose of this exercise is to develop the elements of an algorithm that is similar to the proximal, but uses partial regularization, i.e., a quadratic regularization term that involves only a subset of the coordinates of x. For $c > 0$, let ϕ_c be the real-valued convex function on $\Re^n$ defined by

$$\phi_c(z) = \min_{x \in X} \left\{ f(x) + \frac{1}{2c} \|x - z\|^2 \right\},$$

where f is a convex function over the closed convex set X. Let $x_1, \ldots, x_n$ denote the scalar components of the vector x and let I be a subset of the index set $\{1, \ldots, n\}$. For any $z = (z_1, \ldots, z_n) \in \Re^n$, consider a vector $\bar{z}$ satisfying

$$\bar{z} \in \arg\min_{x \in X} \left\{ f(x) + \frac{1}{2c} \sum_{i \in I} |x_i - z_i|^2 \right\}, \tag{7.39}$$

and let $\tilde{z}$ be the vector with components

$$\tilde{z}_i = \begin{cases} z_i & \forall\, i \in I, \\ \bar{z}_i & \forall\, i \notin I. \end{cases}$$

(a) Show that for a given $z \in \Re^n$, we have

$$\tilde{z} \in \arg\min_{\{x \mid x_i = z_i,\, i \in I\}} \phi_c(x),$$

$$\bar{z} \in \arg\min_{x \in X} \left\{ f(x) + \frac{1}{2c} \|x - \tilde{z}\|^2 \right\}.$$

(b) Interpret $\bar{z}$ as the result of a block coordinate descent step corresponding to the components z_i, $i \notin I$, followed by a proximal minimization step, and show that

$$\phi_c(\bar{z}) \le f(\bar{z}) \le \phi_c(z) \le f(z).$$

Note: Partial regularization, as in iteration (7.39), may yield better approximation to the original problem, and accelerated convergence if f is "well-behaved" with respect to some of the components of x (the components x_i with $i \notin I$), so no regularization is needed for these components.

7.4 ALTERNATING DIRECTION METHODS OF MULTIPLIERS

In this section, we will discuss an algorithm that is related to the method of multipliers, and is particularly well suited for special structures involving among others separability and large sums of component functions. The algorithm uses alternate minimizations to decouple variables that are coupled within the augmented Lagrangian, and is known as the *alternating direction method of multipliers* or ADMM for short. The name comes from the similarity with some methods for solving differential equations, known as alternating direction methods.

The following example, from [BeT89], p. 254, is a continuation of Example 7.3.1 for the method of multipliers. The example illustrates the decoupling process of the ADMM.

Example 7.4.1: (Additive Cost Problems – Continued)

Consider the problem

$$\text{minimize} \quad \sum_{i=1}^{m} f_i(x) \tag{7.40}$$
$$\text{subject to} \quad x \in \cap_{i=1}^{m} X_i,$$

where $f_i : \Re^n \mapsto \Re$ are convex functions and X_i are closed, convex sets with nonempty intersection. We can reformulate this as an equality constrained problem, by introducing additional artificial variables z_i, $i = 1, \ldots, m$, and the equality constraints $x = z_i$:

$$\text{minimize} \quad \sum_{i=1}^{m} f_i(z_i)$$
$$\text{subject to} \quad x = z_i, \quad z_i \in X_i, \quad i = 1, \ldots, m.$$

As motivation for the development of the ADMM for this problem, let us recall the corresponding method of multipliers from Example 7.3.1. There we eliminated the constraints $x = z_i$ by using corresponding multiplier vectors λ_i. At the typical iteration, we find x^{k+1} and z_i^{k+1}, $i = 1, \ldots, m$, that solve the problem

$$\text{minimize} \quad \sum_{i=1}^{m} \left(f_i(z_i) + \lambda_i^{k'}(x - z_i) + \frac{c^k}{2} \|x - z_i\|^2 \right) \tag{7.41}$$
$$\text{subject to} \quad x \in \Re^n, \quad z_i \in X_i, \quad i = 1, \ldots, m,$$

and then update the multipliers according to

$$\lambda_i^{k+1} = \lambda_i^k + c^k(x^{k+1} - z_i^{k+1}), \quad i = 1, \ldots, m.$$

The minimization in Eq. (7.41) can be done by alternating minimizations of x and z_i (a block coordinate descent method), and the multipliers λ_i may be changed only after (typically) many updates of x and z_i (enough to minimize the augmented Lagrangian within adequate precision).

An interesting variation is to perform only a small number of minimizations with respect to x and z_i before changing the multipliers. In the extreme case, where only one minimization is performed, the method takes the form

$$x^{k+1} = \frac{\sum_{i=1}^{m} z_i^k}{m} - \frac{\sum_{i=1}^{m} \lambda_i^k}{mc^k}, \tag{7.42}$$

$$z_i^{k+1} \in \arg\min_{z_i \in X_i} \left\{ f_i(z_i) - \lambda_i^{k'} z_i + \frac{c^k}{2} \|x^{k+1} - z_i\|^2 \right\}, \quad i = 1, \ldots, m, \tag{7.43}$$

followed by the multiplier update

$$\lambda_i^{k+1} = \lambda_i^k + c^k(x^{k+1} - z_i^{k+1}), \quad i = 1, \ldots, m. \tag{7.44}$$

Thus the multiplier iteration is performed after just one block coordinate descent iteration on each of the (now decoupled) variables x and $(z^1, \ldots, z^m)$. The method (7.42)-(7.44) is precisely the ADMM specialized to the problem of this example.

The preceding example also illustrates another advantage of ADMM. Frequently the decoupling process results in computations that are well-suited for parallel and distributed processing (see e.g., [BeT89], [WeO13]). This will be observed in many of the examples to be presented.

We will now formulate the ADMM and discuss its convergence properties. The starting point is the minimization problem of the Fenchel duality context:

$$\begin{aligned} \text{minimize} \quad & f_1(x) + f_2(Ax) \\ \text{subject to} \quad & x \in X, \ Ax \in Z, \end{aligned} \tag{7.45}$$

where A is an $m \times n$ matrix, $f_1 : \Re^n \mapsto \Re$ and $f_2 : \Re^m \mapsto \Re$ are convex functions, X and Z are closed convex sets. We convert this problem to the equivalent constrained minimization problem

$$\begin{aligned} \text{minimize} \quad & f_1(x) + f_2(z) \\ \text{subject to} \quad & x \in X, \ z \in Z, \ Ax = z, \end{aligned} \tag{7.46}$$

We assume throughout this section that *there exists a primal optimal and dual optimal solution pair for the preceding problem, and there is no duality gap*. We introduce its augmented Lagrangian function

$$L_c(x, z, \lambda) = f_1(x) + f_2(z) + \lambda'(Ax - z) + \frac{c}{2}\|Ax - z\|^2.$$

The ADMM, given the current iterates $(x^k, z^k, \lambda^k) \in X \times Z \times \Re^m$, generates a new iterate $(x^{k+1}, z^{k+1}, \lambda^{k+1})$ by first minimizing the augmented Lagrangian with respect to x, then with respect to z, and finally performing a multiplier update:

$$x^{k+1} \in \arg\min_{x \in X} L_c(x, z^k, \lambda^k), \tag{7.47}$$

$$z^{k+1} \in \arg\min_{z \in Z} L_c(x^{k+1}, z, \lambda^k), \tag{7.48}$$

$$\lambda^{k+1} = \lambda^k + c(Ax^{k+1} - z^{k+1}). \tag{7.49}$$

The attainment of the minima above is facilitated by the quadratic penalty term in the augmented Lagrangian. In particular, the minimum in Eq. (7.48) is attained uniquely [cf. Props. 1.1.2 and B.10(b)], while the minimum in Eq. (7.47) is attained if $A'A$ is invertible or if f is coercive [cf. Props. 1.1.2 and B.10(b)]. The penalty parameter c is kept constant in the ADMM. Contrary to the method of multipliers (where c^k is often taken to be increasing with k in order to accelerate convergence), there seems to be no generally good way to adjust c from one iteration to the next. Note that the iteration (7.42)-(7.44), given earlier for the additive cost problem (7.40), is a special case of the preceding iteration, with the notational identification $z = (z_1, \ldots, z_m)$.

The ADMM approach may also be applied to the closely related problem

$$\begin{aligned} \text{minimize} \quad & f_1(x) + f_2(z) \\ \text{subject to} \quad & x \in X, \ z \in Z, \ Ax + Bz = d, \end{aligned} \quad (7.50)$$

where $f_1 : \Re^n \mapsto \Re$, $f_2 : \Re^m \mapsto \Re$ are convex functions, X and Z are closed convex sets, and A, B, and d are given matrices and vector, respectively, of appropriate dimensions [cf. problem (7.1)]. Then the corresponding augmented Lagrangian is

$$L_c(x, z, \lambda) = f_1(x) + f_2(z) + \lambda'(Ax + Bz - d) + \frac{c}{2}\|Ax + Bz - d\|^2, \quad (7.51)$$

and the ADMM iteration takes a similar form [cf. Eqs. (7.47)-(7.49)]:

$$x^{k+1} \in \arg\min_{x \in X} L_c(x, z^k, \lambda^k), \quad (7.52)$$

$$z^{k+1} \in \arg\min_{z \in Z} L_c(x^{k+1}, z, \lambda^k), \quad (7.53)$$

$$\lambda^{k+1} = \lambda^k + c(Ax^{k+1} + Bz^{k+1} - d). \quad (7.54)$$

For some problems, this form may be more convenient than the ADMM of Eqs. (7.47)-(7.49), although the two forms are essentially equivalent.

Example 7.4.2 (Stochastic Programming Using ADMM)

Consider a stochastic programming problem of the form

$$\text{minimize} \quad F(x) + \sum_{j=1}^{m} \pi_j G_j(z_j)$$

$$\text{subject to} \quad x \in X, \ z_j \in Z_j, \ A_j x + B_j z_j = d_j, \ j = 1, \ldots, m;$$

cf. Example 7.2.2. Letting $z = (z_1, \ldots, z_m)$, we see that the problem can be placed in the form (7.50). The augmented Lagrangian is given by

$$L_c(x, z, \lambda) = F(x) + \sum_{j=1}^{m} \Big\{ \pi_j G_j(z_j) + \lambda_j'(A_j x + B_j z_j - d_j)$$

$$+ \frac{c}{2}\|A_j x + B_j z_j - d_j\|^2 \Big\},$$

where $\lambda = (\lambda_1, \ldots, \lambda_m)$. The ADMM (7.52)-(7.54) is written as

$$x^{k+1} \in \arg\min_{x \in X} \Big\{ F(x) + \sum_{j=1}^{m} \Big\{ \lambda_j^{k'} A_j x + \frac{c}{2}\|A_j x + B_j z_j^k - d_j\|^2 \Big\} \Big\},$$

$$z_j^{k+1} \in \arg\min_{z_j \in Z_j} \left\{ \pi_j G_j(z_j) + \lambda_j^{k'} B_j z_j + \frac{c}{2} \|A_j x^{k+1} + B_j z_j - d_j\|^2 \right\},$$

$$\lambda_j^{k+1} = \lambda_j^k + c(A_j x^{k+1} + B_j z_j^{k+1} - d_j), \qquad j = 1, \ldots, m,$$

which is convenient since the minimizations over the variables of the two stages, x and $z_1, \ldots, z_m$, have been decoupled.

Note that the following variant of the stochastic programming problem,

$$\text{minimize} \quad F(x) + \sum_{j=1}^m \pi_j G_j(z_j, A_j x + B_j z_j)$$

$$\text{subject to} \quad x \in X, \quad z_j \in Z_j, \quad j = 1, \ldots, m,$$

admits a similar treatment. Indeed, by introducing variables $u_1, \ldots, u_m$, we can equivalently write the problem as

$$\text{minimize} \quad F(x) + \sum_{j=1}^m \pi_j G_j(z_j, u_j)$$

$$\text{subject to} \quad x \in X, \quad z_j \in Z_j, \quad A_j x + B_j z_j - u_j = 0, \quad j = 1, \ldots, m.$$

With the notational identification $z \sim \{(z_1, u_1), \ldots, (z_m, u_m)\}$, this problem is of the form (7.50) for which the ADMM applies.

Applications of ADMM Involving the ℓ_1-Norm

We will now discuss briefly some examples that involve the ℓ_1-norm, and arise prominently in machine learning and signal processing. In this connection, we note that the ℓ_1 norm finds significant use in regularization contexts, where it tends to induce optimal solutions that are sparse, i.e., have many zero components; see [Ber15a] for further discussion.

Example 7.4.3 (Basis Pursuit)

Consider the problem

$$\text{minimize} \quad \|x\|_1$$

$$\text{subject to} \quad Cx = b,$$

where $\|\cdot\|_1$ is the ℓ_1 norm in $\Re^n$, C is a given $m \times n$ matrix and b is a vector in $\Re^m$. We reformulate it as

$$\text{minimize} \quad f_1(x) + f_2(z)$$

$$\text{subject to} \quad Cx = b, \quad x = z,$$

where $f_1(x) \equiv 0$ and $f_2(z) = \|z\|_1$. The augmented Lagrangian is

$$L_c(x, z, \lambda) = \|z\|_1 + \lambda'(x - z) + \frac{c}{2} \|x - z\|^2.$$

The ADMM iteration (7.47)-(7.49) takes the form

$$x^{k+1} \in \arg\min_{Cx=b} \left\{ \lambda^{k'} x + \frac{c}{2}\|x - z^k\|^2 \right\},$$

$$z^{k+1} \in \arg\min_{z \in \Re^n} \left\{ \|z\|_1 - \lambda^{k'} z + \frac{c}{2}\|x^{k+1} - z\|^2 \right\},$$

$$\lambda^{k+1} = \lambda^k + c(x^{k+1} - z^{k+1}).$$

The iteration for z can also be written as

$$z^{k+1} \in \arg\min_{z \in \Re^n} \left\{ \|z\|_1 + \frac{c}{2} \left\| x^{k+1} - z + \frac{\lambda^k}{c} \right\|^2 \right\}. \qquad (7.55)$$

The type of minimization over z in in the preceding equation arises often in problems involving the ℓ_1-norm. It is straightforward to verify that the solution is given by the so-called *shrinkage operation*, which for any $\alpha > 0$ and $w = (w_1, \ldots, w_m) \in \Re^m$, is defined as

$$S(\alpha, w) \in \arg\min_{z \in \Re^m} \left\{ \|z\|_1 + \frac{1}{2\alpha}\|z - w\|^2 \right\}, \qquad (7.56)$$

and has components given by

$$S_i(\alpha, w) = \begin{cases} w_i - \alpha & \text{if } w_i > \alpha, \\ 0 & \text{if } |w_i| \leq \alpha, \\ w_i + \alpha & \text{if } w_i < -\alpha, \end{cases} \quad i = 1, \ldots, m. \qquad (7.57)$$

Thus the minimization over z in Eq. (7.55) is expressed in terms of the shrinkage operation as

$$z^{k+1} = S\left(\frac{1}{c}, x^{k+1} + \frac{\lambda^k}{c}\right).$$

Example 7.4.4 (ℓ_1 Regularization)

Consider the problem

$$\text{minimize} \quad f(x) + \gamma\|x\|_1$$
$$\text{subject to} \quad x \in X,$$

where $f : \Re^n \mapsto \Re$ is a convex function, X is a closed convex set, and γ is a positive scalar. We reformulate the problem as

$$\text{minimize} \quad f_1(x) + f_2(z)$$
$$\text{subject to} \quad x \in X, \ x = z,$$

where $f_1(x) = f(x)$ and $f_2(z) = \gamma\|z\|_1$. The augmented Lagrangian is

$$L_c(x, z, \lambda) = f(x) + \gamma\|z\|_1 + \lambda'(x - z) + \frac{c}{2}\|x - z\|^2.$$

Sec. 7.4 Alternating Direction Methods of Multipliers

The ADMM iteration (7.47)-(7.49) takes the form

$$x^{k+1} \in \arg\min_{x \in X}\left\{f(x) + \lambda^{k\prime}x + \frac{c}{2}\|x - z^k\|^2\right\},$$

$$z^{k+1} \in \arg\min_{z \in \Re^n}\left\{\gamma\|z\|_1 - \lambda^{k\prime}z + \frac{c}{2}\|x^{k+1} - z\|^2\right\},$$

$$\lambda^{k+1} = \lambda^k + c(x^{k+1} - z^{k+1}).$$

The iteration for z can also be written in closed form, in terms of the shrinkage operation (7.56)-(7.57):

$$z^{k+1} = S\left(\frac{\gamma}{c}, x^{k+1} + \frac{\lambda^k}{c}\right).$$

Example 7.4.5 (Least Absolute Deviations Problem)

Consider the problem

$$\text{minimize} \quad \|Cx - b\|_1$$
$$\text{subject to} \quad x \in \Re^n,$$

where C is an $m \times n$ matrix of rank n, and $b \in \Re^m$ is a given vector. We reformulate the problem as

$$\text{minimize} \quad f_1(x) + f_2(z)$$
$$\text{subject to} \quad Cx - b = z,$$

where $f_1(x) \equiv 0$ and $f_2(z) = \|z\|_1$. Here the augmented Lagrangian function is modified to include the constant vector b [cf. Eq. (7.51)]. It is given by

$$L_c(x, z, \lambda) = \|z\|_1 + \lambda'(Cx - z - b) + \frac{c}{2}\|Cx - z - b\|^2.$$

The ADMM iteration (7.52)-(7.54) takes the form

$$x^{k+1} = (C'C)^{-1}C'\left(z^k + b - \frac{\lambda^k}{c}\right),$$

$$z^{k+1} \in \arg\min_{z \in \Re^m}\left\{\|z\|_1 - \lambda^{k\prime}z + \frac{c}{2}\|Cx^{k+1} - z - b\|^2\right\},$$

$$\lambda^{k+1} = \lambda^k + c(Cx^{k+1} - z^{k+1} - b).$$

Setting $\bar{\lambda}^k = \lambda^k/c$, the iteration can be written in the notationally simpler form

$$x^{k+1} = (C'C)^{-1}C'(z^k + b - \bar{\lambda}^k),$$

$$z^{k+1} \in \arg\min_{z \in \Re^m}\left\{\|z\|_1 + \frac{c}{2}\|Cx^{k+1} - z - b + \bar{\lambda}^k\|^2\right\}, \qquad (7.58)$$

$$\bar{\lambda}^{k+1} = \bar{\lambda}^k + Cx^{k+1} - z^{k+1} - b.$$

The minimization over z in Eq. (7.58) is expressed in terms of the shrinkage operation as

$$z^{k+1} = S\left(\frac{1}{c}, Cx^{k+1} - b - \bar{\lambda}^k\right).$$

Convergence Issues

The convergence properties of ADMM are quite strong. The following proposition gives the main convergence result for the ADMM (7.47)-(7.49). The proof of the proposition is long and not very insightful. It is given in [EcB72] (Th. 8), and for the special case where X and Z are polyhedral, it may be found in Section 3.4 (Prop. 4.2) of [BeT89] (which can be accessed online). A variant of this proof for the ADMM of Eqs. (7.52)-(7.54) is given in [BPC11], and essentially the same convergence result is shown.

Proposition 7.4.1: (ADMM Convergence) Consider problem (7.45), and assume that $A'A$ is invertible. Then the sequence $\{x^k, \lambda^k\}$ generated by the ADMM converges to some optimal primal and dual optimal solution pair (x^*, λ^*), while $\{z^k\}$ converges to Ax^*.

Note that the assumed invertibility of $A'A$ guarantees that the sequence $\{x^k, \lambda^k\}$ is well defined, in the sense that the minima in the iterations (7.47), (7.48), which generate x^{k+1} and z^{k+1}, exist and are unique. To see what can happen when $A'A$ is not invertible, let $m = n = 1$, $X = Z = \Re$, $A = 0$, and f_1 and f_2 be identically 0. Then it can be verified that $\{\lambda^k\}$ and $\{z^k\}$ converge to 0 in one iteration, but $\{x^k\}$ can be any real sequence, and need not even be bounded. In Section 7.4.2 we give a "proximal" variant of the ADMM that involves an extra proximal term in the augmented Lagrangian and is convergent even if $A'A$ is not invertible.

Comparison with the Method of Multipliers

Let us now compare the ADMM with the method of multipliers of Section 7.3. As we have emphasized earlier, its main advantage is that it algorithmically decouples x and z in the penalty term $\|Ax-z\|^2$ or the penalty term $\|Ax + Bz - d\|^2$. In some problem contexts, this is decisively significant.

On the other hand there is a price for the flexibility that the ADMM provides. A major drawback is a much slower practical convergence rate relative to the method of multipliers. Both methods can be shown to have a typically linear convergence rate for the multiplier updates (except under unfavorable circumstances, when the problem is singular, or if $c^k \to \infty$, when the method of multipliers attains a superlinear convergence rate). However, it seems difficult to compare them on the basis of theoretical results alone, because the geometric progression rate at which they converge is different and also because the amount of work between multiplier updates must be properly taken into account. A corollary is that just because the ADMM requires less work between multiplier updates than the method of multipliers, it does not necessarily require less computation time to solve a problem.

A further consideration is that while the ADMM conveniently decouples the minimizations with respect to x and z, the method of multipliers applies to nonconvex problems as well (cf. Section 5.2), and allows for some implementation flexibility that may be exploited by taking advantage of the structure of the given problem:

(a) The minimization of the augmented Lagrangian can be done with a broad variety of methods (not just block coordinate descent). Some of these methods may be well suited for the problem's structure.

(b) The minimization of the augmented Lagrangian need not be done exactly, and its accuracy can be readily controlled through theoretically sound and easily implementable termination criteria.

(c) The adjustment of the penalty parameter c can be used with advantage in the method of multipliers, but there is apparently no general way to do this in ADMM (see Exercise 7.4.3 for a simple illustration). In particular, by taking c to increase to ∞, superlinear or finite convergence can often be achieved in the method of multipliers, while in ADMM, determining a good value of c may require trial and error.

Thus, on balance, it appears that the relative performance merits of ADMM and methods of multipliers are problem-dependent in practice. It all depends on whether the greater ability of ADMM to exploit the problem's structure outweighs its slower convergence properties.

7.4.1 ADMM Applied to Separable Problems

We have often noted that separable problems of the form

$$\begin{aligned} \text{minimize} \quad & \sum_{i=1}^{m} f_i(x_i) \\ \text{subject to} \quad & \sum_{i=1}^{m} A_i x_i = b, \quad x_i \in X_i, \quad i = 1, \ldots, m, \end{aligned} \quad (7.59)$$

where $f_i : \Re^{n_i} \mapsto \Re$ are convex functions, X_i are closed convex sets, and A_i and b are given, have an especially favorable structure that is well-suited for the application of decomposition approaches. Since the primary attractive feature of ADMM is that it decouples the augmented Lagrangian optimization calculations, it is natural to consider its application to this problem.

An idea that readily comes to mind is to form the augmented Lagrangian

$$L_c(x_1, \ldots, x_m, \lambda) = \sum_{i=1}^{m} f_i(x_i) + \lambda' \left(\sum_{i=1}^{m} A_i x_i - b \right) + \frac{c}{2} \left\| \sum_{i=1}^{m} A_i x_i - b \right\|^2,$$

and use an ADMM-like iteration, whereby we minimize L_c sequentially with respect to $x_1,\ldots,x_m$, i.e.,

$$x_i^{k+1} \in \arg\min_{x_i \in X_i} L_c(x_1^{k+1},\ldots,x_{i-1}^{k+1},x_i,x_{i+1}^k,\ldots,x_m^k,\lambda^k), \quad i=1,\ldots,m, \tag{7.60}$$

and follow these minimizations with the multiplier iteration

$$\lambda^{k+1} = \lambda^k + c\left(\sum_{i=1}^m A_i x_i^{k+1} - b\right). \tag{7.61}$$

This approach was proposed in the early days of research on multiplier methods, starting with the paper [StW75], as a means of overcoming the inherent coupling introduced by the quadratic term of the augmented Lagrangian. The context was unrelated to the ADMM, which was unknown at that time.

When there is only one component, $m=1$, we obtain the method of multipliers. When there are only two components, $m=2$, the above method is equivalent to the ADMM of Eqs. (7.52)-(7.54), so it has the corresponding convergence properties. On the other hand, when $m>2$, the method is not a special case of the ADMM that we have discussed and is not covered by similar convergence guarantees. In fact a convergence counterexample has been given for $m=3$ in [CHY16]. The same reference shows that the iteration (7.60)-(7.61) is convergent under additional, but substantially stronger assumptions. Related convergence results are proved in [HoL13] under alternative but also strong assumptions; see also [WHM13].

In what follows we will develop an ADMM (first given in [BeT89], Section 3.4, Example 4.4), which is similar to the iteration (7.60)-(7.61), and is covered by the convergence guarantees of Prop. 7.4.1 without any assumptions other than the ones given in that proposition. We will derive this algorithm, by formulating the separable problem as a special case of the Fenchel framework problem (7.45), and by applying the convergent ADMM (7.47)-(7.49).

We reformulate problem (7.59) by introducing additional variables $z_1,\ldots,z_m$ as follows:

$$\text{minimize} \quad \sum_{i=1}^m f_i(x_i)$$
$$\text{subject to} \quad A_i x_i = z_i, \quad x_i \in X_i, \quad i=1,\ldots,m,$$
$$\sum_{i=1}^m z_i = b.$$

Sec. 7.4 Alternating Direction Methods of Multipliers

We denote $x = (x_1, \ldots, x_m)$, $z = (z_1, \ldots, z_m)$, we view $X = X_1 \times \ldots \times X_m$ as a constraint set for x, we view

$$Z = \left\{ z \; \middle| \; \sum_{i=1}^{m} z_i = b \right\}$$

as a constraint set for z, and we introduce a multiplier vector p_i for each of the equality constraints $A_i x_i = z_i$. The augmented Lagrangian has the separable form

$$L_c(x, z, p) = \sum_{i=1}^{m} \left(f_i(x_i) + (A_i x_i - z_i)' p_i + \frac{c}{2} \|A_i x_i - z_i\|^2 \right), \quad x \in X, \; z \in Z,$$

and the ADMM (7.47)-(7.49) is given by

$$x_i^{k+1} \in \arg\min_{x_i \in X_i} \left\{ f_i(x_i) + (A_i x_i - z_i^k)' p_i^k + \frac{c}{2} \|A_i x_i - z_i^k\|^2 \right\}, \; i = 1, \ldots, m, \tag{7.62}$$

$$z^{k+1} \in \arg\min_{\sum_{i=1}^{m} z_i = b} \left\{ \sum_{i=1}^{m} (A_i x_i^{k+1} - z_i)' p_i^k + \frac{c}{2} \|A_i x_i^{k+1} - z_i\|^2 \right\}, \tag{7.63}$$

$$p_i^{k+1} = p_i^k + c(A_i x_i^{k+1} - z_i^{k+1}). \tag{7.64}$$

We will now show how to simplify this algorithm. We will first obtain the minimization (7.63) for z in closed form by introducing a multiplier vector λ^{k+1} for the constraint $\sum_{i=1}^{m} z_i = b$, and then show that the multipliers p_i^{k+1} obtained from the update (7.64) are all equal to λ^{k+1}. To this end we note that the Lagrangian function corresponding to the minimization (7.63), is given by

$$\sum_{i=1}^{m} \left((A_i x_i^{k+1} - z_i)' p_i^k + \frac{c}{2} \|A_i x_i^{k+1} - z_i\|^2 + \lambda^{k+1'} z_i \right) - \lambda^{k+1'} b.$$

By setting to zero its gradient with respect to z_i, we see that the minimizing vectors z_i^{k+1} are given in terms of λ^{k+1} by

$$z_i^{k+1} = A_i x_i^{k+1} + \frac{p_i^k - \lambda^{k+1}}{c}. \tag{7.65}$$

A key observation is that we can write this equation as

$$\lambda^{k+1} = p_i^k + c(A_i x_i^{k+1} - z_i^{k+1}), \quad i = 1, \ldots, m,$$

so from Eq. (7.64), we have

$$p_i^{k+1} = \lambda^{k+1}, \quad i = 1, \ldots, m.$$

Thus during the algorithm *all the multipliers p_i are updated to a common value*: the multiplier λ^{k+1} of the constraint $\sum_{i=1}^{m} z_i = b$ of problem (7.63).

We now use this fact to simplify the ADMM (7.62)-(7.64). Given z^k and λ^k (which is equal to p_i^k for all i), we first obtain x^{k+1} from the augmented Lagrangian minimization (7.62) as

$$x_i^{k+1} \in \arg\min_{x_i \in X_i} \left\{ f_i(x_i) + (A_i x_i - z_i^k)'\lambda^k + \frac{c}{2}\|A_i x_i - z_i^k\|^2 \right\}, \quad (7.66)$$

for all $i = 1, \ldots, m$. To obtain z^{k+1} and λ^{k+1} from the augmented Lagrangian minimization (7.63), we express z_i^{k+1} in terms of the unknown λ^{k+1} as

$$z_i^{k+1} = A_i x_i^{k+1} + \frac{\lambda^k - \lambda^{k+1}}{c}, \quad i = 1, \ldots, m, \quad (7.67)$$

[cf. Eq. (7.65)], and then obtain λ^{k+1} by requiring that the constraint of the minimization (7.63), $\sum_{i=1}^{m} z_i^{k+1} = b$, is satisfied. Thus, by adding Eq. (7.67) over $i = 1, \ldots, m$, we have

$$\lambda^{k+1} = \lambda^k + \frac{c}{m}\left(\sum_{i=1}^{m} A_i x_i^{k+1} - b \right). \quad (7.68)$$

We then obtain z^{k+1} using Eq. (7.67).

In summary, given (x^k, z^k, λ^k), the iteration of the algorithm obtains $(x^{k+1}, z^{k+1}, \lambda^{k+1})$ by applying the three equations (7.66), (7.68), and (7.67), in that order. The vectors λ^0 and z^0, which are needed at the first iteration, can be chosen arbitrarily. This is a correct ADMM, mathematically equivalent to the algorithm (7.62)-(7.64), which has guaranteed convergence as per Prop. 7.4.1. It is as simple as the iteration (7.60)-(7.61), which, however, is not theoretically sound for $m > 2$ as we noted earlier.

Note that it is possible to eliminate the variables z^k from Eq. (7.66), and write the iteration in terms of (x^k, λ^k). In particular, from Eqs. (7.67) and (7.68), we have

$$z_i^k = A_i x_i^k - \frac{1}{m}\left(\sum_{i=1}^{m} A_i x_i^k - b \right).$$

Substituting this expression in Eq. (7.66), we obtain the equivalent iteration

$$x_i^{k+1} \in \arg\min_{x_i \in X_i} \left\{ f_i(x_i) + (A_i x_i)'\lambda^k \right.$$
$$\left. + \frac{c}{2}\left\| A_i(x_i - x_i^k) + \frac{1}{m}\left(\sum_{i=1}^{m} A_i x_i^k - b \right) \right\|^2 \right\}, \quad (7.69)$$

$$\lambda^{k+1} = \lambda^k + \frac{c}{m}\left(\sum_{i=1}^{m} A_i x_i^{k+1} - b\right). \qquad (7.70)$$

We mention a more refined form of the multiplier iteration (7.70), also from [BeT89], Example 4.4, whereby, the coordinates λ_j^k of the multiplier vector λ^k are updated according to

$$\lambda_j^{k+1} = \lambda_j^k + \frac{c}{m_j}\left(\sum_{i=1}^{m} A_{ji} x_i^{k+1} - b\right), \qquad j = 1, \ldots, r, \qquad (7.71)$$

where r is the row dimension of the matrices A_i, A_{ji} is the jth row of A_i, and m_j is the number of submatrices A_i that have nonzero jth row. Using the j-dependent stepsize c/m_j of Eq. (7.71) in place of the stepsize c/m of Eq. (7.68) may be viewed as a form of diagonal scaling. The derivation of the algorithm of Eqs. (7.66), (7.67), and (7.71) is nearly identical to the one given for the algorithm (7.66)-(7.68). The intuitive idea here is that the components of the vector z_i^k represent estimates of the corresponding components of $A_i x_i$ at the optimum. However, if some of these components are known to be 0 because some of the rows of A_i are 0, then the corresponding values of z_i^k might as well be set to 0 rather than be estimated. If we repeat the preceding derivation of the algorithm (7.66)-(7.68), but without introducing the components z_i that are known to be 0, we obtain by a straightforward calculation the multiplier iteration (7.71).

7.4.2 Connections Between Augmented Lagrangian-Related Methods

We have seen so far several methods that relate to augmented Lagrangians. These are:

(a) The *method of multipliers* and its variations, discussed in Sections 5.2 and 7.3.

(b) The *proximal algorithm*, discussed in Section 3.6, which may be viewed as a dual version of the method of multipliers for convex problems (based on Fenchel duality), as discussed in Section 7.3.

(c) The *ADMM and its special cases*, discussed in the present section.

We will now show that these methods, and some other ones that we have not discussed so far, are all special cases of the ADMM, which thus emerges as a broad framework for methods involving regularization, quadratic penalties, and multiplier iterations. †

---

† A broader and more fundamental unifying framework is an extended form of the proximal algorithm, which can be used to find a zero of a maximal monotone operator; see the paper [Roc76a], and for textbook discussions, [RoW98],

We recall for easy reference the problem solved by the ADMM [cf. Eq. (7.46)]:

$$\text{minimize} \quad f_1(x) + f_2(z)$$
$$\text{subject to} \quad x \in X, \quad z \in Z, \quad Ax = z, \quad (7.72)$$

where A is an $m \times n$ matrix, $f_1 : \Re^n \mapsto \Re$ and $f_2 : \Re^m \mapsto \Re$ are convex functions, X and Z are closed convex sets. The ADMM, given the current iterate $(x^k, z^k, \lambda^k) \in X \times Z \times \Re^m$, generates a new iterate $(x^{k+1}, z^{k+1}, \lambda^{k+1})$ by first minimizing the augmented Lagrangian with respect to x, then with respect to z, and finally performing a multiplier update:

$$x^{k+1} \in \arg\min_{x \in X} L_c(x, z^k, \lambda^k), \quad (7.73)$$

$$z^{k+1} \in \arg\min_{z \in Z} L_c(x^{k+1}, z, \lambda^k), \quad (7.74)$$

$$\lambda^{k+1} = \lambda^k + c(Ax^{k+1} - z^{k+1}), \quad (7.75)$$

where the augmented Lagrangian function is given by

$$L_c(x, z, \lambda) = f_1(x) + f_2(z) + (Ax - z)'\lambda + \frac{c}{2}\|Ax - z\|^2.$$

The Method of Multipliers as a Special Case of ADMM

Consider the preceding ADMM formulation with the notational identifications

$$f_1 = f, \quad f_2(z) \equiv 0, \quad X \sim X, \quad Z = \{z \mid z = b\}, \quad A \sim A,$$

where X is a closed convex subset of $\Re^n$, $f : \Re^n \mapsto \Re$ is a convex function, A is an $m \times n$ matrix, and b is a vector in $\Re^m$. Then the ADMM (7.73)-(7.75) takes the form

$$x^{k+1} \in \arg\min_{x \in X} \left\{ f(x) + (Ax - z^k)'\lambda^k + \frac{c}{2}\|Ax - z^k\|^2 \right\},$$

$$z^{k+1} \in \arg\min_{z=b} \left\{ (Ax^{k+1} - z)'\lambda^k + \frac{c}{2}\|Ax^{k+1} - z\|^2 \right\},$$

$$\lambda^{k+1} = \lambda^k + c(Ax^{k+1} - z^{k+1}).$$

Clearly, we have $z^{k+1} = b$ for all $k \geq 0$, so this ADMM becomes

$$x^{k+1} \in \arg\min_{x \in X} \left\{ f(x) + (Ax - b)'\lambda^k + \frac{c}{2}\|Ax - b\|^2 \right\},$$

$$\lambda^{k+1} = \lambda^k + c(Ax^{k+1} - b),$$

for all $k \geq 1$, and is equivalent to the method of multipliers of Section 7.3.1.

[BaC11], and [Ber15a] (Section 5.1.4), which include many references. This algorithm contains as a special case the ADMM, as shown in [EcB92], and by extension, all the methods discussed in this section.

The Proximal Algorithm as a Special Case of ADMM

Consider the ADMM formulation with the notational identifications

$$f_1 = f, \qquad f_2(z) \equiv 0, \qquad X \sim X, \qquad Z = \Re^n, \qquad A = I,$$

where X is a closed convex subset of $\Re^n$, $f : \Re^n \mapsto \Re$ is a convex function, and I is the identity matrix. Then the ADMM (7.73)-(7.75) takes the form

$$x^{k+1} \in \arg\min_{x \in X} \left\{ f(x) + (x - z^k)' \lambda^k + \frac{c}{2} \|x - z^k\|^2 \right\},$$

$$z^{k+1} \in \arg\min_{z \in \Re^n} \left\{ (x^{k+1} - z)' \lambda^k + \frac{c}{2} \|x^{k+1} - z\|^2 \right\},$$

$$\lambda^{k+1} = \lambda^k + c(x^{k+1} - z^{k+1}).$$

By setting to 0 the gradient of the function minimized in the equation for z^{k+1}, we obtain

$$\lambda^k + c(x^{k+1} - z^{k+1}) = 0,$$

so from the equation for λ^{k+1}, it follows that $\lambda^{k+1} = 0$ for all $k \geq 0$ and $x^{k+1} = z^{k+1}$ for all $k \geq 1$. Thus, the ADMM becomes

$$x^{k+1} \in \arg\min_{x \in X} \left\{ f(x) + \frac{c}{2} \|x - x^k\|^2 \right\},$$

for all $k \geq 2$, and is equivalent to the proximal algorithm.

The Proximal Method of Multipliers as a Special Case of ADMM

The *proximal method of multipliers* is a modified version of the method of multipliers where a proximal term is added to the augmented Lagrangian, which may contribute to greater stability of its minimization. This method, proposed in [Roc76b], applies to the minimization of a convex function $f : \Re^n \mapsto \Re$ subject to a closed convex set constraint $x \in X$, and an equality constraint $Ax = b$, where A is an $m \times n$ matrix [cf. problem (7.25)]. It has the form

$$x^{k+1} \in \arg\min_{x \in X} \left\{ f(x) + \lambda^{k'}(Ax - b) + \frac{c}{2} \|Ax - b\|^2 + \frac{\gamma}{2} \|x - x^k\|^2 \right\}, \tag{7.76}$$

followed by

$$\lambda^{k+1} = \lambda^k + c(Ax^{k+1} - b), \tag{7.77}$$

where $\gamma > 0$ and $c > 0$ are scalar parameters.

Let us consider the ADMM formulation (7.72) with $z = (z_1, z_2)$, where $z_1 \in \Re^m$ and $z_2 \in \Re^n$, and the notational identifications

$$f_1 = f, \quad f_2(z_1, z_2) \equiv 0, \quad X \sim X, \quad Z = \{z_1 \mid z_1 = b\} \times \Re^n, \quad A \sim \begin{pmatrix} A \\ I \end{pmatrix},$$

where I is the identity matrix. Then the ADMM (7.73)-(7.75) takes the form

$$x^{k+1} \in \arg\min_{x \in X} \left\{ f(x) + (Ax - z_1^k)'\lambda_1^k + (x - z_2^k)'\lambda_2^k \right.$$
$$\left. + \frac{c}{2}\|Ax - z_1^k\|^2 + \frac{c}{2}\|x - z_2^k\|^2 \right\},$$

$$z_1^{k+1} \in \arg\min_{z_1 = b} \left\{ (Ax^{k+1} - z_1)'\lambda_1^k + \frac{c}{2}\|Ax^{k+1} - z_1\|^2 \right\},$$

$$z_2^{k+1} \in \arg\min_{z_2 \in \Re^n} \left\{ (x^{k+1} - z_2)'\lambda_1^k + \frac{c}{2}\|x^{k+1} - z_2\|^2 \right\},$$

$$\lambda_1^{k+1} = \lambda_1^k + c(Ax^{k+1} - z_1^{k+1}),$$

$$\lambda_2^{k+1} = \lambda_2^k + c(x^{k+1} - z_2^{k+1}).$$

By using calculations similar to the preceding two cases, we can verify that $z_1^{k+1} = b$ for all $k \geq 0$, while $\lambda_2^{k+1} = 0$ for all $k \geq 0$ and $x^{k+1} = z_2^{k+1}$ for all $k \geq 1$. By setting $\lambda_1^k = \lambda^k$, it follows that the preceding ADMM is equivalent to the proximal method of multipliers (7.76)-(7.77) for the case $\gamma = c$.

The more general form of the proximal method of multipliers where $\gamma \neq c$ is obtained from the preceding algorithm by appropriately scaling A and b. This can be done by starting with both parameters equal to γ in the preceding argument, and then replacing A and b by βA and βb, where $\beta = \sqrt{c/\gamma}$. It then follows that the algorithm (7.76)-(7.77), where $\gamma \neq c$, is also a special case of ADMM.

The Proximal ADMM

We now consider the standard ADMM framework

$$\text{minimize} \quad f_1(x) + f_2(z)$$
$$\text{subject to} \quad x \in X, \quad z \in Z, \quad Ax = z,$$

[cf. Eq. (7.72)], and we introduce a variant of the ADMM, obtained by adding a quadratic regularization term to the augmented Lagrangian. We refer to this method as the *proximal ADMM*, in analogy to the proximal method of multipliers. The method has the form

$$x^{k+1} \in \arg\min_{x \in X} \left\{ L_c(x, z^k, \lambda^k) + \frac{\gamma}{2}\|x - x^k\|^2 \right\}, \qquad (7.78)$$

$$z^{k+1} \in \arg\min_{z \in Z} L_c(x^{k+1}, z, \lambda^k), \qquad (7.79)$$

$$\lambda^{k+1} = \lambda^k + c(Ax^{k+1} - z^{k+1}), \qquad (7.80)$$

where $\gamma > 0$ and $c > 0$ are scalar parameters.

Sec. 7.4   Alternating Direction Methods of Multipliers   707

We consider a reformulation to the ADMM format (7.72), with variables $x \in \Re^n$ and $(z_1, z_2) \in \Re^n \times \Re^m$, and the notational identifications

$$f_1 = f_1, \quad f_2(z_1, z_2) \sim f_2(z_1), \quad X \sim X, \quad Z \sim Z \times \Re^n, \quad A \sim \begin{pmatrix} A \\ \sqrt{\gamma/c}\,I \end{pmatrix}.$$

Similar to the earlier cases, it can be seen that the corresponding ADMM (7.73)-(7.75) takes the form (7.78)-(7.80).

Interestingly, the matrix $\begin{pmatrix} A' & \sqrt{\gamma/c}\,I \end{pmatrix} \begin{pmatrix} A \\ \sqrt{\gamma/c}\,I \end{pmatrix} = A'A + \frac{\gamma}{c}I$ is invertible, so Prop. 7.4.1 can be used to assert the convergence of the proximal ADMM (7.78)-(7.80). By contrast, in the case where $A'A$ is not invertible, Prop. 7.4.1 cannot be used to assert the convergence of the standard ADMM (7.73)-(7.75).

The Proximal ADMM for Separable Problems

Consider the separable problem of Section 7.4.2 [cf. Eq. (7.59)]. By repeating the derivation of that section, we can verify that the proximal ADMM of Eqs. (7.78)-(7.80) contains as a special case the algorithm

$$x_i^{k+1} \in \arg\min_{x_i \in X_i} \left\{ f_i(x_i) + (A_i x_i)'\lambda^k \right.$$
$$\left. + \frac{c}{2}\left\| A_i(x_i - x_i^k) + \frac{1}{m}\left(\sum_{i=1}^m A_i x_i^k - b\right)\right\|^2 + \frac{\gamma}{2}\|x_i - x_i^k\|^2 \right\},$$

for all $i = 1, \ldots, m$, followed by

$$\lambda^{k+1} = \lambda^k + \frac{c}{m}\left(\sum_{i=1}^m A_i x_i^{k+1} - b\right);$$

cf. Eqs. (7.69)-(7.70). This method has the enhanced convergence property noted in connection with the proximal ADMM (7.78)-(7.80).

EXERCISES

7.4.1 (Refined Form of ADMM for Separable Problems [BeT89], Section 3.4)

Complete the details of the derivation given in this section to establish that the algorithm of Eqs. (7.66), (7.67), and (7.71) is a special case of the ADMM for the separable problem (7.59).

7.4.2 (Convergence Rate of ADMM)

Consider the problem
$$\text{minimize} \quad f_1(x) + f_2(z)$$
$$\text{subject to} \quad Ax = z,$$
where f_1 and f_2 are convex quadratic. Assume that $A'A$ is invertible and that the problem has a unique solution (x^*, Ax^*). Let λ^* be the Lagrange multiplier [which exists and is unique since $A'A$ is invertible so (x^*, Ax^*) is a regular point; cf. Prop. 4.1.1]. Show that the ADMM (7.47)-(7.49) converges linearly to (x^*, Ax^*, λ^*) from every starting point (x^0, z^0, λ^0). Use Prop. 5.4.1 to extend this result to the case where f_1 and f_2 are convex and twice continuously differentiable. *Hint*: By Prop. 7.4.1, the ADMM converges to (x^*, Ax^*, λ^*) from every starting point. When f_1 and f_2 are quadratic, the ADMM is a linear stationary iteration. Any such iteration that converges to a unique solution from every starting point must be a contraction, and hence must converge at a linear rate. *Note*: The method of multipliers of Section 7.3.1 [cf. Eqs. (7.29)-(7.30)] also converges at a linear (but typically much faster) rate for the same problem, at the expense of much larger overhead per iteration. The ADMM can also be shown to converge linearly when applied to linear programs; see the paper [EcB90]. It typically does not converge finitely for linear programs, like the method of multipliers for which the finite convergence property of the proximal algorithm applies; cf. Prop. 3.6.5. For analysis of the convergence rate of ADMM and related schemes under unfavorable conditions where the convergence rate is sublinear, see [DaY14a], [DaY14b].

7.4.3 (Role of the Penalty Parameter in ADMM)

The purpose of this exercise is to illustrate by example the difficulty of selecting a good value of c in ADMM. Consider the problem of minimizing $\frac{1}{2}x^2 + \frac{1}{2}(ax)^2$ over $x \in \Re$, which is in the ADMM format with the notational identifications $f_1(x) = \frac{1}{2}x^2$, $f_2(z) = \frac{1}{2}z^2$, and $A = a$. Here a, x, and z are scalars with $a \neq 0$. Consider also the following algorithm with $c > 0$:

$$x^{k+1} = \frac{a(cz^k - \lambda^k)}{1 + ca^2}, \quad z^{k+1} = \frac{\lambda^k + cax^{k+1}}{1+c}, \quad \lambda^{k+1} = \lambda^k + c(ax^{k+1} - z^{k+1}).$$

(a) Verify that this is the ADMM for the problem.

(b) Show that for all $k \geq 1$, we have $\lambda^k = z^k$ and that an equivalent form of the algorithm is

$$x^{k+1} = \frac{a(c-1)}{1+ca^2}z^k, \quad z^{k+1} = \frac{1 + c^2a^2}{(1+ca^2)(1+c)}z^k.$$

(c) Plot the ratio of linear convergence
$$\beta(c) = \frac{1+c^2a^2}{(1+ca^2)(1+c)},$$
as a function of c, and verify that $\beta(c) \in (0,1)$ for all $c > 0$, that $\beta(c) \to 1$ as either $c \downarrow 0$ or $c \uparrow \infty$, and that $\beta(c)$ is minimized over $c \in (0, \infty)$ at $c = 1/|a|$.

7.5 SUBGRADIENT-BASED OPTIMIZATION METHODS

We now return to the convex programming problem

$$\text{minimize } f(x)$$
$$\text{subject to } x \in X, \quad g_j(x) \leq 0, \quad j = 1, \ldots, r,$$

where $f : \Re^n \mapsto \Re$, $g_j : \Re^n \mapsto \Re$ are given convex functions, and X is a convex subset of $\Re^n$. The dual problem is

$$\text{maximize } q(\mu)$$
$$\text{subject to } \mu \in M,$$

where

$$q(\mu) = \inf_{x \in X} L(x, \mu) = \inf_{x \in X} \{f(x) + \mu' g(x)\},$$

is the dual function, and M is the convex set given by

$$M = \{\mu \mid \mu \geq 0, q(\mu) > -\infty\}; \tag{7.81}$$

(cf. Section 6.1).

We consider algorithms for solving the dual problem, which are based on the use of subgradients. We assume throughout this section that for every $\mu \in M$, we can calculate some vector x_μ that minimizes $L(x, \mu)$ over $x \in X$, yielding a subgradient $g(x_\mu)$ of q at μ [cf. Eq. (7.2)]. While the discussion of this section focuses on the case where q is a dual function, *the analysis and the results apply to any function $q : \Re^r \mapsto [-\infty, \infty)$ that is upper-semicontinuous and concave, and for which we can calculate a subgradient at any μ in the set M of Eq. (7.81)*.

We discuss two algorithms, both of which calculate a single subgradient at each iteration: the *subgradient method*, in Section 7.5.1, which at each iteration uses only the current subgradient, and the *cutting plane method*, in Section 7.5.3, which at each iteration uses all the subgradients previously calculated. Despite similarities, the philosophies underlying these two methods are quite different. The subgradient method relates to the gradient and gradient projection methods, and uses subgradients as directions of improvement of the distance to the optimum. On the other hand, the cutting plane method uses subgradients to construct increasingly accurate polyhedral approximations to the dual problem.

7.5.1 Subgradient Methods

The subgradient method generates a sequence of dual feasible points according to the iteration

$$\mu^{k+1} = [\mu^k + s^k g^k]^+,$$

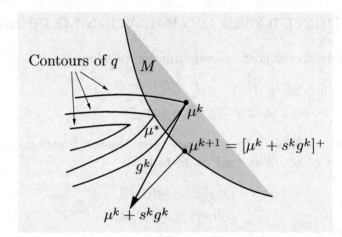

Figure 7.5.1. Illustration of how it may not be possible to improve the dual function using the subgradient iteration

$$\mu^{k+1} = [\mu^k + s^k g^k]^+$$

with a particular choice of subgradient g^k, regardless of the value of s^k.

where g^k is the subgradient $g(x_{\mu^k})$, $[\cdot]^+$ denotes projection on the set M, and s^k is a positive scalar stepsize. To ensure that the method is well-defined, *we assume that M is a closed set*, so that the projection exists and is unique by the projection theorem (Prop. 1.1.4). The iteration looks like the gradient projection method of Section 3.3, except that the subgradient g^k is used in place of the gradient (which may not exist). This difference is significant, however, because in contrast with the gradient projection method, the new iterate may not improve the dual cost for all values of the stepsize; that is, for some k we may have

$$q\bigl([\mu^k + sg^k]^+\bigr) < q(\mu^k), \qquad \forall\ s > 0;$$

see Fig. 7.5.1.

What makes the subgradient method work is that for a sufficiently small stepsize, the distance of the current iterate to the optimal solution set is reduced, as illustrated in Fig. 7.5.2. This is shown in the following proposition, which also provides an estimate for the range of appropriate stepsizes.

Proposition 7.5.1: If μ^k is not optimal, then for every dual optimal solution μ^*, we have

$$\|\mu^{k+1} - \mu^*\| < \|\mu^k - \mu^*\|,$$

Sec. 7.5 Subgradient-Based Optimization Methods

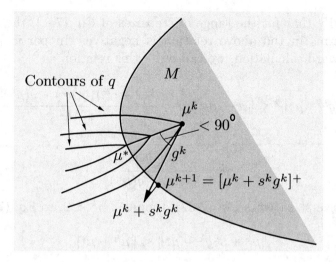

Figure 7.5.2. Illustration of how, given a nonoptimal μ^k, the distance to any optimal solution μ^* is reduced using a subgradient iteration with a sufficiently small stepsize. The crucial fact, which follows from the definition of a subgradient, is that the angle between the subgradient g^k and the vector $\mu^* - \mu^k$ is less than 90 degrees. As a result, the vector

$$\mu^k + s^k g^k$$

is closer to μ^* than μ^k for s^k small enough. Through the projection on M,

$$[\mu^k + s^k g^k]^+$$

gets even closer to μ^*.

---

for all stepsizes s^k such that

$$0 < s^k < \frac{2\big(q(\mu^*) - q(\mu^k)\big)}{\|g^k\|^2}. \tag{7.82}$$

Proof: We have

$$\|\mu^k + s^k g^k - \mu^*\|^2 = \|\mu^k - \mu^*\|^2 - 2s^k(\mu^* - \mu^k)'g^k + (s^k)^2 \|g^k\|^2,$$

and by using the subgradient inequality,

$$(\mu^* - \mu^k)'g^k \geq q(\mu^*) - q(\mu^k),$$

we obtain

$$\|\mu^k + s^k g^k - \mu^*\|^2 \leq \|\mu^k - \mu^*\|^2 - 2s^k\big(q(\mu^*) - q(\mu^k)\big) + (s^k)^2 \|g^k\|^2.$$

We can verify that for the range of stepsizes of Eq. (7.82) the sum of the last two terms in the above relation is negative. In particular, with a straightforward calculation, we can write this relation as

$$\|\mu^k + s^k g^k - \mu^*\|^2 \leq \|\mu^k - \mu^*\|^2 - \frac{\gamma^k(2-\gamma^k)\big(q(\mu^*) - q(\mu^k)\big)^2}{\|g^k\|^2}, \quad (7.83)$$

where

$$\gamma^k = \frac{s^k \|g^k\|^2}{q(\mu^*) - q(\mu^k)}.$$

If the stepsize s^k satisfies Eq. (7.82), then $0 < \gamma^k < 2$, so Eq. (7.83) yields

$$\|\mu^k + s^k g^k - \mu^*\| < \|\mu^k - \mu^*\|.$$

We now observe that since $\mu^* \in M$ and the projection operation is nonexpansive [Prop. 1.1.4(c)], we have

$$\left\| \left[\mu^k + s^k g^k\right]^+ - \mu^* \right\| \leq \|\mu^k + s^k g^k - \mu^*\|.$$

By combining the last two inequalities, the result follows. **Q.E.D.**

The inequality (7.83) can also be used to establish convergence and rate of convergence results for the subgradient method with stepsize rules satisfying Eq. (7.82) (see Exercise 7.5.1). Unfortunately, however, unless we know the dual optimal value $q(\mu^*)$, which is rare, the range of stepsizes (7.82) is unknown. In practice, typically one uses the stepsize formula

$$s^k = \frac{\alpha^k \big(q^k - q(\mu^k)\big)}{\|g^k\|^2}, \quad (7.84)$$

where q^k is an approximation to the optimal dual value and

$$0 < \alpha^k < 2.$$

To estimate the optimal dual value from below, we can use the best current dual value

$$\hat{q}^k = \max_{0 \leq i \leq k} q(\mu^i).$$

As an upper bound, we can use any primal value $f(\bar{x})$ corresponding to a primal feasible solution $\bar{x}$; in many circumstances, primal feasible solutions are naturally obtained in the course of the algorithm. Finally, the special structure of many problems can be exploited to yield improved bounds to the optimal dual value.

Here are two common ways to choose α^k and q^k in the stepsize formula (7.84):

(a) $\alpha^k = 1$ for all k and q^k is given by

$$q^k = \hat{q}^k + \delta^k, \qquad (7.85)$$

where $\hat{q}^k$ is the best current dual value $\hat{q}^k = \max_{0 \le i \le k} q(\mu^i)$, and δ^k is a positive number, which is increased by a certain factor if the previous iteration was a "success", i.e., it improved the best current dual value, and is decreased by some other factor otherwise. Thus, δ^k may be viewed as an "aspiration level" of improvement that we hope to attain in subsequent iterations. If upper bounds $\tilde{q}^k$ to the optimal dual value are available as discussed earlier, then a natural improvement to Eq. (7.85) is

$$q^k = \min\{\tilde{q}^k, \hat{q}^k + \delta^k\}.$$

We refer to the books [BNO03] (Section 8.2.1) and [Ber15a] (Section 3.2) for related convergence analysis. A variant of the method, is to choose

$$q^k = \big(1 + \beta(k)\big)\hat{q}^k,$$

where $\beta(k)$ is a number greater than zero, which is increased or decreased based on algorithmic progress. This variant requires that $\hat{q}^k > 0$.

(b) q^k is the best known upper bound to the optimal dual value at the kth iteration and α^k is a number, which is initially equal to one and is decreased by a certain factor every few iterations. An alternative formula for α^k is

$$\alpha^k = \frac{\beta}{k + \gamma},$$

where β and γ are fixed positive numbers. This formula satisfies the usual conditions for a diminishing stepsize

$$\alpha^k \to 0, \qquad \sum_{k=0}^{\infty} \alpha^k = \infty,$$

and in fact, one can show various convergence results for subgradient methods with a diminishing stepsize (see the exercises).

A number of convergence properties of the subgradient method are developed in the exercises under various assumptions. For example, a convergent variation is given in Exercise 7.5.2. Clearly, there are problems where the subgradient method works very poorly; think of an ill-conditioned differentiable problem, in which case the method becomes a variant of gradient projection.

On the other hand, the subgradient method is simple and works well for many types of problems, yielding good approximate solutions within a few tens or hundreds of iterations. Also, frequently a good primal solution can be obtained thanks to effective heuristics, even with a fairly suboptimal dual solution. There is no clear explanation of these fortuitous phenomena; apparently they are due to the special structure of many of the important types of practical problems.

There are several variations of subgradient methods that aim to accelerate the convergence of the basic method (see e.g., [CFM75], [Sho85], [Min86], [Sho98], [ZLW99], [BLY15]). We develop some of these variations in exercises, and we discuss in some detail one type of variation in the next section.

7.5.2 Approximate and Incremental Subgradient Methods

Similar to their gradient counterparts, discussed in Section 1.2, subgradient methods with errors arise naturally in a variety of contexts. For example, if for a given $\mu \in M$, the dual function value $q(\mu)$ is calculated by minimizing *approximately* $L(x, \mu)$ over $x \in X$, the subgradient obtained [as well as the value of $q(\mu)$] will involve an error.

To analyze such methods, it is useful to introduce a notion of approximate subgradient. In particular, given a scalar $\epsilon \geq 0$ and a vector $\bar{\mu}$ with $q(\bar{\mu}) > -\infty$, we say that g is an ϵ-*subgradient at* $\bar{\mu}$ if

$$q(\mu) \leq q(\bar{\mu}) + \epsilon + g'(\mu - \bar{\mu}), \qquad \forall\, \mu \in \Re^r.$$

The set of all ϵ-subgradients at $\bar{\mu}$ is called the ϵ-*subdifferential at* $\bar{\mu}$ and is denoted by $\partial_\epsilon q(\bar{\mu})$. Note that every subgradient at a given point is also an ϵ-subgradient for all $\epsilon \geq 0$. Generally, however, an ϵ-subgradient need not be a subgradient, unless $\epsilon = 0$:

$$\partial q(\bar{\mu}) = \partial_0 q(\bar{\mu}) \subset \partial_\epsilon q(\bar{\mu}), \qquad \forall\, \epsilon \geq 0.$$

Methods that Use ϵ-Subgradients

An approximate subgradient method is defined by

$$\mu^{k+1} = [\mu^k + s^k g^k]^+, \tag{7.86}$$

where g^k is an ϵ^k-subgradient at μ^k. A convergence analysis for this method is outlined in Exercise 7.5.8, where it is shown that it converges to within 2ϵ of the optimal dual value, assuming that $\epsilon^k \to \epsilon$ and the stepsize is appropriately chosen.

For an interesting context where this method applies, suppose that we minimize *approximately* $L(x, \mu^k)$ over $x \in X$, thereby obtaining a vector $x^k \in X$ with

$$L(x^k, \mu^k) \leq \inf_{x \in X} L(x, \mu^k) + \epsilon^k. \tag{7.87}$$

Sec. 7.5   Subgradient-Based Optimization Methods   715

We claim that the corresponding constraint vector, $g(x^k)$, is an ϵ^k-subgradient at μ^k. Indeed, we have for all $\mu \in \Re^r$

$$\begin{aligned}
q(\mu) &= \inf_{x \in X} \{f(x) + \mu' g(x)\} \\
&\leq f(x^k) + \mu' g(x^k) \\
&= f(x^k) + {\mu^k}' g(x^k) + g(x^k)'(\mu - \mu^k) \\
&\leq q(\mu^k) + \epsilon^k + g(x^k)'(\mu - \mu^k),
\end{aligned}$$

where the last inequality follows from Eq. (7.87). Thus we can view a subgradient method with errors in the Lagrangian minimization, as in Eq. (7.87), as a special case of the approximate subgradient method (7.86).

Incremental Subgradient Methods

Another interesting type of approximate subgradient method is an *incremental* variant that is patterned after the incremental gradient method, discussed in Section 2.4.1. Assume that the dual function has the form

$$q(\mu) = \sum_{i=1}^{m} q_i(\mu),$$

where the q_i are concave and continuous over M (i.e., the structure corresponding to separable problems). The idea of the incremental method is to sequentially take steps along the subgradients of the component functions q_i, with intermediate adjustment of μ after processing each component function. Similar to the incremental approach of Section 2.4, *an iteration is viewed as a cycle of m subiterations*. If μ^k is the vector obtained after k cycles, the vector μ^{k+1} obtained after one more cycle is

$$\mu^{k+1} = \psi_m,$$

where ψ_m is obtained after the m steps

$$\psi_0 = \mu^k, \qquad \psi_i = [\psi_{i-1} + s^k g_i]^+, \quad i = 1, \ldots, m, \qquad (7.88)$$

with g_i being a subgradient of q_i at ψ_{i-1}.

Generally, the characteristics of the incremental subgradient method are similar to the ones we discussed in Section 2.4 for the incremental gradient method. In particular:

(a) When far from the solution, the method can make much faster progress than the nonincremental subgradient method, particularly if the number of component functions is large. The rate of progress also depends on the stepsize.

(b) When close to the solution, the method oscillates and the size of the oscillation (from start to end of a cycle) is proportional to the stepsize. Thus there is a tradeoff between rapid initial convergence (large stepsize) and size of asymptotic oscillation (small stepsize). With a diminishing stepsize the method is capable of attaining convergence (no asymptotic oscillation).

(c) The size of the oscillation depends also on the order in which the component functions q_i are processed within a cycle.

To address the potentially detrimental effect of an unfavorable order of processing the component functions, one may consider randomization. Indeed, a popular technique for incremental gradient methods, discussed in Section 2.4, is to reshuffle randomly the order of the component functions after each cycle. A variation of this method is to pick randomly a function q_i at each step rather than to pick each q_i exactly once in each cycle according to a randomized order.

Let us now focus on the connection between the incremental and the approximate subgradient methods (7.86) and (7.88). An important fact here is that *if two vectors μ and $\bar{\mu}$ are "near" each other, then subgradients at $\bar{\mu}$ can be viewed as ϵ-subgradients at μ, where ϵ is "small."* In particular, if $g \in \partial q(\bar{\mu})$, we have for all $w \in \Re^r$,

$$q(w) \leq q(\bar{\mu}) + g'(w - \bar{\mu})$$
$$\leq q(\mu) + g'(w - \mu) + q(\bar{\mu}) - q(\mu) + g'(\mu - \bar{\mu})$$
$$\leq q(\mu) + g'(w - \mu) + \epsilon,$$

where

$$\epsilon = |q(\bar{\mu}) - q(\mu)| + \|g\| \cdot \|\bar{\mu} - \mu\|.$$

Thus, we have $g \in \partial_\epsilon q(\mu)$, and ϵ is small when $\bar{\mu}$ is near μ.

We now observe from Eq. (7.88) that the ith step within a cycle of the incremental subgradient method involves the direction g_i, which is a subgradient of q_i at the corresponding vector ψ_{i-1}. If the stepsize s^k is small, then ψ_{i-1} is close to the vector μ^k available at the start of the cycle, and hence g_i is an ϵ_i-subgradient at q_i at μ^k, where ϵ_i is small. Let us ignore the projection operation in Eq. (7.88), and let us also use the easily shown relation

$$\partial_{\epsilon_1} q(\mu) + \cdots + \partial_{\epsilon_m} q(\mu) \subset \partial_\epsilon q(\mu), \qquad \forall\ \mu \text{ with } q(\mu) > -\infty,$$

where

$$\epsilon = \epsilon_1 + \cdots + \epsilon_m,$$

to approximate the ϵ-subdifferential of the sum $q = \sum_{i=1}^m q_i$. Then, it can be seen that the incremental subgradient iteration can be viewed as a (nonincremental) ϵ^k-subgradient iteration. The size of ϵ^k depends on the size of s^k, as well as the dual function q, and we generally have $\epsilon^k \to 0$ as $s^k \to 0$.

The view of the incremental subgradient method that we outlined above, parallels the view of the incremental gradient method as a gradient method with errors (see Section 2.4). The convergence analysis of the incremental subgradient method is outlined in the exercises, together with various methods for choosing the stepsize s^k. For an extensive discussion, including an estimate of the convergence rate advantage of methods that randomize the component selection order, we refer to the papers [NeB00], [NeB01], and the books [BNO03], [Ber15a]. There are also aggregated and distributed asynchronous versions of incremental subgradient methods, for which we refer to the papers [NBB01] and [Ber15b].

7.5.3 Cutting Plane Methods

Consider again the dual problem

$$\text{maximize } q(\mu)$$
$$\text{subject to } \mu \in M.$$

The cutting plane method consists of solving at the kth iteration the problem

$$\text{maximize } Q^k(\mu)$$
$$\text{subject to } \mu \in M,$$

where the dual function is replaced by a polyhedral approximation Q^k, constructed using the points μ^i generated so far and their subgradients $g(x_{\mu^i})$, which are denoted by g^i. In particular, for $k = 1, 2, \ldots$,

$$Q^k(\mu) = \min\{q(\mu^0) + (\mu - \mu^0)'g^0, \ldots, q(\mu^{k-1}) + (\mu - \mu^{k-1})'g^{k-1}\} \quad (7.89)$$

and

$$\mu^k \in \arg\max_{\mu \in M} Q^k(\mu); \quad (7.90)$$

see Fig. 7.5.3. We assume that the maximum of $Q^k(\mu)$ above is attained for all k. For those k for which this is not guaranteed, artificial bounds may be placed on the coordinates of μ, so that the maximization will be carried out over a compact set and consequently the maximum will be attained.

The following proposition establishes the convergence properties of the cutting plane method. An important special case arises when the primal problem is a linear program, in which case the dual function is polyhedral and can be put in the form

$$q(\mu) = \min_{i \in I}\{a_i'\mu + b_i\}, \quad (7.91)$$

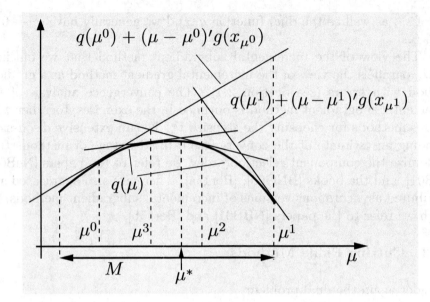

Figure 7.5.3. Illustration of the cutting plane method. With each new iterate μ^i, a new hyperplane

$$q(\mu^i) + (\mu - \mu^i)'g^i$$

is added to the polyhedral approximation of the dual function.

where I is a finite index set, and $a_i \in \Re^r$ and b_i are given vectors and scalars, respectively. As the subgradient g^k in the cutting plane method we select a vector a_{i_k} for which the minimum in Eq. (7.91) is attained (cf. Prop. 7.1.2). In this case, the following proposition shows that the cutting plane method converges finitely; see also Fig. 7.5.4.

Proposition 7.5.2:

(a) Assume that $\{g^k\}$ is a bounded sequence. Then every limit point of a sequence $\{\mu^k\}$ generated by the cutting plane method is a dual optimal solution.

(b) Assume that the dual function q is polyhedral of the form (7.91). Then the cutting plane method with the subgradient selection scheme just described terminates finitely; that is, for some k, μ^k is a dual optimal solution.

Proof: (a) Since for all i, g^i is a subgradient of q at μ^i, we have

$$q(\mu^i) + (\mu - \mu^i)'g^i \geq q(\mu), \qquad \forall\, \mu \in M,$$

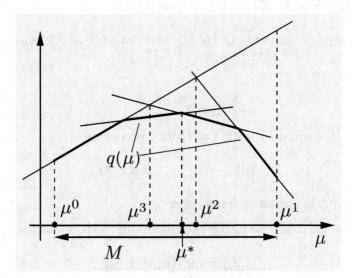

Figure 7.5.4. Illustration of the finite convergence property of the cutting plane method in the case where q is polyhedral. What happens here is that if μ^k is not optimal, a new cutting plane will be added at the corresponding iteration, and there can be only a finite number of cutting planes.

so from the definitions (7.89) and (7.90) of Q^k and μ^k, it follows that

$$Q^k(\mu^k) \geq Q^k(\mu) \geq q(\mu), \qquad \forall \, \mu \in M. \tag{7.92}$$

Suppose that a subsequence $\{\mu^k\}_K$ converges to $\bar{\mu}$. Then, since $\mu^k \geq 0$, we have $\bar{\mu} \geq 0$, and there are two possibilities: (1) $\bar{\mu} \in M$, or (2) $q(\bar{\mu}) = -\infty$. In case (1), by using Eq. (7.92), we obtain

$$Q^k(\mu^k) \geq Q^k(\bar{\mu}) \geq q(\bar{\mu}).$$

In case (2), we also have $Q^k(\mu^k) \geq q(\bar{\mu})$. Thus in both cases (1) and (2), using also the definition (7.89) of Q^k, we have

$$q(\mu^i) + (\mu^k - \mu^i)'g^i \geq Q^k(\mu^k) \geq q(\bar{\mu}). \tag{7.93}$$

We will now take the limit above as $i \to \infty$, $k \to \infty$, $i \in K$, $k \in K$. We have using the upper semicontinuity of q,

$$\limsup_{i \to \infty, \, i \in K} q(\mu^i) \leq q(\bar{\mu}). \tag{7.94}$$

Also by using the assumption that the subgradient sequence $\{g^i\}$ is bounded, we have

$$\lim_{\substack{i \to \infty, \, k \to \infty, \\ i \in K, \, k \in K}} (\mu^k - \mu^i)'g^i = 0. \tag{7.95}$$

By combining Eqs. (7.93)-(7.95), we see that

$$q(\bar{\mu}) \geq \limsup_{k \to \infty,\, k \in K} Q^k(\mu^k) \geq \liminf_{k \to \infty,\, k \in K} Q^k(\mu^k) \geq q(\bar{\mu}),$$

and hence

$$\lim_{k \to \infty,\, k \in K} Q^k(\mu^k) = q(\bar{\mu}).$$

Combining this equation with Eq. (7.92), we obtain

$$q(\bar{\mu}) \geq q(\mu), \qquad \forall\, \mu \in M,$$

showing that $\bar{\mu}$ is a dual optimal solution.

(b) Let i^k be an index attaining the minimum in the equation

$$q(\mu^k) = \min_{i \in I} \{a'_i \mu^k + b_i\},$$

so that a_{i^k} is a subgradient at μ^k. From Eq. (7.92), we see that

$$Q^k(\mu^k) = q(\mu^k) \quad \Longrightarrow \quad \mu^k \text{ is dual optimal.}$$

Therefore, if μ^k is not dual optimal, we must have $Q^k(\mu^k) > q(\mu^k) = a'_{i^k} \mu^k + b_{i^k}$. Since

$$Q^k(\mu^k) = \min_{0 \leq m \leq k-1} \{a'_{i^m} \mu^k + b_{i^m}\},$$

the pair (a_{i^k}, b_{i^k}) is not equal to any of the pairs $(a_{i^0}, b_{i^0}), \ldots, (a_{i^{k-1}}, b_{i^{k-1}})$. It follows that there can be only a finite number of iterations for which μ^k is not dual optimal. **Q.E.D.**

The reader may verify [using Prop. B.21(b) of Appendix B] that the boundedness assumption in Prop. 7.5.2(a) can be replaced by the assumption that $q(\mu)$ is real-valued for all $\mu \in \Re^r$, which can be ascertained if X is a finite set, or alternatively if f and g_j are continuous, and X is a compact set. Despite the finite convergence property shown in Prop. 7.5.2(b), the cutting plane method often tends to converge slowly, even for problems where the dual function is polyhedral. Indeed, typically one should base termination on the upper and lower bounds

$$Q^k(\mu^k) \geq \max_{\mu \in M} q(\mu) \geq \max_{0 \leq i \leq k-1} q(\mu^i),$$

[cf. Eq. (7.92)], rather that wait for finite termination to occur. Nonetheless, the method is often much better suited for solution of particular types of large problems than its competitors; see the Dantzig-Wolfe decomposition method described in Section 7.6.1.

Linearly Constrained Versions

Consider the case where the constraint set M is polyhedral of the form

$$M = \{\mu \mid \gamma_i'\mu + \beta_i \leq 0, \, i \in I\},$$

where I is a finite set, and γ_i and β_i are given vectors and scalars, respectively. Let

$$w(\mu) = \max_{i \in I}\{\gamma_i'\mu + \beta_i\},$$

so the problem is to maximize $q(\mu)$ subject to $w(\mu) \leq 0$. It is then possible to consider a variation of the cutting plane method, where both functions q and w are replaced by polyhedral approximations Q^k and W^k, respectively. The method is

$$\mu^k \in \arg\max_{W^k(\mu) \leq 0} Q^k(\mu).$$

As earlier,

$$Q^k(\mu) = \min\{q(\mu^0) + (\mu - \mu^0)'g^0, \ldots, q(\mu^{k-1}) + (\mu - \mu^{k-1})'g^{k-1}\},$$

with g^i being a subgradient of q at μ^i. The polyhedral approximation W^k is given by

$$W^k(\mu) = \max_{i \in I^k}\{\gamma_i'\mu + \beta_i\},$$

where I^k is a subset of I generated as follows: I^0 is an arbitrary subset of I, and I^k is obtained from I^{k-1} by setting $I^k = I^{k-1}$ if $w(\mu^k) \leq 0$, and by adding to I^{k-1} one or more of the indices $i \notin I^{k-1}$ such that $\gamma_i'\mu^k + \beta_i > 0$ otherwise.

The convergence properties of this method are very similar to the ones of the earlier method. In fact a proposition analogous to Prop. 7.5.2 can be formulated and proved.

Proximal Cutting Plane and Bundle Methods

One of the drawbacks of the cutting plane method is that it can take large steps away from the optimum even when it is close to (or even at) the optimum (cf. Exercise 7.5.7). This phenomenon is referred to as *instability*, and has another undesirable effect, namely, that μ^{k-1} may not be a good starting point for the algorithm that minimizes $Q^k(\mu)$. A way to limit the effects of this phenomenon is to add to the polyhedral function approximation a quadratic regularization term that penalizes large deviations from the current point. Thus in this method, μ^{k+1} is obtained as

$$\mu^{k+1} \in \arg\max_{\mu \in M}\left\{Q^{k+1}(\mu) - \frac{1}{2c^k}\|\mu - \mu^k\|^2\right\}, \tag{7.96}$$

where $\{c^k\}$ is a positive nondecreasing scalar parameter sequence, and as in Eq. (7.89),

$$Q^{k+1}(\mu) = \min\{q(\mu^0) + (\mu - \mu^0)'g^0, \ldots, q(\mu^k) + (\mu - \mu^k)'g^k\}.$$

We recognize this as an approximate version of the proximal algorithm of Section 3.6, where the polyhedral approximation Q^{k+1} is used in place of the dual function q. Accordingly, we refer to it as the *proximal cutting plane method*. One advantage of this method is that the maximum in Eq. (7.96) is guaranteed to be attained. It can also be shown that the method maintains the finite termination property of Prop. 7.5.2(b) in the case where q is polyhedral (see [Ber15a], Section 5.3).

Unfortunately, the proximal cutting plane method may not provide the aimed for increased stability over the ordinary cutting plane method. The reason is that $Q^{k+1}(\mu)$ is polyhedral and if c^k is sufficiently large, the new iterate μ^{k+1} will be the exact minimum of $Q^{k+1}(\mu)$ based on the generic properties of the proximal algorithm, when applied to minimization of polyhedral functions (cf. Prop. 3.6.5). Thus the proximal cutting plane iteration, applied with a large value of c^k will produce the same iterate as the ordinary cutting plane method.

An important variation of the method, which aims to address this difficulty, is *bundle methods*. These have the form

$$\mu^{k+1} \in \arg\max_{\mu \in M}\left\{Q^{k+1}(\mu) - \frac{1}{2c^k}\|\mu - \nu^k\|^2\right\},$$

where the "proximal center" ν^k need not be equal to μ^k, but is rather one of the past iterates μ^i, $i \leq k$. In particular, following the computation of μ^k, the new proximal center ν^k is either set to μ^k, or is left unchanged ($\nu^k = \nu^{k-1}$) depending on whether, according to a certain test, the algorithm has made "sufficient progress" in approximating $q(\mu)$ with $Q^k(\mu)$ or not. The idea is to try to reduce further the effect of instability, by providing additional tendency for the new iterate μ^{k+1} to stay close to the past iterate ν^k. The details of the test scheme for changing ν^k can be complicated, and we refer to the literature for more discussion (e.g., [BGL06] and [Ber15a]).

Central Cutting Plane Methods

These methods maintain a polyhedral approximation

$$Q^k(\mu) = \min\{q(\mu^0) + (\mu - \mu^0)'g^0, \ldots, q(\mu^{k-1}) + (\mu - \mu^{k-1})'g^{k-1}\}$$

to the dual function q, but they generate the next vector μ^k by using a somewhat different mechanism. In particular, instead of maximizing Q^k as in the standard cutting plane method [cf. Eq. (7.90)], they obtain μ^k by finding a "central pair" (μ^k, z^k) within the subset

$$S^k = \{(\mu, z) \mid \mu \in M,\ \tilde{q}^k \leq q(\mu),\ \tilde{q}^k \leq z \leq Q^k(\mu)\},$$

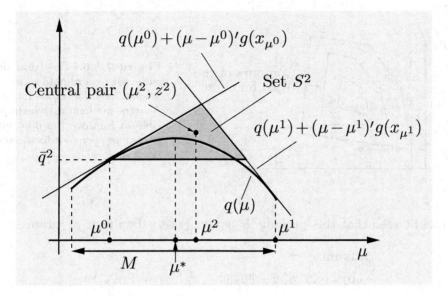

Figure 7.5.5. Illustration of the set

$$S^k = \{(\mu, z) \mid \mu \in M,\ \tilde{q}^k \leq q(\mu),\ \tilde{q}^k \leq z \leq Q^k(\mu)\},$$

in the central cutting plane method.

where $\tilde{q}^k$ is the best lower bound to the optimal dual value that has been found so far,

$$\tilde{q}^k = \max_{i=0,\ldots,k-1} q(\mu^i).$$

The set S^k is illustrated in Fig. 7.5.5.

There are several possible methods for finding the central pair (μ^k, z^k). Roughly, the idea is that the central pair should be "somewhere in the middle" of S. For example, consider the case where S is polyhedral with nonempty interior. Then (μ^k, z^k) could be the *analytic center* of S, where for any polyhedral set

$$P = \{y \mid a_p' y \leq c_p,\ p = 1, \ldots, m\}$$

with nonempty interior, its analytic center is the unique maximizer of $\sum_{p=1}^{m} \ln(c_p - a_p' y)$ over $y \in P$. Another possibility is the *ball center* of S, i.e., the center of the largest inscribed sphere in S; for the generic polyhedral set

$$P = \{y \mid a_p' y \leq c_p,\ p = 1, \ldots, m\}$$

with nonempty interior, the ball center can be obtained by solving the following problem with optimization variables (y, σ):

maximize σ

subject to $a_p'(y + d) \leq c_p,\quad \forall\ \|d\| \leq \sigma,\ p = 1, \ldots, m.$

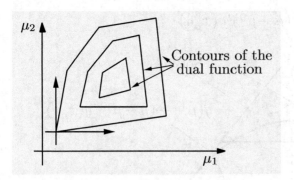

Figure 7.5.6. The basic difficulty with coordinate ascent for a nondifferentiable dual function. At some points it may be impossible to improve the dual function along any coordinate direction.

It can be seen that this problem is equivalent to the linear program

$$\text{maximize } \sigma$$
$$\text{subject to } a'_p y + \|a_p\|\sigma \le c_p, \quad p = 1, \ldots, m.$$

Central cutting plane methods have satisfactory convergence properties, even though they do not terminate finitely in the case of a polyhedral q. Furthermore, these methods have benefited from advances in the implementation of interior point methods; see the references cited at the end of the chapter.

7.5.4 Ascent and Approximate Ascent Methods

The subgradient and cutting plane methods do not guarantee a dual function improvement at each iteration. We will now briefly discuss some of the difficulties in constructing ascent methods, i.e., methods where the dual function is improved at each iterate. For simplicity, we restrict ourselves to the case where the dual function is real-valued, and the dual problem is unconstrained (i.e., $M = \Re^r$). Note, however, that constraints may be lumped into the dual function by using a nondifferentiable penalty function as discussed in Section 5.3.

Coordinate Ascent

It is possible to consider the analog of the coordinate ascent method discussed in Section 2.3.1. Here we maximize the dual function successively along each coordinate direction. Unfortunately there is a fundamental difficulty. It is possible to get stuck at a corner from which it is impossible to make progress along any coordinate direction (see Fig. 7.5.6). This difficulty does not arise in some problem types where the nondifferentiable cost term is separable (cf. the discussion of Section 3.7.1), but may be hard to overcome in general.

In some important special cases, however, such as when the primal problem is a network optimization problem of the type discussed in Sections

Sec. 7.5   Subgradient-Based Optimization Methods   725

6.4.3 and 6.5, it is possible to modify the coordinate ascent method to make it workable. For such problems, two types of approaches have been introduced:

(a) The *relaxation method*, first proposed for the assignment problem in the author's paper [Ber81], and extended to linear cost network flow problems with and without arc gains in the papers [Ber85], [BHT87], and [BeT88]. Here, ascent directions involving several coordinates are constructed, whenever no progress can be made along any coordinate direction.

(b) The *auction algorithm*, first proposed for the assignment problem in the author's paper [Ber79b], and extended to linear cost single commodity network flow problems under the name ϵ-*relaxation* (also known as the *preflow push method*) in the papers [Ber86] and [BeE88]. (See [BeT89], [BeT97], [Ber91], [Ber98] for textbook presentations, [AMO91], [Ber92] for surveys, and the papers [BPT97a], [BPT97b], and [TsB00] for extensions to the convex arc cost case with and without arc gains.) The main idea is to allow a single coordinate to change even if this worsens the dual function. When a coordinate is changed, however, it is set to ϵ plus the value that maximizes the dual function along that coordinate, where ϵ is a positive number. If ϵ is small enough, the algorithm can eventually approach the optimal solution as illustrated in Fig. 7.5.7.

An extensive analysis of these methods can be found in the author's network optimization textbook [Ber98].

Steepest Ascent and ϵ-Ascent Methods

The steepest ascent direction d_μ of a concave function $q : \Re^n \mapsto \Re$ at a vector μ is obtained by solving the problem

$$\text{maximize } q'(\mu; d)$$
$$\text{subject to } \|d\| \leq 1,$$

where

$$q'(\mu; d) = \lim_{\alpha \downarrow 0} \frac{q(\mu + \alpha d) - q(\mu)}{\alpha}$$

is the directional derivative of q at μ in the direction d. Using Prop. B.21 of Appendix B (and making an adjustment for the concavity of q), we have

$$q'(\mu; d) = \min_{g \in \partial q(\mu)} d'g,$$

and by the minimax theorem of Section 6.4.4 [using also the compactness of $\partial q(\mu)$; cf. Prop. B.21(b) in Appendix B], we have

$$\max_{\|d\| \leq 1} q'(\mu; d) = \max_{\|d\| \leq 1} \min_{g \in \partial q(\mu)} d'g = \min_{g \in \partial q(\mu)} \max_{\|d\| \leq 1} d'g = \min_{g \in \partial q(\mu)} \|g\|.$$

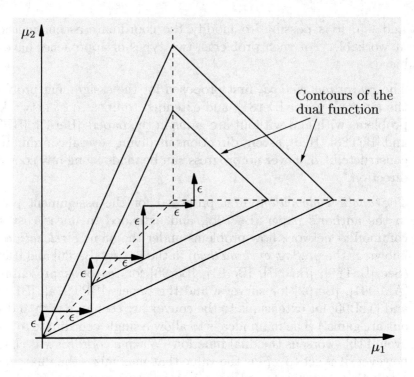

Figure 7.5.7. Path of the ϵ-relaxation method. When ϵ is small, it is possible to approach the optimal solution even if each step does not result in a dual function improvement. The method eventually reaches a small neighborhood of the optimal solution if the problem has a favorable (typically a network) structure.

It follows that the steepest ascent direction is

$$d_\mu = \frac{\bar{g}_\mu}{\|\bar{g}_\mu\|},$$

where $\bar{g}_\mu$ is the unique minimum-norm subgradient at μ:

$$\bar{g}_\mu \in \arg\min_{g \in \partial q(\mu)} \|g\|. \tag{7.97}$$

The steepest ascent method has the form

$$\mu^{k+1} = \mu^k + \alpha^k \bar{g}^k,$$

where $\bar{g}^k$ is the vector of minimum norm in $\partial q(\mu^k)$, and α^k is a positive stepsize that guarantees that $q(\mu^{k+1}) > q(\mu^k)$ if μ^k is not optimal. One disadvantage of this method is that to calculate the steepest ascent direction via Eq. (7.97), one needs to compute the entire subdifferential at the current iterate. It is possible, however, to compute an approximate steepest ascent direction by employing approximations to the subdifferential that involve only a finite number of subgradients; see [Lem74].

A serious theoretical drawback of the steepest ascent method is that, depending on the stepsize rule, it may get stuck far from the optimum. This can happen even if the stepsize is chosen by line maximization (an example is given in [Ber15a], Section 2.1.3). The difficulty in this example is that at the limit, q is nondifferentiable and has subgradients that cannot be approximated by subgradients at any one of the iterates, arbitrarily close to the limit (this would not happen if q were continuously differentiable at the limit).

One way to construct a variant of steepest ascent that is convergent, is to obtain a direction by projecting in the manner of Eq. (7.97) on a larger set than the subdifferential. Such a set should include not only the subgradients at the current iterate, but also the subgradients of neighboring points. One possibility is to replace the subdifferential $\partial q(\mu)$ in the projection of Eq. (7.97) with the ϵ-subdifferential $\partial_\epsilon q(\mu)$ at μ, where ϵ is a positive scalar. The resulting method, called the ϵ-*ascent method*, and first introduced in the papers [BeM71], [BeM73], has the form

$$\mu^{k+1} = \mu^k + \alpha^k \tilde{g}^k,$$

where

$$\tilde{g}^k \in \arg \min_{g \in \partial_\epsilon q(\mu^k)} \|g\|. \tag{7.98}$$

It can be implemented by either calculating explicitly the ϵ-subdifferential $\partial_\epsilon q(\mu)$ (in problems with favorable structure) or by approximating $\partial_\epsilon q(\mu)$ with a finite number of ϵ-subgradients; see [Lem74].

An interesting variation of the ϵ-ascent method applies to the case where the dual function consists of the sum of several concave functions:

$$q(\mu) = q_1(\mu) + \cdots + q_m(\mu).$$

As we have seen, this is very common, particularly for problems with separable structure. Then, we can approximate the ϵ-subdifferential of q with the vector sum

$$\tilde{\partial}_\epsilon q(\mu) = \partial_\epsilon q_1(\mu) + \cdots + \partial_\epsilon q_m(\mu)$$

in the projection problem (7.98). The method (also proposed in [BeM71]) consists of the iteration

$$\mu^{k+1} = \mu^k + \alpha^k \tilde{g}^k,$$

where

$$\tilde{g}^k \in \arg \min_{g \in \tilde{\partial}_\epsilon q(\mu^k)} \|g\|.$$

In many cases, the approximation $\tilde{\partial}_\epsilon q(\mu)$ may be obtained much more easily than the exact ϵ-subdifferential $\partial_\epsilon q(\mu)$, thereby resulting in an algorithm that is much easier to implement. For an application of this algorithm to monotropic programming, we refer to the books [Roc84], [Ber98], and [Ber15a] (Section 6.7). The algorithm draws its validity from the fact

$$\partial_\epsilon q(\mu) \subset \tilde{\partial}_\epsilon q(\mu) \subset \partial_{m\epsilon} q(\mu),$$

in combination with the properties of ϵ-subdifferentials.

EXERCISES

7.5.1 (Convergence of the Subgradient Method [Pol69b]) (www)

Consider the subgradient method $\mu^{k+1} = [\mu^k + s^k g^k]^+$, where

$$s^k = \frac{q^* - q(\mu^k)}{\|g^k\|^2}$$

and q^* is the optimal dual value. Assume that there exists at least one optimal dual solution.

(a) Use Eq. (7.83) to show that $\{\mu^k\}$ is bounded.

(b) Assuming that $\{g^k\}$ is bounded, use Eq. (7.83) to show that $q(\mu^k) \to q^*$, and that $\{\mu^k\}$ converges to some optimal dual solution.

(c) Show that if q is real-valued, $\{g^k\}$ is bounded. *Hint*: Use Prop. B.21 in Appendix B.

7.5.2 (A Convergent Variation of the Subgradient Method) (www)

Consider the subgradient method $\mu^{k+1} = [\mu^k + s^k g^k]^+$, where

$$s^k = \frac{\tilde{q} - q(\mu^k)}{\|g^k\|^2}.$$

(a) Suppose that $\tilde{q}$ is an *underestimate* of the optimal dual value q^* such that $q(\mu^k) < \tilde{q} < q^*$. [Here $\tilde{q}$ is fixed and the algorithm stops at μ^k if $q(\mu^k) \geq \tilde{q}$.] Assuming that $\{g^k\}$ is bounded, show that either for some $\bar{k}$ we have $q(\mu^{\bar{k}}) \geq \tilde{q}$ or else $\{\mu^k\}$ converges to some $\tilde{\mu}$ with $q(\tilde{\mu}) \geq \tilde{q}$. *Hint*: Consider the function $\min\{q(\mu), \tilde{q}\}$ and use the results of Exercise 7.5.1.

(b) Suppose that $\tilde{q}$ is an *overestimate* of the optimal dual value, i.e., $\tilde{q} > q^*$. Assuming that $\{g^k\}$ is bounded, show that the length of the path traveled by the method is infinite, i.e.,

$$\sum_{k=0}^{\infty} s^k \|g^k\| = \sum_{k=0}^{\infty} \frac{\tilde{q} - q(\mu^k)}{\|g^k\|} = \infty.$$

Note: Parts (a) and (b) provide the basis for a method that uses an adjustable level $\tilde{q}$ in the stepsize formula. The method uses two positive scalars δ^0 and B, and operates in cycles during which $\tilde{q}$ is kept constant. Cycle 0 begins with the starting point μ^0. At the beginning of the typical cycle k, we have a vector μ^k and we set $\tilde{q} = q(\mu^k) + \delta^k$. Cycle k consists of successive subgradient iterations that start with μ^k and end when one of the following two occurs:

(1) The dual value exceeds $q(\mu^k) + \delta^k/2$.

(2) The length of the path traveled starting from μ^k exceeds B.

Then cycle $k+1$ begins with μ^{k+1} equal to the vector that has the highest dual value within cycle k, and either $\delta^{k+1} = \delta^k$ or $\delta^{k+1} = \delta^k/2$, depending on whether cycle k terminated with case (1) or case (2), respectively. This method was proposed in the thesis [Brä93], which showed that it has satisfactory convergence properties. For further discussion and analysis, we refer to the book [BNO03], Section 8.2.1.

7.5.3 (Convergence Rate of the Subgradient Method) (www)

Consider the subgradient method of Exercise 7.5.1 and assume that $\{g^k\}$ is a bounded sequence.

(a) Show that
$$\liminf_{k \to \infty} \sqrt{k}\left(q^* - q(\mu^k)\right) = 0.$$

Hint: Use Eq. (7.83) to show that $\sum_{k=0}^{\infty}\left(q^* - q(\mu^k)\right)^2 < \infty$. Assume that $\sqrt{k}\left(q^* - q(\mu^k)\right) \geq \epsilon$ for some $\epsilon > 0$ and arbitrarily large k, and reach a contradiction.

(b) Assume that for some $a > 0$ and all k, we have $q(\mu^*) - q(\mu^k) \geq a\|\mu^* - \mu^k\|$, where μ^* is some dual optimal solution. Use Eq. (7.83) to show that $\{\|\mu^k - \mu^*\|\}$ converges linearly. In particular, for all k we have

$$\|\mu^{k+1} - \mu^*\| \leq r\|\mu^k - \mu^*\|,$$

where $r = \sqrt{1 - a^2/b^2}$ and b is an upper bound for $\|g^k\|$.

7.5.4 (A Variation of the Subgradient Method [CFM75]) (www)

Consider the dual problem and the following variation of the subgradient method

$$\mu^{k+1} = [\mu^k + s^k d^k]^+,$$

where

$$d^k = \begin{cases} g^k & \text{if } k = 0, \\ g^k + \beta^k d^{k-1} & \text{if } k > 0, \end{cases}$$

s^k and β^k are scalars satisfying

$$0 < s^k \leq \frac{q(\mu^*) - q(\mu^k)}{\|d^k\|^2},$$

$$\beta^k = \begin{cases} -\gamma \dfrac{g^{k\prime} d^{k-1}}{\|d^{k-1}\|^2} & \text{if } g^{k\prime} d^{k-1} < 0, \\ 0 & \text{otherwise,} \end{cases}$$

with $\gamma \in [0,2]$, and μ^* is an optimal dual solution. Assuming $\mu^k \neq \mu^*$, show that

$$\|\mu^* - \mu^{k+1}\| < \|\mu^* - \mu^k\|.$$

Furthermore,

$$\frac{(\mu^* - \mu^k)' d^k}{\|d^k\|} \geq \frac{(\mu^* - \mu^k)' g^k}{\|g^k\|},$$

i.e., the angle between d^k and $\mu^* - \mu^k$ is no larger than the angle between g^k and $\mu^* - \mu^k$.

7.5.5

Give an example of a one-dimensional problem where the cutting plane method is started at an optimal solution but does not terminate finitely.

7.5.6 (Coordinate Ascent for Problems with Special Structure)

Consider the maximization of a function $q : \Re^r \mapsto \Re$ subject to $\mu_j \geq 0$, $j = 1, \ldots, r$. Assume that q has the form

$$q(\mu) = Q(\mu) + \sum_{j=1}^{r} q_j(\mu_j),$$

where $Q : \Re^r \mapsto \Re$ is concave and differentiable, and the functions $q_j : \Re \mapsto \Re$ are concave (possibly nondifferentiable). Show that the coordinate ascent method with the line maximization stepsize rule improves the value of q at every μ that is not optimal.

7.5.7 (Solution of Linear Inequalities)

Consider the problem of finding a point in the intersection of m halfspaces

$$C = \{\mu \mid a_i' \mu - b_i \geq 0, \ i = 1, \ldots, m\},$$

which is assumed to be nonempty. Consider the algorithm that stops if $\mu^k \in C$ and otherwise sets

$$\mu^{k+1} = \mu^k + s^k a_{i^k},$$

where i^k is the index of the inequality $a_i' \mu - b_i \geq 0$ that is most violated at μ^k, and

$$s^k = \frac{b_{i^k} - a_{i^k}' \mu^k}{\|a_{i^k}\|^2}.$$

Show that this algorithm is a subgradient method for a suitable function and prove its convergence to a point of C.

7.5.8 (Approximate Subgradient Method) (www)

Let the dual function be real-valued [$q(\mu) > -\infty$ for all $\mu \in \Re^r$], and let μ^* be a dual optimal solution. Consider the subgradient iteration

$$\mu^{k+1} = \left[\mu^k + \frac{q(\mu^*) - q(\mu^k)}{\|g^k\|^2} g^k\right]^+,$$

with the difference that g^k is an ϵ-subgradient at μ^k [i.e., $g^k \in \partial_\epsilon q(\mu^k)$], where $\epsilon > 0$. Show that as long as μ^k is such that $q(\mu^k) < q(\mu^*) - 2\epsilon$, we have $\|\mu^{k+1} - \mu^*\| < \|\mu^k - \mu^*\|$. *Hint*: Similar to the proof of Prop. 7.5.1, show that if g^k is an ϵ-subgradient, then for all $\mu \in M$ we have

$$\|\mu^{k+1} - \mu\|^2 \leq \|\mu^k - \mu\|^2 - 2s^k \left(q(\mu) - q(\mu^k) - \epsilon\right) + (s^k)^2 \|g^k\|^2.$$

7.5.9 (Approximate Subgradient Method with Diminishing Stepsize [CoL94]) (www)

Let the dual function be real-valued [$q(\mu) > -\infty$ for all $\mu \in \Re^r$]. Consider the iteration

$$\mu^{k+1} = \left[\mu^k + s^k g^k\right]^+,$$

where g^k is an ϵ^k-subgradient at μ^k [i.e., $g^k \in \partial_{\epsilon^k} q(\mu^k)$], and we have

$$\sum_{k=0}^\infty s^k = \infty, \qquad s^k \|g^k\|^2 \to 0,$$

$$\epsilon^k \geq 0, \qquad \epsilon^k \to \epsilon.$$

(a) Use the hint of the preceding exercise to show that

$$\sup_{\mu \in M} q(\mu) - \epsilon \leq \limsup_{k \to \infty} q(\mu^k) \leq \sup_{\mu \in M} q(\mu).$$

(b) Assume that $\epsilon^k = 0$ for all k and that

$$\sum_{k=0}^\infty s^k = \infty, \qquad \sum_{k=0}^\infty (s^k)^2 \|g^k\|^2 < \infty.$$

Show that $\limsup_{k \to \infty} q(\mu^k) = \sup_{\mu \in M} q(\mu)$ and that μ^k converges to a maximizing point of q over M if there exists at least one such point.

7.5.10 (Normalized Subgradient Method [Sho85]) (www)

Let the dual function be real-valued [$q(\mu) > -\infty$ for all $\mu \in \Re^r$]. Consider the subgradient iteration

$$\mu^{k+1} = \left[\mu^k + \alpha^k \frac{g^k}{\|g^k\|}\right]^+,$$

where α^k is a positive scalar.

(a) Assume that α^k is fixed ($\alpha^k = \alpha$ for all k). Show that for any $\epsilon > 0$ and any dual optimal solution μ^*, there exists an index $\bar{k}$ and a point $\bar{\mu} \in M$ such that $q(\bar{\mu}) = q(\mu^{\bar{k}})$ and $\|\bar{\mu} - \mu^*\| < \alpha(1+\epsilon)/2$.

(b) Assume that $\alpha^k \to 0$ and $\sum_{k=0}^{\infty} \alpha^k = \infty$, and that the set of dual optimal solutions M^* is nonempty and compact. Show that either $\mu^{\bar{k}} \in M^*$ for some index $\bar{k}$, or else

$$\lim_{k \to \infty} \min_{\mu^* \in M^*} \|\mu^k - \mu^*\| = 0, \qquad \lim_{k \to \infty} q(\mu^k) = \max_{\mu \in M} q(\mu).$$

7.5.11 (Incremental Subgradient Method with Diminishing Stepsize [NeB01a]) (www)

Consider the problem of maximizing over the closed set M the function $q : \Re^r \to \Re$ given by

$$q(\mu) = \sum_{i=1}^{m} q_i(\mu),$$

where each $q_i : \Re^r \mapsto \Re$ is a concave function. The incremental subgradient method for this problem is given by

$$\mu^{k+1} = \psi^{m,k},$$

where $\psi^{m,k}$ is obtained after the m steps

$$\psi^{i,k} = \left[\psi^{i-1,k} + \alpha^k g^{i,k}\right]^+, \qquad g^{i,k} \in \partial q_i(\psi^{i-1,k}), \qquad i = 1, \ldots, m,$$

starting with

$$\psi^{0,k} = \mu^k.$$

Assume the following conditions:

(i) We have $\max_{\mu \in M} q_i(\mu) = q_i^* < \infty$, and the set $\{\mu \mid q_i(\mu) = q_i^*\}$ is nonempty for all i. Furthermore, at least one of the functions q_i has bounded level sets.

(ii) The sequence of subgradients $\{g^{i,k}\}$ is bounded, i.e., for some $C > 0$ we have

$$\|g^{i,k}\| \leq C, \qquad \forall\, i, k.$$

(This is true in particular if each function q_i is polyhedral.)

Show that the set of optimal solutions M^* is nonempty. Furthermore, the following hold:

(a) If the stepsize sequence $\{\alpha^k\}$ is bounded, then a sequence $\{\mu^k\}$ generated by the method is bounded.

(b) If the stepsize sequence $\{\alpha^k\}$ satisfies

$$\lim_{k \to \infty} \alpha^k = 0, \qquad \sum_{k=0}^{\infty} \alpha^k = \infty,$$

then either $\mu^{\bar{k}} \in M^*$ for some $\bar{k}$, or else

$$\lim_{k \to \infty} \min_{\mu^* \in M^*} \|\mu^k - \mu^*\| = 0, \qquad \lim_{k \to \infty} q(\mu^k) = \max_{\mu \in M} q(\mu).$$

(c) If the stepsize sequence $\{\alpha^k\}$ satisfies

$$\sum_{k=0}^{\infty} \alpha^k = \infty, \qquad \sum_{k=0}^{\infty} (\alpha^k)^2 < \infty,$$

then $\{\mu^k\}$ converges to some optimal solution.

7.5.12 (Incremental Subgradient Method with Dynamically Changing Stepsize [NeB01]) (www)

Consider the incremental subgradient method of Exercise 7.5.11 and assume that each $q_i : \Re^r \mapsto \Re$ is a concave function, and the optimal solution set M^* is nonempty. Assume also that for each i, C_i is a scalar satisfying

$$\|g\| \leq C_i, \qquad \forall \, g \in \partial q_i(\mu^k) \cup \partial q_i(\psi^{i-1,k}), \quad k = 0, 1, \ldots,$$

and let $C = \sum_{i=1}^{m} C_i$. Let the stepsize be given by

$$\alpha^k = \frac{\gamma^k \left(q^* - q(\mu^k)\right)}{C^2},$$

where $q^* = \max_{\mu \in M} q(\mu)$, and $\{\gamma^k\}$ satisfies for some scalars γ_l and γ_u

$$0 < \gamma_l \leq \gamma^k \leq \gamma_u < 2, \qquad \forall \, k \geq 0.$$

Show that either $\mu^{\bar{k}} \in M^*$ for some $\bar{k}$, or else $\{\mu^k\}$ converges to some optimal solution.

7.5.13 (ϵ-Complementary Slackness and Approximate Subgradients) (www)

Consider the separable problem

$$\text{minimize} \quad \sum_{i=1}^{n} f_i(x_i)$$

$$\text{subject to} \quad \sum_{i=1}^{n} g_{ij}(x_i) \leq 0, \quad j=1,\ldots,r, \qquad \alpha_i \leq x_i \leq \beta_i, \quad i=1,\ldots,n,$$

where $f_i : \Re \mapsto \Re$, $g_{ij} : \Re \mapsto \Re$ are convex functions. For an $\epsilon > 0$, we say that a pair $(\bar{x}, \bar{\mu})$ satisfies ϵ-*complementary slackness* (see e.g., [BeT89], [Ber98] and the references given there for more on this notion) if $\bar{\mu} \geq 0$, $\bar{x}_i \in [\alpha_i, \beta_i]$ for all i, and

$$0 \leq f_i^+(\bar{x}_i) + \sum_{j=1}^{r} \bar{\mu}_j g_{ij}^+(\bar{x}_i) + \epsilon, \ \forall \, i \in I^-, \qquad f_i^-(\bar{x}_i) + \sum_{j=1}^{r} \bar{\mu}_j g_{ij}^-(\bar{x}_i) - \epsilon \leq 0, \ \forall \, i \in I^+,$$

where

$$I^- = \{i \mid x_i < \beta_i\}, \qquad I^+ = \{i \mid \alpha_i < x_i\},$$

and f_i^-, g_{ij}^- and f_i^+, g_{ij}^+ denote the left and right derivatives of f_i, g_{ij}, respectively. Show that if $(\bar{x}, \bar{\mu})$ satisfies ϵ-complementary slackness, the r-dimensional vector with jth component $\sum_{i=1}^{n} g_{ij}(\bar{x}_i)$ is an $\bar{\epsilon}$-subgradient of the dual function q at $\bar{\mu}$, where $\bar{\epsilon} = \epsilon \sum_{i=1}^{n} (\beta_i - \alpha_i)$. (Based on unpublished joint work with P. Tseng.)

7.5.14 (Subgradient Methods with Small Errors for Sharp Minima [NeB10]) (www)

When using a steepest ascent method with errors to maximize a differentiable function, it is essential for convergence that the errors diminish to 0. The purpose of this exercise is to show that this is not true when using a subgradient method with errors to maximize a nondifferentiable function from a broad class (which includes polyhedral functions). Consider the problem of maximizing a concave function $q : \Re^r \to \Re$ over a closed convex set M, and assume that the optimal solution set, denoted M^*, is nonempty. Consider the iteration

$$\mu^{k+1} = \left[\mu^k + s^k(g^k + r^k)\right]^+,$$

where for all k, g^k is a subgradient of q at μ^k, and r^k is an error such that for all k, we have

$$\|r^k\| \leq \beta, \qquad k = 0, 1, \ldots,$$

where β is some positive scalar. Assume that for some $\gamma > 0$, we have

$$q^* - q(\mu) \geq \gamma \min_{\mu^* \in M^*} \|\mu - \mu^*\|, \qquad \forall \, \mu \in M,$$

where $q^* = \max_{\mu \in M} q(\mu)$, and that for some $\delta > 0$, we have

$$\|g\| \leq \delta, \qquad \forall\, g \in \partial q(\mu^k),\ k = 0, 1, \ldots$$

[these assumptions are satisfied if q is a polyhedral function of the form $q(\mu) = \min_{i=1,\ldots,m}\{a_i'x - b_i\}$, where the a_i and b_i are given vectors and scalars, respectively]. Assuming $\beta < \gamma$, show that if s^k is equal to some constant s for all k, then

$$\limsup_{k \to \infty} q(\mu^k) \geq q^* - \frac{s\gamma(\delta + \beta)^2}{2(\gamma - \beta)},$$

while if $s^k \to 0$ and $\sum_{k=0}^{\infty} s^k = \infty$, then $\limsup_{k \to \infty} q(\mu^k) = q^*$.

7.5.15 (Outer Approximation Methods [GoP79], [PoH91])

Consider the problem

$$\text{minimize} \quad \max_{y \in Y} \phi(x, y) \qquad (7.99)$$
$$\text{subject to} \quad x \in X,$$

where X is a convex and compact subset of $\Re^n$, Y is a compact subset of $\Re^m$, and $\phi : \Re^n \times Y$ is a continuous function such that $\phi(\cdot, y)$ is convex as a function of x for each $y \in Y$. Let x^k be defined as

$$x^k \in \arg\min_{x \in X} \max\{\phi(x, y^0), \phi(x, y^1), \ldots, \phi(x, y^{k-1})\},$$

where y^0 is any vector in Y and

$$y^i \in \arg\max_{y \in Y} \phi(x^i, y), \qquad \forall\, i = 1, 2, \ldots.$$

(a) What is the relation of this method with the cutting plane method?

(b) Show that every limit point of the sequence $\{x^k\}$ is an optimal solution of the original problem (7.99).

(c) Show that the conclusion of part (b) holds also in the case where $Y = [0, 1]$ and y^i is instead defined by

$$y^i \in \arg\max_{y \in \{0, 1/i, 2/i, \ldots, 1\}} \phi(x^i, y).$$

7.6 DECOMPOSITION METHODS

In this section we consider nondifferentiable optimization methods for solving the dual of a separable problem of the form

$$\text{minimize} \quad \sum_{j=1}^{m} f_j(x_j)$$
$$\text{subject to} \quad x_j \in X_j,\ j = 1, \ldots, m, \qquad \sum_{j=1}^{m} A_j x_j = b. \qquad (7.100)$$

Here $f_j : \Re^{n_j} \mapsto \Re$ and X_j are given subsets of $\Re^{n_j}$. The $r \times n_j$ matrices A_j and the vector $b \in \Re^r$ specify the constraint $\sum_{j=1}^{m} A_j x_j = b$, which couples the components x_j. Note that we are making no convexity assumptions on X_j or f_j.

The methodology of this section is based on converting the basic problem with coupling constraints (7.100) into a nondifferentiable optimization problem using two different approaches:

(a) Lagrangian relaxation of the coupling constraints.

(b) Decomposition by right-hand side allocation.

We will apply the subgradient and cutting plane methods of the previous section within the context of each of these two approaches.

7.6.1 Lagrangian Relaxation of the Coupling Constraints

The basic idea here is to solve the dual problem obtained by relaxing the coupling constraints. The dual function is

$$q(\lambda) = \sum_{j=1}^{m} \min_{x_j \in X_j} \{f_j(x_j) + \lambda' A_j x_j\} - \lambda' b.$$

We will assume throughout this subsection that for all j and λ one can find a vector $x_j(\lambda)$ attaining the minimum above. Thus the dual value is

$$q(\lambda) = \sum_{j=1}^{m} \{f_j(x_j(\lambda)) + \lambda' A_j x_j(\lambda)\} - \lambda' b \qquad (7.101)$$

and a subgradient at λ is

$$g_\lambda = \sum_{j=1}^{m} A_j x_j(\lambda) - b, \qquad (7.102)$$

(cf. the discussion of Section 7.1). Thus, both $q(\lambda)$ and g_λ can be calculated by solving m relatively small subproblems (one for each j).

We now discuss the computational solution of the dual problem

$$\begin{aligned} \text{minimize} \quad & q(\lambda) \\ \text{subject to} \quad & \lambda \in \Re^r. \end{aligned} \qquad (7.103)$$

The application of the subgradient method is straightforward, so we concentrate on the cutting plane method.

Cutting Plane Method – Dantzig-Wolfe Decomposition

The cutting plane method consists of the iteration

$$\lambda^k \in \arg\max_{\lambda \in \Re^r} Q^k(\lambda),$$

where $Q^k(\lambda)$ is the piecewise linear approximation of the dual function based on the preceding function values $q(\lambda^0), \ldots, q(\lambda^{k-1})$, and the corresponding subgradients $g^0, \ldots, g^{k-1}$, i.e.,

$$Q^k(\lambda) = \min\bigl\{q(\lambda^0) + (\lambda - \lambda^0)'g^0, \ldots, q(\lambda^{k-1}) + (\lambda - \lambda^{k-1})'g^{k-1}\bigr\}.$$

Let us take a closer look at the subproblem of maximizing $Q^k(\lambda)$. By introducing an auxiliary variable v, we can write this problem as

maximize v
subject to $v \le q(\lambda^i) + (\lambda - \lambda^i)'g^i$, $i = 0, \ldots, k-1$.

This is a linear program in the variables v and λ. Its dual can be verified to have the form

$$\text{minimize} \quad \sum_{i=0}^{k-1} \xi^i\bigl(q(\lambda^i) - \lambda^{i\prime} g^i\bigr)$$

$$\text{subject to} \quad \sum_{i=0}^{k-1} \xi^i = 1, \quad \sum_{i=0}^{k-1} \xi^i g^i = 0,$$

$$\xi^i \ge 0, \quad i = 0, \ldots, k-1,$$

where $\xi^0, \ldots, \xi^{k-1}$ are the dual variables. Using Eqs. (7.101) and (7.102), this problem can be written as

$$\text{minimize} \quad \sum_{j=1}^{m} \left(\sum_{i=0}^{k-1} \xi^i f_j\bigl(x_j(\lambda^i)\bigr) \right)$$

$$\text{subject to} \quad \sum_{i=0}^{k-1} \xi^i = 1, \quad \sum_{j=1}^{m} A_j \left(\sum_{i=0}^{k-1} \xi^i x_j(\lambda^i) \right) = b, \qquad (7.104)$$

$$\xi^i \ge 0, \quad i = 0, \ldots, k-1.$$

The preceding problem is called the *master problem*. It is the dual of the linear subproblem $\max_{\lambda \in \Re^r} Q^k(\lambda)$, which in turn approximates the dual problem $\max_\lambda q(\lambda)$; in short, it is the *dual of the approximate dual*. We have seen in Chapter 6 that duality is often symmetric; that is, the dual problem by appropriate dualization yields the primal. Thus, it is not surprising that the master problem (7.104) and the original primal problem

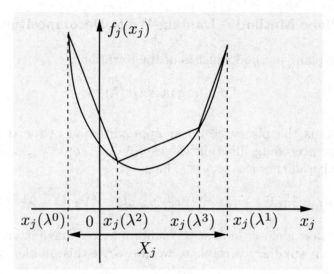

Figure 7.6.1. Viewing the master problem as a piecewise linear inner approximation of the primal problem. Each cost function term $f_j(x_j)$ of the primal is approximated by a piecewise linear function with break points at $x_j(\lambda^0), \ldots, x_j(\lambda^{k-1})$.

(7.100) are closely related. In fact, the master problem may be viewed as a piecewise linear approximation of the primal, as shown in Fig. 7.6.1. In particular, the primal cost function terms $f_j(x_j)$ are approximated by

$$\sum_{i=0}^{k-1} \xi^i f_j\big(x_j(\lambda^i)\big)$$

and the constraint set $X_1 \times \cdots \times X_m$ is approximated by the convex hull of the points $x(\lambda^0), \ldots, x(\lambda^{k-1})$, where $x(\lambda^i) = \big(x_1(\lambda^i), \ldots, x_m(\lambda^i)\big)$, $i = 0, \ldots, k-1$. This type of approximation of the primal problem is called *inner linearization*. It may be viewed as dual to the cutting plane approximation of the dual problem, which is called *outer linearization*. We note that the operations of inner and outer linearization are related by a conjugacy relation, which can be used as a basis for a broad variety of algorithms; see the paper [BeY11], and the book [Ber15a], Chapter 4.

Suppose now that we solve the master problem with a method that yields a geometric multiplier λ^k corresponding to the constraint

$$\sum_{j=1}^{m} A_j \left(\sum_{i=0}^{k-1} \xi^i x_j(\lambda^i) \right) = b;$$

one possibility is to use a method that solves simultaneously both the primal and the dual of a linear program, such as the simplex method. Then, the dual of the master problem [which is the cutting plane subproblem $\max_\lambda Q^k(\lambda)$] is solved by the geometric multiplier λ^k. Therefore, λ^k is the next iterate of the cutting plane problem.

We now summarize the preceding discussion, and state the cutting plane method as applied to the dual problem where only the coupling constraints $\sum_{j=1}^{m} A_j x_j = b$ of the original problem (7.100) are dualized. The method is known as *Dantzig-Wolfe decomposition* or *column generation*, and has played an important role in the development of large-scale optimization.

Cutting Plane – Dantzig-Wolfe Decomposition Method

Start with any λ^0 and for all $j = 1, \ldots, m$, obtain a solution $x_j(\lambda^0)$ of the subproblem

$$\text{minimize} \quad f_j(x_j) + \lambda^{0\prime} A_j x_j$$
$$\text{subject to} \quad x_j \in X_j.$$

Step 1: Given $\lambda^0, \ldots, \lambda^{k-1}$, and the vectors $x_j(\lambda^i)$ for $j = 1, \ldots, m$ and $i = 0, \ldots, k-1$, solve the master problem (7.104) and obtain λ^k, which is a geometric multiplier for the constraint

$$\sum_{j=1}^{m} A_j \left(\sum_{i=0}^{k-1} \xi^i x_j(\lambda^i) \right) = b.$$

Step 2: For all $j = 1, \ldots, m$, obtain a solution $x_j(\lambda^k)$ of the subproblem

$$\text{minimize} \quad f_j(x_j) + \lambda^{k\prime} A_j x_j$$
$$\text{subject to} \quad x_j \in X_j.$$

Step 3: Modify the master problem by adding one more variable ξ^k with cost coefficient $\sum_{j=1}^{m} f_j(x_j(\lambda^k))$ and constraint column

$$\begin{pmatrix} 1 \\ \sum_{j=1}^{m} A_j x_j(\lambda^k) \end{pmatrix},$$

and go to Step 1.

Note that adding one more variable to the master problem (cf. Step 3) amounts to adding one more break point to the piecewise linear approximation of each primal cost term $f_j(x_j)$, and one more extreme point to the convex hull approximations of the sets X_j (cf. Fig. 7.6.1).

7.6.2 Decomposition by Right-Hand Side Allocation

We now consider briefly a different method for decomposing the original primal problem (7.100), which resembles the partitioning methodology of

Section 7.2.2. By introducing auxiliary variables y_j, $j = 1, \ldots, m$, we can write this problem as

$$\text{minimize} \quad \sum_{j=1}^{m} f_j(x_j)$$

$$\text{subject to} \quad x_j \in X_j, \quad j = 1, \ldots, m,$$

$$\sum_{j=1}^{m} y_j = b, \quad A_j x_j = y_j, \quad j = 1, \ldots, m,$$

or equivalently,

$$\text{minimize} \quad \sum_{j=1}^{m} \min_{A_j x_j = y_j, \, x_j \in X_j} f_j(x_j)$$

$$\text{subject to} \quad \sum_{j=1}^{m} y_j = b, \quad y_j \in Y_j, \quad j = 1, \ldots, m,$$
(7.105)

where Y_j is the set of all y_j for which the inner minimization problem

$$\text{minimize} \quad f_j(x_j)$$
$$\text{subject to} \quad A_j x_j = y_j, \quad x_j \in X_j,$$

has at least one feasible solution.

Let us define

$$p_j(y_j) = \min_{A_j x_j = y_j, \, x_j \in X_j} f_j(x_j). \tag{7.106}$$

Then, problem (7.105) can be written as

$$\text{minimize} \quad \sum_{j=1}^{m} p_j(y_j)$$

$$\text{subject to} \quad \sum_{j=1}^{m} y_j = b, \quad y_j \in Y_j, \quad j = 1, \ldots, m.$$
(7.107)

This problem, called the *master problem*, may be solved by nondifferentiable optimization methods. Note from Eq. (7.106), that p_j may be viewed as a primal function of a certain minimization problem. Using this observation and the theory developed in Section 6.4.5, it can be seen that a subgradient of $p_j(y_j)$ at y_j is equal to $-\lambda_j$, where λ_j is the geometric multiplier vector corresponding to the constraint $A_j x_j = y_j$ in the definition (7.106) of p_j. Thus,

$$\text{subgradient of} \sum_{j=1}^{m} p_j(y_j) = - \begin{pmatrix} \lambda_1 \\ \vdots \\ \lambda_m \end{pmatrix}.$$

Sec. 7.6   Decomposition Methods

We can now write explicitly the subgradient and cutting plane methods for solving the master problem. In particular, the subgradient method is given by

$$y^{k+1} = [y^k + s^k \lambda^k]^+,$$

where s^k is a positive stepsize, and $[\cdot]^+$ denotes projection on the set

$$\left\{ y \mid \sum_{j=1}^{m} y_j = b,\ y_j \in Y_j \right\}.$$

Note here that y^k is the $(n_1 + \cdots + n_m)$-dimensional vector with components y_j^k, $j = 1, \ldots, m$, and λ^k is the vector with components λ_j^k, $j = 1, \ldots, m$, where λ_j^k is the geometric multiplier corresponding to the constraint $A_j x_j = y_j^k$ in the inner minimization subproblem

$$\begin{aligned}
\text{minimize}\quad & f_j(x_j) \\
\text{subject to}\quad & A_j x_j = y_j^k, \qquad x_j \in X_j.
\end{aligned}$$

Similarly, the cutting plane method is given by

$$y^k \in \arg \min_{\substack{\sum_{j=1}^{m} y_j = b,\\ y_j \in Y_j,\, j=1,\ldots,m}} P^k(y),$$

where P^k is the piecewise linear function that approximates the cost function $\sum_{j=1}^{m} p_j(y_j)$ of the master problem (7.107), based on previous values of y^i and λ^i, $i = 0, \ldots, k-1$, i.e.,

$$P^k(y) = \max_{i=0,\ldots,k-1} \left\{ \sum_{j=1}^{m} \left(p_j(y_j^i) + (y_j - y_j^i) \lambda_j^i \right) \right\}.$$

EXERCISES

7.6.1 (Linear Problems with Coupling Variables)

Consider the problem with coupling variables of Example 7.2.2 for the case where the cost function and the constraints are linear. Show that its dual is a problem with coupling constraints of the type considered in this section.

7.7 NOTES AND SOURCES

Section 7.1: Extensive accounts of optimality conditions for nondifferentiable (and in some cases nonconvex) optimization are given in the books by Clarke [Cla83], Demjanov and Vasilév [DeV85], Evtushenko [Evt85], Rockafellar and Wets [RoW98], Borwein and Lewis [BoL00], Auslender and Teboulle [AuT03], Bertsekas, Nedic, and Ozdaglar [BNO03], Mordukhovich [Mor06], Bertsekas [Ber09], and Delfour [Del12].

Section 7.2: The material of this section has much in common with Section 3.4 of Bertsekas and Tsitsiklis [BeT89], which describes additional methods and gives many references. The dual coordinate ascent method for quadratic programming dates to Hildreth [Hil57]. See Cryer [Cry71], Censor and Herman [CeH87], Lin and Pang [LiP87], Mangasarian and De Leone [MaD87], [MaD88], and Censor and Zenios [CeZ97] for subsequent work. For analysis and alternative proposals of dual ascent methods for quadratic and other convex problems, see Bertsekas, Hossein, and Tseng [BHT87], Tseng and Bertsekas [TsB87], Hearn and Lawphongpanich [HeL89], Tseng [Tse90], Tseng and Bertsekas [TsB91], Luo and Tseng [LuT91], [LuT92a], [LuT92b], [LuT93c], and Hager and Hearn [HaH93]. The author's textbook [Ber15a] contains a survey of more recent literature.

Section 7.3: The duality relation between the proximal and augmented Lagrangian algorithms was analyzed by Rockafellar [Roc73a], [Roc73b], [Roc74]. A corresponding development for nonquadratic augmented Lagrangian functions was given by Kort and Bertsekas [KoB76] (see also the book account [Ber82a]). For a discussion and recent references on the exponential method of multipliers and the entropy minimization algorithm, see the textbook [Ber15a] and the sources cited for Section 5.2.5.

The incremental augmented Lagrangian method of Section 7.3.3 was derived in the author's book [Ber15a], Section 6.4.3. The author's paper [Ber15b] derived the incremental aggregated versions, both quadratic and exponential, and discussed their relation to the ADMM algorithms of Section 7.4. This paper also gave many references (starting with the paper [StW75]), which are aimed at overcoming the difficulties of the multiplier method related to preservation of separability.

Section 7.4: The ADMM was first proposed by Glowinskii and Morocco [GIM75], and Gabay and Mercier [GaM76], and was further developed by Gabay [Gab79], [Gab83]. It was generalized by Lions and Mercier [LiM79], where the connection with alternating direction methods for solving differential equations was pointed out. The method and its applications in large boundary-value problems were discussed by Fortin and Glowinskii [FoG83].

The recent literature on the ADMM is voluminous and cannot be surveyed here. In our discussion we have followed the analysis of the book by Bertsekas and Tsitsiklis [BeT89] (which among others gave the ADMM for separable problems of Section 7.4.1), and in part the paper by Eckstein and

Bertsekas [EcB92] (which established the connection of the ADMM with the general form of the proximal algorithm of [Roc76a] for finding a zero of a maximal monotone operator, and gave generalizations involving, among others, iterate extrapolation and inexact minimization). In particular, the paper [EcB92] showed that the general form of the proximal algorithm contains as a special case another major method, the Douglas-Ratchford splitting algorithm for finding a zero of the sum of two maximal monotone operators, proposed by Lions and Mercier [LiM79]. The latter algorithm contains in turn as a special case the ADMM, as shown by Gabay [Gab83].

The proximal method of multipliers noted in Section 7.4.3 is due to Rockafellar [Roc76b]. Its convergence proof based on the ADMM seems to have gone unnoticed. A related algorithm, which resembles the proximal ADMM is given by Chen and Teboulle [ChT94].

While there is a fairly complete theory for the method of multipliers applied to nonconvex/twice differentiable problems, this is not so for the ADMM. However, there are papers in the literature that suggest that for certain types of nonconvex problems, the ADMM has reliable convergence properties; this is a subject of ongoing research.

We finally note that there is another algorithm that is related to the ADMM for minimizing the sum of two functions. This is the *proximal gradient algorithm*, summarized in Section 3.6.2, which performs gradient steps on one function followed by a proximal step on the other function. Dual versions of this algorithm produce methods that bear considerable similarity to the ADMM; see [Ber15a], Section 6.3.

Section 7.5: Subgradient methods were first introduced in the Soviet Union in the middle 60s by Shor; the works of Ermoliev and Poljak were also particularly influential. An extensive bibliography for the early period of the subject is given in the edited volume by Balinski and Wolfe [BaW75]. The convergence rate of subgradient methods is discussed by Goffin [Gof77]. Incremental subgradient methods have been considered by several authors; see [Ber10b], [Ber11a] for surveys of the subject with many references. The analysis, is summarized in part in Exercises 7.5.11 and 7.5.12, and is given in detail in the papers by Nedić, Bertsekas, and Borkar [NBB01], and Nedić and Bertsekas [NeB00], [NeB01]; see also [BNO03] and [Ber15a].

The textbooks by Auslender [Aus76], Shapiro [Sha79], Evtushenko [Evt85], Shor [Sho85], Minoux [Min86], Poljak [Pol87], Hiriart-Urruty and Lemarechal [HiL93], Shor [Sho98], Bertsekas, Nedić, and Ozdaglar [BNO03], and Bertsekas [Ber15a] give extensive accounts of subgradient methods that complement our treatment and give many references.

An alternative to subgradient methods, which we have not discussed, is adaptations of quasi-Newton methods to the convex nondifferentiable context. In this connection, we mention subgradient methods with space dilatation, proposed and developed extensively by Shor (see the books [Sho85] and [Sho98]), as well as algorithms based on the ideas of the BFGS method (see e.g., Yu et al. [YVG10], Lewis and Overton [LeO13], and Yousefpour [You15]).

Cutting plane methods were introduced by Cheney and Goldstein [ChG59], and by Kelley [Kel60]. The paper by Bertsekas and Yu [BeY11], and the author's textbook [Ber15a] (Chapter 4) develop the duality relation between cutting plane and simplicial decomposition methods. For analysis of proximal cutting plane and related methods, see Ruszczynski [Rus89], Lemaréchal and Sagastizábal [LeS93], Mifflin [Mif96], Burke and Qian [BuQ98], Mifflin, Sun, and Qi [MSQ98], Bonnans et. al. [BGL06], and Bertsekas [Ber15a].

Central cutting plane methods were introduced by Elzinga and Moore [ElM75]. More recent proposals, some of which relate to interior point methods, are discussed by Goffin and Vial [GoV90], Goffin, Haurie, and Vial [GHV92], Ye [Ye92], Kortanek and No [KoN93], Goffin, Luo, and Ye [GLY94], Atkinson and Vaidya [AtV95], den Hertog et. al. [HKR95], Nesterov [Nes95], and Goffin, Luo, and Ye [GLY96]. For a textbook treatment, see Ye [Ye97], and for a recent survey, see Goffin and Vial [GoV99].

Section 7.6: Three historically important papers on decomposition methods are Dantzig and Wolfe [DaW60], Benders [Ben62], and Everett [Eve63]. The early text by Lasdon [Las70] on large-scale optimization was influential.

The research monograph by Bertsekas and Tsitsiklis [BeT89] develops the connections between decomposition and distributed, possibly asynchronous, computation. The paper by Bertsekas and Yu [BeY11], and the author's textbook [Ber15a] (Sections 4.3-4.6) describe in a more elegant and symmetric form the duality between inner and outer linearization, which we have observed in the context of Dantzig-Wolfe decomposition.

The theoretical and applications literature on large-scale optimization and decomposition is voluminous. We provide a few early references that complement the material we have covered here: Geoffrion [Geo70], [Geo74], [Geo77], Bertsekas [Ber79a], Meyer [Mey79], Cohen [Coh80], Golshtein [Gol85], Tanikawa and Mukai [TaM85], Spingarn [Spi85], Minoux [Min86], Ruszczynski [Rus86], Sen and Sherali [SeS86], Bertsekas and Tsitsiklis [BeT89], Rockafellar [Roc90], Toint and Tuyttens [ToT90], Ferris and Mangasarian [FeM91], Kim and Nazareth [KiN91], Rockafellar and Wets [RoW91], Tseng [Tse91b], [Tse91c], Auslender [Aus92], Fukushima [Fuk92], Pinar and Zenios [PiZ92], Nagurney[Nag93], Patriksson [Pat93], Tseng [Tse93], Eckstein [Eck94b], Migdalas [Mig94], Pinar and Zenios [PiZ94], Mahey, Oualibouch, and Tao [MOT95], Mulvey and Ruszczynski [MuR95], Zhu [Zhu95], Censor and Zenios [1997], Bertsekas [Ber98], Kontogiorgis and Meyer [KoM98], Patriksson [Pat98], Zhao and Luh [ZhL98], Tseng [Tse00], [Tse01a], [Tse01b], [Tse04]. The author's textbook [Ber15a] provides a more extensive account of the currently intense activity in large-scale optimization, and gives additional and more recent references.

APPENDIX A:
Mathematical Background

In this appendix, we collect definitions, notational conventions, and several results from linear algebra and real analysis that are used extensively in nonlinear programming. Only a few proofs are given. Additional proofs can be found in Appendix A of the book by Bertsekas and Tsitsiklis [BeT89], which provides a similar but more extended summary of linear algebra and analysis. Related and additional material can be found in the books by Hoffman and Kunze [HoK71], Lancaster and Tismenetsky [LaT85], and Strang [Str76] (linear algebra), and the books by Ash [Ash72], Ortega and Rheinboldt [OrR70], and Rudin [Rud76] (real analysis).

Set Notation

If X is a set and x is an element of X, we write $x \in X$. A set can be specified in the form $X = \{x \mid x \text{ satisfies } P\}$, as the set of all elements satisfying property P. The union of two sets X_1 and X_2 is denoted by $X_1 \cup X_2$, and their intersection by $X_1 \cap X_2$. The symbols $\exists$ and $\forall$ have the meanings "there exists" and "for all," respectively. The empty set is denoted by $\emptyset$.

The set of real numbers (also referred to as scalars) is denoted by $\Re$. The set $\Re$ augmented with $+\infty$ and $-\infty$ is called the *set of extended real numbers*. We write $-\infty < x < \infty$ for all real numbers x, and $-\infty \leq x \leq \infty$ for all extended real numbers x. We denote by $[a, b]$ the set of (possibly extended) real numbers x satisfying $a \leq x \leq b$. A rounded, instead of square, bracket denotes strict inequality in the definition. Thus $(a, b]$, $[a, b)$, and (a, b) denote the set of all x satisfying $a < x \leq b$, $a \leq x < b$, and $a < x < b$, respectively. Furthermore, we use the natural extensions of the rules of arithmetic: $x \cdot 0 = 0$ for every extended real number x, $x \cdot \infty = \infty$ if $x > 0$, $x \cdot \infty = -\infty$ if $x < 0$, and $x + \infty = \infty$ and $x - \infty = -\infty$ for every scalar x. The expression $\infty - \infty$ is meaningless and is never allowed to occur.

Inf and Sup Notation

The *supremum* of a nonempty set X of scalars, denoted by $\sup X$, is defined as the smallest scalar y such that $y \geq x$ for all $x \in X$. If no such scalar y exists, we say that the supremum of X is ∞. Similarly, the *infimum* of X, denoted by $\inf X$, is defined as the largest scalar y such that $y \leq x$ for all $x \in X$, and is equal to $-\infty$ if no such scalar y exists. For the empty set, we use the convention

$$\sup \emptyset = -\infty, \qquad \inf \emptyset = \infty.$$

If $\sup X$ is equal to a scalar $\bar{x}$ that belongs to the set X, we say that $\bar{x}$ is the *maximum point* of X and we write $\bar{x} = \max X$. Similarly, if $\inf X$ is equal to a scalar $\bar{x}$ that belongs to the set X, we say that $\bar{x}$ is the *minimum point* of X and we write $\bar{x} = \min X$. Thus, when we write $\max X$ (or $\min X$) in place of $\sup X$ (or $\inf X$, respectively), we do so just for emphasis: we indicate that it is either evident, or it is known through earlier analysis, or it is about to be shown that the maximum (or minimum, respectively) of the set X is attained at one of its points.

Function Notation

If f is a function, we use the notation $f : X \mapsto Y$ to indicate the fact that f is defined on a nonempty set X (its *domain*) and takes values in a set Y (its *range*). Thus when using the notation $f : X \mapsto Y$, we implicitly assume that X is nonempty. If $f : X \mapsto Y$ is a function, and U and V are subsets of X and Y, respectively, the set $\{f(x) \mid x \in U\}$ is called the *image* or *forward image of U under f*, and the set $\{x \in X \mid f(x) \in V\}$ is called the *inverse image of V under f*.

A.1 VECTORS AND MATRICES

We denote by $\Re^n$ the set of n-dimensional real vectors. For any $x \in \Re^n$, we use x_i to indicate its ith *coordinate*, also called its ith *component*, and we also write $x = (x_1, \ldots, n)$.

Vectors in $\Re^n$ will be viewed as column vectors, unless the contrary is explicitly stated. For any $x \in \Re^n$, x' denotes the n-dimensional row vector that has the same components as x, arranged in the same order. The *inner product* of two vectors $x, y \in \Re^n$ is defined by $x'y = \sum_{i=1}^{n} x_i y_i$. Two vectors $x, y \in \Re^n$ satisfying $x'y = 0$ are called *orthogonal*.

If x is a vector in $\Re^n$, the notations $x > 0$ and $x \geq 0$ indicate that all components of x are positive and nonnegative, respectively. For any two vectors x and y, the notation $x > y$ means that $x - y > 0$. The notations $x \geq y$, $x < y$, etc., are to be interpreted accordingly.

If X is a set and λ is a scalar, we denote by λX the set $\{\lambda x \mid x \in X\}$. If X_1 and X_2 are two subsets of $\Re^n$, we denote by $X_1 + X_2$ the set

$$\{x_1 + x_2 \mid x_1 \in X_1, x_2 \in X_2\},$$

which is referred to as the *vector sum of X_1 and X_2*. We use a similar notation for the sum of any finite number of subsets. In the case where one of the subsets consists of a single vector $\bar{x}$, we simplify this notation as follows:

$$\bar{x} + X = \{\bar{x} + x \mid x \in X\}.$$

We also denote by $X_1 - X_2$ the set

$$\{x_1 - x_2 \mid x_1 \in X_1, x_2 \in X_2\}.$$

Given sets $X_i \subset \Re^{n_i}$, $i = 1, \ldots, m$, the *Cartesian product* of the X_i, denoted by $X_1 \times \cdots \times X_m$, is the set

$$\{(x_1, \ldots, x_m) \mid x_i \in X_i, i = 1, \ldots, m\},$$

which is a subset of $\Re^{n_1 + \cdots + n_m}$.

Subspaces and Linear Independence

A nonempty subset S of $\Re^n$ is called a *subspace* if $ax + by \in S$ for every $x, y \in S$ and every $a, b \in \Re$. An *affine set* or *linear manifold* in $\Re^n$ is a translated subspace, i.e., a set X of the form $X = \bar{x} + S = \{\bar{x} + x \mid x \in S\}$, where $\bar{x}$ is a vector in $\Re^n$ and S is a subspace of $\Re^n$, called the *subspace parallel to X*. Note that there can be only one subspace S associated with an affine set in this manner. [To see this, let $X = x + S$ and $X = \bar{x} + \bar{S}$ be two representations of the affine set X. Then, we must have $x = \bar{x} + \bar{s}$ for some $\bar{s} \in \bar{S}$ (since $x \in X$), so that $X = \bar{x} + \bar{s} + S$. Since we also have $X = \bar{x} + \bar{S}$, it follows that $S = \bar{S} - \bar{s} = \bar{S}$.] The *span* of a finite collection $\{x_1, \ldots, x_m\}$ of elements of $\Re^n$ is the subspace consisting of all vectors y of the form $y = \sum_{k=1}^{m} \alpha_k x_k$, where each α_k is a scalar.

The vectors $x_1, \ldots, x_m \in \Re^n$ are called *linearly independent* if there exists no set of scalars $\alpha_1, \ldots, \alpha_m$, at least one of which is nonzero, such that $\sum_{k=1}^{m} \alpha_k x_k = 0$. An equivalent definition is that $x_1 \neq 0$, and for every $k > 1$, the vector x_k does not belong to the span of $x_1, \ldots, x_{k-1}$.

If S is a subspace of $\Re^n$ containing at least one nonzero vector, a *basis* for S is a collection of vectors that are linearly independent and whose span is equal to S. Every basis of a given subspace has the same number of vectors. This number is called the *dimension* of S. By convention, the subspace $\{0\}$ is said to have dimension zero. The *dimension of an affine set*

$\bar{x}+S$ is the dimension of the corresponding subspace S. Every subspace of nonzero dimension has a basis that is orthogonal (i.e., any pair of distinct vectors from the basis is orthogonal).

Given any set X, the set of vectors that are orthogonal to all elements of X is a subspace denoted by $X^\perp$:

$$X^\perp = \{y \mid y'x = 0, \, \forall \, x \in X\}.$$

If S is a subspace, $S^\perp$ is called the *orthogonal complement* of S. Any vector x can be uniquely decomposed as the sum of a vector from S and a vector from $S^\perp$. Furthermore, we have $(S^\perp)^\perp = S$.

Matrices

For any matrix A, we use A_{ij}, $[A]_{ij}$, or a_{ij} to denote its ijth element. The *transpose* of A, denoted by A', is defined by $[A']_{ij} = a_{ji}$. For two matrices A and B of compatible dimensions, we have $(AB)' = B'A'$.

If X is a subset of $\Re^n$ and A is an $m \times n$ matrix, then the *image of X under A* is denoted by AX (or $A \cdot X$ if this enhances notational clarity):

$$AX = \{Ax \mid x \in X\}.$$

extrem If Y is a subset of $\Re^m$, the *inverse image of Y under A* is denoted by $A^{-1}Y$ or $A^{-1} \cdot Y$:

$$A^{-1}Y = \{x \mid Ax \in Y\}.$$

If X and Y are subspaces, then AX and $A^{-1}Y$ are also subspaces.

Let A be a square matrix. We say that A is *symmetric* if $A' = A$. We say that A is *diagonal* if $[A]_{ij} = 0$ when $i \neq j$. We say that A is *lower triangular* if $[A]_{ij} = 0$ when $i < j$, and *upper triangular* if $[A]_{ij} = 0$ when $i > j$. We denote by I the identity matrix (the diagonal matrix whose diagonal elements are 1). We denote the *determinant* of A by $\det(A)$.

Let A be an $m \times n$ matrix. The *range space* of A, denoted by $R(A)$, is the set of all $y \in \Re^m$ such that $y = Ax$ for some $x \in \Re^n$. The *nullspace* of A, denoted by $N(A)$, is the set of all $x \in \Re^n$ such that $Ax = 0$. It is seen that $R(A)$ and $N(A)$ are subspaces. The *rank* of A is the dimension of $R(A)$. The rank of A is equal to the maximal number of linearly independent columns of A, and is also equal to the maximal number of linearly independent rows of A. The matrix A and its transpose A' have the same rank. We say that A has *full rank*, if its rank is equal to $\min\{m, n\}$. This is true if and only if either all the rows of A are linearly independent, or all the columns of A are linearly independent.

The range space of an $m \times n$ matrix A is equal to the orthogonal complement of the nullspace of its transpose, i.e.,

$$R(A) = N(A')^\perp.$$

Another way to state this result is that given vectors $a_1, \ldots, a_n \in \Re^m$ (the columns of A) and a vector $x \in \Re^m$, we have $x'y = 0$ for all y such that $a_i'y = 0$ for all i if and only if

$$x = \lambda_1 a_1 + \cdots + \lambda_n a_n$$

for some scalars $\lambda_1, \ldots, \lambda_n$ [compare with Farkas' Lemma (Prop. B.15 in Appendix B)].

A function $f : \Re^n \mapsto \Re$ is said to be *affine* if it has the form $f(x) = a'x + b$ for some $a \in \Re^n$ and $b \in \Re$. Similarly, a function $f : \Re^n \mapsto \Re^m$ is said to be *affine* if it has the form $f(x) = Ax + b$ for some $m \times n$ matrix A and some $b \in \Re^m$. If $b = 0$, f is said to be a *linear function* or *linear transformation*. Sometimes, with slight abuse of terminology, an equation or inequality involving a linear function, such as $a'x = b$ or $a'x \leq b$, is referred to as a *linear equation or inequality*, respectively.

A.2 NORMS, SEQUENCES, LIMITS, AND CONTINUITY

Definition A.1: A *norm* $\|\cdot\|$ on $\Re^n$ is a mapping that assigns a scalar $\|x\|$ to every $x \in \Re^n$ and that has the following properties:

(a) $\|x\| \geq 0$ for all $x \in \Re^n$.

(b) $\|cx\| = |c| \cdot \|x\|$ for every $c \in \Re$ and every $x \in \Re^n$.

(c) $\|x\| = 0$ if and only if $x = 0$.

(d) $\|x + y\| \leq \|x\| + \|y\|$ for all $x, y \in \Re^n$.

The *Euclidean norm* is defined by

$$\|x\| = (x'x)^{1/2} = \left(\sum_{i=1}^{n} |x_i|^2\right)^{1/2}.$$

The space $\Re^n$, equipped with this norm, is called a *Euclidean space*. We will use the Euclidean norm almost exclusively in this book. In particular, in the absence of a clear indication to the contrary, $\|\cdot\|$ will denote the Euclidean norm. Two important results for the Euclidean norm are:

Proposition A.1: (Pythagorean Theorem) If x and y are orthogonal then

$$\|x + y\|^2 = \|x\|^2 + \|y\|^2.$$

> **Proposition A.2: (Schwarz inequality)** For any two vectors x and y, we have
> $$|x'y| \leq \|x\| \cdot \|y\|,$$
> with equality holding if and only if $x = \alpha y$ for some scalar α.

Two other important norms are the *maximum norm* $\|\cdot\|_\infty$ (also called *sup-norm* or ℓ_∞-*norm*), defined by

$$\|x\|_\infty = \max_i |x_i|,$$

and the ℓ_1-*norm* $\|\cdot\|_1$, defined by

$$\|x\|_1 = \sum_{i=1}^n |x_i|.$$

Sequences

We use both subscripts and superscripts in sequence notation. Generally, we prefer subscripts, but we use superscripts whenever we need to reserve the subscript notation for indexing components of vectors and functions. The meaning of the subscripts and superscripts should be clear from the context in which they are used.

A sequence $\{x_k \mid k = 1, 2, \ldots\}$ (or $\{x_k\}$ for short) of scalars is said to *converge* if there exists a scalar x such that for every $\epsilon > 0$ we have $|x_k - x| < \epsilon$ for every k greater than some integer K (that depends on ϵ). The scalar x is said to be the *limit* of $\{x_k\}$, and the sequence $\{x_k\}$ is said to *converge to* x; symbolically, $x_k \to x$ or $\lim_{k \to \infty} x_k = x$. If for every scalar b there exists some K (that depends on b) such that $x_k \geq b$ for all $k \geq K$, we write $x_k \to \infty$ and $\lim_{k \to \infty} x_k = \infty$. Similarly, if for every scalar b there exists some integer K such that $x_k \leq b$ for all $k \geq K$, we write $x_k \to -\infty$ and $\lim_{k \to \infty} x_k = -\infty$. Note, however, that implicit in any of the statements "$\{x_k\}$ converges" or "the limit of $\{x_k\}$ exists" or "$\{x_k\}$ has a limit" is that the limit of $\{x_k\}$ is a scalar. A scalar sequence $\{x_k\}$ is called a *Cauchy sequence* if for every $\epsilon > 0$, there exists some integer K (depending on ϵ) such that $|x_k - x_m| < \epsilon$ for all $k \geq K$ and $m \geq K$.

A scalar sequence $\{x_k\}$ is said to be *bounded above* (respectively, *below*) if there exists some scalar b such that $x_k \leq b$ (respectively, $x_k \geq b$) for all k. It is said to be *bounded* if it is bounded above and bounded below. The sequence $\{x_k\}$ is said to be monotonically *nonincreasing* (respectively, *nondecreasing*) if $x_{k+1} \leq x_k$ (respectively, $x_{k+1} \geq x_k$) for all k. If $x_k \to x$ and $\{x_k\}$ is monotonically nonincreasing (nondecreasing), we also use the notation $x_k \downarrow x$ ($x_k \uparrow x$, respectively).

Proposition A.3: Every bounded and monotonically nonincreasing or nondecreasing scalar sequence converges.

Note that a monotonically nondecreasing sequence $\{x_k\}$ is either bounded, in which case it converges to some scalar x by the above proposition, or else it is unbounded, in which case $x_k \to \infty$. Similarly, a monotonically nonincreasing sequence $\{x_k\}$ is either bounded and converges, or it is unbounded, in which case $x_k \to -\infty$.

Given a scalar sequence $\{x_k\}$, let

$$y_m = \sup\{x_k \mid k \geq m\}, \qquad z_m = \inf\{x_k \mid k \geq m\}.$$

The sequences $\{y_m\}$ and $\{z_m\}$ are nonincreasing and nondecreasing, respectively, and therefore have a limit whenever $\{x_k\}$ is bounded above or is bounded below, respectively (Prop. A.3). The limit of y_m is denoted by $\limsup_{k\to\infty} x_k$, and is referred to as the *upper limit* of $\{x_k\}$. The limit of z_m is denoted by $\liminf_{k\to\infty} x_k$, and is referred to as the *lower limit* of $\{x_k\}$. If $\{x_k\}$ is unbounded above, we write $\limsup_{k\to\infty} x_k = \infty$, and if it is unbounded below, we write $\liminf_{k\to\infty} x_k = -\infty$.

Proposition A.4: Let $\{x_k\}$ and $\{y_k\}$ be scalar sequences.

(a) We have

$$\inf\{x_k \mid k \geq 0\} \leq \liminf_{k\to\infty} x_k \leq \limsup_{k\to\infty} x_k \leq \sup\{x_k \mid k \geq 0\}.$$

(b) $\{x_k\}$ converges if and only if

$$-\infty < \liminf_{k\to\infty} x_k = \limsup_{k\to\infty} x_k < \infty.$$

Furthermore, if $\{x_k\}$ converges, its limit is equal to the common scalar value of $\liminf_{k\to\infty} x_k$ and $\limsup_{k\to\infty} x_k$.

(c) If $x_k \leq y_k$ for all k, then

$$\liminf_{k\to\infty} x_k \leq \liminf_{k\to\infty} y_k, \qquad \limsup_{k\to\infty} x_k \leq \limsup_{k\to\infty} y_k.$$

(d) We have

$$\liminf_{k\to\infty} x_k + \liminf_{k\to\infty} y_k \leq \liminf_{k\to\infty}(x_k + y_k),$$

$$\limsup_{k\to\infty} x_k + \limsup_{k\to\infty} y_k \geq \limsup_{k\to\infty}(x_k + y_k).$$

A sequence $\{x_k\}$ of vectors in $\Re^n$ is said to converge to some $x \in \Re^n$ if the ith component of x_k converges to the ith component of x for every i. We use the notations $x_k \to x$ and $\lim_{k\to\infty} x_k = x$ to indicate convergence for vector sequences as well. The sequence $\{x_k\}$ is called bounded (or Cauchy) if each of its corresponding coordinate sequences is bounded (or Cauchy, respectively). It can be seen that $\{x_k\}$ is bounded if and only if there exists a scalar c such that $\|x_k\| \leq c$ for all k. An infinite subset of a sequence $\{x_k\}$ is called a *subsequence* of $\{x_k\}$. A subsequence can itself be viewed as a sequence, and can be represented as a set $\{x_k \mid k \in \mathcal{K}\}$, where $\mathcal{K}$ is an infinite subset of positive integers (the notation $\{x_k\}_\mathcal{K}$ will also be used).

Definition A.2: We say that a vector $x \in \Re^n$ is a *limit point of a sequence* $\{x_k\}$ in $\Re^n$ if there exists a subsequence of $\{x_k\}$ that converges to x.

Proposition A.5:

(a) A bounded sequence of vectors in $\Re^n$ converges if and only if it has a unique limit point.

(b) A sequence in $\Re^n$ converges if and only if it is a Cauchy sequence.

(c) Every bounded sequence in $\Re^n$ has at least one limit point.

(d) Let $\{x_k\}$ be a scalar sequence. If $\limsup_{k\to\infty} x_k$ ($\liminf_{k\to\infty} x_k$) is finite, then it is the largest (respectively, smallest) limit point of $\{x_k\}$.

$o(\cdot)$ Notation

For a positive integer p and a function $h : \Re^n \mapsto \Re^m$ we write

$$h(x) = o(\|x\|^p)$$

if

$$\lim_{k\to\infty} \frac{h(x_k)}{\|x_k\|^p} = 0,$$

for all sequences $\{x_k\}$ such that $x_k \to 0$ and $x_k \neq 0$ for all k.

Sec. A.2   Norms, Sequences, Limits, and Continuity   753

Closed and Open Sets

We say that x is a *closure point* or *limit point* of a subset X of $\Re^n$ if there exists a sequence $\{x_k\} \subset X$ that converges to x. The *closure* of X, denoted cl(X), is the set of all closure points of X.

Definition A.3: A subset X of $\Re^n$ is called *closed* if it is equal to its closure. It is called *open* if its complement, $\{x \mid x \notin X\}$, is closed. It is called *bounded* if there exists a scalar c such that $\|x\| \leq c$ for all $x \in X$. It is called *compact* if it is closed and bounded. A *neighborhood* of a vector x is an open set containing x. If $X \subset \Re^n$ and $x \in X$, we say that x is an *interior* point of X if there exists a neighborhood of x that is contained in X. A vector $x \in X$ which is not an interior point of X is said to be a *boundary* point of X. The set of all boundary points of X is called the *boundary* of X.

For any norm $\|\cdot\|$ in $\Re^n$, $\epsilon > 0$, and $x^* \in \Re^n$, consider the sets

$$\{x \mid \|x - x^*\| < \epsilon\}, \qquad \{x \mid \|x - x^*\| \leq \epsilon\}.$$

The first set is open and is called an *open sphere* centered at x^*, while the second set is closed and is called a *closed sphere* centered at x^*. Sometimes the terms *open ball* and *closed ball* are used, respectively.

Proposition A.6:

(a) The union of finitely many closed sets is closed.

(b) The intersection of closed sets is closed.

(c) The union of open sets is open.

(d) The intersection of finitely many open sets is open.

(e) A set is open if and only if all of its elements are interior points.

(f) Every subspace of $\Re^n$ is closed.

(g) A subset of $\Re^n$ is compact if and only if it is closed and bounded.

Continuity

Let $f : X \mapsto \Re^m$ be a function, where X is a subset of $\Re^n$, and let x be a vector in X. If there exists a vector $y \in \Re^m$ such that the sequence $\{f(x_k)\}$ converges to y for every sequence $\{x_k\} \subset X$ such that $\lim_{k \to \infty} x_k = x$, we write $\lim_{z \to x} f(z) = y$. If there exists a vector $y \in \Re^m$ such that the sequence $\{f(x_k)\}$ converges to y for every sequence $\{x_k\} \subset X$ such that $\lim_{k\to\infty} x_k = x$ and $x_k \leq x$ (respectively, $x_k \geq x$) for all k, we write $\lim_{z\uparrow x} f(z) = y$ [respectively, $\lim_{z\downarrow x} f(z)$].

Definition A.4: Let X be a subset of $\Re^n$.

(a) A function $f : X \mapsto \Re^m$ is called *continuous* at a vector $x \in X$ if $\lim_{z\to x} f(z) = f(x)$.

(b) A function $f : X \mapsto \Re^m$ is called *right-continuous* (respectively, *left-continuous*) at a vector $x \in X$ if $\lim_{z\downarrow x} f(z) = f(x)$ [respectively, $\lim_{z\uparrow x} f(z) = f(x)$].

(c) A real-valued function $f : X \mapsto \Re$ is called *upper semicontinuous* (respectively, *lower semicontinuous*) at a vector $x \in X$ if $f(x) \geq \limsup_{k\to\infty} f(x_k)$ [respectively, $f(x) \leq \liminf_{k\to\infty} f(x_k)$] for every sequence $\{x_k\} \subset X$ that converges to x.

(d) A function $f : X \mapsto \Re$ is called *coercive* if for every sequence $\{x_k\} \subset X$ such that $\|x_k\| \to \infty$, we have $\lim_{k\to\infty} f(x_k) = \infty$.

If $f : X \mapsto \Re^m$ is continuous at every vector in a subset of its domain X, we say that f *is continuous over that subset*. If $f : X \mapsto \Re^m$ is continuous at every vector in its domain X, we say that f *is continuous*. We say that f is *Lipschitz continuous* if $\|f(x) - f(y)\| \leq L\|x - y\|$ for some scalar L and all $x, y \in X$. We also say that $f : X \mapsto \Re$ is *coercive over a subset* of its domain X if for every sequence $\{x_k\}$ from that subset such that $\|x_k\| \to \infty$, we have $\lim_{k\to\infty} f(x_k) = \infty$. If f is coercive over X, we simply say that f is coercive.

Proposition A.7:

(a) Any vector norm on $\Re^n$ is a continuous function.

(b) Let $f : \Re^m \mapsto \Re^p$ and $g : \Re^n \mapsto \Re^m$ be continuous functions. The composition $f \cdot g : \Re^n \mapsto \Re^p$, defined by $(f \cdot g)(x) = f\bigl(g(x)\bigr)$, is a continuous function.

(c) Let $f : \Re^n \mapsto \Re^m$ be continuous, and let Y be an open (respectively, closed) subset of $\Re^m$. Then the inverse image of Y, $\{x \in \Re^n \mid f(x) \in Y\}$, is open (respectively, closed).

(d) Let $f : \Re^n \mapsto \Re^m$ be continuous, and let X be a compact subset of $\Re^n$. Then the image of X, $\{f(x) \mid x \in X\}$, is compact.

(e) Let X be a closed subset of $\Re^n$ and let $f : X \mapsto \Re$ be lower semicontinuous at all points of X. Then the level set $\{x \in X \mid f(x) \leq \gamma\}$ is closed for all $\gamma \in \Re$.

If X is a nonempty subset of $\Re^n$ and f is a real-valued function whose domain contains X, we say that a vector $x^* \in X$ is a *minimum of f over X* if $f(x^*) = \inf_{x \in X} f(x)$. We also call x^* a *minimizing point* or a *minimizer* or a *minimum* of f over X. Alternatively, we say that f *attains a minimum over X at x^**, and we indicate this by writing

$$x^* \in \arg\min_{x \in X} f(x).$$

If x^* is known to be the unique minimizer of f over X, by slight abuse of notation, we also write

$$x^* = \arg\min_{x \in X} f(x).$$

We use similar notation for maxima. An important property of compactness in connection with optimization problems is the following theorem, which provides conditions for existence of solutions of optimization problems.

Proposition A.8: (Weierstrass' Theorem) Let X be a nonempty subset of $\Re^n$ and let $f : X \mapsto \Re$ be lower semicontinuous at all points of X. Assume that one of the following three conditions holds:

(1) X is compact.

(2) X is closed and f is coercive.

(3) There exists a scalar γ such that the level set

$$\{x \in X \mid f(x) \leq \gamma\}$$

is nonempty and compact.

Then, the set of minima of f over X is nonempty and compact.

Proof: Assume condition (1). Let $\{z_k\} \subset X$ be a sequence such that

$$\lim_{k \to \infty} f(z_k) = \inf_{z \in X} f(z).$$

Since X is bounded, this sequence has at least one limit point x [Prop. A.5(c)]. Since X is closed, x belongs to X, while the lower semicontinuity of f implies that $f(x) \leq \lim_{k \to \infty} f(z_k) = \inf_{z \in X} f(z)$. Therefore, we must have $f(x) = \inf_{z \in X} f(z)$. The set of all minima of f over X is the level set $\{x \in X \mid f(x) \le \inf_{z \in X} f(z)\}$, which is closed by the lower semicontinuity of f [Prop. A.7(e)], and hence compact since X is bounded.

Assume condition (2). Consider a sequence $\{z_k\}$ as in the proof of part (a). Since f is coercive, $\{z_k\}$ must be bounded and the proof proceeds like the proof of part (a).

Assume condition (3). If the given γ is equal to $\inf_{z \in X} f(z)$, the set of minima of f over X is $\{x \in X \mid f(x) \le \gamma\}$, and since by assumption this set is nonempty, we are done. If $\gamma > \inf_{z \in X} f(z)$, consider a sequence $\{z_k\}$ as in the proof of part (a). Then, for all k sufficiently large, z_k must belong to the set $\{x \in X \mid f(x) \le \gamma\}$. Since this set is compact, $\{z_k\}$ must be bounded and the proof proceeds like the proof of part (a). **Q.E.D.**

Note that with appropriate adjustments, the above proposition applies to the existence of maxima of f over X. In particular, if f is upper semicontinuous at all points of X and X is compact, there exists a vector $y \in X$ such that $f(y) = \sup_{z \in X} f(z)$. Note also that under additional convexity assumptions on X and f, there is a more refined theory of existence of optimal solutions, whereby the boundedness assumptions underlying Weierstrass' Theorem are replaced by alternative conditions involving directions of recession (see [BNO03], Section 2.3, [Ber09], Section 3.2).

With an application of Weierstrass' Theorem, we obtain the following *norm equivalence property in* $\Re^n$, which shows that if a sequence converges with respect to one norm, it converges with respect to all other norms.

Proposition A.9: For any two norms $\|\cdot\|$ and $\|\cdot\|'$ on $\Re^n$, there exists some positive constant $c \in \Re$ such that $\|x\| \le c\|x\|'$ for all $x \in \Re^n$.

Proof: Let a be the minimum of $\|x\|'$ over the set of all $x \in \Re^n$ such that $\|x\| = 1$. The latter set is closed and bounded and, therefore, the minimum is attained at some $\tilde{x}$ (Prop. A.8) that must be nonzero since $\|\tilde{x}\| = 1$. For any $x \in \Re^n$, $x \ne 0$, the $\|\cdot\|$ norm of $x/\|x\|$ is equal to 1. Therefore,

$$0 < a = \|\tilde{x}\|' \le \left\|\frac{x}{\|x\|}\right\|' = \frac{\|x\|'}{\|x\|}, \qquad \forall\, x \ne 0,$$

which proves the desired result with $c = 1/a$. **Q.E.D.**

As a corollary, we obtain the following.

Proposition A.10: If a subset of $\Re^n$ is open (respectively, closed, bounded, or compact) for some norm, it is open (respectively, closed, bounded, or compact), for all other norms.

Matrix Norms

A norm $\|\cdot\|$ on the set of $n \times n$ matrices is a real-valued mapping that has the same properties as vector norms do when the matrix is viewed as an element of $\Re^{n^2}$. The norm of an $n \times n$ matrix A is denoted by $\|A\|$.

We are mainly interested in *induced norms*, which are constructed as follows. Given any vector norm $\|\cdot\|$, the corresponding induced matrix norm, also denoted by $\|\cdot\|$, is defined by

$$\|A\| = \max_{\{x \in \Re^n \mid \|x\|=1\}} \|Ax\|. \qquad (A.1)$$

The set over which the maximization takes place above is closed [Prop. A.7(c)] and bounded, while the function being maximized is continuous [Prop. A.7(b)]. Therefore, by Weierstrass' theorem (Prop. A.8) the maximum is attained. It is easily verified that for any vector norm, Eq. (A.1) defines a matrix norm having all the required properties.

Note that by the Schwarz inequality (Prop. A.2), we have

$$\|A\| = \max_{\|x\|=1} \|Ax\| = \max_{\|y\|=\|x\|=1} |y'Ax|.$$

By reversing the roles of x and y in the above relation and by using the equality $y'Ax = x'A'y$, it follows that

$$\|A\| = \|A'\|. \qquad (A.2)$$

A.3 SQUARE MATRICES AND EIGENVALUES

Definition A.5: A square matrix A is called *singular* if its determinant is zero. Otherwise it is called *nonsingular* or *invertible*.

Proposition A.11:
(a) Let A be an $n \times n$ matrix. The following are equivalent:
   (i) The matrix A is nonsingular.
   (ii) The matrix A' is nonsingular.
   (iii) For every nonzero $x \in \Re^n$, we have $Ax \neq 0$.
   (iv) For every $y \in \Re^n$, there exists a unique $x \in \Re^n$ such that $Ax = y$.
   (v) There exists an $n \times n$ matrix B such that $AB = I = BA$.
   (vi) The columns of A are linearly independent.

(vii) The rows of A are linearly independent.

(b) Assuming that A is nonsingular, there is a unique matrix B satisfying $AB = I = BA$, which is called the *inverse* of A and is denoted by A^{-1}.

(c) For any two square invertible matrices A and B of the same dimensions, we have $(AB)^{-1} = B^{-1}A^{-1}$.

Let A and B be square matrices, and let C be a matrix of appropriate dimension. Then we have
$$(A + CBC')^{-1} = A^{-1} - A^{-1}C(B^{-1} + C'A^{-1}C)^{-1}C'A^{-1},$$
provided all the inverses appearing above exist. For a proof, multiply the right-hand side by $A + CBC'$ and show that the product is the identity.

Another useful formula provides the inverse of the partitioned matrix
$$M = \begin{bmatrix} A & B \\ C & D \end{bmatrix}.$$
There holds
$$M^{-1} = \begin{bmatrix} Q & -QBD^{-1} \\ -D^{-1}CQ & D^{-1} + D^{-1}CQBD^{-1} \end{bmatrix},$$
where
$$Q = (A - BD^{-1}C)^{-1},$$
provided all the inverses appearing above exist. For a proof, multiply M with the given expression for M^{-1} and verify that the product is the identity.

Definition A.6: The *characteristic polynomial* ϕ of an $n \times n$ matrix A is defined by $\phi(\lambda) = \det(\lambda I - A)$, where I is the identity matrix of the same size as A. The n (possibly repeated and complex) roots of ϕ are called the *eigenvalues* of A. A vector x (with possibly complex coordinates) such that $Ax = \lambda x$, where λ is an eigenvalue of A, is called an *eigenvector* of A associated with λ.

Proposition A.12: Let A be a square matrix.

(a) A complex number λ is an eigenvalue of A if and only if there exists a nonzero eigenvector associated with λ.

(b) A is singular if and only if it has an eigenvalue that is equal to zero.

Note that the only use of complex numbers in this book is in relation to eigenvalues and eigenvectors. All other matrices or vectors are implicitly assumed to have real components.

Proposition A.13: Let A be an $n \times n$ matrix.

(a) The eigenvalues of a triangular matrix are equal to its diagonal entries.

(b) If S is a nonsingular matrix and $B = SAS^{-1}$, then the eigenvalues of A and B coincide.

(c) The eigenvalues of $cI + A$ are equal to $c + \lambda_1, \ldots, c + \lambda_n$, where $\lambda_1, \ldots, \lambda_n$ are the eigenvalues of A.

(d) The eigenvalues of A^k are equal to $\lambda_1^k, \ldots, \lambda_n^k$, where $\lambda_1, \ldots, \lambda_n$ are the eigenvalues of A.

(e) If A is nonsingular, then the eigenvalues of A^{-1} are the reciprocals of the eigenvalues of A.

(f) The eigenvalues of A and A' coincide.

Definition A.7: The *spectral radius* $\rho(A)$ of a square matrix A is defined as the maximum of the magnitudes of the eigenvalues of A.

It can be shown that the roots of a polynomial depend continuously on the coefficients of the polynomial. For this reason, the eigenvalues of a square matrix A depend continuously on A, and we obtain the following.

Proposition A.14: The eigenvalues of a square matrix A depend continuously on the elements of A. In particular, $\rho(A)$ is a continuous function of A.

The next two propositions are fundamental for the convergence theory of linear iterative methods.

Proposition A.15: For any induced matrix norm $\|\cdot\|$ and any square matrix A we have

$$\lim_{k\to\infty} \|A^k\|^{1/k} = \rho(A) \leq \|A\|.$$

Furthermore, given any $\epsilon > 0$, there exists an induced matrix norm $\|\cdot\|$ such that

$$\|A\| = \rho(A) + \epsilon.$$

Proposition A.16: Let A be a square matrix. We have

$$\lim_{k\to\infty} A^k = 0$$

if and only if $\rho(A) < 1$.

A corollary of the above proposition is that the iteration $x_{k+1} = Ax_k$ converges to 0 for every initial condition x_0 if and only if $\rho(A) < 1$. From this it also follows that if $\rho(A) < 1$, the iteration $x_{k+1} = Ax_k + b$ converges to the vector $x^* = (I-A)^{-1}b$ for every initial condition x_0 and every vector b. To see this, note that the iteration $x_{k+1} = Ax_k + b$ can equivalently be written as $y_{k+1} = Ay_k$, where $y_k = x_k - x^*$.

A.4 SYMMETRIC AND POSITIVE DEFINITE MATRICES

Symmetric matrices have several special properties, particularly with respect to their eigenvalues and eigenvectors. In this section, $\|\cdot\|$ denotes the Euclidean norm throughout.

Proposition A.17: Let A be a symmetric $n \times n$ matrix. Then:

(a) The eigenvalues of A are real.

(b) The matrix A has a set of n mutually orthogonal, real, and nonzero eigenvectors $x_1, \ldots, x_n$.

(c) Suppose that the eigenvectors in part (b) have been normalized so that $\|x_i\| = 1$ for each i. Then

$$A = \sum_{i=1}^{n} \lambda_i x_i x_i',$$

where λ_i is the eigenvalue corresponding to x_i.

Sec. A.4 Symmetric and Positive Definite Matrices

Proposition A.18: Let A be a symmetric $n \times n$ matrix, let $\lambda_1 \leq \cdots \leq \lambda_n$ be its (real) eigenvalues, and let $x_1, \ldots, x_n$ be associated orthogonal eigenvectors, normalized so that $\|x_i\| = 1$ for all i. Then:

(a) $\|A\| = \rho(A) = \max\{|\lambda_1|, |\lambda_n|\}$, where $\|\cdot\|$ is the matrix norm induced by the Euclidean norm.

(b) $\lambda_1 \|y\|^2 \leq y'Ay \leq \lambda_n \|y\|^2$ for all $y \in \Re^n$.

(c) (*Courant-Fisher Minimax Principle*) For all $i = 1, \ldots, n$, and for all i-dimensional subspaces $\overline{S}_i$ and all $(n - i + 1)$-dimensional subspaces $\underline{S}_i$, there holds

$$\min_{\|y\|=1,\, y \in \underline{S}_i} y'Ay \leq \lambda_i \leq \max_{\|y\|=1,\, y \in \overline{S}_i} y'Ay.$$

Furthermore, equality on the left (right) side above is attained if $\underline{S}_i$ is the subspace spanned by $x_i, \ldots, x_n$ ($\overline{S}_i$ is the subspace spanned by $x_1, \ldots, x_i$, respectively).

(d) (*Interlocking Eigenvalues Lemma*) Let $\tilde{\lambda}_1 \leq \tilde{\lambda}_2 \leq \cdots \leq \tilde{\lambda}_n$ be the eigenvalues of $A + bb'$, where b is a vector in $\Re^n$. Then,

$$\lambda_1 \leq \tilde{\lambda}_1 \leq \lambda_2 \leq \tilde{\lambda}_2 \leq \cdots \leq \lambda_n \leq \tilde{\lambda}_n.$$

Proof: (a) We know that $\|A\| \geq \rho(A)$ (Prop. A.15), so we need to show the reverse inequality. We express an arbitrary vector $y \in \Re^n$ in the form $y = \sum_{i=1}^n \xi_i x_i$, where each ξ_i is a suitable scalar. Using the orthogonality of the vectors x_i and the Pythagorean Theorem (Prop. A.1), we obtain $\|y\|^2 = \sum_{i=1}^n |\xi_i|^2 \cdot \|x_i\|^2$. Using the Pythagorean Theorem again, we obtain

$$\|Ay\|^2 = \left\| \sum_{i=1}^n \lambda_i \xi_i x_i \right\|^2 = \sum_{i=1}^n |\lambda_i|^2 \cdot |\xi_i|^2 \cdot \|x_i\|^2 \leq \rho^2(A) \|y\|^2.$$

Since this is true for every y, we obtain $\|A\| \leq \rho(A)$ and the desired result follows.

(b) As in part (a), we express the generic $y \in \Re^n$ as $y = \sum_{i=1}^n \xi_i x_i$. We have, using the orthogonality of the vectors x_i, $i = 1, \ldots, n$, and the fact $\|x_i\| = 1$,

$$y'Ay = \sum_{i=1}^n \lambda_i |\xi_i|^2 \|x_i\|^2 = \sum_{i=1}^n \lambda_i |\xi_i|^2$$

and

$$\|y\|^2 = \sum_{i=1}^n |\xi_i|^2 \|x_i\|^2 = \sum_{i=1}^n |\xi_i|^2.$$

These two relations prove the desired result.

(c) Let $\underline{X}_i$ be the subspace spanned by $x_1, \ldots, x_i$. The subspaces $\underline{X}_i$ and $\underline{S}_i$ must have a common vector x_0 with $\|x_0\| = 1$, since the sum of their dimensions is $n + 1$ [if there was no common nonzero vector, we could take sets of basis vectors for $\underline{X}_i$ and S_i (a total of $n + 1$ in number), which would have to be linearly independent, yielding a contradiction]. The vector x_0 can be expressed as a linear combination $x_0 = \sum_{j=1}^{i} \xi_j x_j$, and since $\|x_0\| = 1$ and $\|x_i\| = 1$ for all $i = 1, \ldots, n$, we must have

$$\sum_{j=1}^{i} \xi_j^2 = 1.$$

We also have using the expression

$$A = \sum_{j=1}^{n} \lambda_j x_j x_j'$$

[cf. Prop. A.17(c)],

$$x_0' A x_0 = \sum_{j=1}^{i} \lambda_j \xi_j^2 \leq \lambda_i \left(\sum_{j=1}^{i} \xi_j^2 \right).$$

Combining the last two relations, we obtain $x_0' A x_0 \leq \lambda_i$, which proves the left-hand side of the desired inequality. The right-hand side is proved similarly. Furthermore, we have $x_i' A x_i = \lambda_i$, so equality is attained as in the final assertion.

(d) From part (c) we have

$$\lambda_i = \max_{\underline{S}_i} \min_{\|y\|=1,\, y \in \underline{S}_i} y'Ay \leq \max_{\underline{S}_i} \min_{\|y\|=1,\, y \in \underline{S}_i} y'(A + bb')y \leq \tilde{\lambda}_i,$$

so that $\lambda_i \leq \tilde{\lambda}_i$ for all i. Furthermore, from part (c), for some $(n - i + 1)$-dimensional subspace $\underline{\tilde{S}}_i$ we have

$$\tilde{\lambda}_i = \min_{\|y\|=1,\, y \in \underline{\tilde{S}}_i} y'(A + bb')y.$$

Using this relation and the left-hand side of the inequality of part (c), applied to the subspace $\{y \mid y \in \underline{\tilde{S}}_i,\, b'y = 0\}$, whose dimension is at least $(n - i)$, we obtain

$$\tilde{\lambda}_i \leq \min_{\|y\|=1,\, y \in \underline{\tilde{S}}_i,\, b'y=0} y'(A + bb')y = \min_{\|y\|=1,\, y \in \underline{\tilde{S}}_i,\, b'y=0} y'Ay \leq \lambda_{i+1},$$

Sec. A.4   Symmetric and Positive Definite Matrices   763

and the proof is complete.   **Q.E.D.**

Proposition A.19: Let A be a square matrix, and let $\|\cdot\|$ be the matrix norm induced by the Euclidean norm. Then:

(a) If A is symmetric, then $\|A^k\| = \|A\|^k$ for any positive integer k.

(b) $\|A\|^2 = \|A'A\| = \|AA'\|$.

(c) If A is symmetric and nonsingular, then $\|A^{-1}\|$ is equal to the reciprocal of the smallest of the absolute values of the eigenvalues of A.

Proof: (a) If A is symmetric then A^k is symmetric. Using Prop. A.18(a), we have $\|A^k\| = \rho(A^k)$. Using Prop. A.13(d), we obtain $\rho(A^k) = \rho(A)^k$, which is equal to $\|A\|^k$ by Prop. A.18(a).

(b) For any vector x such that $\|x\| = 1$, we have, using the Schwarz inequality (Prop. A.2),

$$\|Ax\|^2 = x'A'Ax \le \|x\| \cdot \|A'Ax\| \le \|x\| \cdot \|A'A\| \cdot \|x\| = \|A'A\|.$$

Thus, $\|A\|^2 \le \|A'A\|$. On the other hand,

$$\|A'A\| = \max_{\|y\|=\|x\|=1} |y'A'Ax| \le \max_{\|y\|=\|x\|=1} \|Ay\| \cdot \|Ax\| = \|A\|^2.$$

Therefore, $\|A\|^2 = \|A'A\|$. The equality $\|A\|^2 = \|AA'\|$ is obtained by replacing A by A' and using Eq. (A.2).

(c) This follows by combining Prop. A.13(e) with Prop. A.18(a).   **Q.E.D.**

Definition A.8: A symmetric $n \times n$ matrix A is called *positive definite* if $x'Ax > 0$ for all $x \in \Re^n$, $x \ne 0$. It is called *nonnegative definite* or *positive semidefinite* if $x'Ax \ge 0$ for all $x \in \Re^n$.

Throughout this book, the notion of positive and negative definiteness applies exclusively to symmetric matrices. Thus *whenever we say that a matrix is positive or negative (semi)definite, we implicitly assume that the matrix is symmetric.*

Proposition A.20:

(a) For any $m \times n$ matrix A, the matrix $A'A$ is symmetric and nonnegative definite. The matrix $A'A$ is positive definite if and only if A has rank n. In particular, if $m = n$, $A'A$ is positive definite if and only if A is nonsingular.

(b) A square symmetric matrix is nonnegative definite (respectively, positive definite) if and only if all of its eigenvalues are nonnegative (respectively, positive).

(c) The inverse of a symmetric positive definite matrix is symmetric and positive definite.

Proof: (a) Symmetry is obvious. For any vector $x \in \Re^n$, we have $x'A'Ax = \|Ax\|^2 \geq 0$, which establishes nonnegative definiteness. Positive definiteness is obtained if and only if the inequality is strict for every $x \neq 0$, which is the case if and only if $Ax \neq 0$ for every $x \neq 0$. This is equivalent to A having rank n.

(b) Let λ and x be an eigenvalue and a corresponding real nonzero eigenvector of a symmetric nonnegative definite matrix A. Then $0 \leq x'Ax = \lambda x'x = \lambda \|x\|^2$, which proves that $\lambda \geq 0$. For the converse result, let y be an arbitrary vector in $\Re^n$. Let $\lambda_1, \ldots, \lambda_n$ be the eigenvalues of A, assumed to be nonnegative, and let $x_1, \ldots, x_n$ be a corresponding set of nonzero, real, and orthogonal eigenvectors. Let us express y in the form $y = \sum_{i=1}^{n} \xi_i x_i$. Then $y'Ay = (\sum_{i=1}^{n} \xi_i x_i)'(\sum_{i=1}^{n} \xi_i \lambda_i x_i)$. From the orthogonality of the eigenvectors, the latter expression is equal to $\sum_{i=1}^{n} \xi_i^2 \lambda_i \|x_i\|^2 \geq 0$, which proves that A is nonnegative definite. The proof for the case of positive definite matrices is similar.

(c) The eigenvalues of A^{-1} are the reciprocal of the eigenvalues of A [Prop. A.13(e)], so the result follows using part (b). **Q.E.D.**

Proposition A.21: Let A be a square symmetric nonnegative definite matrix.

(a) There exists a symmetric matrix Q with the property $Q^2 = A$. Such a matrix is called a *symmetric square root* of A and is denoted by $A^{1/2}$.

(b) A symmetric square root $A^{1/2}$ is invertible if and only if A is invertible. Its inverse is denoted by $A^{-1/2}$.

(c) There holds $A^{-1/2}A^{-1/2} = A^{-1}$.

Sec. A.5   Derivatives

> (d) There holds $AA^{1/2} = A^{1/2}A$.

Proof: (a) Let $\lambda_1, \ldots, \lambda_n$ be the eigenvalues of A and let $x_1, \ldots, x_n$ be corresponding nonzero, real, and orthogonal eigenvectors normalized so that $\|x_k\| = 1$ for each k. We let

$$A^{1/2} = \sum_{k=1}^{n} \lambda_k^{1/2} x_k x_k',$$

where $\lambda_k^{1/2}$ is the nonnegative square root of λ_k. We then have

$$A^{1/2} A^{1/2} = \sum_{i=1}^{n} \sum_{k=1}^{n} \lambda_i^{1/2} \lambda_k^{1/2} x_i x_i' x_k x_k' = \sum_{k=1}^{n} \lambda_k x_k x_k' = A.$$

Here the second equality follows from the orthogonality of distinct eigenvectors; the last equality follows from Prop. A.17(c). We now notice that each one of the matrices $x_k x_k'$ is symmetric, so $A^{1/2}$ is also symmetric.

(b) This follows from the fact that the eigenvalues of A are the squares of the eigenvalues of $A^{1/2}$ [Prop. A.13(d)].

(c) We have $(A^{-1/2} A^{-1/2})A = A^{-1/2}(A^{-1/2} A^{1/2}) A^{1/2} = A^{-1/2} I A^{1/2} = I$.

(d) We have $AA^{1/2} = A^{1/2} A^{1/2} A^{1/2} = A^{1/2} A$.   **Q.E.D.**

A symmetric square root of A is not unique. For example, let $A^{1/2}$ be as in the proof of Prop. A.21(a) and notice that the matrix $-A^{1/2}$ also has the property $(-A^{1/2})(-A^{1/2}) = A$. However, if A is positive definite, it can be shown that the matrix $A^{1/2}$ we have constructed is the only symmetric and positive definite square root of A.

A.5 DERIVATIVES

Let $f : \Re^n \mapsto \Re$ be some function, fix some $x \in \Re^n$, and consider the expression

$$\lim_{\alpha \to 0} \frac{f(x + \alpha e_i) - f(x)}{\alpha},$$

where e_i is the ith unit vector (all components are 0 except for the ith component which is 1). If the above limit exists, it is called the ith *partial derivative* of f at the vector x and it is denoted by $(\partial f / \partial x_i)(x)$ or $\partial f(x)/\partial x_i$ (x_i in this section will denote the ith component of the vector

x). Assuming all of these partial derivatives exist, the *gradient* of f at x is defined as the column vector

$$\nabla f(x) = \begin{pmatrix} \frac{\partial f(x)}{\partial x_1} \\ \vdots \\ \frac{\partial f(x)}{\partial x_n} \end{pmatrix}.$$

For any $y \in \Re^n$, we define the one-sided *directional derivative* of f in the direction y to be

$$f'(x;y) = \lim_{\alpha \downarrow 0} \frac{f(x + \alpha y) - f(x)}{\alpha},$$

provided that the limit exists.

If the directional derivative of f at a vector x exists in all directions y and $f'(x;y)$ is a linear function of y, we say that f is *differentiable* at x. This type of differentiability is also called *Gateaux differentiability*. It is seen that f is differentiable at x if and only if the gradient $\nabla f(x)$ exists and satisfies

$$\nabla f(x)'y = f'(x;y), \qquad \forall \, y \in \Re^n.$$

The function f is called *differentiable over a subset U of $\Re^n$* if it is differentiable at every $x \in U$. The function f is called *differentiable* (without qualification) if it is differentiable at all $x \in \Re^n$.

If f is differentiable over an open set U and $\nabla f(\cdot)$ is continuous at all $x \in U$, f is said to be *continuously differentiable over U*. It can then be shown that

$$\lim_{y \to 0} \frac{f(x+y) - f(x) - \nabla f(x)'y}{\|y\|} = 0, \qquad \forall \, x \in U, \qquad (A.3)$$

where $\|\cdot\|$ is an arbitrary vector norm. If f is continuously differentiable over $\Re^n$, then f is also called a *smooth* function. If f is not smooth, it is called *nonsmooth*.

The preceding equation can also be used as an alternative definition of differentiability. In particular, f is called *Frechet differentiable* at x if there exists a vector g satisfying Eq. (A.3) with $\nabla f(x)$ replaced by g. If such a vector g exists, it can be seen that all the partial derivatives $(\partial f/\partial x_i)(x)$ exist and that $g = \nabla f(x)$. Frechet differentiability implies (Gateaux) differentiability but not conversely (see for example Ortega and Rheinboldt [OrR70] for a detailed discussion). In this book, when dealing with a differentiable function f, we will always assume that f is continuously differentiable over some open set [$\nabla f(\cdot)$ is a continuous function over that set], in which case f is both Gateaux and Frechet differentiable, and the distinctions made above are of no consequence.

The definitions of differentiability of f at a vector x only involve the values of f in a neighborhood of x. Thus, these definitions can be used for functions f that are not defined on all of $\Re^n$, but are defined instead in a neighborhood of the vector at which the derivative is computed. In particular, for functions $f : X \mapsto \Re$, where X is a strict subset of $\Re^n$, we use the above definition of differentiability of f at a vector x, *provided x is an interior point of the domain X*. Similarly, we use the above definition of continuous differentiability of f over a subset U, *provided U is an open subset of the domain X*. Thus any mention of continuous differentiability of a function over a subset implicitly assumes that this subset is open.

Differentiation of Vector-Valued Functions

A function $f : \Re^n \mapsto \Re^m$, with component functions $f_1, \ldots, f_m$, is called differentiable (or smooth) if each component is differentiable (or smooth, respectively). The *gradient matrix* of f, denoted $\nabla f(x)$, is the $n \times m$ matrix whose ith column is the gradient $\nabla f_i(x)$ of f_i:

$$\nabla f(x) = \begin{bmatrix} \nabla f_1(x) \cdots \nabla f_m(x) \end{bmatrix}.$$

The transpose of ∇f is called the *Jacobian* of f and is the matrix whose ijth entry is equal to the partial derivative $\partial f_i / \partial x_j$.

Now suppose that each one of the partial derivatives of a function $f : \Re^n \mapsto \Re$ is a smooth function of x. We use the notation $(\partial^2 f / \partial x_i \partial x_j)(x)$ to indicate the ith partial derivative of $\partial f / \partial x_j$ at a vector $x \in \Re^n$. The *Hessian* of f is the matrix whose ijth entry is equal to $(\partial^2 f / \partial x_i \partial x_j)(x)$, and is denoted by $\nabla^2 f(x)$. We have $(\partial^2 f / \partial x_i \partial x_j)(x) = (\partial^2 f / \partial x_j \partial x_i)(x)$ for every x, which implies that $\nabla^2 f(x)$ is symmetric.

If $f : \Re^{m+n} \mapsto \Re$ is a function of (x, y), where $x \in \Re^m$ and $y \in \Re^n$, and $x_1, \ldots, x_m$ and $y_1, \ldots, y_n$ denote the components of x and y, respectively, we write

$$\nabla_x f(x,y) = \begin{pmatrix} \frac{\partial f(x,y)}{\partial x_1} \\ \vdots \\ \frac{\partial f(x,y)}{\partial x_m} \end{pmatrix}, \quad \nabla_y f(x,y) = \begin{pmatrix} \frac{\partial f(x,y)}{\partial y_1} \\ \vdots \\ \frac{\partial f(x,y)}{\partial y_n} \end{pmatrix}.$$

We denote by $\nabla^2_{xx} f(x,y)$, $\nabla^2_{xy} f(x,y)$, and $\nabla^2_{yy} f(x,y)$ the matrices with components

$$\left[\nabla^2_{xx} f(x,y)\right]_{ij} = \frac{\partial^2 f(x,y)}{\partial x_i \partial x_j}, \quad \left[\nabla^2_{xy} f(x,y)\right]_{ij} = \frac{\partial^2 f(x,y)}{\partial x_i \partial y_j},$$

$$\left[\nabla^2_{yy} f(x,y)\right]_{ij} = \frac{\partial^2 f(x,y)}{\partial y_i \partial y_j}.$$

If $f : \Re^{m+n} \mapsto \Re^r$, and $f_1, f_2, \ldots, f_r$ are the component functions of f, we write

$$\nabla_x f(x,y) = \big[\nabla_x f_1(x,y) \cdots \nabla_x f_r(x,y)\big],$$

$$\nabla_y f(x,y) = \big[\nabla_y f_1(x,y) \cdots \nabla_y f_r(x,y)\big].$$

Let $f : \Re^k \mapsto \Re^m$ and $g : \Re^m \mapsto \Re^n$ be smooth functions, and let h be their composition, i.e.,

$$h(x) = g\big(f(x)\big).$$

Then, the *chain rule* for differentiation states that

$$\nabla h(x) = \nabla f(x) \nabla g\big(f(x)\big), \qquad \forall\, x \in \Re^k.$$

Some examples of useful relations that follow from the chain rule are:

$$\nabla\big(f(Ax)\big) = A'\nabla f(Ax), \qquad \nabla^2\big(f(Ax)\big) = A'\nabla^2 f(Ax) A,$$

where A is a matrix,

$$\nabla_x \Big(f\big(h(x), y\big)\Big) = \nabla h(x) \nabla_h f\big(h(x), y\big),$$

$$\nabla_x \Big(f\big(h(x), g(x)\big)\Big) = \nabla h(x) \nabla_h f\big(h(x), g(x)\big) + \nabla g(x) \nabla_g f\big(h(x), g(x)\big).$$

Differentiation Theorems

We now state some theorems relating to differentiable functions that will be useful for our purposes.

Proposition A.22: (**Mean Value Theorem**) If $f : \Re \mapsto \Re$ is continuously differentiable over an open interval I, then for every $x, y \in I$, there exists some $\xi \in [x, y]$ such that

$$f(y) - f(x) = \nabla f(\xi)(y - x).$$

Proposition A.23: (Second Order Expansions) Let $f: \Re^n \mapsto \Re$ be twice continuously differentiable over an open sphere S centered at a vector x.

(a) For all y such that $x + y \in S$,
$$f(x+y) = f(x) + y'\nabla f(x) + \tfrac{1}{2}y'\left(\int_0^1 \left(\int_0^t \nabla^2 f(x+\tau y)d\tau\right)dt\right)y.$$

(b) For all y such that $x + y \in S$, there exists an $\alpha \in [0,1]$ such that
$$f(x+y) = f(x) + y'\nabla f(x) + \tfrac{1}{2}y'\nabla^2 f(x+\alpha y)y.$$

(c) For all y such that $x + y \in S$ there holds
$$f(x+y) = f(x) + y'\nabla f(x) + \tfrac{1}{2}y'\nabla^2 f(x)y + o(\|y\|^2).$$

Proposition A.24: (Descent Lemma) Let $f: \Re^n \mapsto \Re$ be continuously differentiable, and let x and y be two vectors in $\Re^n$. Suppose that
$$\|\nabla f(x+ty) - \nabla f(x)\| \le Lt\|y\|, \qquad \forall\, t \in [0,1],$$
where L is some scalar. Then
$$f(x+y) \le f(x) + y'\nabla f(x) + \frac{L}{2}\|y\|^2.$$

Proof: Let t be a scalar parameter and let $g(t) = f(x+ty)$. The chain rule yields $(dg/dt)(t) = y'\nabla f(x+ty)$. Now
$$f(x+y) - f(x) = g(1) - g(0) = \int_0^1 \frac{dg}{dt}(t)\,dt = \int_0^1 y'\nabla f(x+ty)\,dt$$
$$\le \int_0^1 y'\nabla f(x)\,dt + \left|\int_0^1 y'(\nabla f(x+ty) - \nabla f(x))\,dt\right|$$
$$\le \int_0^1 y'\nabla f(x)\,dt + \int_0^1 \|y\|\cdot\|\nabla f(x+ty) - \nabla f(x)\|\,dt$$
$$\le y'\nabla f(x) + \|y\|\int_0^1 Lt\|y\|\,dt = y'\nabla f(x) + \frac{L}{2}\|y\|^2.$$

Q.E.D.

> **Proposition A.25: (Implicit Function Theorem)** Let $f: \Re^{n+m} \mapsto \Re^m$ be a function of $x \in \Re^n$ and $y \in \Re^m$ such that:
>
> (1) $f(\bar{x}, \bar{y}) = 0$.
>
> (2) f is continuous, and has a continuous and nonsingular gradient matrix $\nabla_y f(x, y)$ in an open set containing $(\bar{x}, \bar{y})$.
>
> Then there exist open sets $S_{\bar{x}} \subset \Re^n$ and $S_{\bar{y}} \subset \Re^m$ containing $\bar{x}$ and $\bar{y}$, respectively, and a continuous function $\phi: S_{\bar{x}} \mapsto S_{\bar{y}}$ such that $\bar{y} = \phi(\bar{x})$ and $f(x, \phi(x)) = 0$ for all $x \in S_{\bar{x}}$. The function ϕ is unique in the sense that if $x \in S_{\bar{x}}$, $y \in S_{\bar{y}}$, and $f(x, y) = 0$, then $y = \phi(x)$. Furthermore, if for some integer $p > 0$, f is p times continuously differentiable the same is true for ϕ, and we have
>
> $$\nabla \phi(x) = -\nabla_x f(x, \phi(x)) \big(\nabla_y f(x, \phi(x))\big)^{-1}, \qquad \forall\, x \in S_{\bar{x}}.$$

As a final word of caution to the reader, let us mention that one can easily get confused with gradient notation and its use in various formulas, such as for example the order of multiplication of various gradients in the chain rule and the Implicit Function Theorem. Perhaps the safest guideline to minimize errors is to remember our conventions:

(a) A vector is viewed as a column vector (an $n \times 1$ matrix).

(b) The gradient ∇f of a scalar function $f: \Re^n \mapsto \Re$ is also viewed as a column vector.

(c) The gradient matrix ∇f of a vector function $f: \Re^n \mapsto \Re^m$ with components $f_1, \ldots, f_m$ is the $n \times m$ matrix whose columns are the (column) vectors $\nabla f_1, \ldots, \nabla f_m$.

With these rules in mind one can use "dimension matching" as an effective guide to writing correct formulas quickly.

A.6 CONVERGENCE THEOREMS

Many iterative algorithms can be written as

$$x_{k+1} = T(x_k), \qquad k = 0, 1, \ldots,$$

where $T: X \mapsto X$ is a mapping from a set $X \subset \Re^n$ into itself, and has the property

$$\|T(x) - T(y)\| \leq \rho \|x - y\|, \qquad \forall\, x, y \in X. \tag{A.4}$$

Here $\|\cdot\|$ is some norm, and ρ is a scalar with $0 \leq \rho < 1$. Such a mapping is called a *contraction mapping*, or simply a *contraction*. The scalar ρ is called the *contraction modulus* of T. Note that a mapping T may be a contraction for some choice of the norm $\|\cdot\|$ and fail to be a contraction under a different choice of norm.

Any vector $x^* \in X$ satisfying $T(x^*) = x^*$ is called a *fixed point* of T and the iteration $x_{k+1} = T(x_k)$ is an important algorithm for finding such a fixed point. The following is the central result regarding contraction mappings.

Proposition A.26: (Contraction Mapping Theorem) Suppose that $T : X \mapsto X$ is a contraction of modulus $\rho \in [0, 1)$ and that X is a closed subset of $\Re^n$. Then:

(a) (*Existence and Uniqueness of Fixed Point*) The mapping T has a unique fixed point $x^* \in X$.

(b) (*Convergence*) For every initial vector $x_0 \in X$, the sequence $\{x_k\}$ generated by $x_{k+1} = T(x_k)$ converges to x^*. In particular,

$$\|x_k - x^*\| \leq \rho^k \|x_0 - x^*\|, \qquad \forall\, k \geq 0.$$

Proof: (a) Fix some $x_0 \in X$ and consider the sequence $\{x_k\}$ generated by $x_{k+1} = T(x_k)$. We have, from the contraction property [cf. Eq. (A.4)],

$$\|x_{k+1} - x_k\| \leq \rho \|x_k - x_{k-1}\|,$$

for all $k \geq 1$, which implies

$$\|x_{k+1} - x_k\| \leq \rho^k \|x_1 - x_0\|, \qquad \forall\, k \geq 0.$$

It follows that for every $k \geq 0$ and $m \geq 1$, we have

$$\|x_{k+m} - x_k\| \leq \sum_{i=1}^{m} \|x_{k+i} - x_{k+i-1}\|$$

$$\leq \rho^k (1 + \rho + \cdots + \rho^{m-1}) \|x_1 - x_0\|$$

$$\leq \frac{\rho^k}{1-\rho} \|x_1 - x_0\|.$$

Therefore, $\{x_k\}$ is a Cauchy sequence and must converge to a limit, denoted x^* (Prop. A.5). Furthermore, since X is closed, x^* belongs to X. We have for all $k \geq 1$,

$$\|T(x^*) - x^*\| \leq \|T(x^*) - x_k\| + \|x_k - x^*\| \leq \rho \|x^* - x_{k-1}\| + \|x_k - x^*\|$$

and since x_k converges to x^*, we obtain $T(x^*) = x^*$. Therefore, the limit x^* of x_k is a fixed point of T. It is a unique fixed point because if y^* were another fixed point, we would have

$$\|x^* - y^*\| = \|T(x^*) - T(y^*)\| \le \rho\|x^* - y^*\|,$$

which implies that $x^* = y^*$.

(b) We have

$$\|x_{k'} - x^*\| = \|T(x_{k'-1}) - T(x^*)\| \le \rho\|x_{k'-1} - x^*\|,$$

for all $k' \ge 1$, so by applying this relation successively for $k' = k, k-1, \ldots, 1$, we obtain the desired result. **Q.E.D.**

The type of convergence demonstrated in part (b) of the preceding proposition is referred to as *linear convergence*. More precisely, given a sequence $\{x_k\}$ that converges to some $x^* \in \Re^n$, and a continuous (error) function $e : \Re^n \mapsto \Re$ such that $e(x^*) = 0$, we say that $\{e(x^k)\}$ *converges linearly or geometrically*, if there exist $q > 0$ and $\beta \in (0, 1)$ such that for all k

$$e(x^k) \le q\beta^k.$$

Typical examples of error functions that we use are $e(x) = \|x_k - x^*\|$ and $e(x) = f(x) - f(x^*)$, where f is the cost function of an optimization problem.

We note that the convergence of contraction iterations is maintained when there are additional decaying perturbations in $T(x_k)$, i.e.,

$$x_{k+1} = T(x_k) + w_k, \qquad (A.5)$$

where $T : \Re^n \mapsto \Re^n$ is a contraction and $\{w_k\}$ is a sequence in $\Re^n$ such that $w_k \to 0$ (see the discussion following the subsequent Prop. A.30). A related useful fact is that when $\{\|w_k\|\}$ is linearly decaying, then the linear convergence of $\{x_k\}$ is maintained. In particular, consider the iteration (A.5), and assume that T is a contraction of modulus $\rho \in [0, 1)$ and for some scalars $q > 0$ and $\sigma \in (0, 1)$ we have $\|w_k\| \le q\sigma^k$, for all k. Then we claim that $\{x_k\}$ converges to x^*, the unique fixed point of T, and for every scalar γ with $\max\{\rho, \sigma\} < \gamma < 1$, there exits a scalar $p > 0$ such that

$$\|x_k - x^*\| \le p\gamma^k, \qquad \forall\ k \ge 0. \qquad (A.6)$$

To see this, we note that for all k, we have

$$\|x_k - x^*\| = \|T(x_{k-1}) - x^* + w_{k-1}\| \le \|T(x_{k-1}) - x^*\| + \|w_{k-1}\|,$$

so that by using the contraction property,

$$\|x_k - x^*\| \le \rho\|x_{k-1} - x^*\| + q\sigma^{k-1}.$$

Sec. A.6 Convergence Theorems

Replacing k with $k-1$, we have

$$\|x_{k-1} - x^*\| \leq \rho\|x_{k-2} - x^*\| + q\sigma^{k-2},$$

and by combining the preceding two relations,

$$\|x_k - x^*\| \leq \rho^2\|x_{k-2} - x^*\| + q(\sigma^{k-1} + \rho\sigma^{k-2}).$$

Proceeding similarly, we obtain for all k,

$$\|x_k - x^*\| \leq \rho^k\|x_0 - x^*\| + q(\sigma^{k-1} + \rho\sigma^{k-2} + \cdots + \rho^{k-2}\sigma + \rho^{k-1})$$
$$\leq \rho^k\|x_0 - x^*\| + kq\big(\max\{\rho, \sigma\}\big)^{k-1}$$
$$\leq \gamma^k\|x_0 - x^*\| + \bar{q}\gamma^k,$$

where for a given $\gamma \in \big(\max\{\rho, \sigma\}, 1\big)$, $\bar{q}$ is such that $kq\big(\max\{\rho, \sigma\}\big)^{k-1} \leq \bar{q}\gamma^k$ for all k. This shows Eq. (A.6).

In the case of a linear mapping

$$T(x) = Ax + b,$$

where A is an $n \times n$ matrix and $b \in \Re^n$, it can be shown that T is a contraction mapping with respect to some norm (but not necessarily all norms) if and only if all the eigenvalues of A lie strictly within the unit circle. For a proof, see [OrR70], or [Ber12], Example 1.5.1.

Contractions with Respect to a Weighted Maximum Norm

Given a vector $\xi = (\xi_1, \ldots, \xi_n)' \in \Re^n$, with positive components $\xi_i > 0$, the weighted maximum norm corresponding to ξ is defined by

$$\|x\|_\xi = \max_{i=1,\ldots,n} \frac{|x_i|}{\xi_i}, \qquad x \in \Re^n.$$

Consider the linear mapping

$$T(x) = Ax + b, \tag{A.7}$$

where A is an $n \times n$ matrix with components a_{ij} and b is a vector in $\Re^n$. The following proposition gives useful criteria for T to be a weighted maximum norm contraction.

Proposition A.27: Consider the mapping T of Eq. (A.7).

(a) T is a contraction with respect to $\|\cdot\|_\xi$ with modulus ρ if and only if

$$\frac{\sum_{j=1}^n |a_{ij}|\xi_j}{\xi_i} \leq \rho, \qquad \forall\, i = 1, \ldots, n.$$

(b) Let P be a stochastic $n \times n$ matrix P (i.e., its components p_{ij} satisfy $p_{ij} \geq 0$ for all $i, j = 1, \ldots, n$, and $\sum_{j=1}^{n} p_{ij} = 1$ for all $i = 1, \ldots, n$), and assume that

$$|a_{ij}| \leq p_{ij}, \qquad \forall\, i, j = 1, \ldots, n,$$

and that for some row index $\bar{i} \in \{1, \ldots, n\}$,

$$|a_{\bar{i}j}| < p_{\bar{i}j}, \qquad \forall\, j = 1, \ldots, n.$$

Assume further that P corresponds to an irreducible Markov chain (one with a single recurrent class and no transient states) and that $\xi = (\xi_1, \ldots, \xi_n)' \in \Re^n$ is its invariant distribution, i.e.,

$$\xi_i > 0, \quad i = 1, \ldots, n, \qquad \sum_{i=1}^{n} \xi_i = 1, \qquad \xi' = \xi' P.$$

Then T is a contraction with respect to the norm $\|\cdot\|_\xi$.

Part (a) of the preceding proposition is given as Prop. 1.5.2(a) of [Ber12], while part (b) is given as Prop. 1 of [BeY09].

Convergence of Iterations with Delays

The following two propositions deal with iterations that involve delayed iterates.

Proposition A.28: (Iterations with Delays I) Let $\{\alpha_k\}$ be a scalar sequence such that

$$|\alpha_k| \leq \sum_{i=1}^{n} \beta_i |\alpha_{k-i}|, \qquad \forall\, k = 0, 1, \ldots,$$

where $\beta_i > 0$, $i = 1, \ldots, n$, are some scalars with $\sum_{i=1}^{n} \beta_i < 1$, and n is a positive integer. Then the sequence $\{\gamma_k\}$, where

$$\gamma_k = \max_{i=1,\ldots,n} \frac{|a_{k-i}|}{\xi_i},$$

converges to 0 linearly, where $\xi = (\xi_1, \ldots, \xi_n)'$ is the unique solution of the system of equations

Sec. A.6   Convergence Theorems   775

$$\sum_{i=1}^{n} \xi_i = 1, \quad \xi_j = \frac{\beta_j}{\sum_{i=1}^{n} \beta_i}\xi_1 + \xi_{j+1}, \; j=1,\ldots,n-1, \quad \xi_n = \frac{\beta_n}{\sum_{i=1}^{n} \beta_i}\xi_1.$$

Proof: The given system of equations can be seen to have a unique solution by successively expressing $\xi_n, \xi_{n-1}, \ldots, \xi_2$ in terms of ξ_1, and then determining ξ_1 from the equation $\sum_{i=1}^{n} \xi_i = 1$. Furthermore, we can easily verify the equation $\xi' = \xi' P$ for ξ to be the invariant distribution of the irreducible matrix P given by

$$P = \begin{pmatrix} \beta_1/\sum_{i=1}^{n}\beta_i & \beta_2/\sum_{i=1}^{n}\beta_i & \cdots & \beta_{n-1}/\sum_{i=1}^{n}\beta_i & \beta_n/\sum_{i=1}^{n}\beta_i \\ 1 & 0 & \cdots & 0 & 0 \\ \vdots & \vdots & \cdots & \vdots & \vdots \\ 0 & 0 & \cdots & 1 & 0 \end{pmatrix}.$$

The proof follows by using Prop. A.27(b).   **Q.E.D.**

The preceding proposition can be used to show that an iteration of the form

$$\alpha_k = \gamma + \sum_{i=1}^{n} \beta_i \alpha_{k-i},$$

where γ is a scalar, and $\beta_1, \ldots, \beta_n$ are scalars satisfying $\sum_{i=1}^{n} |\beta_i| < 1$, converges to

$$\frac{\gamma}{1 - \sum_{i=1}^{n} \beta_i}.$$

The following proposition is due to [FAJ14], whose proof we follow closely. Additional related results are given in [Fey16].

Proposition A.29: (Iterations with Delays II) Let $\{\alpha_k\}$ be a nonnegative sequence satisfying

$$\alpha_{k+1} \leq p\alpha_k + q \max_{\max\{0,k-d\} \leq \ell \leq k} \alpha_\ell, \quad \forall \; k=0,1,\ldots, \quad (A.8)$$

for some positive integer d and nonnegative scalars p and q such that $p+q < 1$. Then we have

$$\alpha_k \leq \rho^k \alpha_0, \quad \forall \; k = 0, 1, \ldots, \quad (A.9)$$

where $\rho = (p+q)^{\frac{1}{1+d}}$.

Proof: We first show a preliminary relation. Since $p + q < 1$, we have

$$1 \leq (p+q)^{-\frac{b}{1+b}},$$

which implies that

$$\begin{aligned}
p + q\rho^{-b} &= p + q(p+q)^{-\frac{b}{1+b}} \\
&\leq (p+q)(p+q)^{-\frac{b}{1+b}} \\
&= (p+q)^{\frac{1}{1+b}} \\
&= \rho.
\end{aligned} \qquad (A.10)$$

We now show Eq. (A.9) by induction. It clearly holds for $k = 0$. Assume that it holds for all k up to some $\bar{k}$. Then

$$\alpha_k \leq \rho^k \alpha_0, \qquad \forall\ k = \max\{0, \bar{k} - b\}, \ldots, \bar{k}.$$

From this relation and Eq. (A.8), we have

$$\begin{aligned}
\alpha_{\bar{k}+1} &\leq p\rho^{\bar{k}}\alpha_0 + q\left(\max_{\max\{0,\bar{k}-b\}\leq \ell \leq \bar{k}} \rho^\ell \alpha_0\right) \\
&\leq p\rho^{\bar{k}}\alpha_0 + q\rho^{\max\{0,\bar{k}-b\}}\alpha_0 \\
&\leq p\rho^{\bar{k}}\alpha_0 + q\rho^{\bar{k}-b}\alpha_0 \\
&= (p + q\rho^{-b})\rho^{\bar{k}}\alpha_0.
\end{aligned}$$

Using also Eq. (A.10), we have $\alpha_{\bar{k}+1} \leq \rho^{\bar{k}+1}\alpha_0$, and this completes the induction. **Q.E.D.**

Nonstationary Iterations

For nonstationary iterations of the form $x_{k+1} = T_k(x_k)$, where the function T_k depends on k, the ideas of the preceding propositions may apply but with modifications. The following proposition is often useful in this respect.

Proposition A.30: Let $\{\alpha_k\}$ be a nonnegative scalar sequence such that

$$\alpha_{k+1} \leq (1 - \gamma_k)\alpha_k + \beta_k, \qquad \forall\ k = 0, 1, \ldots,$$

where $0 \leq \beta_k$, $0 < \gamma_k \leq 1$ for all k, and

$$\sum_{k=0}^{\infty} \gamma_k = \infty, \qquad \frac{\beta_k}{\gamma_k} \to 0.$$

Then $\alpha_k \to 0$.

Sec. A.6 Convergence Theorems

Proof: We first show that given any $\epsilon > 0$, we have $\alpha_k < \epsilon$ for infinitely many k. Indeed, if this were not so, by letting $\bar{k}$ be such that $\alpha_k \geq \epsilon$ and $\beta_k/\gamma_k \leq \epsilon/2$ for all $k \geq \bar{k}$, we would have for all $k \geq \bar{k}$

$$\alpha_{k+1} \leq \alpha_k - \gamma_k \alpha_k + \beta_k \leq \alpha_k - \gamma_k \epsilon + \frac{\gamma_k \epsilon}{2} = \alpha_k - \frac{\gamma_k \epsilon}{2}.$$

Therefore, for all $m \geq \bar{k}$,

$$\alpha_{m+1} \leq \alpha_{\bar{k}} - \frac{\epsilon}{2} \sum_{k=\bar{k}}^{m} \gamma_k.$$

Since $\{\alpha_k\}$ is nonnegative and $\sum_{k=0}^{\infty} \gamma_k = \infty$, we obtain a contradiction.

Thus, given any $\epsilon > 0$, there exists $\bar{k}$ such that $\beta_k/\gamma_k < \epsilon$ for all $k \geq \bar{k}$ and $\alpha_{\bar{k}} < \epsilon$. We then have

$$\alpha_{\bar{k}+1} \leq (1 - \gamma_k)\alpha_{\bar{k}} + \beta_k < (1 - \gamma_k)\epsilon + \gamma_k \epsilon = \epsilon.$$

By repeating this argument, we obtain $\alpha_k < \epsilon$ for all $k \geq \bar{k}$. Since ϵ can be arbitrarily small, it follows that $\alpha_k \to 0$. **Q.E.D.**

As an example, consider the iteration

$$x_{k+1} = T(x_k) + w_k,$$

where $T : \Re^n \mapsto \Re^n$ is a contraction of modulus $\rho \in (0, 1)$ and $\{w_k\}$ is a sequence in $\Re^n$ such that $w_k \to 0$. Then we have

$$\|x_{k+1} - x^*\| \leq \|T(x_k) - x^*\| + \|w_k\| \leq \rho\|x_k - x^*\| + \|w_k\|,$$

and Prop. A.30 applies with $\alpha_k = \|x_k - x^*\|$, $\gamma_k = 1 - \rho$, and $\beta_k = \|w_k\|$, showing that $x_k \to x^*$.

As another example, consider a sequence of "approximate" contraction mappings $T_k : \Re^n \mapsto \Re^n$, satisfying

$$\|T_k(x) - T_k(y)\| \leq (1 - \gamma_k)\|x - y\| + \beta_k, \qquad \forall\ x, y \in \Re^n,\ k = 0, 1, \ldots,$$

where $\gamma_k \in (0, 1]$, for all k, and

$$\sum_{k=0}^{\infty} \gamma_k = \infty, \qquad \frac{\beta_k}{\gamma_k} \to 0.$$

Assume also that all the mappings T_k have a common fixed point x^*. Then

$$\|x_{k+1} - x^*\| = \|T_k(x_k) - T_k(x^*)\| \leq (1 - \gamma_k)\|x_k - x^*\| + \beta_k,$$

and from Prop. A.30, it follows that the sequence $\{x_k\}$ generated by the iteration $x_{k+1} = T_k(x_k)$ converges to x^* starting from any $x_0 \in \Re^n$.

Supermartingale Convergence

We next give a convergence theorem relating to deterministic sequences. It is a special case of a fundamental theorem, known as the *supermartingale convergence theorem*, which relates to convergence of sequences of random variables. We will not need this more general theorem in our analysis, and we refer to [Ber15a] and [WaB13] for some of its applications in incremental optimization methods with randomized order of component selection.

Proposition A.31: Let $\{Y_k\}$, $\{Z_k\}$, $\{W_k\}$, and $\{V_k\}$ be four scalar sequences such that

$$Y_{k+1} \leq (1+V_k)Y_k - Z_k + W_k, \qquad k = 0, 1, \ldots, \qquad (A.11)$$

$\{Z_k\}$, $\{W_k\}$, and $\{V_k\}$ are nonnegative, and

$$\sum_{k=0}^{\infty} W_k < \infty, \qquad \sum_{k=0}^{\infty} V_k < \infty.$$

Then either $Y_k \to -\infty$, or else $\{Y_k\}$ converges to a finite value and $\sum_{k=0}^{\infty} Z_k < \infty$.

Proof: We first give the proof assuming that $V_k \equiv 0$, and then generalize. In this case, using the nonnegativity of $\{Z_k\}$, we have

$$Y_{k+1} \leq Y_k + W_k.$$

By writing this relation for the index k set to $\bar{k}, \ldots, k$, where $k \geq \bar{k}$, and adding, we have

$$Y_{k+1} \leq Y_{\bar{k}} + \sum_{\ell=\bar{k}}^{k} W_\ell \leq Y_{\bar{k}} + \sum_{\ell=\bar{k}}^{\infty} W_\ell.$$

Since $\sum_{k=0}^{\infty} W_k < \infty$, it follows that $\{Y_k\}$ is bounded above, and by taking upper limit of the left hand side as $k \to \infty$ and lower limit of the right hand side as $\bar{k} \to \infty$, we have

$$\limsup_{k \to \infty} Y_k \leq \liminf_{\bar{k} \to \infty} Y_{\bar{k}} < \infty.$$

This implies that either $Y_k \to -\infty$, or else $\{Y_k\}$ converges to a finite value. In the latter case, by writing Eq. (A.11) for the index k set to $0, \ldots, k$, and adding, we have

$$\sum_{\ell=0}^{k} Z_\ell \leq Y_0 + \sum_{\ell=0}^{k} W_\ell - Y_{k+1}, \qquad \forall\, k = 0, 1, \ldots,$$

Sec. A.6 Convergence Theorems

so by taking the limit as $k \to \infty$, we obtain $\sum_{\ell=0}^{\infty} Z_\ell < \infty$.

We now extend the proof to the case of a general nonnegative sequence $\{V_k\}$. We first note that

$$\log \prod_{\ell=0}^{k}(1 + V_\ell) = \sum_{\ell=0}^{k} \log(1 + V_\ell) \le \sum_{k=0}^{\infty} V_k,$$

since we generally have $(1+a) \le e^a$ and $\log(1+a) \le a$ for any $a \ge 0$. Thus the assumption $\sum_{k=0}^{\infty} V_k < \infty$ implies that

$$\prod_{\ell=0}^{\infty}(1 + V_\ell) < \infty. \tag{A.12}$$

Define

$$\bar{Y}_k = Y_k \prod_{\ell=0}^{k-1}(1+V_\ell)^{-1}, \quad \bar{Z}_k = Z_k \prod_{\ell=0}^{k}(1+V_\ell)^{-1}, \quad \bar{W}_k = W_k \prod_{\ell=0}^{k}(1+V_\ell)^{-1}.$$

Multiplying Eq. (A.11) with $\prod_{\ell=0}^{k}(1+V_\ell)^{-1}$, we obtain

$$\bar{Y}_{k+1} \le \bar{Y}_k - \bar{Z}_k + \bar{W}_k.$$

Since $\bar{W}_k \le W_k$, the hypothesis $\sum_{k=0}^{\infty} W_k < \infty$ implies $\sum_{k=0}^{\infty} \bar{W}_k < \infty$, so from the special case of the result already shown, we have that either $\bar{Y}_k \to -\infty$ or else $\{\bar{Y}_k\}$ converges to a finite value and $\sum_{k=0}^{\infty} \bar{Z}_k < \infty$. Since

$$Y_k = \bar{Y}_k \prod_{\ell=0}^{k-1}(1+V_\ell), \qquad Z_k = \bar{Z}_k \prod_{\ell=0}^{k}(1+V_\ell),$$

and $\prod_{\ell=0}^{k-1}(1+V_\ell)$ converges to a finite value by the nonnegativity of $\{V_k\}$ and Eq. (A.12), it follows that either $Y_k \to -\infty$ or else $\{Y_k\}$ converges to a finite value and $\sum_{k=0}^{\infty} Z_k < \infty$. **Q.E.D.**

Fejér Monotonicity

Supermartingale convergence theorems can be applied in a variety of contexts. One such context, the so called *Fejér monotonicity* theory, deals with iterations that "almost" decrease the distance to *every* element of some given set X^*. We may then often show that such iterations are convergent to a (unique) element of X^*. Applications of this idea arise when X^* is the set of optimal solutions of an optimization problem or the set of fixed points of a certain mapping. Examples are various gradient and subgradient projection methods with a diminishing stepsize that arise in various contexts in this book.

Proposition A.32: (Fejér Convergence Theorem) Let X^* be a nonempty subset of $\Re^n$, and let $\{x_k\} \subset \Re^n$ be a sequence satisfying for some $p > 0$ and for all k,

$$\|x_{k+1} - x^*\|^p \leq (1+\beta_k)\|x_k - x^*\|^p - \gamma_k \phi(x_k; x^*) + \delta_k, \qquad \forall\, x^* \in X^*,$$

where $\{\beta_k\}$, $\{\gamma_k\}$, and $\{\delta_k\}$ are nonnegative sequences satisfying

$$\sum_{k=0}^{\infty} \beta_k < \infty, \qquad \sum_{k=0}^{\infty} \gamma_k = \infty, \qquad \sum_{k=0}^{\infty} \delta_k < \infty,$$

$\phi : \Re^n \times X^* \mapsto [0, \infty)$ is some nonnegative function, and $\|\cdot\|$ is some norm. Then:

(a) The minimum distance sequence $\inf_{x^* \in X^*} \|x_k - x^*\|$ converges, and in particular, $\{x_k\}$ is bounded.

(b) If $\{x_k\}$ has a limit point $\bar{x}$ that belongs to X^*, then the entire sequence $\{x_k\}$ converges to $\bar{x}$.

(c) Suppose that for some $x^* \in X^*$, $\phi(\cdot; x^*)$ is lower semicontinuous and satisfies

$$\phi(x; x^*) = 0 \qquad \text{if and only if} \qquad x \in X^*. \tag{A.13}$$

Then $\{x_k\}$ converges to a point in X^*.

Proof: (a) Let $\{\epsilon_k\}$ be a positive sequence such that $\sum_{k=0}^{\infty}(1+\beta_k)\epsilon_k < \infty$, and let x_k^* be a point of X^* such that

$$\|x_k - x_k^*\|^p \leq \inf_{x^* \in X^*} \|x_k - x^*\|^p + \epsilon_k.$$

Then since ϕ is nonnegative, we have for all k,

$$\inf_{x^* \in X^*} \|x_{k+1} - x^*\|^p \leq \|x_{k+1} - x_k^*\|^p \leq (1+\beta_k)\|x_k - x_k^*\|^p + \delta_k,$$

and by combining the last two relations, we obtain

$$\inf_{x^* \in X^*} \|x_{k+1} - x^*\|^p \leq (1+\beta_k) \inf_{x^* \in X^*} \|x_k - x^*\|^p + (1+\beta_k)\epsilon_k + \delta_k.$$

The result follows by applying Prop. A.31 with

$$Y_k = \inf_{x^* \in X^*} \|x_k - x^*\|^p, \quad Z_k = 0, \quad W_k = (1+\beta_k)\epsilon_k + \delta_k, \quad V_k = \beta_k.$$

Sec. A.6   Convergence Theorems

(b) Following the argument of the proof of Prop. A.31, define for all k,

$$\bar{Y}_k = \|x_k - \bar{x}\|^p \prod_{\ell=0}^{k-1}(1+\beta_\ell)^{-1}, \qquad \bar{\delta}_k = \delta_k \prod_{\ell=0}^{k}(1+\beta_\ell)^{-1}.$$

Then from our hypotheses, we have $\sum_{k=0}^{\infty} \bar{\delta}_k < \infty$ and

$$\bar{Y}_{k+1} \leq \bar{Y}_k + \bar{\delta}_k, \qquad \forall\, k = 0, 1, \ldots, \tag{A.14}$$

while $\{\bar{Y}_k\}$ has a limit point at 0, since $\bar{x}$ is a limit point of $\{x_k\}$. For any $\epsilon > 0$, let $\bar{k}$ be such that

$$\bar{Y}_{\bar{k}} \leq \epsilon, \qquad \sum_{\ell=\bar{k}}^{\infty} \bar{\delta}_\ell \leq \epsilon,$$

so that by adding Eq. (A.14), we obtain for all $k > \bar{k}$,

$$\bar{Y}_k \leq \bar{Y}_{\bar{k}} + \sum_{\ell=\bar{k}}^{\infty} \bar{\delta}_\ell \leq 2\epsilon.$$

Since ϵ is arbitrarily small, it follows that $\bar{Y}_k \to 0$. We now note that as in Eq. (A.12),

$$\prod_{\ell=0}^{\infty}(1+\beta_\ell)^{-1} < \infty,$$

so that $\bar{Y}_k \to 0$ implies that $\|x_k - \bar{x}\|^p \to 0$, and hence $x_k \to \bar{x}$.

(c) From Prop. A.31, it follows that

$$\sum_{k=0}^{\infty} \gamma_k \phi(x_k; x^*) < \infty.$$

Thus $\lim_{k\to\infty,\, k\in\mathcal{K}} \phi(x_k; x^*) = 0$ for some subsequence $\{x_k\}_\mathcal{K}$. By part (a), $\{x_k\}$ is bounded, so the subsequence $\{x_k\}_\mathcal{K}$ has a limit point $\bar{x}$, and by the lower semicontinuity of $\phi(\cdot; x^*)$, we must have

$$\phi(\bar{x}; x^*) \leq \lim_{k\to\infty,\, k\in\mathcal{K}} \phi(x_k; x^*) = 0,$$

which in view of the nonnegativity of ϕ, implies that $\phi(\bar{x}; x^*) = 0$. Using the hypothesis (A.13), it follows that $\bar{x} \in X^*$, so by part (b), the entire sequence $\{x_k\}$ converges to $\bar{x}$.   **Q.E.D.**

APPENDIX B:
Convex Analysis

Convexity is central in nonlinear programming, and has a rich mathematical theory. In this appendix, we selectively collect the definitions, notational conventions, and results that we will need. For detailed textbook accounts of convex analysis and its connections with optimization, see Rockafellar [Roc70], Ekeland and Teman [EkT76], Hiriart-Urruty and Lemarechal [HiL93], Rockafellar and Wets [RoW98], Borwein and Lewis [BoL00], Bonnans and Shapiro [BoS00], Zalinescu [Zal02], Auslender and Teboulle [AuT03], Bertsekas, Nedić, and Ozdaglar [BNO03], and Bertsekas [Ber09].

A discussion of generalized notions of convexity, including quasiconvexity and pseudoconvexity, and their applications in optimization can be found in the books by Avriel [Avr76], Bazaraa, Sherali, and Shetty [BSS93], Mangasarian [Man69], and the references quoted therein.

The author's convex optimization theory textbook [Ber09] is consistent with the notation and content of this appendix, but develops the subject in much greater depth and detail. Proofs of the results quoted are generally given in this textbook, and on some occasions, in the author's convex optimization algorithms textbook [Ber15a]. In a few cases of important convex optimization-related results, a proof is included here.

B.1 CONVEX SETS AND FUNCTIONS

A subset C of $\Re^n$ is called *convex* if

$$\alpha x + (1-\alpha)y \in C, \qquad \forall\ x, y \in C,\ \forall\ \alpha \in [0,1]. \qquad (\text{B.1})$$

The following proposition provides some means for verifying convexity of a set.

Proposition B.1:

(a) For any collection $\{C_i \mid i \in I\}$ of convex sets, the set intersection $\cap_{i \in I} C_i$ is convex.

(b) The vector sum of two convex sets C_1 and C_2 is convex.

(c) The image of a convex set under a linear transformation is convex.

(d) If C is a convex set and $f : C \mapsto \Re$ is a convex function, the level sets $\{x \in C \mid f(x) \leq \alpha\}$ and $\{x \in C \mid f(x) < \alpha\}$ are convex for all scalars α.

Proof: See Prop. 1.1.1 and Section 1.1.1 of [Ber09]. **Q.E.D.**

Let C be a convex subset of $\Re^n$. A function $f : C \mapsto \Re$ is called *convex* if

$$f\big(\alpha x + (1-\alpha)y\big) \leq \alpha f(x) + (1-\alpha) f(y), \qquad \forall \, x, y \in C, \, \forall \, \alpha \in [0, 1]. \quad \text{(B.2)}$$

The function f is called *concave* if $-f$ is convex. The function f is called *strictly convex* if the above inequality is strict for all $x, y \in C$ with $x \neq y$, and all $\alpha \in (0, 1)$. For a function $f : \Re^n \mapsto \Re$, we also say that f is *convex over the convex set* C if Eq. (B.2) holds.

We occasionally deal with functions $f : C \mapsto [-\infty, \infty]$ that can take infinite values. The *epigraph* of such a function f is the subset of $\Re^{n+1}$ given by

$$\text{epi}(f) = \big\{(x, w) \mid x \in C, \, w \in \Re, \, f(x) \leq w\big\}.$$

We say that $f : C \mapsto (-\infty, \infty]$ is convex if C is convex and epi(f) is a convex set. Note that a function $f : C \mapsto (-\infty, \infty]$ is convex if Eq. (B.2) holds (here the rules of arithmetic are extended to include $\infty + \infty = \infty$, $0 \cdot \infty = 0$, and $\alpha \cdot \infty = \infty$, for all $\alpha > 0$).

The *effective domain* of f is the set

$$\text{dom}(f) = \big\{x \in C \mid f(x) < \infty\big\},$$

which is convex if f is convex. The function f is called *closed* if epi(f) is a closed set, and it is called *proper* if dom(f) is nonempty and $f(x) > -\infty$ for all $x \in C$.

By restricting the definition of a convex function to its effective domain we can avoid calculations with ∞, and we will often do this. However, in some analyses it is more economical to use convex functions that can take the value of infinity.

A useful property, obtained by repeated application of the definition of convexity [cf. Eq. (B.2)], is that if $x_1, \ldots, x_m \in C$, $\alpha_1, \ldots, \alpha_m \geq 0$, and

$\sum_{i=1}^{m} \alpha_i = 1$, then

$$f\left(\sum_{i=1}^{m} \alpha_i x_i\right) \leq \sum_{i=1}^{m} \alpha_i f(x_i).$$

This is a special case of *Jensen's inequality* and can be used to prove a number of interesting inequalities in applied mathematics and probability theory.

The following proposition provides some means for recognizing convex functions.

Proposition B.2:

(a) A linear function is convex.

(b) Any vector norm is convex.

(c) The weighted sum of convex functions, with positive weights, is convex.

(d) If I is an index set, C is a convex subset of $\Re^n$, and $f_i : C \mapsto (-\infty, \infty]$ is convex for each $i \in I$, then the function $h : C \mapsto (-\infty, \infty]$ defined by

$$h(x) = \sup_{i \in I} f_i(x)$$

is also convex.

(e) If $F : \Re^{n+m} \mapsto \Re$ is a convex function of the pair (x, z) where $x \in \Re^n$, $z \in \Re^m$, and Z is a convex set such that $\inf_{z \in Z} F(x, z) > -\infty$ for all $x \in \Re^n$, then the function $f : \Re^n \to \Re$ defined by

$$f(x) = \inf_{z \in Z} F(x, z), \qquad \forall\, x \in \Re^n,$$

is convex.

Proof: For parts (a)-(d), see Props. 1.1.4-1.1.6 and Section 1.1.3 of [Ber09]. For part (e), see Prop. 3.3.1 of [Ber09]. **Q.E.D.**

Characterizations of Differentiable Convex Functions

For differentiable functions, there is an alternative characterization of convexity, given in the following proposition.

Proposition B.3: (First Derivative Characterizations) Let C be a convex subset of $\Re^n$ and let $f : \Re^n \mapsto \Re$ be differentiable over $\Re^n$.

(a) f is convex over C if and only if
$$f(z) \geq f(x) + (z - x)'\nabla f(x), \qquad \forall\, x, z \in C.$$

(b) f is strictly convex over C if and only if the above inequality is strict whenever $x \neq z$.

(c) Let f be convex. For a scalar $L > 0$ the following five properties are equivalent:

(i) $\|\nabla f(x) - \nabla f(y)\| \leq L\,\|x - y\|$, for all $x, y \in \Re^n$.

(ii) $f(x) + \nabla f(x)'(y - x) + \frac{1}{2L}\|\nabla f(x) - \nabla f(y)\|^2 \leq f(y)$, for all $x, y \in \Re^n$.

(iii) $\bigl(\nabla f(x) - \nabla f(y)\bigr)'(x - y) \geq \frac{1}{L}\|\nabla f(x) - \nabla f(y)\|^2$, for all $x, y \in \Re^n$.

(iv) $f(y) \leq f(x) + \nabla f(x)'(y - x) + \frac{L}{2}\|y - x\|^2$, for all $x, y \in \Re^n$.

(v) $\bigl(\nabla f(x) - \nabla f(y)\bigr)'(x - y) \leq L\|x - y\|^2$, for all $x, y \in \Re^n$.

Proof: For parts (a) and (b), see Prop. 1.1.7 and Section 1.1.4 of [Ber09]. For part (c), see [Ber15a], Exercise 6.1 (with solution included). **Q.E.D.**

For twice differentiable convex functions, there is another characterization of convexity, which is given in the following proposition.

Proposition B.4: (Second Derivative Characterizations) Let C be a convex subset of $\Re^n$ and let $f : \Re^n \mapsto \Re$ be twice continuously differentiable over $\Re^n$.

(a) If $\nabla^2 f(x)$ is positive semidefinite for all $x \in C$, then f is convex over C.

(b) If $\nabla^2 f(x)$ is positive definite for every $x \in C$, then f is strictly convex over C.

(c) If C is open and f is convex over C, then $\nabla^2 f(x)$ is positive semidefinite for all $x \in C$.

(d) If $f(x) = x'Qx$, where Q is a symmetric matrix, then f is convex if and only if Q is positive semidefinite. Furthermore, f is strictly convex if and only if Q is positive definite.

Proof: See Prop. 1.1.10 and Section 1.1.4 of [Ber09]. **Q.E.D.**

The conclusion of Prop. B.4(c) can also be proved if C is assumed to have nonempty interior instead of being open. We now consider a strengthened form of strict convexity for a continuously differentiable function $f : \Re^n \mapsto \Re$. We say that f is *strongly convex* if for some $\sigma > 0$, we have

$$f(y) \geq f(x) + \nabla f(x)'(y-x) + \frac{\sigma}{2}\|x-y\|^2, \qquad \forall\, x, y \in \Re^n. \tag{B.3}$$

It can be shown that an equivalent definition is that

$$\bigl(\nabla f(x) - \nabla f(y)\bigr)'(x-y) \geq \sigma\|x-y\|^2, \qquad \forall\, x, y \in \Re^n. \tag{B.4}$$

A proof of this may be found in several sources, including the on-line exercises of Chapter 1 of [Ber09]. By fixing x in the definition (B.3), we see that a strongly convex function majorizes a coercive function, so it is itself coercive. It is also strictly convex, as shown among other properties by the following proposition.

Proposition B.5: (Strong Convexity) Let $f : \Re^n \mapsto \Re$ be a function that is continuously differentiable. Then:

(a) If f strongly convex in the sense that it satisfies Eq. (B.4) for some $\sigma > 0$, then f is strictly convex. If in addition, ∇f satisfies the Lipschitz condition

$$\|\nabla f(x) - \nabla f(y)\| \leq L\,\|x-y\|, \qquad \forall\, x, y \in \Re^n, \tag{B.5}$$

for some $L > 0$, then we have for all $x, y \in \Re^n$

$$\bigl(\nabla f(x) - \nabla f(y)\bigr)'(x-y) \geq \frac{\sigma L}{\sigma + L}\|x-y\|^2 + \frac{1}{\sigma + L}\|\nabla f(x) - \nabla f(y)\|^2. \tag{B.6}$$

(b) If f is twice continuously differentiable over $\Re^n$, then f satisfies Eq. (B.4) if and only if the matrix $\nabla^2 f(x) - \sigma I$, where I is the identity, is positive semidefinite for every $x \in \Re^n$.

Proof: (a) Fix some $x, y \in \Re^n$ such that $x \neq y$, and define the function $h : [0,1] \mapsto \Re$ by

$$h(t) = f\bigl(x + t(y-x)\bigr).$$

Consider some $t, \bar{t} \in [0, 1]$ such that $t < \bar{t}$. Using the chain rule and Eq. (B.4), we have

$$\left(\frac{dh(\bar{t})}{dt} - \frac{dh(t)}{dt}\right)(\bar{t} - t)$$
$$= \Big(\nabla f\big(x + \bar{t}(y - x)\big) - \nabla f\big(x + t(y - x)\big)\Big)'(y - x)(\bar{t} - t)$$
$$\geq \sigma(\bar{t} - t)^2 \|x - y\|^2 > 0.$$

Thus, dh/dt is strictly increasing, and for any $t \in (0, 1)$

$$\frac{h(t) - h(0)}{t} = \frac{1}{t}\int_0^t \frac{dh(\tau)}{d\tau}\, d\tau < \frac{1}{1-t}\int_t^1 \frac{dh(\tau)}{d\tau}\, d\tau = \frac{h(1) - h(t)}{1-t}.$$

Equivalently, we have $th(1) + (1 - t)h(0) > h(t)$, so from the definition of h, we obtain

$$tf(y) + (1 - t)f(x) > f\big(ty + (1 - t)x\big).$$

Since this inequality was proved for arbitrary $t \in (0, 1)$ and $x \neq y$, it follows that f is strictly convex.

We now assume that the Lipschitz condition (B.5) holds, and show Eq. (B.6). From Eqs. (B.4) and (B.5), we have $\sigma \leq L$. If $\sigma = L$, the result follows by combining the relation (iii) of Prop. B.3(c) and the relation

$$\|\nabla f(x) - \nabla f(y)\| \geq \sigma\|x - y\|, \qquad \forall\ x, y \in \Re^n,$$

which is a consequence of the strong convexity assumption (B.4). For $\sigma < L$ consider the function

$$\phi(x) = f(x) - \frac{\sigma}{2}\|x\|^2.$$

We will show that $\nabla\phi$, which is given by

$$\nabla\phi(x) = \nabla f(x) - \sigma x, \tag{B.7}$$

is Lipschitz continuous with constant $L - \sigma$. To this end, based on the equivalence of statements (i) and (v) of Prop. B.3(c), it is sufficient to show that

$$\big(\nabla\phi(x) - \nabla\phi(y)\big)'(x - y) \leq (L - \sigma)\|x - y\|^2, \qquad \forall\ x, y \in \Re^n,$$

or, using the expression (B.7) for $\nabla\phi$,

$$\big(\nabla f(x) - \nabla f(y) - \sigma(x - y)\big)'(x - y) \leq (L - \sigma)\|x - y\|^2, \qquad \forall\ x, y \in \Re^n.$$

This relation is equivalently written as

$$\big(\nabla f(x) - \nabla f(y)\big)'(x - y) \leq L\|x - y\|^2, \qquad \forall\ x, y \in \Re^n,$$

Sec. B.1 Convex Sets and Functions

and is true by the equivalence of statements (i) and (v) of Prop. B.3(c).

Having shown that $\nabla \phi$ is Lipschitz continuous with constant $L - \sigma$, we use the equivalence of statements (i) and (iii) of Prop. B.3(c) to the function ϕ and obtain

$$\big(\nabla\phi(x) - \nabla\phi(y)\big)'(x-y) \geq \frac{1}{L-\sigma}\big\|\nabla\phi(x) - \nabla\phi(y)\big\|^2.$$

Using the expression (B.7) for $\nabla \phi$ in this relation, we have

$$\big(\nabla f(x) - \nabla f(y) - \sigma(x-y)\big)'(x-y) \geq \frac{1}{L-\sigma}\big\|\nabla f(x) - \nabla f(y) - \sigma(x-y)\big\|^2,$$

which after expanding the quadratic and collecting terms, can be verified to be equivalent to the desired relation.

(b) Suppose that f satisfies Eq. (B.4). We fix some $x \in \Re^n$, let d be any vector in $\Re^n$, and let γ be a scalar in $(0,1]$. We use the second order expansion of Prop. A.23(b) twice to obtain

$$f(x + \gamma d) = f(x) + \gamma d'\nabla f(x) + \frac{\gamma^2}{2} d'\nabla^2 f(x + t\gamma d)d,$$

and

$$f(x) = f(x + \gamma d) - \gamma d'\nabla f(x + \gamma d) + \frac{\gamma^2}{2} d'\nabla^2 f(x + s\gamma d)d,$$

for some t and s belonging to $[0,1]$. By adding these two equations and using Eq. (B.4), we obtain

$$\frac{\gamma^2}{2} d'\big(\nabla^2 f(x+s\gamma d) + \nabla^2 f(x+t\gamma d)\big)d = \big(\nabla f(x+\gamma d) - \nabla f(x)\big)'(\gamma d) \geq \sigma \gamma^2 \|d\|^2.$$

We divide both sides by γ^2 and then take the limit as $\gamma \to 0$ to conclude that $d'\nabla^2 f(x)d \geq \sigma\|d\|^2$. Since this inequality was proved for every $d \in \Re^n$, it follows that $\nabla^2 f(x) - \sigma I$ is positive semidefinite.

Conversely, assume that $\nabla^2 f(x) - \sigma I$ is positive semidefinite for all $x \in \Re^n$. Fix some $x, y \in \Re^n$ such that $x \neq y$, and consider the function $g : [0, 1] \mapsto \Re$ defined by

$$g(t) = \nabla f\big(tx + (1-t)y\big)'(x-y).$$

Using the Mean Value Theorem (Prop. A.22 in Appendix A), we have

$$\big(\nabla f(x) - \nabla f(y)\big)'(x-y) = g(1) - g(0) = \frac{dg(t)}{dt}$$

for some $t \in [0,1]$. Since $\nabla^2 f\big(tx + (1-t)y\big) - \sigma I$ is positive semidefinite, we have

$$\frac{dg(t)}{dt} = (x-y)'\nabla^2 f\big(tx + (1-t)y\big)(x-y) \geq \sigma\|x-y\|^2.$$

By combining the preceding two relations, we obtain Eq. (B.4). **Q.E.D.**

Convex and Affine Hulls

Let X be a subset of $\Re^n$. A *convex combination* of elements of X is a vector of the form $\sum_{i=1}^m \alpha_i x_i$, where $x_1, \ldots, x_m$ belong to X and $\alpha_1, \ldots, \alpha_m$ are scalars such that

$$\alpha_i \geq 0, \quad i = 1, \ldots, m, \qquad \sum_{i=1}^m \alpha_i = 1.$$

The *convex hull* of X, denoted $\mathrm{conv}(X)$, is the set of all convex combinations of elements of X. In particular, if X consists of a finite number of vectors $x_1, \ldots, x_m$, its convex hull is

$$\mathrm{conv}(\{x_1, \ldots, x_m\}) = \left\{ \sum_{i=1}^m \alpha_i x_i \;\Big|\; \alpha_i \geq 0,\, i = 1, \ldots, m,\, \sum_{i=1}^m \alpha_i = 1 \right\}.$$

It is straightforward to verify that $\mathrm{conv}(X)$ is a convex set, and using this, to assert that $\mathrm{conv}(X)$ is the intersection of all convex sets containing X.

We recall that a linear manifold M is a set of the form $x + S = \{z \mid z - x \in S\}$, where S is a subspace, called the subspace parallel to M. If S is a subset of $\Re^n$, the *affine hull* of S, denoted $\mathrm{aff}(S)$, is the intersection of all linear manifolds containing S. Note that $\mathrm{aff}(S)$ is itself a linear manifold and that it contains $\mathrm{conv}(S)$. It can be seen that the affine hull of S and the affine hull of $\mathrm{conv}(S)$ coincide.

Given a nonempty subset X of $\Re^n$, a *nonnegative combination* of elements of X is a vector of the form $\sum_{i=1}^m \alpha_i x_i$, where m is a positive integer, $x_1, \ldots, x_m$ belong to X, and $\alpha_1, \ldots, \alpha_m$ are nonnegative scalars. If the scalars α_i are all positive, $\sum_{i=1}^m \alpha_i x_i$ is said to be a *positive combination*. A set $C \subset \Re^n$ is said to be a *cone* if $ax \in C$ for all $a > 0$ and $x \in C$. The *cone generated by* X, denoted $\mathrm{cone}(X)$, is the set of all nonnegative combinations of elements of X. It is easily seen that $\mathrm{cone}(X)$ is a convex cone containing the origin, although it need not be closed even if X is compact.

The following is a fundamental characterization of convex hulls.

Proposition B.6: (Caratheodory's Theorem) Let X be a nonempty subset of $\Re^n$.

(a) Every nonzero vector from $\mathrm{cone}(X)$ can be represented as a positive combination of linearly independent vectors from X.

(b) Every vector from $\mathrm{conv}(X)$ can be represented as a convex combination of no more than $n + 1$ vectors from X.

Proof: See Prop. 1.2.1 and Section 1.2 of [Ber09]. **Q.E.D.**

Sec. B.1   Convex Sets and Functions   791

Closure and Continuity Properties

We now explore some topological properties of convex sets and functions. Let C be a convex subset of $\Re^n$. We say that x is a *relative interior point* of C, if $x \in C$ and there exists a neighborhood N of x such that $N \cap \text{aff}(C) \subset C$, i.e., if x is an interior point of C relative to $\text{aff}(C)$. The *relative interior of* C, denoted $\text{ri}(C)$, is the set of all relative interior points of C. For example, if C is a line segment connecting two distinct points in the plane, then $\text{ri}(C)$ consists of all points of C except for the end points.

Proposition B.7: Let C be a nonempty convex set.

(a) (*Line Segment Principle*) If $x \in \text{ri}(C)$ and $\bar{x} \in \text{cl}(C)$, then all points on the line segment connecting x and $\bar{x}$, except possibly $\bar{x}$, belong to $\text{ri}(C)$.

(b) (*Nonemptiness of Relative Interior*) $\text{ri}(C)$ is a nonempty convex set, and has the same affine hull as C. In fact, if m is the dimension of $\text{aff}(C)$ and $m > 0$, there exist vectors $x_0, x_1, \ldots, x_m \in \text{ri}(C)$ such that $x_1 - x_0, \ldots, x_m - x_0$ span the subspace parallel to $\text{aff}(C)$.

(c) (*Prolongation Lemma*) $x \in \text{ri}(C)$ if and only if every line segment in C having x as one endpoint can be prolonged beyond x without leaving C [i.e., for every $\bar{x} \in C$, there exists a $\gamma > 1$ such that $x + (\gamma - 1)(x - \bar{x}) \in C$].

Proof: See Props. 1.3.1-1.3.3 and Section 1.3 of [Ber09].   **Q.E.D.**

An important property of the closure of a convex set C is that it does not "differ" much from C, in the sense that $\text{cl}(C)$ and C have the same relative interior. (This is not true for a nonconvex set; take for example the set of rational numbers.) The next proposition proves this property, together with some additional related facts.

Proposition B.8: (**Properties of Closure and Relative Interior**)

(a) The closure $\text{cl}(C)$ and the relative interior $\text{ri}(C)$ of a convex set C are convex. Furthermore $\text{ri}(\text{cl}(C)) = \text{ri}(C)$.

(b) For a convex set C, we have $\text{cl}(C) = \text{cl}(\text{ri}(C))$.

(c) Let C and $\bar{C}$ be nonempty convex sets. Then the following three conditions are equivalent:

(i) C and $\bar{C}$ have the same relative interior.

(ii) C and $\bar{C}$ have the same closure.

(iii) $\mathrm{ri}(C) \subset \bar{C} \subset \mathrm{cl}(C)$.

(d) The vector sum of two closed convex sets at least one of which is compact, is a closed convex set.

(e) The image of a convex and compact set under a linear transformation is a convex and compact set.

(f) The convex hull of a compact set is compact.

(g) If C_1 and C_2 are convex sets then

$$\mathrm{ri}(C_1 \times C_2) = \mathrm{ri}(C_1) \times \mathrm{ri}(C_2).$$

Moreover, if $\mathrm{ri}(C_1)$ and $\mathrm{ri}(C_2)$ have a nonempty intersection, then

$$\mathrm{ri}(C_1 + C_2) = \mathrm{ri}(C_1) + \mathrm{ri}(C_2), \quad \mathrm{ri}(C_1 \cap C_2) = \mathrm{ri}(C_1) \cap \mathrm{ri}(C_2).$$

Proof: See Section 1.3.1 of [Ber09]. **Q.E.D.**

An important property of real-valued convex functions over $\Re^n$ is that they are continuous. Extended real-valued convex functions also have interesting continuity properties; see [Ber09], Sections 1.3.2, 1.3.3, for a fuller account. We have the following proposition.

Proposition B.9: (Continuity of a Convex Function) If $f : \Re^n \mapsto \Re$ is convex, then it is continuous. More generally, if $C \subset \Re^n$ is convex and $f : C \mapsto \Re$ is convex, then f is continuous in the relative interior of C.

Proof: See Section 1.3.2 of [Ber09]. **Q.E.D.**

Another important fact is that in order for all of the level sets of a closed convex function to be compact, it is sufficient that one of its nonempty level sets be compact. This follows from the theory of directions of recession (the specialization to convex functions of the notions of asymptotic sequences and asymptotic directions of Section 3.1.2). This theory is developed in Sections 1.4 and 3.2 of [Ber09], but will not be needed in this book. The following proposition is sufficient for our purposes.

Sec. B.2   Hyperplanes

Proposition B.10: (Nonemptiness and Compactness of the Set of Minimizing Points)

(a) The set of minimizing points of a convex function $f : \Re^n \mapsto \Re$ over a closed convex set X is nonempty and compact if and only if all its level sets,

$$L_a = \{x \in X \mid f(x) \leq a\}, \qquad a \in \Re,$$

are compact.

(b) The set of minimizing points over a closed convex set X of a sum $f_1 + \cdots + f_m$, where $f_1, \ldots, f_m$ are real-valued convex functions on $\Re^n$, is nonempty and compact if either X is compact, or if one of the functions is coercive (for example it is positive definite quadratic).

Proof: See Section 1.4 and Prop. 3.2.3 of [Ber09]. **Q.E.D.**

B.2 HYPERPLANES

A *hyperplane* in $\Re^n$ is a set of the form $\{x \mid a'x = b\}$, where a is nonzero vector in $\Re^n$ and b is a scalar. If $\bar{x}$ is any vector in a hyperplane $H = \{x \mid a'x = b\}$, then we must have $a'\bar{x} = b$, so the hyperplane can be equivalently described as

$$H = \{x \mid a'x = a'\bar{x}\},$$

or

$$H = \bar{x} + \{x \mid a'x = 0\}.$$

Thus, H is an affine set that is parallel to the subspace $\{x \mid a'x = 0\}$. The vector a is orthogonal to this subspace, and consequently, a is called the *normal* vector of H; see Fig. B.1.

The sets

$$\{x \mid a'x \geq b\}, \qquad \{x \mid a'x \leq b\},$$

are called the *closed halfspaces* associated with the hyperplane (also referred to as the *positive and negative halfspaces*, respectively). The sets

$$\{x \mid a'x > b\}, \qquad \{x \mid a'x < b\},$$

are called the *open halfspaces* associated with the hyperplane.

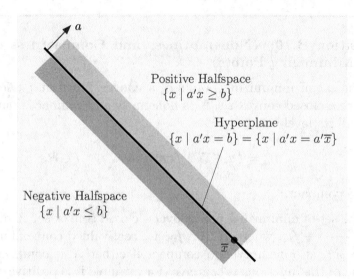

Figure B.1. Illustration of the hyperplane $H = \{x \mid a'x = b\}$. If $\bar{x}$ is any vector in the hyperplane, then the hyperplane can be equivalently described as

$$H = \{x \mid a'x = a'\bar{x}\} = \bar{x} + \{x \mid a'x = 0\}.$$

The hyperplane divides the space into two halfspaces as illustrated.

Proposition B.11: (Supporting Hyperplane Theorem) If $C \subset \Re^n$ is a convex set and $\bar{x}$ is a point that does not belong to the interior of C, there exists a vector $a \neq 0$ such that

$$a'x \geq a'\bar{x}, \qquad \forall\, x \in C.$$

Proof: See Prop. 1.5.1 of [Ber09]. **Q.E.D.**

Proposition B.12: (Separating Hyperplane Theorem) If C_1 and C_2 are two nonempty and disjoint convex subsets of $\Re^n$, there exists a hyperplane that separates them, i.e., a vector $a \neq 0$ such that

$$a'x_1 \leq a'x_2, \qquad \forall\, x_1 \in C_1,\ x_2 \in C_2.$$

Proof: See Prop. 1.5.2 of [Ber09]. **Q.E.D.**

> **Proposition B.13: (Strict Separation Theorem)** If C_1 and C_2 are two nonempty and disjoint convex sets such that C_1 is closed and C_2 is compact, there exists a hyperplane that strictly separates them, i.e., a vector $a \neq 0$ and a scalar b such that
>
> $$a'x_1 < b < a'x_2, \qquad \forall \; x_1 \in C_1, \; x_2 \in C_2.$$

Proof: See Prop. 1.5.3 of [Ber09]. **Q.E.D.**

The preceding proposition may be used to provide a fundamental characterization of closed convex sets, namely that *every closed convex set is the intersection of the halfspaces that contain it*. To see this, let C be the set at issue, and note that C is contained in the intersection of the halfspaces that contain C. To show the reverse inclusion, let $x \notin C$. Applying the Strict Separation Theorem (Prop. B.13) to the sets C and $\{x\}$, we see that there exists a halfspace containing C but not containing x. Hence, if $x \notin C$, then x cannot belong to the intersection of the halfspaces containing C, proving the result.

We finally provide a special type of separation theorem that is particularly useful in convex optimization. The proof is somewhat complicated, and can be found in [Roc70] (Ths. 11.3 and 20.2), in [BNO03] (Props. 2.4.6 and 3.5.1), and in [Ber09] (Props 1.5.6 and 1.5.7).

> **Proposition B.14: (Proper Separation)**
>
> (a) Let C_1 and C_2 be two nonempty convex subsets of $\Re^n$. There exists a hyperplane that separates C_1 and C_2, and does not contain both C_1 and C_2 if and only if
>
> $$\mathrm{ri}(C_1) \cap \mathrm{ri}(C_2) = \emptyset.$$
>
> (b) Let C and P be two nonempty convex subsets of $\Re^n$ such that P is the intersection of a finite number of closed halfspaces. There exists a hyperplane that separates C and P, and does not contain C if and only if
>
> $$\mathrm{ri}(C) \cap P = \emptyset.$$

Proof: See Props 1.5.6 and 1.5.7 of [Ber09]. **Q.E.D.**

B.3 CONES AND POLYHEDRAL CONVEXITY

We now develop some basic results regarding cones and polyhedral sets, in the context of the objectives of this book. A much broader discussion is found in Ch. 2 of [Ber09]. We introduce three important types of cones.

Given a cone C, the cone given by

$$C^\perp = \{y \mid y'x \leq 0, \ \forall \ x \in C\},$$

is called the *polar cone* of C. Note that the polar cone of a subspace is the orthogonal complement, illustrating that the notion of polarity may be viewed as a generalization of the notion of orthogonality.

A cone C is said to be *finitely generated*, if it has the form

$$C = \left\{ x \ \Big| \ x = \sum_{j=1}^{r} \mu_j a_j, \ \mu_j \geq 0, \ j = 1, \ldots, r \right\},$$

where $a_1, \ldots, a_r$ are some vectors.

A cone C is said to be *polyhedral*, if it has the form

$$C = \{x \mid a_j' x \leq 0, \ j = 1, \ldots, r\},$$

where $a_1, \ldots, a_r$ are some vectors.

It is straightforward to show that the polar cone of any cone, as well as all finitely generated and polyhedral cones are convex, by verifying the definition of convexity of Eq. (B.1). Furthermore, polar and polyhedral cones are closed, since they are intersections of closed halfspaces. Finitely generated cones are also closed as shown in part (b) of the following proposition, which also provides some additional important results.

Proposition B.15:

(a) (*Polar Cone Theorem*) For any nonempty closed convex cone C, we have $(C^\perp)^\perp = C$.

(b) Let $a_1, \ldots, a_r$ be vectors of $\Re^n$. Then the finitely generated cone

$$C = \left\{ x \ \Big| \ x = \sum_{j=1}^{r} \mu_j a_j, \ \mu_j \geq 0, \ j = 1, \ldots, r \right\}$$

is closed and its polar cone is the polyhedral cone given by

$$C^\perp = \{x \mid x' a_j \leq 0, \ j = 1, \ldots, r\}.$$

(c) (*Minkowski-Weyl Theorem*) A cone is polyhedral if and only if it is finitely generated.

Sec. B.3   Cones and Polyhedral Convexity

(d) (*Farkas' Lemma*) Let $x, e_1, \ldots, e_m$, and $a_1, \ldots, a_r$ be vectors of $\Re^n$. We have $x'y \leq 0$ for all vectors $y \in \Re^n$ such that

$$y'e_i = 0, \quad \forall\, i = 1, \ldots, m, \qquad y'a_j \leq 0, \quad \forall\, j = 1, \ldots, r,$$

if and only if x can be expressed as

$$x = \sum_{i=1}^{m} \lambda_i e_i + \sum_{j=1}^{r} \mu_j a_j,$$

where λ_i and μ_j are some scalars with $\mu_j \geq 0$ for all j.

Proof: See Props. 2.2.1, 2.3.1, and 2.3.2 of [Ber09].   **Q.E.D.**

Polyhedral Sets

A subset of $\Re^n$ is said to be a *polyhedral set* (or *polyhedron*) if it is nonempty and it is the intersection of a finite number of closed halfspaces, i.e., if it is of the form

$$P = \{x \mid a_j' x \leq b_j,\; j = 1, \ldots, r\},$$

where a_j are some vectors and b_j are some scalars.

The following is a fundamental result, showing that a polyhedral set can be represented as the sum of a finitely generated cone and the convex hull of a finite set of points. The proof is based on an interesting construction that can be used to translate results about polyhedral cones to results about polyhedral sets.

Proposition B.16: A set P is polyhedral if and only if there exist a nonempty and finite set of vectors $\{v_1, \ldots, v_m\}$, and a finitely generated cone C such that

$$P = \left\{ x \;\middle|\; x = y + \sum_{j=1}^{m} \mu_j v_j,\; y \in C,\; \sum_{j=1}^{m} \mu_j = 1,\; \mu_j \geq 0,\; j = 1, \ldots, m \right\}.$$

Proof: See Prop. 2.3.3 of [Ber09].   **Q.E.D.**

B.4 EXTREME POINTS AND LINEAR PROGRAMMING

A vector x is said to be an *extreme point* of a convex set C if x belongs to C and there do not exist vectors $y, z \in C$, and a scalar $\alpha \in (0, 1)$ such that

$$y \neq x, \qquad z \neq x, \qquad x = \alpha y + (1 - \alpha)z.$$

An equivalent definition is that x cannot be expressed as a convex combination of some vectors of C, all of which are different from x.

An important fact that forms the basis for the simplex method of linear programming, is that if a linear function f attains a minimum over a polyhedral set C having at least one extreme point, then f attains a minimum at some extreme point of C (as well as possibly at some other nonextreme points). We will prove this fact after considering the more general case where f is concave, and C is closed and convex. We first show a preliminary result.

Proposition B.17: Let C be a nonempty, closed, convex set in $\Re^n$.

(a) If H is a hyperplane that passes through a boundary point of C and contains C in one of its halfspaces, then every extreme point of $C \cap H$ is also an extreme point of C.

(b) C has at least one extreme point if and only if it does not contain a line, i.e., a set L of the form $L = \{x + \alpha d \mid \alpha \in \Re\}$ with $d \neq 0$.

Proof: (a) Let $\bar{x}$ be an element of T which is not an extreme point of C. Then we have $\bar{x} = \alpha y + (1 - \alpha)z$ for some $\alpha \in (0, 1)$, and some $y \in C$ and $z \in C$, with $y \neq x$ and $z \neq x$. Since $\bar{x} \in H$, $\bar{x}$ is a boundary point of C, and the halfspace containing C is of the form $\{x \mid a'x \geq a'\bar{x}\}$, where $a \neq 0$. Then $a'y \geq a'\bar{x}$ and $a'z \geq a'\bar{x}$, which in view of $\bar{x} = \alpha y + (1 - \alpha)z$, implies that $a'y = a'\bar{x}$ and $a'z = a'\bar{x}$. Therefore, $y \in T$ and $z \in T$, showing that $\bar{x}$ cannot be an extreme point of T.

(b) Assume that C has an extreme point x and contains a line $L = \{\bar{x} + \alpha d \mid \alpha \in \Re\}$, where $d \neq 0$. We will arrive at a contradiction. For each integer $n > 0$, the vector

$$x_n = \left(1 - \frac{1}{n}\right)x + \frac{1}{n}(\bar{x} + nd) = x + d + \frac{1}{n}(\bar{x} - x)$$

lies in the line segment connecting x and $\bar{x} + nd$, so it belongs to C. Since C is closed, $x + d = \lim_{n \to \infty} x_n$ must also belong to C. Similarly, we show that $x - d$ must belong to C. Thus $x - d$, x, and $x + d$ all belong to C, contradicting the hypothesis that x is an extreme point.

Conversely, we use induction on the dimension of the space to show that if C does not contain a line, it must have an extreme point. This is true in the real line $\Re^1$, so assume it is true in $\Re^{n-1}$. If a nonempty, closed, convex subset C of $\Re^n$ contains no line, it must have some boundary point $\bar{x}$. Take any hyperplane H passing through $\bar{x}$ and containing C in one of its halfspaces. Then, since H is an $(n-1)$-dimensional manifold, the set $C \cap H$ lies in an $(n-1)$-dimensional space and contains no line, so by the induction hypothesis, it must have an extreme point. By part (a), this extreme point must also be an extreme point of C. **Q.E.D.**

Proposition B.18: Let C be a convex subset of $\Re^n$, and let C^* be the set of minima of a concave function $f : C \mapsto \Re$ over C.

(a) If C^* contains a relative interior point of C, then f must be constant over C, i.e., $C^* = C$.

(b) If C is closed and contains at least one extreme point, and C^* is nonempty, then C^* contains some extreme point of C.

Proof: (a) Let x^* belong to $C^* \cap \mathrm{ri}(C)$, and let x be any vector in C. By the prolongation lemma of Prop. B.7(c), there exists a $\gamma > 1$ such that the vector
$$\hat{x} = x^* + (\gamma - 1)(x^* - x)$$
belongs to C, implying that
$$x^* = \frac{1}{\gamma}\hat{x} + \frac{\gamma - 1}{\gamma}x.$$
By the concavity of the function f, we have
$$f(x^*) \geq \frac{1}{\gamma}f(\hat{x}) + \frac{\gamma - 1}{\gamma}f(x),$$
and since $f(\hat{x}) \geq f(x^*)$ and $f(x) \geq f(x^*)$, we obtain
$$f(x^*) \geq \frac{1}{\gamma}f(\hat{x}) + \frac{\gamma - 1}{\gamma}f(x) \geq f(x^*).$$
Hence $f(x) = f(x^*)$.

(b) Let x^* minimize f over C. If $x^* \in \mathrm{ri}(C)$, by part (a), f must be constant over C, so it attains a minimum at an extreme point of C (since C has at least one extreme point by assumption). If $x^* \notin \mathrm{ri}(C)$, then by Prop. B.14(a), there exists a hyperplane H_1 properly separating x^* and C. Since $x^* \in C$, H_1 must contain x^*, so by the proper separation property, H_1

cannot contain C, and it follows that the intersection $C \cap H_1$ has dimension smaller than the dimension of C.

If $x^* \in \text{ri}(C \cap H_1)$, then f must be constant over $C \cap H_1$, so it attains a minimum at an extreme point of $C \cap H_1$ [since C contains an extreme point, it does not contain a line by Prop. B.17(b), and hence $C \cap H_1$ does not contain a line, which implies that $C \cap H_1$ has an extreme point]. By Prop. B.17(a), this optimal extreme point is also an extreme point of C. If $x^* \notin \text{ri}(C \cap H_1)$, there exists a hyperplane H_2 properly separating x^* and $C \cap H_1$. Again, since $x^* \in C \cap H_1$, H_2 contains x^*, so it cannot contain $C \cap H_1$, and it follows that the intersection $C \cap H_1 \cap H_2$ has dimension smaller than the dimension of $C \cap H_1$.

If $x^* \in \text{ri}(C \cap H_1 \cap H_2)$, then f must be constant over $C \cap H_1 \cap H_2$, etc. Since with each new hyperplane, the dimension of the intersection of C with the generated hyperplanes is reduced, this process will be repeated at most n times, until x^* is a relative interior point of some set $C \cap H_1 \cap \cdots \cap H_k$, at which time an extreme point of $C \cap H_1 \cap \cdots \cap H_k$ will be obtained. Through a reverse argument, repeatedly applying Prop. B.17(a), it follows that this extreme point is an extreme point of C. **Q.E.D.**

As a corollary we have the following:

Proposition B.19: Let C be a closed convex set and let $f : C \mapsto \Re$ be a concave function. Assume that for some invertible $n \times n$ matrix A and some $b \in \Re^n$ we have

$$Ax \geq b, \qquad \forall\, x \in C.$$

Then if f attains a minimum over C, it attains a minimum at some extreme point of C.

Proof: Consider the transformation $x = A^{-1}y$ and the problem of minimizing

$$h(y) = f(A^{-1}y)$$

over $Y = \{y \mid A^{-1}y \in C\}$. The function h is concave over the closed convex set Y. Furthermore, $y \geq b$ for all $y \in Y$, implying that Y does not contain a line, so that by Prop. B.17(b), Y contains an extreme point. If follows from Prop. B.18(b) that h attains a minimum at some extreme point y^* of Y. Then f attains its minimum over C at $x^* = A^{-1}y^*$, while x^* is an extreme point of C, since it can be verified that invertible transformations of sets map extreme points to extreme points. **Q.E.D.**

Extreme Points of Polyhedral Sets

We now consider a polyhedral set P and we characterize the set of its extreme points (also called *vertices*). By Prop. B.16, P can be represented as
$$P = C + \hat{P},$$
where C is a finitely generated cone C and $\hat{P}$ is the convex hull of some vectors $v_1, \ldots, v_m$:
$$\hat{P} = \left\{ x \,\Big|\, x = \sum_{j=1}^{m} \mu_j v_j, \ \sum_{j=1}^{m} \mu_j = 1, \ \mu_j \geq 0, \ j = 1, \ldots, m \right\}.$$
We note that an extreme point $\bar{x}$ of P cannot be of the form $\bar{x} = c + \hat{x}$, where $c \neq 0$, $c \in C$, and $\hat{x} \in \hat{P}$, since in this case $\bar{x}$ would be the midpoint of the line segment connecting the distinct vectors $\hat{x}$ and $2c + \hat{x}$. Therefore, an extreme point of P must belong to $\hat{P}$, and since $\hat{P} \subset P$, it must also be an extreme point of $\hat{P}$. An extreme point of $\hat{P}$ must be one of the vectors $v_1, \ldots, v_m$, since otherwise this point would be expressible as a convex combination of $v_1, \ldots, v_m$. Thus the set of extreme points of P is either empty or finite. Using Prop. B.17(b), it follows that *the set of extreme points of P is nonempty and finite if and only if P contains no line.*

If P is bounded, then we must have $P = \hat{P}$, and it can be shown that *P is equal to the convex hull of its extreme points* (not just the convex hull of the vectors $v_1, \ldots, v_m$). For a sketch of the proof note that if P is represented as
$$P = \operatorname{conv}(\{v_1, \ldots, v_m\}) + C,$$
where $v_1, \ldots, v_m$ are some vectors and C is a finitely generated cone (cf. Prop. B.16), then the set of extreme points of P is a subset of $\{v_1, \ldots, v_m\}$. The reason is that an extreme point $\bar{x}$ cannot be of the form $\bar{x} = \tilde{x} + y$, where $\tilde{x} \in \operatorname{conv}(\{v_1, \ldots, v_m\})$ and $y \neq 0$, $y \in C$, since in this case $\bar{x}$ would be the midpoint of the line segment connecting the distinct vectors $\tilde{x}$ and $\tilde{x} + 2y$. It thus follows that an extreme point must belong to $\operatorname{conv}(\{v_1, \ldots, v_m\})$.

The following proposition gives another and more specific characterization of extreme points of polyhedral sets, and is central in the theory of linear programming.

Proposition B.20: Let P be a polyhedral set in $\Re^n$.

(a) If P has the form
$$P = \{x \mid a_j'x \leq b_j, \ j = 1, \ldots, r\},$$
where a_j and b_j are given vectors and scalars, respectively, then a vector $v \in P$ is an extreme point of P if and only if the set

$$A_v = \{a_j \mid a_j'v = b_j, \, j \in \{1,\ldots,r\}\}$$

contains n linearly independent vectors.

(b) If P has the form

$$P = \{x \mid Ax = b, \, x \geq 0\},$$

where A is a given $m \times n$ matrix and b is a given vector, then a vector $v \in P$ is an extreme point of P if and only if the columns of A corresponding to the nonzero coordinates of v are linearly independent.

(c) *(Fundamental Theorem of Linear Programming)* Assume that P has at least one extreme point. Then if a linear function attains a minimum over P, it attains a minimum at some extreme point of P.

Proof: (a) If the set A_v contains fewer than n linearly independent vectors, then the system of equations

$$a_j'w = 0, \qquad \forall \, a_j \in A_v$$

has a nonzero solution $\bar{w}$. For sufficiently small $\gamma > 0$, we have $v + \gamma\bar{w} \in P$ and $v - \gamma\bar{w} \in P$, thus showing that v is not an extreme point. Thus, if v is an extreme point, A_v must contain n linearly independent vectors.

Conversely, suppose that A_v contains a subset $\bar{A}_v$ consisting of n linearly independent vectors. Suppose that for some $y \in P$, $z \in P$, and $\alpha \in (0,1)$, we have $v = \alpha y + (1-\alpha)z$. Then for all $a_j \in \bar{A}_v$, we have

$$b_j = a_j'v = \alpha a_j'y + (1-\alpha)a_j'z \leq \alpha b_j + (1-\alpha)b_j = b_j.$$

Thus v, y, and z are all solutions of the system of n linearly independent equations

$$a_j'w = b_j, \qquad \forall \, a_j \in \bar{A}_v.$$

Hence $v = y = z$, implying that v is an extreme point.

(b) Let k be the number of zero coordinates of v, and consider the matrix $\bar{A}$, which is the same as A except that the columns corresponding to the zero coordinates of v are set to zero. We write P in the form

$$P = \{x \mid Ax \leq b, \, -Ax \leq -b, \, -x \leq 0\},$$

and apply the result of part (a). We obtain that v is an extreme point if and only if $\bar{A}$ contains $n - k$ linearly independent rows, which is equivalent to the $n - k$ nonzero columns of $\bar{A}$ (corresponding to the nonzero coordinates of v) being linearly independent.

(c) Since P is polyhedral, it has a representation

$$P = \{x \mid Ax \geq b\},$$

for some $m \times n$ matrix A and some $b \in \Re^m$. If A had rank less than n, then its nullspace would contain some nonzero vector $\bar{x}$, so P would contain a line parallel to $\bar{x}$, contradicting the existence of an extreme point [cf. Prop. B.17(b)]. Thus A has rank n and hence it must contain n linearly independent rows that constitute an $n \times n$ invertible submatrix $\hat{A}$. If $\hat{b}$ is the corresponding subvector of b, we see that every $x \in P$ satisfies $\hat{A}x \geq \hat{b}$. The result then follows using Prop. B.19. **Q.E.D.**

B.5 DIFFERENTIABILITY ISSUES

Convex functions have interesting differentiability properties, which we discuss in this section. We first consider real-valued functions. Recall that the directional derivative of a function $f : \Re^n \mapsto \Re$ at a point $x \in \Re^n$ in the direction $y \in \Re^n$ is given by

$$f'(x;y) = \lim_{\alpha \downarrow 0} \frac{f(x + \alpha y) - f(x)}{\alpha},$$

provided that the limit exists, in which case we say that f is *directionally differentiable at x in the direction y*, and we call $f'(x;y)$ the *directional derivative of f at x in the direction y*. We say that f is *directionally differentiable at x* if it is directionally differentiable at x in all directions. Recall also that f is differentiable at x if it is directionally differentiable at x and $f'(x;y)$ is linear, as a function of y, of the form

$$f'(x;y) = \nabla f(x)'y,$$

where $\nabla f(x)$ is the gradient of f at x. It can be shown that if f is differentiable, then its gradient is continuous over $\Re^n$ (see [Ber15a], Exercise 3.4).

Given a convex function $f : \Re^n \mapsto \Re$, we say that a vector $d \in \Re^n$ is a *subgradient* of f at a point $x \in \Re^n$ if

$$f(z) \geq f(x) + (z - x)'d, \qquad \forall\, z \in \Re^n. \tag{B.8}$$

If instead f is a concave function, we say that d is a subgradient of f at x if $-d$ is a subgradient of the convex function $-f$ at x. The set of all subgradients of a convex (or concave) function f at $x \in \Re^n$ is called the *subdifferential* of f at x, and is denoted by $\partial f(x)$.

The next proposition clarifies the relationship between the directional derivative and the subdifferential, and provides some basic properties of subgradients.

Proposition B.21: Let $f : \Re^n \mapsto \Re$ be a convex function. For every $x \in \Re^n$, the following hold:

(a) A vector d is a subgradient of f at x if and only if

$$f'(x;y) \geq y'd, \quad \forall\, y \in \Re^n.$$

(b) The subdifferential $\partial f(x)$ is a nonempty, convex, and compact set, and there holds

$$f'(x;y) = \max_{d \in \partial f(x)} y'd, \quad \forall\, y \in \Re^n.$$

Furthermore, if X is a bounded set, the set $\cup_{x \in X} \partial f(x)$ is bounded.

(c) f is differentiable at x with gradient $\nabla f(x)$, if and only if it has $\nabla f(x)$ as its unique subgradient at x. Moreover, if f is differentiable over $\Re^n$, then $\nabla f(\cdot)$ is a continuous function.

(d) If a sequence $\{x_k\}$ converges to x and $d_k \in \partial f(x_k)$ for all k, the sequence $\{d_k\}$ is bounded and each of its limit points is a subgradient of f at x.

(e) If f is equal to the sum $f_1 + \cdots + f_m$ of convex functions $f_j : \Re^n \mapsto \Re$, $j = 1, \ldots, m$, then $\partial f(x)$ is equal to the vector sum $\partial f_1(x) + \cdots + \partial f_m(x)$.

(f) If f is equal to the composition of a convex function $h : \Re^m \mapsto \Re$ and an $m \times n$ matrix A [$f(x) = h(Ax)$], then $\partial f(x)$ is equal to $A'\partial h(Ax) = \{A'g \mid g \in \partial h(Ax)\}$.

(g) A vector $x^* \in X$ minimizes f over a convex set $X \subset \Re^n$ if and only if there exists a subgradient $d \in \partial f(x^*)$ such that

$$d'(z - x^*) \geq 0, \quad \forall\, z \in X.$$

Proof: See Props. 3.1.1-3.1.4, and Exercise 3.4 of [Ber15a]. **Q.E.D.**

Note that the necessary condition for optimality of part (g) of the preceding proposition generalizes the optimality condition of Section 1.1 for the case where f is differentiable:

$$\nabla f(x^*)'(z - x^*) \geq 0, \qquad \forall \ z \in X.$$

In the special case where $X = \Re^n$, we obtain a basic necessary and sufficient condition for unconstrained optimality of x^*, namely $0 \in \partial f(x^*)$. This optimality condition is also evident from the subgradient inequality (B.8).

A case of great interest in optimization involves functions of the form

$$f(x) = \max_{z \in Z} \phi(x, z).$$

The directional derivative and the subdifferential of f can be described in terms of the directional derivative and the subdifferential of ϕ, evaluated at points $\bar{z}$ where the maximum is attained, as shown by the following proposition.

Proposition B.22: (Danskin's Theorem) Let $Z \subset \Re^m$ be a compact set, and let $\phi : \Re^n \times Z \mapsto \Re$ be continuous and such that $\phi(\cdot, z) : \Re^n \mapsto \Re$ is convex for each $z \in Z$.

(a) The function $f : \Re^n \mapsto \Re$ given by

$$f(x) = \max_{z \in Z} \phi(x, z) \tag{B.9}$$

is convex and has directional derivative given by

$$f'(x; y) = \max_{z \in Z(x)} \phi'(x, z; y),$$

where $\phi'(x, z; y)$ is the directional derivative of the function $\phi(\cdot, z)$ at x in the direction y, and $Z(x)$ is the set of maximizing points in Eq. (B.9)

$$Z(x) = \left\{ \bar{z} \ \Big| \ \phi(x, \bar{z}) = \max_{z \in Z} \phi(x, z) \right\}.$$

In particular, if $Z(x)$ consists of a unique point $\bar{z}$ and $\phi(\cdot, \bar{z})$ is differentiable at x, then f is differentiable at x, and $\nabla f(x) = \nabla_x \phi(x, \bar{z})$, where $\nabla_x \phi(x, \bar{z})$ is the vector with coordinates

$$\frac{\partial \phi(x, \bar{z})}{\partial x_i}, \qquad i = 1, \ldots, n.$$

(b) If $\phi(\cdot, z)$ is differentiable for all $z \in Z$ and $\nabla_x \phi(x, \cdot)$ is continuous on Z for each x, then

$$\partial f(x) = \mathrm{conv}\{\nabla_x \phi(x, z) \mid z \in Z(x)\}, \qquad \forall\, x \in \Re^n. \qquad (\mathrm{B}.10)$$

In particular, if ϕ is linear in x for all $z \in Z$, i.e.,

$$\phi(x, z) = a_z' x + b_z, \qquad \forall\, z \in Z,$$

then

$$\partial f(x) = \mathrm{conv}\{a_z \mid z \in Z(x)\}.$$

Proof: See Prop. 4.5.1 of [BNO03] or Exercise 3.5 of [Ber15a] (with solution included). **Q.E.D.**

The preceding proposition derives its origin from a theorem by Danskin [Dan67] that provides a formula for the directional derivative of the maximum of a (not necessarily convex) directionally differentiable function. When adapted to a convex function f, this formula yields the expression (B.10) for $\partial f(x)$.

Another important subdifferential formula relates to the subgradients of an expected value function

$$f(x) = E\{F(x, \omega)\},$$

where ω is a random variable taking values in a set Ω, and $F(\cdot, \omega) : \Re^n \mapsto \Re$ is a real-valued convex function such that f is real-valued (note that f is easily verified to be convex). If ω takes a finite number of values with probabilities $p(\omega)$, then the formulas

$$f'(x; d) = E\{F'(x, \omega; d)\}, \qquad \partial f(x) = E\{\partial F(x, \omega)\}, \qquad (\mathrm{B}.11)$$

hold because they can be written in terms of finite sums as

$$f'(x; d) = \sum_{\omega \in \Omega} p(\omega) F'(x, \omega; d), \qquad \partial f(x) = \sum_{\omega \in \Omega} p(\omega) \partial F(x, \omega),$$

so Prop. B.21(e) applies. However, the formulas (B.11) hold even in the case where Ω is uncountably infinite, with appropriate mathematical interpretation of the integral of set-valued functions $E\{\partial F(x, \omega)\}$ as the set of integrals

$$\int_{\omega \in \Omega} g(x, \omega)\, dP(\omega), \qquad (\mathrm{B}.12)$$

where $g(x,\omega) \in \partial F(x,\omega)$, $\omega \in \Omega$ (measurability issues must be addressed in this context). For a formal proof and analysis, see the author's papers [Ber72], [Ber73], which also provide a necessary and sufficient condition for f to be differentiable, even when $F(\cdot,\omega)$ is not. In this connection, it is important to note that the integration over ω in Eq. (B.12) may smooth out the nondifferentiabilities of $F(\cdot,\omega)$ if ω is a "continuous" random variable. This property can be used in turn in algorithms, including schemes that bring to bear the methodology of differentiable optimization.

Subgradients of Extended Real-Valued Convex Functions

The notion of a subdifferential and a subgradient of a convex extended real-valued function $f : \Re^n \mapsto (-\infty, \infty]$ can be developed along the lines of the present section. In particular, a vector d is a subgradient of f at a vector x such that $f(x) < \infty$ if the subgradient inequality holds, i.e.,

$$f(z) \geq f(x) + (z-x)'d, \qquad \forall\, z \in \Re^n. \tag{B.13}$$

The subdifferential $\partial f(x)$ is the set of all subgradients of the convex function f. By convention, $\partial f(x)$ is considered empty for all x with $f(x) = \infty$.

Note that $\partial f(x)$ is always a closed set, since for any x with $f(x) < \infty$, it is the set of all d that lie in the intersection of the infinite collection of closed halfspaces defined by Eq. (B.13). However, contrary to the case of real-valued functions, $\partial f(x)$ may be empty, or closed but unbounded, even if $f(x) < \infty$. For example, the subdifferential of the extended real-valued convex function

$$f(x) = \begin{cases} -\sqrt{x} & \text{if } 0 \leq x \leq 1, \\ \infty & \text{otherwise,} \end{cases}$$

is given by

$$\partial f(x) = \begin{cases} -\frac{1}{2\sqrt{x}} & \text{if } 0 < x < 1, \\ [-1/2, \infty) & \text{if } x = 1, \\ \emptyset & \text{if } x \leq 0 \text{ or } 1 < x. \end{cases}$$

Thus, $\partial f(x)$ can be empty and can be unbounded at points x that belong to the effective domain of f (as in the cases $x = 0$ and $x = 1$, respectively, of the above example). However, it can be shown that $\partial f(x)$ is nonempty and compact at points x that are *interior* points of the effective domain of f, as also illustrated by the above example. Also $\partial f(x)$ is nonempty at points x that are *relative interior* points of the effective domain of f. These facts are shown in [Ber09], Prop. 5.4.1.

There are generalized versions of some of the preceding results within the context of extended real-valued convex functions, but with appropriate adjustments and additional assumptions to deal with cases where $\partial f(x)$ may be empty or noncompact. For example the sum differentiation formula

$$\partial(f_1 + \cdots + f_m)(x) = \partial f_1(x) + \cdots + \partial f_m(x)$$

[cf. Prop. B.21(e)] may fail even for x in the effective domain of $f_1+\cdots+f_m$; a condition such as that the relative interiors of the effective domains of the extended real-valued convex functions $f_1,\ldots,f_m$ have a point in common is necessary for the formula to hold for all $x \in \Re^n$ (see the books [Roc70] and [Ber09]). There is a similar result for the subdifferential of the composition $f(x) = h(Ax)$ [cf. Prop. B.21(f)], for the case where h is extended real-valued convex and A is a matrix: we have

$$\partial f(x) = A'\partial h(Ax), \qquad \forall\, x \in \Re^n,$$

if the range of A contains a point in the relative interior of $\mathrm{dom}(h)$.

Danskin's Theorem for Extended Real-Valued Convex Functions

Let us finally note an extension of Danskin's Theorem [Prop. B.22(b)], which provides a more general formula for the subdifferential $\partial f(x)$ of the function

$$f(x) = \sup_{z \in Z} \phi(x, z), \tag{B.14}$$

where Z is a compact set. This version of the theorem does not require that $\phi(\cdot, z)$ is differentiable. Instead it assumes that $\phi(\cdot, z)$ is an extended real-valued closed proper convex function for each $z \in Z$, that $\mathrm{int}\big(\mathrm{dom}(f)\big)$ [the interior of the set $\mathrm{dom}(f) = \{x \mid f(x) < \infty\}$] is nonempty, and that ϕ is continuous on the set $\mathrm{int}\big(\mathrm{dom}(f)\big) \times Z$. Then for all $x \in \mathrm{int}\big(\mathrm{dom}(f)\big)$, we have

$$\partial f(x) = \mathrm{conv}\big\{\partial \phi(x, z) \mid z \in Z(x)\big\},$$

where $\partial \phi(x, z)$ is the subdifferential of $\phi(\cdot, z)$ at x for any $z \in Z$, and $Z(x)$ is the set of maximizing points in Eq. (B.14); for a formal statement and proof of this result, see Prop. A.22 of the author's Ph.D. thesis, which may be found on-line [Ber71].

APPENDIX C:
Line Search Methods

In this appendix we describe algorithms for one-dimensional minimization. These are iterative algorithms, used to implement (approximately) the line minimization stepsize rules.

We briefly present three practical methods. The first two use polynomial interpolation, one requiring derivatives, the second only function values. The third, the Golden Section method, also requires just function values. By contrast with the interpolation methods, it does not depend on the existence of derivatives of the minimized function and may be applied even to discontinuous functions. Its validity depends, however, on a certain unimodality assumption.

In our presentation of the interpolation methods, we consider minimization of the function

$$g(\alpha) = f(x + \alpha d),$$

where f is continuously differentiable. By the chain rule, we have

$$g'(\alpha) = \frac{dg(\alpha)}{d\alpha} = \nabla f(x + \alpha d)'d.$$

We assume that

$$g'(0) = \nabla f(x)'d < 0,$$

i.e., that d is a descent direction at x. We give no convergence or rate of convergence results, but under some fairly natural assumptions, it can be shown that the interpolation methods converge superlinearly.

C.1 CUBIC INTERPOLATION

The cubic interpolation method successively determines at each iteration an appropriate interval $[a, b]$ within which a local minimum of g is guaranteed to exist. It then fits a cubic polynomial to the values $g(a)$, $g(b)$, $g'(a)$, $g'(b)$. The minimizing point $\bar{\alpha}$ of this cubic polynomial lies within $[a, b]$ and replaces one of the two points a or b for the next iteration.

Cubic Interpolation

Step 1: (Determination of the Initial Interval) Let $s > 0$ be some scalar. (Note: If d "approximates well" the Newton direction, then we take $s = 1$.) Evaluate $g(\alpha)$ and $g'(\alpha)$ at the points $\alpha = 0$, s, $2s$, $4s$, $8s$, ..., until two successive points a and b are found such that either $g'(b) \geq 0$ or $g(b) \geq g(a)$. Then, it can be seen that a local minimum of g exists within the interval $(a, b]$. [Note: If $g(s)$ is "much larger" than $g(0)$, it is advisable to replace s by βs, where $\beta \in (0, 1)$, for example $\beta = \frac{1}{2}$ or $\beta = \frac{1}{5}$, and repeat this step.] One can show that this step can be carried out if $\lim_{\alpha \to \infty} g(\alpha) > g(0)$.

Step 2: (Updating of the Current Interval) Given the current interval $[a, b]$, a cubic polynomial is fitted to the four values $g(a)$, $g'(a)$, $g(b)$, $g'(b)$. The cubic can be shown to have a unique minimum $\bar{\alpha}$ in the interval $(a, b]$ given by

$$\bar{\alpha} = b - \frac{g'(b) + w - z}{g'(b) - g'(a) + 2w}(b - a),$$

where

$$z = \frac{3(g(b) - g(a))}{b - a} + g'(a) + g'(b),$$

$$w = \sqrt{z^2 - g'(a)g'(b)}.$$

If $g'(\bar{\alpha}) \geq 0$ or $g(\bar{\alpha}) \geq g(a)$ replace b by $\bar{\alpha}$. If $g'(\bar{\alpha}) < 0$ and $g(\bar{\alpha}) < g(a)$ replace a by $\bar{\alpha}$. (Note: In practice the computation is terminated once the length of the current interval becomes smaller than a prespecified tolerance or else we obtain $\bar{\alpha} = b$.)

C.2 QUADRATIC INTERPOLATION

This method uses three points a, b, and c such that $a < b < c$, and $g(a) > g(b)$ and $g(b) < g(c)$. Such a set of points is referred to as a *three-point pattern*. It can be seen that a local minimum of g must lie between the extreme points a and c of a three-point pattern a, b, c. At each iteration, the method fits a quadratic polynomial to the three values $g(a)$, $g(b)$, and $g(c)$, and replaces one of the points a, b, and c by the minimizing point of this quadratic polynomial (see Fig. C.1).

Sec. C.2 Quadratic Interpolation

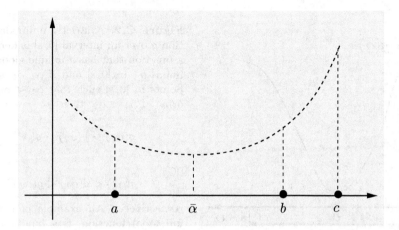

Figure C.1. A three-point pattern and the associated quadratic polynomial. If $\bar{\alpha}$ minimizes the quadratic, a new three point pattern is obtained using $\bar{\alpha}$ and two of the three points a, b, and c ($\bar{\alpha}$, and a, b in the example of the figure).

Quadratic Interpolation

Step 1: (Determination of Initial Three-Point Pattern) We search along the line as in the cubic interpolation method until we find three successive points a, b, and c with $a < b < c$ such that $g(a) > g(b)$ and $g(b) < g(c)$. As for the cubic interpolation method, we assume that this stage can be carried out, and we can show that this is guaranteed if $\lim_{\alpha \to \infty} g(\alpha) > g(0)$.

Step 2: (Updating the Current Three-Point Pattern) Given the current three-point pattern a, b, c, we fit a quadratic polynomial to the values $g(a)$, $g(b)$, and $g(c)$, and we determine its unique minimum $\bar{\alpha}$. It can be shown that $\bar{\alpha} \in (a, c)$ and that

$$\bar{\alpha} = \frac{1}{2} \frac{g(a)(c^2 - b^2) + g(b)(a^2 - c^2) + g(c)(b^2 - a^2)}{g(a)(c - b) + g(b)(a - c) + g(c)(b - a)}.$$

Then, we form a new three-point pattern as follows. If $\bar{\alpha} > b$, we replace a or c by $\bar{\alpha}$ depending on whether $g(\bar{\alpha} < g(b)$ or $g(\bar{\alpha}) > g(b)$, respectively. If $\bar{\alpha} < b$, we replace c or a by $\bar{\alpha}$ depending on whether $g(\bar{\alpha}) < g(b)$ or $g(\bar{\alpha}) > g(b)$, respectively. [Note: If $g(\bar{\alpha}) = g(b)$ then a special local search near $\bar{\alpha}$ should be conducted to replace $\bar{\alpha}$ by a point $\bar{\alpha}'$ with $g(\bar{\alpha}') \neq g(b)$. The computation is terminated when the length of the three-point pattern is smaller than a certain tolerance.]

An alternative possibility for quadratic interpolation is to determine the minimum $\bar{a}$ of the quadratic polynomial that has the same value as g at the points 0 and a, and the same first derivative as g at 0. It can be

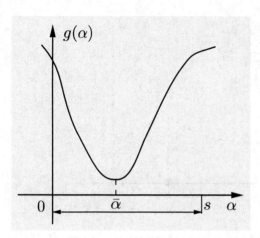

Figure C.2. A strictly unimodal function g over an interval $[0, s]$ is defined as a function that has a unique global minimum α^* in $[0, s]$ and if α_1, α_2 are two points in $[0, s]$ such that $\alpha_1 < \alpha_2 < \alpha^*$ or $\alpha^* < \alpha_1 < \alpha_2$, then

$$g(\alpha_1) > g(\alpha_2) > g(\alpha^*)$$

or

$$g(\alpha^*) < g(\alpha_1) < g(\alpha_2),$$

respectively. An example of a strictly unimodal function, is a function which is strictly convex over $[0, s]$.

verified that this minimum is given by

$$\bar{a} = \frac{g'(0)a^2}{2\bigl(g'(0)a + g(0) - g(a)\bigr)}.$$

C.3 THE GOLDEN SECTION METHOD

Here, we assume that $g(\alpha)$ is *strictly unimodal* in the interval $[0, s]$, as defined in Fig. C.2. The Golden Section method minimizes g over $[0, s]$ by determining at the kth iteration an interval $[\alpha_k, \bar{\alpha}_k]$ containing α^*. These intervals are obtained using the number

$$\tau = \frac{3 - \sqrt{5}}{2},$$

which satisfies $\tau = (1-\tau)^2$ and is related to the Fibonacci number sequence. The significance of this number will be seen shortly.

Initially, we take

$$[\alpha_0, \bar{\alpha}_0] = [0, s].$$

Given $[\alpha_k, \bar{\alpha}_k]$, we determine $[\alpha_{k+1}, \bar{\alpha}_{k+1}]$ so that $\alpha^* \in [\alpha_{k+1}, \bar{\alpha}_{k+1}]$ as follows. We calculate

$$b_k = \alpha_k + \tau(\bar{\alpha}_k - \alpha_k)$$
$$\bar{b}_k = \bar{\alpha}_k - \tau(\bar{\alpha}_k - \alpha_k)$$

and $g(b_k)$, $g(\bar{b}_k)$. Then:

(1) If $g(b_k) < g(\bar{b}_k)$ we set

$$\alpha_{k+1} = \alpha_k, \quad \bar{\alpha}_{k+1} = b_k \quad \text{if} \quad g(\alpha_k) \leq g(b_k)$$

Sec. C.3 The Golden Section Method 813

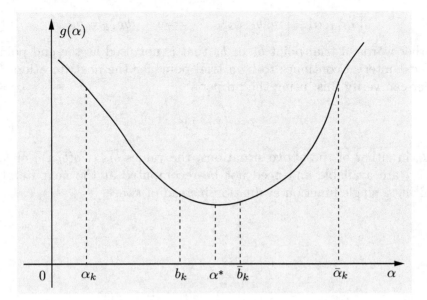

Figure C.3. Golden Section search. Given the interval $[\alpha_k, \bar{\alpha}_k]$ containing the minimum α^*, we calculate

$$b_k = \alpha_k + \tau(\bar{\alpha}_k - \alpha_k)$$

and

$$\bar{b}_k = \bar{\alpha}_k - \tau(\bar{\alpha}_k - \alpha_k).$$

The new interval $[\alpha_{k+1}, \bar{\alpha}_{k+1}]$ has either b_k or $\bar{b}_k$ as one of its endpoints.

$$\alpha_{k+1} = \alpha_k, \quad \bar{\alpha}_{k+1} = \bar{b}_k \quad \text{if} \quad g(\alpha_k) > g(b_k).$$

(2) If $g(b_k) > g(\bar{b}_k)$ we set

$$\alpha_{k+1} = \bar{b}_k, \quad \bar{\alpha}_{k+1} = \bar{\alpha}_k \quad \text{if} \quad g(\bar{b}_k) \geq g(\bar{\alpha}_k)$$

$$\alpha_{k+1} = b_k, \quad \bar{\alpha}_{k+1} = \bar{a}_k \quad \text{if} \quad g(\bar{b}_k) < g(\alpha_k).$$

(3) If $g(b_k) = g(\bar{b}_k)$ we set

$$\alpha_{k+1} = b_k, \quad \bar{\alpha}_{k+1} = \bar{b}_k.$$

Based on the definition of a strictly unimodal function it can be shown (see Fig. C.3) that the intervals $[\alpha_k, \bar{\alpha}_k]$ contain α^* and their lengths converge to zero. In practice, the computation is terminated once $(\bar{\alpha}_k - \alpha_k)$ becomes smaller than a prespecified tolerance.

An important fact, which rests on the choice of the particular number τ is that

$$[\alpha_{k+1}, \bar{\alpha}_{k+1}] = [\alpha_k, \bar{b}_k] \quad \Longrightarrow \quad \bar{b}_{k+1} = b_k,$$

$$[\alpha_{k+1}, \bar{\alpha}_{k+1}] = [b_k, \bar{\alpha}_k] \quad \Longrightarrow \quad b_{k+1} = \bar{b}_k.$$

In other words, a trial point b_k or $\bar{b}_k$ that is not used as the end point of the next interval continues to be a trial point for the next iteration. The reader can verify this, using the property

$$\tau = (1 - \tau)^2.$$

Thus, in either of the above situations, the values $\bar{b}_{k+1}$, $g(\bar{b}_{k+1})$ or b_{k+1}, $g(b_{k+1})$ are available and need not be recomputed at the next iteration, requiring a single function evaluation instead of two.

APPENDIX D:
Implementation of Newton's Method

In this appendix we describe a globally convergent version of Newton's method based on the modified Cholesky factorization approach discussed in Section 1.4. A computer code implementing the method can be freely obtained from the author's web page or through the book's web page.

D.1 CHOLESKY FACTORIZATION

We will give an algorithm for factoring a positive definite symmetric matrix A as

$$A = LL',$$

where L is lower triangular. This is the *Cholesky factorization*. Let a_{ij} be the elements of A and let A_i be the ith leading principal submatrix of A, i.e., the submatrix

$$A_i = \begin{bmatrix} a_{11} & a_{12} & \cdots & a_{1i} \\ a_{21} & a_{22} & \cdots & a_{2i} \\ \vdots & \vdots & \ddots & \vdots \\ a_{i1} & a_{i2} & \cdots & a_{ii} \end{bmatrix}.$$

It is seen that this submatrix is positive definite, since for any $y \in \Re_i$, $y \neq 0$, we have by the positive definiteness of A

$$y' A_i y = \begin{bmatrix} y' & 0 \end{bmatrix} A \begin{bmatrix} y \\ 0 \end{bmatrix} > 0.$$

The factorization of A is obtained by successive factorization of $A_1, A_2, \ldots$. Indeed we have $A_1 = L_1 L_1'$, where $L_1 = [\sqrt{a_{11}}]$. Suppose we have the Cholesky factorization of A_{i-1},

$$A_{i-1} = L_{i-1} L_{i-1}'. \tag{D.1}$$

Let us write

$$A_i = \begin{bmatrix} A_{i-1} & \beta_i \\ \beta_i' & a_{ii} \end{bmatrix}, \tag{D.2}$$

where β_i is the column vector

$$\beta_i = \begin{bmatrix} a_{1i} \\ \vdots \\ a_{i-1,i} \end{bmatrix}. \tag{D.3}$$

Based on Eqs. (D.1)-(D.3), it can be verified that

$$A_i = L_i L_i',$$

where

$$L_i = \begin{bmatrix} L_{i-1} & 0 \\ l_i' & \lambda_{ii} \end{bmatrix}, \tag{D.4}$$

and

$$l_i = L_{i-1}^{-1} \beta_i, \qquad \lambda_{ii} = \sqrt{a_{ii} - l_i' l_i}. \tag{D.5}$$

The scalar λ_{ii} is well defined because it can be shown that $a_{ii} - l_i' l_i > 0$. This is seen by defining $b = A_{i-1}^{-1} \beta_i$, and by using the positive definiteness of A_i to write

$$0 < \begin{bmatrix} b' & -1 \end{bmatrix} A_i \begin{bmatrix} b \\ -1 \end{bmatrix} = b' A_{i-1} b - 2 b' \beta_i + a_{ii}$$
$$= b' \beta_i - 2 b' \beta_i + a_{ii} = a_{ii} - b' \beta_i$$
$$= a_{ii} - \beta_i' A_{i-1}^{-1} \beta_i = a_{ii} - \beta_i' (L_{i-1} L_{i-1}')^{-1} \beta_i$$
$$= a_{ii} - (L_{i-1}^{-1} \beta_i)' (L_{i-1}^{-1} \beta_i) = a_{ii} - l_i' l_i.$$

The preceding construction can also be used to show that the Cholesky factorization is unique among factorizations involving lower triangular matrices with positive elements along the diagonal. Indeed, A_1 has a unique such factorization, and if A_{i-1} has a unique factorization $A_{i-1} = L_{i-1} L_{i-1}'$, then L_i is uniquely determined from the requirement $A_i = L_i L_i'$ with the diagonal elements of L_i positive, and Eqs. (D.4) and (D.5).

Cholesky Factorization by Columns

In the preceding algorithm, we calculate L by rows, i.e., we first calculate the first row of L, then the second row, etc. An alternative and equivalent method is to calculate L by columns, i.e., first calculate the first column of L, then the second column, etc. To see how this can be done, we note that the first column of A is equal to the first column of L multiplied with l_{11}, i.e.,

$$a_{i1} = l_{11} l_{i1}, \quad i = 1, \ldots, n,$$

from which we obtain

$$l_{11} = \sqrt{a_{11}},$$

$$l_{i1} = \frac{a_{i1}}{l_{11}}, \quad i = 2, \ldots, n.$$

Similarly, given columns $1, 2, \ldots, j-1$ of L, we equate the elements of the jth column of A with the corresponding elements of LL' and we obtain the elements of the jth column of L as follows:

$$l_{jj} = \sqrt{a_{jj} - \sum_{m=1}^{j-1} l_{jm}^2},$$

$$l_{ij} = \frac{a_{ij} - \sum_{m=1}^{j-1} l_{jm} l_{im}}{l_{jj}}, \quad i = j+1, \ldots, n.$$

D.2 APPLICATION TO A MODIFIED NEWTON METHOD

Consider now adding to A a diagonal correction E and simultaneously factoring the matrix

$$F = A + E,$$

where E is such that F is positive definite. The elements of E are introduced sequentially during the factorization process as some diagonal elements of the triangular factor are discovered, which are either negative or are close to zero, indicating that A is either not positive definite or is nearly singular. As discussed in Section 1.4, this is a principal method by which Newton's method is modified to enhance its global convergence properties. The precise mechanization is as follows:

We first fix positive scalars μ_1 and μ_2, where $\mu_1 < \mu_2$. We calculate the first column of the triangular factor L of F by

$$l_{11} = \begin{cases} \sqrt{a_{11}} & \text{if } \mu_1 < a_{11}, \\ \sqrt{\mu_2} & \text{otherwise}, \end{cases}$$

$$l_{i1} = \frac{a_{i1}}{l_{11}}, \qquad i = 2, \ldots, n.$$

Similarly, given columns $1, 2, \ldots, j-1$ of L, we obtain the elements of the jth column from the equations

$$l_{jj} = \begin{cases} \sqrt{a_{jj} - \sum_{m=1}^{j-1} l_{jm}^2} & \text{if } \mu_1 < a_{11} - \sum_{m=1}^{j-1} l_{jm}^2, \\ \sqrt{\mu_2} & \text{otherwise,} \end{cases}$$

$$l_{ij} = \frac{a_{ij} - \sum_{m=1}^{j-1} l_{jm} l_{im}}{l_{jj}}, \qquad i = j+1, \ldots, n.$$

In words, if the diagonal element of LL' comes out less than μ_1, we bring it up to μ_2.

Note that the jth diagonal element of the correction matrix E is equal to zero if $\mu_1 < a_{jj} - \sum_{m=1}^{j-1} l_{jm}^2$ and is equal to

$$\mu_2 - \left(a_{jj} - \sum_{m=1}^{j-1} l_{jm}^2 \right)$$

otherwise.

The preceding scheme can be used to modify Newton's method, where at the kth iteration, we add a diagonal correction Δ^k to the Hessian $\nabla^2 f(x^k)$ and simultaneously obtain the Cholesky factorization $L^k L^{k'}$ of $\nabla^2 f(x^k) + \Delta^k$ as described above. A modified Newton direction d^k is then obtained by first solving the triangular system

$$L^k y = -\nabla f(x^k),$$

and then solving the triangular system

$$L^{k'} d^k = y.$$

Solving the first system is called *forward elimination* and is accomplished in $O(n^2)$ arithmetic operations using the equations

$$y_1 = -\frac{\partial f(x^k)/\partial x_1}{l_{11}},$$

$$y_i = -\frac{\partial f(x^k)/\partial x_i + \sum_{m=1}^{i-1} l_{im} y^m}{l_{ii}}, \qquad i = 2, \ldots, n,$$

where l_{im} is the imth element of L^k. Solving the second system is called *back substitution* and is accomplished again in $O(n^2)$ arithmetic operations using the equations

$$d^n = \frac{y^n}{l_{nn}},$$

$$d_i = \frac{y_i - \sum_{m=i+1}^{n} l_{mi} d_m}{l_{ii}}, \qquad i = 1, \ldots, n-1.$$

The next point x^{k+1} is obtained from

$$x^{k+1} = x^k + \alpha^k d^k,$$

where α^k is chosen by the Armijo rule with unity initial step whenever the Hessian is not modified ($\Delta^k = 0$) and by means of a line minimization otherwise.

Assuming fixed values of μ_1 and μ_2, the following may be verified for the modified Newton's method just described:

(a) The algorithm is globally convergent in the sense that every limit point of $\{x^k\}$ is a stationary point of f. This can be shown using Prop. 1.2.1 in Section 1.2.

(b) For each local minimum x^* with positive definite Hessian, there exist scalars $\mu > 0$ and $\epsilon > 0$ such that if $\mu_1 \leq \mu$ and $\|x^0 - x^*\| \leq \epsilon$, then $x^k \to x^*$, $\Delta^k = 0$, and $\alpha^k = 1$ for all k. In other words if μ_1 is not chosen too large, the Hessian will never be modified near x^*, the method will be reduced to the pure form of Newton's method, and the convergence to x^* will be superlinear. The theoretical requirement that μ_1 be sufficiently small can be eliminated by making μ_1 dependent on the norm of the gradient (e.g. $\mu_1 = c\|\nabla f(x^k)\|$, where c is some positive scalar).

Practical Choice of Parameters and Stepsize Selection

We now address some practical issues. As discussed earlier, one should try to choose μ_1 small in order to avoid detrimental modification of the Hessian. Some trial and error with one's particular problem may be required here. As a practical matter, we recommend choosing initially $\mu_1 = 0$ and increasing μ_1 only if difficulties arise due to roundoff error or extremely large norm of calculated direction. (Choosing $\mu_1 = 0$, runs counter to our convergence theory because the generated directions are not guaranteed to be gradient related, but the practical consequences of this are typically insignificant.)

The parameter μ_2 should generally be chosen considerably larger than μ_1. It can be seen that choosing μ_2 very small can make the modified Hessian matrix $L^k L^{k'}$ nearly singular. On the other hand, choosing μ_2 very large has the effect of making nearly zero the coordinates of d^k that correspond to nonzero diagonal elements of the correction matrix Δ^k. Generally, some trial and error is necessary to determine a proper value of μ_2. A good guideline is to try a relatively small value of μ_2 and to increase μ_2 if the stepsize generated by the line minimization algorithm is substantially smaller than unity. The idea here is that small values of μ_2 tend to produce directions d^k with large value of norm and hence small values of stepsize. Thus a small value of stepsize indicates that μ_2 is chosen smaller than appropriate, and suggests that an increase of μ_2 is desirable. It is also possible to construct along these lines an adaptive scheme that changes the values of μ_1 and μ_2 in the course of the algorithm.

The following scheme to set and adjust μ_1 and μ_2 has worked well for the author. At each iteration k, we determine the maximal absolute diagonal element of the Hessian, i.e.,

$$w^k = \max\left\{\left|\frac{\partial^2 f(x^k)}{(x_1)^2}\right|, \ldots, \left|\frac{\partial^2 f(x^k)}{(x_n)^2}\right|\right\},$$

and we set μ_1 and μ_2 to

$$\mu_1 = r_1 w^k, \qquad \mu_2 = r_2 w^k.$$

The scalar r_1 is set at some "small" (or zero) value. The scalar r_2 is changed each time the Hessian is modified; it is multiplied by 5 if the stepsize obtained by the minimization rule is less than 0.2, and it is divided by 5 each time the stepsize is larger than 0.9.

Finally, regarding stepsize selection, any of a large number of possible line minimization algorithms can be used for those iterations where the Hessian is modified (in other iterations the Armijo rule with unity initial stepsize is used). One possibility is to use quadratic interpolation based on function values; see Section C.2 in Appendix C.

It is worth noting that if the cost function is quadratic, then it can be shown that a unity stepsize results in cost reduction for any values of μ_1 and μ_2. In other words if f is quadratic (not necessarily positive definite), we have

$$f\bigl(x^k - (F^k)^{-1}\nabla f(x^k)\bigr) \leq f(x^k),$$

where $F^k = \nabla^2 f(x^k) + \Delta^k$ and Δ^k is any positive definite matrix such that F^k is positive definite. As a result, a stepsize near unity is appropriate for initiating the line minimization algorithm. This fact can be used to guide the implementation of the line minimization routine.

References

[AFB06] Ahn, S., Fessler, J., Blatt, D., and Hero, A. O., 2006. "Convergent Incremental Optimization Transfer Algorithms: Application to Tomography," IEEE Transactions on Medical Imaging, Vol. 25, pp. 283-296.

[AHR93] Anstreicher, K. M., den Hertog, D., Roos, C., and Terlaky, T., 1993. "A Long Step Barrier Method for Convex Quadratic Programming," Algorithmica, Vol. 10, pp. 365-382.

[AHR97] Auslender, A., Cominetti, R., and Haddou, M., 1997. "Asymptotic Analysis for Penalty and Barrier Methods in Convex and Linear Programming," Math. Operations Res., Vol. 22, pp. 43-62.

[AHU58] Arrow, K. J., Hurwicz, L., and Uzawa, H., (Eds.), 1958. Studies in Linear and Nonlinear Programming, Stanford Univ. Press, Stanford, CA.

[AHU61] Arrow, K. J., Hurwicz, L., and Uzawa, H., 1961. "Constraint Qualifications in Maximization Problems," Naval Research Logistics Quarterly, Vol. 8, pp. 175-191.

[AMO91] Ahuja, R. K., Magnanti, T. L., and Orlin, J. B., 1991. "Some Recent Advances in Network Flows," SIAM Review, Vol. 33, pp. 175-219.

[AaL97] Aarts, E., and Lenstra, J. K., 1997. Local Search in Combinatorial Optimization, Wiley, N. Y.

[Aba67] Abadie, J., 1967. "On the Kuhn-Tucker Theorem," in Nonlinear Programming, Abadie, J., (Ed.), North Holland, Amsterdam.

[Ali92] Alizadeh, F., 1992. "Optimization over the Positive-Definite Cone: Interior Point Methods and Combinatorial Applications," in Pardalos, P., (Ed.), Advances in Optimization and Parallel Computing, North Holland, Amsterdam.

[Ali95] Alizadeh, F., 1995. "Interior-Point Methods in Semidefinite Programming with Applications in Combinatorial Applications," SIAM J. on Optimization, Vol. 5, pp. 13-51.

[AnH13] Andersen, M. S., and Hansen, P. C., 2013. "Generalized Row-Action Methods for Tomographic Imaging," Numerical Algorithms, Vol. 67, pp. 1-24.

[AnV94] Anstreicher, K. M., and Vial, J.-P., 1994. "On the Convergence of an Infeasible Primal-Dual Interior-Point Method for Convex Programming," Optimization Methods and Software, Vol. 3, pp. 273-283.

[Arm66] Armijo, L., 1966. "Minimization of Functions Having Continuous Partial Derivatives," Pacific J. Math., Vol. 16, pp. 1-3.

[Ash72] Ash, R. B., 1972. Real Analysis and Probability, Academic Press, N. Y.

[AtV95] Atkinson, D. S., and Vaidya, P. M., 1995. "A Cutting Plane Algorithm for Convex Programming that Uses Analytic Centers," Math. Programming, Vol. 69, pp. 1-44.

[AuC90] Auslender, A., and Cominetti, R., 1990. "First and Second Order Sensitivity Conditions," Optimization, Vol. 21, pp. 1-13.

[AuE76] Aubin, J. P., and Ekeland, I., 1976. "Estimates of the Duality Gap in Nonconvex Optimization," Math. Operations Res., Vol. 1, pp. 225-245.

[AuT03] Auslender, A., and Teboulle, M., 2003. Asymptotic Cones and Functions in Optimization and Variational Inequalities, Springer, N. Y.

[Aus76] Auslender, A., 1976. Optimization: Methodes Numeriques, Mason, Paris.

[Aus92] Auslender, A., 1992. "Asymptotic Properties of the Fenchel Dual Functional and Applications to Decomposition Properties," Vol. 73, pp. 427-449.

[Aus96] Auslender, A., 1996. "Non Coercive Optimization Problems," Math. of Operations Research, Vol. 21, pp. 769-782.

[Aus97] Auslender, A., 1997. "How to Deal with the Unbounded in Optimization: Theory and Algorithms," Math. Programing, Vol. 79, pp. 3-18.

[Avr76] Avriel, M., 1976. Nonlinear Programming: Analysis and Methods, Prentice-Hall, Englewood Cliffs, N. J.

[BCS99] Bonnans, J. F., Cominetti, R., and Shapiro, A., 1999. "Second Order Optimality Conditions Based on Parabolic Second Order Tangent Sets," SIAM J. on Optimization, Vol. 9, pp. 466-492.

[BGG84] Bertsekas, D. P., Gafni, E. M., and Gallager, R. G., 1984. "Second Derivative Algorithms for Minimum Delay Distributed Routing in Networks," IEEE Trans. on Communications, Vol. 32, pp. 911-919.

[BGI95] Burachik, R., Grana Drummond, L. M., Iusem, A. N., and Svaiter, B. F., 1995. "Full Convergence of the Steepest Descent Method with Inexact Line Searches," Optimization, Vol. 32, pp. 137-146.

[BGL06] Bonnans, J. F., Gilbert, J. C., Lemaréchal, C., and Sagastizábal, S. C., 2006. Numerical Optimization: Theoretical and Practical Aspects, Springer, N. Y.

[BGS72] Bazaraa, M. S., Goode, J. J., and Shetty, C. M., 1972. "Constraint Qualifications Revisited," Management Science, Vol. 18, pp. 567-573.

[BGT81] Bland, R. G., Goldfarb, D., and Todd, M. J., 1981. "The Ellipsoid Method: A Survey," Operations Research, Vol. 29, pp. 1039-91.

[BHG08] Blatt, D., Hero, A. O., Gauchman, H., 2008. "A Convergent Incremental Gradient Method with a Constant Step Size," SIAM J. Optimization, Vol. 18, pp. 29-51.

[BHT87] Bertsekas, D. P., Hossein, P., and Tseng, P., 1987. "Relaxation Methods for Network Flow Problems with Convex Arc Costs," SIAM J. on Control and Optimization, Vol. 25, pp. 1219-1243.

[BJS90] Bazaraa, M. S., Jarvis, J. J., and Sherali, H. D., 1990. Linear Programming and Network Flows, 2nd edition, Wiley, N. Y.

[BLY15] Bragin, M. A., Luh, P. B., Yan, J. H., Yu, N., and Stern, G. A., 2015. "Convergence of the Surrogate Lagrangian Relaxation Method," J. of Optimization Theory and Applications, Vol. 164, pp. 173-201.

[BMM95a] Ball, M. O., Magnanti, T. L., Monma, C. L., and Nemhauser, G. L., 1995. Network Models, Handbooks in OR and MS, Vol. 7, North-Holland, Amsterdam.

[BMM95b] Ball, M. O., Magnanti, T. L., Monma, C. L., and Nemhauser, G. L., 1995. Network Routing, Handbooks in OR and MS, Vol. 8, North-Holland, Amsterdam.

[BMR00] Birgin, E. G., Martinez, J. M., and Raydan, M., 2000. "Nonmonotone Spectral

Projected Gradient Methods on Convex Sets," SIAM J. on Optimization, Vol. 10, pp. 1196-1211.

[BMS99] Boltyanski, V., Martini, H., and Soltan, V., 1999. Geometric Methods and Optimization Problems, Kluwer, Boston.

[BMT90] Burke, J. V., Moré, J. J., and Toraldo, G., 1990. "Convergence Properties of Trust Region Methods for Linear and Convex Constraints," Math. Programming, Vol. 47, pp. 305-336.

[BNO03] Bertsekas, D. P., Nedić, A., and Ozdaglar, A. E., 2003. Convex Analysis and Optimization, Athena Scientific, Belmont, MA.

[BOT06] Bertsekas, D. P., Ozdaglar, A. E., and Tseng, P., 2006 "Enhanced Fritz John Optimality Conditions for Convex Programming," SIAM J. on Optimization, Vol. 16, pp. 766-797.

[BPC11] Boyd, S., Parikh, N., Chu, E., Peleato, B., and Eckstein, J., 2011. Distributed Optimization and Statistical Learning via the Alternating Direction Method of Multipliers, Now Publishers Inc, Boston, MA.

[BPP13] Bhatnagar, S., Prasad, H., and Prashanth, L. A., 2013. Stochastic Recursive Algorithms for Optimization, Lecture Notes in Control and Information Sciences, Springer, N. Y.

[BPT92] Bonnans, J. F., Panier, E. R., Tits, A. L., and Zhou, J. L., 1992. "Avoiding the Maratos Effect by Means of a Nonmonotone Line Search II. Inequality Constrained Problems – Feasible Iterates," SIAM J. Numer. Anal., Vol. 29, pp. 1187-1202.

[BPT97a] Bertsekas, D. P., Polymenakos, L. C., and Tseng, P., 1997. "An ϵ-Relaxation Method for Separable Convex Cost Network Flow Problems," SIAM J. on Optimization, Vol. 7, pp. 853-870.

[BPT97b] Bertsekas, D. P., Polymenakos, L. C., and Tseng, P., 1997. "Epsilon-Relaxation and Auction Methods for Separable Convex Cost Network Flow Problems," in Network Optimization, Pardalos, P. M., Hearn, D. W., and Hager, W. W., (Eds.), Lecture Notes in Economics and Mathematical Systems, Springer-Verlag, N. Y., pp. 103-126.

[BSL14] Bergmann, R., Steidl, G., Laus, F., and Weinmann, A., 2014. "Second Order Differences of Cyclic Data and Applications in Variational Denoising," arXiv preprint arXiv:1405.5349.

[BSS93] Bazaraa, M. S., Sherali, H. D., and Shetty, C. M., 1993. Nonlinear Programming Theory and Algorithms, 2nd edition, Wiley, N. Y.

[BST14] Bolte, J., Sabach, S., and Teboulle, M., 2014. "Proximal Alternating Linearized Minimization for Nonconvex and Nonsmooth Problems," Math. Programming, Vol. 146, pp. 1-36.

[BTW82] Boggs, P. T., Tolle, J. W., and Wang, P., 1982. "On the Local Convergence of Quasi-Newton Methods for Constrained Optimization," SIAM J. on Control and Optimization, Vol. 20, pp. 161-171.

[BaB88] Barzilai, J., and Borwein, J. M., 1988. "Two-Point Step Size Gradient Methods," IMA J. Numerical Analysis, Vol. 8, pp. 141-148.

[BaC11] Bauschke, H. H., and Combettes, P. L., 2011. Convex Analysis and Monotone Operator Theory in Hilbert Spaces, Springer, NY.

[BaD93] Barzilai, J., and Dempster, M. A. H., 1993. "Measuring Rates of Convergence of Numerical Algorithms," J. Opt. Theory and Appl., Vol. 78, pp. 109-125.

[BaL89] Bayer, D. A., and Lagarias, J. C., 1989. "The Nonlinear Geometry of Linear Programming. I. Affine and Projective Scaling Trajectories. II. Legendre Transform

Coordinates and Central Trajectories. III. Projective Legendre Transform Coordinates and Hilbert Geometry," Trans. Amer. Math. Soc., Vol. 314, pp. 499-581.

[BaT85] Balas, E., and Toth, P., 1985. "Branch and Bound Methods," in The Traveling Salesman Problem, Lawler, E., Lenstra, J. K., Rinnoy Kan, A. H. G., and Shmoys, D. B., (Eds.), Wiley, N. Y., pp. 361-401.

[BaW75] Balinski, M., and Wolfe, P., (Eds.), 1975. Nondifferentiable Optimization, Math. Programming Study 3, North-Holland, Amsterdam.

[Bac14] Bacak, M., 2014. "Computing Medians and Means in Hadamard Spaces," arXiv preprint arXiv:1210.2145v3.

[Bac16] Bacak, M., 2016. "A variational Approach to Stochastic Minimization of Convex Functionals," arXiv preprint arXiv:1605.03289.

[BeE88] Bertsekas, D. P., and Eckstein, J., 1988. "Dual Coordinate Step Methods for Linear Network Flow Problems," Math. Programming, Vol. 42, pp. 203-243.

[BeG82] Bertsekas, D. P., and Gafni, E., 1982. "Projection Methods for Variational Inequalities with Application to the Traffic Assignment Problem," Math. Programming Studies, Vol. 17, pp. 139-159.

[BeG83] Bertsekas, D. P., and Gafni, E., 1983. "Projected Newton Methods and Optimization of Multicommodity Flows," IEEE Trans. Automat. Control, Vol. AC-28, pp. 1090-1096.

[BeG92] Bertsekas, D. P., and Gallager, R. G., 1992. Data Networks, 2nd edition, Prentice-Hall, Englewood Cliffs, N. J.

[BeM71] Bertsekas, D. P, and Mitter, S. K., 1971. "Steepest Descent for Optimization Problems with Nondifferentiable Cost Functionals," Proc. 5th Annual Princeton Confer. Inform. Sci. Systems, Princeton, N. J., pp. 347-351.

[BeM73] Bertsekas, D. P., and Mitter, S. K., 1973. "A Descent Numerical Method for Optimization Problems with Nondifferentiable Cost Functionals," SIAM J. on Control, Vol. 11, pp. 637-652.

[BeN01] Ben-Tal, A., and Nemirovski, A., 2001. Lectures on Modern Convex Optimization: Analysis, Algorithms, and Engineering Applications, SIAM, Philadelphia.

[BeO02] Bertsekas, D. P., and Ozdaglar, A. E., 2002. "Pseudonormality and a Lagrange Multiplier Theory for Constrained Optimization," J. Opt. Th. and Appl., Vol. 114, pp. 287-343.

[BeS15] Beck, A., and Shtern, S., 2015. "Linearly Convergent Away-Step Conditional Gradient for Non-Strongly Convex Functions," arXiv preprint arXiv:1504.05002.

[BeT88] Bertsekas, D. P., and Tseng, P., 1988. "Relaxation Methods for Minimum Cost Ordinary and Generalized Network Flow Problems," Operations Research, Vol. 36, pp. 93-114.

[BeT89] Bertsekas, D. P., and Tsitsiklis, J. N., 1989. Parallel and Distributed Computation: Numerical Methods, Prentice-Hall, Englewood Cliffs, N. J; republished by Athena Scientific, Belmont, MA, 1997.

[BeT91] Bertsekas, D. P., and Tsitsiklis, J. N., 1991. "Some Aspects of Parallel and Distributed Iterative Algorithms - A Survey," Automatica, Vol. 27, pp. 3-21.

[BeT94] Bertsekas, D. P., and Tseng, P., 1994. "Partial Proximal Minimization Algorithms for Convex Programming," SIAM J. on Optimization, Vol. 4, pp. 551-572.

[BeT96] Bertsekas, D. P., and Tsitsiklis, J. N., 1996. Neuro-Dynamic Programming, Athena Scientific, Belmont, MA.

[BeT97] Bertsimas, D., and Tsitsiklis, J. N., 1997. Introduction to Linear Optimization, Athena Scientific, Belmont, MA.

[BeT00] Bertsekas, D. P., and Tsitsiklis, J. N., 2000. "Gradient Convergence of Gradient Methods with Errors," SIAM J. on Optimization, Vol. 36, pp. 627-642.

[BeT07] Bertsekas, D. P., and Tseng, P., 2007. "Set Intersection Theorems and Existence of Optimal Solutions," Mathematical Programming, Vol. 110, pp. 287-314.

[BeT08] Bertsekas, D. P., and Tsitsiklis, J. N., 2008. Introduction to Probability, 2nd Edition, Athena Scientific, Belmont, MA.

[BeT13] Beck, A., and Tetruashvili, L., 2013. "On the Convergence of Block Coordinate Descent Type Methods," SIAM J. on Optimization, Vol. 23, pp. 2037-2060.

[BeY09] Bertsekas, D. P., and Yu, H., 2009. "Projected Equation Methods for Approximate Solution of Large Linear Systems," J. of Computational and Applied Mathematics, Vol. 227, pp. 27-50.

[BeY10] Bertsekas, D. P., and Yu, H., 2010. "Asynchronous Distributed Policy Iteration in Dynamic Programming," Proc. of Allerton Conf. on Communication, Control and Computing, Allerton Park, Ill, pp. 1368-1374.

[BeY11] Bertsekas, D. P., and Yu, H., 2011. "A Unifying Polyhedral Approximation Framework for Convex Optimization," SIAM J. on Optimization, Vol. 21, pp. 333-360.

[BeZ82] Ben-Tal, A., and Zowe, J., 1982. "A Unified Theory of First and Second-Order Conditions for Extremum Problems in Topological Vector Spaces," Math. Programming Studies, Vol. 19, pp. 39-76.

[BeZ97] Ben-Tal, A., and Zibulevsky, M., 1997. "Penalty/Barrier Multiplier Methods for Convex Programming Problems," SIAM J. on Optimization, Vol. 7, pp. 347-366.

[Ben62] Benders, J. F., 1962. "Partitioning Procedures for Solving Mixed Variables Programming Problems," Numer. Math., Vol. 4, pp. 238-252.

[Ber71] Bertsekas, D. P., 1971. "Control of Uncertain Systems With a Set-Membership Description of the Uncertainty," Ph.D. Thesis, Dept. of EECS, MIT; may be downloaded from http://web.mit.edu/dimitrib/www/publ.html.

[Ber72] Bertsekas, D. P., 1972. "Stochastic Optimization Problems with Nondifferentiable Cost Functionals with an Application in Stochastic Programming," Proc. 1972 IEEE Conf. Decision and Control, pp. 555-559.

[Ber73] Bertsekas, D. P., 1973. "Stochastic Optimization Problems with Nondifferentiable Cost Functionals," J. of Optimization Theory and Applications, Vol. 12, pp. 218-231.

[Ber74] Bertsekas, D. P., 1974. "Partial Conjugate Gradient Methods for a Class of Optimal Control Problems," IEEE Trans. Automat. Control, Vol. 19, pp. 209-217.

[Ber75a] Bertsekas, D. P., 1975. "Necessary and Sufficient Conditions for a Penalty Method to be Exact," Math. Programming, Vol. 9, pp. 87-99.

[Ber75b] Bertsekas, D. P., 1975. "Combined Primal-Dual and Penalty Methods for Constrained Optimization, SIAM J. on Control, Vol. 13, pp. 521-544.

[Ber75c] Bertsekas, D. P., 1975. "Nondifferentiable Optimization Via Approximation," Math. Programming Study 3, Balinski, M., and Wolfe, P., (Eds.), North-Holland, Amsterdam, pp. 1-25.

[Ber75d] Bertsekas, D. P., 1975. "On the Method of Multipliers for Convex Programming," IEEE Transactions on Aut. Control, Vol. 20, pp. 385-388.

[Ber76a] Bertsekas, D. P., 1976. "On Penalty and Multiplier Methods for Constrained Optimization," SIAM J. on Control and Optimization, Vol. 14, pp. 216-235.

[Ber76b] Bertsekas, D. P., 1976. "Multiplier Methods: A Survey," Automatica, Vol. 12, pp. 133-145.

[Ber76c] Bertsekas, D. P., 1976. "On the Goldstein-Levitin-Poljak Gradient Projection Method," IEEE Trans. Automat. Control, Vol. 21, pp. 174-184.

[Ber77] Bertsekas, D. P., 1977. "Approximation Procedures Based on the Method of Multipliers," J. Opt. Th. and Appl., Vol. 23, pp. 487-510.

[Ber78] Bertsekas, D. P., 1978. "Local Convex Conjugacy and Fenchel Duality," Preprints of Triennial World Congress of IFAC, Helsinki, Vol. 2, pp. 1079-1084.

[Ber79a] Bertsekas, D. P., 1979. "Convexification Procedures and Decomposition Algorithms for Large-Scale Nonconvex Optimization Problems," J. Opt. Th. and Appl., Vol. 29, pp. 169-197.

[Ber79b] Bertsekas, D. P., 1979. "A Distributed Algorithm for the Assignment Problem," Lab. for Information and Decision Systems Working Paper, M.I.T.

[Ber80a] Bertsekas, D. P., 1980. "A Class of Optimal Routing Algorithms for Communication Networks," Proc. of the Fifth International Conference on Computer Communication, Atlanta, Ga., pp. 71-76.

[Ber80b] Bertsekas, D. P., 1980. "Variable Metric Methods for Constrained Optimization Based on Differentiable Exact Penalty Functions," Proc. Allerton Conference on Communication, Control, and Computation, Allerton Park, Ill., pp. 584-593.

[Ber81] Bertsekas, D. P., 1981. "A New Algorithm for the Assignment Problem," Math. Programming, Vol. 21, pp. 152-171.

[Ber82a] Bertsekas, D. P., 1982. Constrained Optimization and Lagrange Multiplier Methods, Academic Press, N. Y.; republished by Athena Scientific, Belmont, MA, 1997.

[Ber82b] Bertsekas, D. P., 1982. "Projected Newton Methods for Optimization Problems with Simple Constraints," SIAM J. on Control and Optimization, Vol. 20, pp. 221-246.

[Ber82c] Bertsekas, D. P., 1982. "Enlarging the Region of Convergence of Newton's Method for Constrained Optimization," J. Opt. Th. and Appl., Vol. 36, pp. 221-252.

[Ber82d] Bertsekas, D. P., 1982. "Distributed Dynamic Programming," IEEE Trans. Aut. Control, Vol. AC-27, pp. 610-616.

[Ber82e] Bertsekas, D. P., 1982. "Notes on Nonlinear Programming and Discrete-Time Optimal Control," Lab. for Information and Decision Systems Report LIDS-P-919, MIT.

[Ber83] Bertsekas, D. P., 1983. "Distributed Asynchronous Computation of Fixed Points," Math. Programming, Vol. 27, pp. 107-120.

[Ber85] Bertsekas, D. P., 1985. "A Unified Framework for Minimum Cost Network Flow Problems," Math. Programming, Vol. 32, pp. 125-145.

[Ber86] Bertsekas, D. P., 1986. "Distributed Relaxation Methods for Linear Network Flow Problems," Proceedings of 25th IEEE Conference on Decision and Control, pp. 2101-2106.

[Ber91] Bertsekas, D. P., 1991. Linear Network Optimization: Algorithms and Codes, M.I.T. Press, Cambridge, MA.

[Ber92] Bertsekas, D. P., 1992. "Auction Algorithms for Network Problems: A Tutorial Introduction," Computational Optimization and Applications, Vol. 1, pp. 7-66.

[Ber96a] D. P. Bertsekas, 1996. "Thevenin Decomposition and Network Optimization," J. Opt. Theory and Appl., Vol. 89, pp. 1-15.

[Ber96b] Bertsekas, D. P., 1996. "Incremental Least Squares Methods and the Extended Kalman Filter," SIAM J. on Optimization, Vol. 6, pp. 807-822.

[Ber97] Bertsekas, D. P., 1997. "A New Class of Incremental Gradient Methods for Least Squares Problems," SIAM J. on Optimization, Vol. 7, pp. 913-926.

[Ber98] Bertsekas, D. P., 1998. Network Optimization: Continuous and Discrete Models, Athena Scientific, Belmont, MA.

[Ber99] Bertsekas, D. P., 1999. "A Note on Error Bounds for Convex and Nonconvex Problems," Computational Optimization and Applications, Vol. 12, pp. 41-51.

[Ber05a] Bertsekas, D. P., 2005. Dynamic Programming and Optimal Control, 3rd Edition, Vol. I, Athena Scientific, Belmont, MA.

[Ber05b] Bertsekas, D. P., 2005. "Dynamic Programming and Suboptimal Control: A Survey from ADP to MPC," Fundamental Issues in Control, Special Issue for the CDC-ECC-05, European J. of Control, Vol. 11, Nos. 4-5.

[Ber05c] Bertsekas, D. P., 2005. "Lagrange Multipliers with Optimal Sensitivity Properties in Constrained Optimization," Lab. for Information and Decision Systems Report 2632, MIT; in Proc. of the 2004 Erice Workshop on Large Scale Nonlinear Optimization, Erice, Italy, Kluwer.

[Ber09] Bertsekas, D. P., 2009. Convex Optimization Theory, Athena Scientific, Belmont, MA.

[Ber10a] Bertsekas, D. P., 2010. "Extended Monotropic Programming and Duality," Lab. for Information and Decision Systems Report LIDS-P-2692, MIT, March 2006, corrected in Feb. 2010.

[Ber10b] Bertsekas, D. P., 2010. "Incremental Gradient, Subgradient, and Proximal Methods for Convex Optimization: A Survey," Lab. for Information and Decision Systems Report LIDS-P-2848, MIT.

[Ber11a] Bertsekas, D. P., 2011. "Incremental Proximal Methods for Large Scale Convex Optimization," Math. Programming, Vol. 129, pp. 163-195.

[Ber11b] Bertsekas, D. P., 2011. "Centralized and Distributed Newton Methods for Network Optimization and Extensions," Lab. for Information and Decision Systems Report LIDS-P-2866, MIT.

[Ber12] Bertsekas, D. P., 2012. Dynamic Programming and Optimal Control: Approximate Dynamic Programming, 4th Edition, Vol. II, Athena Scientific, Belmont, MA.

[Ber13] Bertsekas, D. P., 2013. Abstract Dynamic Programming, Athena Scientific, Belmont, MA.

[Ber15a] Bertsekas, D. P., 2015. Convex Optimization Algorithms, Athena Scientific, Belmont, MA.

[Ber15b] Bertsekas, D. P., 2015. "Incremental Aggregated Proximal and Augmented Lagrangian Algorithms," Lab. for Information and Decision Systems Report LIDS-P-3176, MIT, September 2015.

[BiL97] Birge, J. R., and Louveaux, 1997. Introduction to Stochastic Programming, Springer-Verlag, New York, N. Y.

[Bia15] Bianchi, P., 2015. "Ergodic Convergence of a Stochastic Proximal Point Algorithm," arXiv preprint arXiv:1504.05400.

[Bis95] Bishop, C. M, 1995. Neural Networks for Pattern Recognition, Oxford University Press, N. Y.

[BoL00] Borwein, J. M., and Lewis, A. S., 2000. Convex Analysis and Nonlinear Optimization, Springer-Verlag, N. Y.

[BoL05] Bottou, L., and LeCun, Y., 2005. "On-Line Learning for Very Large Datasets," Applied Stochastic Models in Business and Industry, Vol. 21, pp. 137-151.

[BoS00] Bonnans, J. F., and Shapiro, A., 2000. Perturbation Analysis of Optimization Problems, Springer-Verlag, N. Y.

[BoT80] Boggs, P. T., and Tolle, J. W., 1980. "Augmented Lagrangians which are Quadratic in the Multiplier," J. Opt. Th. and Appl., Vol. 31, pp. 17-26.

[BoV04] Boyd, S., and Vandenbergue, L., 2004. Convex Optimization, Cambridge Univ. Press, Cambridge, U.K.

[Bog90] Bogart, K. P., 1990. Introductory Combinatorics, Harcourt Brace Jovanovich, Inc., New York, N. Y.

[Bon89a] Bonnans, J. F., 1989. "A Variant of a Projected Variable Metric Method for Bound Constrained Optimization Problems," Report, INRIA, France.

[Bon89b] Bonnans, J. F., 1989. "Asymptotic Admissibility of the Unit Stepsize in Exact Penalty Methods," SIAM J. on Control and Optimization, Vol. 27, pp. 631-641.

[Bon92] Bonnans, J. F., 1992. "Directional Derivatives of Optimal Solutions in Smooth Nonlinear Programming," J. Opt. Theory and Appl., Vol. 73, pp. 27-45.

[Bon94] Bonnans, J. F., 1994. "Local Analysis of Newton Type Methods for Variational Inequalities and Nonlinear Programming," J. Applied Math. Optimization, Vol. 29, pp. 161-186.

[Bor08] Borkar, V. S., 2008. Stochastic Approximation: A Dynamical Systems Viewpoint, Cambridge Univ. Press.

[Brä93] Brännlund, U., 1993. "On Relaxation Methods for Nonsmooth Convex Optimization," Doctoral Thesis, Royal Institute of Technology, Stockholm, Sweden.

[Bro70] Broyden, C. G., 1970. "The Convergence of a Class of Double Rank Minimization Algorithms," J. Inst. Math. Appl., Vol. 6, pp. 76-90.

[BuM88] Burke, J. V., and Moré, J. J., 1988. "On the Identification of Active Constraints," SIAM J. Numer. Anal., Vol. 25, pp. 1197-1211.

[BuQ98] Burke, J. V., and Qian, M., 1998. "A Variable Metric Proximal Point Algorithm for Monotone Operators," SIAM J. on Control and Optimization, Vol. 37, pp. 353-375.

[CCP70] Canon, M. D., Cullum, C. D., and Polak, E., 1970. Theory of Optimal Control and Mathematical Programming, McGraw-Hill, N. Y.

[CCP98] Cook, W., Cunningham, W., Pulleyblank, W., and Schrijver, A., 1998. Combinatorial Optimization, Wiley, N. Y.

[CFM75] Camerini, P. M., Fratta, L., and Maffioli, F., 1975. "On Improving Relaxation Methods by Modified Gradient Techniques," Math. Programming Studies, Vol. 3, pp. 26-34.

[CGT91] Conn, A. R., Gould, N. I. M., and Toint, P. L., 1991. "A Globally Convergent Augmented Lagrangian Algorithm for Optimization with General Constraints and Simple Bounds," SIAM J. Numer. Anal., Vol. 28, pp. 545-572.

[CGT92] Conn, A. R., Gould, N. I. M., and Toint, P. L., 1992. "LANCELOT: A FORTRAN Package for Large-Scale Nonlinear Optimization," Springer-Verlag, N. Y.

[CGT00] Conn, A. R., Gould, N. I. M., and Toint, P. L., 2000. Trust Region Methods, SIAM, Philadelphia, PA.

[CHY16] Chen, C., He, B., Ye, Y., and Yuan, X., 2016. "The Direct Extension of ADMM for Multi-Block Convex Minimization Problems is not Necessarily Convergent," Math. Programming, Series A, Vol. 155, pp. 57-79.

[CPS92] Cottle, R., Pang, J. S., and Stone, R. E., 1992. The Linear Complementarity Problem, Academic Press, Boston.

[CPS11] Choi, S. C. T., Paige, C. C., and Saunders, M. A., 2011. "MINRES-QLP: A Krylov Subspace Method for Indefinite or Singular Symmetric Systems," SIAM Journal on Scientific Computing, Vol. 33, pp. 1810-1836.

[CaC68] Canon, M. D., and Cullum, C. D., 1968. "A Tight Upper Bound on the Rate of Convergence of the Frank-Wolfe Algorithm," SIAM J. on Control, Vol. 6, pp. 509-516.

[CaF97] Caprara, A., and Fischetti, M., 1997. "Branch and Cut Algorithms," in Annotated Bibliographies in Combinatorial Optimization, Dell'Amico, M., Maffioli, F., and Martello, S., (Eds.), Wiley, Chisester, Chapter 4.

[CaG74] Cantor, D. G., Gerla, M., 1974. "Optimal Routing in Packet Switched Computer Networks," IEEE Trans. on Computing, Vol. C-23, pp. 1062-1068.

[CaM87] Calamai, P. H., and Moré, J. J., 1987. "Projected Gradient Methods for Linearly Constrained Problems," Math. Programming, Vol. 39, pp. 98-116.

[Cam94] Cameron. (book on integer progr) ADD

[Car61] Carroll, C. W., 1961. "The Created Response Surface Technique for Optimizing Nonlinear Restrained Systems," Operations Research, Vol. 9, pp. 169-184.

[Cau47] Cauchy, M. A., 1847. "Analyse Mathématique–Méthode Générale Pour La Résolution des Systémes d' Equations Simultanées," Comptes Vendus Acad. Sc., Paris.

[CeH87] Censor, Y., and Herman, G. T., 1987. "On Some Optimization Techniques in Image Reconstruction from Projections," Applied Numer. Math., Vol. 3, pp. 365-391.

[CeZ92] Censor, Y., and Zenios, S. A., 1992. "The Proximal Minimization Algorithm with D-Functions," J. Opt. Theory and Appl., Vol. 73, pp. 451-464.

[CeZ97] Censor, Y., and Zenios, S. A., 1997. Parallel Optimization: Theory, Algorithms, and Applications, Oxford University Press, N. Y.

[ChG59] Cheney, E. W., and Goldstein, A. A., 1959. "Newton's Method for Convex Programming and Tchebycheff Approximation," Numer. Math., Vol. I, pp. 253-268.

[ChT93] Chen, G., and Teboulle, M., 1993. "Convergence Analysis of a Proximal-Like Minimization Algorithm Using Bregman Functions," SIAM J. on Optimization, Vol. 3, pp. 538-543.

[ChT94] Chen, G., and Teboulle, M., 1994. "A Proximal-Based Decomposition Method for Convex Minimization Problems," Math. Programming, Vol. 64, pp. 81-101.

[Cla83] Clarke, F. H., 1983. Nonsmooth Analysis and Optimization, Wiley-Interscience, N. Y.

[CoC82a] Coleman, T. F., and Conn, A. R., 1982. "Nonlinear Programming Via an Exact Penalty Function: Asymptotic Analysis," Math. Progamming, Vol. 24, pp. 123-136.

[CoC82b] Coleman, T. F., and Conn, A. R., 1982. "Nonlinear Programming Via an Exact Penalty Function: Global Analysis," Math. Programming, Vol. 24, pp. 137-161.

[CoL94] Correa, R., and Lemarechal, C., 1994. "Convergence of Some Algorithms for Convex Minimization," Math. Programming, Vol. 62, pp. 261-276.

[CoT13] Couellan, N. P., and Trafalis, T. B., 2013. "On-line SVM Learning via an Incremental Primal-Dual Technique," Optimization Methods and Software, Vol. 28, pp. 256-275.

[Coh80] Cohen, G., 1980. "Auxiliary Problem Principle and Decomposition of Optimization Problems," J. Opt. Theory and Appl., Vol. 32, pp. 277-305.

[Cro58] Croes, G. A., 1958. "A Method for Solving Traveling Salesman Problems," Operations Research, Vol. 6, pp. 791-812.

[Cry71] Cryer, C. W., 1971. "The Solution of a Quadratic Programming Problem Using Systematic Overrelaxation," SIAM J. on Control, Vol. 9, pp. 385-392.

[Cul71] Cullum, J., 1971. "An Explicit Procedure for Discretizing Continuous Optimal Control Problems," J. Opt. Theory and Appl., Vol. 8, pp. 15-34.

[Cyb89] Cybenko, 1989. "Approximation by Superpositions of a Sigmoidal Function," Math. of Control, Signals, and Systems, Vol. 2, pp. 303-314.

[DCD14] Defazio, A. J., Caetano, T. S., and Domke, J., 2014. "Finito: A Faster, Permutable Incremental Gradient Method for Big Data Problems," Proceedings of the 31st ICML, Beijing.

[DCR15] Duchi, J. C., Chaturapruek, S., and Re, R., 2015. "Asynchronous Stochastic Convex Optimization," arXiv preprint arXiv:1508.00882.

[DES82] Dembo, R. S., Eisenstadt, S. C., and Steihaug, T., 1982. "Inexact Newton Methods," SIAM J. Numer. Anal., Vol. 19, pp. 400-408.

[DFJ54] Dantzig, G. B., Fulkerson, D. R., and Johnson, S. M., 1954. "Solution of a Large-Scale Traveling-Salesman Problem," Operations Research, Vol. 2, pp. 393-410.

[DHS06] Dai, Y. H., Hager, W. W., Schittkowski, K., and Zhang, H., 2006. "The Cyclic Barzilai-Borwein Method for Unconstrained Optimization," IMA J. of Numerical Analysis, Vol. 26, pp. 604-627.

[DKK83] Decker, D. W., Keller, H. B., and Kelley, C. T., 1983. "Convergence Rates for Newton's Method at Singular Points," SIAM J. Numer. Anal., Vol. 20, pp. 296-314.

[DaW60] Dantzig, G. B., and Wolfe, P., 1960. "Decomposition Principle for Linear Programs," Operations Research, Vol. 8, pp. 101-111.

[DaY14a] Davis, D., and Yin, W., 2014. "Convergence Rate Analysis of Several Splitting Schemes," arXiv preprint arXiv:1406.4834.

[DaY14b] Davis, D., and Yin, W., 2014. "Convergence Rates of Relaxed Peaceman-Rachford and ADMM Under Regularity Assumptions, arXiv preprint arXiv:1407.5210.

[Dan67] Danskin, J. M., 1967. The Theory of Max-Min and its Application to Weapons Allocation Problems, Springer, NY.

[Dan71] Daniel, J. W., 1971. The Aproximate Minimization of Functionals, Prentice-Hall, Englewood Cliffs, N. J.

[Dav59] Davidon, W. C., 1959. "Variable Metric Method for Minimization," Argonne National Lab., Report ANL-5990 (Rev.), Argonne, Ill. Reprinted with a new preface in SIAM J. on Optimization, Vol. 1, 1991, pp. 1-17.

[Dav76] Davidon, W. C., 1976. "New Least Squares Algorithms," J. Opt. Theory and Appl., Vol. 18, pp. 187-197.

[DeK80] Decker, D. W., and Kelley, C. T., 1980. "Newton's Method at Singular Points, Parts I and II," SIAM J. Numer. Anal., Vol. 17, pp. 66-70, 465-471.

[DeM77] Dennis, J. E., and Moré, J. J., 1977. "Quasi-Newton Methods: Motivation and Theory," SIAM Review, Vol. 19, pp. 46-89.

[DeR70] Demjanov, V. F., and Rubinov, A. M., 1970. Approximate Methods in Optimization Problems, American Elsevier, N. Y.

[DeS83] Dennis, J. E., and Schnabel, R. E., 1983. Numerical Methods for Unconstrained Optimization and Nonlinear Equations, Prentice-Hall, Englewood Cliffs, N. J.

[DeT83] Dembo, R. S., and Tulowitzki, U., 1983. "On the Minimization of Quadratic Functions Subject to Box Constraints," Working Paper Series B No. 71, School of Organization and Management, Yale Univ., New Haven, Conn.

[DeT91] Dennis, J. E., and Torczon, V., 1991. "Direct Search Methods on Parallel Machines," SIAM J. on Optimization, Vol. 1, pp. 448-474.

[DeT93] De Angelis, P. L., and Toraldo, G., 1993. "On the Identification Property of a Projected Gradient Method," SIAM J. Numer. Anal., Vol. 30, pp. 1483-1497.

[DeV85] Demjanov, V. F., and Vasilév, L. V., 1985. Nondifferentiable Optimization, Optimization Software, N. Y.

[Del12] Delfour, M. C., 2012. Introduction to Optimization and Semidifferential Calculus, SIAM, Phila.

[Deu12] Deuflhard, P., 2012. "A Short History of Newton's Method," Documenta Mathematica, Optimization Stories, pp. 25-30.

[DiG79] DiPillo, G., and Grippo, L., 1979. "A New Class of Augmented Lagrangians in Nonlinear Programming," SIAM J. on Control and Optimization, Vol. 17, pp. 618-628.

[DiG89] DiPillo, G., and Grippo, L., 1989. "Exact Penalty Functions in Constrained Optimization," SIAM J. on Control and Optimization, Vol. 27, pp. 1333-1360.

[Dix72a] Dixon, L. C. W., 1972. "Quasi-Newton Algorithms Generate Identical Points," Math. Programming, Vol. 2, pp. 383-387.

[Dix72b] Dixon, L. C. W., 1972. "Quasi-Newton Algorithms Generate Identical Points. II. The Proofs of Four New Theorems," Math. Programming, Vol. 3, pp. 345-358.

[DoJ86] Dontchev, A. L., and Jongen, H. Th., 1986. "On the Regularity of the Kuhn-Tucker Curve," SIAM J. on Control and Optimization, Vol. 24, pp. 169-176.

[DoR09] Dontchev, A. L., and Rockafellar, R. T., 2009. Implicit Functions and Solution Mappings," 2nd edition, Springer, N. Y.

[DuB89] Dunn, J. C., and Bertsekas, D. P., 1989. "Efficient Dynamic Programming Implementations of Newton's Method for Unconstrained Optimal Control Problems," J. Opt. Theory and Appl., Vol. 63, pp. 23-38.

[DuM65] Dubovitskii, M. D., and Milyutin, A. A., 1965. "Extremum Problems in the Presence of Restriction," USSR Comp. Math. and Math. Phys., Vol. 5, pp. 1-80.

[DuS83] Dunn, J. C., and Sachs, E., 1983. "The Effect of Perturbations on the Convergence Rates of Optimization Algorithms," Appl. Math. Optim., Vol. 10, pp. 143-157.

[DuZ89] Du, D.-Z., and Zhang, X.-S., 1989. "Global Convergence of Rosen's Gradient Projection Method," Math. Programming, Vol. 44, pp. 357-366.

[Dun79] Dunn, J. C., 1979. "Rates of Convergence for Conditional Gradient Algorithms Near Singular and Nonsingular Extremals," SIAM J. on Control and Optimization, Vol. 17, pp. 187-211.

[Dun80a] Dunn, J. C., 1980. "Convergence Rates for Conditional Gradient Sequences Generated by Implicit Step Length Rules," SIAM J. on Control and Optimization, Vol.

18, pp. 473-487.

[Dun80b] Dunn, J. C., 1980. "Newton's Method and the Goldstein Step Length Rule for Constrained Minimization Problems," SIAM J. on Control and Optimization, Vol. 18, pp. 659-674.

[Dun81] Dunn, J. C., 1981. "Global and Asymptotic Convergence Rate Estimates for a Class of Projected Gradient Processes," SIAM J. on Control and Optimization, Vol. 19, pp. 368-400.

[Dun87] Dunn, J. C., 1987. "On the Convergence of Projected Gradient Processes to Singular Critical Points," J. Opt. Theory and Appl., Vol. 55, pp. 203-216.

[Dun88a] Dunn, J. C., 1988. "Gradient Projection Methods for Systems Optimization Problems," Control and Dynamic Systems, Vol. 29, pp. 135-195.

[Dun88b] Dunn, J. C., 1988. "A Projected Newton Method for Minimization Problems with Nonlinear Inequality Constraints," Numer. Math., Vol. 53, pp. 377-409.

[Dun91a] Dunn, J. C., 1991. "Scaled Gradient Projection Methods for Optimal Control Problems and Other Structured Nonlinear Programs," in New Trends in Systems Theory, Conte, G., et al, (Eds.), Birkhöuser, Boston, MA.

[Dun91b] Dunn, J. C., 1991. "A Subspace Decomposition Principle for Scaled Gradient Projection Methods: Global Theory," SIAM J. on Control and Optimization, Vol. 29, pp. 219-246.

[Dun93a] Dunn, J. C., 1993. "A Subspace Decomposition Principle for Scaled Gradient Projection Methods: Local Theory," SIAM J. on Control and Optimization, Vol. 31, pp. 219-246.

[Dun93b] Dunn, J. C., 1993. "Second-Order Multiplier Update Calculations for Optimal Control Problems and Related Large Scale Nonlinear Programs," SIAM J. on Optimization, Vol. 3, pp. 489-502.

[Dun93c] Dunn, J. C., 1993. Private Communication.

[Dun94] Dunn, J. C., 1994. "Gradient-Related Constrained Minimization Algorithms in Function Spaces: Convergence Properties and Computational Implications," in Large Scale Optimization: State of the Art, Hager, W. W., Hearn, D. W., and Pardalos, P. M., (Eds.), Kluwer, Boston.

[Eas58] Eastman, W. L., 1958. Linear Programming with Pattern Constraints, Ph.D. Thesis, Harvard University, Cambridge, MA.

[EcB90] Eckstein, J., and Bertsekas, D. P., 1990. "An Alternating Direction Method for Linear Programming," Report LIDS-P-1967, Lab. for Info. and Dec. Systems, M.I.T.

[EcB92] Eckstein, J., and Bertsekas, D. P., 1992. "On the Douglas-Rachford Splitting Method and the Proximal Point Algorithm for Maximal Monotone Operators," Math. Programming, Vol. 55, pp. 293-318.

[Eck94a] Eckstein, J., 1994. "Nonlinear Proximal Point Algorithms Using Bregman Functions, with Applications to Convex Programming," Math. Operations Res., Vol. 18, pp. 202-226.

[Eck94b] Eckstein, J., 1994. "Parallel Alternating Direction Multiplier Decomposition of Convex Programs," J. Opt. Theory and Appl., Vol. 80, pp. 39-62.

[EkT76] Ekeland, I., and Teman, R., 1976. Convex Analysis and Variational Problems, North-Holland Publ., Amsterdam.

[ElM75] Elzinga, J., and Moore, T. G., 1975. "A Central Cutting Plane Algorithm for the Convex Programming Problem," Math. Programming, Vol. 8, pp. 134-145.

[Eve63] Everett, H., 1963. "Generalized Lagrange Multiplier Method for Solving Problems of Optimal Allocation of Resources," Operations Research, Vol. 11, pp. 399-417.

[Evt85] Evtushenko, Y. G., 1985. Numerical Optimization Techniques, Optimization Software, N. Y.

[FAJ14] Feyzmahdavian, H. R., Aytekin, A., and Johansson, M., 2014. "A Delayed Proximal Gradient Method with Linear Convergence Rate," in Prop. of 2014 IEEE International Workshop on Machine Learning for Signal Processing (MLSP), pp. 1-6.

[FGK73] Fratta, L., Gerla, M., and Kleinrock, L., 1973. "The Flow Deviation Method: An Approach to Store-and-Forward Communication Network Design," Networks, Vol. 3, pp. 97-133.

[FGW02] Forsgren, A., Gill, P. E., and Wright, M. H., 2002. "Interior Methods for Nonlinear Optimization," SIAM Review, Vol. 44, pp. 525-597.

[FaF63] Fadeev, D. K., and Fadeeva, V. N., 1963. Computational Methods of Linear Algebra, Freeman, San Francisco, CA.

[FaP03] Facchinei, F., and Pang, J.-S., 2003. Finite-Dimensional Variational Inequalities and Complementarity Problems, Springer Verlag, N. Y.

[Fab73] Fabian, V., 1973. "Asymptotically Efficient Stochastic Approximation: The RM Case," Ann. Statist., Vol. 1, pp. 486-495.

[FeM91] Ferris, M. C., and Mangasarian, O. L., 1991. "Parallel Constraint Distribution," SIAM J. on Optimization, Vol. 1, pp. 487-500.

[Fen49] Fenchel, W., 1949. "On Conjugate Convex Functions," Canad. J. Math., Vol. 1, pp. 73-77.

[Fen51] Fenchel, W., 1951. "Convex Cones, Sets, and Functions," Mimeographed Notes, Princeton Univ.

[Fey16] Feyzmahdavian, H. R., 2016. Performance Analysis of Positive Systems and Optimization Algorithms with Time-Delays, Doctoral Thesis, KTH, Sweden.

[FiM68] Fiacco, A. V., and McCormick, G. P., 1968. Nonlinear Programming: Sequential Unconstrained Minimization Techniques, Wiley, N. Y.

[Fia83] Fiacco, A. V., 1983. Introduction to Sensitivity and Stability Analysis in Nonlinear Programming, Academic Press, N. Y.

[FlH95] Florian, M. S., and Hearn, D., 1995. "Network Equilibrium Models and Algorithms," Handbooks in OR and MS, Ball, M. O., Magnanti, T. L., Monma, C. L., and Nemhauser, G. L., (Eds.), Vol. 8, North-Holland, Amsterdam, pp. 485-550.

[FlP63] Fletcher, R., and Powell, M. J. D., 1963. "A Rapidly Convergent Descent Algorithm for Minimization," Comput. J., Vol. 6, pp. 163-168.

[FlP95] Floudas, C., and Pardalos, P. M., (Eds.), 1995. State of the Art in Global Optimization: Computational Methods and Applications, Kluwer, Boston.

[Fla92] Flam, S. D., 1992. "On Finite Convergence and Constraint Identification of Subgradient Projection Methods," Math. Programming, Vol. 57, pp 427-437.

[Fle70a] Fletcher, R., 1970. "A New Approach to Variable Metric Algorithms," Computer J., Vol. 13, pp. 317-322.

[Fle70b] Fletcher, R., 1970. "A Class of Methods for Nonlinear Programming with Termination and Convergence Properties," in Integer and Nonlinear Programming, Abadie, J., (Ed.), pp. 157-173, North-Holland Publ., Amsterdam.

[Fle00] Fletcher, R., 2000. Practical Methods of Optimization, 2nd edition, Wiley, NY.

[Flo95] Floudas, C. A., 1995. Nonlinear and Mixed-Integer Optimization: Fundamentals and Applications, Oxford University Press, N. Y.

[FoG83] Fortin, M., and Glowinski, R., (Eds.), 1983. Augmented Lagrangian Methods: Applications to the Numerical Solution of Boundary-Value Problems, North-Holland, Amsterdam.

[FoS11] Fong, D. C. L., and Saunders, M., 2011. "LSMR: An Iterative Algorithm for Sparse Least-Squares Problems," SIAM Journal on Scientific Computing, Vol. 33, pp. 2950-2971.

[FoS12] Fong, D. C. L., and Saunders, M., 2012. "CG Versus MINRES: An Empirical Comparison," SQU Journal for Science, Vol. 17, pp. 44-62.

[FrG16] Freund, R., and Grigas, P., 2016. "New Analysis and Results for the Frank-Wolfe Method," Math. Programming, Series A, Vol. 155, pp. 199-230.

[FrW56] Frank, M., and Wolfe, P., 1956. "An Algorithm for Quadratic Programming," Naval Research Logistics Quarterly, Vol. 3, pp. 95-110.

[Fra12] Frauendorfer, K., 2012. Stochastic Two-Stage Programming, Springer, N. Y.

[Fre91] Freund, R. M., 1991. "Theoretical Efficiency of a Shifted Barrier Function Algorithm in Linear Programming," Linear Algebra and Appl., Vol. 152, pp. 19-41.

[Fri56] Frisch, M. R., 1956. "La Resolution des Problemes de Programme Lineaire par la Methode du Potential Logarithmique," Cahiers du Seminaire D'Econometrie, Vol. 4, pp. 7-20.

[Fuk92] Fukushima, M., 1992. "Application of the Alternating Direction Method of Multipliers to Separable Convex Programming," Comp. Opt. and Appl., Vol. 1, pp. 93-111.

[GFJ15] Ghadimi, E., Feyzmahdavian, H. R., and Johansson, M., 2015. "Global Convergence of the Heavy-Ball Method for Convex Optimization," in European Control Conference (ECC), pp. 310-315.

[GHV92] Goffin, J. L., Haurie, A., and Vial, J. P., 1992. "Decomposition and Nondifferentiable Optimization with the Projective Algorithm," Management Science, Vol. 38, pp. 284-302.

[GKT51] Gale, D., Kuhn, H. W., and Tucker, A. W., 1951. "Linear Programming and the Theory of Games," in Activity Analysis of Production and Allocation, Koopmans, T. C., (Ed.), Wiley, N. Y.

[GKX10] Gupta, M. D., Kumar, S., and Xiao, J. 2010. "L1 Projections with Box Constraints," arXiv preprint arXiv:1010.0141.

[GLL91] Grippo, L., Lampariello, F., and Lucidi, S., 1991. "A Class of Nonmonotone Stabilization Methods in Unconstrained Minimization," Numer. Math., Vol. 59, pp. 779-805.

[GLY94] Goffin, J. L., Luo, Z.-Q., and Ye, Y., 1994. "On the Complexity of a Column Generation Algorithm for Convex or Quasiconvex Feasibility Problems," in Large Scale Optimization: State of the Art, Hager, W. W., Hearn, D. W., and Pardalos, P. M., (Eds.), Kluwer, Boston.

[GLY96] Goffin, J. L., Luo, Z.-Q., and Ye, Y., 1996. "Complexity Analysis of an Interior Cutting Plane Method for Convex Feasibility Problems," SIAM J. on Optimization, Vol. 6, pp. 638-652.

[GMW81] Gill, P. E., Murray, W., and Wright, M. H., 1981. Practical Optimization, Academic Press, N. Y.

[GMW91] Gill, P. E., Murray, W., and Wright, M. H., 1991. Numerical Linear Algebra and Optimization, Vol. I, Addison-Wesley, Redwood City, CA.

[GOP15a] Gurbuzbalaban, M., Ozdaglar, A., and Parrilo, P., 2015. "On the Convergence Rate of Incremental Aggregated Gradient Algorithms," arXiv preprint arXiv:1506.02081.

[GOP15b] Gurbuzbalaban, M., Ozdaglar, A., and Parrilo, P., 2015. "Convergence Rate of Incremental Gradient and Newton Methods," arXiv preprint arXiv:1510.08562.

[GOP15c] Gurbuzbalaban, M., Ozdaglar, A., and Parrilo, P., 2015. "Why Random Reshuffling Beats Stochastic Gradient Descent," arXiv preprint arXiv:1510.08560.

[GaB82] Gafni, E. M., and Bertsekas, D. P., 1982. "Convergence of a Gradient Projection Method," Report LIDS-P-1201, Lab. for Info. and Dec. Systems, M.I.T.

[GaB84] Gafni, E. M., and Bertsekas, D. P., 1984. "Two-Metric Projection Methods for Constrained Optimization," SIAM J. on Control and Optimization, Vol. 22, pp. 936-964.

[GaD88] Gawande, M., and Dunn, J. C., 1988. "Variable Metric Gradient Projection Processes in Convex Feasible Sets Defined by Nonlinear Inequalities," Appl. Math. Optim., Vol. 17, pp. 103-119.

[GaJ88] Gauvin, J., and Janin, R., 1988. "Directional Behavior of Optimal Solutions in Nonlinear Mathematical Programming," Math. of Operations Res., Vol. 13, pp. 629-649.

[GaM76] Gabay, D., and Mercier, B., 1976. "A Dual Algorithm for the Solution of Nonlinear Variational Problems via Finite-Element Approximations," Comp. Math. Appl., Vol. 2, pp. 17-40.

[GaM92] Gaudioso, M., and Monaco, M. F, 1992. "Variants to the Cutting Plane Approach for Convex Nondifferentiable Optimization," Optimization, Vol. 25, pp. 65-75.

[Gab79] Gabay, D., 1979. Methodes Numeriques pour l'Optimization Non Lineaire, These de Doctorat d'Etat et Sciences Mathematiques, Univ. Pierre at Marie Curie (Paris VI).

[Gab82] Gabay, D., 1982. "Reduced Quasi-Newton Methods with Feasibility Improvement for Nonlinearly Constrained Optimization," Math. Programming Studies, Vol. 16, pp. 18-44.

[Gab83] Gabay, D., 1983. "Applications of the Method of Multipliers to Variational Inequalities," in M. Fortin and R. Glowinski, eds., Augmented Lagrangian Methods: Applications to the Solution of Boundary-Value Problems, North-Holland, Amsterdam.

[Gai94] Gaivoronski, A. A., 1994. "Convergence Analysis of Parallel Backpropagation Algorithm for Neural Networks," Optimization Methods and Software, Vol. 4, pp. 117-134.

[Gal77] Gallager, R. G., 1977. "A Minimum Delay Routing Algorithm Using Distributed Computation," IEEE Trans. on Communications, Vol. 25, pp. 73-85.

[Gau77] Gauvin, J., 1977. "A Necessary and Sufficient Condition to Have Bounded Multipliers in Convex Programming," Math. Programming., Vol. 12, pp. 136-138.

[Geo70] Geoffrion, A. M., 1970. "Elements of Large-Scale Mathematical Programming, I, II," Management Science, Vol. 16, pp. 652-675, 676-691.

[Geo74] Geoffrion, A. M., 1974. "Lagrangian Relaxation for Integer Programming," Math. Programming Studies, Vol. 2, pp. 82-114.

[Geo77] Geoffrion, A. M., 1977. "Objective Function Approximations in Mathematical Programming," Math. Programming, Vol. 13, pp. 23-27.

[GiK95] Gilmore, P., and Kelley, C. T., 1995. "An Implicit Filtering Algorithm for Optimization of Functions with Many Local Minima," SIAM J. on Optimization, Vol.

5, pp. 269-285.

[GiM74] Gill, P. E., and Murray, W., (Eds.), 1974. Numerical Methods for Constrained Optimization, Academic Press, N. Y.

[GlL97] Glover, F., and Laguna, M., 1997. Tabu Search, Kluwer, Boston.

[GlM75] Glowinski, R. and Marrocco, A., 1975. "Sur l' Approximation par Elements Finis d' Ordre un et la Resolution par Penalisation-Dualite d'une Classe de Problemes de Dirichlet Non Lineaires" Revue Francaise d'Automatique Informatique Recherche Operationnelle, Analyse Numerique, R-2, pp. 41-76.

[GlP79] Glad, T., and Polak, E., 1979. "A Multiplier Method with Automatic Limitation of Penalty Growth," Math. Programming, Vol. 17, pp. 140-155.

[Gla79] Glad, T., 1979. "Properties of Updating Methods for the Multipliers in Augmented Lagrangians," J. Opt. Th. and Appl., Vol. 28, pp. 135-156.

[GoP79] Gonzaga, C., and Polak, E., 1979. "On Constraint Dropping Schemes and Optimality Functions for a Class of Outer Approximations Algorithms," SIAM J. on Control and Optimization, Vol. 17, pp.477-493.

[GoT71] Gould, F. J., and Tolle, J., 1971. "A Necessary and Sufficient Condition for Constrained Optimization," SIAM J. Applied Math., Vol. 20, pp. 164-172.

[GoT72] Gould, F. J., and Tolle, J., 1972. "Geometry of Optimality Conditions and Constraint Qualifications," Math. Programming, Vol. 2, pp. 1-18.

[GoV90] Goffin, J. L., and Vial, J. P., 1990. "Cutting Planes and Column Generation Techniques with the Projective Algorithm," J. Opt. Th. and Appl., Vol. 65, pp. 409-429.

[GoV99] Goffin, J. L., and Vial, J. P., 1999. "Convex Nondifferentiable Optimization: A Survey Focussed on the Analytic Center Cutting Plane Method," Logilab Technical Report, Department of Management Studies, University of Geneva, Switzerland; also GERAD Tech. Report G-99-17, McGill Univ., Montreal, Canada.

[Gof77] Goffin, J. L., 1977. "On Convergence Rates of Subgradient Optimization Methods," Math. Programming, Vol. 13, pp. 329-347.

[Gol62] Goldstein, A. A., 1962. "Cauchy's Method of Minimization," Numer. Math., Vol. 4, pp. 146-150.

[Gol64] Goldstein, A. A., 1964. "Convex Programming in Hibert Space," Bull. Amer. Math. Soc., Vol. 70, pp. 709-710.

[Gol67] Goldstein, A. A., 1967. Constructive Real Analysis, Harper and Row, N. Y.

[Gol70] Goldfarb, D., 1970. "A Family of Variable-Metric Methods Derived by Variational Means," Math. Comp., Vol. 24, pp. 23-26.

[Gol85] Golshtein, E. G., 1985. "A Decomposition Method for Linear and Convex Programming Problems," Matecon, Vol. 21, pp. 1077-1091.

[Gol89] Goldberg, D. E., 1989. Genetic Algorithms in Search, Optimization, and Machine Learning, Addison Wesley, Reading, MA.

[Gom58] Gomory, R. E., 1958. "Outline of an Algorithm for Integer Solutions to Linear Programs," Bulletin of the American Mathematical Society, Vol. 64, pp. 275-278.

[Gon91] Gonzaga, C. C., 1991. "Large Step Path-Following Methods for Linear Programming, Part I: Barrier Function Method," SIAM J. on Optimization, Vol. 1, pp. 268-279.

[Gon92] Gonzaga, C. C., 1992. "Path Following Methods for Linear Programming," SIAM Review, Vol. 34, pp. 167-227.

[Gon00] Gonzaga, C. C., 2000. "Two Facts on the Convergence of the Cauchy Algorithm," J. of Optimization Theory and Applications, Vol. 107, pp. 591-600.

[GrS00] Grippo, L., and Sciandrone, M., 2000. "On the Convergence of the Block Nonlinear Gauss-Seidel Method Under Convex Constraints," Operations Research Letters, Vol. 26, pp. 127-136.

[GrW08] Griewank, A., and Walther, A., 2008. Evaluating Derivatives: Principles and Techniques of Algorithmic Differentiation, SIAM.

[Gri94] Grippo, L., 1994. "A Class of Unconstrained Minimization Methods for Neural Network Training," Optimization Methods and Software, Vol. 4, pp. 135-150.

[GuM86] Guelat, J., and Marcotte, P., 1986. "Some Comments on Wolfe's 'Away Step'," Math. Programming, Vol. 35, pp. 110-119.

[Gui69] Guignard, M., 1969. "Generalized Kuhn-Tucker Conditions for Mathematical Programming Problems in a Banach Space," SIAM J. on Control, Vol. 7, pp. 232-241.

[Gul92] Guler, O., 1992. "New Proximal Point Algorithms for Convex Minimization," SIAM J. on Optimization, Vol. 2, pp. 649-664.

[Gul94] Guler, O., 1994. "Limiting Behavior of Weighted Central Paths in Linear Programming," Math. Programming, Vol. 65, pp. 347-363.

[HDY12] Hinton, G., Deng, L., Yu, D., Dahl, G. E., Mohamed, A., Jaitly, N., Senior A., et al. 2012. "Deep Neural Networks for Acoustic Modeling in Speech Recognition: The Shared Views of Four Research Groups," Signal Processing Magazine, IEEE, Vol. 29, pp. 82-97.

[HKR95] den Hertog, D., Kaliski, J., Roos, C., and Terlaky, T., 1995. "A Path-Following Cutting Plane Method for Convex Programming," Annals of Operations Research, Vol. 58, pp. 69-98.

[HLV87] Hearn, D. W., Lawphongpanich, S., and Ventura, J. A., 1987. "Restricted Simplicial Decomposition: Computation and Extensions," Math. Programming Studies, Vol. 31, pp. 119-136.

[HPT00] Horst, R., Pardalos, P. M., and Thoai, N. V., 2000. Introduction to Global Optimization, 2nd Edition, Kluwer, Boston.

[HDR13] Hong, M., Wang, X., Razaviyayn, M., and Luo, Z. Q., 2013. "Iteration Complexity Analysis of Block Coordinate Descent Methods," arXiv preprint arXiv:1310.6957.

[HaB70] Haarhoff, P. C., and Buys, J. D, 1970. "A New Method for the Optimization of a Nonlinear Function Subject to Nonlinear Constraints," Computer J., Vol. 13, pp. 178-184.

[HaH93] Hager, W. W., and Hearn, D. W., 1993. "Application of the Dual Active Set Algorithm to Quadratic Network Optimization," Computational Optimization and Applications, Vol. 1, pp. 349-373.

[HaM79] Han, S. P., and Mangasarian, O. L., 1979. "Exact Penalty Functions in Nonlinear Programming," Math. Programming, Vol. 17, pp. 251-269.

[Hag99] Hager, W. W., 1999. "Stabilized Sequential Quadratic Programming," Computational Optimization and Applications, Vol. 12.

[Han77] Han, S. P., 1977. "A Globally Convergent Method for Nonlinear Programming," J. Opt. Th. and Appl., Vol. 22, pp. 297-309.

[Hay11] Haykin, S., 2011. Neural Networks and Learning Machines, (3rd Ed.), Pearson Education, Upper Saddle River, N. J.

[HeK70] Held, M., and Karp, R. M., 1970. "The Traveling Salesman Problem and Minimum Spanning Trees," Operations Research, Vol. 18, pp. 1138-1162.

[HeK71] Held, M., and Karp, R. M., 1971. "The Traveling Salesman Problem and Minimum Spanning Trees: Part II," Math. Programming, Vol. 1, pp. 6-25.

[HeL89] Hearn, D. W., and Lawphongpanich, S., 1989. "Lagrangian Dual Ascent by Generalized Linear Programming," Operations Res. Letters, Vol. 8, pp. 189-196.

[HeS52] Hestenes, M. R., and Stiefel, E. L., 1952. "Methods of Conjugate Gradients for Solving Linear Systems," J. Res. Nat. Bur. Standards Sect. B, Vol. 49, pp. 409-436.

[Hei96] Heinkenschloss, M., 1996. "Projected Sequential Quadratic Programming Methods," SIAM J. on Optimization, Vol. 6, pp. 373-417.

[Her94] den Hertog, D., 1994. Interior Point Approach to Linear, Quadratic, and Convex Programming, Kluwer, Dordrecht, The Netherlands.

[Hes69] Hestenes, M. R., 1969. "Multiplier and Gradient Methods," J. Opt. Th. and Appl., Vol. 4, pp. 303-320.

[Hes75] Hestenes, M. R., 1975. Optimization Theory: The Finite Dimensional Case, Wiley, N. Y.

[Hes80] Hestenes, M. R., 1980. Conjugate Direction Methods in Optimization, Springer-Verlag, Berlin and N. Y.

[HiL93] Hiriart-Urruty, J.-B., and Lemarechal, C., 1993. Convex Analysis and Minimization Algorithms, Vols. I and II, Springer-Verlag, Berlin and N. Y.

[Hil57] Hildreth, C., 1957. "A Quadratic Programming Procedure," Naval Res. Logist. Quart., Vol. 4, pp. 79-85. See also "Erratum," Naval Res. Logist. Quart., Vol. 4, p. 361.

[HoJ61] Hooke, R., and Jeeves, T. A., 1961. "Direct Search Solution of Numerical and Statistical Problems," J. Assoc. Comp. Mach., Vol. 8, pp. 212-221.

[HoK71] Hoffman, K., and Kunze, R., 1971. Linear Algebra, Prentice-Hall, Englewood Cliffs, N. J.

[HoL13] Hong, M., and Luo, Z. Q., 2013. "On the Linear Convergence of the Alternating Direction Method of Multipliers," arXiv preprint arXiv:1208.3922.

[Hoh77] Hohenbalken, B. von, 1977. "Simplicial Decomposition in Nonlinear Programming," Math. Programming, Vol. 13, pp. 49-68.

[Hol74] Holloway, C. A., 1974. "An Extension of the Frank and Wolfe Method of Feasible Directions," Math. Programming, Vol. 6, pp. 14-27.

[HuD84] Hughes, G. C., and Dunn, J. C., 1984. "Newton-Goldstein Convergence Rates for Convex Constrained Minimization Problems with Singular Solutions," Appl. Math. Optim., Vol. 12, pp. 203-230.

[HuL88] Huang, C., and Litzenberger, R. H., 1988. Foundations of Financial Economics, Prentice-Hall, Englewood Cliffs, N. J.

[IJT15a] Iusem, A., Jofre, A., and Thompson, P., 2015. "Approximate Projection Methods for Monotone Stochastic Variational Inequalities," Math. Programming, to appear.

[IJT15b] Iusem, A., Jofre, A., and Thompson, P., 2015. "Incremental Constraint Projection Methods for Monotone Stochastic Variational Inequalities," Math. Operations Res., to appear.

[IST94] Iusem, A. N., Svaiter, B., and Teboulle, M., 1994. "Entropy-Like Proximal Methods in Convex Programming," Math. Operations Res., Vol. 19, pp. 790-814.

[IbF96] Ibaraki, S., and Fukushima, M., 1996. "Partial Proximal Method of Multipliers for Convex Programming Problems," J. of Operations Research Society of Japan, Vol. 39, pp. 213-229.

[IbK88] Ibaraki, T., and Katoh, N., 1988. Resource Allocation Problems: Algorithmic Approaches, M.I.T. Press, Cambridge, MA.

[Iof94] Ioffe, A., 1994. "On Sensitivity Analysis of Nonlinear Programs in Banach Spaces: The Approach via Composite Unconstrained Minimization," SIAM J. on Optimization, Vol. 4, pp. 1-43.

[IuT95] Iusem, A. N., and Teboulle, M., 1995. "Convergence Rate Analysis of Nonquadratic Proximal Methods for Convex and Linear Programming," Math. Operations Res., Vol. 20.

[Ius99] Iusem, A. N., 1999. "Augmented Lagrangian Methods and Proximal Point Methods for Convex Minimization," Investigacion Operativa, Vol. 8, pp. 11-49 .

[JRT95] Junger, M., Reinelt, G., and Rinaldi, G., 1995. "Practical Problem Solving with Cutting Plane Algorithms in Combinatorial Oprimization," in Combinatorial Optimization, Cook, W. J., Lovasz, L., and Seymour, P., (Eds.), DIMACS Series in Discrete Mathematics and Computer Science, AMS, pp. 11-152.

[JaS95] Jarre, F., and Saunders, M. A., 1995. "A Practical Interior-Point Method for Convex Programming," SIAM J. Optimization, Vol. 5, pp. 149-171.

[Jag13] Jaggi, M., 2013. "Revisiting Frank-Wolfe: Projection-Free Sparse Convex Optimization," Proc. of ICML 2013.

[JeW92] Jeyakumar, V., and Wolkowicz, H., 1992. "Generalizations of Slater's Constraint Qualification for Infinite Convex Programs," Math. Programming, Vol. 57, pp. 85-101.

[Joh48] John, F., 1948. "Extremum Problems with Inequalities as Subsidiary Conditions," in Studies and Essays: Courant Anniversary Volume, K. O. Friedrichs, Neugebauer, O. E., and Stoker, J. J., (Eds.), Wiley-Interscience, N. Y., pp. 187-204.

[KAK89] Korst, J., Aarts, E. H., and Korst, A., 1989. Simulated Annealing and Boltzmann Machines: A Stochastic Approach to Combinatorial Optimization and Neural Computing, Wiley, N. Y.

[KLT03] Kolda, T. G., Lewis, R. M., and Torczon, V., 2003. "Optimization by Direct Search: New Perspectives on Some Classical and Modern Methods," SIAM Review, Vol. 45, pp. 385-482.

[KMN91] Kojima, M., Meggido, N., Noma, T., and Yoshise, A., 1991. A Unified Approach to Interior Point Algorithms for Linear Complementarity Problems, Springer-Verlag, Berlin.

[KMY89] Kojima, M., Mizuno, S., and Yoshise, A., 1989. "A Primal-Dual Interior Point Algorithm for Linear Programming," in Progress in Mathematical Programming, Interior Point and Related Methods, Meggido, N., (Ed.), Springer-Verlag, N. Y., pp. 29-47.

[KaW94] Kall, P., and Wallace, S. W., 1994. Stochastic Programming, Wiley, Chichester, UK.

[Kan39] Kantorovich, L. V., 1939. "The Method of Successive Approximations for Functional Equations," Acta Math., Vol. 71, pp. 63-97.

[Kan45] Kantorovich, L. V., 1945. "On an Effective Method of Solution of Extremal Problems for a Quadratic Functional," Dokl. Akad. Nauk SSSR, Vol. 48, pp. 483-487.

[Kan49] Kantorovich, L. V., 1949. "On Newton's Method," Trudy Mat. Inst. Steklov, Vol. 28, pp. 104-144. Translated in Selected Articles in Numerical Analysis by C. D. Benster, 104.1-144.2.

[Kar39] Karush, W., 1939. "Minima of Functions of Several Variables with Inequalities as Side Conditions," M.S. Thesis, Department of Math., University of Chicago.

[Kar84] Karmarkar, N., 1984. "A New Polynomial-Time Algorithm for Linear Programming," Combinatorica, Vol. 4, pp. 373-395.

[KeG88] Keerthi, S. S., and Gilbert, E. G., 1988. "Optimal, Infinite Horizon Feedback Laws for a General Class of Constrained Discete Time Systems: Stability and Moving-Horizon Approximations," J. Optimization Theory Appl., Vo. 57, pp. 265-293.

[Kel60] Kelley, J. E., 1960. "The Cutting-Plane Method for Solving Convex Programs," J. Soc. Indust. Appl. Math., Vol. 8, pp. 703-712.

[Kel99] Kelley, C. T., 1999. Iterative Methods for Optimization, SIAM, Philadelphia.

[Kha79] Khachiyan, L. G., 1979. "A Polynomial Algorithm for Linear Programming," Soviet Math. Doklady, Vol. 20, pp. 191-194.

[KiN91] Kim, S., and Nazareth, J. L., 1991. "The Decomposition Principle and Algorithms for Linear Programming," Linear Algebra and its Applications, Vol. 152, pp. 119-133.

[KiR92] King, A., and Rockafellar, R. T., 1992. "Sensitivity Analysis for Nonsmooth Generalized Equations," Math. Programming, Vol. 55, pp. 193-212.

[KlH98] Klatte, D., and Henrion, R., 1998. "Regularity and Stability in Nonlinear Semi-Infinite Optimization," in Semi-Infinite Programming, Reemtsen, R., and Ruckman, J. J., (Eds.), Kluwer, Boston, pp. 69-102.

[KoB72] Kort, B. W., and Bertsekas, D. P., 1972. "A New Penalty Function Method for Constrained Minimization," Proc. 1972 IEEE Confer. Decision Control, New Orleans, LA, pp. 162-166.

[KoB76] Kort, B. W., and Bertsekas, D. P., 1976. "Combined Primal-Dual and Penalty Methods for Convex Programming," SIAM J. on Control and Optimization, Vol. 14, pp. 268-294.

[KoM98] Kontogiorgis, S., and Meyer, R. R., 1998. "A Variable-Penalty Alternating Directions Method for Convex Optimization," Math. Programming, Vol. 83, pp. 29-53.

[KoN93] Kortanek, K. O., and No, H., 1993. "A Central Cutting Plane Algorithm for Convex Semi-Infinite Programming Problems," SIAM J. on Optimization, Vol. 3, pp. 901-918.

[KoZ93] Kortanek, K. O., and Zhu, J., 1993. "A Polynomial Barrier Algorithm for Linearly Constrained Convex Programming Problems," Math. Operations Res., Vol. 18, pp. 116-127.

[KoZ95] Kortanek, K. O., and Zhu, J., 1995. "On Controlling the Parameter in the Logarithmic Barrier Term for Convex Programming Problems," J. Opt. Th. and Appl., Vol. 84, pp. 117-143.

[Koh74] Kohonen, T., 1974. "An Adaptive Associative Memory Principle," IEEE Trans. on Computers, Vol. C-23, pp. 444-445.

[Kor75] Kort, B. W., 1975. "Combined Primal-Dual and Penalty Function Algorithms for Nonlinear Programming," Ph.D. Thesis, Dept. of Engineering-Economic Systems, Stanford Univ., Stanford, Ca.

[Kor76] Korpelevich, G. M., 1976. "The Extragradient Method for Finding Saddle Points and Other Problems," Matecon, Vol. 12, pp. 747-756.

[KuC78] Kushner, H. J., and Clark, D. S., 1978. Stochastic Approximation Methods for Constrained and Unconstrained Systems, Springer-Verlag, N. Y.

[KuT51] Kuhn, H. W., and Tucker, A. W., 1951. "Nonlinear Programming," in Proc. of the Second Berkeley Symposium on Math. Statistics and Probability, Neyman, J., (Ed.), Univ. of California Press, Berkeley, CA, pp. 481-492.

[KuY97] Kushner, H. J., and Yin, G., 1997. Stochastic Approximation Methods, Springer-Verlag, N. Y.

[Kuh76] Kuhn, H. W., 1976. "Nonlinear Programming: A Historical View," in Nonlinear Programming, Cottle, R. W., and Lemke, C. E., (Eds.), SIAM-AMS Proc., Vol. IX, American Math. Soc., Providence, RI, pp. 1-26.

[LJS12] Lacoste-Julien, S., Jaggi, M., Schmidt, M., and Pletscher, P., 2012. "Block-Coordinate Frank-Wolfe Optimization for Structural SVMs," arXiv preprint arXiv:1207-.4747.

[LMS92] Lustig, I. J., Marsten, R. E., and Shanno, D. F., 1992. "On Implementing Mehrotra's Predictor-Corrector Interior-Point Method for Linear Programming," SIAM J. on Optimization, Vol. 2, pp. 435-449.

[LPW92] Ljung, L., Pflug, G., and Walk, H., 1992. Stochastic Approximation and Optimization of Random Systems, Birkhauser, Boston.

[LRW98] Lagarias, J. C., Reeds, J. A., Wright, M. H., and Wright, P. E., 1998. "Convergence Properties of the Nelder-Mead Simplex Method in Low Dimensions," SIAM J. on Optimization, Vol. 9, pp. 112-147.

[LST01] Lucidi, S., Sciandrone, M., and Tseng, P., 2001. "Objective-Derivative-Free Methods for Constrained Optimization," Math. Programming, Vol. 92, pp. 37-59.

[LVB98] Lobo, M. S., Vandenberghe, L., Boyd, S., and Lebret, H., 1998. "Applications of Second-Order Cone Programming," Linear Algebra and Applications, Vol. 284, pp. 193-228.

[LaD60] Land, A. H., and Doig, A. G., 1960. "An Automatic Method for Solving Discrete Programming Problems," Econometrica, Vol. 28, pp. 497-520.

[LaJ13] Lacoste-Julien, S., and Jaggi, M., 2013. "An Affine Invariant Linear Convergence Analysis for Frank-Wolfe Algorithms," arXiv preprint arXiv:1312.7864.

[LaT85] Lancaster, P., and Tismenetsky, M., 1985. The Theory of Matrices, Academic Press, N. Y.

[LaW78] Lasdon, L. S., and Waren, A. D., 1978. "Generalized Reduced Gradient Software for Linearly and Nonlinearly Constrained Problems," in Design and Implementation of Optimization Software, Greenberg, H. J., (Ed.), Sijthoff and Noordhoff, Holland, pp. 335-362.

[Lan15] Landi, G., 2015. "A Modified Newton Projection Method for ℓ_1-Regularized Least Squares Image Deblurring," J. of Mathematical Imaging and Vision, Vol. 51, pp. 195-208.

[Las70] Lasdon, L. S., 1970. Optimization Theory for Large Systems, Macmillian, N. Y.; republished by Dover Pubs, N. Y., 2002.

[Law76] Lawler, E., 1976. Combinatorial Optimization: Networks and Matroids, Holt, Reinhart, and Winston, N. Y.

[LeL10] Leventhal, D., and Lewis, A. S., 2010. "Randomized Methods for Linear Constraints: Convergence Rates and Conditioning," Math. of Operations Research, Vol. 35, pp. 641-654.

[LeO16] Lewis, A. S., and Overton, M. L., 2013. "Nonsmooth Optimization via Quasi-Newton Methods," Math. Programming, Vol. 141, pp. 135-163.

[LeP65] Levitin, E. S., and Poljak, B. T., 1965. "Constrained Minimization Methods," Ž. Vyčisl. Mat. i Mat. Fiz., Vol. 6, pp. 787-823.

[LeS93] Lemaréchal, C., and Sagastizábal, C., 1993. "An Approach to Variable Metric Bundle Methods," in Systems Modelling and Optimization, Proc. of the 16th IFIP-TC7 Conference, Compiègne, Henry, J., and Yvon, J.-P., (Eds.), Lecture Notes in Control and Information Sciences 197, pp. 144-162.

[Lem74] Lemarechal, C., 1974. "An Algorithm for Minimizing Convex Functions," in Information Processing '74, Rosenfeld, J. L., (Ed.), pp. 552-556, North-Holland, Amsterdam.

[LiM79] Lions, P. L., and Mercier, B., 1979. "Splitting Algorithms for the Sum of Two Nonlinear Operators," SIAM J. on Numerical Analysis, Vol. 16, pp. 964-979.

[LiP87] Lin, Y. Y., and Pang, J.-S., 1987. "Iterative Methods for Large Convex Quadratic Programs: A Survey," SIAM J. on Control and Optimization, Vol. 18, pp. 383-411.

[Lin07] Lin, C. J., 2007. "Projected Gradient Methods for Nonnegative Matrix Factorization," Neural Computation, Vol. 19, pp. 2756-2779.

[LuB96] Lucena, A., and Beasley, J. E., 1996. "Branch and Cut Algorithms," in Advances in Linear and Integer Programming, Beasley, J. E., (Ed.), Oxford University Press, N. Y., Chapter 5.

[LuT91] Luo, Z. Q., and Tseng, P., 1991. "On the Convergence of a Matrix-Splitting Algorithm for the Symmetric Monotone Linear Complementarity Problem," SIAM J. on Control and Optimization, Vol. 29, pp. 1037-1060.

[LuT92a] Luo, Z. Q., and Tseng, P., 1992. "On the Convergence of the Coordinate Descent Method for Convex Differentiable Minimization," J. Opt. Th. and Appl., Vol. 72, pp. 7-35.

[LuT92b] Luo, Z. Q., and Tseng, P., 1992. "On the Linear Convergence of Descent Methods for Convex Essentially Smooth Minimization," SIAM J. on Control and Optimization, Vol. 30, pp. 408-425.

[LuT93a] Luo, Z. Q., and Tseng, P., 1993. "Error Bounds and Convergence Analysis of Feasible Descent Methods: A General Approach," Annals of Operations Res., Vol. 46, pp. 157-178.

[LuT93b] Luo, Z. Q., and Tseng, P., 1993. "Error Bound and Reduced Gradient Projection Algorithms for Convex Minimization over a Polyhedral Set," SIAM J. on Optimization, Vol. 3, pp. 43-59.

[LuT93c] Luo, Z. Q., and Tseng, P., 1993. "On the Convergence Rate of Dual Ascent Methods for Linearly Constrained Convex Minimization," Math. of Operations Res., Vol. 18, pp. 846-867.

[LuT94a] Luo, Z. Q., and Tseng, P., 1994. "Analysis of an Approximate Gradient Projection Method with Applications to the Backpropagation Algorithm," Optimization Methods and Software, Vol. 4, pp. 85-101.

[LuT94b] Luo, Z. Q., and Tseng, P., 1994. "On the Rate of Convergence of a Distributed Asynchronous Routing Algorithm," IEEE Transactions on Automatic Control, Vol. 39, pp. 1123-1129.

[LuX15] Lu, Z., and Xiao, L., 2015. "On the Complexity Analysis of Randomized Block-Coordinate Descent Methods," Math. Programming, Vol. 152, pp. 615-642.

[LuY16] Luenberger, D. G., and Ye, Y., 1916. Introduction to Linear and Nonlinear Programming, 4th edition, Springer, N. Y.

[Lue69] Luenberger, D. G., 1969. Optimization by Vector Space Methods, Wiley, N. Y.

[Lue84] Luenberger, D. G., 1984. Introduction to Linear and Nonlinear Programming, 2nd edition, Addison-Wesley, Reading, MA.

[Lue98] Luenberger, D. G., 1998. Investment Science, Oxford University Press, N. Y.

[Luo91] Luo, Z. Q., 1991. "On the Convergence of the LMS Algorithm with Adaptive Learning Rate for Linear Feedforward Networks," Neural Computation, Vol. 3, pp. 226-245.

[Luq84] Luque, F.J., 1984. "Asymptotic Convergence Analysis of the Proximal Point Algorithm," SIAM J. on Control and Optimization, Vol. 22, pp. 277-293.

[MMS91] McShane, K. A., Monma, C. L., and Shanno, D., 1991. "An Implementation of a Primal-Dual Interior Point Method for Linear Programming," ORSA J. on Computing, Vol. 1, pp. 70-83.

[MMZ95] McKenna, M. P., Mesirov, J. P., and Zenios, S. A., 1995. "Data Parallel Quadratic Programming on Box-Constrained Problems," SIAM J. on Optimization, Vol. 5, pp. 570-589.

[MOT95] Mahey, P., Oualibouch, S., and Tao, P. D., 1995. "Proximal Decomposition on the Graph of a Maximal Monotone Operator," SIAM J. on Optimization, Vol. 5, pp. 454-466.

[MRR00] Mayne, D. Q., Rawlings, J. B., Rao, C. V., and Scokaert, P. O. M., 2000. "Constrained Model Predictive Control: Stability and Optimality," Automatica, Vol. 36, pp. 789-814.

[MSQ98] Mifflin, R., Sun, D., and Qi, L., 1998. "Quasi-Newton Bundle-Type Methods for Nondifferentiable Convex Optimization," SIAM J. on Optimization, Vol. 8, pp. 583-603.

[MTW93] Monteiro, R. D. C., Tsuchiya, T., and Wang, Y., 1993. "A Simplified Global Convergence Proof of the Affine Scaling Algorithm," Annals of Operations Res., Vol. 47, pp. 443-482.

[MYF03] Moriyama, H., Yamashita, N., and Fukushima, M., 2003. "The Incremental Gauss-Newton Algorithm with Adaptive Stepsize Rule," Computational Optimization and Applications, Vol. 26, pp. 107-141.

[MaD87] Mangasarian, O. L., and De Leone, R., 1987. "Parallel Successive Overelaxation Methods for Symmetric Linear Complementarity Problems and Linear Programs," J. of Optimization Th. and Appl., Vol. 54, pp. 437-446.

[MaD88] Mangasarian, O. L., and De Leone, R., 1988. "Parallel Gradient Projection Successive Overrelaxation for Symmetric Linear Complementarity Problems," Annals of Operations Res., Vol. 14, pp. 41-59.

[MaF67] Mangasarian, O. L., and Fromovitz, S., 1967. "The Fritz John Necessary Optimality Conditions in the Presence of Equality and Inequality Constraints," J. Math. Anal. and Appl., Vol. 17, pp. 37-47.

[MaP82] Mayne, D. Q., and Polak, E., 1982. "A Superlinearly Convergent Algorithm for Constrained Optimization Problems," Math. Programming Studies, Vol. 16, pp. 45-61.

[MaS94] Mangasarian, O. L., and Solodov, M. V., 1994. "Serial and Parallel Backpropagation Convergence Via Nonmonotone Perturbed Minimization," Optimization Methods and Software, Vol. 4, pp. 103-116.

[Mai13] Mairal, J., 2013. "Optimization with First-Order Surrogate Functions," arXiv preprint arXiv:1305.3120.

[Mai14] Mairal, J., 2014. "Incremental Majorization-Minimization Optimization with Application to Large-Scale Machine Learning," arXiv preprint arXiv:1402.4419.

[Man69] Mangasarian, O. L., 1969. Nonlinear Programming, Prentice-Hall, Englewood Cliffs, N. J.; also SIAM, Classics in Applied Mathematics 10, Phila., PA., 1994.

[Mar70] Martinet, B., 1970. "Regularisation d'Inequations Variationnelles par Approximations Successives," Rev. Francaise Inf. Rech. Oper., pp. 154-159.

[Mar72] Martinet, B., 1972. "Determination Approchee d'un Point Fixe d'une Application Pseudo-Contractante," C. R. Acad. Sci. Paris, 274A, pp. 163-165.

[Mar78] Maratos, N., 1978. "Exact Penalty Function Algorithms for Finite Dimensional and Control Optimization Problems," Ph.D. Thesis, Imperial College Sci. Tech, Univ. of London.

[McK98] McKinnon, K. I. M., 1998. "Convergence of the Nelder-Mead Simplex Method to a Non-Stationary Point," SIAM J. on Optimization, Vol. 9, pp. 148-158.

[McL80] McLinden, L., 1980. "The Complementarity Problem for Maximal Monotone Multifunctions," in Variational Inequalities and Complementarity Problems, Cottle, R., Giannessi, F., and Lions, J.-L., (Eds.), Wiley, N. Y., pp. 251-270.

[McS73] McShane, E. J., 1973. "The Lagrange Multiplier Rule," American Mathematical Monthly, Vol. 80, pp. 922-925.

[Meg88] Megiddo, N., 1988. "Pathways to the Optimal Set in Linear Programming," in Progress in Mathematical Programming, Megiddo, N., (Ed.), Springer-Verlag, N. Y., pp. 131-158.

[Meh92] Mehrotra, S., 1992. "On the Implementation of a Primal-Dual Interior Point Method," SIAM J. on Optimization, Vol. 2, pp. 575-601.

[Mey79] Meyer, R. R., 1979. "Two-Segment Separable Programming," Management Science, Vol. 25, pp. 385-395.

[Mey07] Meyn, S., 2007. Control Techniques for Complex Networks, Cambridge Univ. Press, N. Y.

[Mif96] Mifflin, R., 1996. "A Quasi-Second-Order Proximal Bundle Algorithm," Math. Programming, Vol. 73, pp. 51-72.

[Mig94] Migdalas, A., 1994. "A Regularization of the Frank-Wolfe Method and Unification of Certain Nonlinear Programming Methods," Math. Programming, Vol. 65, pp. 331-345.

[Min60] Minty, G. J., 1960. "Monotone Networks," Proc. Roy. Soc. London, A, Vol. 257, pp. 194-212.

[Min86] Minoux, M., 1986. Mathematical Programming: Theory and Algorithms, Wiley, N. Y.

[Mit66] Mitter, S. K., 1966. "Successive Approximation Methods for the Solution of Optimal Control Problems," Automatica, Vol. 3, pp. 135-149.

[MoA89a] Monteiro, R. D. C., and Adler, I., 1989. "Interior Path Following Primal-Dual Algorithms, Part I: Linear Programming," Math. Programming, Vol. 44, pp. 27-41.

[MoA89b] Monteiro, R. D. C., and Adler, I., 1989. "Interior Path Following Primal-Dual Algorithms, Part II: Convex Quadratic Programming," Math. Programming, Vol. 44, pp. 43-66.

[MoL99] Morari, M., and Lee, J. H., 1999. "Model Predictive Control: Past, Present, and Future," Computers and Chemical Engineering, Vol. 23, pp. 667-682.

[MoS83] Moré, J. J., and Sorensen, D. C., 1983. "Computing a Trust Region Step," SIAM J. on Scientific and Statistical Computing, Vol. 4, pp. 553-572.

[MoT89] Moré, J. J., and Toraldo, G., 1989. "Algorithms for Bound Constrained Quadratic Programming Problems," Numer. Math., Vol. 55, pp. 377-400.

[MoW93] Moré, J. J., and Wright, S. J., 1993. Optimization Software Guide, SIAM, Frontiers in Applied Mathematics 14, Phila., PA.

[Mor88] Mordukhovich, B. S., 1988. Approximation Methods in Problems of Optimization and Control, Nauka, Moscow.

[Mor06] Mordukhovich, B. S., 2006. Variational Analysis and Generalized Differentiation I: Basic Theory, Springer, N. Y.

[MuP75] Mukai, H., and Polak, E., 1975. "A Quadratically Convergent Primal-Dual Algorithm with Global Convergence Properties for Solving Optimization Problems with Equality Constraints," Math. Programming, Vol. 9, pp. 336-349.

[MuR95] Mulvey, J. M., and Ruszcynski, A., 1995. "A New Scenario Decomposition Method for Large Scale Stochastic Optimization," Operations Research, Vol. 43, pp. 477-490.

[MuS87] Murtagh, B. A., and Saunders, M. A., 1987. "MINOS 5.1 User's Guide," Technical Report SOL-83-20R, Stanford Univ.

[Mur92] Murty, K. G., 1992. Network Programming, Prentice-Hall, Englewood Cliffs, N. J.

[NBB01] Nedić, A., Bertsekas, D. P., and Borkar, V. S., 2001. "Distributed Asynchronous Incremental Subgradient Methods," in Inherently Parallel Algorithms in Feasibility and Optimization and Their Applications, Butnariu, D., Censor, Y., and Reich, S., (Eds.), Elsevier Science, Amsterdam, Netherlands.

[NSL15] Nutini, J., Schmidt, M., Laradji, I. H., Friedlander, M., and Koepke, H., 2015. "Coordinate Descent Converges Faster with the Gauss-Southwell Rule Than Random Selection," arXiv preprint arXiv:1506.00552.

[NaQ96] Nazareth, J. L., and Qi, L., 1996. "Globalization of Newton's Method for Solving Nonlinear Equations," Numerical Linear Algebra with Applications, Vol. 3, pp. 239-249.

[NaS89] Nash, S. G., and Sofer, 1989. "Block Truncated-Newton Methods for Parallel Optimization," Math. Programming, Vol. 45, pp. 529-546.

[NaT02] Nazareth, L., and Tseng, P., 2002. "Gilding the Lily: A Variant of the Nelder-Mead Algorithm Based on Golden-Section Search," Computational Optimization and Applications, Vol. 22, pp. 133-144.

[Nag93] Nagurney, A., 1993. Network Economics: A Variational Inequality Approach, Kluwer, Dordrecht, The Netherlands.

[Nas85] Nash, S. G., 1985. "Preconditioning of Truncated-Newton Methods," SIAM J. on Scientific and Statistical Computing, Vol. 6, pp. 599-616.

[Naz94] Nazareth, J. L., 1994. The Newton-Cauchy Framework: A Unified Approach to Unconstrained Nonlinear Minimization, Lecture Notes in Computer Science No. 769, Springer-Verlag, Berlin and New York.

[Naz96] Nazareth, J. L., 1996. "Lagrangian Globalization: Solving Nonlinear Equations via Constrained Optimization," in Mathematics of Numerical Analysis, Renegar, J., Shub, M., and Smale, S., (Eds.), Lectures in Applied Mathematics, Vol. 32, The American Mathematical Society, Providence, RI, pp. 533-542.

[NeB00] Nedić, A., and Bertsekas, D. P., 2000. "Convergence Rate of Incremental Subgradient Algorithms," Stochastic Optimization: Algorithms and Applications, S. Uryasev and P. M. Pardalos, Eds., Kluwer, pp. 263-304.

[NeB01] Nedić, A., and Bertsekas, D. P., 2001. "Incremental Subgradient Methods for Nondifferentiable Optimization," SIAM J. on Optimization, Vol. 12, pp. 109-138.

[NeB10] Nedić, A., and Bertsekas, D. P., 2010. "The Effect of Deterministic Noise in Subgradient Methods," Math. Programming, Ser. A, Vol. 125, pp. 75-99.

[NeN94] Nesterov, Y., and Nemirovskii, A., 1994. Interior Point Polynomial Algorithms in Convex Programming, SIAM, Studies in Applied Mathematics 13, Phila., PA.

[NeW88] Nemhauser, G. L., and Wolsey, L. A., 1988. Integer and Combinatorial Optimization, Wiley, N. Y.

[NeY83] Nemirovsky, A., and Yudin, D. B., 1983. Problem Complexity and Method Efficiency, Wiley, N. Y.

[Ned11] Nedić, A., 2011. "Random Algorithms for Convex Minimization Problems," Math. Programming, Vol. 129, pp. 225?253.

[Nes83] Nesterov, Y., 1983. "A Method for Unconstrained Convex Minimization Problem with the Rate of Convergence $O(1/k^2)$," Doklady AN SSSR, Vol. 269, pp. 543-547; translated as Soviet Math. Dokl.

[Nes95] Nesterov, Y., 1995. "Complexity Estimates of Some Cutting Plane Methods Based on Analytic Barrier," Math. Programming, Vol. 69, pp. 149-176.

[Nes04] Nesterov, Y., 2004. Introductory Lectures on Convex Optimization, Kluwer Academic Publisher, Dordrecht, The Netherlands.

[Nes12] Nesterov, Y., 2012. "Efficiency of Coordinate Descent Methods on Huge-Scale Optimization Problems," SIAM J. on Optimization, Vol. 22, pp. 341-362.

[Neu28] Neumann, J. von, 1928. "Zur Theorie der Gesellschaftsspiele," Math. Ann., Vol. 100, pp. 295-320.

[NgS79] Nguyen, V. H., and Strodiot, J. J., 1979. "On the Convergence Rate of a Penalty Function Method of Exponential Type," J. Opt. Th. and Appl., Vol. 27, pp. 495-508.

[NoW06] Nocedal, J., and Wright, S. J., 2006. Numerical Optimization, 2nd Edition, Springer, NY.

[Noc80] Nocedal, J., 1980. "Updating Quasi-Newton Matrices with Limited Storage," Math. of Computation, Vol. 35, pp. 773-782.

[OLR85] O'hEigeartaigh, M., Lenstra, S. K., and Rinnoy Kan, A. H. G., (Eds.), 1985. Combinatorial Optimization: Annotated Bibliographies, Wiley, N. Y.

[OrL74] Oren, S. S., and Luenberger, D. G., 1974. "Self-Scaling Variable Metric Algorithm, Part I," Management Science, Vol. 20, pp. 845-862.

[OrR70] Ortega, J. M., and Rheinboldt, W. C., 1970. Iterative Solution of Nonlinear Equations in Several Variables, Academic Press, N. Y.

[Ore73] Oren, S. S., 1973. "Self-Scaling Variable Metric Algorithm, Part II," Management Science, Vol. 20, pp. 863-874.

[OzB03] Ozdaglar, A. E., and Bertsekas, D. P., 2003. "Routing and Wavelength Assignment in Optical Networks," IEEE Trans. on Networking, pp. 259-272.

[OzB04] Ozdaglar, A. E., and Bertsekas, D. P., 2004. "The Relation Between Pseudonormality and Quasiregularity in Constrained Optimization," Optimization Methods and Software, Vol. 19, pp. 493–506.

[PCC93] Pulleyblank, W., Cook, W., Cunningham, W., and Schrijver, A., 1993. An Introduction to Combinatorial Optimization, Wiley, N. Y.

[PaM89] Pantoja, J. F. A. D., and Mayne, D. Q., 1989. "Sequential Quadratic Programming Algorithm for Discrete Optimal Control Problems with Control Inequality Constraints," Intern. J. on Control, Vol. 53, pp. 823-836.

[PaR87] Pardalos, P. M., and Rosen, J. B., 1987. Constrained Global Optimization: Algorithms and Applications, Springer-Verlag, N. Y.

[PaS82] Papadimitriou, C. H., and Steiglitz, K., 1982. Combinatorial Optimization: Algorithms and Complexity, Prentice-Hall, Englewood Cliffs, N. J.

[PaT91] Panier, E. R., and Tits, A. L., 1991. "Avoiding the Maratos Effect by Means of a Nonmonotone Line Search. I.," SIAM J. on Numer. Anal., Vol. 28, pp. 1183-1195.

[Pan84] Pang, J.-S., 1984. "On the Convergence of Dual Ascent Methods for Large-Scale Linearly Constrained Optimization Problems," Unpublished Manuscript, School of Management, Univ. of Texas, Dallas, Texas.

[Pap81] Papavassilopoulos, G., 1981. "Algorithms for a Class of Nondifferentiable Problems," J. Opt. Th. and Appl., Vol. 34, pp. 41-82.

[Pap82] Pappas, T. N., 1982. "Solution of Nonlinear Equations by Davidon's Least Squares Method," M.S. Thesis, Dept. of Electrical Engineering and Computer Science, M.I.T., Cambridge, MA.

[Pat93] Patriksson, M., 1993. "Partial Linearization Methods for Nonlinear Programming," J. Opt. Th. and Appl., Vol. 78, pp. 227-246.

[Pat98] Patriksson, M., 1998. Nonlinear Programming and Variational Inequalities: A Unified Approach, Kluwer, Dordtrecht, The Netherlands.

[Per78] Perry, A., 1978. "A Modified Conjugate Gradient Algorithm," Operations Research, Vol. 26, pp. 1073-1078.

[Pfl96] Pflug, G. C., 1996. Optimization of Stochastic Models, Kluwer, Boston.

[PiP73] Pironneau, O., and Polak, E., 1973. "Rate of Convergence of a Class of Methods of Feasible Directions," SIAM J. Numer. Anal., Vol. 10, pp. 161-173.

[PiZ92] Pinar, M. C., and Zenios, S. A., 1992. "Parallel Decomposition of Multicommodity Network Flows Using a Linear-Quadratic Penalty Algorithm," ORSA J. on Computing, Vol. 4, pp. 235-249.

[PiZ94] Pinar, M. C., and Zenios, S. A., 1994. "On Smoothing Exact Penalty Functions for Convex Constrained Problems," SIAM J. on Optimization, Vol. 4, pp. 486-511.

[PoH91] Polak, E., and He, L., 1991. "Finite-Termination Schemes for Solving Semi-infinite Satisfycing Problems," J. Opt. Theory and Appl., Vol. 70, pp. 429-442.

[PoR69] Polak, E., and Ribiere, G., 1969. "Note sur la Convergence de Methodes de Directions Conjugees," Rev. Fr. Inform. Rech. Oper., Vol. 16-R1, pp. 35-43.

[PoT73a] Poljak, B. T., and Tsypkin, Y. Z., 1973. "Pseudogradient Adaptation and Training Algorithms," Automation and Remote Control, pp. 45-68.

[PoT73b] Poljak, B. T., and Tretjakov, N. V., 1973. "The Method of Penalty Estimates for Conditional Extremum Problems," Z. Vyčisl. Mat. i Mat. Fiz., Vol. 13, pp. 34-46.

[PoT80a] Polak, E., and Tits, A. L., 1980. "A Globally Convergent, Implementable Multiplier Method with Automatic Penalty Limitation," Applied Math. and Optimization, Vol. 6, pp. 335-360.

[PoT80b] Poljak, B. T., and Tsypkin, Y. Z., 1980. "Adaptive Estimation Algorithms (Convergence, Optimality, Stability)," Automation and Remote Control, Vol. 40, pp. 378-389.

[PoT81] Poljak, B. T., and Tsypkin, Y. Z., 1981. "Optimal Pseudogradient Adaptation Algorithms," Automation and Remote Control, Vol. 41, pp. 1101-1110.

[PoT97] Polyak, R., and Teboulle, M., 1997. "Nonlinear Rescaling and Proximal-Like Methods in Convex Optimization," Math. Programming, Vol. 76, pp. 265-284.

[Pol64] Poljak, B. T., 1964. "Some Methods of Speeding up the Convergence of Iteration Methods," Z. VyČisl. Mat. i Mat. Fiz., Vol. 4, pp. 1-17.

[Pol69a] Poljak, B. T., 1969. "The Conjugate Gradient Method in Extremal Problems," Z. Vyčisl. Mat. i Mat. Fiz., Vol. 9, pp. 94-112.

[Pol69b] Poljak, B. T., 1969. "Minimization of Unsmooth Functionals," Z. Vyčisl. Mat. i Mat. Fiz., Vol. 9, pp. 14-29.

[Pol70] Poljak, B. T., 1970. "Iterative Methods Using Lagrange Multipliers for Solving Extremal Problems with Constraints of the Equation Type," Z. VyČisl. Mat. i Mat. Fiz., Vol. 10, pp. 1098-1106.

[Pol71] Polak, E., 1971. Computational Methods in Optimization: A Unified Approach, Academic Press, N. Y.

[Pol73] Polak, E., 1973. "A Historical Survey of Computational Methods in Optimal Control," SIAM Review, Vol. 15, pp. 553-584.

[Pol79] Poljak, B. T., 1979. "On Bertsekas' Method for Minimization of Composite Functions," Internat. Symp. Systems Opt. Analysis, Benoussan, A., and Lions, J. L., (Eds.), pp. 179-186, Springer-Verlag, Berlin and N. Y.

[Pol87] Poljak, B. T., 1987. Introduction to Optimization, Optimization Software Inc., N. Y.

[Pol92] Polyak, R., 1992. "Modified Barrier Functions (Theory and Methods)," Math. Programming, Vol. 54, pp. 177-222.

[Pol97] Polak, E., 1997. Optimization: Algorithms and Consistent Approximations, Springer-Verlag, N. Y.

[Pot94] Potra, F. A., 1994. "A Quadratically Convergent Predictor-Corrector Method for Solving Linear Programs from Infeasible Starting Points," Math. Programming, Vol. 67, pp. 383–406.

[Pow64] Powell, M. J. D., 1964. "An Efficient Method for Finding the Minimum of a Function of Several Variables without Calculating Derivatives," The Computer Journal, Vol. VII, pp. 155-162.

[Pow69] Powell, M. J. D., 1969. "A Method for Nonlinear Constraints in Minimizing Problems," in Optimization, Fletcher, R., (Ed.), Academic Press, N. Y, pp. 283-298.

[Pow73] Powell, M. J. D., 1973. "On Search Directions for Minimization Algorithms," Math. Programming, Vol. 4, pp. 193-201.

[Pre95] Prekopa, A., 1995. Stochastic Programming, Kluwer, Boston.

[PsD75] Pschenichny, B. N., and Danilin, Y. M., 1975. "Numerical Methods in Extremal Problems," MIR, Moscow, (Engl. trans., 1978).

[Psc70] Pschenichny, B. N., 1970. "Algorithms for the General Problem of Mathematical Proramming," Kibernetika (Kiev), Vol. 6, pp. 120-125.

[Pyt98] Pytlak, R., 1998. "An Efficient Algorithm for Large-Scale Nonlinear Programming Problems with Simple Bounds on the Variables," SIAM J. on Optimization, Vol. 8, pp. 532-560.

[RGV14] Richard, E., Gaiffas, S., and Vayatis, N., 2014. "Link Prediction in Graphs with Autoregressive Features," J. of Machine Learning Research, Vol. 15, pp. 565-593.

[RHL13] Razaviyayn, M., Hong, M., and Luo, Z. Q., 2013. "A Unified Convergence Analysis of Block Successive Minimization Methods for Nonsmooth Optimization," SIAM J. on Optimization, Vol. 23, pp. 1126-1153.

[RSP16] Reddi, S. J., Sra, S., Poczos, B., and Smola, A., 2016. "Fast Incremental Method for Nonconvex Optimization," arXiv preprint arXiv:1603.06159.

[Ray93] Raydan, M., 1993. "On the Barzilai and Borwein Choice of Steplength for the Gradient Method," IMA J. Num. Anal., Vol. 13, pp. 321-326.

[Ray97] Raydan, M., 1997. "The Barzilai and Borwein Gradient Method for the Large Scale Unconstrained Minimization Problem," SIAM J. on Optimization, Vol. 7, pp. 26-33.

[ReR98] Reemtsen, R., and Ruckman, J. J., (Eds.), 1998. Semi-Infinite Programming, Kluwer, Boston.

[Ren01] Renegar, J., 2001. A Mathematical View of Interior-Point Methods in Convex Optimization, SIAM, Phila.

[RiT14] Richtarik, P., and Takac, M., 2014. "Iteration Complexity of Randomized Block-Coordinate Descent Methods for Minimizing a Composite Function," Math. Programming, Vol. 144, pp. 1-38.

[RoW91] Rockafellar, R. T., and Wets, R. J.-B., 1991. "Scenarios and Policy Aggregation in Optimization under Uncertainty," Math. of Operations Res., Vol. 16, pp. 119-147.

[RoW98] Rockafellar, R. T., and Wets, R. J.-B., 1998. Variational Analysis, Springer-Verlag, Berlin.

[Rob74] Robinson, S. M., 1974. "Perturbed Kuhn-Tucker Points and Rates of Convergence for a Class of Nonlinear Programming Algorithms," Math. Programming, Vol. 7, pp. 1-16.

[Rob87] Robinson, S. M., 1987. "Local Structure of Feasible Sets in Nonlinear Programming, Part III. Stability and Sensitivity," Math. Programming Studies, Vol. 30, pp. 45-66.

[Roc67] Rockafellar, R. T., 1967. "Convex Programming and Systems of Elementary Monotonic Relations," J. of Math. Analysis and Applications, Vol. 19, pp. 543-564.

[Roc70] Rockafellar, R. T., 1970. Convex Analysis, Princeton Univ. Press, Princeton, N. J.

[Roc73a] Rockafellar, R. T., 1973. "A Dual Approach to Solving Nonlinear Programming Problems by Unconstrained Optimization," Math. Programming, pp. 354-373.

[Roc73b] Rockafellar, R. T., 1973. "The Multiplier Method of Hestenes and Powell Applied to Convex Programming," J. Opt. Th. and Appl., Vol. 12, pp. 555-562.

[Roc74] Rockafellar, R. T., 1974. "Augmented Lagrange Multiplier Functions and Duality in Nonconvex Programming," SIAM J. on Control, Vol. 12, pp. 268-285.

[Roc76a] Rockafellar, R. T., 1976. "Monotone Operators and the Proximal Point Algorithm," SIAM J. on Control and Optimization, Vol. 14, pp. 877-898.

[Roc76b] Rockafellar, R. T., 1976. "Augmented Lagrangians and Applications of the Proximal Point Algorithm in Convex Programming," Math. Operations Res., Vol. 1, pp. 97-116.

[Roc76c] Rockafellar, R. T., 1976. "Solving a Nonlinear Programming Problem by Way of a Dual Problem," Symp. Matematica, Vol. 27, pp. 135-160.

[Roc81] Rockafellar, R. T., 1981. "Monotropic Programming: Descent Algorithms and Duality," in Nonlinear Programming 4, by Mangasarian, O. L., Meyer, R. R., and Robinson, S. M., (Eds.), Academic Press, N. Y., pp. 327-366.

[Roc84] Rockafellar, R. T., 1984. Network Flows and Monotropic Optimization, Wiley, N. Y.; republished by Athena Scientific, Belmont, MA, 1998.

[Roc90] Rockafellar, R. T., 1990. "Computational Schemes for Solving Large-Scale Problems in Extended Linear-Quadratic Programming," Math. Programming, Vol. 48, pp. 447-474.

[Roc93] Rockafellar, R. T., 1993. "Lagrange Multipliers and Optimality," SIAM Review, Vol. 35, pp. 183-238.

[Ros60a] Rosenbrock, H. H., 1960. "An Automatic Method for Finding the Greatest or Least Value of a Function," Computer J., Vol. 3, pp. 175-184.

[Ros60b] Rosen, J. B., 1960. "The Gradient Projection Method for Nonlinear Programming, Part I, Linear Constraints," SIAM J. Applied Math., Vol. 8, pp. 514-553.

[RuK04] Rubinstein, R. Y., and Kroese, D. P., 2004. The Cross-Entropy Method: A Unified Approach to Combinatorial Optimization, Springer, N. Y.

[Rud76] Rudin, W., 1976. Real Analysis, McGraw-Hill, N. Y.

[Rus86] Ruszczynski, A., 1986. "A Regularized Decomposition Method for Minimizing a Sum of Polyhedral Functions," Math. Programming, Vol. 35, pp. 309-333.

[Rus89] Ruszczynski, A., 1989. "An Augmented Lagrangian Decomposition Method for Block Diagonal Linear Programming Problems," Operations Res. Letters, Vol. 8, pp. 287-294.

[Rus95] Ruszczynski, A., 1995. "On Convergence of an Augmented Lagrangian Decomposition Method for Sparse Convex Optimization," Math. of Operations Res., Vol. 20, pp. 634-656.

[Rus97] Ruszczynski, A., 1997. "Decomposition Methods in Stochastic Programming," Math. Programming, Vol. 79, pp. 333-353.

[Rus06] Ruszczynski, A., 2006. Nonlinear Optimization, Princeton Univ. Press, Princeton, N. J.

[SBC93] Saarinen, S., Bramley, R., and Cybenko, G., 1993. "Ill-Conditioning in Neural Network Training Problems," SIAM J. Sci. Comput., Vol. 14, pp. 693-714.

[SBK64] Shah, B., Buehler, R., and Kempthorne, O., 1964. "Some Algorithms for Minimizing a Function of Several Variables," J. Soc. Indust. Appl. Math., Vol. 12, pp. 74-92.

[SFR09] Schmidt, M., Fung, G., and Rosales, R., 2009. "Optimization Methods for ℓ_1-Regularization," Univ. of British Columbia, Technical Report TR-2009-19.

[SHH62] Spendley, W. G., Hext, G. R., and Himsworth, F. R., 1962. "Sequential Application of Simplex Designs in Optimisation and Evolutionary Operation," Technometrics, Vol. 4, pp. 441-461.

[SHM16] Silver, D., Huang, A., Maddison, C. J., Guez, A., Sifre, L., Van Den Driessche, G., Schrittwieser J., et al. 2016. "Mastering the Game of Go with Deep Neural Networks and Tree Search," Nature, Vol. 529, pp. 484-489.

[SKS12] Schmidt, M., Kim, D., and Sra, S., 2012. "Projected Newton-Type Methods in Machine Learning," in Optimization for Machine Learning, by Sra, S., Nowozin, S., and Wright, S. J., (eds.), MIT Press, Cambridge, MA, pp. 305-329.

[SLB13] Schmidt, M., Le Roux, N., and Bach, F., 2013. "Minimizing Finite Sums with the Stochastic Average Gradient," arXiv preprint arXiv:1309.2388.

[SaS86] Saad, Y., and Schultz, M. H., 1986. "GMRES: A Generalized Minimal Residual Algorithm for Solving Nonsymmetric Linear Systems," SIAM J. Sci. Statist. Comput., Vol. 7, pp. 856-869.

[Sah96] Sahinidis, N. V., 1996. "BARON: A General Purpose Global Optimization Software Package," Journal of Global Optimization, Vol. 8, pp. 201-205.

[Sah04] Sahinidis, N. V., 2004. "Optimization Under Uncertainty: State-of-the-Art and Opportunities," Computers and Chemical Engineering, Vol. 28, pp. 971-983.

[Sak66] Sakrison, D. T., 1966. "Stochastic Approximation: A Recursive Method for Solving Regression Problems," in Advances in Communication Theory and Applications, 2, A. V. Balakrishnan, ed., Academic Press, NY, pp. 51-106.

[ScF14] Schmidt, M., and Friedlander, M. P., 2014. "Coordinate Descent Converges Faster with the Gauss-Southwell Rule than Random Selection," Advances in Neural Information Processing Systems 27 (NIPS 2014).

[Sch82] Schnabel, R. B., 1982. "Determining Feasibility of a Set of Nonlinear Inequality Constraints," Math. Programming Studies, Vol. 16, pp. 137-148.

[Sch86] Schrijver, A., 1986. Theory of Linear and Integer Programming, Wiley, N. Y.

[Sch93] Schrijver, A., 1993. Combinatorial Optimization: Polyhedra and Efficiency, Springer, N. Y.

[Sch10] Schmidt, M., 2010. "Graphical Model Structure Learning with L1-Regularization," PhD Thesis, Univ. of British Columbia.

[Sch12] Schittkowski, K., 2012. Nonlinear Programming Codes: Information, Tests, Performance, Springer Science and Business Media.

[SeS86] Sen, S., and Sherali, H. D., 1986. "A Class of Convergent Primal-Dual Subgradient Algorithms for Decomposable Convex Programs," Math. Programming, Vol. 35, pp. 279-297.

[ShA99] Sherali, H. D., and Adams, W. P., 1999. A Reformulation-Linearization Technique for Solving Discrete and Continuous Nonconvex Problems, Kluwer, Boston.

[ShD14] Shapiro, A., and Dentcheva D., 2014. Lectures on Stochastic Programming: Modeling and Theory, SIAM, Phila.

[Sha70] Shanno, D. F., 1970. "Conditioning of Quasi-Newton Methods for Function Minimization," Math. Comput., Vol. 27, pp. 647-656.

[Sha78] Shanno, D. F., 1978. "Conjugate Gradient Methods with Inexact Line Searches," Math. of Operations Res., Vol. 3, pp. 244-256.

[Sha79] Shapiro, J. E., 1979. Mathematical Programming Structures and Algorithms, Wiley, N. Y.

[Sha88] Shapiro, A., 1988. "Sensitivity Analysis of Nonlinear Programs and Differentiability Properties of Metric Projections," SIAM J. on Control and Optimization, Vol. 26, pp. 628-645.

[Sho85] Shor, N. Z., 1985. Minimization Methods for Nondifferentiable Functions, Springer-Verlag, Berlin.

[Sho98] Shor, N. Z., 1998. Nondifferentiable Optimization and Polynomial Problems, Kluwer, Dordrecht, the Netherlands.

[Sla50] Slater, M., 1950. "Lagrange Multipliers Revisited: A Contribution to Non-Linear Programming," Cowles Commission Discussion Paper, Math. 403.

[Sol98] Solodov, M. V., 1998. "Incremental Gradient Algorithms with Stepsizes Bounded Away from Zero," Computational Optimization and Applications, Vol. 11, pp. 23-35.

[Son86] Sonnevend, G., 1986. "An "Analytical Centre" for Polyhedrons and New Classes of Global Algorithms for Linear (Smooth, Convex) Programming," Lecture Notes in Control and Information Sciences, Vol. 84, pp. 866-878.

[Spa03] Spall, J. C., 2003. Introduction to Stochastic Search and Optimization: Estimation, Simulation, and Control, J. Wiley, Hoboken, N. J.

[Spa12] Spall, J. C., 2012. "Cyclic Seesaw Process for Optimization and Identification," J. of Optimization Theory and Applications, Vol. 154, pp. 187-208.

[Spi85] Spingarn, J. E., 1985. "Applications of the Method of Partial Inverses to Convex Programming: Decomposition," Math. Programming, Vol. 32, pp. 199-223.

[StW75] Stephanopoulos, G., and Westerberg, A. W., 1975. "The Use of Hestenes' Method of Multipliers to Resolve Dual Gaps in Engineering System Optimization," J. Opt. Th. and Applications, Vol. 15, pp. 285-309.

[Str76] Strang, G., 1976. Linear Algebra and Its Applications, Academic Press, N. Y.

[TBA86] Tsitsiklis, J. N., Bertsekas, D. P., and Athans, M., 1986. "Distributed Asynchronous Deterministic and Stochastic Gradient Optimization Algorithms," IEEE Trans. on Aut. Control, Vol. AC-31, pp. 803-812.

[TBT90] Tseng, P., Bertsekas, D. P., and Tsitsiklis, J. N., 1990. "Partially Asynchronous Algorithms for Network Flow and Other Problems," SIAM J. on Control and Optimization, Vol. 28, pp. 678-710.

[TTA15] Toulis, P., Tran, D., and Airoldi, E. M., 2015. "Stability and Optimality in Stochastic Gradient Descent," arXiv preprint arXiv:1505.02417.

[TZY95] Tapia, R. A., Zhang, Y., and Ye, Y., 1995. "On the Convergence of the Iteration Sequence in Primal-Dual Interior-Point Methods," Math. Programming, Vol. 68, pp. 141-154.

[TaM85] Tanikawa, A., and Mukai, H., 1985. "A New Technique for Nonconvex Primal-Dual Decomposition," IEEE Trans. on Aut. Control, Vol. AC-30, pp. 133-143.

[TaP13] Talischi, C., and Paulino, G. H., 2013. "A Consistent Operator Splitting Algorithm and a Two-Metric Variant: Application to Topology Optimization," arXiv preprint arXiv:1307.5100.

[Tap77] Tapia, R. A., 1977. "Diagonalized Multiplier Methods and Quasi-Newton Methods for Constrained Minimization," J. Opt. Th. and Applications, Vol. 22, pp. 135-194.

[Teb92] Teboulle, M., 1992. "Entropic Proximal Mappings with Applications to Nonlinear Programming," Math. Operations Res., Vol. 17, pp. 1-21.

[ToT90] Toint, P. L., and Tuyttens, D., 1990. "On Large Scale Nonlinear Network Optimization," Math. Programming, Vol. 48, pp. 125-159.

[ToV67] Topkis, D. M., and Veinott, A. F., 1967. "On the Convergence of Some Feasible Directions Algorithms for Nonlinear Programming," SIAM J. on Control, Vol. 5, pp. 268-279.

[Tor91] Torczon, V., 1991. "On the Convergence of the Multidimensional Search Algorithm," SIAM J. on Optimization, Vol. 1, pp. 123-145.

[TrW80] Traub, J. F., and Wozniakowski, H., 1980. A General Theory of Optimal Algorithms, Academic Press, N. Y.

[TsB86] Tsitsiklis, J. N., and Bertsekas, D. P., 1986. "Distributed Asynchronous Optimal Routing in Data Networks," IEEE Trans. on Automatic Control, Vol. 31, pp. 325-331.

[TsB87] Tseng, P., and Bertsekas, D. P., 1987. "Relaxation Methods for Problems with Strictly Convex Separable Costs and Linear Constraints," Math. Programming, Vol. 38, pp. 303-321.

[TsB90] Tseng, P., and Bertsekas, D. P., 1990. "Relaxation Methods for Monotropic Programs," Math. Programming, Vol. 46, pp. 127-151.

[TsB91] Tseng, P., and Bertsekas, D. P., 1991. "Relaxation Methods for Problems with Strictly Convex Costs and Linear Constraints," Math. Operations Res., Vol. 16, pp. 462-481.

[TsB93] Tseng, P., and Bertsekas, D. P., 1993. "On the Convergence of the Exponential Multiplier Method for Convex Programming," Math. Programming, Vol. 60, pp. 1-19.

[TsB00] Tseng, P., and Bertsekas, D. P., 2000. "An Epsilon-Relaxation Method for Separable Convex Cost Generalized Network Flow Problems," Math. Programming, Vol. 88, pp. 85-104.

[Tse89] Tseng, P., 1989. "A Simple Complexity Proof for a Polynomial-Time Linear Programming Algorithm," Operations Res. Letters, Vol. 8, pp. 155-159.

[Tse90] Tseng, P., 1990. "Dual Ascent Methods for Problems with Strictly Convex Costs and Linear Constraints: A Unified Approach," SIAM J. on Control and Optimization, Vol. 28, pp. 214-242.

[Tse91a] Tseng, P., 1991. "On the Rate of Convergence of a Partially Asynchronous Gradient Projection Algorithm," SIAM J. on Optimization, Vol. 4, pp. 603-619.

[Tse91b] Tseng, P., 1991. "Relaxation Method for Large Scale Linear Programming using Decomposition," Math. of Operations Res., Vol. 17, pp. 859-880.

[Tse91c] Tseng, P., 1991. "Decomposition Algorithm for Convex Differentiable Minimization," J. Opt. Theory and Appl., Vol. 70, pp. 109-135.

[Tse92] Tseng, P., 1992. "Complexity Analysis of a Linear Complementarity Algorithm Based on a Lyapunov Function," Math. Programming, Vol. 53, pp. 297-306.

[Tse93] Tseng, P., 1993. "Dual Coordinate Ascent Methods for Non-Strictly Convex Minimization," Math. Programming, Vol. 59, pp. 231-247.

[Tse95a] Tseng, P., 1995. "Fortified-Descent Simplicial Search Method," Report, Dept. of Math., University of Washington, Seattle, Wash.; also in SIAM J. on Optimization, Vol. 10, 2000, pp. 269-288.

[Tse95b] Tseng, P., 1995. "Simplified Analysis of an $O(nL)$-Iteration Infeasible Predictor-Corrector Path Following Method for Monotone LCP," in Recent Trends in Optimization Theory and Applications, Agarwal, R. P., (Ed.), World Scientific, pp. 423-434.

[Tse98] Tseng, P., 1998. "Incremental Gradient(-Projection) Method with Momentum Term and Adaptive Stepsize Rule," SIAM J. on Optimization, Vol. 8, pp. 506-531.

[Tse00] Tseng, P., 2000. "A Modified Forward-Backward Splitting Method for Maximal Monotone Mappings," SIAM J. on Control and Optimization, Vol. 38, pp. 431-446.

[Tse01a] Tseng, P., 2001. "Convergence of Block Coordinate Descent Methods for Non-differentiable Minimization," J. Optim. Theory Appl., Vol. 109, pp. 475-494.

[Tse01b] Tseng, P., 2001. "An Epsilon Out-of-Kilter Method for Monotropic Programming," Math. of Operations Research, Vol. 26, pp. 221-233.

[Tse04] Tseng, P., 2004. "An Analysis of the EM Algorithm and Entropy-Like Proximal Point Methods," Math. Operations Research, Vol. 29, pp. 27-44.

[Tse08] Tseng, P., 2008. "On Accelerated Proximal Gradient Methods for Convex-Concave Optimization," Report, Math. Dept., Univ. of Washington.

[VaB95] Vandenberghe, L., and Boyd, S., 1995. "A Primal-Dual Potential Reduction Method for Problems Involving Matrix Inequalities," Math. Programming, Vol. 69, pp. 205-236.

[VaW89] Varaiya, P., and Wets, R. J.-B., 1989. "Stochastic Dynamic Optimization Approaches and Computation," Mathematical Programming: State of the Art, M. Iri and K. Tanabe (eds.), Kluwer, Boston, pp. 309-332.

[VeH93] Ventura, J. A., and Hearn, D. W., 1993. "Restricted Simplicial Decomposition for Convex Constrained Problems," Math. Programming, Vol. 59, pp. 71-85.

[Ven67] Venter, J. H., 1967. "An Extension of the Robbins-Monro Procedure," Ann. Math. Statist., Vol. 38, pp. 181-190.

[WDS13] Weinmann, A., Demaret, L., and Storath, M., 2013. "Total Variation Regularization for Manifold-Valued Data," arXiv preprint arXiv:1312.7710.

[WHM13] Wang, X., Hong, M., Ma, S., Luo, Z. Q., 2013. "Solving Multiple-Block Separable Convex Minimization Problems Using Two-Block Alternating Direction Method of Multipliers," arXiv preprint arXiv:1308.5294.

[WQB98] Wei, Z., Qi, L., and Birge, J. R., 1998. "New Method for Nonsmooth Convex Optimization," J. of Inequalities and Applications, Vol. 2, pp. 157-179.

[WSK14] Wytock, M., Sra, S., and Kolter, J. K., 2014. "Fast Newton Methods for the Group Fused Lasso," Proc. of 2014 Conf. on Uncertainty in Artificial Intelligence.

[WaB13a] Wang, M., and Bertsekas, D. P., 2013. "Incremental Constraint Projection-Proximal Methods for Nonsmooth Convex Optimization," Lab. for Information and Decision Systems Report LIDS-P-2907, MIT.

[WaB13b] Wang, M., and Bertsekas, D. P., 2013. "Convergence of Iterative Simulation-Based Methods for Singular Linear Systems," Stochastic Systems, Vol. 3, pp. 38-95.

[WaB15] Wang, M., and Bertsekas, D. P., 2015. "Incremental Constraint Projection Methods for Variational Inequalities." Math. Programming," Vol. 150.2, pp. 321-363.

[WaB16] Wang, M., and Bertsekas, D. P., 2016. "Stochastic First-Order Methods with Random Constraint Projection." SIAM J. on Optimization, Vol. 26, pp. 681-717.

[WeO13] Wei, E., and Ozdaglar, A., 2013. "On the $O(1/k)$ Convergence of Asynchronous Distributed Alternating Direction Method of Multipliers," arXiv preprint arXiv:1307.8254.

[Web29] Weber, A., 1929. Theory of Location of Industries, (Engl. Transl. by C. J. Friedrich), Univ. of Chicago Press, Chicago, Ill.

[WiH60] Widrow, B., and Hoff, M. E., 1960. "Adaptive Switching Circuits," Institute of Radio Engineers, Western Electronic Show and Convention, Convention Record, Part 4, pp. 96-104.

[Wil63] Wilson, R. B., 1963. "A Simplicial Algorithm for Concave Programming," Ph.D. Thesis, Grad. Sch. Business Admin., Harvard Univ., Cambridge, MA.

[Wol98] Wolsey, L. A., 1998. Integer Programming, Wiley, N. Y.

[Wri92] Wright, S. J., 1992. "An Interior Point Algorithm for Linearly Constrained Optimization," SIAM J. on Optimization, Vol. 2, pp. 450-473.

[Wri93a] Wright, S. J., 1993. "Identifiable Surfaces in Constrained Optimization," SIAM J. on Control and Optimization, Vol. 31, pp. 1063-1079.

[Wri93b] Wright, S. J., 1993. "Interior Point Methods for Optimal Control of Discrete Time Systems," J. Opt. Theory and Appl., Vol. 77, pp. 161-187.

[Wri93c] Wright, S. J., 1993. "A Path-Following Infeasible-Interior-Point Algorithm for

Linear Complementarity Problems," Optimization Methods and Software, Vol. 2, pp. 79-106.

[Wri94] Wright, S. J., 1994. "An Infeasible-Interior-Point Algorithm for Linear Complementarity Problems," Math. Programming, Vol. 67, pp. 29-52.

[Wri96] Wright, S. J., 1996. "A Path-Following Interior-Point Algorithm for Linear and Quadratic Problems," Annals of Operations Res., Vol. 62, pp. 103-130.

[Wri97a] Wright, S. J., 1997. Primal-Dual Interior Point Methods, SIAM, Phila., PA.

[Wri97b] Wright, S. J., 1997. "Applying New Optimization Algorithms to Model Predictive Control," Chemical Process Control-V, CACHE, AIChE Symposium Series No. 316, Vol. 93, pp. 147-155.

[Wri98] Wright, S. J., 1998. "Superlinear Convergence of a Stabilized SQP Method to a Degenerate Solution," Computational Optimization and Applications, Vol. 11, pp. 253-275.

[WuB01] Wu, C., and Bertsekas, D. P., 2001. "Distributed Power Control Algorithms for Wireless Networks," IEEE Trans. on Vehicular Technology, Vol. 50, pp. 504-514.

[YSQ14] You, K., Song, S., and Qiu, L., 2014. "Randomized Incremental Least Squares for Distributed Estimation Over Sensor Networks," Preprints of the 19th World Congress The International Federation of Automatic Control Cape Town, South Africa.

[YVG10] Yu, J., Vishwanathan, S. V. N., Gunter, S., and Schraudolph, N. N., 2010. "A Quasi-Newton Approach to Nonsmooth Convex Optimization Problems in Machine Learning," J. of Machine Learning Research, Vol. 11, pp. 1145-1200.

[Ye92] Ye, Y., 1992. "A Potential Reduction Algorithm Allowing Column Generation," SIAM J. on Optimization, Vol. 2, pp. 7-20.

[Ye97] Ye, Y., 1997. Interior Point Algorithms: Theory and Analysis, Wiley Interscience, N. Y.

[Ypm95] Ypma, T. J., 1995. "Historical Development of the Newton-Raphson Method," SIAM Review, Vol. 37, pp. 531-551.

[You15] Yousefpour, R., 2015. "Combination of Steepest Descent and BFGS Methods for Nonconvex Nonsmooth Optimization," Numerical Algorithms, pp. 1-34.

[ZLW99] Zhao, X., Luh, P. B., and Wang, J., 1999. "Surrogate Gradient Algorithm for Lagrangian Relaxation," J. Opt. Theory and Appl., Vol. 100, pp. 699-712.

[ZTP93] Zhang, Y., Tapia, R. A., and Potra, F., 1993. "On the Superlinear Convergence of Interior-Point Algorithms for a General Class of Problems," SIAM J. on Optimization, Vol. 3 pp. 413-422.

[Zal02] Zalinescu, C., 2002. Convex Analysis in General Vector Spaces, World Scientific, Singapore.

[Zan67a] Zangwill, W. I., 1967. "Minimizing a Function Without Calculating Derivatives," The Computer Journal, Vol. X, pp. 293-296.

[Zan67b] Zangwill, W. I., 1967. "Nonlinear Programming via Penalty Functions," Management Science, Vol. 13, pp. 344-358.

[Zan69] Zangwill, W. I., 1969. Nonlinear Programming, Prentice-Hall, Englewood Cliffs, N. J.

[ZhT92] Zhang, Y., and Tapia, R. A., 1992. "Superlinear and Quadratic Convergence of Primal-Dual Interior-Point Algorithms for Linear Programming Revisited," J. Opt. Theory and Appl., Vol. 73, pp. 229-242.

[ZhT93] Zhang, Y., and Tapia, R. A., 1993. "A Superlinearly Convergent Polynomial Primal-Dual Interior-Point Algorithm for Linear Programming," SIAM J. on Optimization, Vol. 3, pp. 118-133.

[Zho93] Zhou, L., 1993. "A Simple Proof of the Shapley-Folkman Theorem," Economic Theory, Vol. 3, pp. 371-372.

[Zhu95] Zhu, C., 1995. "On the Primal-Dual Steepest Descent Algorithm for Extended Linear-Quadratic Programming," SIAM J. on Optimization, Vol. 5, pp. 114-128.

[Zou60] Zoutendijk, G., 1960. Methods of Feasible Directions, Elsevier Publ. Co., Amsterdam.

[Zou76] Zoutendijk, G., 1976. Mathematical Programming Methods, North Holland, Amsterdam.

INDEX

A

Active constraint identification 276, 340
Additive cost function 4, 158, 320
Adjoint equation 213, 214
Affine hull 790
Alternating direction method of multipliers (ADMM) 666, 692, 742
Analytic center 451, 723
Arithmetic-geometric mean inequality 13, 18, 362
Armijo rule 36, 49, 95, 260, 283, 285, 339
Assignment problem 632, 635, 725
Asymptotic direction 247, 338, 792
Asymptotic sequence 247, 338, 792
Asynchronous gradient method 202, 203, 676
Auction algorithm 725
Augmented Lagrangian 367, 469, 531, 683
Automatic differentiation 229

B

BFGS method 139, 145, 743
Backpropagation 165, 218
Ball center 723
Barrier function 447
Barrier method 447
Basis 747
Boundary point 753
Bounded sequence 750
Bounded set 753
Branch-and-bound 564, 639, 664
Branch-and-cut 647
Bundle methods 722

C

Capture theorem 40, 87
Caratheodory's theorem 790
Cauchy sequence 750
Central cutting plane methods 722, 744
Central difference formula 148
Central path 450, 469
Certainty equivalent control 234
Chain rule 768, 804, 808
Characteristic polynomial 759
Cholesky factorization 101, 818
Classification 110, 159, 685
Closed function 784
Closed set 753

Closure point 753
Coercive 754
Column generation 739
Compact set 753
Complementary slackness (CS) 380
Complementary violation (CV) 389
Concave function 784
Condition number 72, 91
Conditional gradient method 262, 338
Cone 790
Cone generated by a set 790
Cone of first order feasible variations 404
Conic programming 575, 604, 661
Conjugate concave function 598
Conjugate convex function 598
Conjugate direction method 120
Conjugate directions 120
Conjugate gradient method 34, 125, 221, 334
Constant stepsize rule 36, 51, 152, 198, 260
Constraint qualifications 392, 421
Continuous function 754
Continuously differentiable function 766
Contraction mapping 772
Contraction modulus 772
Convergence of iterations with delays 774, 775
Convergence of nonstationary iterations 776
Convergence rate 69, 204
Convex combination 790
Convex function 784
Convex hull 790
Convex set 783
Coordinate descent methods 34, 149, 195, 323, 724
Costate vector 212, 214
Courant-Fisher minimax principle 762
Craig's method 131
Cubic interpolation 812
Cutting plane methods 717, 739, 744

D

DFP method 139
Danskin's theorem 806, 808

Dantzig-Wolfe decomposition 739
Deep neural networks 109
Delays in iterations 195, 774
Descent lemma 770
Diagonal matrix 748
Diagonal scaling 33, 189, 230, 703
Dimension 747
Diminishing stepsize rule 38, 54, 165, 205
Direct search methods 154, 232
Directional convexity 595
Directional derivative 767, 803, 806
Directionally differentiable function 803
Discrete optimization 2, 564, 576, 630, 664
Discretized Newton's method 34
Distributed algorithms 150, 194, 535, 693
Dual function 432, 561
Dual problem 432, 562
Duality gap 564, 583, 597, 661, 670
Duality for linear programming 435, 439, 441, 468, 584
Duality for quadratic programming 437, 584, 673

E

ϵ-ascent method 727
ϵ-complementary slackness 734
ϵ-relaxation method 725
ϵ-subdifferential 714
ϵ-subgradient 714
ϵ-subgradient method 714
Effective domain 784
Eigenvalue 759
Eigenvector 759
Ellipsoid method 68
Entropic descent 323
Entropy minimization algorithm 687
Epigraph 784
Euclidean norm 749
Euclidean space 749
Exact differentiable penalty 503, 520, 542, 552
Exact nondifferentiable penalty 442, 503, 540, 549, 552, 576
Exponential multiplier method 496, 687
Extended Kalman filter 181
Extended monotropic programming 576, 614, 662, 727
Extragradient method 342
Extrapolation 80, 81, 94, 136
Extreme point 798
Extreme points of polyhedra 801

F

Facility location problem 636
Farkas' lemma 402, 429, 584, 797
Feasible direction 257
Feasible direction method 258
Feasible vector 6
Fejér convergence 780
Fejér monotonicity 779
Fenchel duality 598
Fenchel duality theorem 602
Fenchel optimality conditions 600
Finite differences 148
Finitely generated cone 796
Forward difference formula 148
Frank-Wolfe method 262
Frank-Wolfe theorem 250
Frechet differentiability 766
Fritz John conditions 387, 415, 424, 425, 661
Full rank matrix 748

G

GMRES method 131
Game theory 440, 620
Gateaux differentiability 767
Gauss-Newton method 34, 112
Gauss-Seidel method 324
Generalized assignment 636
Generated cone 790
Geometric convergence rate 69, 772
Geometric multiplier 556
Global maximum 6
Global minimum 6
Golden section method 812
Gordan's theorem 423, 597
Gradient 767
Gradient method 30, 195, 221
Gradient method with errors 44, 166, 190, 198
Gradient projection 272, 322
Gradient related directions 41, 49, 51, 58, 260
Gram-Schmidt procedure 123

H

Hamiltonian function 218
Heavy ball method 81, 94
Hessian matrix 768
Hierarchical decomposition 326, 676
Hoffman's bound 580
Hyperplane 793
Hyperplane normal 793

Index 859

I

Ill-conditioning 72, 475
Implicit function theorem 15, 441, 489, 544, 770
Incremental Gauss-Newton method 178, 194
Incremental Newton method 185
Incremental aggregated gradient method 172, 196, 232
Incremental aggregated proximal algorithm 320, 322, 689
Incremental augmented Lagrangian method 687
Incremental gradient method 161, 190, 192, 195, 680
Incremental proximal algorithm 320, 341, 688
Incremental subgradient method 232, 715, 732, 733
Induced norm 757
Infimum 746
Informative Lagrange mutiplier 419
Inner linearization 738
Inner product 746
Integer programming 2, 564, 662
Interior point 753
Interior point methods 449
Interlocking eigenvalues lemma 761
Inverse matrix 758
Invertible matrix 757
Iterations with delays 774

J

Jacobi method 158
Jacobian matrix 767
Jensen's inequality 785

K

Kalman filter 179
Kantorovich inequality 83
Karush-Kuhn-Tucker conditions 344, 379, 381, 386, 392, 424, 430, 442
Knapsack problem 579, 636

L

ℓ_1 regularization 160, 696
Lagrange multiplier 344, 557
Lagrangian function 356, 379
Lagrangian methods 527
Lagrangian relaxation 648
Least squares problems 34, 108
Left-continuous 754

Levenberg-Marquardt method 113
Limit point 753
Limited memory quasi-Newton method 146
Limited minimization rule 35, 259
Line segment principle 447, 791
Linear convergence rate 69, 772
Linear independence/interior point constraint qualification 394, 398, 428
Linear manifold 747, 790
Linear programming 248, 306, 435, 439, 441, 450, 458, 584, 798, 802
Linearization algorithm 513, 539
Linearly independent 747
Lipschitz continuity 754
Local convergence 40, 96, 529, 545
Local maximum 6
Local minimum 6
Local proximal algorithm 321
Location theory 24, 596
Logarithmic barrier method 450
Logistic regression 159
Long-step interior point method 458
Lower limit 751
Lower semicontinuous 754
Lower triangular matrix 748

M

Manifold suboptimization method 283, 298, 340
Maratos' effect 526, 542, 551
Maximum a posteriori (MAP) rule 111
Max crossing point problem 555, 660
Maximum norm 750
Mean value theorem 768
Mirror descent 319, 323
Min common point problem 554, 660
Minimax problems 396, 552, 620
Minimax theorem 441, 621, 629
Minimization rule 35, 260
Minkowski-Weyl theorem 796
Model predictive control 234
Modified Newton's method 33, 817
Modified barrier method 496
Momentum term 80
Monotropic programming 576, 612, 617, 661, 672, 727
Multicommodity flow problem 242, 263, 291, 331
Multilayer perceptron 108
Multiplier method 398, 480, 493, 544, 546, 548, 682, 741

N

Necessary conditions 9, 15, 252, 346, 379, 430, 567
Neighborhood 753
Nelder-Mead method 154
Nesterov's method 83
Network optimization 3, 242, 263, 291, 331, 618, 631
Neural networks 108, 111, 217
Newton's method 32, 85, 95, 222, 289, 454, 535
Nonderivative methods 148
Nonexpansive mapping 20
Nonmonotone stepsize rule 39
Nonnegative combination 790
Nonnegative definite matrix 763
Nonsingular local minimum 10
Nonsingular matrix 757
Nonstationary iterations 776
Norm 749
Normal cone 415
Nullspace 748

O

Open set 753
Optimal control 4, 144, 210, 253, 595
Orthogonal basis 748
Orthogonal vectors 746
Outer linearization 738

P

PARTAN 136
Parallel computation 150
Parallel projections algorithm 329
Parallel subspace 747
Parallel tangents method 136
Pareto optimality 594
Partial elimination of constraints 488
Partially asynchronous algorithms 197, 204, 233
Partitioning 665, 677, 681
Path following method 450
Penalty method 471
Point of attraction 529
Polar cone 796
Polar cone theorem 796
Polyhedral cone 796
Polyhedral set 797
Positive definite matrix 763
Positive semidefinite matrix 763
Preconditioned conjugate gradient method 127, 137, 226
Preflow push method 725
Primal-dual method 458, 502
Primal function 373, 504, 623, 665, 740
Primal problem 432, 504, 555
Projection theorem 20, 244
Proper function 784
Proper hyperplane separation 795
Proximal ADMM 706, 707, 743
Proximal algorithm 307, 340, 682, 690
Proximal cutting plane method 722, 744
Proximal gradient algorithm 319, 743
Proximal method of multipliers 705, 743
Pseudoconvexity 783
Pseudonormality 397, 407, 419, 427, 442
Pythagorean theorem 749

Q

Quadratic convergence 70, 98
Quadratic interpolation 811
Quadratic programming 244, 248, 250, 303, 437, 458, 584, 673
Quasi-Newton methods 34, 138, 548, 743
Quasiconvexity 783
Quasinormality 427, 443
Quasiregularity 406, 427, 443

R

Range space 748
Rank 748
Receding horizon control 234
Recession cone 792
Recession directions 338, 756, 792
Recession function 792
Reduced gradient method 340
Regression problems 118, 159, 685
Regular point 347, 378
Regularization 160, 696
Relative interior 791
Relaxation method 725
Reshuffling in incremental methods 161, 171, 174
Retractive asymptotic sequence 247, 338
Riccati equation 220, 222, 225
Right-continuous 754
Rollout algorithms 234
Routing problems 228, 241, 263, 291, 331, 338, 342

S

Saddle point theorem 441, 567, 621, 629
Scaling 72, 230, 290, 531
Schwarz inequality 750
Second order cone programming 608
Second order expansion 769
Self-scaling quasi-Newton method 143, 146

Semi-infinite programming 407, 442
Semidefinite programming 610
Sensitivity 14, 26, 370, 384, 579, 625, 638, 662, 699, 707
Separable problems 534, 570, 661, 675, 699, 707
Separating hyperplane theorem 794
Sequential quadratic programming 503, 513, 539, 545
Shapley-Folkman theorem 571, 582
Sharp minimum 315, 734
Short-step interior point method 458, 468
Shortest path problem 632
Shrinkage operator 696
Simplex method for direct search 154
Simplex method for linear programming 306
Simplicial decomposition method 269, 339, 341, 744
Singular local minimum 10
Singular matrix 757
Singular problems 79, 87, 118
Slater constraint qualification 395, 398, 575, 589, 661
Smoothing 229, 627, 807
Spacer steps 43, 58
Span 747
Spectral radius 759
Sphere 753
Square root of a matrix 764
Stationary point 8, 238
Steepest descent 31, 725
Steepest descent with errors 44, 64, 66, 92
Stochastic approximation 46, 170
Stochastic gradient descent 170
Stochastic gradient methods 46, 170, 197
Stochastic optimal control 234
Stochastic programming 680, 694
Strict complementary slackness 383, 423
Strict hyperplane separation theorem 795
Strictly convex function 784
Strong duality theorem 583, 589, 592
Strongly convex function 787
Subdifferential 804
Subgradient 803
Subgradient method 709, 743
Sufficient conditions 10, 22, 252, 364, 383, 385, 423, 430, 567
Superlinear convergence rate 69
Supermartingale convergence theorem 778
Support vector machines 160
Supporting hyperplane theorem 794
Supremum 746

Symmetric matrix 748

T

Tangent cone 400
Termination of algorithms 42, 267, 720
Theorems of the alternative 403, 423, 429, 597, 661
Total unimodularity 634
Totally asynchronous algorithms 197, 208, 233
Traffic assignment 243
Training of neural networks 109, 218, 232
Transportation problem 440
Transpose matrix 748
Traveling salesman problem 636
Truncated Newton method 106, 332
Trust region methods 103
Two-metric projection method 282, 292, 340

U

Unimodal function 812
Unimodular matrix 634
Unimodularity 634
Upper limit 751
Upper semicontinuous 754
Upper triangular matrix 748

V

Valid inequalities 647
Value iteration algorithm 200
Varignon frame 24, 597

W

Weak duality theorem 563
Weber problem 24, 596
Weierstrass' theorem 12, 755
Weighted maximum norm 773
Wolfe conditions 63

Z

Zero sum games 440, 621, 661